AF615580

Basic Principles of

# LIGAND FIELD THEORY

Basic Principles of

# LIGAND FIELD THEORY

Hans L. Schläfer

*and*

Günter Gliemann

*Translated from the German by*

*David F. Ilten*

WILEY–INTERSCIENCE

a division of John Wiley & Sons Ltd.

LONDON NEW YORK SYDNEY TORONTO

Library of Congress catalog card no. 77-83220

SBN 471 76100 1

Set on Monophoto Filmsetter and
printed by J. W. Arrowsmith Ltd., Bristol, England

Dem Altmeister der Komplexspektroskopie
**Prof. Dr. Martin Linhard**
in freundschaftlicher Verbundenheit

# Preface

It is not necessary to expand upon the great importance of ligand field theory in the chemistry of transition metal compounds. With the present treatment we have attempted to give an easily-readable introduction to this theory. The necessary concepts are for the most part introduced on the basis of experimental results, and we hope this will prove useful to the experimental chemist.

The book represents a revised and expanded form of lectures and problem sessions held in the physical chemical institute in Frankfurt as well as in the International Summer School in Constance (Lake Constance) in 1962 and in courses on ligand field theory held in Buckow (DDR) in 1964 and in other places.

We wish to thank Dr. G. Herzog and Dr. O. Steinborn, who kindly assisted with the reading of proofs.

*Frankfurt am Main, January, 1966.*

HANS LUDWIG SCHLÄFER
GÜNTER GLIEMANN

# General introduction

During the last ten years a theoretical concept, the so-called crystal field or ligand field theory, has proved of great value in discussions of the properties of transition metal compounds and especially of metal complexes. With the aid of ligand field theoretical concepts it is possible to understand a whole series of physical and chemical properties of transition metal complexes from a unified point of view, and to obtain at least semiquantitative information. On this basis the optical properties of the vast majority of coordination compounds of transition metal ions can be interpreted. Other problems, e.g. the description of magnetic properties, of stereochemical irregularities and questions of kinetic and thermodynamic stability can also be treated effectively.

It is hardly possible today to discuss the chemistry of the transition metals, as offered in general lectures on inorganic chemistry, without employing ligand field theory. It is difficult to find a better example of how useful meaningfully chosen models can be for understanding a large body of exceedingly varied experimental results. Nevertheless, only comparatively few basic concepts are required for a first qualitative understanding of the theory.

In this book an attempt will be made to offer the chemist access to the ligand field theory. We thus intentionally make no attempt to present a complete and self-contained systematic treatment. We refer to the excellent monographs by Ballhausen* and Griffith* as well as to numerous review articles in which those interested in a systematic presentation will find everything necessary. The books by Jørgensen* give a good general survey, especially of the interpretation of absorption spectra of transition metal complexes. Cotton* offers a comparable treatment of group theoretical methods.

According to the impression gained from numerous discussions with colleagues and students, and especially from the experience of introductory lectures held at the Frankfurt institute and elsewhere, understanding ligand field theory presents the chemist with certain difficulties. Not the least cause of these is a lack of familiarity with group and quantum theoretical concepts, often coupled with a lack of the necessary mathematical background. Therefore the emphasis has been placed on didactical

* These references are listed in the bibliography at the end of the book, p. 499 ff.

questions in the following presentation. Prerequisites for understanding this book are simply a basic knowledge of complex chemistry, physical chemistry, physics and mathematics, which the student should have normally obtained by the 'Diplom' examination.

The book consists of two main parts. The first part gives a qualitative introduction to the basic concepts, using selected examples from the field of transition metal complex spectroscopy. These concepts are then used in a study of the magnetic behaviour and thermodynamic stability of complex ions in solutions, the kinetic behaviour in ligand exchange reactions, and other questions from the physical chemistry of coordination compounds. This is done almost completely without the use of mathematical formalism. Rather we have attempted to give a phenomenological presentation on the basis of experimental findings. In doing this it was not possible to avoid making certain compromises, which certainly could be open to much criticism.

The material of this first part is intended to give the reader an impression of the underlying model and of its capabilities of interpreting the properties of transition metal complexes. Among other things it is shown what concrete information can be obtained in special cases, e.g. in the identification of *cis–trans* isomers of $[MA_4B_2]$ complexes ($M = Cr^{3+}, Co^{3+}$) on the basis of absorption spectra, the question of the appearance of low-spin complex ions, the connection between stereochemistry and the orbital moment contribution to the magnetic moment for $Co^{2+}$ and $Ni^{2+}$ compounds, the problem of the pronounced stability of $Co^{3+}$ and $Cr^{3+}$ complexes in ligand exchange reactions, and so on.

In the second part an introduction to the mathematical treatment is offered to those who are interested in a deeper understanding. Here we have utilized experience gained in holding seminars and problem sessions. The material of this more demanding part requires rather intensive concentration, especially from the chemist not fully acquainted with the thinking patterns of the theoretical physicist. Also, the knowledge of certain mathematical principles is essential. As an aid we have collected in an appendix important elements of matrix and determinant theory, as well as explanations of the terms operator, eigenvalue, normalization, orthogonality, etc. in condensed form.

It is recommended that the reader not familiar with the elementary basis of quantum theory should study the appropriate sections in one of the introductory texts cited before beginning the material of part B. A presentation of these additional points would have produced a book of unmanageable length. The derivation of the necessary quantum mechanical relations and group theoretical theorems is not given; neither is the pertinent literature. The second part rather attempts to explain the necessary calculatory methods by using selected examples. The chapter on the theory of the states of free atoms and ions is followed by a section on the basic principles of group theory, in so far as these are necessary for application in ligand field theory. After the treatment of the splitting of the one-electron *d* states in ligand fields of various symmetries, a presentation of the methods of determining the term systems of complex ions whose central ions contain several *d* electrons follows. The molecular orbital method is also treated (and in the appendix the valence bond method). Sections on spin-orbit coupling as well as magnetic susceptibility follow. In all cases the problems are explained on the basis of typical examples, for which the method of calculation is presented.

Finally a collection of important original literature, arranged in order of topics, is given, which is intended to make it easier for the reader to study particular questions in more detail.

The following book was arranged so as to give the chemist an easily-readable introduction to the ligand field theory. After studying the first part he should be in a position to understand the important concepts of the theory and to have an idea of their applicability. After completing the second part the reader should be capable of grasping the quantitative side as well, at least in its elementary form, so that he can treat simple problems and can read the original literature. The material given in the review articles and monographs cited in the bibliography should then offer no further difficulties.

For the majority of the symmetries which usually appear for complex ions, especially for the cubic microsymmetry realized at least approximately in the largest number of cases, (symmetry groups $O_h$ or $T_d$), the term systems for the complexes with central ion having 1 to 9 *d* electrons have been calculated and the matrix elements are given in the original papers (cf. bibliography). Generally, therefore the chemist today fortunately does not need to carry out calculations himself. He will, using term diagrams or tabulated matrix elements, simply obtain the parameters characterizing the ligand field, e.g. from the absorption spectra, and assign the observed bands to specific electronic transitions. Also in those cases where spin-orbit coupling plays an important rôle, e.g. in magnetic susceptibility or the *g* factor, the theory is so far developed that for many problems one can refer to original papers in which the needed matrix elements are given.

# Contents

**General introduction** . . . . . . . . **ix**

PART A

**1 Optical properties of transition metal complexes**

1.1 Historical survey and description of the problem . . 2
1.2 Survey of the experimental findings; absorption spectra of selected metal complexes . . . . . . . 4
1.3 The complex models of Kossel–Magnus and of Pauling . 13
1.4 The extended ionic model: The ligand field . . . 15
1.5 Term splitting under cubic symmetry . . . . 24
1.6 Analysis of absorption spectra . . . . . 28
1.7 The strong- and weak-field methods . . . . 51
1.8 Term diagrams . . . . . . . . . 70
1.9 Dq values, Racah parameter B, spectrochemical and nephelauxetic series . . . . . . . . 76
1.10 Relative intensities and half widths . . . . 84
1.11 Spin-orbit coupling . . . . . . . . 93
1.12 Treatment of complexes by the molecular orbital theory. 97
1.13 Charge-transfer spectra . . . . . . . 109

**2 Magnetic properties of transition metal complexes**

2.1 General introduction. Survey of experimental findings . 118
2.2 Interpretation of low-spin complexes according to Pauling 120
2.3 Interpretation of the appearance of low-spin complexes on the basis of the ligand field theory . . . . 126
2.4 Orbital moment contribution and stereochemistry . . 131
2.5 The refined theory of the paramagnetic susceptibility of transition metal complexes; consideration of spin-orbit coupling . . . . . . . . . . 137

**3 Stabilization of ions**

3.1 Ligand field stabilization of cubic complexes . . . 142
3.2 Stabilization and tendency of formation of low-spin complexes . . . . . . . . . . 147
3.3 Influence of ligand field stabilization on the thermodynamic properties of transition metal complexes . . 150
3.4 The relative stability of tetrahedral and octahedral high-spin complexes . . . . . . . . 158
3.5 Ionic radii of transition metals of the iron group . . 160

**4 Consequences of the Jahn–Teller theorem for transition metal complexes** . . . 164

**5 The kinetic stability of transition metal complexes** . . . 176

**6 Bonding** . . . 186

PART B

**1 The theory of free atoms and ions**

1.1 One-electron systems . . . 207

*1.1.1 The energy* . . . 207

*1.1.2 The orbital angular momentum* . . . 215

*1.1.3 The electron spin* . . . 220

1.2 Many-electron systems . . . 222

*1.2.1 The Schrödinger equation* . . . 222

*1.2.2 The central-field approximation* . . . 223

*1.2.3 Electron configurations* . . . 226

*1.2.4 Total angular momentum and total spin* . . . 232

*1.2.5 The states* $^{2S+1}L$ . . . 235

*1.2.6 The functions* $\psi(L, M_L, S, M_s)$ . . . 238

*1.2.7 Electron interaction* . . . 243

*1.2.8 The matrix elements* $(\psi, H_1, \psi)$ . . . 247

**2 Group theory**

2.1 Symmetry operations and symmetry groups . . . 255

*2.1.1 Definition of a symmetry operation and of a group* . 255

*2.1.2 Classes and subgroups* . . . 260

*2.1.3 Special symmetry groups* . . . 261

2.2 Representations . . . 270

*2.2.1 Definition of a representation* . . . 270

*2.2.2 Equivalent representations: similarity transformations* . . . 276

*2.2.3 Characters* . . . 277

*2.2.4 Reduction of representations* . . . 278

2.3 Coordinate transformations . . . 286

2.4 Group theory and quantum mechanics . . . 293

*2.4.1 The symmetry group of the Hamiltonian* . . . 293

*2.4.2 Degenerate systems* . . . 296

*2.4.3 Perturbation theory* . . . 309

2.5 Product representations . . . 312

2.6 Double groups . . . 317

**3 Ligand field theory**
3.1 The model . . . . . . . . . . 321
3.2 Ligand fields . . . . . . . . . . 324
3.3 $d^1$ systems . . . . . . . . . . 333
*3.3.1 Ligand fields of $O_h$ symmetry* . . . . . 336
*3.3.2 Ligand fields of symmetry $T_d$* . . . . . 342
*3.3.3 Ligand fields of symmetry $D_{4h}$* . . . . . 343
3.4 $d^N$ systems . . . . . . . . . . 347
*3.4.1 The weak-field case* . . . . . . . 348
*3.4.2 Strong-field case* . . . . . . . 362
*3.4.3 Rigorous treatment* . . . . . . . 386

**4 The molecular orbital theory as an extension of the ionic ligand field theory**
4.1 The method . . . . . . . . . . 407
4.2 Symmetry-adapted functions . . . . . . 410
4.3 Energies . . . . . . . . . . 419

**5 Spin–orbit coupling and magnetic properties**
5.1 Spin–orbit coupling in free atoms and atomic ions . . 429
5.2 Spin–orbit coupling of $d^1$ ions in ligand fields of octahedral symmetry . . . . . . . . . . 438
5.3 Zeeman effect in free atoms and atomic ions . . . . 448
5.4 Zeeman effect on octahedrally bound $d^1$ ions . . . . 453
5.5 Magnetic susceptibility . . . . . . . . 459

PART C
**Appendix**
**1 Mathematical terms and aids** . . . . . . . 465

**2 Character tables of the point groups of most importance for the ligand field theory** . . . . . . . . . 479

**3 The most important symmetries of complex ions** . . . . 481

**4 Hybridization of central ion orbitals in the valence bond theory** . 484

**5 Selection rules** . . . . . . . . . . 494

**6 Atomic units** . . . . . . . . . . 498

**Bibliography** . . . . . . . . . . 499

**Subject Index** . . . . . . . . . . 521

PART A

# 1. Optical properties of transition metal complexes

In the following it is planned to present the underlying concepts of the ligand field theory on the basis of a discussion of selected absorption spectra of transition metal ions in the visible and ultraviolet regions. We have chosen this method because the interpretation of spectra represents an especially impressive example of the application of the theory. Also, such a phenomenological treatment is well suited to the experimental background of the chemist. Certainly it would have been simpler to introduce the principles of the ligand field theory by considering the magnetic moments of the transition metal complex ions. In that case one is generally interested in only the energetically lowest-lying (most stable) state of a compound—the ground state. On the other hand, a treatment of the optical properties requires the consideration of the energetically higher-lying (excited) states because the maxima of the absorption bands are to the first approximation associated with transitions from the ground state to the respective excited states. It is well known that in most cases it is possible to explain magnetic properties—at least in so far as is necessary to rationalize the existence of low-spin complexes—with the help of *Pauling's valence bond theory*. This theory is of course familiar to the chemist. However, the valence bond theory does not provide a suitable interpretation of absorption spectra, whereas the crystal field or ligand field theory* does. When considering magnetic properties one is concerned with comparing only a single value with experiment, namely the multiplicity of the ground state and thereby the effective magnetic moment at room temperature of the compound in question. However, when one is interested in interpreting absorption spectra, the theory must provide the number and relative location of all the various absorption bands (e.g. the $[Mn(H_2O)_6]^{2+}$ ion has eight bands). Certain effects such as the splitting of energy levels accompanying a descent in symmetry are sometimes reflected directly in the visible absorption spectrum. Therefore, the range

* Concerning the use of the terms crystal field or ligand field, cf. p. 17 (footnote).

of phenomena embraced by the theory is greater than for the case of magnetic moments.

## 1.1 Historical survey and description of the problem

A collection of brilliantly coloured, beautifully crystallized salts of metal complexes is aesthetically pleasing to the casual observer; for the spectroscopist it presents an immediate challenge. The prefixes such as violeo-, praseo-, luteo-, purpureo-, roseo-,* refer to the most variegated visual impressions given by such compounds. When one considers their underlying chemical structures, the remarkable colour changes which accompany a change of ligand or of steric arrangement are brought to mind. The transition metal complexes are especially well suited for investigations on the relationship between colour and composition. They show an amazing variety of colours and at the same time variations of the basic chemical structure are easily obtained. It is therefore not surprising that directly after the introduction of the coordination theory by Alfred Werner[1] several of his students began to study the absorption spectra of solution complexes. The results of these early investigations conducted in Werner's laboratory in Zürich are to be found only in doctoral dissertations.

Starting about 1915 spectroscopic studies were undertaken by such Japanese investigators as Shibata and co-workers[2] and later by Tsushida

---

* These notations, which were introduced by Fremy (*Ann. Chim. Phys.*, **35**, 257 (1852)), give the colours of the appropriate complexes, e.g. $[Co(NH_3)_5Br]^{2+}$ is also called bromopurpureo-cobalt (III) ion. Today one occasionally uses these notations to describe classes of compounds.

| Colour | Name | Structure |
|---|---|---|
| green | praseo- | trans-$[Co(NH_3)_4Cl_2]^+$ |
| violet | violeo- | cis-$[Co(NH_3)_4Cl_2]^+$ |
| purple red | purpureo- | $[Co(NH_3)_5Cl]^{2+}$ |
| rose | roseo- | $[Co(NH_3)_5H_2O]^{3+}$ |
| yellow | luteo- | $[Cr(NH_3)_6]^{3+}$ bzw. $[Co\,en_3]^{3+}$ |
| yellow | croceo- | trans-$[Co(NH_3)_4(NO_2)_2]^+$ |
| brown | flavo- | cis-$[Co(NH_3)_4(NO_2)_2]^+$ |

1. Cf. A. Werner: *Ber. dt. chem. Ges.*, **40**, 15 (1907) A Werner: *Neuere Anschauungen auf dem Gebiet der anorganischen Chemie*, revised by P. Pfeiffer, Friedr. Vieweg and Sohn, Braunschweig, 1923. R. Weinland: *Einführung in die Chemie der Komplexverbindungen*, 2nd edn, Ferdinand Enke Verlag, Stuttgart, 1924.
2. Y. Shibata: cf. contribution in *J. Coll. Sci. imp. Univ., Tokyo*, 1915 to 1921, also *J. chem. Soc. Japan*, 1915; *Compt. Rend.*, 1913.

*et al.*[3]. Also, von Kiss and co-workers[4] and Samuel *et al.*[5] have measured numerous absorption spectra of complexes in solution. On the basis of these findings the authors attempted to derive phenomenological rules, for example, relating the number of bands to the nature of the central ion, or describing the position of the bands as a function of the nature of the ligands for complexes having similar structures. Joos and Deutschbein[6] and Bose and Datta[7] attempted to give interpretations using the theory of electrons. It was soon recognized that *d* electrons are important in determining the colour of transition metal ion complexes. Further, concepts of polarization as introduced by Fajans proved to be useful[8].

None of these efforts, however, led to a satisfactory solution of the problem cited. Certainly not the least cause of this failure was the fact that the published spectra could not stand up to a critical examination. First of all in the years before 1940 the technique and apparatus for measuring absorption spectra had not been developed far enough to give results of the quality which can be obtained today almost effortlessly using commercial spectrophotometers. Secondly, far too little attention was paid to the state of the solution for the majority of the spectral studies on solutions. Frequently, depending upon the conditions of concentration and acidity, the most diverse ionic species can appear (consecutive equilibria between various complex species, hydrolysed ions, etc.). These species can all contribute to the absorption. That is, one cannot assume that the absorbing ion in the solution is the same ion which exists in the solid salt. Furthermore, in many cases the substances studied were not sufficiently pure; they can seldom be prepared without chemically-similar side products being present.

Only after about 1940 did systematic investigations of the absorption spectra of dissolved complexes take all necessary factors into account.

---

3. R. Tsuchida: *Bull. chem. Soc. Japan*, **13**, 388, 436, 471 (1938); **15**, 427 (1940); and other works.
4. A. v. Kiss and others: *Acta chem. mineralog. physica* (*Szeged*), **7**, 119 (1939); *Z. phys. Chem.* **A 180**, 117 (1937); **A 186**, 239 (1940); *Z. anorg. allg. Chem.* **235**, 407 (1937); **244**, 98 (1940); **245**, 355 (1940); **246**, 28 (1941); other works, e.g. *Acta chim. hung.*
5. R. Samuel and others: *Trans. Faraday Soc.*, **31**, 423 (1935); *Z. phys. Chem.*, **B 22**, 424 (1933); *Z. Phys.*, **80**, 395 (1933); and other works.
6. G. Joos: *Ann. Phys.*, **81**, 1076 (1926); **85**, 641 (1928); *Ergebn. exakt. Naturw.*, **18**, 78 (1939); *Z. phys. Chem.*, **B 20**, 1 (1933); *Phys. Z.*, **29**, 117 (1928); and other works. O. Deutschbein: *Z. Phys.*, **77**, 489 (1932); *Ann. Phys.*, **14**, 753 (1932); **20**, 828 (1934); and other works.
7. D. M. Bose and S. Datta: *Z. Phys.*, **80**, 376 (1932).
8. K. Fajans: *Z. Kristallogr. Mineralogr. Petrogr.*, **A 61**, 18 (1925); **A 66**, 321 (1928); *Naturwiss.*, **11**, 165 (1923). K. Fajans and G. Joos: *Z. Phys.*, **23**, 1 (1924).

Of these, the work of Linhard and co-workers[9] on $Co^{3+}$ and $Cr^{3+}$ complexes stands out as an example of utmost precision. With this work the necessary material for a theoretical investigation became available. Because many compounds are stable only in the solid state, the spectroscopy of crystalline compounds, especially the determination of reflectance spectra of finely pulverized salts, as has been made possible by the critical and pioneering work of Kortüm[10], has developed into a very important tool of the complex spectroscopist.

## 1.2. Survey of the experimental findings; absorption spectra of selected metal complexes

The absorption spectra of complexes of transition metals, with the exception of certain examples which will be treated later, have in general an appearance such as is schematically indicated in Figure A.1. In the

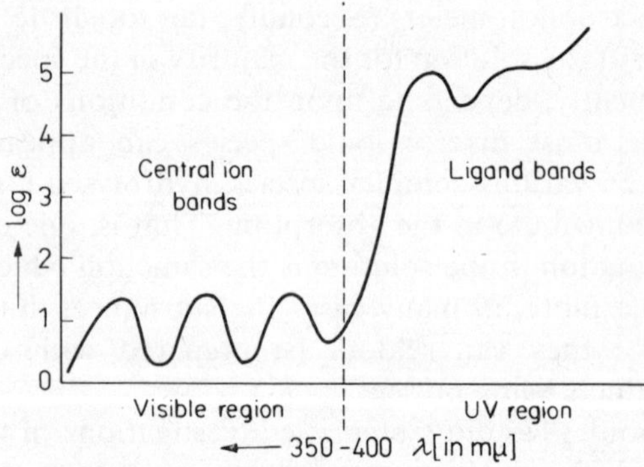


Figure A.1. Absorption spectrum of a transition metal ion complex (schematic). $\lambda <$ ca. 350–400 m$\mu$ ligand bands. (*a*) Charge transfer bands (Transitions: I Central ion → ligand; II ligand → central ion). (*b*) Transitions between states of the ligands. Inner-ligand transitions. e.g. $\pi \rightarrow \pi^*$, $n \rightarrow \pi^*$; $\lambda >$ about 350–400 m$\mu$ central ion bands ($d \rightarrow d$ bands).

---

9. M. Linhard and others: *Z. Electrochem.*, **50**, 224 (1944); *Z. anorg. allg. Chem.*, **262**, 328 (1950); **264**, 321 (1951); **266**, 49, 73 (1951); **267**, 113, 121 (1951); **271**, 101, 131 (1952); **278**, 287 (1955); *Z. phys. Chem.* (Frankfurt), **5**, 20 (1955); **11**, 308 (1957); and additional works.
10. G. Kortüm and J. Vogel: *Z. phys. Chem.* (Frankfurt), **18**, 110, 230 (1958). G. Kortüm and G. Schreyer: *Angew. Chem.*, **67**, 694 (1955); *Z. Naturf.*, **11a**, 1018 (1956). G. Kortüm: *Spectrochim. Acta. Coll. spectrosc. internat.* VI, 534 (1957). G. Kortüm and D. Oelkrug: *Naturwiss*, **23**, 600 (1966).

region between 200 and 1000 m$\mu$ (50,000–10,000 cm$^{-1}$) bands of varying intensity are found. In the long wavelength spectral region, that is, with $\lambda >$ 350–400 m$\mu$, one finds one or more bands with values of the logarithm of the molar decadic extinction coefficient (log $\varepsilon$) for the band maxima which lie between roughly 0 and 2. In the short wavelength portion of the spectrum, that is, for wavelengths $<$350–400 m$\mu$, intense absorption bands with log $\varepsilon$ values of the order of 4–5 appear. They are approximately as intense as similar bands in organic dyes.

For the case of metal complexes where the central ion is not a transition metal ion, e.g. $Al^{3+}$ or $Zn^{2+}$, the weak bands at long wavelengths do not appear. However, the intense absorption bands at long wavelengths remain. (Compare Figures A.2 and A.3, the spectra of $[Al(C_2O_4)_3]^{3-}$ and

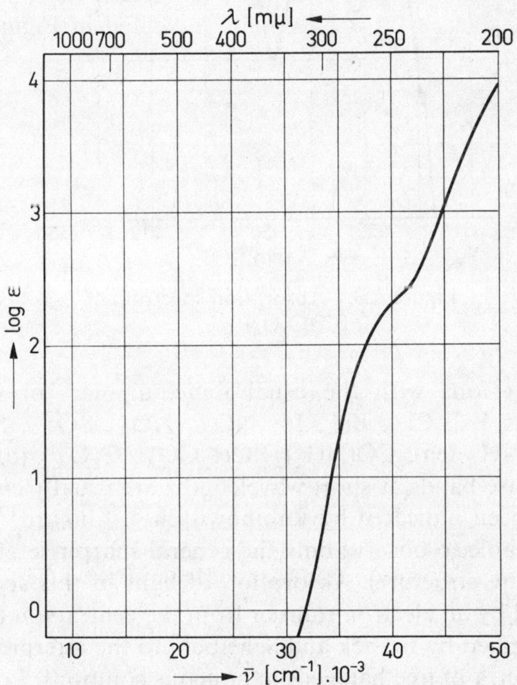


Figure A.2. Absorption spectrum of $[Al(C_2O_4)_3]^{3-}$.

$[Cr(C_2O_4)_3]^{3-}$). The weak bands in the long wavelength region were already assigned to the *d* electrons of the central ion by Shibata. They originate from transitions within the *d* shell, so-called $d \rightarrow d$ transitions. We shall refer to them summarily as *central ion bands* ($d \rightarrow d$ bands).

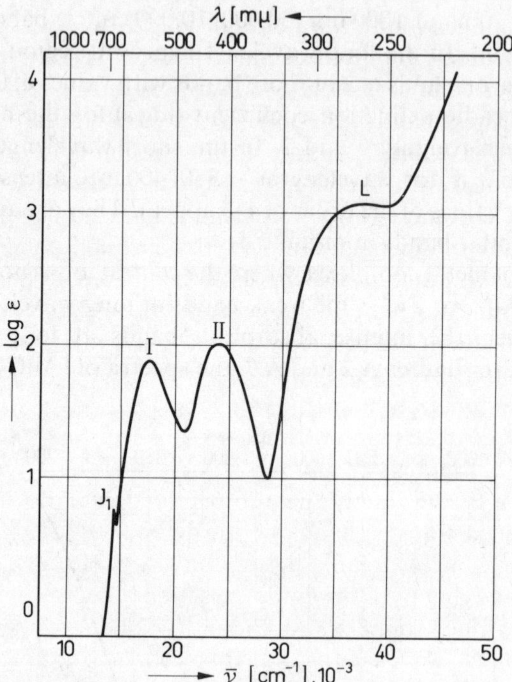


Figure A.3. Absorption spectrum of $[Cr(C_2O_4)_3]^{3-}$.

For complex ions with the usual ionic, dipolar- or dipolar-ionic ligands such as $F^-$, $Cl^-$, $Br^-$, $I^-$, $NO_3^-$, $NO_2^-$, $SO_4^{2-}$, $SO_3^{2-}$, $NH_3$, $NH_2CH_2CH_2NH_2$ (en), $CO(NH_2)_2$, $CH_3CO_2^-$, $C_2O_4^{2-}$ (ox), $H_2O$ and ROH, the intense bands at short wavelengths are usually *charge transfer bands*, as has been indicated by Rabinowitsch[11], Linhard[12] and others. One usually is able to observe only the general sharp rise of these bands (and not the fine structure). Absorption of light in this spectral region is accompanied by an electron transfer from the central ion to the ligand, an idea first applied by Franck and Scheibe[13] to the interpretation of the absorption spectra of free halogens in aqueous solutions. For the case of the free halogens $Cl^-$, $Br^-$, and $I^-$ the electron transfer probably takes

11. E. Rabinowitsch: *Rev. mod. Phys.*, **14**, 112 (1942); cf. also L. E. Orgel: *Q. Rev. chem. Soc.*, **8**, 422 (1954).
12. M. Linhard and others: *Z. anorg. allg. Chem.*, **262**, 328 (1950); **266**, 49 (1951); and additional works.
13. J. Franck and G. Scheibe: *Z. phys. Chem.*, **A 139**, 22 (1928). J. Franck and F. Haber: *Sitz. Ber. preuss. Akad. Wiss., phys.-math. Kl.*, 1931, 250.

place from the halogen ion to a water molecule of the hydrate sphere*.

For chloro-, bromo-, and iodo-complexes, the most intense absorption bands of the inner complex halogen ions appear in the short wavelength spectral region. These bands generally have a form similar to that of free halogen ions, but have undergone a red shift. Similar facts are true for the complex ions formed with the other ligands which were discussed earlier. For the case of chromium(III) trisethylenediamine molecules one finds in the u.v. the band of the inner-complex ethylenediamine molecule. This, however, appears at longer wavelengths as compared to the band of the free ethylenediamine molecule. It is in general observed that the absorption of the free ligand, after it has been complexed, undergoes a small red shift†.

In Figure A.4 the absorption bands of various free ions and molecules which frequently function as complex ligands are presented. (Compare the band of the free oxalate ion with the u.v. absorption of $[Cr(C_2O_4)_3]^{3-}$ or $[Al(C_2O_4)_3]^{3-}$, Figures A.2 and A.3.)

We shall refer to the intense u.v. absorption bands as ligand bands. Besides the charge–transfer bands of the ligand, other bands, such as those of complexes with organic ligands having $\pi$-electron systems, can appear. When studying pyridine complexes, one observes the u.v. absorption band of the inner complex of pyridine, which has undergone only a slight shift to longer wavelengths in comparison to that of free pyridine (Figure A.5). Frequently traces of the vibrational fine structure of the pyridine band are found in the spectra of pyridine complexes (Figure A.9). This is a case of $\pi \rightarrow \pi^*$ transitions (*inner ligand transitions*) of the pyridine molecule which are modified slightly by the polarizing influence of the central ion as the complex is formed. In certain cases it is also to be expected that electron transfers between essentially ligand states can appear.

---

* Interpretation of the spectra of dissolved halogen ions $X^-$ as charge-transfer spectra is supported by the fact that in these spectra in general two intense bands are found. Their difference in wave-number $\Delta\bar{\nu}$ corresponds approximately to that of the ground doublet ${}^2P_{3/2}$–${}^2P_{1/2}$ of the free halogen atoms equation

$$X^-(H_2O)+h\nu_1 \rightarrow X\,{}^2P_{1/2}\,(H_2O^-)$$

$$X^-(H_2O)+h\nu_2 \rightarrow X\,{}^2P_{3/2}\,(H_2O^-)$$

$\Delta\bar{\nu}$ for the free halogen atoms I, Br, and Cl is 7600, 3600 and 880 $cm^{-1}$ respectively. Schiebe, Fromherz and Menschick (*Z. Phys. Chem.*, Haber Band, 22(1928); *Z. Elektrochem.*, **34**, 497 (1928); *Z. Phys. Chem.*, **B5**, 355 (1929); *Z. Phys. Chem.*, **B7**, 439 (1930); **B3** 1 (1929) found the following separations of the band maxima for aqueous alkali halogenide solutions: for $I^-$ 7300 $cm^{-1}$, for $Br^{-1}$ 2000 $cm^{-1}$, for $Cl^-$ only one band which can be interpreted as being the superposition of two maxima of the same height lying close to one another. In recent years these suggestions have been refined. (R. Platzman and J. Franck, *Farkas Memorial Volume*, Weizman Press, Jerusalem 1952; *Z. Physik*, **138**, 411 (1954)).

† Complexes having central ions of lower valence are an exception, e.g. the tris-dipyridyl compounds described by Herzog and coworkers. $[V\text{dip}]^+$, $[V\text{dip}_3]$, $[V\text{dip}_3]^-$ (dip = $\alpha, \alpha'$ dipyridyl).

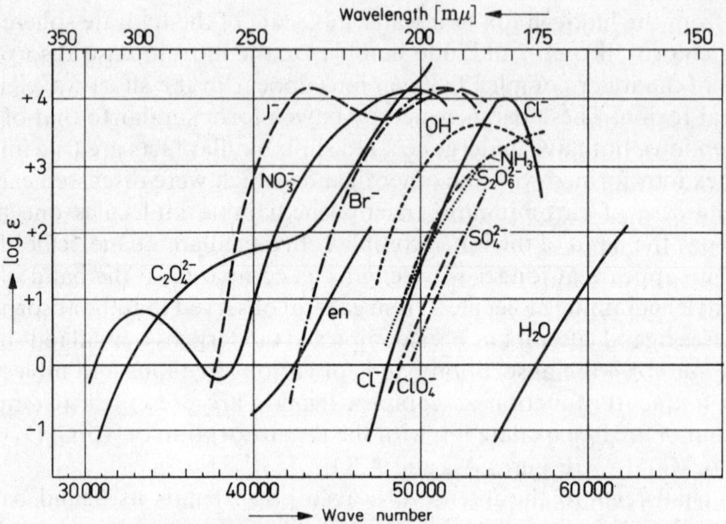


Figure A.4. Absorption spectra of various free ions and molecules which appear as complex ligands. ($ClO_4^-$ and $S_2O_6^{2-}$ are only in very rare cases coordinated in the first sphere.)

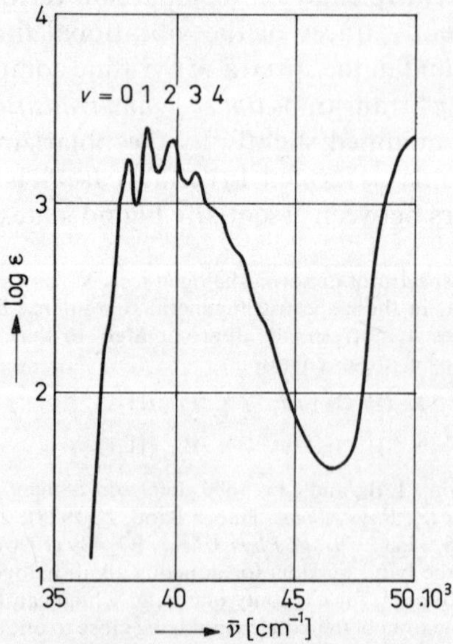


Figure A.5. The absorption spectrum of the free pyridine molecule, showing vibrational fine-structure (Pyridine dissolved in hexane).

As further examples of selected spectra the absorption spectra of the ions $[Ti(H_2O)_6]^{3+}$ and $[V(H_2O)_6]^{3+}$ can be seen in Figures A.6 and A.7. The spectrum of the hexaquotitanium(III) ion was taken under suitable conditions on solutions of titanium(III) sulphate, while the spectrum of the hexaquovanadium(III) ion was measured on plane-parallel sliced plates of crystals of the ammonium or caesium vanadium alum.

In alums the trivalent metal ion is surrounded by six $H_2O$ molecules in an almost octahedral array. $[Ti(H_2O)_6]^{3+}$ has one and $[V(H_2O)_6]^{3+}$ two central ion bands at long wavelengths. Reflectance* spectra of the complexes $[K_3MoCl_6]$ as well as $(NH_4)_2$ $[MoCl_5H_2O]$ are shown in Figure A.8. Besides the weak long wavelength bands one recognizes the sharp rise of the intense short wavelength electron transfer bands of the chloride ions of the inner complex. Finally in Figure A.9 the reflectance spectra of $[CrCl_3py_3]$, $[MoCl_3py_3]$ and $[TlCl_3py_3]$ (py = pyridine) can be seen. In addition to two long wavelength bands for the chromium and molybdenum compounds one observes in the u.v. the ligand band of the inner complex pyridine showing evidence of vibrational structure. The thallium compound has only the u.v. band of the inner-complex pyridine, because $Tl^{3+}$ is not a transition metal ion.

Such a differentiation between central ion and ligand bands can be carried out for the majority of the cases of complexes having transition metal central ions. It is especially applicable to complexes with the ligands

---

* In absorption spectra $\varepsilon$ or $\log \varepsilon$ is plotted as the ordinate. The molar decadic extinction coefficient is obtained with the aid of the Lambert–Beer Law

$$E = \log \frac{J_0}{J} = \varepsilon c d,$$

where $c$ is the concentration of the absorbing material in mole/l and $d$ is the length of the cuvette (optical path) in cm, $J_0$ is the intensity of the incident (monochromatic) light, $J$ the intensity of the light after being transmitted through the solution the distance $d$, $E$ the absorbency (cf. B. G. Kortüm, *Kolorimetrie, Photometrie und Spektrometrie*, 4th edition, Springer-Verlag, Berlin, 1962).

In reflectance spectra the ordinate is the Kubelka–Munk Function $f(R)$ or its logarithm:

$$f(R) \equiv \frac{(1-R)^2}{2R} = \frac{\varepsilon c}{0{\cdot}4343 s}$$

$\varepsilon$ is the molar decadic extinction coefficient, $c$ the concentration $s$ the average scattering coefficient, which depends upon the particle size and $R$ the relative diffuse reflectance power

$$R = \frac{J_{\text{refl.}}}{J_{\text{refl. stand.}}}$$

$J_{\text{refl.}}$ is the intensity of the diffuse reflected light. Cf. P. Kubelka and F. Munk, *Z. Techn. Physik*, **12**, 593 (1931); *J. Opt. Soc. America*, **38**, 448 (1948).

One can, as has been shown by G. Kortüm[10], obtain the typical colour curves of the materials in question from the reflection power, measured in the usual manner. Measurements are carried out on finely pulverized samples which have been diluted with a white standard. No quantitative statements concerning the intensities can be made from such measurements, however.

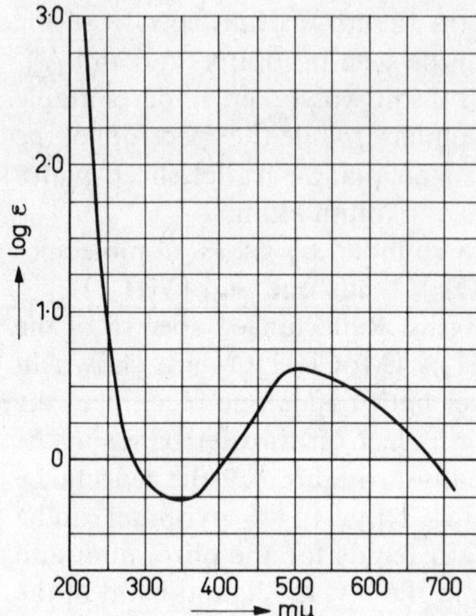


Figure A.6. Absorption spectrum of $[Ti(H_2O)_6]^{3+}$. Ti(III)-sulphate $c = 0{\cdot}117$ m in dilute $H_2SO_4$.

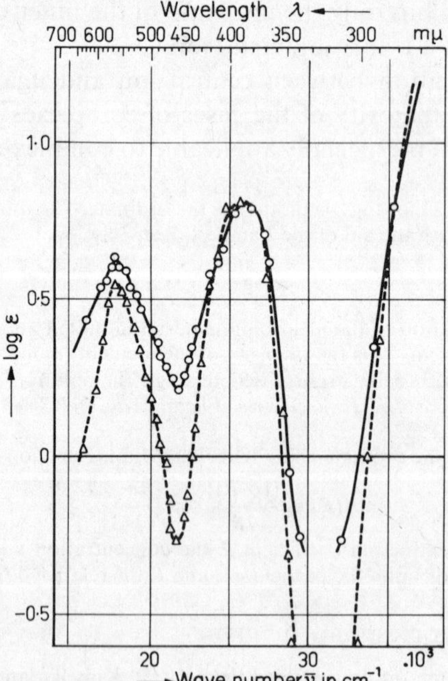


Figure A.7. Crystalline spectrum of $[V(H_2O)_6]^{3+}$.

—— $CsV(SO_4)_2 \cdot 12H_2O$
--- $NH_4V(SO_4)_2 \cdot 12H_2O$
plane parallel crystalline plates. d = 0·5 and 1 mm.

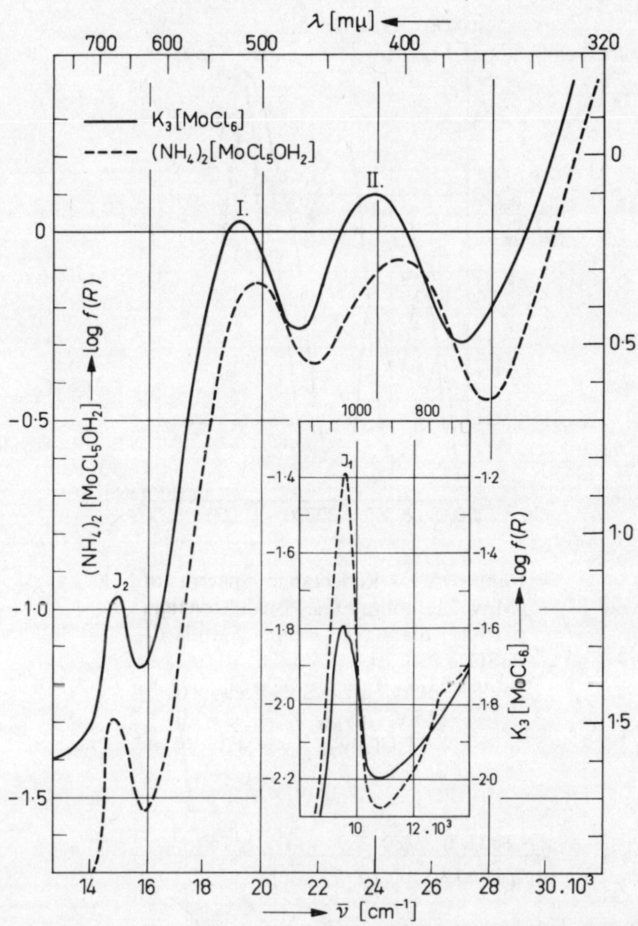


Figure A.8. Reflectance spectra of $K_3[MoCl_6]$ and $(NH_4)_2[MoCl_5OH_2]$. Powder, average particle diameter $< 5\mu$, undiluted. Standard: $BaSO_4$.

discussed above. Also, the spectra of complexes having ligands such as $SCN^-$ and $CN^-$ correspond in many cases to the scheme shown in Figure A.1. Such a division of the spectrum into these two band types is no longer possible for certain cyano-complexes and for complexes with CO and NO as ligands; that is, for carbonyl and nitrosyl complexes or $\alpha,\alpha'$-dipyridyl complexes with central ions of a low valency state. Also for such ions as $CrO_4^{2-}$ or $MnO_4^-$ other conditions dominate, as can be seen from the spectrum of the permanganate ion in Figure A.10.

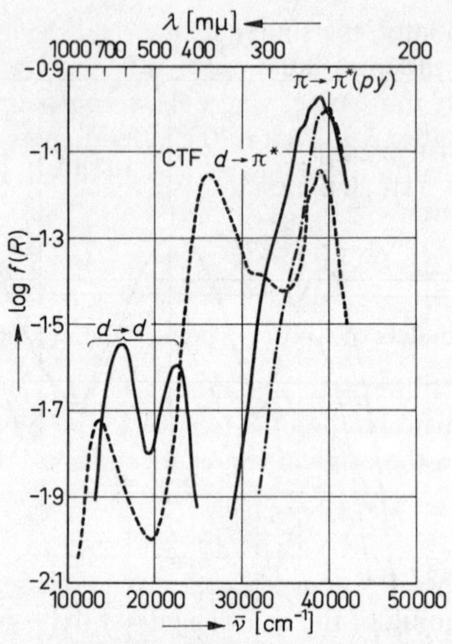


Figure A.9. Reflectance spectra of $[Mpy_3Cl_3]$ complexes. Powder, average particle diameter $< 5\,\mu$. Standard: $BaSO_4$.

——$[CrCl_3py_3]$, $\gamma = 5{\cdot}9 \times 10^{-3}$;
– – –$[MoCl_3py_3]$, $\gamma = 5{\cdot}3 \times 10^{-3}$;
–·–·–$[TlCl_3py_3]$, $\gamma = 4{\cdot}3 \times 10^{-3}$.

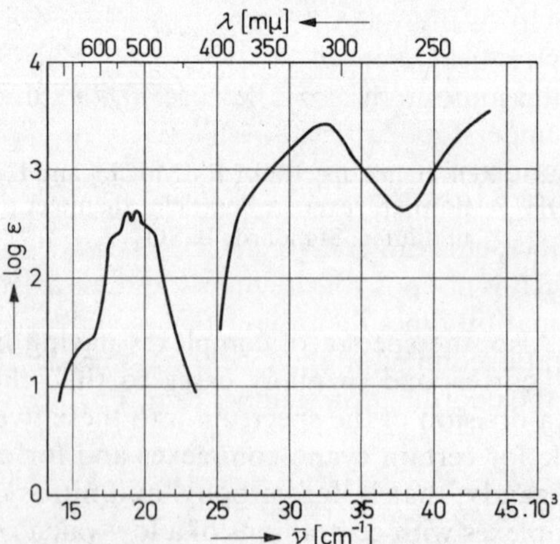


Figure A.10, Absorption spectrum of $[MnO_4]^-$.

As will be shown later, the simple ligand field theory provides a useful approximation for the interpretation of the long wavelength portion of the spectrum in only those cases where the complexes have spectra of the type indicated in Figure A.1. That is, the individual absorption properties of the components, although more or less modified, must still be recognizable in the complex.

## 1.3. The complex models of Kossel–Magnus and of Pauling

Two, at first sight alternative, points of view have been developed to explain the mechanism binding the ligands to the central ion, a problem not touched upon in the original classical coordination theory.

### (a) *The ionic model*

According to Kossel[14] and Magnus[15], whose ideas were developed about 1920, the binding of the components of the complex is, at least to the first approximation, determined by purely electrostatic forces. The components (central ion and ligands) were considered to be rigid, impenetrable spheres, having a well-defined radius. A complex ion can be described in terms of interactions between dipoles and point charges, point dipoles or combined charges, all envisaged as lying at the centre of these classical spheres. In the extended theory the polarization is considered by taking into account not only the permanent dipoles (moment $\mu_p$) but also the induced dipoles (moment $\mu_i$). The induced dipole moment can be estimated with the help of the polarizability $\alpha$ using the relation $\mu_i = \alpha F$, where $F$ is the electric field strength.

With the aid of this ionic model, a whole series of phenomena in complex chemistry can be understood, at least qualitatively. It is possible to give the heats of formation of the complexes as a function of charge, dipole moment and radius. Also such questions as to the maximum size of the coordination number, etc., can be handled successfully.

Using this model it is not possible to understand the existence of certain anomalous coordination types on a purely energetic basis. For example, the planar quadratic configuration appears alongside the tetrahedral configuration for the coordination number four. Further the appearance

14. W. Kossel: *Ann. Phys.*, **49**, 229 (1916); *Naturwiss.*, **7**, 339, 360 (1919); **11**, 589, 1923; *Z. Elektrochem.*, **26**, 314 (1920); *Z. Physik*, **1**, 395 (1920).
15. A. Magnus: *Z. anorg. allg. Chem.*, **124**, 289 (1922); *Phys. Z.*, **23**, 241 (1922); cf. also F. J. Garrik: *Philos. Mag.*, **9**, 131 (1930); **10**, 71, 76 (1930); **11**, 741 (1932) as well as E. J. W. Verwey: *Chem. Weekbl.*, **25**, 250 (1928).

of certain low-spin complexes, e.g. diamagnetic cobalt(III) cannot be explained.

(b) *The covalent model*

According to Sidgwick[16] and Pauling[17] the binding between the central ion and the ligand, at least for a series of strong complexes, such as $[Fe(CN)_6]^{3-}$ or $[Co(NH_3)_6]^{3+}$, is to be considered to be essentially covalent.

In such typical complex ions ['Durchdringungskomplexe' (penetration complexes) after Biltz] *electron pair bonds* as are known from the hydrogen molecule are assumed to exist between the central ion and the ligand. Sidgwick[16] made the central point of his discussion the attaining of noble gas configurations by the formation of electron pairs with electrons of the ligands. (In some cases, as when NO acts as the ligand, electron triplets were considered.) The concept of the effective number of electrons[18] plays an important rôle. According to the original treatment of Pauling, a certain number of electrons will be transferred from the ligand to the central ion. These can then form covalent bonds with the unpaired electrons remaining on the ligands. In 1931 Pauling carried out studies to determine which states of the central ion are available to accept transferred electrons and in which way equivalent bonds to the ligands can be generated by mixing (*hybridizing*) these states.

The results show that four equivalent bonds ($dsp^2$) can be generated from one $d$ state, two $p$ states and an $s$ state. Since their bonding directions lie in a plane, quadratic planar geometry should be expected. On the other hand, such a configuration would not be possible if an $s$ state and three $p$ states ($sp^3$) were available for hybridization. However, in this case four equivalent bonds to the corners of a tetrahedron can be produced. Two $d$, one $s$ and three $p$ states are required to give an octahedral configuration ($d^2sp^3$).

Pauling's concepts, which are based on the valence bond theory, explain the appearance of anomalous coordination types. Kimball[19] discussed the corresponding equivalent bonds for coordination numbers

16. N. V. Sidgwick: *J. chem. Soc.*, **123**, 725 (1923); *Trans. Faraday Soc.*, **19**, 469 (1923); *Chemy Ind.*, **42**, 316 (1923); cf. also: *The Chemical Elements and their Compounds*, Oxford University Press, London–New York 1950.
17. L. Pauling: *The Nature of the Chemical Bond*, 3rd edn., pp. 145 ff., Ithaca, New York, Cornell University Press, 1960; cf. also *J. Am. chem. Soc.*, **53**, 1367 (1931); **54**, 988 (1932); *Proc. natn. Acad. Sci. U.S.A.*, **14**, 359 (1928).
18. N. V. Sidgwick: *Trans. Faraday Soc.*, **19**, 472 (1923).
19. G. E. Kimball: *J. chem. Phys.*, **8**, 188 (1940).

from two to eight for the most diverse geometrical arrangements. For the example of coordination number 8, $d^4sp^3$ is found for the case that the ligands are located at the corners of an archimedian tetragonal antiprism (symmetry $D_{4d}$) or a dodecahedron (symmetry $D_{2d}$)* [20]. When ligands are located on the corners of a cube[19,21] $f$ states are needed in addition so that equivalent bonds may be formed. Even the simple ionic treatment shows that the cubic orientation is energetically less favourable than the antiprismatic configuration. The ion $[Mo(CN)_8]^{4-}$ is an example of a complex having antiprismatic or dodecahedral structure.

Using the concepts of Pauling many low-spin complexes (e.g. the appearance of diamagnetism in $[Co(NH_3)_6]^{3+}$) can be explained without difficulty (cf. section 2.2). Unfortunately, neither the ionic model nor the covalent model of Pauling can help us with our problem of interpreting optical properties. To solve this problem it is necessary to determine the term system of the complex ion in question, i.e. the position of the energy states on the energy scale must be ascertained. Not only the ground states, but also the higher excited states must be considered because the absorption bands correspond to electronic transitions from the ground states to these excited states. It is apparent that the ionic model can offer no help. Also, it is virtually impossible to acquire more information with the aid of Pauling's model because of the mathematical complexity of the quantum mechanical covalent scheme. The calculation of the many energy terms which are necessary for the analysis of optical spectra appears to be an almost completely hopeless task.

Note. The application of the valence bond theory to octahedral complexes is explicitly treated in Appendix 4, pp. 484 ff.

## 1.4. The extended ionic model: The ligand field

We have seen that for a large number of complexes one can differentiate between the individual absorption properties of the components of the complex (central ion and ligands)—even though these bands may be slightly changed from their original form (cf. section 1.2). When treating such cases theoretically, it is therefore useful to employ a model which in the first rough approximation assumes separated electron clouds for the central ion and the ligands. Such a model should also be applicable to typical (strong) complexes such as $[Co(NH_3)_6]^{3+}$ when the absorption spectra are of the type shown in Figure A.1. For the case of diamagnetic

---

* The symmetry notation is that of Schönflies, cf. Part B, pp. 269 ff.

20. G. H. Duffey: *J. chem. Phys.*, **18**, 1444, 746 (1950).
21. G. Racah: *J. chem. Phys.*, **11**, 214 (1943).

$[Co(NH_3)_6]^{3+}$ or $[Coen_3]^{3+}$ ion two weak bands in the long wavelength region are observed (Figure A.11) as well as an intense band at short wavelengths corresponding to the inner complex $NH_3$ or ethylenediamine, which has undergone a shift to longer wavelengths in comparison to the free $NH_3$ or ethylenediamine molecule.

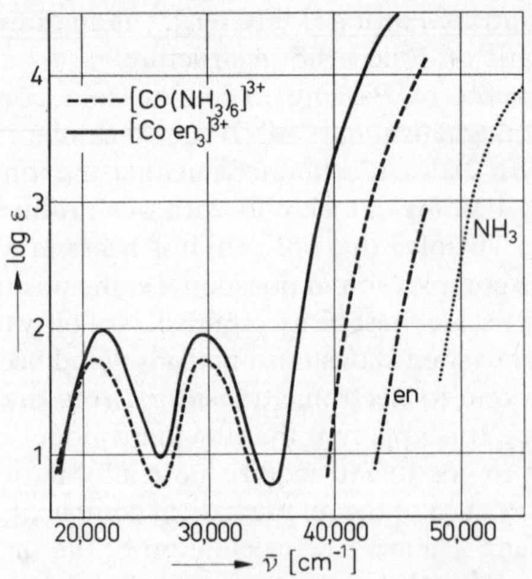


Figure A.11. Absorption spectra of $[Co(NH_3)_6]^{3+}$ and $[Coen_3]^{3+}$ together with those of the free ligands ethylenediamine and ammonia. $[Co(NH_3)_6]$ $(ClO_4)_3$ and $[Coen_3]_3$ $(ClO_4)_3$ in aqueous solution.

Kossel and Magnus (section 1.3) represented the components of the complex as rigid impenetrable spheres. The interaction between the spheres results from charges or dipoles which are assumed to lie at the centre of the ligand or the central ion. This ionic model is shown schematically in Figure A.12*a*. Contrasted to this is Pauling's model (sections 1.3 or 2.2) in which the electronic structure of the central ion and the ligands is explicitly used in the calculation. The states of the central ion are combined with one another in a suitable fashion to form equivalent (hybridized) bonds to the ligands. Figure A.12*c* shows this covalent model schematically. One can speak only of an electron cloud for the entire complex.

The *extended ionic theory* no longer considers the components of the complex as being structureless entities, as was done by Kossel and Magnus.

Rather, the influence of the electric field produced by the remaining components on the internal structure of one specific component is treated.

First of all, one is interested in the changes the electronic system of the central ion undergoes under the influence of the electric field produced by the ligands—the so-called *ligand field*. Therefore, the extended theory, i.e. the *crystal field* or *ligand field theory*, differs from the classical theory in that the electronic structure of the central ion is considered. The ligands are, however, to the first approximation still represented by point charges or by point dipoles*. A schematic representation of the extended ionic model is shown in Figure A.12*b*.

The electronic structure of most ions which can function as central ions is known for the *free-ion* case, that is for the case where these ions are not influenced by any sort of interaction with neighbours. Besides the energetically lowest-lying normal or ground state of the free ions, the energy of the excited terms of the particular ion is also known from the analysis of the corresponding emission spectra[22]. The term systems of the free ions can also be determined approximately theoretically by using the

---

* The designation ligand field theory is used by several authors when the electronic structure of the ligands is taken into consideration in a certain manner and the molecular orbital theory (cf. section 1.12) is combined with the model systems employed. For example one can discuss the contributions of $\pi$ bonds in this manner.

The theory denoted above as the extended ionic theory is called 'crystal field theory' by the aforementioned authors. Were this terminology to be followed, a 'crystal field theory' would be a theory where the electric field operating on the electrons of the metal ion would be assumed to be produced by point charges or by point dipoles.

According to the opinion of the present authors this terminology is unfortunate. For 'crystal field' the lattice field of Madelung is usually understood. As is well known, besides the crystal component under immediate consideration, all other crystal components contribute to this field as well. While it is meaningful to speak of a crystal field when treating crystallized complexes, this is no longer justified for the case of the isolated dissolved complex ion. Here the field of the ligands coordinated about the central ion, that is the ligand field, must be considered.

As detailed investigations have shown, however, it is to a good approximation permissible in the treatment of crystallized transition metal ion complexes to consider only those neighbours located immediately about the metal ion in a discussion of the electric field. More remote lattice components make only a negligible contribution to the electric field. As long as one is interested in the lowest-lying energy states, as is the case in a discussion of the optical absorption spectra, it suffices to consider only those components coordinated in the first sphere. Solely the microsymmetry at the site of the metal ion, which is determined by the geometrical arrangement of the nearest neighbours, is of importance.

We shall, therefore, speak of the ligand field theory and not of the crystal field theory. The ligand field is the electric field generated at the site of the transition metal ion by its nearest neighbours. Therefore, to the first approximation, one can consider the ligands to be point charges or point dipoles (extended ionic model) as is shown in Figure A.12*b*, or one can also take the electronic structure of the ligands into consideration as is discussed in section 1.12.

22. Cf. e.g. C. E. Moore: Atomic Energy Levels. *Natn. Bur. Stand. Civ.*, **467**, Vol. 1 (1949); Vol. 2 (1952); Vol. 3 (1958). Landolt-Börnstein: *Phys. chem. Tabellen*.

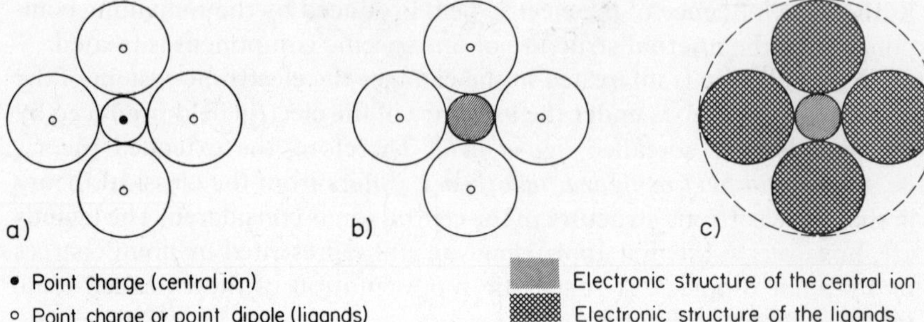


Figure A.12. Various complex models (schematic). (*a*) Ionic model (after W. Kossel and A. Magnus). Central ion and ligands treated as point charges or point dipoles. (*b*) Extended ionic model. Electron system of the central ion under the influence of the electric field of the ligands, which are considered to be point charges or point dipoles. (*c*) Covalent model (after L. Pauling). Complex ion treated as a single electronic system. Electronic structure of the central ion and the ligands is taken into account.

method* of Slater[23] (cf. Part B, Chapter 1). This is of course always necessary when an analysis of the emission spectra is not available.

Every term of the free ion can be characterized first by the energy content of the ion for the case that it exists in a state of this term and secondly by the orbital angular momentum and the spin of the electronic system. These parameters are denoted by the quantum numbers $L$ and $S$ or by the multiplicity which is conventionally calculated from $S$ using $M = 2S+1$. For the numerical values $L = 0, 1, 2, 3, \ldots$ it is conventional to use the letters $S, P, D, F, \ldots$. For the case of *Russell–Saunders coupling*, a state is characterized by the symbol $^{2S+1}L_J$, whereby the quantum number $J$ denoting the total angular momentum is often written as a lower right subscript. As an example, the well-known sodium D doublet corresponds to the transition $^2S_{1/2} \rightarrow {}^2P_{3/2}$. The term system explaining the Na D line is shown in Figure A.13.

Russell–Saunders coupling ($LS$ coupling) is valid to a rather good approximation for atoms which are not too heavy. The coupling of all of

* The term systems of the free central ions with configuration $d^n$ ($2 \leqq n \leqq 8$) can be expressed by *Racah* parameters $B$ and $C$. (G. Racah, *Phys. Rev.*, **62**, 438 (1942): **63**, 367 (1943); cf. also J. S. Griffith, *The Theory of Transition Metal Ions*, Cambridge University Press, 1961, pp. 83 ff.). These parameters are in turn sums of *Slater* radial integrals $F_2$ and $F_4$.

23. J. C. Slater: *Phys. Rev.*, **34**, 1293 (1929). Complete summarized statement in E. U. Condon and G. H. Shortley: *The Theory of Atomic Spectra*, 2nd edn., Cambridge University Press, London–New York 1953; cf. also C. W. Ufford: *Phys. Rev.*, **44**, 732 (1933). G. H. Shortley: *Phys. Rev.*, **43**, 451 (1933). M. H. Johnson, Jr.: *Phys. Rev.*, **43**, 632 (1933). D. R. Inglis and N. Ginsburg: *Phys. Rev.*, **43**, 194 (1933).

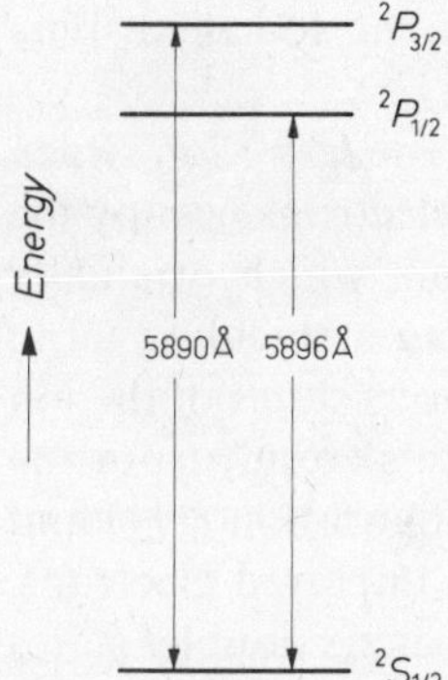


Figure A.13. The sodium D line.

the orbital angular momenta $\vec{l}_i$ of the electrons with one another and the coupling of the spins $\vec{s}_i$ with one another is large in comparison to the coupling between the orbital angular momentum and spin for the individual electrons*. The orbital angular momenta $\vec{l}_i$ of all the electrons are added together to give the resultant total orbital angular momentum $\vec{L}$ which determines the term character ($S, P, D, F, \ldots$). Similarly the spin angular momentum $\vec{S}$, which has an integral value when an even number of electrons are coupled and a half-integral value for an odd number of electrons. Finally $\vec{L}$ and $\vec{S}$ are added vectorially to give the total angular momentum $\vec{J}$ of the atom.

$$\vec{L} = \sum_i \vec{l}_i \qquad \vec{S} = \sum_i \vec{s}_i \qquad \vec{J} = \vec{L} + \vec{S}$$

The following expressions hold for the absolute values of the angular momenta:

$$|\vec{l}| = \sqrt{l(l+1)}\,\hbar \qquad |\vec{s}| = \sqrt{s(s+1)}\,\hbar$$

$$|\vec{L}| = \sqrt{L(L+1)}\,\hbar \qquad |\vec{S}| = \sqrt{S(S+1)}\,\hbar \qquad |\vec{J}| = \sqrt{J(J+1)}\,\hbar\,,$$

where $\hbar = h/2\pi$ ($h$ = *Planck's* constant). $s$, $l$ are the quantum numbers of a single electron and $S$, $L$ and $J$ are the quantum numbers of the many electron system. (For more details concerning the term symbolism for atomic structures, the reader is referred to Finkelnburg, *Introduction to Atomic Physics*, 10th Edition, Springer-Verlag, Berlin, 1964.)

In the following discussion we shall, except where otherwise stated, neglect the so-called multiplet splitting and denote with a dash in the term

* The use of an arrow over a quantity indicates that it is a vector quantity, e.g. $\vec{L}$ is the total orbital angular momentum vector.

system states belonging to different $J$-values. That is, the so-called spin–orbit coupling will not be considered*.

As an example, let us consider the term system of the $Cr^{3+}$ ion which arises from the electron configuration $3d^3$. That is, 3 electrons occupy the $3d$ subshell of the ion. When the electron interaction, which originally had been completely neglected or approximated using a shielding term, is taken into account, for the case of $LS$ coupling various states of the ion with multiplicities 2 and 4 (doublet and quartet terms) result from the $d^3$ configuration. The former belong to the spin configuration (↑↓↑) and the latter to the spin configuration with three (↑↑↑) unpaired electrons. In total there are two quartet states $^4P$ and $^4F$, as well as six doublet states $^2P$, $a^2D$, $b^2D$, $^2F$, $^2G$ and $^2H$. Their relative order on the energy scale is shown in Figure A.14. In agreement with *Hund's* rules†[24] the ground term is the $^4F$ term.

For the configuration $d^2$, for example, the terms $^3P$, $^3F$, $^1S$, $^1D$ and $^1G$ are present, whereby the $^3F$ term is the energetically lowest-lying.

Every state in the term system of an atom or an ion has a particular energy which appears as an eigenvalue of the Schrödinger equation for the problem in question. One or more eigenfunctions $\Psi$ belong to every energy eigenvalue. The number of linearly independent eigenfunctions gives the degeneracy of the particular state. A $P$ state has three eigenfunctions and is, therefore, a threefold orbitally degenerate state. The eigenfunctions $\Psi$ of a many-electron system can be expressed as antisymmetrized products of one electron eigenfunctions. Every atomic one-electron eigenfunction can be written as the product of a function of $r$,

---

* This is based on the fact that for transition metal ions the energy splitting associated with the spin-orbit coupling (cf. section 1.11) is usually of the order of a few $cm^{-1}$ to one thousand $cm^{-1}$. The *Stark* effect splittings, which lead to partial or complete removal of orbital degeneracy (cf. p. 22), lie between ca. 10,000–20,000 $cm^{-1}$. For complex ions having central ions of higher transition metal series the spin-orbit coupling generally can no longer be neglected.

† The rules given by F. Hund for determining the ground state are:

(1) $S$ assumes the maximum value allowed by the *Pauli* exclusion principle.

(2) Likewise $L$ assumes the maximum allowed value.

(3) If a restriction of the possibilities occurs because of the *Pauli* principle, and a choice must be made between reducing the value of $S$ or of $L$, then $L$ is to be reduced in value, so that $S$ can be maximized.

(4) The value of $J$ is to be chosen as small as possible in the first half of a subgroup and as large as possible in the second half. ($J_0 = L - S$ when less than half a shell is filled, $J_0 = L + S$ when more than half the shell is filled.

24. F. Hund: *Z. Physik*, **33**, 345 (1925); cf. also B. A. Eucken: *Lehrbuch* (*Textbook*) *der chemischen Physik*., Vol. 1, 3rd edn. S. 288, Akad. Verlagsges. Geest and Portig K. G., Leipzig 1949.

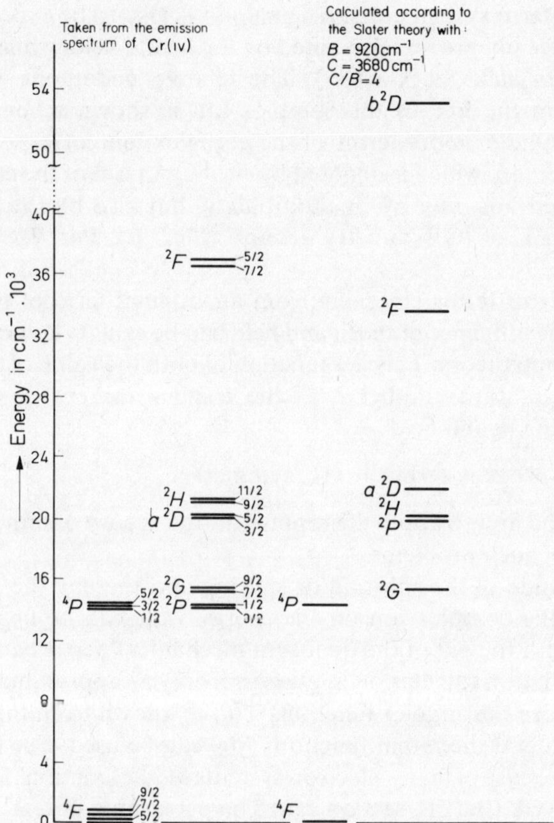


Figure A.14. Term system of the free $Cr^{3+}$ ion.

the distance of the electron from the nucleus and a function of the angular coordinates $\theta$ and $\varphi$.

$$\psi = R(r) \cdot Y(\theta, \varphi).$$

The radial function $R(r)$ is an analytic function for the one-electron system. For the case of a many-electron system it can be approximated by analytic functions. The angular function $Y(\theta, \varphi)$ is independent of the particular atom under consideration. For $d$ electrons ($l = 2$) there are five linearly independent functions $d_{xy}$, $d_{xz}$, $d_{yz}$, $d_{x^2-y^2}$ and $d_{z^2}$ (cf. Part B, section 1.1), for $p$ electrons ($l = 1$) three, $p_x$, $p_y$, $p_z$ and for $s$ electrons ($l = 0$) one such function (Part B, section 1.1).

Upon forming a complex the central ion is influenced by the electric field of the ligands and the states of its electrons are changed. A similar change of the term system of the original ion is also to be expected. Some terms of the free ion are merely shifted by the ligand field, while others are split (*inner complex Stark effect*). The change undergone during the transition from the free to the complex ion is shown schematically in Figure A.15. One or more terms of the complex ion correspond to each term of the free ion, which is denoted by ${}^{2S+1}L$. A term of the complex ion is characterized not only by its multiplicity, but also by the irreducible representation $\Gamma$ of its symmetry group: ${}^{2S+1}\Gamma$ (cf. Part B, Chapters 2 and 3).

The number of terms emerging from an original term of the free ion when under the influence of the ligand field can be *exactly* determined with the help of group theory. This is a function of both the value of the angular momentum quantum number $L$ of the original (parent) term and the symmetry of the ligand field:

$$\text{Number of progeny terms} = f(L, \text{symmetry}).$$

Bethe[25] was the first to apply the representation theory of finite groups to the solution of such problems.

The magnitude of the splitting or shifting and thereby the position of the states of the complex ion on the energy scale can be *approximately* determined with the aid of the quantum mechanical perturbation theory. Such a perturbation calculation is, of course, only an approximate method. For this purpose the angular functions $Y(\theta, \varphi)$ known from the hydrogen atom problem and the radial functions $R(r)$ can be used. For these latter functions in the case of many-electron systems *Slater* functions for example can be employed. (Part B, section 1.2.) The perturbing potential contains as parameters the central ion-ligand separations ($R$), the charge ($Ze$), as well as the dipole moments ($\mu$) of the ligands:

$$\text{Magnitude of the splitting or shifting} = F(R, Ze, \mu).$$

Such calculations were first carried out by Bethe[25], Kramers[26] and Van Vleck[27], although magnetochemical problems were in the foreground. Van Vleck[28] had already interpreted a spectroscopic effect, the so-called chromium doublet in chrome alum using the concepts sketched above. However, because of the influence of

25. H. Bethe: *Ann Physik* [5] **3**, 133 (1929); *Z. Physik*, **60**, 218 (1930).
26. H. A. Kramers: *Proc. Acad. Sci.* (*Amsterdam*), **32**, 1176 (1929); **33**, 953 (1930); *Compt. rend.*, **191**, 784 (1930).
27. J. H. Van Vleck: *Theory of Electric and Magnetic Susceptibilities*, Oxford University Press, Oxford–New York 1932; *Phys. Rev.*, **41**, 208 (1932); *J. chem. Phys.*, **3**, 803, 807 (1935); cf. also C. J. Gorter: *Phys. Rev.*, **42**, 437 (1932) as well as W. G. Penney and R. Schlapp: *Phys. Rev.*, **42**, 666 (1932); **41**, 194 (1932); **43**, 485 (1933).
28. R. Finkelstein and J. H. Van Vleck: *J. chem. Phys.*, **8**, 790 (1940).

*Pauling's* valence bond method no one thought to apply such considerations to interprete other properties of transition metal ion complexes, such as their absorption spectra. After 1950, beginning with the papers of Hartmann and Ilse[29], a rapid development took place, which showed that the ligand field theory is of great importance in understanding the chemistry of the transition metal compounds.

Generally speaking the separations $R$ in the complex ions are not known exactly. Even when they have been determined x-ray crystallographically for a crystallized compound, the values for the effective charges and effective dipole moments of the ligands remain speculative. The magnitude of the induced moment $\mu_i$, which together with the permanent dipole moment $\mu_p$ (the latter can usually be determined experimentally for the free ligand), gives the effective moment $\mu_{eff}$, is to a large extent undetermined. Estimating the induced moment with the aid of the relation $\mu_i = \alpha F$ ($\alpha$ = polarizability and $F$ = electric field strength) is not justified, as this relation is only valid for weak homogeneous fields. The fields at the ligand sites are assuredly strongly inhomogeneous. Determining the position of the states of the complex ion on the energy scale is also complicated by the indeterminacy of the field parameters. For this reason an absolute calculation of the magnitude of the splittings and shifts and thereby of the band positions is virtually impossible. In practice the magnitudes ($R$, $Ze$ and $\mu$) are kept as parameters in the calculation and the splitting is determined from experimental data (cf. pp. 71 ff., 77). Such parameters characterizing the field are implicitly contained in the quantities $\Delta$ or $Dq$ (cf. sections 1.7a and 1.8). These are used to measure the field for the case of cubic symmetry. Likewise the magnitudes of the splittings and shifts can be given as functions of these latter parameters (cf. Part B, Chapter 3, p. 340).

The change of position of the ground term of the complex ion in comparison to the free ion implies in general a change in the binding energy of the complex ion in addition to that obtained from the simple ionic model. This contribution to the energy $E_s$ (Figure A.15) will be denoted as the *ligand (crystal) field stabilization energy*.

The change in the ground term together with the changes of the higher terms is the determining factor for the *optical behaviour* of the complex ion because the maxima of the long wavelength absorption bands correspond to transitions from the ground state to higher states (as indicated in Figure A.15 with arrows).

The multiplicity of the ground term ($M = 2S+1$) determines roughly the magnetic behaviour of the complex ion*.

* The magnetic moment of the transition metal ion complexes is approximately dependent upon the value of $S$ or upon the number of unpaired electrons. (Cf. section 2.1).

29. F. E. Ilse: Dissertation Frankfurt a.M. 1946. F. E. Ilse and H. Hartmann: *Z. phys. Chem.*, **197**, 239 (1951); *Z. Naturf.*, **6a**, 751 (1951).

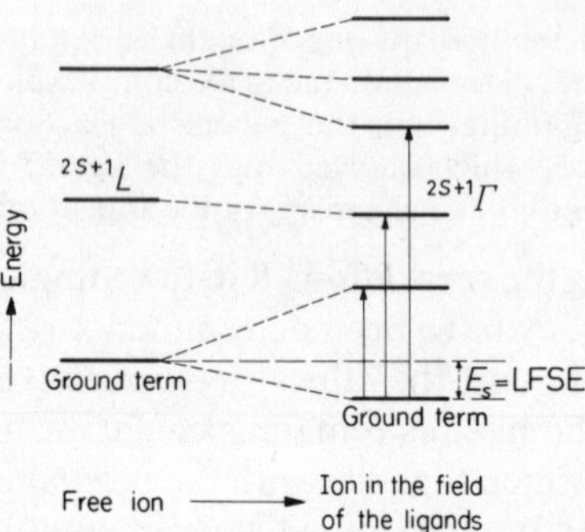


Figure A.15. The term system of the free ion and the term system of the ion in the ligand field (schematic). The shift produced by the classical electrostatic interaction (spherically symmetrical portion of the potential) is to the first approximation the same for all terms and is not shown. This shift does not influence the optical and magnetic properties.

The theory of free atoms and atomic ions is exhaustively treated in chapter 1 of Part B.

## 1.5. Term splitting under cubic symmetry

In this section the term splitting for the case of cubic symmetry of the ligand field will be considered. It is precisely this symmetry which is frequently, at least approximately realized for numerous complexes where the ligands are located at the vertices of an octahedron, $[MA_6]$, or a tetrahedron, $[MA_4]$.

A group theoretical analysis shows (Part B, chapters 2, pp. 255 ff. and 3, pp. 321 ff.) that an $S$ state remains unchanged in the cubic field, a $P$ state does not split, a $D$ state is split into two, an $F$ state into three and a $G$ state into four states (Figure A.16). This holds for octahedral ($O_h$) as well as for tetrahedral ($T_d$) symmetry.

A spherically symmetric ion has, when only the orbital angular momentum is considered, one linearly independent eigenfunction $\Psi$ for the $S$ state, three for the $P$ state, five for the $D$ state and seven for the $F$ state. In this sense, e.g. the $D$ state is fivefold orbitally degenerate. The degeneracy is partially or completely removed by the presence of the ligand field. When the symmetry is cubic, a fivefold degenerate $D$ state splits into two

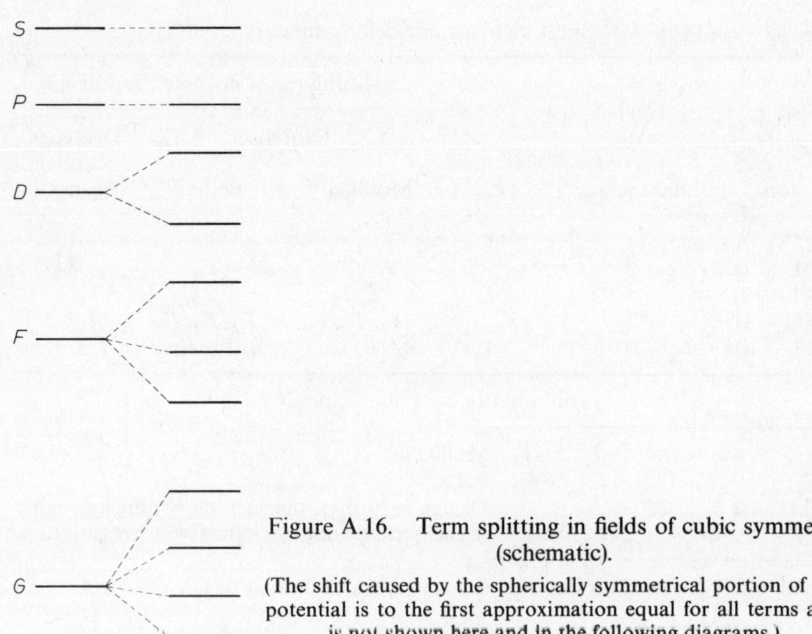


Figure A.16. Term splitting in fields of cubic symmetry (schematic).

(The shift caused by the spherically symmetrical portion of the potential is to the first approximation equal for all terms and is not shown here and in the following diagrams.)

states, a triply degenerate one and a doubly-degenerate state. Correspondingly, the sevenfold degenerate *F* state splits into two triply degenerate states and one nondegenerate state.

Table A.1 shows the splitting for cubic symmetry and gives the *degrees of degeneracy*. Every state in the ligand field is characterized by its symmetry species, that is by its irreducible representation $\Gamma$. In the table the notation is given according to Bethe[25] as well as according to Mulliken[30]. For the triply-degenerate states $T_1$ and $T_2$ the symbols $F_1$ and $F_2$ are sometimes used. For octahedral complexes, because of the presence of a centre of symmetry which is lacking for the tetrahedral case, the states are characterized by the additional index *g* (*gerade*, or even), e.g. $T_{2g}$ instead of $T_2$.

If the perturbation calculations are carried out to determine the energetic positioning of the states, it is seen that the order of the terms is always inverted for tetrahedral symmetry as compared to octahedral symmetry. (A qualitative explanation for the case of $d^1$ is given in section 1.71). For example, if the energetic order of the *F* state of the free ion is $A_{2g}$, $T_{2g}$, $T_{1g}$ for the octahedral ligand field, then it is $T_1$, $T_2$, $A_2$ for the case of tetrahedral coordination (Figure A.17).

30. R. S. Mulliken: *J. chem. Phys.*, **3**, 375, 506, 517, 586 (1935).

Table A.1. Splitting in a cubic field (symmetry $O_h$ or $T_d$)

| State of the free ion | Degree of orbital degeneracy | Splitting terms in cubic ligand field | | | |
|---|---|---|---|---|---|
| | | Number | Notation: Mulliken[30] | Notation: Bethe[25] | Degree of orbital degeneracy |
| $S(L = 0)$ | 1 | 1 | $A_1$ | $\Gamma_1$ | 1 |
| $P(L = 1)$ | 3 | 1 | $T_1$ | $\Gamma_4$ | 3 |
| $D(L = 2)$ | 5 | 2 | $E, T_2$ | $\Gamma_3, \Gamma_5$ | 2, 3 |
| $F(L = 3)$ | 7 | 3 | $A_2, T_1, T_2$ | $\Gamma_2, \Gamma_4, \Gamma_5$ | 1, 3, 3 |
| $G(L = 4)$ | 9 | 4 | $A_1, E, T_1, T_2$ | $\Gamma_1, \Gamma_3, \Gamma_4, \Gamma_5$ | 1, 2, 3, 3 |

Term notation for cubic symmetry:

| | | | | | |
|---|---|---|---|---|---|
| $A_1$ | $A_2$ | $E$ | $T_1$ | $T_2$ | Mulliken[30] |
| $\Gamma_1$ | $\Gamma_2$ | $\Gamma_3$ | $\Gamma_4$ | $\Gamma_5$ | Bethe[25] |
| 1 | 1 | 2 | 3 | 3 | Degree of orbital degeneracy = dimensionality of the corresponding irreducible representation |

(Small letters are used for one-electron states, e.g. $\gamma_3$ or $e$, $\gamma_5$ or $t_2$).

The ground states of the ions of the first transition metal series which come into question as central ions and arise from the configuration $3d^1$ to $3d^9$ have $L = 2$ or 3, except for $3d^5$ with $L = 0$. They are ordered symmetrically about the configuration $3d^5S$, which corresponds to the half-filled shell. The splitting diagrams are also symmetrical about $d^5$,

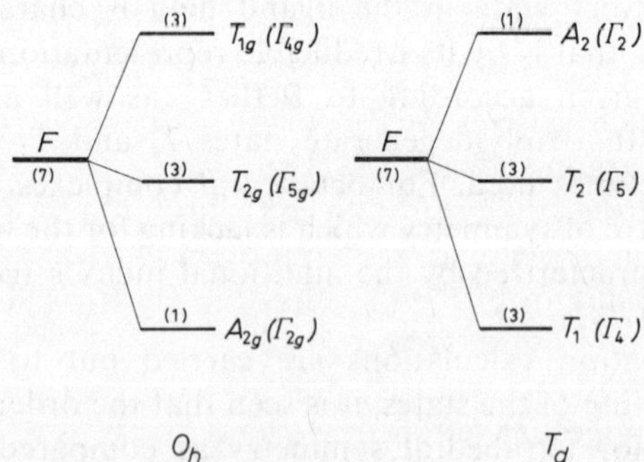


Figure A.17. The splitting of an $F$ state in octahedral ($O_h$) and tetrahedral ($T_d$) fields (schematic).

(The numbers in parentheses here and in the following diagrams give the orbital degeneracy of the corresponding states.)

as is schematically shown in Figure A.18 for an octahedral ligand field. $d^1$ and $d^9$ have equivalent splitting diagrams, but the energetic order of the states is inverted. (Similar arguments apply to the $d^2$, $d^8$ ; $d^3$, $d^7$ and $d^4$, $d^6$ pairs.) For $d^1$ the $T_{2g}$ state is the lowest lying state and the $E_g$ state is the excited state. For $d^9$ the $E_g$ state is the ground state and the $T_{2g}$ state the excited state. The splitting is therefore analogous for $d^{5-n}$ and $d^{5+n}$ ($n = 1, 2, 3, 4$), although their term order is inverted. These regularities can be coordinated with the *reciprocity between electrons and 'holes'*, which is known from atomic spectroscopy*.

| $d^1$ | $d^2$ | $d^3$ | $d^4$ | $d^5$ | $d^6$ | $d^7$ | $d^8$ | $d^9$ |
|---|---|---|---|---|---|---|---|---|
| $Ti^{3+}$ | $V^{3+}$ | $Cr^{3+}$ | $Mn^{3+}$ | $Fe^{3+}$ | $Co^{3+}$ | | | |
| | | $V^{2+}$ | $Cr^{2+}$ | $Mn^{2+}$ | $Fe^{2+}$ | $Co^{2+}$ | $Ni^{2+}$ | $Cu^{2+}$ |
| $^2D$ | $^3F$ | $^4F$ | $^5D$ | $^6S$ | $^5D$ | $^4F$ | $^3F$ | $^2D$ |

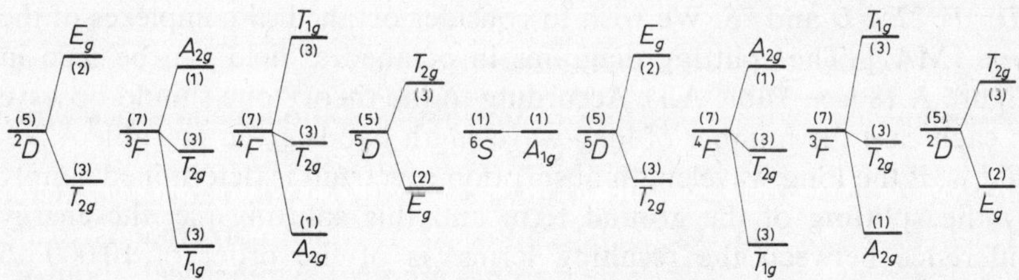


Figure A.18. Term splitting of the ground states of the ions $Ti^{3+}$ to $Cu^{2+}$ inclusive in an octahedral field (schematic).

The states which arise from a configuration with $n\,(\leq 5)$ $d$ electrons outside a closed argon shell are the same as arise from the configuration of $(10-n)$ $d$ electrons, that is, from a configuration of $n$ holes in the filled $d^{10}$ shell. The term inversion of configurations which are symmetric about $d^5$ is understandable from the fact that the electric field of the ligands affects electrons and holes oppositely.

If the term order in Figure A.18 is inverted (e.g. for $d^9$: $(E_g, T_{2g}) \rightarrow (T_2, E)$) it then applies for tetrahedral symmetry.

For an explanation of group theoretical terminology and methods and the theory of term splitting see Part B, chapter 2.

* The interaction between two holes gives the same Coulomb repulsion as that between two electrons. The magnetic moment corresponding to the orbiting of a positron has, however, the opposite sign as that of an electron, so that the multiplet structure is inverted (cf. Part B, chapter 1, p. 232).

## 1.6. Analysis of absorption spectra

### (a) *Simplified theory, consideration of the splitting of the ground state of the free ion*

In attempting to interpret the long wavelength splitting of the ground state of the free ion we shall first of all proceed according to the scheme developed above. That is, we shall attempt to utilize the rough approximation of studying the splitting and shifting of the lowest-lying terms of the free central ion in the electric field of the ligands.

When the separation of the first excited state of the free ion from the ground state is sufficiently large, it should be adequate to consider only the splitting of the ground term of the free ion. In this manner it should also be possible to understand the observed spectra at least qualitatively—that is, without dealing with finer details.

As an example, consider the trivalent ions $Ti^{3+}$, $V^{3+}$, $Cr^{3+}$, $Mn^{3+}$ and $Fe^{3+}$ [31] in the first half of the first transition metal series with the electronic configuration $d^1$ to $d^5$. The ground terms of the free central ions are $^2D$, $^3F$, $^4F$, $^5D$ and $^6S$. We wish to consider octahedral complexes of the type [$MA_6$]. The splitting diagrams in octahedral field can be seen in Figure A.18 (see Table A.1). According to the theory one should observe in each case the number of long wavelength bands given in Table A.2—that is, if the long wavelength absorption spectrum is determined simply by the splitting of the ground term and this splitting (i.e. the energy difference between the resulting terms) is of the order of 10,000 to 20,000 $cm^{-1}$*.

Table A.2. Number of bands expected from the theory when only the splitting of the ground state of the free ion is considered

| Ion | Ground state | Number of bands expected for $O_h$ symmetry |
|---|---|---|
| $Ti^{3+}$ | ($d^1$) $^2D$ | 1 |
| $V^{3+}$ | ($d^2$) $^3F$ | 2 |
| $Cr^{3+}$ | ($d^3$) $^4F$ | 2 |
| $Mn^{3+}$, $Cr^{2+}$ | ($d^4$) $^5D$ | 1 |
| $Fe^{3+}$, $Mn^{2+}$ | ($d^5$) $^6S$ | – |

* The assumption that the separation between the ground term and the higher terms of the free ion is large compared to the splitting in the ligand field is justified only for $Ti^{3+}$ and $Cu^{2+}$ ($d^1$ and $d^9$). In all other cases the excited states must be considered as well, as we shall show later.

31. Cf. e.g. H. Hartmann and H. L. Schläfer: *Z. Naturf.*, **6a**, 764 (1951).

Studies on complexes such as $[Ti(H_2O)_6]^{3+}$ [32] (Figure A.6) and $[Mn(H_2O)_6]^{3+}$ [31] (on crystalline caesium manganese alum) show in agreement with the theory *one* long wavelength band, while two long wavelength bands are observed for $[V(H_2O)_6]^{3+}$ [33] (Fig. A.7) and for $[Cr(NH_3)]^{3+}$ (Figure A.23). Finally the aqueous complex of iron(III) should be almost colourless in the long wavelength spectral region because the S state does not split. This is in qualitative agreement with experiment. For such a solution of iron(III) an absorption is found with an $\varepsilon$ value smaller by the factor 100–1000 than is the case for the other ions of the series. This applies to $Mn^{2+}$ acqueous complexes as well. These show an exceedingly weak rose colour. Finally one finds for the aqueous complexes of $Cr^{2+}$ exactly as for $[Mn(H_2O)_6]^{3+}$ only one long wavelength band[34].

These qualitative statements concerning the spectra (when applied to compounds having the configurations $d^4$ and $d^5$) are valid only for high spin complexes; that is, those where the ground state has the same multiplicity as the free ion (see section 1.6 f).

The positions of the individual bands in the spectrum are obtained correctly to the right order of magnitude for the respective complexes when ligand field calculations are carried out using plausible values for the separations and moments. For example, one calculates for $[Ti(H_2O)_6]^{3+}$ ion assuming a reasonable value of the Ti–$H_2O$ distance of 3 au* and a moment of 0·715 au* an absorption band ($^2T_{2g} \rightarrow {}^2E_g$) at 570 m$\mu$[29], and the experimental value[32] for the band maximum is 492 m$\mu$.

It is also understandable that the observed absorption bands are weak. The values of the molar decadic extinction coefficient $\varepsilon$ for the band maxima i.e. between 5 and 100–200, depending upon the complex. All of the transitions are between states of the same parity ($d$ electrons, $l = 2$, gerade $\nleftrightarrow$ gerade). On the basis of the *Laporte* rule such transitions between states of the same parity are forbidden as pure electronic transitions in systems having a centre of inversion.

They are possible, however, as can be shown by group theory, in combination with normal modes of the proper symmetry species (irreducible representation) of the complex ion (see also section 1.10). With the following examples it will be shown that the theory is also capable of explaining finer details of the spectra.

---

| * 1 au(atomic unit) | $= 4{\cdot}770 \times 10^{-10}$ e.s.u. | (charge) |
| --- | --- | --- |
| | $= 2{\cdot}52 \times 10^{-18}$ e.s.u. cm | (dipole moment) |
| | $= 0{\cdot}529 \times 10^{18}$ cm | (length) |

(cf. Appendix VI, p. 503).

32. H. Hartmann and H. L. Schläfer: *Z. phys. Chem.*, **197**, 116 (1951).
33. H. Hartmann and H. L. Schläfer: *Z. Naturf.*, **6a**, 754 (1951).
34. H. L. Schläfer and H. Skoludek: *Z. phys. Chem.* (Frankfurt), **11**, 277 (1957).

(b) *Spectra of* $[TiA_6]^{3+}$ *complexes* (A = $H_2O$, $CO(NH_2)_2$)

According to what has been outlined above, complexes of $Ti^{3+}$ having octahedral symmetry should show a weak band at long wavelengths because the $^2D$ term of the free ion splits into two terms in an octahedral field. Only the $^2D$ term arises from the $3d^1$ configuration. As the next higher term a $^2S$ term comes into consideration. This arises from the configuration $4s^1$ of the free $Ti^{3+}$ ion. This, however, lies 80,000 $cm^{-1}$ above the $^2D$ term, so that it is permissible to restrict a discussion of the long wavelength portion of the spectrum of $Ti^{3+}$ in actuality to only the $^2D$ ground state.

In spite of this, after a more careful study of the spectra of octahedral $Ti^{3+}$ complexes, one finds that either the absorption bands are strongly asymmetric, as is the case for hexaquo titanium(III) ion [32,35]. (Solutions of $TiCl_3 \cdot 6H_2O$ or titanium(III) sulphate in $H_2O$ contain this ion preferentially under appropriate conditions of concentration and pH) (cf. Figure A.6), or that a double structure is evidenced, as is the case for $[Ti(CO(NH_2)_2)_6]^{3+}$. Figure A.19 shows the spectrum of a thin section of crystalline $[Ti(Co(NH_2)_2)_6]I_3$ [35] and Figure A.20 the long wavelength band in the reflectance spectrum.

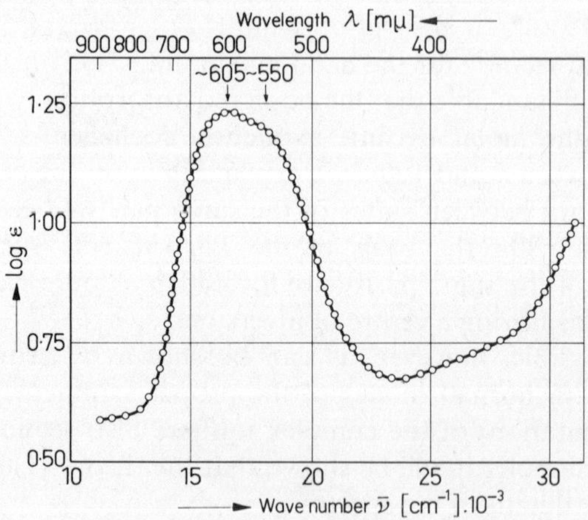


Figure A.19. Crystalline spectrum of $[Ti(CO(NH_2)_2)_6]I_3$. (Section d = 0·2 mm).

35. H. Hartmann, H. L. Schläfer and K. H. Hansen: *Z. anorg. allg. Chem.*, **289**, 40 (1957).

While solutions of $TiCl_3 \cdot 6H_2O$ show one broad asymmetric absorption band at about 20,000 cm$^{-1}$, similar to Figure A.6, one observes quite clearly in the reflectance spectra of the solid material two separate bands (Figure A.21) at 15,000 and 18,300 cm$^{-1}$. It is probable that no $[Ti(H_2O)_6]^{3+}$ units exist in crystalline Ti(III) chloride hexahydrate. The spectral position of both bands seems to indicate that $Cl^-$ is coordinated next to $H_2O$. That is, that $[Ti(H_2O)_4Cl_2]^+$ units are possibly present.

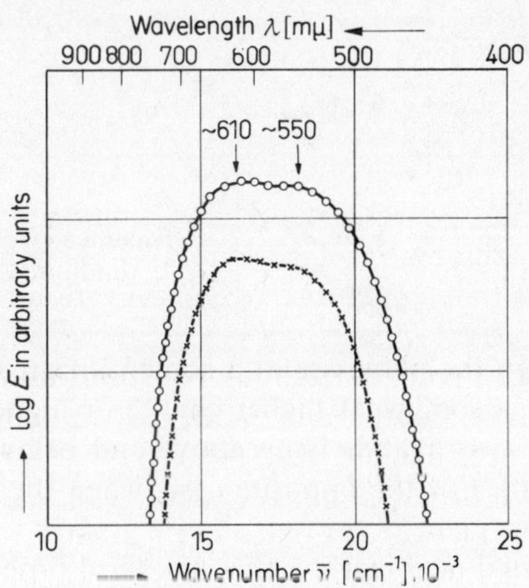


Figure A.20. Reflectance spectrum of $[Ti(CO(NH_2)_2)_6]I_3$ showing the long wave band only. (MgO-standard).

It follows from the findings on hexaquo and hexaurea Ti(III) ion that a strictly $O_h$ symmetrical ligand field cannot be present. According to the theory, the results of which are shown in Figure A.22, two bands appear only when at least tetragonal if not rhombic symmetry is present. With the transition from octahedral ($O_h$) to tetragonal ($D_{4h}$) symmetry of the ligand field, the degeneracy of the excited $^2E_g$ state is removed. It is split into two simple states: $(^2E_g(O_h) \rightarrow {}^2A_{1g}(D_{4h}) + {}^2B_{1g}(D_{4h}))$. The ground state $^2T_{2g}$ is likewise split into two states, a simple and a doubly degenerate state. $(^2T_{2g}(O_h) \rightarrow {}^2B_{2g}(D_{4h}) + {}^2E_g(D_{4h}))$. With the transition to rhombohedral ($D_{2h}$) symmetry, the degeneracy of the ground state is also completely removed $(^2E_g(D_{4h}) \rightarrow {}^2B_{1g}(D_{2h}) + {}^2B_{2g}(D_{2h});\ {}^2B_{2g}(D_{4h}) \rightarrow {}^2B_{3g}(D_{2h}))$*.

The order of the terms in Figure A.22 is valid for the case that the tetragonal symmetry arises from the fact that for the octahedron the

* Cf. Part B, p. 231 for term notation for $D_{4h}$ symmetry and Figure A.22 for $D_{2h}$ symmetry.

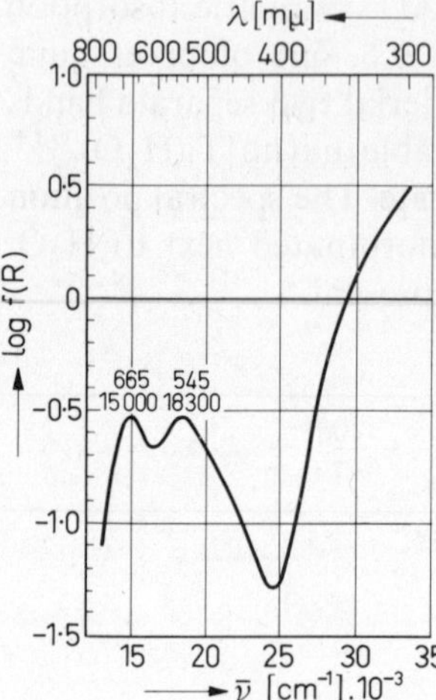


Figure A.21. Reflectance spectrum of $TiCl_3 . 6H_2O$ (undiluted).

four ligands lying in the plane are at a somewhat greater distance from the central ion (or a somewhat higher charge or higher effective dipole moment) than the two ligands lying above and below the plane (compressed octahedron). For the opposite case, when the two ligands lying above and below the plane of the four show a greater separation from the central ion (elongated octahedron), then the term order is $^2E_g$, $^2B_{2g}$ and $^2A_{1g}$, $^2B_{1g}$. For trigonal symmetry ($D_3$) (left side of Figure A.22) only the ground state splits into two states; the excited state remains doubly degenerate. When the octahedron is compressed in the direction of a threefold axis, the $^2A_1$ state is the lowest-lying one. For an octahedron elongated in the direction of a threefold axis, the order of the two lowest terms in Figure A.22 is inverted ($^2E$, $^2A_1$, $^2E$).

The experimentally observed asymmetry or double structure of the $Ti^{3+}$ band can therefore be explained on the basis of a splitting of the $^2E_g$ state through a tetragonal or rhombic component of the ligand field, where the magnitude of the splitting is ~2000–3000 $cm^{-1}$. (Separation of the two band maxima, which can be determined by an analysis of the asymmetric band*.)

Unfortunately no crystallographic data are available for the corresponding $[TiA_6]^{3+}$ complexes, so that nothing can be said about the positions of the ligands in the crystallized compounds and about possible deviations from the regular octahedral configuration. On the basis of

* Concerning the separation of asymmetric bands into the single bands from which they can be constructed by superposition the reader is referred to, e.g. C. K. Jørgensen, *Acta Chem. Scand.*, **8**, 1495 (1954); H. Siebert and M. Linhard, *Z. Phys. Chem.* (Frankfurt), **11**, 318 (1957).

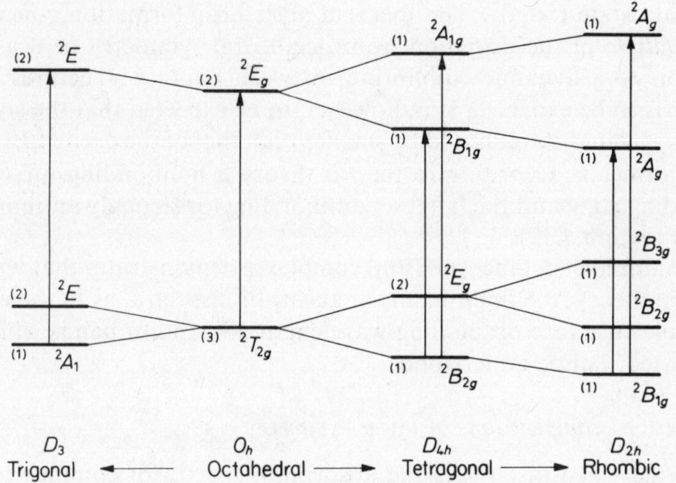


Figure A.22. Splitting of the $^2D$ state of $Ti^{3+}$ in fields of various symmetries (schematic).

spectroscopic findings, one is led to expect such a deviation from octahedral symmetry in the direction of tetragonal distortion for $[TiA_6]^{3+}$ complexes, as we have seen.

The *Theorem of Jahn and Teller* (see chapter 4), which has completely general validity, leads to the same result for $[TiA_6]^{3+}$ complexes. According to this theorem, any polyatomic entity, with the exception of linear molecules, is stable only when its ground state is orbitally nondegenerate. Generally speaking, according to Jahn and Teller (again excepting linear molecules) the totally-symmetric nuclear configuration of an electronic state of a molecular entity which is orbitally degenerate is unstable.

For stabilization to occur a descent in symmetry must take place until the degeneracy of the ground term is completely lifted. This can result either from a transition of the system to another nuclear orientation in the sense of a static reduction of the symmetry, or the entity can exist in several equivalent limiting forms which can be easily transformed into one another by vibrations. It is further required that the pertinent electronic states of each of these limiting forms be orbitally nondegenerate.

It follows that regular octahedral Ti(III) complexes cannot be stable because of the threefold orbital degeneracy of the $^2T_{2g}$ ground state. A distortion of the octahedron in the sense of a tetragonal orientation of the ligands is required so that (cf. Figure A.22) the ground state is a non-

36. H. A. Jahn and E. Teller: *Proc. R. Soc.*, **A 161**, 220 (1937). H. A. Jahn: *Proc. R. Soc.*, **A 164**, 117 (1938).

degenerate state $(^2B_{2g})$*. The theorem gives no information concerning the *magnitude* of the deviation from octahedral symmetry or if a static distortion or a dynamic equilibrium between limiting structures of the complex is to be expected. It is, however, to be expected that the splitting of the $T_{2g}$ ground state should be smaller than that of the excited $E_g$ state* for the former is according to the MO theory a nonbonding (or weakly antibonding state and the latter an antibonding (or strongly antibonding) state (cf. section 1.12).

The example of octahedral Ti(III) complexes demonstrates that with the aid of the theory, the finer details of absorption spectra, as in the case of the double structure of the long wavelength absorption bands which we have treated, can be understood.

### (c) *Spectra of complex ions of lower symmetry*

As a rule when discussing the absorption spectra of complex ions of lower symmetry one can, to begin with, always proceed from the central ion to octahedral symmetry $(O_h)$. By this means in most cases results are obtained which roughly describe the observed optical properties. Generally the largest effect is caused by the transition from the spherical symmetry of the free ion to octahedral symmetry. The supplementary effect of the fields of lower symmetry, which usually leads to a further splitting of the states, manifests itself often in the spectra through a *broadening* of the absorption bands or by their becoming *asymmetric*. For the usual absorption bands of the transition metal complexes, which have a relatively large half-width† ($\delta \sim 3500$–$4000\ \mathrm{cm}^{-1}$) the term splitting through the reduction of symmetry compared to the octahedral case must generally be of the order of several thousand $\mathrm{cm}^{-1}$ if the appearance of *separate* bands is to be discernible in the spectrum. One finds, e.g. in the spectrum of $[CrCl_3py_3]$ (cf. Figure A.9) only two long wavelength Cr bands, as in the

---

* Paramagnetic resonance measurements or studies of the accurate temperature dependence of the paramagnetic susceptibility give for the case of tetragonal splitting of the ground state values of ca. 200–900 $\mathrm{cm}^{-1}$ and it is also found that the simple $^2B_{2g}$ state is the ground state.

† The half-width, $\delta$, is the width of the absorption band in $\mathrm{cm}^{-1}$ when $\varepsilon$, the molar decadic extinction coefficient is half so great as $\varepsilon_{max}$, the value for the band maximum. For symmetrical bands the following expression holds:

$$(\bar{\nu}_{max} - \bar{\nu}) = \pm\frac{\delta}{2} \text{ for } \varepsilon = \frac{\varepsilon_{max}}{2}.$$

where $\bar{\nu}_{max}$ is the wavenumber at the band maximum, $\nu$ is the wavenumber at $\varepsilon_{max/2}$ (cf. G. Kortüm, *Kolorimetrie, Photometrie und Spektrometrie*, 4th Ed., Springer-Verlag, Berlin. 1962).

case of $[CrA_6]^{3+}$ complexes, although the symmetry here is by no means octahedral. In Figure A.23 the spectrum of octahedral Cr(III) hexammine ion $[Cr(NH_3)_6]^{3+}$ and that of trisethylenediamine complex ion $[Cren_3]^{3+}$

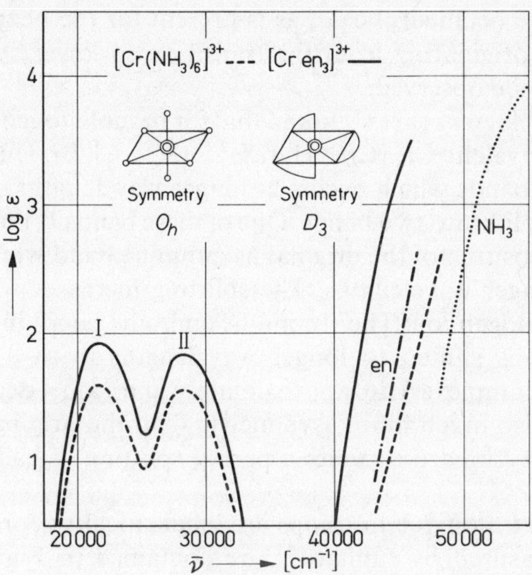


Figure A.23. Absorption spectra of $[Cr(NH_3)_6]^{3+}$ and $[Cren_3]^{3+}$. $[Cr(NH_3)_6]$ $(ClO_4)_3$ and $[Cren_3]$ $(ClO_4)_3$ in $H_2O$. The absorption of the free ligands ammonia and ethylenediamine is also shown.

can be seen. The latter ion, for which case the ethylenediamine molecule occupies two coordination positions and forms an inner complex pentagon, displays $D_3$ symmetry. One sees that the absorption spectrum of $[Cren_3]^{3+}$ shows a very small broadening of the two long wavelength chromium bands compared to the spectrum of $[Cr(NH_3)_6]^{3+}$. Also, the intensity of both bands is somewhat greater than that of the hexammine bands.

Likewise for other $M(A\text{–}A)_3$ complexes in which A–A is a bidentate chelating agent, e.g. $NH_2CH_2CH_2CH_2NH_2$ (trimethylenediamine) or the oxalate or malonate ion, one can understand the spectra by assuming octahedral ($O_h$) symmetry of the field for the first approximation, although the actual symmetry of the compounds is $D_3$ (cf. Figure A.3, *Spectrum of* $[Crox_3]^{3-}$, where likewise only two long wavelength chromium bands are to be seen). In all of these cases the additional splitting caused by the $D_3$ symmetry as compared to the octahedral symmetry $[A_{2g}(O_h) \rightarrow A_2(D_3)$; $T_{2g}(O_h) \rightarrow E(D_3)+A_1(D_3)$; $T_{1g}(O_h) \rightarrow E(D_3)+A_2(D_3)]$ is almost too small to be seen in the spectra.

For the *monoacidopentammine* and *diacidotetrammine* Cr(III) and Co(III) *complexes*, on the other hand, characteristic differences in the absorption spectra are observed, as will be shown for Cr(III) complexes. These can be interpreted with the aid of the additional splitting and shifting of the terms compared to the octahedral case, as is present for the hexammines. The term splittings originating from the transition to lower symmetries are large enough to be observed.

Linhard and his coworkers[37] found that for monohalogenopentammine complexes of trivalent Cr, $[Cr(NH_3)_5X]^{2+}$ (X = Cl, Br, I) the one of the two chromium bands which occurs at longer wavelengths, which we will denote with $I$, splits into two bands. One of these bands $I_b$ remains approximately at the position of the original hexammine band while the other $I_a$ is shifted to longer wavelengths. The splitting increases with increasing radius of the halogen ion. The second hexammine band, in the following denoted by $II$, is shifted to longer wavelengths (with a transition to monohalogen complexes) to approximately the same degree as is the band $I_a$. It is also much more asymmetric than the original hexammine band $II$, so that Linhard assumes a partial splitting of band $II$ into two bands.

In Figure A.24 these relationships are schematically represented, while the spectra measured by Linhard[37] are contained in Figure A.25. The

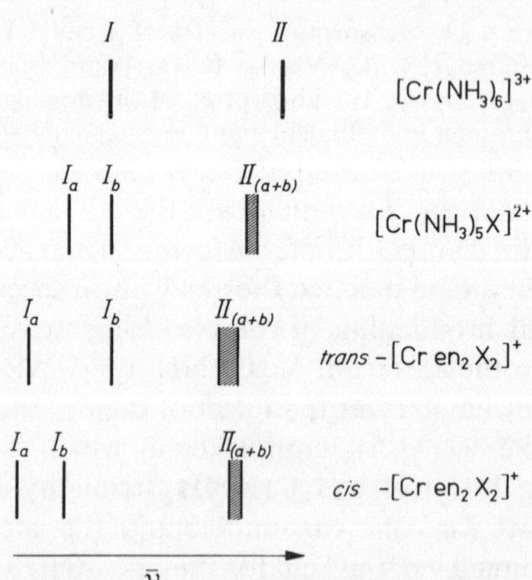


Figure A.24. Experimental band positions and splittings for Cr(III) complexes of various symmetries (schematic). (The band maxima are indicated by solid lines.)

37. M. Linhard and M. Weigel: *Z. anorg. allg. Chem.*, **266,** 49 (1951).

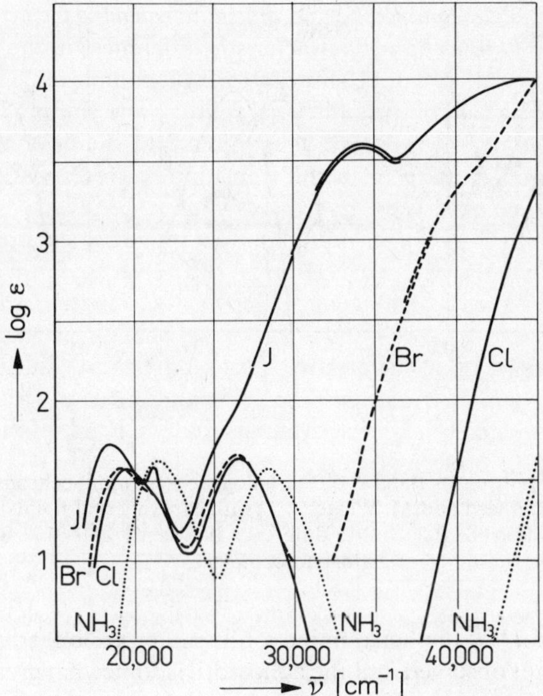


Figure A.25. Absorption of $[Cr(NH_3)_5X]^{2+}$ complexes (X = Cl, Br, I) after Linhard and M. Weigel[37]. The absorption of $[Cr(NH_3)_6]^{3+}$ is indicated by dotted curves for comparison.

splittings of band *I* can be best seen for the iodocomplexes. Band *II* is, however, covered to a large extent by the intense (charge-) electron-transfer band of the inner complex $I^-$. In Figure A.26 the splitting of band *I* is displayed once again on an expanded scale.

The spectrum of the *trans*-dihalogentetrammine chromium(III) ion is similar to that of the monohalogenpentammine ion. Band *I* of the hexammine ion splits into two bands $I_a$ and $I_b$, whereby the one lies approximately at the position of the original hexammine band *I* and the other is shifted to longer wavelengths.

Band *II* of the hexammine upon undergoing a transition to the *trans*-dihalogen complex evidences a red shift to about the same degree as band $I_a$ and has a shoulder, indicating the splitting into $II_a$ and $II_b$.

For *cis*-dihalogenotetramminechromium(III) ion one observes a shifting of band *I* to longer wavelengths. It is also split into bands $I_a$ and $I_b$.

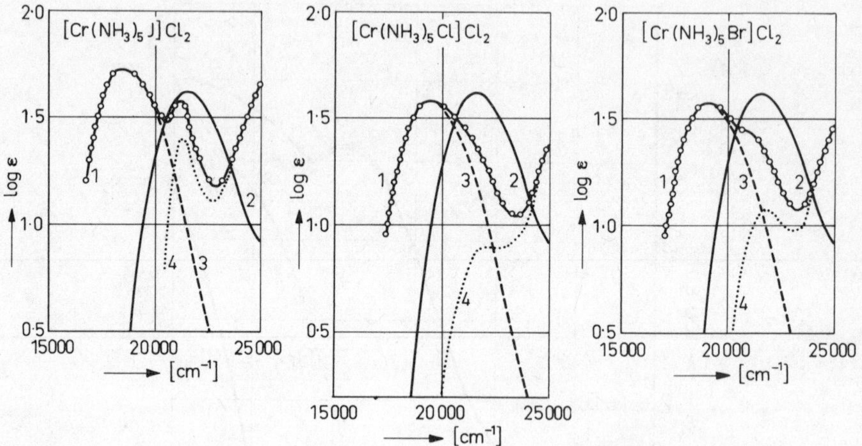


Figure A.26. Splitting of Band I of the halogenopentamminechromium(III) complexes after M. Linhard and M. Weigel[37]. 1 (with experimental points) = halogeno-complexes; 2 = $[Cr(NH_3)_6]^{3+}$; 2 (3 difference between 1 and 3) = Band $I_b$ of the halogeno-complexes.

Likewise band *II* of the hexammine ion is shifted to longer wavelengths. A splitting is not observed, but the half-width is somewhat greater than for the hexammine band. In Figure A.27 the spectra taken by Linhard[38] are given. One recognizes further that the spectrum of the *cis*-dihalogeno compound shows a greater intensity than does the *trans*-compound.

The spectra displayed were taken on the complexes *cis*-$[Cren_2Cl_2]^+$ and *trans*-$[Cren_2Cl_2]^+$ * and the spectrum of $[Cren_3]^{3+}$ is also presented for comparison.

These observations can be understood with the aid of the ligand field theory, as was shown by Hartmann and Kruse[39]. Figure A.28 shows schematically the term system of the ions $[CrA_6]$ (symmetry $O_h$), $[CrA_5B]$ (symmetry $C_{4v}$), *trans*-$[CrA_4B_2]$ (symmetry $D_{4h}$) and *cis*-$[CrA_4B_2]$ (symmetry $C_{2v}$).

According to the group-theoretical analysis, the two excited states split into a singly and a doubly orbitally degenerate state, respectively, with the transition from octahedral symmetry to $C_{4v}$ symmetry $T_{2g}(O_h) \rightarrow$

* The compounds *trans*- and *cis*-$[Cr(NH_3)_4X_2]$* are difficult to obtain (in contrast to the corresponding Co(III) complexes). Therefore, for Cr(III) the ethylenediammine compounds were used, for which cases the substitution of one $NH_2CH_2CH_2NH_2$ molecule for two $NH_3$ molecules is hardly discernible in the spectrum, as can be seen from the previous discussion:

38. M. Linhard and M. Weigel: *Z. phys. Chem.* (Frankfurt), **5**, 20 (1955).
39. H. Hartmann and H. H. Kruse: *Z. phys. Chem.* (Frankfurt), **5**, 9 (1955).

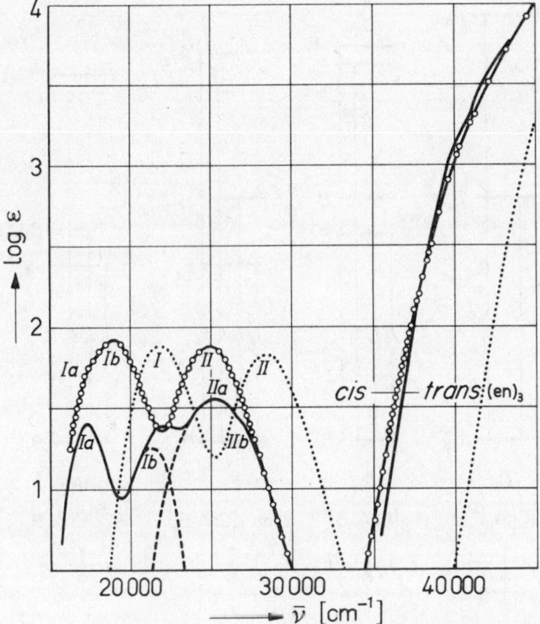


Figure A.27. Absorption of *cis* and *trans* $[Cren_2Cl_2]^+$ complexes after M. Linhard and M. Weigel[38]. The absorption curve for $[Cren_3]^{3+}$ is indicated (dotted) for comparison.

$B_2(C_{4v})+E(C_{4v})$; $T_{1g}(O_h) \rightarrow A_2(C_{4v})+E(C_{4v})$. It follows from a perturbation calculation that the $B_2$ state remains virtually at the position of the $T_{2g}$ state. However, the $E$ state, which arises from the $T_{2g}$ state, is shifted to lower energies, as is also the case for the progeny. Likewise, the two products of the $T_{1g}$ state are shifted to lower energies. Their separation is much smaller than the separation of the two states which arise from $T_{2g}$. The shift to lower energies is approximately equal to the amount the $E$ state, which arises from the $T_{2g}$ state, sinks. The transitions shown with arrows correspond to the maxima of the observed absorption bands (Figures A.24, A.25, A.26 and A.27), whereby $I_a$ and $I_b$ are observed as separate bands and $II_a$ and $II_b$, because of the small separation between $E$ and $A_2$ as one asymmetrical band. The absorption spectra of the complexes having symmetries $D_{4h}$ and $C_{2v}$ are interpreted in a similar manner. Also, the fact that the shifting of the terms of the *trans*-complexes as compared to the position of the octahedral terms is approximately twice as great as is the case for monochloropentammine complexes is correctly shown by the theory.

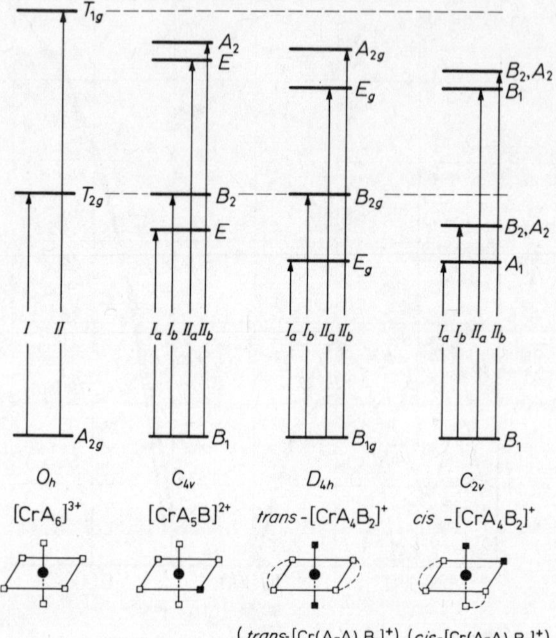


Figure A.28. Terms system of $Cr^{3+}$ ion in fields of various symmetries (schematic). (Only terms arising from $^4F$ without consideration of term interaction are shown.)

The difference in intensity between the spectra of the *cis*-$[Cren_2Cl_2]^{2+}$ and the *trans*-$[Cren_2Cl_2]^+$ ion can be qualitatively understood when one considers that the *trans*-compound has a centre of symmetry, while the *cis*-compound does not (cf. section 1.10).

The example cited shows how it is sometimes possible to distinguish between stereoisomeric forms of Cr(III) complexes by using absorption spectra in connection with group-theoretical considerations.

This is possible in an exactly analogous manner for $Co^{3+}$ complexes having the stated symmetries. These spectra have also been studied by Linhard[37,38,40], who ascertained that their general appearance is very similar to that of the corresponding Cr(III) complexes. The term systems are in this case similar, only that the term order for $[Co(NH_3)_6]^{3+}$ is $^1A_{1g}$, $^1T_{1g}$, $^1T_{2g}$ and the term notations for the compounds of lower

40. M. Linhard and H. Flygare: *Z. anorg. allg. Chem.*, **262**, 328 (1950). M. Linhard and M. Weigel: *Z. anorg. allg. Chem.*, **264**, 321 (1951); **271**, 101 (1952).

symmetry are correspondingly different[41]. One is then in the position to conclude from the spectrum if a compound of the type $[Co^{III}A_4B_2]$ exists in the *cis* or in the *trans* form.

### (d) *The complete theory: Consideration of the splitting of all the energetically low-lying states of the free ion*

To this point we have considered merely the splitting of the ground term of the free ion. This is only justified when the separation between the ground term and the higher terms of the free ion is large enough. As we shall see, this is truly the case for $Ti^{3+}$ and $Cu^{2+}$, but not for the remaining metal ions such as $V^{3+}$, $Cr^{3+}$, $Mn^{3+}$, $Fe^{3+}$, $Co^{2+}$ and $Ni^{2+}$, that is, for ions with the configurations $d^2$ to $d^8$.

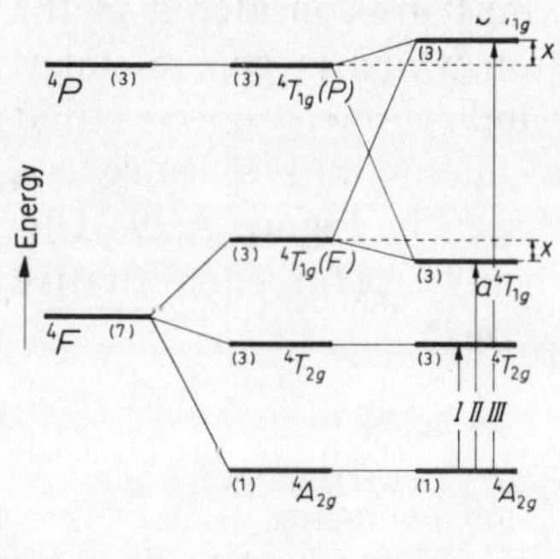


Figure A.29. Term system of $Cr^{3+}$ in a ligand field of octahedral symmetry. Only the quartet terms are considered (schematic).

As an illustrative example we shall discuss the situation for $Cr^{3+}$ complexes of octahedral symmetry. Only the splitting of the $^4F$ ground term of the free $Cr^{3+}$ ion in a ligand field of symmetry $O_h$ has been considered until now. In this approximation the appearance of both observed long wavelength chromium bands ($^4A_{2g} \rightarrow {}^4T_{2g}$; $^4A_{2g} \rightarrow {}^4T_{1g}$) could be explained (Figure A.17). Such a consideration represents, however, only the first rough approximation because as can be seen from the term system of the free $Cr^{3+}$ ion (Figure A.14), at least the $^4P$ terms, which lies at about

41. F. Basolo, C. J. Ballhausen and J. Bjerrum: *Acta chem. scand.*, **9**, 810 (1955). J. S. Griffith and L. E. Orgel: *J. chem. Soc.*, **1956**, 4981. C. J. Ballhausen and C. K. Jørgensen: *Kgl. Danske Vidensk. Selsk. mat. fysiske Medd.*, **29**, No. 14 (1955).

14,300 $cm^{-1}$ plays an additional rôle. Also, there are a number of doublet terms of which the $^2G$ and $^2P$ are the lowest lying. A complete treatment of the problem must take all these states ($^4F$, $^4P$, as well as $^2P$, $a^2D$, $b^2D$, $^2F$, $^2G$ and $^2H$) into consideration.

To begin with, let us consider only the $^4P$ term which has the same multiplicity as the ground term. It is not split by an octahedral field but is merely shifted. One should then observe another chromium band *III* in the near u.v. For most $Cr^{3+}$ complexes, this band is covered by the ligand absorption. However, for the $[Cr(H_2O)_6]^{3+}$ ion the band of the inner complex $H_2O$ rises at extremely short wavelengths and band *III* (Figure A.30) can easily be seen at 37,000 $cm^{-1}$. In order to bring the position of the chromium bands into agreement with experiment, the interaction of the two $^4T_{1g}$ terms, one of which arises from the $^4F$ and the other from the $^4P$ state of the free ion, must be considered in the perturbation calculation. This is an interaction which in general takes place between terms of the same multiplicity and symmetry species (that is, the same irreducible representation) (term interaction). It leads to a *repulsion* of the terms concerned, as can be seen in Figure A.29. This interaction must be taken into consideration in any case for a quantitative determination of the term systems of complex ions*.

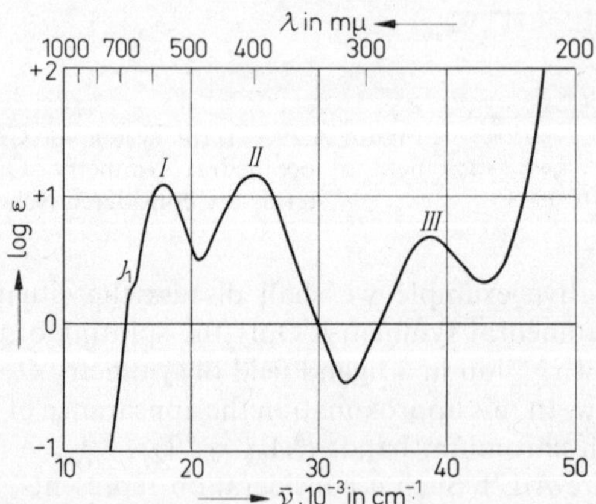


Figure A.30. Absorption spectrum of $[Cr(H_2O)_6]^{3+}$. $[Cr(H_2O)_6]$ $(ClO_4)_3$ in perchloric acid aqueous solution.

* Consideration of the term interaction means that in the determinant derived from the perturbation theory those nondiagonal elements which come from different terms of the same multiplicity and symmetry species must be included in the calculation.

With a more careful consideration of the spectra of octahedral or quasi-octahedral Cr(III) complexes, one recognizes a shoulder of small half-width and intensity on the declining portion of band *I* in the red, in case it does not lie at too long wavelengths. For trisoxalatochromium(III) ion (cf. Figure A.3) this can be seen as a distinct peak ($J_1$) at 14,340 cm$^{-1}$ (692 m$\mu$) and likewise for trismalonatocomplex at 14,450 cm$^{-1}$ (692 m$\mu$). For the hexaquo ion (Figure A.30) the band ($J_1$) is recognizable only as a shoulder on band *I* because this lies at a relatively long wavelength. In some cases, e.g. for the trisethylenediamine chromium (III) ion (Figure A.31) the band ($J_1$) shows a clear doublet structure (15,100 and 15,600 cm$^{-1}$; 665 and 640 m$\mu$*). This band is much less sensitive to a change of the ligand than are the chromium bands *I* and *II* (or *III*) and always appears in the region between about 12,800 and 15,600 cm$^{-1}$. For crystalline chromium compounds, e.g. as in the case of alum or ruby ($Cr^{3+}$ in $Al_2O_3$) one likewise observes absorption in this region—especially at low temperatures. In solutions, e.g. for the trisoxalato complex, the half-width of the sharp band is 300 cm$^{-1}$ (cf. Figure A.3)[42].

The doublet terms of the free ion ($^2P$, $a^2D$, $b^2D$, $^2F$, $^2G$ and $^2H$) also must be brought into consideration in the theoretical treatment. This was first done by Finkelstein and van Vleck[28]. The result of such a complete

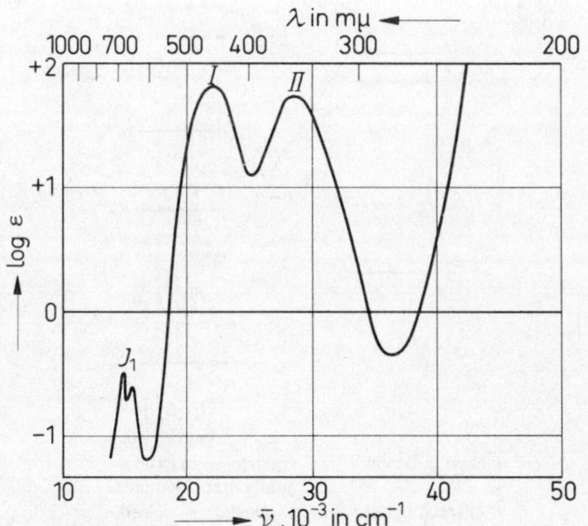


Figure A.31. Absorption spectrum of $[Cr\,en_3]^{3+}$. $[Cr\,en_3](ClO_4)_3$ in $H_2O$.

* With greater resolution of the spectrophotometer the vibrational structure of the band can clearly be seen.

42. Cf. e.g. H. L. Schläfer: *Z. phys. Chem.* (Frankfurt), **11**, 65 (1957).

treatment, whereby the interaction between the doublet states and interactions between the quartet states of the same irreducible representation is considered as schematically shown in Figure A.32. For the sake of clarity only the $^2G$ term, the lowest-lying of the doublet terms, is shown.

The sharp band at 12,800–15,600 cm$^{-1}$, a so-called *intercombination band*, results from the $^4A_{2g} \rightarrow {}^2E_g$ transition, an electronic transition between terms of different multiplicities (quartet → doublet). Besides being parity-forbidden, it is also spin-forbidden*. It is therefore quite understandable that the intensity of the intercombination band ($J_1$ in Figures A.3, A.30 and A.31) is down in intensity by a factor of approximately 100 in ε as compared with the intensity of the normal chromium bands *I*, *II* (or *III*).

The correct position below the $^4T_{2g}$ term (Figure A.32) of the $^2E_g$ state, arising from $^2G$, is obtained only when the total interaction with all $^2E_g$ states arising from the remaining doublet states of the free ion ($^2H$, $a^2D$, $b^2D$) is taken into consideration. Correspondingly, one must take account of the interaction of the $^2T_{2g}$ states with one another (they arise from $^2H$, $^2G$, $^2F$, $a^2D$, $b^2D$ states) and with the $^2T_{1g}$ states. (These arise in the ligand field from $^2H$, $^2G$, $^2F$ and $^2P$ states.)

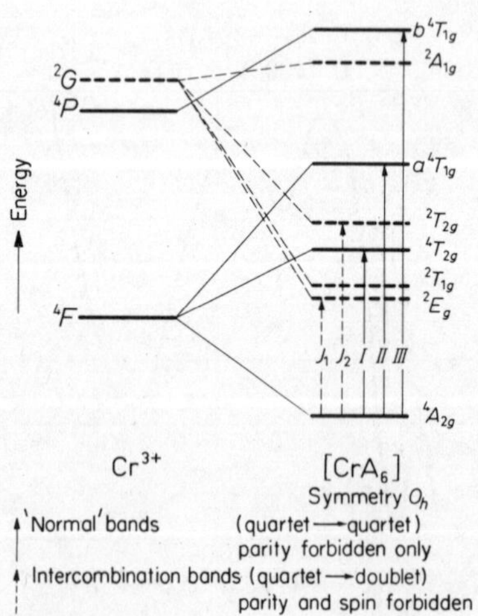


Figure A.32. $Cr^{3+}$ in a ligand field of $O_h$ symmetry, complete term system (schematic). ($^2G$ is the only one of the doublet terms shown.)

* See section 1.10.

In some cases one also observes a second intercombination band ($J_2$) corresponding to the $^2A_{2g} \rightarrow {}^2T_{2g}$ transition (Figure A.32). It is a sharp peak lying in the minimum between the bands *I* and *II* and is found, e.g., for crystalline chromalum at 21,000 cm$^{-1}$ [43] and for ruby[44] at essentially the same position. Figure A.33 shows absorption spectra of ruby single crystals. The intercombination band $J_2$ is clearly discernible.

From the example of $Cr^{3+}$ complexes it is seen that the details of the absorption spectra are satisfactorily described by the theory*.

When, as is usually the case, the terms of the free ion have a spacing comparable to the magnitude of the term splittings in the ligand field, then all of the energetically lower-lying terms of the free ion must be taken into account. In addition the interaction between terms of the same symmetry species and multiplicity must be considered. The restriction to considering only the splitting of the ground term of the free ion, with which for didactical purposes we began our discussion of optical properties, is only rarely justified, e.g. for $Ti^{3+}$ and $Cu^{2+}$. Otherwise it leads to erroneous results.

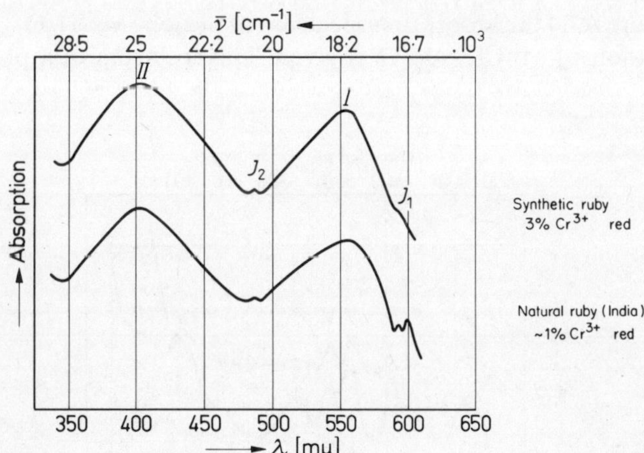


Figure A.33. Absorption spectra of single crystals of ruby (After A. Neuhaus[44]).

* The double structure of the intercombination band $^4A_{2g} \rightarrow {}^2E_g$ 'Chromium doublet' can perhaps be interpreted as the John–Teller splitting of the excited state. Both ruby lines $R_1$ and $R_2$ can be understood according to Tanabe and Sugano [*Disc. Faraday Soc.*, **26** (1958)] by taking into account trigonal symmetry and spin-orbit coupling.

43. F. H. Spedding and G. C. Nutting: *J. chem. Phys.*, **2**, 421 (1934); **3**, 369 (1935).
44. A. Neuhaus: *Z. Kristallogr.*, **113**, 195 (1960).

(e) *Pure intercombination spectra*, $[Mn(H_2O)_6]^{2+}$ *and* $[Fe(H_2O)_6]^{3+}$

During our discussion of the spectra of $Cr^{3+}$ ions we have become acquainted with intercombination bands, which, because they are spin-forbidden, have a low intensity. Such intercombination bands can occur when we exclude complexes having central ions with configurations $d^1$ and $d^9$ because terms of different multiplicity arise from the configurations $d^2$ to $d^8$. Frequently the weak intercombination band will be covered by the more intense bands corresponding to transitions between terms of the same multiplicity. Nevertheless, the theory allows predictions to be made in these cases as to if intercombination bands appear and as to in what spectral region they should lie.

There are also cases where the long wavelength portion of the spectrum shows only intercombination bands. Such *pure intercombination spectra* are observed for high-spin (see pp. 119 ff.) complexes of central ions with configuration $d^5$, e.g. $[Mn(H_2O)_6]^{2+}$ and $[Fe(H_2O)_6]^{3+}$, which have a sextet ground state.

The weak colouration of solutions containing these ions could be made qualitatively plausible by noting that the $^6S$ ground term of the free $Mn^{2+}$ or $Fe^{3+}$ does not split. (cf. section 1.6a.)

In Figure A.34 the absorption spectrum of the ion $[Mn(H_2O)_6]^{2+}$ taken in a solution of $MnCl_2 \cdot 4H_2O$ is given. The spectrum of a plate of the

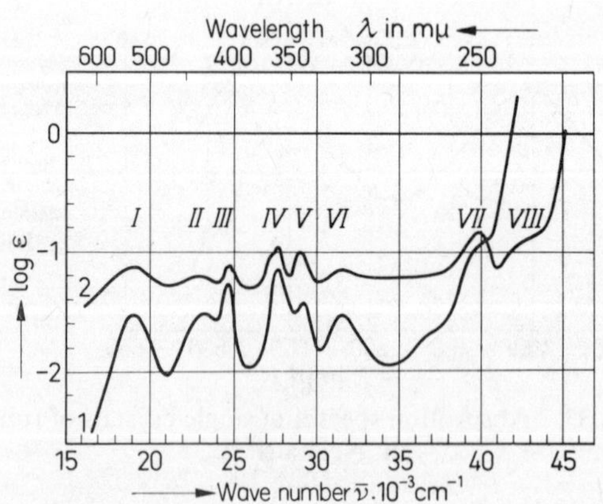


Figure A.34. Absorption spectra of $[Mn(H_2O)_6]^{2+}$.

1 = $MnCl_2$,$4H_2O$ in aqueous solution (c = 4.8 M).
2 = Crystalline spectrum $MnCl_2 . 4H_2O$ (crystalline plates $d$ = 1–7 mm).

crystalline tetrahydrate is shown as well. One recognizes eight exceedingly weak absorption bands having maxima log ε values less than −1.

If one considers the term system of free $Mn^{2+}$ (or $Fe^{3+}$) ion (Figure A.35, left side), it is seen that besides the $^6S$ ground term a series of quartet terms exists, in the order of increasing energy the states $^4G$, $^4P$, $^4D$ and $^4F$. Furthermore, doublet terms also belong to the configuration $d^5$. However, they lie energetically so high that we shall leave them out of consideration for the following discussion—especially when we do not consider very large ligand field strengths.

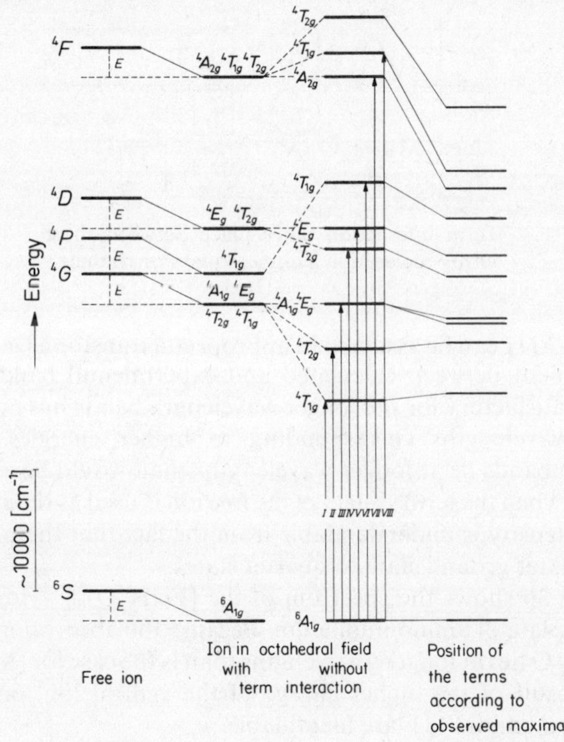


Figure A.35. Term system of an ion of $d^5$ configuration in an octahedral field. ($[Mn(H_2O)_6]^{2+}$) (schematic).

When the calculation is carried out without considering term interaction, the result is simply an equal shifting of all terms of the free ion. This is a singularity of the $d^5$ system with a half-filled spherically symmetric $d$ shell. If term interaction between the quartet terms of the same irreducible representation (symmetry species) is considered, one obtains the term system given in Figure A.35 after carrying out the perturbation calculation.

Table A.3 contains the terms of the free ion and the splitting terms for $O_h$ symmetry which are to be expected according to group theory.

Table A.3. Splitting of the sextet and quartet terms of a $d^5$ ion in an octahedral field

| | | | | |
|---|---|---|---|---|
| $^6S$ | $^6A_{1g}$ | | | |
| $^4P$ | $^4T_{1g}$ (circle) | | | |
| $^4D$ | $^4E_g$ (triangle) | $^4T_{2g}$ (square) | | |
| $^4F$ | $^4A_{2g}$ | $^4T_{1g}$ (circle) | $^4T_{2g}$ (square) | |
| $^4G$ | $^4A_{1g}$ | $^4E_g$ (triangle) | $^4T_{1g}$ (circle) | $^4T_{2g}$ (square) |

Term interaction takes place between those terms enclosed in a circle, square or triangle.

Bands *I–VIII* can be assigned to appropriate transitions in this manner. The agreement between calculated and experimental band positions is relatively satisfactory for the longer wavelength bands but poor for bands at short wavelengths corresponding to higher energies. The short-wavelength bands lie at longer wavelengths than would be expected from the theory, when the term system of the free ion is used as the starting point. The low intensity is understandable from the fact that the transitions are from the sextet ground state to quartet states.

Figure A.36 shows the spectrum of the $[Fe(H_2O)_6]^{3+}$ ion taken on a crystalline plate of ammonium alum. Because the absorption of the inner complex $H_2O$ lies at longer wavelengths than is the case for $[Mn(H_2O)_6]^{2+}$ ion as a result of the higher charge of the central ion, only the inter-combination bands *I–IV* are identifiable.

### (f) *Term crossing: critical field strength*

During our consideration of the complete term system of octahedral $Cr^{3+}$ complexes (Figure A.32) we have seen that *term crossing* can occur in the presence of the ligand field. Two progeny terms of the $^2G$ term, which lies energetically higher in the free ion than the $^4F$ ground term, lie lower than the second progeny term $^4T_{2g}$ of the $^4F$ term. These are the terms $^2E_g$ and $^2T_{1g}$. Likewise the $^2T_{2g}$ term, which also arises from the $^2G$ term

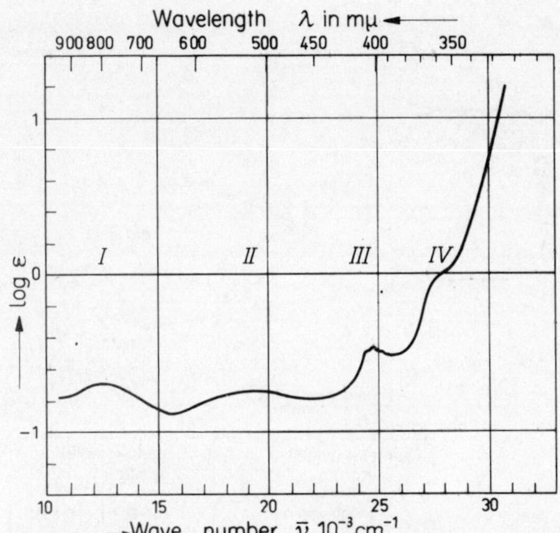


Figure A.36. Crystalline spectrum of $NH_4Fe(SO_4)_2 \cdot 12H_2O$ (crystalline plates $d$ = 1–5 mm.)

of the free ion, lies under the $^4T_{1g}$ term, the energetically highest term resulting from $^4F$.

Such term crossings can occur when the splitting is of the order of the separation of the terms of the free ion from one another. In certain cases it can happen that a term arising from an energetically higher-lying term of a different multiplicity in the free ion can become the ground term of the complex ion.

As an example we shall consider the situation for octahedral complexes of trivalent cobalt. The ground term of the free ion is a $^5D$ term, which in an octahedral ligand field splits into an energetically lower-lying $^5T_{2g}$ and a higher-lying $^5E_g$ state (Figure A.18). Besides the quintet terms (↑↑↑↑↑↓), there are also the triplet (↑↑↑↓↑↓) and the singlet terms (↑↓↑↓↑↓) arising from the configuration $d^6$ of the free ion. The calculation shows that for a certain *critical ligand field strength* on term $^1A_{1g}$ resulting from a higher-lying singlet term of the free ion crosses the $^5T_{2g}$ splitting term of the $^5D$ state (Figure A.37). For field strengths of the ligand field smaller than the critical value at point P, one observes in the absorption spectrum, mainly bands arising from transitions from the quintet ground term to higher quintet terms. The weak intercombination bands which may possibly be present are not taken into consideration. For higher field strengths, that is, to the right of the *cross-over point* P, one finds a spectrum that is essentially determined by singlet → singlet transitions.

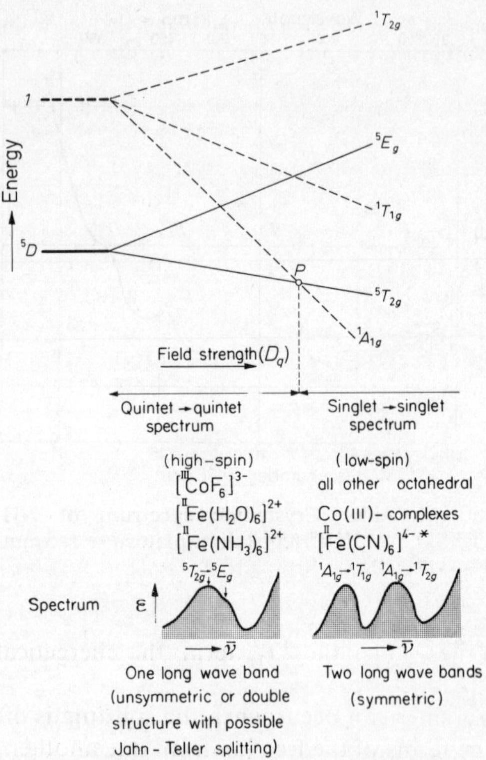


Figure A.37. Term system of octahedral cobalt (III) and iron(III) complexes (simplified, schematic).

The $[CoF_6]^{3-}$ ion has a quintet–quintet spectrum (a long-wavelength band $^5T_{2g} \rightarrow {}^5E_g$ that, similarly to the case for $Ti^{3+}$ (cf. section 1.6b), consists of two partial bands[45] as a result of the pronounced *Jahn–Teller* splitting). The remaining octahedral $Co^{3+}$ complexes, as $[Co(H_2O)_6]^{3+}$, $[Co(NH_3)_6]^{3+}$ (cf. Figure A.11) all show two long-wavelength bands $^1A_{1g} \rightarrow {}^1T_{1g}$; $^1A_{1g} \rightarrow {}^1T_{2g}$, that is, singlet → singlet spectra. In the first case we are to the left of the cross-over point P and in the second case to the right. With a continuous change of the strength of the ligand field, which can be thought of as the successive application of increasingly smaller ligands or of ligands with increasing electric dipole moments, the appearance of the absorption spectrum changes enormously when the point P is crossed.

45. F. A. Cotton and M. D. Meyers: *J. Am. chem. Soc.*, **82**, 5023 (1960).

The consequences of term crossing for the magnetic behaviour of the complexes will be discussed in Part A, Chapter 2.3.

Analogous term crossings can occur for octahedral complexes of the configurations $d^4$($Mn^{3+}$, $Cr^{2+}$), $d^5$($Fe^{3+}$, $Mn^{2+}$) as well as $d^7$($Co^{2+}$). Likewise, they are found for complexes of bivalent iron, which belong to the configuration $d^6$ of the central ion, as do the complexes of trivalent cobalt. For example for $Mn^{2+}$ and $Fe^{3+}$ complexes for suitable ligand field strengths a splitting product of a doublet term can become the ground term. While to the left of the crossing point P sextet → quartet intercombination spectra of low intensity are observed, more intense doublet → doublet spectra are found to the right of this point.

## 1.7. The strong- and weak-field methods

To this point we have always used the following procedure to determine the term system of a transition metal ion complex: One begins with the free central ion having the configuration $d^N$ ($N = 1$ to 9). For the example of an ion of an atom of the first transition series $N$ $d$ electrons ($l = 2$) occupy the $3d$ states of the free ion. For the case of $LS$ coupling (see p. 18) one obtains the term system of the free ion under the influence of the electron–electron interaction. This was at first neglected or approximated by a shielding field. Every state in the term system is denoted by the absolute value of the orbital angular momentum and the spin angular momentum or by the corresponding quantum numbers $L$ and $S$.

In a succeeding step the influence of the electric field of the ligands on the terms of the free ion is considered. The terms resulting from this method can be assigned to the original terms of the free ion from which they arise under the influence of the field. If the interaction between those terms of the same multiplicity and which belong to the same irreducible representation of the symmetry group of the ligand system is considered (cf. section 1.6d), an energetic change of the terms interacting with one another is found compared to the case without term interaction. Terms of this type can no longer be assigned to only one original term of the free ion having a particular $L$ value. They belong rather to the totality of the original terms taking part in the term interaction.

For the $Cr^{3+}$ ion in an octahedral environment an interaction takes place between the two $^4T_{1g}$ states (see Figure A.29). The states arising when this term interaction is taken into account can no longer be unambiguously assigned to the $^4P$ or $^4F$ state of the free ion. The lower-lying state (a $^4T_{1g}$) has predominantly $F$ and the higher-lying $b^4T_{1g}$) predominantly $P$ character.

$Mn^{2+}$ or $Fe^{3+}$ for the case of octahedral symmetry of the field (see section 1.6e) show three ${}^4T_{2g}$ states arising from ${}^4G$, ${}^4F$ and ${}^4D$ states of the free ion. In addition there are two terms ${}^4E_g$ which arise from the original states ${}^4G$ and ${}^4D$ (cf. Table A.3). As we have seen in this case it is only when term interaction is taken into account that a splitting occurs (cf. Figure A.35).

When attempting to determine the term system of the complex ion, it is assumed as a starting point that the *electron interaction is large compared to the effect of the ligand field.*

$$d^N \xrightarrow[\text{Electron interaction}]{\text{Step 1}} {}^{2S+1}L \xrightarrow[\text{Ligand field (term interaction)}]{\text{Step 2}} {}^{2S+1}\Gamma$$

This is the *weak-field case* and exists in various degrees of approximation as is explained in sections 1.6a and 1.6f (considering only the splitting of the ground term of the free ion, consideration of the higher terms as well, with and without term interaction).

A second possible procedure for determining the term system of a complex ion is obtained when it is assumed that the *ligand field is large compared to the interaction* between the single $d$ electrons; the $LS$ coupling in the free ion will be destroyed. In the first step the splitting of a *one-electron d state* in the electric field of the ligands is considered. The occupation numbers of the one-electron states appear formally in the ligand field theory in place of the quantum numbers $L$ and $S$. For the octahedral ligand field the $d$ electrons are distributed between the threefold orbitally degenerate $d\varepsilon(t_{2g})$ and the doubly degenerate $d\gamma(e_g)$ state. Taking account of the Pauli-principle, from the $N = 1$ to 9 $d$ electrons, $n \leqq 6$ can occupy the $d\varepsilon$ and $N-n \leqq 4$ the $d\gamma$ state.

In a second step the electron interaction as well as the so-called configuration interaction is taken into consideration

$$d^N \xrightarrow[\text{Ligand field (octahedral)}]{\text{Step 1}} d\varepsilon^n\, d\gamma^{N-n} \xrightarrow[\text{Electron interaction (configuration interaction)}]{\text{Step 2}} {}^{2S+1}\Gamma$$

This is the strong-field case, in which various approximation steps can be carried out (consideration of only the ligand field splitting, consideration of the electron interaction with and without taking into account interaction of the states of the same multiplicity and species arising from various ligand field configurations $d\varepsilon^n\, d\gamma^{N-n}$).

In the following section the behaviour of a one-electron $d$ state in a ligand field will be discussed in detail. In conjunction with this the many-electron problem will be treated from the standpoint of the strong-field case.

(a) *One-electron d states in a ligand field*

The $d$ electrons outside the closed shells play an important rôle in determining the physical and chemical properties of transition metal complexes. A *one-electron d state* ($l = 2$) is for the free ion, i.e., the spherically symmetrical case, a fivefold degenerate state with regard to the orbital angular momentum. It possesses five linearly independent eigenfunctions $\psi$. Each of these eigenfunctions may be written as the product of a function $R(r)$, dependent upon the distance of the electron from the nucleus and a function $Y(\theta, \phi)$, containing the angular coordinates of the electron (see Part B, pp. 219 ff.). The angular dependent part of the eigenfunction, the *angular function*, is of particular interest for the following discussion and will, therefore, be considered more thoroughly.

Figure A.38 shows a sketch of the angular dependent functions for the five $d$ states. The lines denote boundary surfaces * on which these angular functions have the same value. The signs in Figure A.38 are those of the functions.

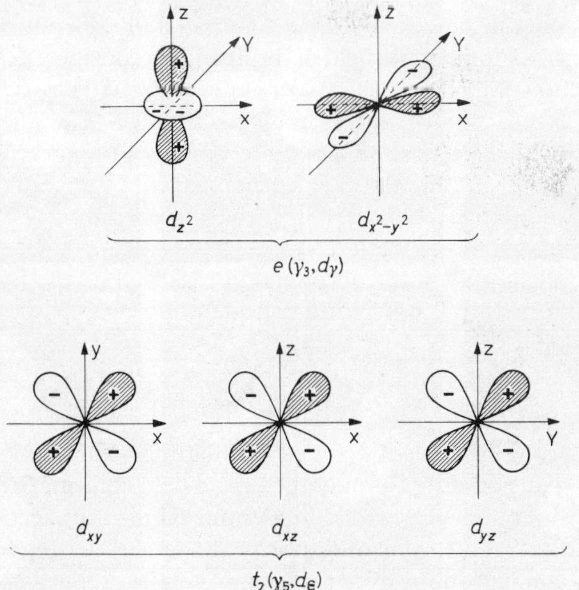


Figure A.38. The five $d$ functions.

* Cf. C. A. Coulson: *Valence*, pp. 20 ff., Oxford University Press, 1953.

One can understand the situation especially well from Figure A.39, in which photographs of spatial models of the functions, together with the *s* and the three $p_x$, $p_y$ and $p_z$ functions, are shown.

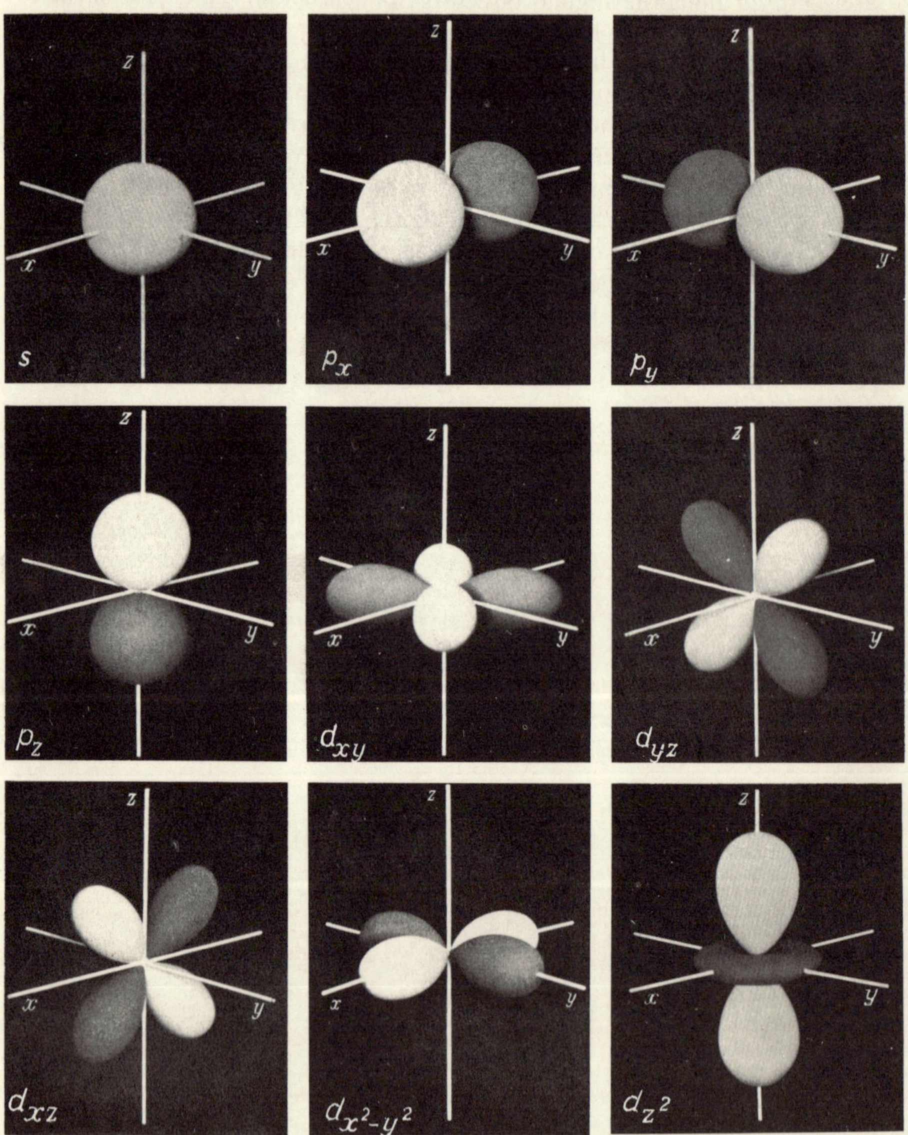


Figure A.39. Models of the angular functions (light areas indicate positive values of the functions, dark areas negative values). (From L. E. Sutton, trans. by E. Fluck: *Chemische Bindung und Molekülstruktur. Anorganische und allgemeine Chemie in Einzeldarstellungen*, Bd. III, Springer-Verlag, Berlin, Göttingen-Heidelberg, 1961.

On the basis of their symmetry properties, one distinguishes between two types of $d$ orbitals; these are denoted by the symbols $d\varepsilon$ and $d\gamma$. For $d\varepsilon$ $\gamma_5$ or $t_2$ are sometimes used and for $d\gamma$ $\gamma_3$ and $e$. The three $d\varepsilon$ functions are sketched along the angular bisectors of the cartesian axes. The two $d\gamma$ functions, on the other hand, have their maximal amplitudes along the cartesian axes*:

$$(\gamma_5, t_2)\; d\varepsilon: d_{xy}, d_{xz}, d_{yz}$$
$$(\gamma_3, e)\; d\gamma: d_{x^2-y^2}, d_{z^2}.$$

$|\psi|^2$ is a measure of the position probability density of the electron. From the shape and extension of the amplitude functions one sees that the position probability for a $d\varepsilon$ electron is the greatest in the regions between the cartesian axes. For a $d\gamma$ electron, however, it is maximal in the direction of the cartesian axes.

The analytical expressions for the five real $d$ functions[46] are:

$$d\gamma\begin{cases} d_{z^2} = d_0 = R_{nd}(r)\cdot\sqrt{5/\pi}/4\cdot(2z^2-x^2-y^2)/r^2 = \phi_1 \\ d_{x^2-y^2} = \frac{1}{\sqrt{2}}(d_2+d_{-2}) = R_{nd}(r)\cdot\sqrt{15/\pi}/4\cdot(x^2-y^2)/r^2 = \phi_2 \end{cases}$$

$$d\varepsilon\begin{cases} d_{xy} = \frac{1}{i\sqrt{2}}(d_2-d_{-2}) = R_{nd}(r)\cdot\sqrt{15/\pi}/2\cdot xy/r^2 = \phi_3 \\ d_{xz} = -\frac{1}{\sqrt{2}}(d_1-d_{-1}) = R_{nd}(r)\cdot\sqrt{15/\pi}/2\cdot xz/r^2 = \phi_4 \\ d_{yz} = -\frac{1}{i\sqrt{2}}(d_1+d_{-1}) = R_{nd}(r)\cdot\sqrt{15/\pi}/2\cdot yz/r^2 = \phi_5 \end{cases}$$

The indices 0, $\pm 1$, and $\pm 2$ indicate that the one-electron $d$ orbitals have the values for the magnetic orbital angular momentum quantum number $m_l = 0, \pm 1, \pm 2$. The $d_{m_l}$ functions have the form†:

$$d_{m_l} = R_{nl}(r)\cdot Y_{l,m_l}(\theta,\phi)$$
$$= R_{nl}(r)\cdot\Theta_{l,m_l}(\theta)\cdot\Phi_{m_l}(\phi)$$

Only two of the three possible functions $d_{x^2-y^2}$, $d_{z^2-x^2}$ and $d_{z^2-y^2}$ can be linearly independent at one time.

*Although each pair of functions could be used, conventionally $d_{x^2-y^2}$ and a hybrid of the other two denoted by $d_{z^2}$ are employed.

$$d_{z^2} = \frac{1}{\sqrt{2}}[d_{z^2-x^2}+d_{z^2-y^2}].$$

From Figure A.38 or A.39 it is seen that the form of $d_{z^2}$ is different from that of the other functions.

†Cf. Part B, pp. 206 ff.

46. Cf. e.g. L. Pauling and L. B. Wilson: *Introduction to Quantum Mechanics*, pp. 132 ff., McGraw-Hill, New York–London, 1935, 935.

with

$$\Theta_{l,m_l} = (-1)^l \sqrt{\frac{2l+1}{2} \frac{(l+m_l)!}{(l-m_l)!}} \frac{1}{2^l l!} \frac{1}{\sin^{m_l}\theta} \frac{d^{l-m_l}}{(d\cos\theta)^{l-m_l}} \sin^{2l}\theta$$

and

$$\Phi_{m_l}(\phi) = \frac{1}{\sqrt{2\pi}} e^{im_l\phi}.$$

For $d$ electrons the orbital angular momentum quantum number is $l = 3$. $Y_{l,m_l}$ are the spherical harmonics. For the radial functions $R_n(r)$ for example *Slater* functions[47] having the form

$$R_{nl}(r) = N \cdot r^{n'-1} e^{-\frac{Z-\sigma}{n'} r}$$

may be used. $N$ is a normalization factor, $n'$ the effective principal quantum number, $Z$ the nuclear charge and $\sigma$ the shielding constant.

For the spherically symmetric free ion all five $d$ states have the same energy. That is in general no longer the case when going from spherical symmetry to lower symmetries, for example when the free ion is brought into an electric field that is generated by ions or dipolar molecules having a specific geometric arrangement.

In order to obtain a qualitative survey of the situation for octahedral ($O_h$) ligand field symmetry, we shall represent the ligands by negative charges (or electric dipoles whose negative ends point to the central ion) located along the Cartesian axes and concentrated about the central ion situated at the origin of the coordinate system. Each Cartesian axis connects a pair of vertices of the octahedron lying opposite to each other. They are therefore identical with the three fourfold axes (Figure A.40). If one considers the interaction of the negative charge of the ligands with the negative $d$ electrons of the central ion, it is seen that the electrostatic repulsion for an electron in the $d\gamma$ states must be greater than that of an electron located in the $d\varepsilon$ states. The electron density is the greatest in the direction of the cartesian axes in the former case (Figure A.38, A.39 and A.41), while in the latter it is maximal in the region between the ligands (Figures A.38, A.39, and A.42). In this manner a splitting of the $d\varepsilon$ and $d\gamma$ states, which are degenerate in the spherically symmetric case, occurs, after which the progeny of the $d\varepsilon$ states are the energetically lower lying.

Discussing this in somewhat more detail, we see that the two ligands located on the $z$ axis at $+z$ and $-z$ exert the same influence on an electron in the $d_{x^2-y^2}$ or $d_{xy}$ state. The four other ligands in the $xy$ plane show a

47. J. C. Slater: *Phys. Rev.*, **36**, 57 (1930); cf. also B. H. Eyring, J. Walter and G. E. Kimball: *Quantum Chemistry*, p. 162, John Wiley, London 1954.

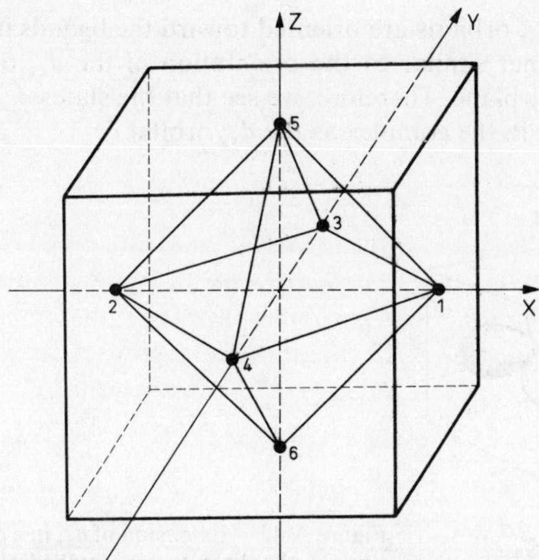


Figure A.40. Octahedron with its coordinate system inscribed in a cube.

stronger interaction with $d_{x^2-y^2}$ than with $d_{xy}$. An electron in the $d_{x^2-y^2}$ state has a charge distribution with a maximum in the direction of the bonds to the ligands located at $+x$, $-x$ and $+y$, $-y$. For a $d_{xy}$ electron the maximum electron density lies in the direction of the angular bisectors between the bonds. It follows that the $d_{x^2-y^2}$ state and the $d_{xy}$ state of the octahedral complex ion have different energies. The latter is the energetically lower-lying because of the lower electrostatic repulsion energy.

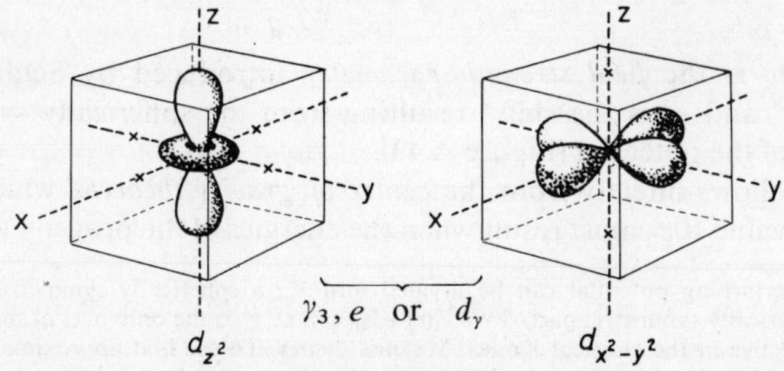


Figure A.41. Extension of $d_{z^2}$ and $d_{x^2-y^2}$ in a cube. × = position of the ligands for octahedral configuration.

The $d_{xz}$ and $d_{yz}$ orbitals are oriented toward the ligands in the $xz$ or $yz$ plane in a manner similar to the orientation of the $d_{xy}$ orbitals to the ligands in the $xy$ plane. Therefore, we see that the states $d_{yz}$ and $d_{xz}$ have the same energy in the complex as the $d_{xy}$ orbital.

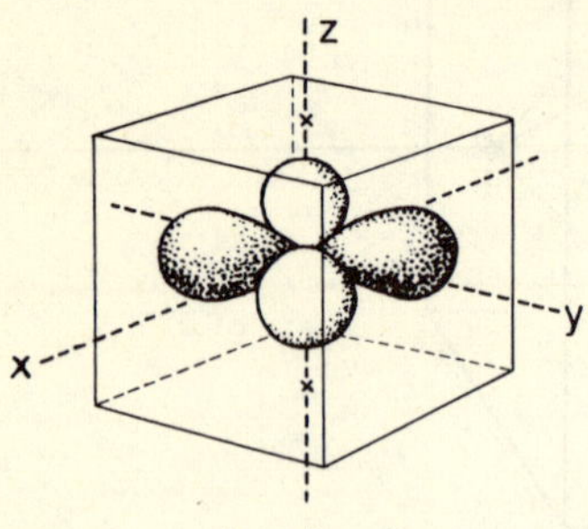


$\underbrace{d_{xy}, d_{xz}, d_{yz}}$
$\gamma_5, t_2$ or $d_\varepsilon$

Figure A.42. Extension of $d_{xy}$ in a cube. X = position of the ligands for octahedral configuration.

The $d_{z^2}$ orbital has a different form than the other $d$ functions (Figure A.38, cf. footnote, p. 53). An electron in this state interacts strongly with the ligands on the $z$ axis, so that it cannot be concluded that the $d_{z^2}$ orbitals should lie energetically higher than the three states of equal energy $d_{xy}, d_{xz}, d_{yz}$. That the state $d_{z^2}$ is degenerate with $d_{x^2-y^2}$ cannot be seen directly even qualitatively. Group theoretical arguments, however, as well as calculations lead to this result.

The energies of the progeny terms $E(t_{2g})$ and $E(e_g)$ as compared to the energy of the $d$ state in the free ion are obtained by carrying out perturbation calculations (Part B, pp. 336 ff.).

$$E(t_{2g}) = \varepsilon_0 - 4Dq$$
$$E(e_g) = \varepsilon_0 + 6Dq,$$

where $Dq$ is the *field strength parameter* introduced by Schlapp and Penney[48] and $\varepsilon_0$ is the shift* resulting from the spherically symmetric portion of the potential (Figure A.43).

This follows directly from the *centre of gravity theorem*, which states that the value $10\varepsilon_0$ must result when the energies of the progeny terms are

* The perturbing potential can be divided into $V_k$, a spherically symmetric, and $V_0$, a non-spherically symmetric part: $V = V_k(r) + V_0(x, y, z)$. $V_k$ is the only part of the potential that is effective in the classical Kossel–Magnus theory. To the first approximation it has virtually no influence on the optical and magnetic properties. It is $V_0$ that produces that splitting.

48. R. Schlapp and W. G. Penney, *Phys. Rev.*, **42**, 666 (1932).

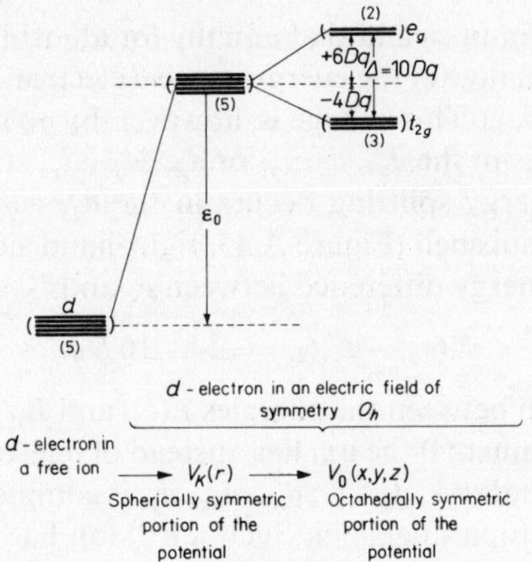


Figure A.43. Splitting of a one-electron $d$ state in an octahedral ligand field. $\varepsilon_0$ is of the order of 20–40 eV, $\Delta \equiv 10\ Dq$ on the other hand lies between 1 and 3 eV.

multiplied by their degrees of degeneracy* and these values added together.

Consider in this connection the following argument: We assume a free ion with one outer $d$ electron having the energy $E$. If this ion is brought into the centre of a sphere having a negative surface charge of magnitude $6q$ ($q$ is a unit of charge) homogeneously distributed, then the energy of the $d$ electron is labilized by an amount $\varepsilon_0$ by virtue of the electrostatic repulsive forces. $\varepsilon_0$ is not dependent upon in which orbital $d_{z^2}, d_{x^2-y^2}, d_{xy}, d_{xz}, d_{yz}$ the electron is found because the $d$ electron charge distributions of these orbitals differ only in their angular parts. These are, of course, immaterial for the interaction with a spherically symmetric environment (Figure A.43, left side). Let us now consider that the negative charge on the surface of the sphere is so distributed that the point charges $q$ are located at the vertices of an octahedron which is circumscribed by the sphere; the vertices of the octahedron lie on the surface of the sphere. The electrostatic potential $V$ generated by the octahedral charge distribution is made up additively of a spherically symmetric part $V_k(r)$, which is equal to the potential that is produced by the homogeneous charge distribution described above, and an angular-dependent part $V_0(x, y, z)$. For the case of an octahedral charge distribution by virtue of the spherically symmetric portion of the potential $V_k(r)$ the energy of the $d$ electron is first of all

* The degeneracies for $d\varepsilon$ and $d\gamma$ are $3 \times 2 = 6$ or $2 \times 2 = 4$ because every orbital has two possible spin states.

increased by an amount $\varepsilon_0$ and that equally for all orbitals $d_{z^2}$, $d_{x^2-y^2}$, $d_{xy}$, $d_{xz}$, $d_{yz}$. A further change of the energy of the $d$ electron arises as a result of the potential $V_0(x, y, z)$. This change is, however, by no means independent of if the $d$ electron is in the $d_{z^2}$, $d_{x^2-y^2}$ or $d_{xy}$, $d_{yz}$, $d_{yz}$ state. As has already been shown, an energy splitting occurs in the $d\gamma$, $e_g(d_{z^2}, d_{x^2-y^2})$ and the $d\varepsilon$, $t_{2g}(d_{xy}, d_{xz}, d_{yz})$ subshell (Figure A.43, right-hand side).

We denote the energy difference between $e_g$ and $t_{2g}$ with $\Delta \equiv 10\,Dq$.

$$E(e_g) - E(t_{2g}) = \Delta \equiv 10\,Dq. \tag{1}$$

A further relation between the energies $E(e_g)$ and $E(t_{2g})$ can be obtained in the following manner: If the ion has, instead of one, ten $d$ electrons, then each of the five orbitals $d_{z^2}$, $d_{x^2-y^2}$, $d_{xy}$, $d_{xz}$, $d_{yz}$ is completely filled with two electrons having antiparallel spin. Such a $d^{10}$ ion has a spherically symmetric charge distribution. The energy change caused by the interaction with the outer octahedral charge distribution for the 10 $d$ electrons must total $10\varepsilon_0$.

$$4E(e_g) + 6E(t_{2g}) = 10\varepsilon_0. \quad \textit{(centre of gravity theorem)} \tag{2}$$

As has already been shown (cf. p. 56) the $t_{2g}$ state is stabilized by the transition from spherical symmetry to octahedral symmetry and the $e_g$ state is labilized with reference to the energy change $\varepsilon_0$ arising from $V_k(r)$. This stabilization or labilizaton is denoted by

$$x = \varepsilon_0 - E(t_{2g}) \quad \text{and} \quad y = E(e_g) - \varepsilon_0.$$

With this equation (1) becomes

$$x + y = \Delta \equiv 10\,Dq. \tag{3}$$

$$4y - 6x = 0,$$

$$2y = 3x$$

and equation (2)

$$4y - 6x = 0,$$

that is

$$2y = 3x$$

or, considering equation (3);

$$x = \frac{2}{5}\Delta = 4\,Dq$$

$$y = \frac{3}{5}\Delta = 6\,Dq.$$

The positions of $d\varepsilon$ and $d\gamma$ as compared with $\varepsilon_0$ are given by the integrals (Part B, section 3.3, pp. 338 ff.).

$$-4Dq = \int \phi_n^* V_o \phi_n \, d\tau \qquad n = 3,4,5$$
$$+6Dq = \int \phi_m^* V_o \phi_m \, d\tau \qquad m = 1,2$$

where $V_0$ is the octahedral perturbation potential and $\phi_n$ and $\phi_m$ are the real $d$ functions given on p. 55. For the case of point charges having the separation $R$ from the central ion $\Delta \equiv 10\,Dq$ is given by

$$10Dq = \frac{5\,q\overline{r^4}}{3R^5},$$

and for the case of point dipoles with the separation $R$ from the central ion by

$$10Dq = \frac{25\,\mu\overline{r^4}}{3R^6}$$

$q$ is the charge of the ligand, $\mu$ is its dipole moment; $r^4$ is the expectation value of the fourth power of the distance of the $d$ electron from the nucleus of the central ion:

$$\overline{r^4} = \int_0^\infty R_{nd}^2(r)\, r^4 \cdot r^2 dr.$$

For tetrahedral symmetry ($T_d$) of the ligand field it is seen from similar arguments with the aid of the functions shown in Figures A.38 and A.39 that analogously to the octahedral case a splitting in $d\gamma(d_{x^2-y^2}, d_{z^2})$ and $d\varepsilon(d_{xy}, d_{xz}, d_{yz})$ takes place. Here, however, the $d\gamma$ states are energetically lower lying than the $d\varepsilon$ states.

For the tetrahedron the cartesian axes are identical with the twofold axes through the midpoints of the edges of the tetrahedron, as can be seen from Figure A.44. Comparing the octahedral with the tetrahedral case, one sees (Figure A.40 and A.44) on the basis of the spatial orientation of the angular functions not only the *inversion of terms*, but also that the magnitude of the splitting for the tetrahedron must be smaller than for the octahedron. This assumes identical ligands and identical ligand–central ion separations. In Figure A.44 the vertices of the octahedron coincide with the midpoints of the surface planes of the cube.

The $d_{xy}$, $d_{xz}$, and $d_{yz}$ orbitals are spatially so oriented that an electron located in one of these states has its maximum electron density in the direction of the angular bisectors of the coordinate axes pairs sketched in Figure A.44. The $d_{x^2-y^2}$ and $d_{z^2}$ functions are oriented symmetrically about the angular bisectors between the tetrahedral axes. They therefore point in the direction of the midpoints of the edges of the tetrahedron. It is also

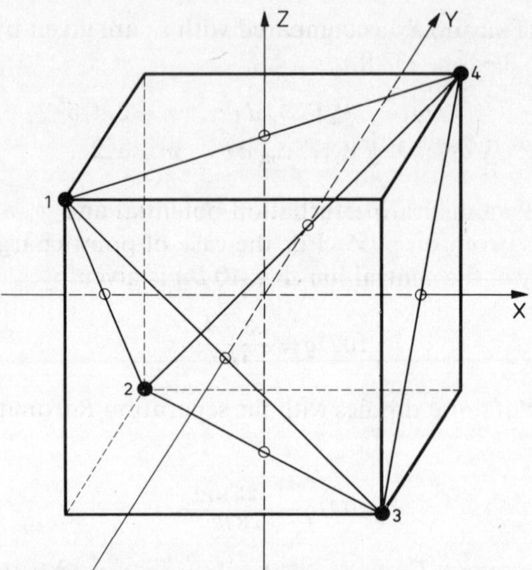


Figure A.44. Tetrahedron with its coordinate system inscribed in a cube. (The points ○ correspond to the vertices of the octahedron, cf. Figure A.40 or A.41, A.42.)

seen that the $d\varepsilon$ functions lie nearer to the ligands than do the $d\gamma$ functions. Thereby the states corresponding to the latter are stabilized compared to those of the former. The smaller splitting $d\gamma - d\varepsilon$ for the tetrahedron compared with the octahedron arises obviously from the fact that the amplitude functions for the octahedron approach the ligands more closely than they do in the tetrahedral case, i.e., the electrostatic repulsion for the octahedron is greater. Further, the number of ligands is reduced by two for the tetrahedron compared with the octahedron.

The splitting pattern for an arrangement of eight ligands on the vertices of a cube (coordination number 8, symmetry $O_h$) is similar to that of the tetrahedron. $d\gamma$ is stabilized compared to $d\varepsilon$. For equal nuclear separations ligands-central ion and equal charges or dipole moments the splitting, however, is greater than for the tetrahedron, as we have seen from Figure A.44.

In Figure A.45 the situation for the symmetries $O_h$ (octahedron), $T_d$ (tetrahedron) and $O_h$ (cube) are shown. As follows from the theory the splitting in the tetrahedral case is $\frac{4}{9}$ of that of the octahedral case. For the cube it is twice as great as for the tetrahedron, again assuming that the distances central ion-ligand remain unchanged.

$$\Delta_{\text{tetrahedron}} = -\tfrac{4}{9}\Delta_{\text{octahedron}}; \quad \Delta_{\text{cube}} = 2\Delta_{\text{tetrahedron}} = -\tfrac{8}{9}\Delta_{\text{octahedron}}$$

The minus sign indicates that the order of the terms is inverted.

In Figure A.46 the schematic splitting diagrams for a one-electron $d$ state for quadratic planar (symmetry $D_{4h}$) and for tetragonal antiprismatic (symmetry $D_{4d}$) ligand orientation are given. The qualitative term order can be determined through similar arguments as were applied for cubic symmetry under consideration of the spatial extension of the angular functions.

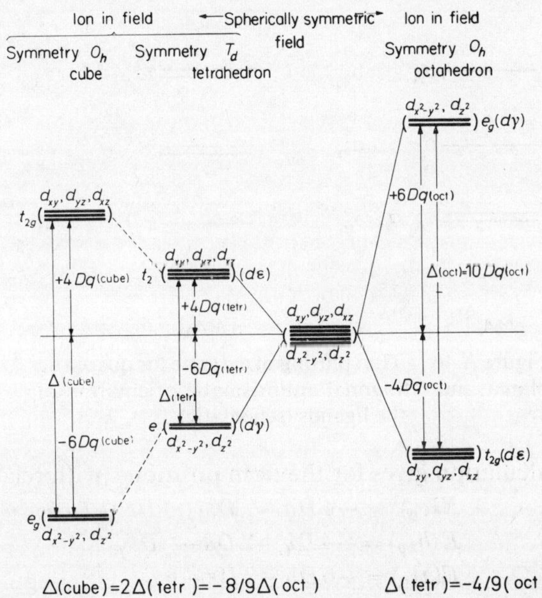


Figure A.45. Splitting of a one-electron $d$ state in fields of cubic symmetry.

Figure A.47 shows qualitatively the situation as one goes from octahedral to *tetragonal symmetry*. The term energies are given here by three parameters, the octahedral field strength parameter $Dq$, as well as by $Ds$ and $Dt$. The two latter parameters give the magnitude of the tetragonal field components. The quadratic planar orientation is obtained from the octahedral in that the ligands at $+z$ and $-z$ are removed to $\infty$.

For $Ds$ the centre of gravity theorem is valid for the $t_{2g}$ and $e_g$ states individually. For $Dt$, however, it is valid only for the total configuration, that is, in relation to the *d* state which is fivefold orbitally degenerate in a spherically symmetric field.

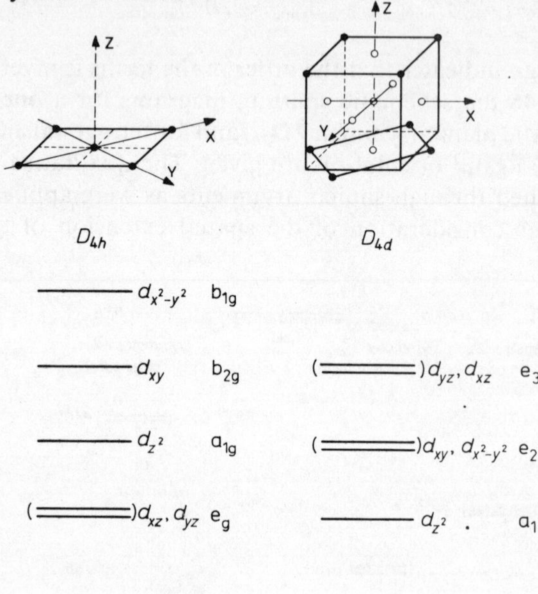


Figure A.46. The splitting of a *d* state for quadratic planar and tetragonal antiprismatic orientation of the ligands (schematic).

The exact calculation gives for the term positions (with relation to $\varepsilon_0$):

$$\begin{aligned} E(e_g) &= -4Dq - Ds + 4Dt \\ E(b_{2g}) &= -4Dq + 2Ds - Dt \\ E(a_{1g}) &= +6Dq - 2Ds - 6Dt \\ E(b_{1g}) &= +6Dq + 2Ds - Dt \end{aligned}$$

Figure A.47 shows the splitting for an *elongated* octahedron ($Ds > 0$, $Dt > 0$) for a *compressed* octahedron ($Ds < 0$, $Dt < 0$) the term order is $b_{2g}, e_g; b_{1g}, a_{1g}$ (cf. Part B, pp. 343 ff.).

### (b) *The many electron problem*[49]

In the strong field case a state of a complex ion is associated with the specific field configuration from which it arises upon the introduction of

49. Cf. e.g. H. Hartmann: *Z. phys. Chem.* (Frankfurt), 4, 376 (1955); **11**, 209 (1957); as well as *Proc. Int. Conf. Co-ord. Chem., Rome*, 1 (1958); *Suppl. to Scientific Research*, **28**°, 1 (1958); *J. inorg. nucl. Chem.*, **8**, 1 (1958); or D. S. McClure, in: *Solid State Physics*, pp. 408 ff., Academic Press, New York–London, **9** (1959).

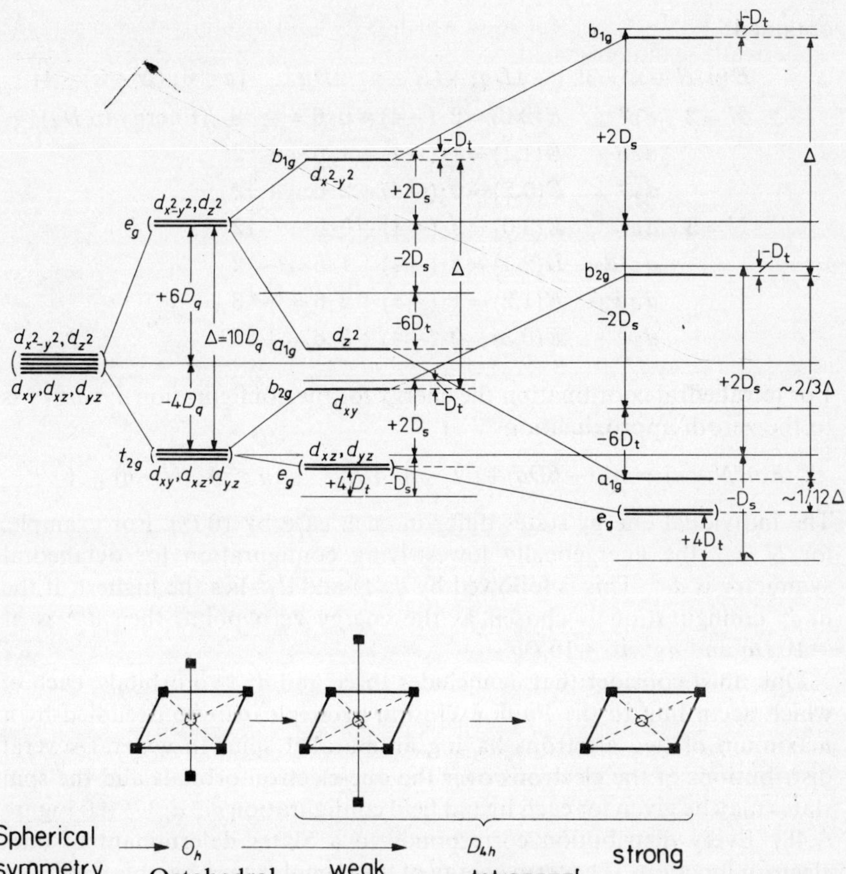


Figure A.47. Splitting diagram for a $d$ state during the transition from octahedral to tetragonal symmetry.

electron interaction. The ligand field configuration is denoted by the occupation numbers of the one-electron states. When the central ion has several $d$ electrons, then the states of the complex, e.g. for the octahedral ligand field, can be associated with various electron configurations $d\varepsilon^n\, d\gamma^{N-n}$, where $N$ is the number of $d$ electrons of the central ion. Each such electron configuration is given an energy value in the zero*th* approximation (without consideration of electron interaction) which lies lower the greater $n$ and the smaller $N-n$ is.

The *energy of the ligand field configuration* $d\varepsilon^n\, d\gamma^{N-n}$ is obtained in the zero*th* approximation by multiplying the number of electrons $n$ and $N-n$ in the $d\varepsilon(t_{2g})$ or $d\gamma(e_g)$ state by $-4Dq$ or $+4Dq$ and adding the values thus obtained:

$$E(n,N-n)=n\cdot(-4Dq)+(N-n)\cdot 6Dq \qquad [n\leq 6;\ (N-n)\leq 4]$$

$$\begin{array}{lll}
N=2 & d\varepsilon^2 & E(2,0)=2\cdot(-4)+0\cdot 6=-\ 8 \quad \text{(Energy in } Dq\text{)} \\
 & d\varepsilon d\gamma & E(1,1)=1\cdot(-4)+1\cdot 6=+\ 2 \\
 & d\gamma^2 & E(0,2)=0\cdot(-4)+2\cdot 6=+12 \\
N=3 & d\varepsilon^3 & E(3,0)=3\cdot(-4)+0\cdot 6=-12 \\
 & d\varepsilon^2 d\gamma & E(2,1)=2\cdot(-4)+1\cdot 6=-\ 2 \\
 & d\varepsilon d\gamma^2 & E(1,2)=1\cdot(-4)+2\cdot 6=+\ 8 \\
 & d\gamma^3 & E(0,3)=0\cdot(-4)+3\cdot 6=+18
\end{array}$$

For tetrahedral coordination the energy for the configuration $d\gamma^n\, d\varepsilon^{N-n}$ is to the zeroth approximation:*

$$E(n, N-n) = n\cdot(-6Dq)+(N-n)\cdot 4Dq. \qquad [n \leqq 4; (N-n) \leqq 6].$$

The individual energy states differ in each case by 10 $Dq$. For example, for $N = 2$ the energetically lowest-lying configuration for octahedral symmetry is $d\varepsilon^2$. This is followed by $d\varepsilon\, d\gamma$ and $d\gamma^2$ lies the highest. If the $d\varepsilon\, d\gamma$ configuration is chosen as the energy zero point, then $d\varepsilon^2$ is at $-10\,Dq$ and $d\gamma^2$ at $+10\,Dq$.

One must consider that $d\varepsilon$ includes three and $d\gamma$ two orbitals, each of which according to the Pauli exclusion principle, can be occupied by a maximum of two electrons having antiparallel spin. In general several distributions of the electrons over the one-electron orbitals and the spin states may be given for each ligand field configuration $d\varepsilon^n\, d\gamma^{N-n}$ (cf. Figure A.48). Every distribution corresponds to a Slater determinant of one-electron functions. The term energy of the complex may be obtained from a perturbation calculation using these determinants as basis functions and treating the electron interaction as the perturbation.

For example the following states of the complex ion arise from the three ligand field configurations $d\varepsilon^2$, $d\varepsilon\, d\gamma$ and $d\gamma^2$ when the electron interaction is included:

| | | | | |
|---|---|---|---|---|
| $d\varepsilon^2$: | multiplicity 3 | $^3T_{1g}$ | multiplicity 1 | $^1A_{1g}$, $^1E_g$, $^1T_{2g}$ |
| $d\varepsilon\, d\gamma$: | multiplicity 3 | $^3T_{1g}$, $^3T_{2g}$ | multiplicity 1 | $^1T_{1g}$, $^1T_{2g}$ |
| $d\gamma^2$: | multiplicity 3 | $^3A_{2g}$ | multiplicity 1 | $^1A_{1g}$, $^1E_g$ |

The number of states of the complex ion which arise from a given configuration $d\varepsilon^n\, d\gamma^{N-n}$ as well as the symmetry species of each of these states, i.e. its irreducible representation, is determined by reducing the

* According to the results obtained in the preceding section, we have $Dq$ (tetrahedron) = $-4/9$ $Dq$ (octahedron).

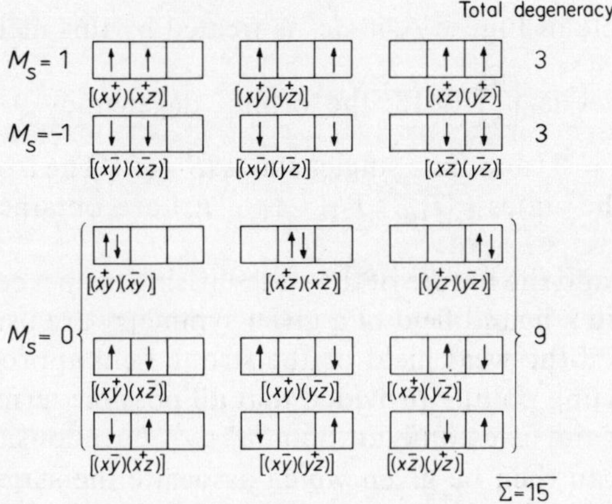


Figure A48. The possible distributions of two electrons over the $d\varepsilon(t_2)$ states. It can be seen from the figure that there are 15 different distributions possible. Therefore the total degeneracy (orbital + spin is 15. $9 = 3 \times 3$ of these occupations belong to triplet states ($M_s = +1, 0, -1$). The remaining 6 belonging to $M_s = 0$ must then represent singlet states. It must be emphasized that in general the occupations given in the figure can by no means be identified directly with the states. These are as a rule describable by superpositions of the distributions corresponding to a given $M_s$ value and correspond to linear combinations of the functions representing the given occupations.

direct product of the representations of the occupied one-electron states, always taking into account the Pauli principle (Part B, pp. 362 ff.).

The total number of states is also obtained when the possible electron distributions over the one-electron $d\varepsilon$ and $d\gamma$ states are considered using the Pauli exclusion principle.*

* The number of states for $q$ electrons which fill a subshell having $e$ orbitals, each of which can be occupied by a maximum of two electrons with antiparallel spin, is obtained from the relation:

$$\binom{2e}{q} = \frac{(2e)!}{q!(2e-q)!}.$$

Correspondingly one obtains for the configuration $d\varepsilon^n\, d\gamma^{N-n}$ ($d\varepsilon$ subshell with $e = 3$, $d\gamma$ subshell with $e = 2$) the total number of states:

$$\binom{6}{n} \cdot \binom{4}{N-n} = \frac{6!}{n!(6-n)!} \cdot \frac{4!}{(N-n)!\,(4-(N-n))!}.$$

e.g. $d\varepsilon^2\, d\gamma^1$

$$\binom{6}{2} \cdot \binom{4}{1} = \frac{6!}{2!(6-2)!} \cdot \frac{4!}{1!(4-1)!} = 15 \cdot 4 = 60.$$

As an example in Figure A.48 $d\varepsilon^2$ is treated by this method. The total number of states is $\binom{6}{2} = 15$, the orbital degeneracy 9. There are six triplet states ($S = 1$) and nine singlet states ($S = 0$). The irreducible representations of the states ($^3T_{1g}$, $^1T_{2g}$, $^1A_{1g}$, $^1E_g$) are obtained using group theory.

The number and the nature of the states arising from a configuration $d^N$ of the free ion in a ligand field of a given symmetry are usually the same, independent of if the weak field or the strong field approximation were used as the starting point—provided that all possible terms $^{2S+1}L$ of the free ion or all ligand field configurations $d\varepsilon^n\, d\gamma^{N-n}$ are considered. Correlation *diagrams* can then be given which associate the states obtained for the weak field case with those states obtained for the strong field case. For states of maximum multiplicity (quartet states) of the configuration $d^3$ ($Cr^{3+}$) such a correlation diagram is shown schematically in Figure A.49.

To the left are the two quartet-states $^4F$ and $^4P$, which arise from the configuration $d^3$ of the free ion when the electron interaction is considered. These split in an octahedral field into $^4A_{2g}(F)$, $^4T_{2g}(F)$, $^4T_{1g}(F)$ and $^4T_{1g}(P)$. If one considers the term interaction between the two $^4T_{1g}$ states, the desired term system of the complex ion is obtained.

To the right the three states of the ligand configurations $d\varepsilon^3$, $d\varepsilon^2\, d\gamma$, and $d\varepsilon\, d\gamma^2$ from which quartet states can arise are given. These are obtained starting with the configuration $d^3$ of the free ion, when the distribution of

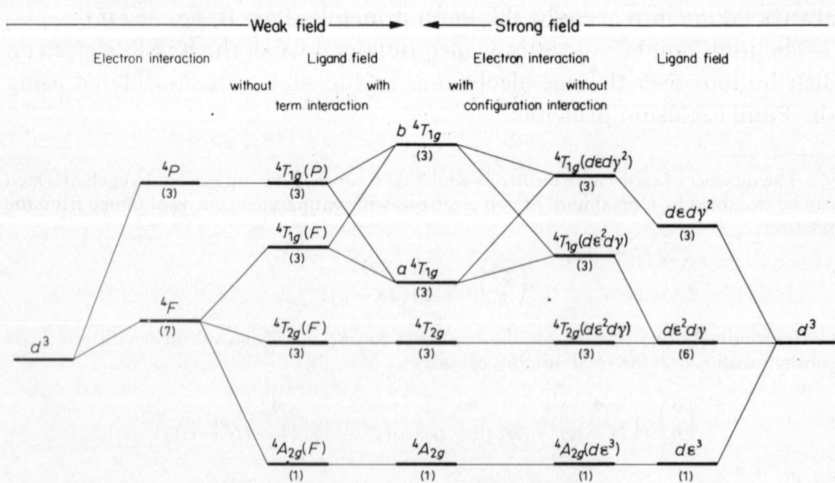


Figure A.49. Correlation diagram for $d^3$ in an octahedral field (schematic, quartet terms only).

$N = 3$ electrons over the $d\varepsilon$ and $d\gamma$ one-electron states arising when a fivefold orbitally degenerate $d$ state is split by a field of octahedral symmetry is considered. With the introduction of the electron interaction the states ${}^4A_{2g}(d\varepsilon^3)$, ${}^4T_{2g}(d\varepsilon^2\,d\gamma)$, ${}^4T_{1g}(d\varepsilon^2\,d\gamma)$ and ${}^4T_{1g}(d\varepsilon\,d\gamma^2)$ result. If the configuration interaction between the two ${}^4T_{1g}$ states arising from the different ligand field configurations, namely $d\varepsilon^2\,d\gamma$ and $d\varepsilon\,d\gamma^2$, is considered, the desired term system of the complex ion is finally obtained.

The complete quantitative treatment leads to the same result according to both the weak-field case and the strong-field case, as long as the same one-electron functions are used (e.g. hydrogen-like functions or Slater functions, p. 56). This is, however, no longer the case if certain higher-lying states of the free ion, are omitted or if higher configurations $d\varepsilon^n\,d\gamma^{N-n}$ are neglected, as often occurs to allow the simplification of the calculation. For a complete treatment all term and configuration interactions must be taken into account.

The term system of a complex ion can be determined in the following manner (Part B, section 3.4):

One proceeds according to the *weak-field method*. With the help of a perturbation calculation the splittings and shifts of the states ${}^{2S+1}L$ of the free ions in the ligand field are determined. To the first approximation the term interaction can be neglected. In a second step the interaction between terms of the same symmetry species (irreducible representation) and multiplicity is taken into account.

Analogously one can proceed according to the *strong-field method* and for a rough discussion consider only the energies of several ligand field configurations in the zero*th* approximation (cf. chapter 2: Discussion of magnetic behaviour, section 3a and chapter 3: The question of stabilization energies). For a more accurate description of the situation all of the ligand field functions must be considered. It must further be ascertained which states arise from them under the influence of electron interaction. One obtains the desired term system of the complex ion by considering the interaction between terms of the same symmetry species and multiplicity which arise from different ligand field configurations.

An advantage in treating a specific problem according to the strong-field method as opposed to the weak-field method, when one is not interested in the absorption spectra, i.e. the higher (excited) states, is that effects can be understood on the basis of a simplified approximation. No explicit calculations are necessary. Such calculations must almost always be carried out when using the weak-field approach.

If the configuration interaction is not taken into account when using the strong-field case, the values obtained for the term energies are better than when term interaction is neglected for the weak-field case. The reason for this is that the off-diagonal elements give only a comparatively small contribution in the former case, while for the latter procedure the off-diagonal elements give large contributions to the term energy. (In the first case the off-diagonal elements are small compared to the diagonal elements and in the latter case they are large.)

## 1.8. Term diagrams

The results of quantitative calculations of the term systems of a complex ion are conveniently presented in the form of *diagrams* in which the term energies are given as functions of the parameters characterizing the ligand field.

For the case of *cubic* symmetry of the ligand field the positions of the terms on the energy scale depend to the first approximation upon only a *single* parameter $\Delta \equiv 10Dq$ (cf. p. 60). Therefore, the term energies (in $cm^{-1}$) are given as a function of $Dq$ (also in $cm^{-1}$). Figure A.50 shows such a diagram calculated by Orgel for $Mn^{2+}$ in an octahedral environment. For $Dq = 0$ one has the terms of the free ion as they are obtained from emission spectra, that is, the case of vanishing ligand field. The value of the ligand field strength increases to the right in the diagram.

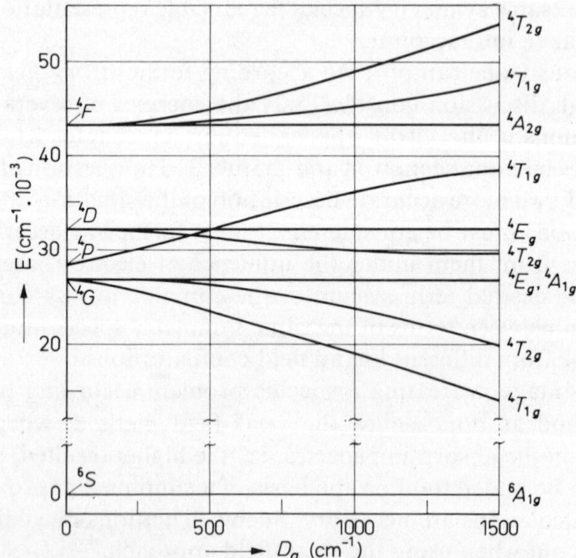


Figure A.50. Term diagram for ($d^5$)$Mn^{2+}$ for octahedral symmetry according to Orgel[50] (sextet and quartet terms only).

This term diagram should be compared with Figure A.35, in which the positions of the terms for a given value of *Dq*, that is for a given field strength, are drawn for $Mn^{2+}$ for octahedral field symmetry.

As has already been mentioned, the field strength parameter *Dq* is in the extended ionic model (crystal field model) a function of the separations, charges and dipole moments, i.e., therefore of parameters having very uncertain values.

For the $[Ti(H_2O)_6]^{3+}$ ion (p. 30) a calculation using plausible values for the distance $Ti^{2+}$—$H_2O$ and for the effective dipole moment of $H_2O$ leads to ${}^2T_{2g}$—${}^2E_g \sim 17{,}600\ cm^{-1}$, while the experimental value, which can be obtained from the band maximum of the long-wavelength weak band is $20{,}400\ cm^{-1}$.

It is not possible to carry out absolute calculations of *Dq* values for specific complexes. Rather they are obtained from absorption spectra in that, by using a term diagram as described above, as good an agreement as possible between the experimental band positions and the corresponding terms in the $E = f(Dq)$ diagram is obtained. In this procedure one particular band is used to determine *Dq* and it can then be tested if the other bands lie as would be predicted by the theory. To calculate *Dq* the corresponding matrix elements are used which have been determined (cf., e.g., Orgel[50]) for octahedral complexes of configurations $d^1$–$d^9$ of the central ion for the weak field case.

Tanabe and Sugano[51] have carried out analogous calculations for the strong field case considering term interaction and have given the corresponding matrix elements and term diagrams as well. Figure A.51 shows as an example such a diagram for $d^3(Cr^{3+}, V^{2+}, Mo^{3+})$. Here the term energy $E/B$ is plotted against $Dq/B$. *B* is a value introduced by Racah[52] (cf. p. 18 or 74, or Part B, p. 251) as a measure of the electron interaction and can be expressed in terms of *Slater** integrals. The term energies in the *Tanabe–Sugano* diagrams depend upon two parameters (*Dq* and *B*). *B* lies between 700 and $12{,}000\ cm^{-1}$ for the ions of interest and can be obtained approximately from the emission spectra, i.e., from the energy differences between terms of the free ions†. The second electron interaction parameter which also appears is according to experience set equal to $\sim 4B$.

---

* $B = F_2 - 5F_4$, $C = 35F_4$. $F_2$ and $F_4$ are those radial integrals defined by Slater (*Phys. Rev.*, **34**, 123 (1929: cf. E. U. Condon and G. H. Shortley: *The Theory of Atomic Spectra*, Cambridge University Press, 1955) (cf. Part B, p. 259).

† The *B* values for the complexes must be reduced in value as compared to the appropriate *B* values of the free metal ions (cf. p. 73, and section 1.9b).

50. L. E. Orgel: *J. chem. Phys.*, **23**, 1004 (1955).
51. Y. Tanabe and S. Sugano: *J. phys. Soc. Japan*, **9**, 753, 766 (1954).
52. G. Racah: *Phys. Rev.*, **62**, 438 (1942); **63**, 367 (1943); **76**, 1352 (1949).

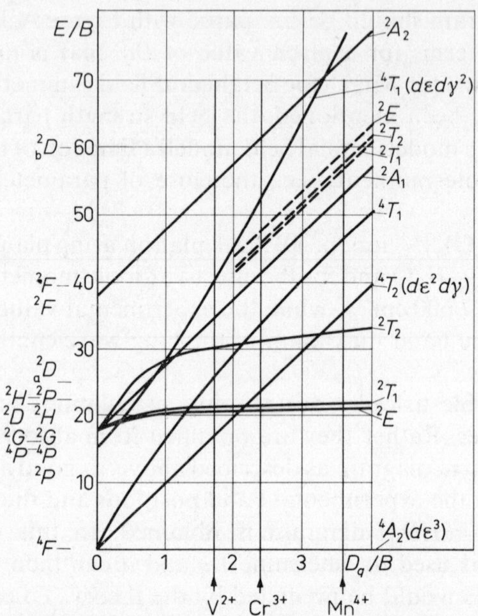


Figure A.51. Term diagram for octahedral $d^3$ complexes according to Tanabe and Sugano[51]. In the diagram the term energies normalized to $B$ values for $[V(H_2O)_6]^{2+}$, $[Cr(H_2O)_6]^{3+}$ and which the ions $V^{2+}$, $Cr^{3+}$ and $Mn^{4+}$ are found are indicated by perpendicular lines. The $Dq$ and $B$ values for $[V(H_2O)_6]^{2+}$, $[Cr(H_2O)_6]^{3+}$ and $[MnF_6]^{2-}$ (C. K. Jørgensen; *Advances in Chemical Physics* **5**, 33 ff.) (1963) were used:

$$V^{2+}\ Dq/B \approx 1{\cdot}8$$
$$Cr^{3+} \qquad \approx 2{\cdot}4$$
$$Mn^{4+} \qquad \approx 3{\cdot}6$$
$$(Mo^{3+} \qquad \approx 4{\cdot}3)$$

$Mo^{3+}$ complexes of octahedral symmetry lie at $Dq/B \approx 4{\cdot}3$, i.e., the intercombination band $^4A_{2g} \rightarrow {}^4T_{2g}$ should also lie before the first spin-allowed band $^4A_{2g} \rightarrow {}^4T_{2g}$. Cf. Figure A.8, where the two intercombination bands $J_1$ ($^4A_{2g} \rightarrow {}^2E_g, {}^2T_{2g}$) and $J_2$($^4A_{2g} \rightarrow {}^2T_{2g}$) can be identified in the reflectance spectrum of $[MoCl_6]^{3-}$. For Cr(III) complexes $J_2$ lies in the minimum between the two spin-allowed bands (*I* and *II*) $^4A_{2g} \rightarrow {}^4T_{2g}$ and $^4A_{2g} \rightarrow {}^4aT_1{}^g$ and can be observed in favourable cases (cf. Figure A.33).

To give an example of the capabilities of the theory the experimental bands of $[Cr(H_2O)_6]^{3+}$ ion are compared with those obtained theoretically using the method of Tanabe and Sugano:

$KCr(SO_4)_2 \cdot 12H_2O$ Crystal

| Electronic transition | calculated | observed | | |
|---|---|---|---|---|
| $^4A_{2g}(d\varepsilon^3) \rightarrow {}^4T_{2g}(d\varepsilon^2\, d\gamma)$ | 17,200 $cm^{-1}$ | 18,000 $cm^{-1}$ | *I* | broad bands |
| $\rightarrow {}^4T_{1g}(d\varepsilon^2\, d\gamma)$ | 24,600 $cm^{-1}$ | 24,600 $cm^{-1}$ | *II* | broad bands |
| $\rightarrow {}^4T_{1g}(d\varepsilon\, d\gamma^2)$ | 38,200 $cm^{-1}$ | 36,600 $cm^{-1}$ | *III* | broad bands |
| $\rightarrow {}^2E_g(d\varepsilon^3)$ | 14,900 $cm^{-1}$ | 14,900 $cm^{-1}$ | $J_1$ | Lines |
| $\rightarrow {}^2T_{1g}(d\varepsilon^3)$ | 15,300 $cm^{-1}$ | 15,100 $cm^{-1}$ | $J_1$ | Lines |
| $\rightarrow {}^2T_{2g}(d\varepsilon^3)$ | 21,800 $cm^{-1}$ | 21,000 $cm^{-1}$ | $J_2$ | Lines |

The absorption band lying at 24,600 $cm^{-1}$ (*II* in Figure A.30) is used as the reference band in the diagram. With $B = 765$ $cm^{-1}$ the positions of the remaining broad bands, as well as of the intercombination bands, can be satisfactorily described. The $Dq$ value obtained is 1720 $cm^{-1}$. (The energy difference $^4A_{2g} - {}^4T_{2g}$ corresponding to the maximum of the first band is $10Dq$.)

In Figure A.52 the *Tanabe-Sugano* diagram for an octahedral complex ion of the configuration $d^6$ ($Co^{3+}$, $Fe^{2+}$) is depicted, showing a *term crossing* for a critical field strength (cf. section 1.6f and Figure A.37). For field strengths smaller than $Dq/B = 2$ the $^5T_{2g}$ state is the ground state. For field strengths greater than $Dq/B = 2$ a $^1A_{1g}$ state is the ground state.

For Tanabe–Sugano diagrams the ground state is always sketched so as to coincide with the $y$ axis.

The *Orgel* diagrams $E = f(Dq)$ give the term energies for cubic symmetry as a function of the *single* parameter $Dq$. For the electron interaction a very specialized fixed value of the parameter $B$ is assumed, approximately that of the respective free ion. The term system of the free ion is the starting point of the calculation as usual (cf. section 1.6). For $Dq = 0$ the term positions are identical with those of the free ions, or when a smaller $B$ value is used, which always proves to be necessary, the terms lie at somewhat lower energies compared with those of the free ion. Such an Orgel diagram is with regard to the term positions valid for only a particular central ion; for example for a $d^5$ ion either for $Mn^{2+}$ or for $Fe^{3+}$.

For Tanabe–Sugano diagrams $E/B = \phi(Dq/B)$ the positions of the terms are given as a function of two parameters, the ligand field strength parameter $Dq$ and the electron interaction parameter $B$ (with $C \sim 4B$). Such a diagram is valid for all central ions of a particular configuration

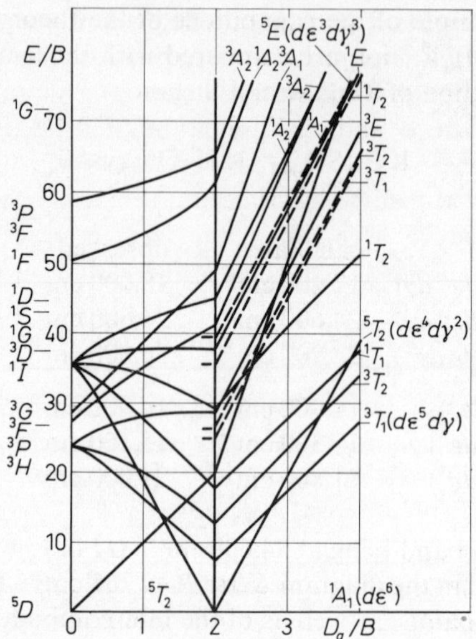


Figure A.52. Term diagram for octahedral $d^6$ complexes after Tanabe and Sugano[51].

$d^N$, e.g. a $d^5$ diagram is valid for both $Mn^{2+}$ and $Fe^{3+}$, where for each central ion the corresponding $B$ value is used.

The parameter $B$ is not identical with that obtained from the term system of the appropriate free ion, but is smaller (cf. section 1.9b). The $B$ value which was used for the interpretation of the $[Cr(H_2O)_6]^{3+}$ spectrum ($B = 765\ \mathrm{cm}^{-1}$) is smaller than that of the free $Cr^{3+}$ ($B = 918\ \mathrm{cm}^{-1}$) obtained from the emission spectrum of $Cr^{IV}$. Depending upon the nature of the ligands somewhat different $B$ values are assumed for the same central ion.

The following rules hold for term diagrams;

1. States having the same multiplicity and irreducible representation cannot cross (*non-crossing rule*).

2. The energies of states which appear only once are linearly dependent upon the strength of the ligand field. If states $^{2S+1}\Gamma$ appear several times, their energy as a rule *no longer depends linearly* upon the field strength. Rather, one finds in this case curves in the term diagram whose deviations from a straight line arise from term or configuration interaction (cf. sections 1.6d and 1.6e, and pp. 73 ff.).

3. For states arising in the ligand field from the same initial state $^{2S+1}L$ of the free ion the *centre of gravity rule* is valid for the case that no term interaction appears with states of the same symmetry species (irreducible representation) and multiplicity that arise from other initial states of the free ion. In Table A.4 the energies of the progeny terms arising in an octahedral field from the ground terms of the ions $d^1$ to $d^9$ are given (cf. section 1.5, Figure A.18). Here the term interaction with splitting terms from higher states of the free ion which influence the energy of some of these states are not considered. (Weak-field case, first approximation, diagonal elements only.) The validity of the centre of gravity theorem is seen from the above.

Table A.4. States and term energies for the case of octahedral symmetry (using the weak field approximation without consideration of term interaction between excited states of the same irreducible representation and multiplicity)

| Ground state of the free ion | | Energy of the states for $O_h$ symmetry in units of $Dq$. |
|---|---|---|
| $d^1$ | $^2D$ | $^2T_{2g}(-4)$, $^2E_g(+6)$ |
| $d^2$ | $^3F$ | $^3T_{1g}(-6)$, $^3T_{2g}(+2)$, $^3A_{2g}(+12)$ |
| $d^3$ | $^4F$ | $^4A_{2g}(-12)$, $^4T_{2g}(-2)$, $^4T_{1g}(+6)$ |
| $d^4$ | $^5D$ | $^5E_g(-6)$, $^5T_{2g}(+1)$ |
| $d^5$ | $^6S$ | $^6A_{1g}(0)$ |
| $d^6$ | $^5D$ | $^5T_{2g}(-4)$, $^5E_g(+6)$ |
| $d^7$ | $^4F$ | $^4T_{1g}(-6)$, $^4T_{2g}(+2)$, $^4A_{2g}(+12)$ |
| $d^8$ | $^3F$ | $^3A_{2g}(-12)$, $^3T_{2g}(-2)$, $^3T_{1g}(+6)$ |
| $d^9$ | $^2D$ | $^2E_g(-6)$, $^2T_{2g}(+4)$ |

4. A term diagram can be extended to include negative $Dq$ values. This occurs in Orgel diagrams in that the $E = f(Dq)$ curves are extended into the region of negative $Dq$ values, as is shown in Figure A.53 for the example of the quartet terms of a $d^3$ ion. For positive $Dq$ values the diagram for octahedral symmetry of the ligand field holds; for negative $Dq$ values, the diagram for tetrahedral symmetry of the ligand field. The distance $^4F - {}^4P$ is chosen* depending upon which $d^8$ or $d^9$ central ions one considers. The energy of both $^4T_1$ states without consideration of term interaction is given by the dashed curve as a function of $Dq$. The right side holds also for tetrahedral $d^7$ and the left side for octahedral $d^7$ compounds. With consideration of term interaction the hyperbolic branches are

* The diagram is also valid for the triplet terms of $d^2$ or $d^8$ ions. The right side corresponds to $d^2(T_d)$ or $d^8(O_h)$ and the left to $d^2(O_h)$ or $d^8(T_d)$.

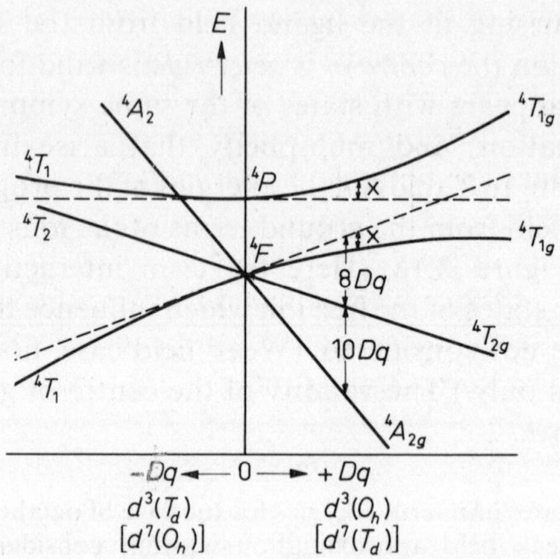


Figure A.53. $d^3(d^7)$ ion in a cubic field (schematic). The effects of term- or configuration interaction in term diagrams are shown. ---- without consideration of the interaction between $^4T_1(F)$ and $^4T_1(F)$.

obtained. The centre of gravity theorem is valid for the progeny terms of $^4F$ when the term interaction with $^4T_{1g}(P)$ is not considered.

$$\left.\begin{aligned} E(^4A_{2g}) &= -12Dq \\ E(^4T_{2g}) &= -\ \ 2Dq \\ E(^4T_{1g}) &= +\ \ 6Dq \end{aligned}\right\} \quad \text{without term interaction}$$

$$\left.\begin{aligned} E(^4A_{2g}) &= -12Dq \\ E(^4T_{2g}) &= -\ \ 2Dq \\ E(^4T_{1g}) &= +\ \ 6Dq - X \end{aligned}\right\} \quad \text{with term interaction}$$

where $X$ symbolically stands for the deviation from the straight line $E = 6Dq$ produced by the term interaction*.

## 1.9. *Dq* values, Racah parameter *B*, spectrochemical and nephelauxetic series

### (a) *The spectrochemical series*

As was shown in the last section from the absorption spectra of octahedral or quasioctahedral transition metal ion complexes it is possible to

*Compare with the diagram Figure A.29 which gives the term positions for $d^3$ for the case of $O_h$ symmetry and a given field strength.

determine the values of the *field strength parameter Dq* corresponding to various central ions and ligands. This can be done *graphically by* using term diagrams. The measured spectrum is fitted as closely as possible to the diagram and the $Dq$ or $Dq/B$ value is read from the abscissa. Somewhat more accurate values are obtained *numerically* by using tabulated matrix elements[50,51]. In some cases ($d^3$, $d^6$ and $d^8$) the maximum of the first spin-allowed band yields directly the value $\Delta \equiv 10Dq$. In other cases ($d^2$, $d^6$ (diamag.) and $d^7$) $\Delta$ cannot be obtained directly, rather corrections for the configuration or term interaction must be considered. The value of $\Delta \equiv 10Dq$ is in these cases approximately 20 per cent different from the energy value of the first band maximum. The determination of $Dq$ for high-spin $d^5$ complexes is comparatively complicated because of the intercombination bands.

When $Dq$ can be obtained directly from the spectra, then the electron interaction parameters $B$ can also be determined, together with $B/C$.* In all other cases $Dq$ and $B$ are determined using the matrix elements according to a trial and error method of the least mean square deviation. An upper bound on the $B$ value is obtained from the emission spectra, that is, from the term systems of the corresponding free ion (Table A.5). The $B$ values for the complexes must however, be reduced compared with those of the free ions, as we have seen, e.g. for $[Cr(H_2O)_6]^{3+}$ ion. For complexes with central ions of the first series of transition metals this reduction is between 5 and 40 per cent (cf. section 1.9b).

The following general rules for the size of the parameter $Dq$ are obtained from a consideration of the spectroscopic material:

1. Complexes with the same ligands which belong to the same series of transition metals and have central ions with the same nuclear charge show very similar $Dq$ values. For the first series of transition metals $Dq$

Table A.5. $B$-values of various free metal ions in $cm^{-1}$

| | | | | | | | | |
|---|---|---|---|---|---|---|---|---|
| $3d^3$ | $V^{2+}$ | 757 | $4d^3$ | $Mo^{3+}$ | ~610 | $5d^3$ | $Re^{4+}$ | ~780 |
| | $Cr^{3+}$ | 918 | | | | | $Ir^{6+}$ | ~650 |
| | $Mn^{4+}$ | 1064 | | | | | | |
| $3d^6$ | $Co^{3+}$ | ~1100 | $4d^6$ | $Rh^{3+}$ | ~720 | $5d^6$ | $Ir^{3+}$ | ~660 |
| $3d^8$ | $Ni^{2+}$ | 1041 | $4d^8$ | $Pd^{2+}$ | 683 | $5d^8$ | $Pt^{2+}$ | ~600 |

The values denoted with ~ have been estimated by extrapolation from known values for isoelectronic ions. The other values were taken from the appropriate emission spectra (cf. Y. Tanabe and S. Sugano, *J. Phys. Soc.* (*Japan*), **9**, 766 (1954)).

* Experience shows that a useful approximation is $C \sim 4B$ (cf. p. 73).

values of between 2040 $cm^{-1}$ ($M = Ti^{3+}$) and 1370 $cm^{-1}$ ($M = Fe^{3+}$) are found for trivalent central ions such as $[M(H_2O)_6]^{3+}$.

Table A.6. *Dq* values for aquocomplexes (in $cm^{-1}$)

| | Ti | V | Cr | Mn | Fe | Co | Ni | Cu |
|---|---|---|---|---|---|---|---|---|
| bivalent | — | 1220 | (1260) | 780 | 1050 | 970 | 850 | (1250) |
| trivalent | 2040 | 1900 | 1770 | (2100) | 1400 | 2000 | — | — |

( ) tetragonally distorted complexes, therefore the *Dq* values are uncertain.

2. Complexes with identical ligands show larger *Dq* values when the nuclear charge of the central ion is greater. Thus one finds for $[\overset{III}{M}(H_2O)_6]^{3+}$ ions *Dq* values of the order of about 2000 $cm^{-1}$, for $[\overset{II}{M}(H_2O)_6]^{2+}$ ions about 1000 $cm^{-1}$ (see Table A.6). In general *Dq* for a complex with a trivalent central ion is between 40 and 80 per cent greater than for a corresponding complex with a bivalent central ion.

(For complexes with very low valencies of the central ion (anomalous valencies), e.g. $[V^{I}dip_3]^+$ (dip = 2,2′-dipyridyl), this is no longer the case.)

3. If one considers complexes with the same ligands but with central ions belonging to different transition metal series, then one finds an increase of 30–40 per cent in *Dq* when going from the first to the second series of transition metal ions or from the second to the third series.

4. The most common ligands can be arranged in a series according to increasing *Dq* values, i.e., increasing ligand field strength, where the order of the ligands within such a scheme is independent of the central ion of the complexes*:

$$J^- < Br^- < Cl^- \sim \underline{S}CN^- \sim N_3^- < (C_2H_5O)_2PS_2^- < F^- < (C_2H_5)_2NCS_2^- <$$
$$< (NH_2)_2C\underline{O} < OH^- < (COO)_2^{--} \sim H_2O < \underline{N}CS^- < NH_2CH_2COO^- <$$
$$< NCSHg^+ \sim NH_3 \sim C_5H_5N < NH_2 \cdot CH_2 \cdot CH_2 \cdot NH_2 \sim \underline{S}O_3^{--} < NH_2OH <$$
$$< \underline{N}O_2^- < H^- \sim CH_3^- < \underline{C}N^-.$$

The given series of the ligands ordered according to increasing *Dq* values is identical with the spectrochemical series determined empirically by

* This series corresponds to that given by C. K. Jørgensen at the International Summer School on Ligand Field Theory, Konstanz, 1962. 2,2′-dipyridyl and *o*-phenanthroline come after ethylenediamine and before $NO_2^-$ in the series. The series holds only for complexes with normal valencies. An underscored atomic symbol indicates that the ligand is coordinated with that atom.

Tsuchida[53] on the basis of spectroscopic investigations on appropriate compounds. If a ligand in a complex ion is replaced by the ligand to the right in the series, then a shift of the spectrum to shorter wavelengths (hypsochromic effect), i.e. to higher wave numbers, is seen.

The $Cu^{2+}$ aquo ion $[Cu(H_2O)_6]^{2+}$, e.g. in aqueous solutions of $CuSO_4 \cdot 5H_2O$ is weakly blue-coloured. The absorption band lies here at $\approx 12{,}600\ cm^{-1}$. With the addition of ammonia, copper ammine complexes are formed in which $H_2O$ has been replaced by ammonia. These complexes have the familiar intense blue-violet colour. The absorption maximum now lies at $13{,}100\ cm^{-1}$, because the substitution of $H_2O$ by $NH_3$ is associated with an increase of the ligand field strength. Anhydrous $CuSO_4$ is as is well-known colorless. The ligand field of $SO_4^{2-}$ ions is in this case so weak that the absorption band moves to the near infrared.

Ordering according to with which atom the various ligands coordinate with the metal ions we find:

$$J < Br < Cl < S < F < O < N < C,$$

a series going in the direction of decreasing radii of the atoms. This series, arranged according to the coordinated atoms of the ligands, was already recognized in principle by Fajans[54].

5. *Dq*, as mentioned by Jørgensen[55], can be written approximately as a function *f*, dependent only upon the ligands, multiplied by a function *g*, dependent upon only the central ion:

$$Dq \approx f\,(\text{ligand}) \times g\,(\text{central ion}).$$

If one considers $[MA_6]$ complexes with the same ligands A, a series of central ions arranged according to inreasing *Dq* values can be given

$$\text{Mn(II)} < \text{Co(II)} \approx \text{Ni(II)} < \text{V(II)} < \text{Fe(III)} < \text{Cr(III)} < \text{Co(III)} \lessapprox \text{Mn(IV)} < \\ < \text{Mo(III)} < \text{Rh(III)} < \text{Ir(III)} < \text{Re(IV)} < \text{Pt(IV)}.$$

6. A *rule of 'average environment'* can be determined empirically which states that the *Dq* values of mixed complexes $[MA_nB_{6-n}]$ can be found approximately through linear interpretation between *Dq* values for $[MA_6]$ and $[MB_6]$:

$$Dq([MA_n B_{6-n}]) \approx \frac{n}{6} Dq([MA_6]) + \frac{6-n}{6} Dq([MB_6]).$$

53. R. Tsuchida: *Bull. chem. soc. Japan*, **13**, 388, 436, 471 (1938). R. Tsuchida and M. Kobayashi: *Bull. chem. Soc. Japan*, **13**, 47 (1938).
54. K. Fajans: *Naturwiss.*, **11**, 165 (1922).
55. C. K. Jørgensen: *Absorption Spectra and Chemical Bonding in Complexes*, pp. 107 ff., Pergamon Press, Oxford–London–New York–Paris 1962.

In some cases of ligands lying widely separated in the spectrochemical series one observes that the absorption bands of the complexes are split upon going from a type with six identical ligands to mixed complexes. As an example in section 1.6c we became acquainted with the complexes of trivalent chromium and trivalent cobalt of type $[MA_5B]$, $[MA_4B_2]$ ($A{=}NH_3$ ; B=Cl, Br, I), where the spectral splitting could be utilized to distinguish between the stereoisomeric forms *cis* and *trans*-$[MA_4B_2]$. Generally, however, the term splitting accompanying a descent in symmetry is so small that it cannot be directly observed because of the large half width of the bands.

Table A.7 contains the values of $\Delta \equiv 10Dq$[56] for a series of ligands and central ions. The $B$ values for the free central ions as well as the corresponding complex ions are likewise given.

It is to be taken into account in a discussion of $Dq$ values that as a result of Jahn–Teller distortion (pp. 33 and 164) octahedral symmetry is by no means present for the high-spin $d^4$ and $d^9$ complexes $[MA_6]$. Rather such complexes are more or less strongly tetragonally distorted. These are complexes with doubly orbitally degenerate $E_g$ ground states. Such a Jahn–Teller distortion appears for complexes with the triply orbitally degenerate ground states $T_{1g}$ or $T_{2g}$ as well, but to such a small degree that the deviations from octahedral symmetry are comparatively small. One observes in those cases where excited $E_g$ states are present (e.g. $Ti^{3+}$ ; see p. 33) a Jahn–Teller splitting of the excited state, whereby asymmetrical absorption bands or ones with a double structure can arise.

Although for $Cr^{2+}$, $Mn^{3+}$ and $Cu^{2+}$ complexes with six identical ligands the determination of $Dq$ is problematical because of the tetragonal field symmetry, for $Ti^{3+}$ the determination of the $Dq$ values from the maximum of a long wave length band is more easily realized, however, because of the reasons mentioned, with a certain degree of uncertainty.

From the arrangement of the ligands according to increasing $Dq$ values, one sees that monatomic ionic ligands in general produce weaker splittings than polyatomic ions or molecules. It is remarkable that the strongly polarizable $I^-$ ion causes a smaller effect than the much more weakly polarizable $F^-$ ion. The rôle played by polarization is therefore exactly the opposite of what would be expected. Likewise the fact that $OH^-$ and $(COO)_2^{2-}$ are before $H_2O$ in the series is not readily understandable. From an electrostatic standpoint it is also scarcely explainable that $Dq$ increases so markedly when one goes from divalent to trivalent ions. The capability of certain molecules and polyatomic ions to form $\pi$ bonds (p. 106) with the

56. C. K. Jørgensen: *Energy levels of Complex and Gaseous Ions*, J. Gjellerups Forlag, Copenhagen 1957.

Table A.7. Summary of $\Delta$ and $B$ values for various transition metal complexes after Jørgensen[56]

| | $6\,Br^-$ | $6\,Cl^-$ | $3\,ox^{--}$ | $6\,H_2O$ | $enta^{-4}$ | $6\,NH_3$ | 3 en | $6\,CN^-$ |
|---|---|---|---|---|---|---|---|---|
| $3d$ Titanium(III) | — | — | — | 20,300 | 18,400 | — | — | — |
| $3d^2$ Vanadium(III) | — | — | 16,500 | 17,700 | — | — | — | — |
| $3d^3$ Vanadium(III) | — | — | — | 12,600 | — | — | — | — |
| Chromium(III) | — | 13,600 | 17,400 | 17,400 | 18,400 | 21,600 | 21,900 | 26,300 |
| $B = 950$ | — | 510 | 640 | 750 | 720 | 670 | 620 | 520 |
| $4d^3$ Molybdenum | — | 19,200 | — | — | — | — | — | — |
| $3d^7$ Chromium(II) | — | — | — | 13,900 | — | — | — | — |
| Magnesium(III) | — | — | 20,100 | 21,000 | — | — | — | — |
| $3d^5$ Magnesium(II) | — | — | — | 7,800 | 6,800 | — | 9,100 | — |
| $B = 850$ | — | — | — | 790 | 760 | — | 750 | — |
| Iron(III) | — | — | — | 13,700 | — | — | — | — |
| $B \sim 1000$ | — | — | — | 770 | — | — | — | — |
| $3d^6$ Iron(II) | — | — | — | 10,400 | 9,700 | — | — | 33,000 |
| Cobalt(III) | — | — | 18,000 | 18,600 | 20,400 | 23,000 | 23,300 | 34,000 |
| $B \sim 1050$ | — | — | 560 | 720 | 660 | 660 | 620 | 440 |
| $4d^6$ Rhodium(III) | 18,900 | 20,300 | 26,300 | 27,000 | — | 33,900 | 34,400 | — |
| $B \sim 800$ | 300 | 400 | — | 500 | — | 460 | 460 | — |
| $5d^6$ Iridium(III) | 23,100 | 24,900 | — | — | — | — | 41,200 | — |
| $B = 660$ | 250 | 300 | — | — | — | — | — | — |
| Platinum(IV) | 24,000 | 29,000 | — | — | — | — | — | — |
| $3d^7$ Cobalt(II) | — | — | — | 9,300 | 10,200 | 10,100 | 11,000 | — |
| $B = 1030$ | — | — | — | ~970 | ~940 | — | — | — |
| $3d^8$ Nickel(II) | 7,000 | 7,300 | — | 8,500 | 10,100 | 10,800 | 11,600 | — |
| $B = 1130$ | 760 | 780 | — | 940 | 870 | 890 | 840 | — |
| $3d^9$ Copper(II) | — | — | — | 12,600 | 13,600 | 15,100 | 16,400 | — |

enta = ethylenediamine tetraacetic acid.

central ion is apparently of importance to the value of $Dq$. The extension of such $\pi$ bonds can, as will be shown later, lead to an increase in $Dq$. Increasing $Dq$ values cannot be equated with a decreasing degree of covalence. Chemical experience would indicate an increase of covalent binding in the direction $F^- < Cl^- < Br^- < I^-$. However, the order of the halogens is exactly inverted in the spectrochemical series.

The given series of atoms coordinated by the ligands goes in the direction of decreasing atomic radii, so that it is in qualitative agreement with a purely electrostatic interpretation of the change in $Dq$.

The value $Dq$ determined empirically from the spectra is a complicated quantity, composed of various contributions (cf. p.118 ).

1. Purely electrostatic effects. These were taken into account in the extended ionic model. The distances, charges and dipole moments (cf. p. 61 for explicit expressions for $Dq$) are parameters of the perturbing potential.
2. Effects arising from $\sigma$ bonds between the central ion and lone electron pairs of the ligands.
3. Influence of dative $d_\pi \rightarrow p_\pi$ bonds (metal $\rightarrow$ ligands).
4. Influence of dative $p_\pi \rightarrow d_\pi$ bonds (ligands $\rightarrow$ metal).

The influences mentioned under 1, 2 and 3 lead to an increase in $Dq$, while the influence 4 tends to produce a decrease of the field strength parameter. The influence from 1 which was solely considered in the original electrostatic theory gives only a moderately large contribution to $Dq$. The effects 2 and 3 can no longer be separated from one another in cases of strong interaction, as for example for certain cyano complexes.

For these reasons an absolute calculation of $Dq$ values is impracticable. On the other hand, the semi-empirical theory in which the $Dq$ values are taken from the measured spectra is of great practical importance for a quantitative discussion of the optical and other physical and chemical properties of transition metal ion complexes.

### (b) *The nephelauxetic series*

It has already been mentioned that the values of the Racah parameter $B_{\text{fr. ion}}$ obtained from the emission spectra of the free ions are not in agreement with those values $B_{\text{compl. ion}}$, determined from the absorption spectra of the complex ions. The latter values are always smaller than the former (see Table A.7). $B$ contains the Slater radial integrals $F_2$ and $F_4$ (cf. p. 71). Differing $B$ values for free central ions and for central ions in the complex mean that the radial distribution of the $d$ electrons is different for different environments. In the term diagrams this is evident in that for $\Delta = 0$ the terms $^{2S+1}L$ do not show the same distances as are obtained from the emission spectra. Rather, smaller term differences appear. The lower values of $B_{\text{compl. ion}}$ compared with $B_{\text{fr. ion}}$ refer to the fact that the charge cloud of the *d* electrons is most likely more spread out in the complex than in the free ion as a result of the covalent bond effects (*d electron delocalization*). These observations of Owen[57], Orgel[50] and Tanabe and Sugano[51] caused Schaffer and Jørgensen[58] to undertake an ordering of the various ligands according to decreasing values of the relation

$$\beta = \frac{B_{\text{compl. ion}}}{B_{\text{fr. ion}}}.$$

One obtains in this manner the following 'nephelauxetic' (cloud expanding) series of the ligands, which expresses fairly well their tendency to form a covalent bond between the central ion and the ligands. For a specific central ion in a specific oxidation state one finds:

$$F^- > H_2O > (NH_2)_2C\underline{O} > NH_3 > (COO)_2^{--} \sim NH_2CH_2CH_2NH_2 > \underline{N}CS^- >$$
$$> Cl^- \sim CN^- > Br^- > (C_2H_5O)_2PS_2^- \sim S^{--} \sim J^- > (C_2H_5O)_2PSe_2^-.$$

This series goes for the most part parallel with the degree of electronegativity of the atoms of the ligands which coordinate with the central ion:

$$F > O > N > Cl > Br > S \sim J > Se.$$

A nephelauxetic series for the central ions can also be given, although with specific uncertainties:

$$Mn(II) > Ni(II) \sim Co(II) \sim Mo(III) > Cr(III) >$$
$$> Fe(III) > Rh(III) \sim Ir(III) > Co(III) > Mn(IV) > Pt(IV).$$

It shows a decrease of $\beta$ with increasing oxidation number.

The nephelauxetic effect, as measured by the quantity $\beta$, can be expressed approximately as a product of two functions of in each case one variable[55]:

$$\beta = \varphi\,(\text{ligands}) \times \eta\,(\text{central ion})$$

The nephelauxetic series corresponds much more to the chemist's conception of the appearance of covalent bonds than does the spectrochemical series. One sees that, for example, $F^-$ and $H_2O$ are at the left end of the nephelauxetic series, i.e. their $\beta$ values lie very close to 1. Complexes with more covalently bound ligands as e.g. $Br^-$ and $I^-$ lie more in the

57. J. Owen: *Proc. R. Soc.*, **A 227**, 183 (1955).
58. C. E. Schäffer and C. K. Jørgensen: *J. inorg. nucl. Chem.*, **8**, 143 (1958); *Suppl. to Scientific Research*, **28°**, 143 (1958); *Proc. Int. Symp. Co-ord. Comp.*, Rome, 1958.

middle or to the right end of the series. That is, they have $\beta$ values which are much smaller than 1. In the spectrochemical series $I^-$ and $Br^-$ were on the left end. Many covalent complexes lie in the spectrochemical series close to $H_2O$; they have $Dq$ values which are not very different from those of the corresponding aquo-complexes. For a thorough discussion of the nephelauxetic effect, the reader is referred to Jørgensen (e.g. *Advances in Chemical Physics*, **5**, 67 (1963).

Jørgensen distinguishes between the following two reasons for the reduction of the $B$ values for complexes as compared with those of the corresponding free ions:

1. Electrons are transferred to a greater or lesser degree from the ligands to the central ion. The effective charge of the central ion is thereby lowered; the charge is no longer $Z = +1, +2, +3 \ldots$, rather $Z^* < Z$ (*central field covalency*).

2. A delocalization of the d electrons in the direction of the ligands ($\sigma$ antibonding $d\gamma$ and $\pi$ antibonding $d\varepsilon$ electrons (*symmetry restricted covalency*).

The description of this situation can be carried out effectively within the framework of the MO Theory (cf. pp. 97 ff.).

## 1.10. Relative intensities and half widths

(a) *Intensity of the $d \rightarrow d$ absorption bands*

In a discussion of the absorption spectra of metal complexes we have seen that besides the *normal* bands corresponding to transitions between states of the *same* multiplicity, frequently less intense *intercombination bands* are found. These are associated with transitions between states of different multiplicity.

All of the absorption bands corresponding to $d \rightarrow d$ transitions are weak; they must to the first approximation then be forbidden transitions. The normal bands have values of the molar decadic extinction coefficient $\varepsilon$ which seldom exceed 200. The intercombination bands, as we have seen from the example of $Cr^{3+}$ and $Mn^{2+}$ complexes (pp. 42 and 46), are a factor of 100–1000 weaker ($\varepsilon \sim 10^{-2}$–1).

Because we are concerned with transitions between states, all of which arise from the $d^N$ configuration, all of the states show the same parity (even ($g$))*. For the case of a very weak ligand field spectral selection rules are

---

* The parity of a state gives the behaviour of its eigenfunction upon reflection through the origin of the coordinate system (centre of inversion), that is, when the position coordinates $x$, $y$ and $z$ are replaced respectively by $-x$, $-y$, and $-z$. If $\Psi$ transforms into $\Psi$, the parity is even (*gerade*, $g$): if into $-\Psi$ odd (*ungerade*, $u$) (cf. Part C, pp. 494 ff.).

valid which correspond for the most part to those of the free atoms or ions. The following types of transitions in principle come into discussion:

(a) Electric dipole radiation transition,
(b) Magnetic dipole radiation transitions,
(c) Electric quadrupole radiation transitions.

For free ions only electric quadrupole and magnetic dipole transitions are allowed between states of the same parity which arise from the same configuration. According to the Laporte rule electric dipole transitions between states of the same parity are forbidden ($g \nleftrightarrow g$; $u \nleftrightarrow u$) (cf. Appendix 5, p. 494).

An asymmetric electric field can produce a sufficiently great perturbation of the inversion symmetry, so that the electric dipole transitions are no longer rigorously forbidden. This situation applies to complexes.* The parity-forbidden selection rule is removed for a metal ion in a complex. The two states combining with one another must show different symmetry properties in at least one component of a translation vector.† The prohibition is removed when states of different parity (e.g. *p* states) are mixed in with *d* states. Such a mixing can, as has been noted by Van Vleck[59], result from:

1. Absence of a symmetry centre and
2. Perturbation of the inversion symmetry present through vibrations of the proper irreducible representation.

The first case is present for a large number of complexes, i.e. static hemihedral ligand fields should be the principal cause for the intensity of the observed bands. Tetrahedral complexes, e.g. $[CoCl_4]^{2-}$ or $[CuCl_4]^{2-}$, do not have a centre of symmetry Ballhausen and Liehr[60] have theoretically estimated the intensity for such cases by considering a mixture of 3*d* with 4*p* states.

For octahedral complexes such as $[Ti(H_2O)_6]^{3+}$ or $[Cr(NH_3)_6]^{3+}$ the second case is present. Here the static field is virtually holohedral; a centre of symmetry is present. The parity-forbidden condition can only be

* That the $d \rightarrow d$ bands of complexes in general cannot be magnetic dipole or electric quadrupole transitions follows for the normal bands from a comparison of the experimentally-determined with the theoretically-estimated oscillator strengths of these transitions. The values obtained for the oscillator strengths under the assumption of magnetic dipole and electric quadrupole transitions are too small[51]. Only electric dipole radiation leads to oscillator strengths $> \sim 10^{-6}$.

†A transition is allowed when the product of the representations of the corresponding symmetry group of the two combining states contains the representation of the transition moment operator. The components of the electric dipole radiation operator transform as translations, i.e. as the components of spatial vectors.

59. J. H. Van Vleck: *J. phys. Chem.*, **41**, 67 (1937).
60. C. J. Ballhausen and A. D. Liehr: *J. molec. Spectros.*, **2**, 342 (1958).

removed by asymmetric vibrations of the suitable species, i.e. in this manner dynamic hemihedral perturbations are superimposed upon the static holohedral field.

For *cis–trans* isomers $[MA_2B_4]$ as *trans*- and *cis*-$[Cren_2Cl_2]^+$ the *cis* isomers always show a somewhat higher intensity than the *trans* complexes (cf. Figure A.27). It can therefore be concluded that the static hemihedral fields are a contributing factor to intensity differences because for the *cis* complex the centre of symmetry present for the *trans* isomers is missing.

The hemihedral static fields do not appear to be principally responsible for the intensity of the parity-forbidden $d \rightarrow d$ band. This follows from the fact that complexes such as $[MA_5B)$ or *cis*-$[MA_4B_2]$ without an inversion centre have only slightly less intense bands than the holohedric compounds $[MA_6]$ or *trans*-$[MA_4B_2]$, (see the corresponding chromium(III) complexes, Figures A.25 and A.27).

In general, tetrahedral complexes absorb much more strongly than octahedral complexes. For the example of cobalt(II)-tetrachloroion $[CoCl_4]^{2-}$ the observed intensities are about 100 times stronger than for the hexaquo-ion $[Co(H_2O)_6]^{2+}$ or other octahedral complexes like $[Co(NH_3)_6]^{2+}$. The band lying from $\sim$20,000–22,000 cm$^{-1}$ corresponding to the transition ${}^4T_{1g}(F) \rightarrow {}^4T_{1g}(P)$ has $\varepsilon_{max}$ values of $\sim$10. For the tetrachlorocomplex a value of 600 is found (see Figure A.54) for the band at $\sim$15,000 cm$^{-1}$, ${}^4A_2(F) \rightarrow {}^4T_1(P)$. This is visually apparent because octahedral cobalt(II) complexes are usually weakly rose-coloured, while tetrahedral complexes have an intense blue colour.

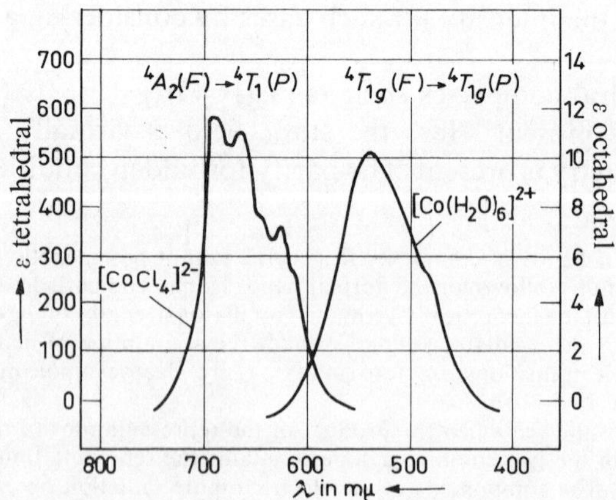


Figure A.54. Spectra of $[Co(H_2O)_6]^{2+}$ and $[CoCl_4]^{2-}$ in the region between 800 and 400 m$\mu$.

The spectra of complexes with trivalent central ions are generally a factor of 10 more intense in $\varepsilon$ than the spectra of corresponding complexes with bivalent central ions. Further one observes a general increase in intensity when going to complexes having ligands with greater $Dq$ values. (The spectrum of a hexammine complex is more intense than a hexaquo ion). Therefore, in general, the stronger the ligand field, the more intense is the absorption.

The very low intensity of intercombination bands is understandable because the transitions are not only parity-forbidden but also spin-forbidden, because the transitions are between states of different multiplicity. Spin–orbit coupling is responsible for the appearance of such bands (cf. section 1.11).

To the spin-allowed normal bands ($\varepsilon_{max} \sim 10$–$10^2$; $\delta \sim 3000\ \mathrm{cm}^{-1}$) correspond oscillator strengths* of $f \sim 10^{-3}$–$10^{-5}$. For the narrow intercombination bands, as are found in $Cr^{3+}$ complexes ($\varepsilon_{max} \sim 10^{-2}$–$10^{-1}$; $\delta \sim 300\ \mathrm{cm}^{-1}$), the $f$ values range from $\sim 10^{-7}$ to $10^{-8}$ and for the broad intercombination bands, e.g. $[Fe(H_2O)_6]^{3+}$ ion $f$ is $\sim 10^{-6}$.

It is expected that the intensity of the absorption bands corresponding to transitions which are forbidden as pure electronic transitions should be temperature dependent. At sufficiently low temperatures the vibrations necessary for such a transition are partially frozen in. Holmes and McClure[61] have made such observations for various crystalline hydrates, e.g. for $NiSO_4 \cdot 7H_2O$. In going from room temperature to 77°K, a decrease of the oscillator strength of 40 per cent is observed.

For complexes having less than cubic symmetry the assignment of bands to specific transitions can be tested by carrying out investigations with polarized light on suitably oriented crystals. Such experiments have been undertaken for the most part by Japanese investigators.[62]

*Trans*-$[Coen_2X_2]^+$(X=Cl, Br) has for example tetragonal ($D_{4h}$) symmetry. The spectra of these ions are virtually identical with those of the corresponding tetramine *trans*-$[Co(NH_3)_4X_2]^+$. The perchlorates crystallize in a monoclinic system, with the *a* axis almost superimposed upon the fourfold axes of the complex ion and the *b* axis nearly perpendicular to these axes. As with chromium(III) complexes (cf. Figure A.55) one observes with the transition from $[Co(NH_3)_6]^{3+}$ or $[Coen_3]^{3+}$ ion

---

* Calculated from the measured absorption bands using $f = 4{\cdot}32 \times 10^{-9} \int_{-\infty}^{\infty} \varepsilon(\bar{\nu})\, d\bar{\nu}$, or when the band assumes a gaussian form, $F = 4{\cdot}6 \times 10^{-9} \varepsilon_{max} \delta$ ($\varepsilon_{max}$ = value of the molar decadic extinction coefficient $\varepsilon$ for the band maximum, $\delta$ = half-width in $\mathrm{cm}^{-1}$ (see footnote, p. 34) $\bar{\nu}$ = wave-number in $\mathrm{cm}^{-1}$.

61. O. Holmes and D. S. McClure; *J. chem. Phys.*, **26**, 1686 (1957).
62. S. Yamada, A. Nakahara, Y. Sjimura and R. Tsuchida: *Bull. chem. Soc. Japan*, **28**, 222 (1955).

to the tetragonal dichlorocomplex a splitting of the band at longest wavelengths ($^1A_{1g} \rightarrow {}^1T_{1g}$) into two bands. The one at longer wavelengths corresponds to the $^1A_{1g} \rightarrow {}^1E_g$ transition and the one at shorter wavelengths to $^1A_{1g} \rightarrow {}^1A_{2g}$. ($^1T_{1g}(O_h) \rightarrow {}^1A_{2g}(D_{4h}) + {}^1E_g(D_{4h})$; $^1A_g(O_h) \rightarrow {}^1A_g(D_{4h})$).

Within the longer wavelength band at 620 m$\mu$ the strongest absorption is found for light with its electric vector vibrating along the *a* axis. Light with its vector oriented in the direction of the *b* axis is also appreciably, but less strongly, absorbed. The shorter wavelength band at 430 m$\mu$ disappears almost completely for light polarized in the direction of the *a* axis, while light polarized in the direction of the *b* axis shows strong absorption (cf. Figure A.55).

This behaviour agrees with the assignment of both bands. The electric vector along the *a* axis transforms according to $A_{2u}$, while the perpendicular vector is transformed as $E_u$ of the symmetry group $D_{4h}$. The transition according to $^1A_{2g}$ can take

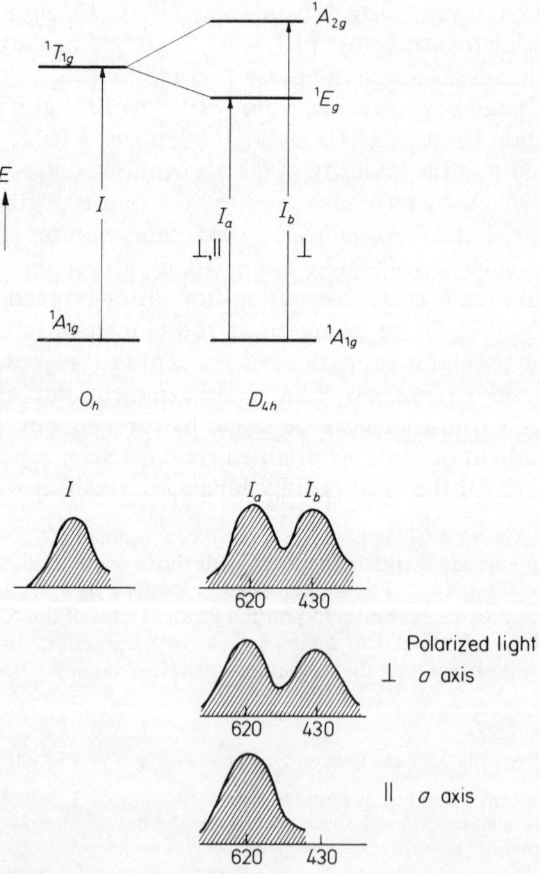


Figure A.55. Investigations on *trans*-[Co $en_2X_2]^+$, X = Cl, Br, with polarized light (schematic).

place only in the presence of vibrations of the irreducible representation $\alpha_{1u}$ (*a* axis) and $\varepsilon_u$ (*b* axis), which are simultaneously excited. Transitions according to $^1E_g$ can occur only with $\varepsilon_u$ (*a* axis) or $\alpha_{1u}, \alpha_{2u}, \beta_{1u}$ or $\beta_{2u}$ (*b* axis)*. Therefore the transition $^1A_{1g} \rightarrow {}^1A_{2g}$ can only result when the light vector vibrates in the direction of the *b* axis, while $^1A_{1g} \rightarrow {}^1E_g$ is possible for light of both orientations. It therefore follows from experimental investigations with polarized light that the excited $^1A_{2g}$ state is associated with the shorter wavelength band at 430 m$\mu$.

(b) *Halfwidth of d → d bands*

Another problem deserving discussion is the question of the half width $\delta$ of the absorption bands. In the spectra of transition metal complexes one finds, e.g., for chromium(III) complexes, besides broad bands ($\delta \sim 3000$ cm$^{-1}$), intercombination bands of very small half-width ($\delta \sim 300$ cm$^{-1}$). For crystalline complexes, especially at low temperature, sharp absorption lines are observed. We have seen such sharp bands not only for chromium(III) complexes (see p. 43 ff. Figures A.31, A.33) but also for $[Mn(H_2O)_6]^{2+}$ (see p. 46 Figure A.34).

An explanation of the simultaneous appearance of broad, indistinct bands and sharp bands or lines was first given by Tanabe and Sugano[51] and later by Orgel[63].

We shall consider the situation for a diatomic molecule. In Figure A.56 the potential curves for the ground and the excited states together with their vibrational levels are shown schematically. If both potential curves have, as indicated in the left-hand side of the figure, minima at approximately the same value of the nuclear separation ($r_0'' = r_0'$), then according to the Franck–Condon principle† a spectrum is to be expected corresponding to essentially the 0–0 transition or the 1–1 transition (in case the first excited vibrational level is already occupied at room temperature). Only very few lines should be expected, or if modification through thermal and other effects is present, a relatively sharp band. If, on the other hand, the equilibrium separations in the ground state and in the first excited state do not coincide ($r_0' \neq r_0''$), as shown in the right-hand portion of the figure for the case of greater nuclear separation in the excited state, then transitions to a larger number of vibrational levels of the excited electronic state take place. This leads to a broader absorption band.

---

* The vibrations are denoted by $\alpha_1, \alpha_2, \beta_1, \beta_2$ and $\varepsilon$ corresponding to the irreducible representations $A_1, A_2, B_1, B_2$ and $E$ of the symmetry group $D_4$.

† The *Franck–Condon* principle states that because of the small mass of the electrons as compared with that of the nuclei the change in the electronic orientation of the molecule connected with the excitation of the electron takes place so rapidly that the position and velocity of the heavy nuclei remains essentially unchanged during the transition.

63. L. E. Orgel: *J. chem. Phys.*, **23**, 1824 (1955).

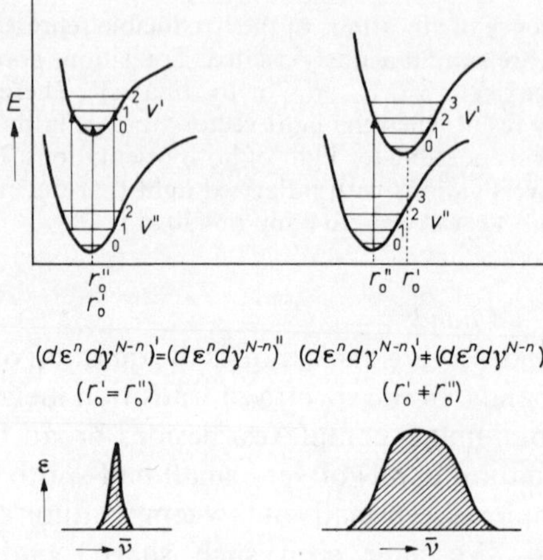


Figure A.56. An explanation of the half-width of absorption bands. The following relations hold for the values of the nuclear separation $r_0'$ and $r_0''$ corresponding to the minima of the potential curves:

1. $n' > n''$ i.e., $(N-n)' < (N-n)''$ $\quad r_0' < r_0''$
2. $n' < n''$ i.e., $(N-n)' > (N-n)''$ $\quad r_0' > r_0''$

Case 1. e.g., $Mn^{2+}$ $O_h$ symmetry

$^6A_{1g}(d\varepsilon^3\, d\gamma^2)$ $\quad ^4T_{1g}(d\varepsilon^4\, d\gamma)$

$$n' = 4, \quad (N-n)' = 1$$

$$n' = 3, \quad (N-n)' = 2$$

Case 2, e.g., $Cr^{3+}$ $O_h$ symmetry

$^4A_{2g}(d\varepsilon^3)$ $\quad ^4T_{1g}(d\varepsilon^2\, d\gamma)$

$$n' = 2, \quad (N-n)' = 1$$

$$n' = 3, \quad (N-n)' = 0$$

For every state of the complex ion arising, e.g., from an octahedral ligand configuration $d\varepsilon^n d\gamma^{N-n}$ the number of electrons ($n$ and $N-n$) in the two electronic states $t_{2g}$ and $e_g$ determines the slope of the curve, which gives the term energy as a function of the field strength in the Orgel diagram. To the first approximation the relation

$$\frac{d(\Delta E)}{d(Dq)} = -4n + 6(N-n).$$

is valid.

For an electronic transition between states of different slope the electronic distribution between the $(t_{2g})\, d\varepsilon$ and $(e_g)\, d\gamma$ states is changed. This means that the equilibrium distance metal ion–ligand is changed in going from the ground state to the excited state. The potential curves of both states have minima which are shifted relative to one another (Figure A.56, right side). In such a case $(d\varepsilon^n\, d\gamma^{N-n})' \neq (d\varepsilon^n\, d\gamma^{N-n})''$ one observes broad absorption bands. The greater the difference between the slopes $d(\Delta E)/d(Dq)$ of the two states between which the transition takes place is, the greater the half-width of the band in question should be. If the transition occurs between states of the complex ion arising from the same ligand field configuration $(d\varepsilon^n d\gamma^{N-n})' = (d\varepsilon^n\, d\gamma^{N-n})''$, that is between states having the same distribution of electrons between the $d\varepsilon$ and $d\gamma$ levels, then the nuclear separation in the lower and the upper state is essentially identical (Figure A.56, left side). One observes sharp lines or bands having a small half-width.

The first case is present for normal spin-allowed bands of the complex ions, while the second case arises for certain spin-forbidden bands.

As an example let us consider the octahedral chromium(III) complexes. The ground state ${}^4A_{2g}$ arises from the configuration $d\varepsilon^3$ and the two excited states ${}^4T_{2g}$ and $a^4T_{1g}$ from $d\varepsilon^2\, d\gamma$ or the higher $b^4T_{1g}$ term from $d\varepsilon\, d\gamma^2$ (cf. Figure A.32, p. 44). The transitions ${}^4A_{2g} \rightarrow {}^4T_{2g}$ or ${}^4A_{2g} \rightarrow {}^4T_{1g}$ lead to broad bands having half widths of $\sim$3000–4000 cm$^{-1}$. Also in the crystal essentially no decrease in the half-width is observed when going to low temperatures.

The weak spin-forbidden bands at 650–680 m$\mu$ correspond to transitions according to ${}^2E_g$ (or ${}^2T_{1g}$). These states also arise from the configuration $d\varepsilon^3$. The bands have a very small half-width ($\sim$200–300 cm$^{-1}$). In crystalline chromium(III) complexes one observes, especially at low temperatures, lines (cf. Figures A.31, A.33 p. 43 ff.) which are sharp enough to allow study of Zeeman splittings.

For octahedral cobalt(III) complexes one finds two broad bands associated with the transitions ${}^1A_{1g}(d\varepsilon^6) \rightarrow {}^1T_{1g}(d\varepsilon^5\, d\gamma)$ or ${}^1T_{2g}(d\varepsilon^5\, d\gamma)$. In the long wavelength spectral region, similar to the chromium(III) complexes, a spin-forbidden intercombination band corresponding to a singlet → triplet transition, appears in the declining region of the long wavelength first band. It has a half-width comparable to that of the two spin-allowed bands. This is understandable because its appearance is ascribable to the ${}^1A_{1g}(d\varepsilon^6) \rightarrow {}^3T_{1g}(d\varepsilon^5\, d\gamma)$ transition. Figure A.57 shows the spectra of $[Co(NH_3)_6]^{3+}$ and $[Coen_3]^{3+}$ in the region of the intercombination band.

On this basis the differences in the half-widths of the spin-forbidden bands in the spectrum of $[Mn(H_2O)_6]^{2+}$ ion can be qualitatively

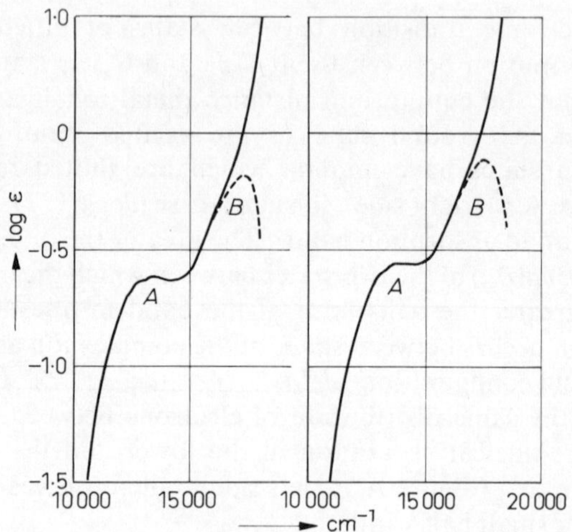


Figure A.57. The intercombination bands $^1A_{1g}(d\varepsilon^6) \rightarrow {}^3T_{1g}(d\varepsilon^5\, d\gamma)$ of the complex ions $[Co(NH_3)_6]^{3+}$ and $[Co\, en_3]^{3+}$.

understood[63]. From the spectrum the following ordering of the bands according to decreasing half-width is obtained, when the appropriate excited state is given (see Figure A.58):

$$^4G(A_{1g}) \approx {}^4G(E_g) < {}^4D(E_g) < {}^4D(T_{2g}) < {}^4G(T_{2g}) < {}^4G(T_{1g}).$$

With the aid of the term diagram for $Mn^{2+}$ (Figure A.50) it is seen that the increase of the half-width runs parallel to the increase of the slope of the $E = f(Dq)$ curves for the $Dq$ value of ca. 780 cm$^{-1}$, which is suitable for the hexaquo ion.

The band at longest wavelengths is the broadest at 18,800 cm$^{-1}$ $^6A_{1g}(d\varepsilon^3\, d\gamma^2) \rightarrow {}^4T_{1g}(d\varepsilon^4\, d\gamma)$; then follows the band at 23,000 cm$^{-1}$ $^6A_{1g}(d\varepsilon^3\, d\gamma^2) \rightarrow {}^4T_{2g}(d\varepsilon^4\, d\gamma)$. The very sharp bands at 24,900 and 25,150 cm$^{-1}$ correspond to the transitions $^6A_{1g}(d\varepsilon^3\, d\gamma^2) \rightarrow {}^4E_g$, $^4A_{1g}(d\varepsilon^3\, d\gamma^2)$. Also sharply peaked is the band lying at 29,700 cm$^{-1}$ $^6A_{1g}(d\varepsilon^3\, d\gamma^2) \rightarrow {}^4E_g(d\varepsilon^3\, d\gamma^2)$. The band lying at 28,000 cm$^{-1}$ also has identical configurations for the ground and the excited state and is associated with the transition to $^4T_{2g}(d\varepsilon^3\, d\gamma^2)$.

Finally it should also be noted that the same factor is decisive for the spectral shifting of the spin-allowed bands, as that which determines the half-width. Both are dependent upon the quantity $d(\Delta E)/d(Dq)$. Consequently the spectrochemical series of the ligands and the band widths are closely related. According to what has been said it is clear that the broader the band, the greater is the shift that is seen when the ligands are replaced

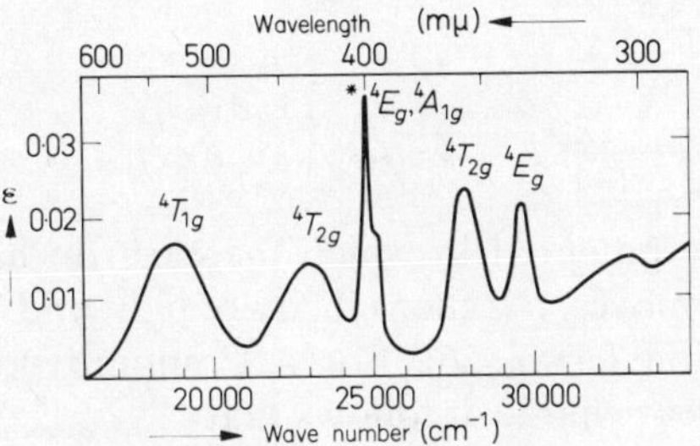


Figure A.58. Spectrum of $[Mn(H_2O)_6]^{2+}$.

by neighbours in the spectrochemical series. This is applicable only for spin-allowed bands for which the direction of the shift is always the same. For spin-forbidden bands $d(\Delta E)/d(Dq)$ can be zero or even negative.

## 1.11. Spin–orbit coupling

In the discussion of the absorption spectra of the complex ions of the first (and second) transition metal series we have considered until now only the orbital degeneracy and its partial or complete removal in fields of lower symmetry. A $d$ state of the free ion is in this approximation a fivefold orbitally degenerate state. Because each of the five one-electron $d$ states ($d_{xy}$, $d_{xz}$, $d_{yz}$; $d_{x^2-y^2}$ and $d_{z^2}$) can be occupied according to the Pauli principle by a maximum of two electrons (with antiparallel spin), such a $d$ state is in total tenfold degenerate.

If we wish to treat the magnetic properties of the compounds as well as other fine details of the spectra more accurately, especially for the case of complexes with central ions of the higher series of transition metals, then the interaction of the magnetic spin moment of the electron with the magnetic field arising from its orbital motion, the so-called *spin–orbit coupling*, must be considered.

The spin–orbit interaction is described by means of a *spin–orbit coupling constant* $\zeta_{nl}$ or $\lambda$, the magnitude of which can be obtained for the free ion from the multiplet splitting of the emission spectra. $\zeta_{nl}$ is a *one-electron parameter* and $\lambda$ a many-electron parameter, based upon one term of configuration $nd^N$ ($N = 1$–$9$).

The spin–orbit coupling constant $\zeta_{nl}$ for one electron is always positive*.

$$\zeta_{nl} = \hbar^2 \int_0^\infty R_{nl}^2 \, \xi(r) \, \mathrm{d}r$$

* Cf. Part B, chapter 5.

with

$$\xi(r) = \frac{-e}{2m^2c^2}\left(\frac{1}{r}\frac{\partial V(r)}{\partial r}\right),$$

whereby $V(r)$ is the potential in which the electron having the mass $m$ and the charge $-e$ moves. $c$ is the velocity of light and $R_{nl}$ the radial part of the wave function (cf. p. 56). For a Coulomb potential $V = Ze/r$, where $Z$ is the nuclear charge, it follows that

$$\zeta_{nl} = \frac{e^2\hbar^2}{2m^2c^2a_0^3} \cdot \frac{Z^4}{n^3 l(l+\frac{1}{2})(l+1)},$$

when $a_0$ is the *Bohr radius.*

For the ground term (maximal spin and maximal orbital angular momentum quantum number) which arises from the configuration $d^N$

$$\lambda = \pm\zeta_{nd}/2S,$$

where the positive sign is for $N < 5$ and the negative for $N > 5$. For $N = 5$, $\lambda = 0$. (See, for example, E. U. Condon and G. H. Shortley: *The Theory of Atomic Spectra*, Cambridge University Press, 1951, pp. 120 ff.).

In the first series of transition metals the $\zeta_{3d}$ values lie between 154 $cm^{-1}$ for $Ti^{3+}$ and 829 $cm^{-1}$ for $Cu^{2+}$. In the second transition series the $\zeta_{4d}$ values lie higher (e.g. 500 $cm^{-1}$ for $Zr^{3+}$, 1500 $cm^{-1}$ for $Ru^{+}$) (cf. Table A.8).

In the absence of an external field a $^2D$ term corresponds to a single *nd* electron, as we have seen for the $Ti^{3+}$ ion (cf. section 1.6b). When the spin–orbit interaction is taken into consideration, this state consists of two multiplet components $^2D_{3/2}$ and $^2D_{5/2}$ whose separation on the energy scale is $\frac{5}{2}\zeta_{3d}$ (Figure A.59, left side). For $Ti^{3+}$ the distance between the two multiplets is 384 $cm^{-1}$ (cf. Part B, p. 438).

If the spin is not considered, a fivefold orbitally degenerate one-electron $d$ state or a $^2D$ state splits in $O_h$ symmetry into a triply orbitally degenerate state $^2T_{2g}$ and a doubly degenerate $^2E_g$ state.

If the spin–orbit coupling is considered, the $^2T_{2g}$ state, which is in total sixfold degenerate, splits into two states $(^2T_{2g}^7)\Gamma_7$ and $(^2T_{2g}^8)\Gamma_8$, the first of which is doubly and the second fourfold degenerate. The $^2E_g$ state, denoted by $(^2E_g^8)\Gamma_8$ is not split. It has the same symmetry as $^2T_{2g}^8$ and is likewise fourfold degenerate. These relations are schematically depicted in Figure A.59 (right side)*.

* For the notation of the states used for ambiguous representations, cf. Part B, pp. 315, 319.

Table A.8. Values of the one-electron spin–orbit coupling constants $\zeta_{nd}$ for several free ions

| Ion | $\zeta_{3d}$[cm$^{-1}$] | Ion | $\zeta_{4d}$[cm$^{-1}$] | Number of *d*-electrons |
|---|---|---|---|---|
| $Ti^{3+}$ | 154 | $Zr^{3+}$ | 500 | 1 |
| $Ti^{2+}$ | 123 | $Zr^{2+}$ | 468 | 2 |
| $V^{3+}$ | 217 | $Nb^{3+}$ | 800 | |
| $V^{2+}$ | 169 | $Nb^{2+}$ | 560 | 3 |
| $Cr^{3+}$ | 275 | $Mo^{3+}$ | 800 | |
| $Cr^{2+}$ | 229 | $Ru^{4+}$ | 1600 | 4 |
| $Mn^{3+}$ | 352 | | | |
| $Mn^{2+}$ | 395 | | | 5 |
| $Fe^{3+}$ | 440 | $Ru^{3+}$ | 1500 | |
| $Fe^{2+}$ | 408 | | | 6 |
| $Co^{3+}$ | 500 | | | |
| $Co^{2+}$ | 530 | | | 7 |
| $Ni^{2+}$ | 630 | | | 8 |
| $Cu^{2+}$ | 829 | $Ag^{2+}$ | 1843 | 9 |

A summary of spin–orbit coupling constants $\zeta_{nd}$ for the transition metal ions up to oxidation state M(VI) for the 3*d*- and 4*d*-series is given by T. M. Dunn: *Trans. Faraday Soc.*, **57**, 1441 (1961).

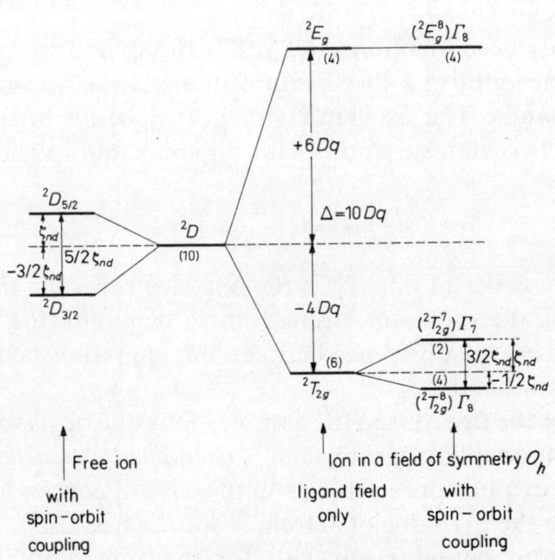


Figure A.59. Splitting diagram for a $^2D$ state in a field of $O_h$ symmetry with consideration of the spin–orbit interaction (schematic).

If the magnitudes of the coupling constants $\zeta_{nd}$ (Table A.8) are compared with the values for $\Delta \equiv 10Dq$, which lie at $\approx$ 10,000–20,000 cm$^{-1}$, it is seen that it is in general permissible to neglect spin–orbit coupling for a discussion of optical properties. That is, to proceed as we have thus far and to consider only the orbital degeneracy.

Because the splitting for equal distances between the central ion and the ligands for the case of tetrahedral coordination is only $\frac{4}{9}$ of the splittings for octahedral coordination (cf. p. 63), it is to be expected that under certain conditions the spin–orbit coupling energy could be of the order of the energy values for the ligand field splitting. This should be the case for e.g. $Co^{2+}$ and $Cu^{2+}$ compounds, because the $\zeta_{3d}$ values are here approximately equally large. In the absorption spectra of such complex ions it should be possible to find indications of structures attributable to spin–orbit coupling.

The intense band found in the 14,300–16,400 cm$^{-1}$ region of the absorption spectrum of $[CoCl_4]^{2-}$ ion can be assigned to the transition $^4F(A_2) \rightarrow {}^4P(T_1)$. It shows a well-developed, characteristic structure (five or six maxima (Figure A.54)), which results from the splitting of the $^4P(T_1)$ state by spin–orbit interaction. In a rigorously cubic field the $^4P(T_1)$ state should split into four states, two of which are doubly and two fourfold degenerate. For sufficiently low symmetry a maximum of six doubly degenerate states (Kramers doublets) can arise. $\zeta_{3d}$ is for $Co^{2+}$ 530 cm$^{-1}$ and $\Delta \approx 3300$ cm$^{-1}$.

As has already been mentioned (cf. p. 87), the spin–orbit coupling must be taken into account in a discussion of the relative intensities of intercombination bands. The oscillator strength $f_{ab}$ of an intercombination band between two states $a$ and $b$ with different multiplicity is given by

$$f_{ab} = f_0 \left( \frac{k \zeta_{nl}}{\Delta E} \right)^2$$

Here $k$ is of the order of one, $\zeta_{nl}$ is the one-electron spin–orbit coupling constant and $f_0$ the oscillator strength for a neighbouring spin-allowed band. $\Delta E$ is the energy difference between this spin-allowed band and the intercombination band.

In going from the free ions to the complex ions one observes a reduction of the spin–orbit coupling parameters $\zeta_{nd}$ (*relativistic nephelauxetic effect*). This reduction can be appreciable as in the case of copper hexaquo ions. Compared with the $\zeta_{3d}$ value of 830 cm$^{-1}$ for the free $Cu^{2+}$ ion the value for the spin–orbit coupling constant for the aquo complexes is only $\approx$ 620 cm$^{-1}$. From the results of paramagnetic resonance measurements it is found that, compared to the free ions in general, a decrease of the $\zeta_{nd}$ values of 10–40 per cent in the complexes in question appears. The situation is analogous to that of the electronic interaction parameter $B$ (cf. p. 83). A quantity

$$\beta^* = \frac{\zeta_{nd\,(\text{compl. ion})}}{\zeta_{nd\,(\text{free ion})}}$$

can be defined.

The spin–orbit coupling constant should become smaller, the more the electron cloud is spread out. This can be caused either from a radial extension of the wave function (delocalization of $d$ electrons) or by charge transfer from the central ion to the ligands[64,65].

The spin–orbit coupling is of critical importance for the treatment of absorption spectra of compounds of the rare earth elements. The multiplet distances are of the order of 1000 cm$^{-1}$, while the ligand field splitting of a given $^{2S+1}L_J$ state of the free ion is only about 100 cm$^{-1}$. The $4f$ electrons are shielded from the environment by the $5s$ and $5p$ shells, so that sharp lines or bands of small half-width are observed in the spectra.

## 1.12. Treatment of complexes by the molecular orbital theory*

Another approximation method besides the ligand field theory, which has occupied our attention until now, is the molecular orbital theory. This was originally developed for diatomic molecules and aromatic hydrocarbons[66,67] and was applied first by Van Vleck[68] to complexes. The electronic structure of the central ion and the ligands is considered and it is assumed that the electrons move in orbitals extending over the entire molecule (central ion and ligands). One can only speak meaningfully of an electron cloud for the total molecular entity (cf. Figure A.12$c$, p. 18). The molecular orbitals (MO's) can be built up from the atomic orbitals (AO's) through a suitable linear combination; hence LCAO = linear combination of atomic orbitals. Besides the eigenfunctions of the ligands, which are occupied in the zeroth approximation by the lone pair electrons, the five $d$ eigenfunctions, the three $p$ functions and a $s$ eigenfunction of the central ion are available for constructing these molecular eigenfunctions.

---

* For details of the MO method, see, for example, C. A. Coulson: *Valence*, Chapter IV, Oxford University Press, 1952.

64. J. Owen: *Proc. R. Soc.*, **A 227**, 183 (1955).
65. T. M. Dunn: *J. chem. Soc.*, **1959**, 623.
66. R. S. Mulliken: *Rev. mod. Physics*, **4**, 1 (1932); *Phys. Rev.*, **41**, 49 (1932); **43**, 279 (1933); *J. chem. Phys.*, **3**, 375 (1935).
67. F. Hund: *Z. Physik*, **63**, 719 (1930); *Manual of Physics*, Vol. 24/I, 561 (1933).
68. J. H. Van Vleck; *J. chem. Phys.*, **3**, 803, 809 (1935).

The five $d$ angular functions have already been introduced (cf. p. 53) and the corresponding representation of the $s$ and three $p$ functions $p_x$, $p_y$, $p_z$ are given in Figures A.60 and A.39.

The fundamental ideas of the LCAO method will be discussed using the example of the hydrogen molecule ion $H_2^+$.

If the electron belongs either to the one hydrogen nucleus A or to the other hydrogen nucleus B (nuclear separation is very large compared with the Bohr radius), then the ground state for the first case is described by a function $\phi_A$ and in the second case by a function $\phi_B$. If the electron belongs to both nuclei, as in the actual molecule $H_2^+$, its position can be described

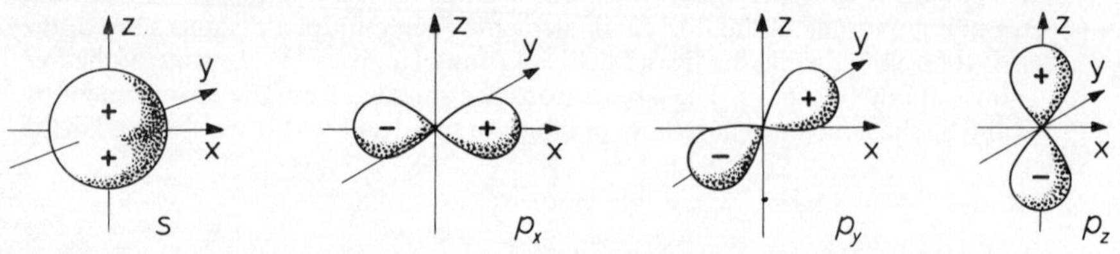


Figure A.60. Sketches of the angular part of the $s$ and $p$ orbitals.

by a superposition of both functions. If $\phi_A$ and $\phi_B$ are orthogonal and normalized, one obtains for the two normalized LCAO functions the linear combinations (assuming that the overlap is vanishingly small):

$$\psi_b = 1/\sqrt{2}\cdot(\phi_A + \phi_B)$$
$$\psi_a = 1/\sqrt{2}\ (\phi_A - \phi_B).$$

The physical significance of the LCAO function is seen when $\phi_A^2$ and $\phi_B^2$ are graphically compared with $\psi_b^2$ or $\psi_a^2$ as is done in Figure A.61.

An electron occupying the molecular orbital described by the function $\psi_b$ has an appreciable position probability density between the nuclei A and B ($\psi^2\,d\tau$ is a measure of the position probability density.) A bonding orbital is associated with $\psi_b$. An electron occupying the molecular orbital associated with $\psi_a$ has a smaller position probability density in the region between the nuclei, as for the case $\psi_b$. A nodal plane bisects the A–B axis. The state $\psi_a$ is an antibonding state. Figure A.62 shows the energy of the two MO's relative to the energy of an isolated hydrogen atom. The AO's $\phi_A$ and $\phi_B$ have the same energy, while the bonding molecular orbital $\psi_b$ is stabilized by an amount $E_{AB}$ compared with $\phi_A$ or $\phi_B$ and the antibonding molecular orbital $\psi_a$ is destablized by the amount $E_{AB}$*.

* In addition there is the labellizing energy (arising from the Coulomb integral) which is equally large for the two molecular orbitals $\psi_a$ and $\psi_b$ and is in general very small compared with $E_{AB}$. It is not considered in Figure A.62.

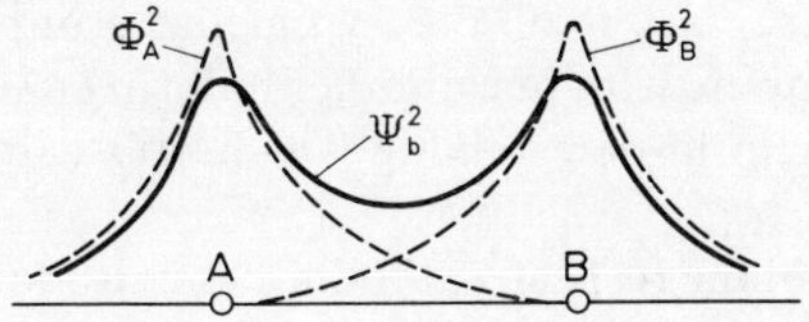


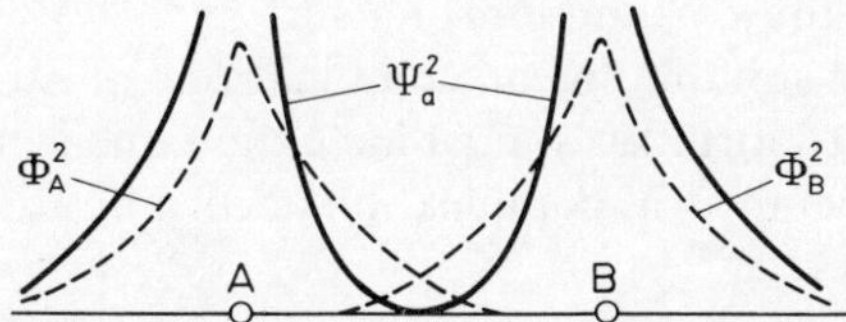


Figure A.61. $\Psi_a^2$ and $\Psi_b^2$ for a diatomic molecule A-B, e.g., $H_2^+$.

A qualitatively similar situation is found for all MO energy diagrams resulting from two atoms, each of which contributes one atomic orbital. If $\phi_A$ and $\phi_B$ do not have the same energy (heteronuclear molecules) then $\psi_b$ and $\psi_a$ lie approximately equidistant from the energetic mean of the two AO's. For all MO diagrams, even for complicated cases, the energies of the AO's of the atoms constituting the molecule in question are arranged on the left and right side, while in the middle the energies of the resulting MO's are given*.

The determination of the MO's is greatly simplified when the molecule considered is highly symmetrical. Only orbitals which transform according

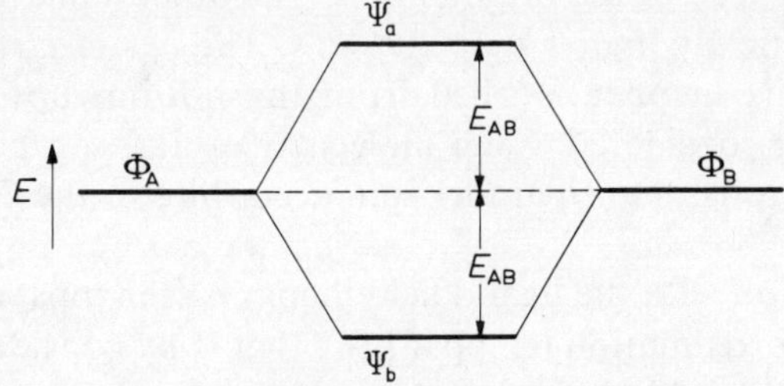


Figure A.62. Energy diagram for the bonding and antibonding molecular orbitals of a homonuclear diatomic molecule (cf. footnote*.)

*For the AO's linear combinations from *s*, *p*, *d*, ... functions are also possible.

to the same representation of the symmetry group of the molecular framework can combine with one another to form LCAO–MO's. The following three criteria help in determining which AO's can combine with one another:

(a) The energies of the respective atomic orbitals must be of comparable magnitude.

(b) The electron clouds of the AO's must show the greatest possible overlap (the degree of overlap can be obtained qualitatively from the diagram of the angular functions, as they are shown in Figure A.60 or A.38 and A.39 for the *s*, *p*, and *d* orbitals.)

(c) The AO's (or suitable linear combinations of AO's) which combine with one another to form MO's must have the same symmetry properties.

In treating an octahedral complex ion according to the MO theory one proceeds as follows:

1. The octahedral group ($O_h$) is, as is every other point group, a subgroup of the rotation–reflection group of the sphere. Therefore all AO's of the central ion (or suitable linear combinations of these AO's) can be divided into families which transform according to the irreducible representations of the group $O_h$.

2. The orbitals of the ligands are next considered. We can differentiate $\sigma$, $\pi$, etc. states according to their orientation with respect to the line-of-centres ligand–central ion, that is, according to the bond direction. $\sigma$ orbitals have rotational symmetry with reference to the bond direction, $\pi$ states have a nodal plane containing the bond axis (see Figure A.63).

3. The MO's of the complex ion are generated through suitable linear combinations of the orbitals obtained according to 1 and 2 with consideration of (a), (b) and (c).

The number of MO's obtained is the same as the number of AO's brought into consideration. Half of the MO's are in general bonding; they are stabilized compared to the original AO's. The other half are antibonding; they lie energetically higher than the AO's. The *one-electron term system* obtained in this manner is filled from the bottom upwards with the electrons being considered. Each molecular orbital can be filled with at most two electrons of antiparallel spin according to the Pauli exclusion principle.

The connection with the ligand field theory is seen most easily from the strong-field approximation (cf. pp. 51 ff.), that is by considering the effect of the ligand field upon the *d* electron states (without introducing electron interaction).

The five *d* orbitals belong in an octahedral field to the representations $T_{2g}$ and $E_g$ of the group $O_h$. The orbitals of the ligands can likewise be classified according to the irreducible representations of this group.

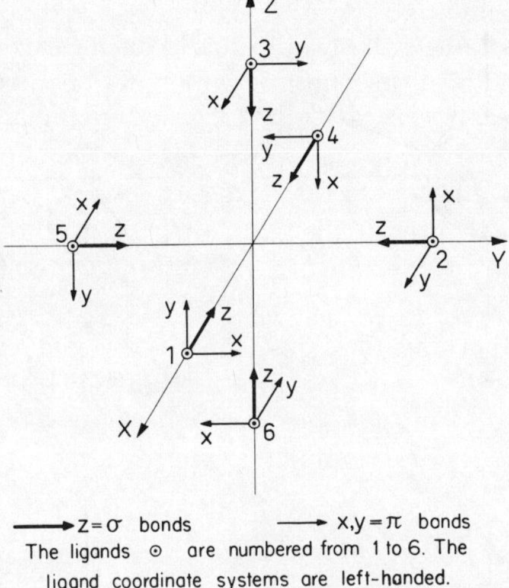


Figure A.63. σ and π bonds in an octahedral complex ion.

From each ligand a $p$ state having an angular function in the bond direction ($\sigma$ bond) and two perpendicular $p$ orbitals ($\pi$ bonds) are considered. Figure A.63 shows schematically the orientation of the ligand orbitals for all six ligands. Table A.10 gives the ligand or central ion states which together form $\sigma$ or $\pi$ bonds. Figure A.64 shows the appropriate orbitals for the case of $\sigma$ bonds.

As an example, the $d_{x^2-y^2}$ function overlaps in general the $\sigma$ orbitals of the ligands 1,4 and 2,5, oriented at $x$, $-x$ and $y$, $-y$. The $\sigma$ orbitals of the ligands are composed of linear combinations of AO's as shown in Table A.10, column 3. The $d_{z^2}$ orbitals overlap with $\sigma$ orbitals at $x$, $-x$; $y$, $-y$, preferentially though at $z$, $-z$ (ligands 1,4; 2,5 and 3,6). The wave functions for $d\,e_g$, $\sigma\,e_g$ are*

$$\left.\begin{aligned}\phi_1 &= \alpha\, d_{z^2} + \{(1-\alpha^2)/12\}^{\frac{1}{2}}\,[z_1+z_2-2z_3+z_4+z_5-2z_6]\\ \phi_2 &= \alpha\, d_{x^2-y^2} + \{(1-\alpha^2)/4\}^{\frac{1}{2}}\,[z_1-z_2+z_4-z_5]\end{aligned}\right\}\ \text{bonding functions}$$

$$\left.\begin{aligned}\phi_1' &= (1-\alpha^2)^{\frac{1}{2}}\, d_{z^2} - \left(\frac{\alpha^2}{12}\right)^{\frac{1}{2}}\,[z_1+z_2-2z_3+z_4+z_5-2z_6]\\ \phi_2' &= (1-\alpha^2)^{\frac{1}{2}}\, d_{x^2-y^2} - \left(\frac{\alpha^2}{4}\right)^{\frac{1}{2}}\,[z_1-z_2+z_4-z_5]\end{aligned}\right|\ \text{antibonding functions}$$

---

* This formulation is valid only when the overlap integrals between all central ion and ligand functions are zero (cf. Part B, p. 427).

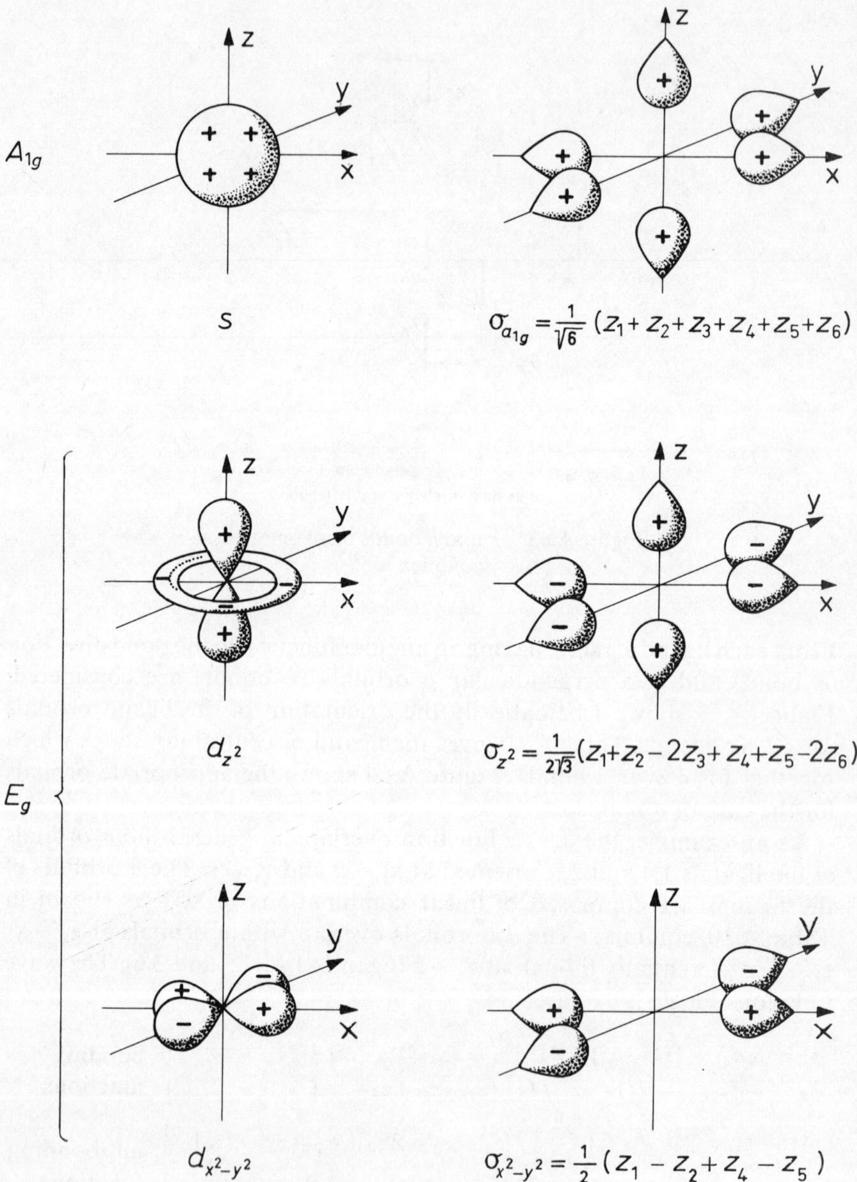


Figure A.64. Orbitals of the central ion and ligand orbitals for the case of pure $\sigma$ bonding (cf. Table A.10 and Figure A.65).

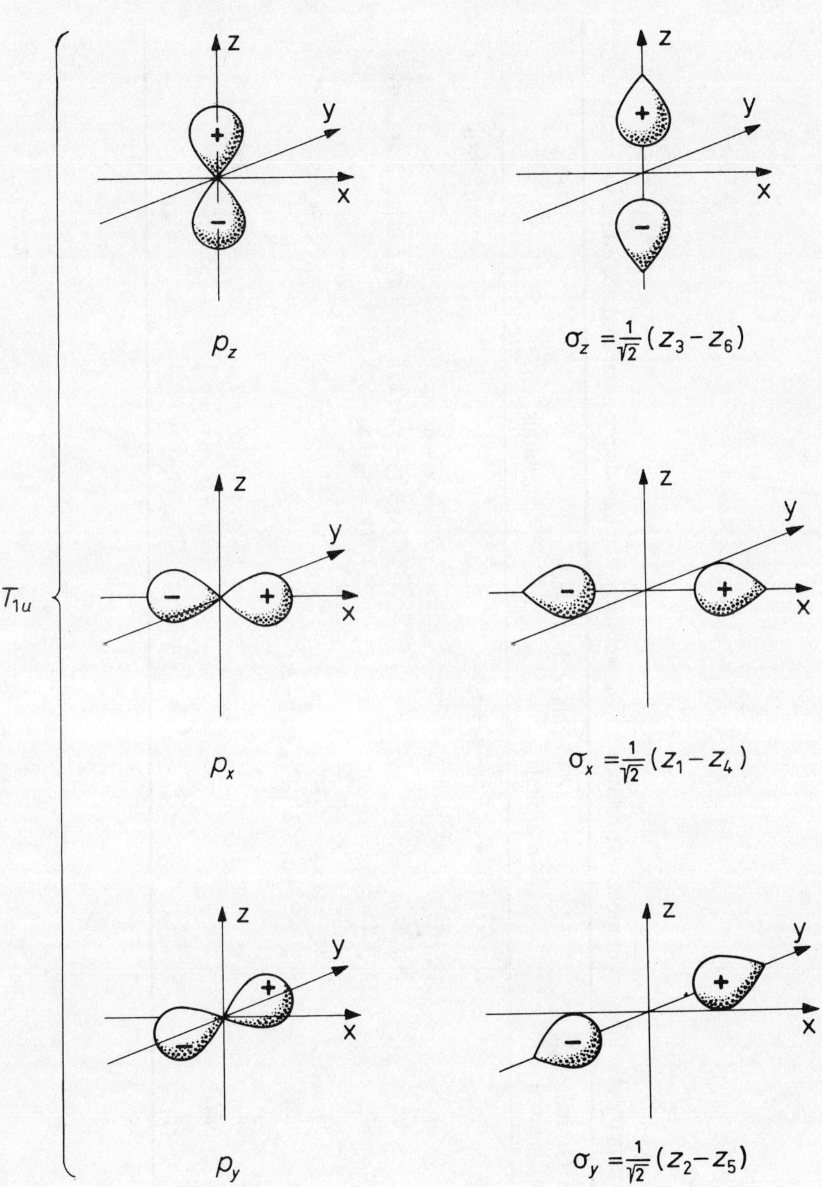


Continuation of Figure A.64.

Table A.10. Symmetry classification of molecular orbitals for $O_h$ symmetry (the ligands are numbered according to Figure A.63)

| Symmetry | Orbital of the central ion | Orbitals of the ligands: $\sigma$ orbitals | Orbitals of the ligands: $\pi$ orbitals | MO's |
|---|---|---|---|---|
| $a_{1g}$ | $4s$ | $1/\sqrt{6}(z_1+z_2+z_3+z_4+z_5+z_6)$ | | $sa_{1g}, \sigma a_{1g}$ |
| $t_{1u}$ | $4p_x$ | $1/\sqrt{2}(z_1-z_4)$ | $1/2(y_2+x_3-x_5-y_6)$ | $pt_{1u}, \sigma t_{1u}$ ⋮ $pt_{1u}, \pi t_{1u}$ |
| | $4p_y$ | $1/\sqrt{2}(z_2-z_5)$ | $1/2(x_1+y_3-y_4-x_6)$ | |
| | $4p_z$ | $1/\sqrt{2}(z_3-z_6)$ | $1/2(y_1+x_2-x_4-y_5)$ | |
| $e_g$ | $3d_{x^2-y^2}$ | $1/2(z_1-z_2+z_4-z_5)$ | | $de_g, \sigma e_g$ |
| | $3d_{z^2}$ | $1/2\sqrt{3}(z_1+z_2-2z_3+z_4+z_5-2z_6)$ | | |
| $t_{2g}$ | $3d_{xy}$ | | $1/2(x_1+y_2-y_4-x_5)$ | $dt_{2g}, \pi t_{2g}$ |
| | $3d_{xz}$ | | $1/2(y_1+x_3+x_4+y_6)$ | |
| | $3d_{yz}$ | | $1/2(x_2+y_3+y_5+x_6)$ | |

$\alpha$ is a coefficient giving the degree of interaction of the states. For a purely electrovalent complex $\alpha^2$ has the value 1; for a purely covalent complex the value 0·5. The first condition corresponds to the limiting case for which the extended ionic model was conceived (see note p. 17), the second condition corresponds to the limiting case of Pauling's model. The deviation from the extended ionic model is expressed by the value of $\alpha$.

Four other bonding and antibonding functions can be constructed from the $s$, $p_x$ and $p_y$ and $p_z$ orbitals of the central ion with the $\sigma$ orbitals of the ligands, as shown in column 3 of Table A.10. The $d_{xy}$, $d_{xz}$, $d_{yz}$ orbitals of the central ion can no longer combine with $\sigma$ states of the ligands; they are nonbonding.

Figure A.65 shows a schematic energy diagram for the MO's of an octahedral complex ion, where the $\pi$ bond portions are neglected. The order of the bonding $a_{1g}$, $t_{1u}$ and $e_g$ states is not known, likewise for the antibonding $a_{1g}^*$ and $t_{1u}^*$ states. It is essential that the $(d\varepsilon)t_{2g}$ states of the central ion are not influenced by the $\sigma$ bonds (nonbonding states), while

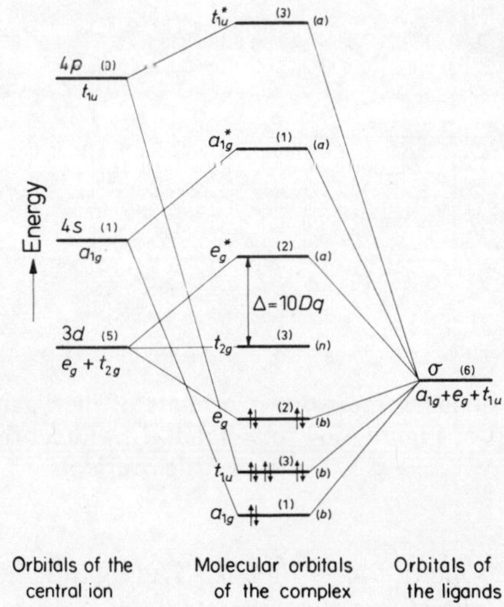


Figure A.65. Schematic representation of the molecular orbitals of an octahedral complex ($\sigma$-bonds only). The numbers in parentheses give the degeneracy of the corresponding orbitals. The electrons contributed by the ligands[12] are denoted by arrows. $b$ = bonding, $a$ = antibonding, $n$ = nonbonding state. The antibonding states are, as is conventional, starred.

the $(d\gamma)e_g$ states combine with the ligand states, where an antibonding $e_g^*$ state arises alongside a bonding $e_g$ state. The twelve electrons contributed by the ligands (in Figure A.65 shown by arrows) occupy the bonding states $a_{1g}$, $t_{1u}$ and $e_g$. The $d$ electrons of the central ion occupy the $t_{2g}$ and $e_g^*$ states. The distance between the nonbonding $t_{2g}$ and the antibonding $e_g^*$ state corresponds to the value of $\Delta \equiv 10\,Dq$.

With inclusion of $\pi$ bond contributions it can be qualitatively understood on the basis of overlap, that the $\pi$ bonds usually should be much weaker than the $\sigma$ bonds (see Figure A.66). The $(d\varepsilon)t_{2g}$ central ion states $(d_{xy}, d_{xz}, d_{yz})$ can combine with the $\pi$ states of the ligands having symmetry $t_{2g}$ (see Table A.10, column 4). Thereby a bonding and an antibonding molecular state arise, which are each threefold orbitally degenerate. Two cases can be distinguished for ligands which enter into $\pi$ bonds with the central ion. These are schematically depicted in Figure A.67.

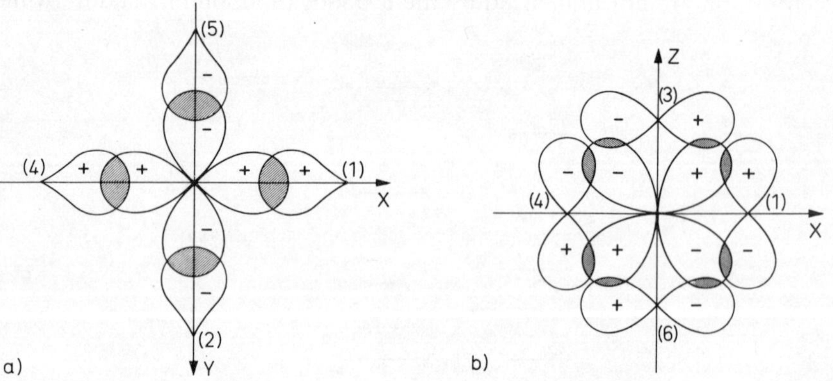


Figure A.66. (a) $\sigma$ bonds $d_{x^2-y^2}$ with $\sigma$ orbitals of the ligands at $x$, $-x$, $y$, $-y$ strongly overlapping), cf. Figure A.64. (b) $\pi$ bonds $d_{xz}$ with $\pi$ orbitals of the ligands at $x$, $-x$, $z$, $-z$ (very little overlap).

1. *Dative $\pi$ bonds metal ion $\rightarrow$ ligands.* The ligands have $\pi$ states, which are energetically higher lying than the states of the metal ions. They arise from $3d$ states and are unoccupied in the ground state. The bonding $t_{2g}$-MO contains essentially portions of the $d_{xy}, d_{xz}, d_{yz}$ orbitals of the central ion. It decreases, compared with the case where only a purely electrostatic interaction is assumed. This is an immediate consequence of the electron delocalization in connection with the formation of $\pi$ bonds. The energetically higher lying antibonding $t_{2g}^*$ state has essentially ligand character. It remains unoccupied. Similar conditions are present for phosphine or arsine-complexes. The value of $\Delta$ the energy difference $e_g^* - t_{2g}$ is increased

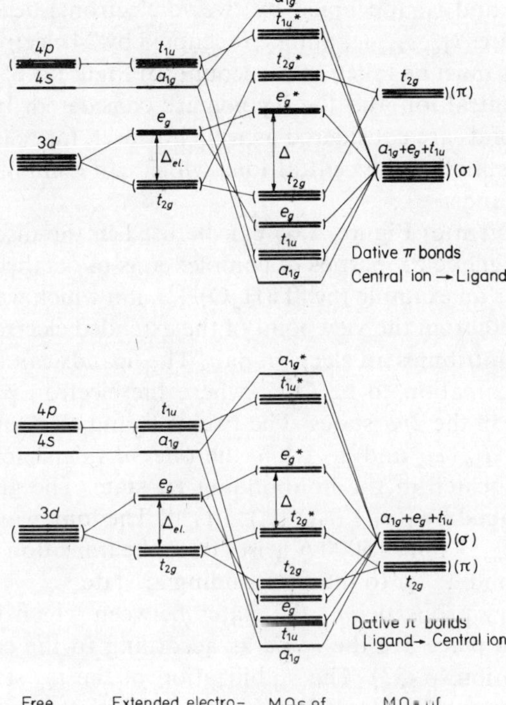


Figure A.67. Schematic MO diagrams for octahedral complex ions with consideration of $\pi$ bonds of the $t_{2g}$ states of the metal ion with the $t_{2g}$ states of the ligands.

through $\pi$ bonding, compared with the value $\Delta_{el}$ to be expected from the purely electrostatic consideration.

2. *Dative $\pi$ bonds ligands $\rightarrow$ central ion.* In certain cases, e.g. $OH^-$, the ligands have $\pi$ states lying at low energies which are occupied in the ground state. The interaction with the $d\varepsilon$ states leads to a bonding $t_{2g}$ molecular orbital having essentially ligand character and a weakly anti-bonding $t_{2g}^*$ molecular orbital having essentially metallic character. $t_{2g}^*$ lies consequently higher than for pure electrostatic interaction. $\Delta$, the energy difference $e_g^* - t_{2g}^*$, is consequently smaller than in case 1, nevertheless in general still larger than $\Delta_{el}$.

When one considers an octahedral complex ion and brings besides $\sigma$ bonds $\pi$ bonds into the discussion as well, then besides the states of the ligands $a_{1g}$, $e_g$ and $t_{1u}$ (occupied by twelve electrons) necessary for the $\sigma$ bonds, the states $t_{1g}$, $t_{2g}$, $t_{1u}$ and $t_{2u}$ (occupied by 24 electrons) necessary for the $\pi$ bonds must be taken into account. In Figure A.65 only $\sigma$ bonds between the central ion and the ligands are considered. In Figure A.67 additional $\pi$ bonds are considered as well, however, for reasons of clarity only those $t_{2g}$ states of the central ions which can combine with the $t_{2g}$ states of the ligands.

The term system of Figure A.65 can be used in the discussion of the optical and magnetic properties of complex ions of octahedral structure. We shall take as an example the $[Ti(H_2O)_6]^{3+}$ ion which we have already treated (see p. 30) from the viewpoint of the extended electrostatic theory. Every ligand contributes an electron pair. The ligands can be assumed in the first approximation to be $O^{2-}$, where the electron pair with antiparallel spin is in the $2p\sigma$ states. The twelve ligand electrons occupy the bonding states $a_{1g}$, $t_{1u}$ and $e_g$, while the one $3d$ valence electron of the central ion is located in the nonbonding $t_{2g}$ state. The ground state is therefore described by $[(a_{1g})^2(t_{1u})^6(e_g)^4]\ (t_{2g})^1$. The long wavelength band of the $[Ti(H_2O)_6]^{3+}$ ion at 5000 Å arises from the transition of an electron from the nonbonding $t_{2g}$ to the antibonding $e_g^*$ state.

The symmetry properties of the states between which the electronic transitions take place are the same as according to the extended ionic theory (description, p. 33). The stabilization of the $t_{2g}$ state compared with the $e_g$ state arises in the MO theory because the latter has antibonding character. According to the extended electrostatic theory the splitting and thereby the stabilization of the $t_{2g}$ state as compared with the $e_g$ state arises only from the differing magnitudes of the electrostatic interaction (see p. 55 ff.).

$\Delta \equiv 10Dq$ can, as already indicated, be obtained from the absorption spectra of suitable octahedral complexes. This experimental quantity can be used to discuss the optical and magnetic properties, the stereochemistry, the heats of formation, etc. $\Delta$ is accordingly, not as in the extended ionic model the result of a purely electrostatic interaction. Rather it becomes clear in a consideration according to the molecular orbital theory, that besides the electrostatic contribution $\Delta_{el}$ at least three other contributions must be considered: a positive contribution from $\sigma$ bonds, a positive contribution from dative $\pi$ (M → I) bonds as well as a negative contribution from dative $\pi$ (L → M) bonds.

$$\Delta \approx \Delta_{el} + \Delta\sigma + \Delta\pi(M \rightarrow L) - \Delta\pi(L \rightarrow M).$$

The additivity is of course only a very rough simplification.

For the many-electron case the states of the complex ion arise from the corresponding configurations (occupations of the MO's) only after the electron interaction has been introduced. For example, the occupation $[(a_{1g})^2(t_{1u})^6(e_g)^4](t_{2g})^3$ is valid for the ground state of the $[Cr(H_2O)_6]^{3+}$ ion. Excited configurations are $[(a_{1g})^2(t_{1u})^6(e_g)^4]\,(t_{2g})^2(e_g^*)$ and $[(a_{1g})^2(t_{1u})^6(e_g)^4]$ $(t_{2g})(e_g^*)^2$ (see p. 68, Figure A.49). The states of the complex ion belonging to these occupations of the MO's are, when only those of multiplicity 4 are considered, $[(a_{1g})^2(t_{1u})^6(e_g)^4]\,(t_{2g})^3 \rightarrow {}^4A_{2g}$; $[(a_{1g})^2(t_{1u})^6(e_g)^4]\,(t_{2g})^2(e_g^*) \rightarrow$ ${}^4T_{2g}+a^4T_{1g}$; $[(a_{1g})^2(t_{1u})^6(e_g)^4]\,(t_{2g})(e_g^*)^2 \rightarrow b^4T_{1g}$.

A description of the complexes of the transition metal ions with the help of the MO theory gives a more satisfactory and complete picture of the situation than the extended ionic theory in so far as the covalent parts of the coordination bonds are taken into consideration. This means, however, that exactly the advantage of the extended ionic theory, namely its simplicity and lucidity, is lost. Besides this, in the realm of the MO theory, calculations with consideration of electron interaction are impracticable. One must rely upon discussions in the one-electron term scheme. In the treatment of concrete problems, in so far as this is possible, practically and for reasons of economy, the extended ionic theory is utilized. However, it must always be remembered that the electrostatic interaction gives only one contribution to the experimentally determined field parameter. A combination of the extended ionic theory with the MO theory allows for qualitative speculations concerning the rôle of the ligands, the $\pi$ bond contributions etc. and also allows the discussion of electron-transfer bands, as will be shown in the next section.

## 1.13. Charge–transfer spectra

We have seen in the last section that in the treatment of a complex ion according to the MO theory more energy states for the complexes considered are obtained than by the use of the extended ionic model as described in sections 1 to 10 (the electrostatic ligand field model). Here only the $d$ states of the central ion in the electric field of the ligands were considered and were represented as point charges or point dipoles, while the MO theory considers the electronic states of the ligands as well.

The consideration of the ligand states allows the clarification of the electron transfer processes between the central ion and the ligands. A transfer of charge from the ligand to the central ion or the reverse process can take place. In the parlance of the MO theory this means, for example for the first case, that the electronic transitions from states which are located essentially on the central ion, and therefore have $d$ character, to those having essentially ligand character can appear. These transitions are spectroscopically observable as *charge–transfer bands.* Depending upon the direction of the electronic transition one speaks of *metal oxidation* or *metal reduction bands*. They appear predominantly in the ultraviolet spectral region or for some complexes in the visible. The charge–transfer bands have, in the rule, a factor of 100–1000 greater intensity than the bands corresponding to $d \rightarrow d$ transitions, which have been discussed in the preceding sections. The oscillator strengths of the charge–transfer bands lie at $f \sim 0{\cdot}1$ and the values of the molar decadic extinction coefficient $\varepsilon_{max}$ for the band maxima at $10^4$–$10^5$. It can be concluded that likewise the majority of the charge–transfer transitions should be allowed as pure electronic transitions. That is they take place between states of different parity ($g \leftrightarrow u$). Likewise the participating states should have the same multiplicity.

Metal oxidation bands which can be approximately assigned to the transitions $t_2 \rightarrow \pi_1^*$ or $\pi_2^*$ or $e \rightarrow \pi_1^*$ between states on the central ion and empty antibonding ligand states of the pyridine molecules are found, e.g. $[M^{II}X_2py_2]$ and $[M^{III}X_3py_3]$ ($M = Cu^{2+}$, $Pt^{2+}$; $Mo^{3+}$, $Ir^{3+}$)[69,70] in the region from 24,000–36,000 $cm^{-1}$) see Figure A.9: $[MoCl_3py_3]$, band CTF at about 390 m$\mu$). Similar bands appear, e.g. also for *o*-phenanthroline complexes of the bivalent iron.

For $[Co(NH_3)_5X]^{2+}$ complexes metal reduction bands are observed, which are associated with a charge transfer of an electron essentially located on a $X^-$ ion to a *d* state of the central ion. In the series Cl, Br, I = X a shifting of the charge-transfer bands to longer wavelengths takes place (Figure A.68). The band at longer wavelengths CTF 1 (Figure A.68) has a lower intensity than the shorter wavelength band CTF 2, which has the same intensity $\varepsilon_{max} \sim 20{,}000$ for monohalogenopentammine complexes. The distance between the two bands lies for the various $[Co(NH_3)_5X]^{2+}$ complexes between ca. 8000–9000 $cm^{-1}$. The intensity of CTF 1 increases in the series Cl, Br, I proportional to the spin–orbit coupling constants $\zeta_{np}$ of the halogen ($\zeta_{np}$: F = 250, Cl = 600, Br = 2500, and I = 5000 $cm^{-1}$). Orgel[71] and Yamatera[72] assign CTF 1 to the transitions $\pi \rightarrow d_{z^2}$ and CTF 2 to the transitions $\sigma \rightarrow d_{z^2}$ (with the ligand X oriented in the *xy* plane).

Also the spectra of anions as $SO_4^{2-}$, $MnO_4^-$, $CrO_4^{2-}$ are classified as charge–transfer spectra. They can be understood when the transfer of a

69. C. K. Jørgensen: *Acta. chem. scand.*, **11**, 166 (1957).
70. H. L. Schläfer and E. König: *Z. phys. Chem.* (Frankfurt), **30**, 145 (1961); **26**, 371 (1960).
71. L. E. Orgel: *Report to the 10th Solvay Council, Quelques problèmes de chimie minérale*, p. 289, Stoop, Brussels 1956.
72. J. Yamatera: *Naturwiss.*, **44**, 632 (1957).

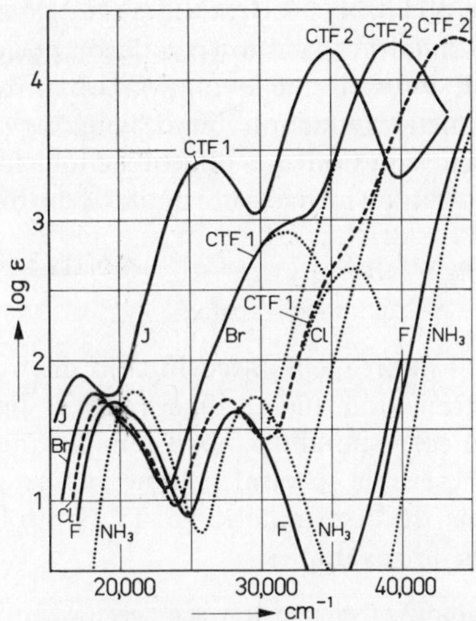


Figure A.68. Absorption spectra of monohalogeno-pentammine-cobalt(III) complex ions$[Co(NH_3)_5X]^{2+}$, X = F, Cl, Br, I (after Linnard and Weigel). The spectrum of $[Co(NH_3)_6]^{3+}$ denoted by $NH_3$ is also sketched for comparison.

$\pi$ electron from an oxygen ligand to a state localized on a central ion is assumed[73].

Metal reduction bands of a complex ion lie in the rule at longer wavelengths the easier the ligands are to oxidize. One observes that metal oxidation bands appear at longer wavelengths the more easily oxidizable the metal ion is. Finally one finds when comparing corresponding complexes of central ions from various transition metal series that the bands are shifted to shorter wavelengths when going from the first to the higher series. The direction of this effect is qualitatively in agreement with the fact that for complexes of metal ions of the higher transition series an increasing tendency to stabilization of the higher valence states is observed.

Charge–transfer bands are closely related to the *photochemical redox-reactions* observed when solutions of corresponding complexes are irradiated with light of suitable wavelength. Systems showing spontaneous redox reactions by daylight frequently have charge–transfer bands, which extend into the visible spectral region. As an example the azidocobalt(III)

73. M. Wolfsberg and L. Helmholz: *J. chem. Phys.*, **20**, 837 (1952).

ammines of type $[Co(NH_3)_5N_3^{++}$, $[Co(NH_3)_4(N_3)_2]^+$ or $[Co(NH_3)_3(N_3)_3]$ are introduced. As Linhard[74] has shown, a decomposition of the aqueous solutions of complexes with the evolution of nitrogen follows upon irradiation in the metal reduction band lying at about 30,000 cm$^{-1}$ ($\log \varepsilon_{max} \sim 4$). In acid solution the photoreaction, for example for the monoazidocomplex, can be formally formulated as follows:

$$[Co(NH_3)_5N_3]^{++} \xrightarrow{h\nu} Co^{++} + 5NH_3 + N_3$$

$$2N_3 \longrightarrow 3N_2$$

A large portion of the reagents used in colorimetry for the analytical determination of transition metal ions contain ligands which form complexes with the corresponding metal ions having intensive charge–transfer bands in the visible spectral region. The determination of $Fe^{3+}$ with thiocyanate ions, the determination of $Ti^{4+}$ with $H_2O_2$ or from $Fe^{2+}$ with *o*-phenathroline are examples.

Certain general features of charge–transfer spectra can also be qualitatively understood on the basis of an ionic picture[75]. The energy required to transfer an electron from the ligand to the central ion can be estimated as $\Delta U = -\Delta I + I_L + \Delta A$, whereby $\Delta I$ is the difference between the *nth* and the $(n-1)$*th* ionization energy of the central ion, $I_L$ the ionization energy of the ligand and $\Delta A$ the difference of the electrostatic heat of formation for the complex ion in the ground state (central ion charge $n$, ligand charge $-1$) and in the excited state (central ion charge $n-1$, ligand charge 0). This primitive model explains e.g. qualitatively the observed red shift of the charge–transfer bands for iron(III) halogeno-complexes with the increasing number of halogen ligands.

The concept 'charge transfer' implies that an electronic transition between a MO which in the first approximation is associated with a ligand and another MO, associated in first approximation with the central ion, takes place. The stronger the coupling between the states of the central ion place. The stronger the coupling between the states of the central ion which combine with one another, and the ligands, the more difficult it is to distinguish between corresponding molecular states of the complex ion having essentially ligand character and those essentially localized on the central ion.

The term 'charge–transfer spectrum' loses its meaning to the degree that the MO's of the complex ion between which the electronic transition takes place are no longer localized. The appearance of charge–transfer bands has been shown for many inorganic complexes. An interpretation

74. M. Linhard and M. Weigel: *Z. anorg. allg. Chem.*, **266**, 64 (1951); **267**, 121 (1951).
75. E. Rabinowitsch: *Rev. mod. Phys.*, **14**, 112 (1942). H. L. Schläfer, *Z. phys. Chem.* (Frankfurt), **3**, 222, 263 (1955).

of these bands on the basis of the MO theory has, however, been carried through in only a few cases. Essentially two reasons are responsible for this: On the one hand there are insufficient reliable experimental results and secondly the identification of the energy states is complicated because the MO theory yields a one-electron term system from which the states of the complex ion arise only after the electron interaction has been included. But such a consideration meets great difficulty.

Figure A.69 gives greatly simplified one-electron term systems[76c] of an octahedral $MX_6$ complex for three different cases. The possible electronic

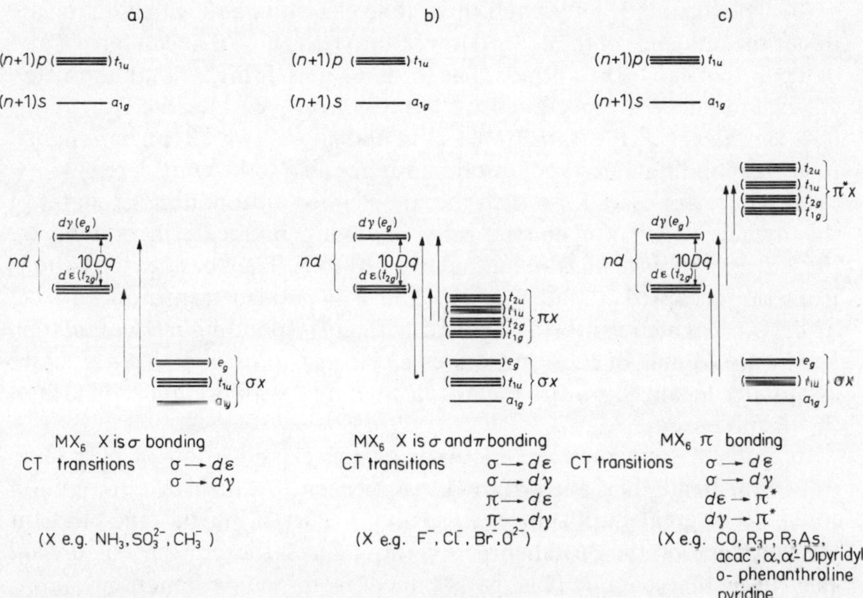


Figure A.69. Possible cases of charge-transfer bands for complexes having octahedral microsymmetry (schematic). In the figure only those states of the ligands X and the central ion M which are available for bonding are shown. From these MO's result which must be considered in a more exact discussion. Compare e.g., Figure A.69*a* with Figure A.65, in which the MO-scheme for the case of pure σ bonds is given. It must always be kept in mind that the localization of states on the ligands or on the central ion is only a very rough approximation. In particular, the degree to which delocalization of the electrons plays a rôle depends upon the values of the mixing coefficients α (cf. p. 105).

---

76. C. K. Jørgensen: (a) *Proceedings 8, ICCC*, Wien, 1964, p. 67 ff., Springer-Verlag, Wien–New York 1964. (b) *Orbitals in Atoms and Molecules*, p. 84, Academic Press, London–New York, 1962. (c) Essays in Coordination Chemistry, *Experientia Suppl.* IX, p. 98, Birkhäuser Verlag, Basel–Stuttgart, 1964. (d) *Molec. Physics*, **6**, 43 (1963); **2**, 309 (1959).

transitions corresponding to charge–transfer bands are indicated with arrows. The states essentially localized on the central ion *nd* (with the subshells $d\varepsilon$ and $d\gamma$) as well as the higher-lying states $(n+1)s$ and $(n+1)p$ are always on the left-hand side, while to the right the states essentially localized on the ligand, which are important for the case in question, are shown.

(a) Each of the ligands X has one lone electron pair that is capable of forming $\sigma$ bonds. The six $\sigma$ states of the ligands are each occupied by two electrons. Transitions $\sigma \rightarrow d\varepsilon$ and $\sigma \rightarrow d\gamma$ are possible which can be associated with appropriate *metal reduction bands.*

(b) The ligands X have each three lone electron pairs which dissociate under the influence of the central ion M into two pairs, that can form $\pi$ and one pair that can form $\sigma$ bonds. The former states lie higher than the latter. All ligand states are occupied by two electrons. Besides the transitions $\sigma \rightarrow d\varepsilon$ and $\sigma \rightarrow d\gamma$ the transition $\pi \rightarrow d\varepsilon$ and $\pi \rightarrow d\gamma$, which can be assigned to corresponding metal reduction bands are also to be considered*.

(c) There are cases for which the unoccupied antibonding $\pi^*$ states of the ligand, which are energetically low-lying in the term system are present and allow a '*back-donation*' from M to X. Then besides transitions from the occupied $\sigma$ states $\sigma \rightarrow d\varepsilon$ and $\sigma \rightarrow d\gamma$ also transition $d\varepsilon \rightarrow \pi^*$ or $d\gamma \rightarrow \pi^*$, which can be associated with corresponding *metal oxidation bands,* are found. In these processes electronic transitions from *d* states essentially localized on the metal ion to empty antibonding $\pi^*$ states of the ligand ensue.

Recently a semiempirical method for the classification of electronic transition bands has been given by Jørgensen[76], which in our opinion could be of great importance. Jørgensen argues using the one-electron term system of the MO theory and introduces the concept of *optical electronegativity* $x_{\text{opt.}}$. This is obtained from the wavenumbers $\bar{\nu}_{\text{corr.}}$ (measured in kK (kiloayser), $1\,\text{kK} = 1000\,\text{cm}^{-1}$) (corrected by the spin pairing energy *P*) of the corresponding bands $\pi \rightarrow d\varepsilon$ (for $O_h$ symmetry) or $\pi \rightarrow d\gamma$ (for $T_d$ symmetry) according to

$$\bar{\nu}_{\text{corr.}} = [x_{\text{opt.}}(\text{X}) - x_{\text{opt.}}(\text{M})] \cdot 30\,\text{kK}.$$

X denotes the ligands, M the metal ion. The relation is valid for complex ions having cubic microsymmetry. The value of the spin *S* of the central ion is changed through an electronic transition. *P* lies at 4000–6000 $\text{cm}^{-1}$ for compounds with central ions of the 3*d* series and for those of the 4*d* and 5*d* series at 3000 $\text{cm}^{-1}$.

---

* There are also ligands where two lone electron pairs must be considered (e.g. $H_2O$ or $NH_2^-$), one yielding $\sigma$ and one yielding $\pi$ bonds.

Bands $\pi \to d\gamma$ ($O_h$ symmetry) or $\pi \to d\varepsilon$ ($T_d$ symmetry), i.e. bands corresponding to transitions from the $\pi$ states of the ligands to the $\sigma$ antibonding $d$ states, must differ from the former bands in their spectral positions by $\Delta \equiv 10Dq$ when besides this the spin pairing energy is brought into consideration. They must appear at larger wavenumbers (cf. Figure A.69b).

Using the method of Jørgensen a table of values of the optical electronegativities can be obtained from the spectra. These values are of interest for the discussion of chemical problems and are closely related to Mulliken's concept of electronegativity. For complexes having $\pi$ electron systems as ligands, as, e.g. pyridine, 2,2′-dipyridyl or *o*-phenanthroline, one observes in the spectrum the absorption band of the free ligand, slightly shifted to longer wavelengths (cf. Figure A.9; Spectra of $[CrCl_3py_3]$, $[MoCl_3py_3]$; indications of vibrational structure of the band $\pi \to \pi^*$ of the inner complex pyridine are recognizable). Here we are concerned with transitions, which take place between MO's which are localized to a large extent on the ligands (*intra-ligand spectra, inner-ligand transitions*). They are changed only to a small degree compared with the states of the free ligand.

Qualitatively these changes experienced by the term system of the ligand when the ligand enters a complex ion can be understood when the electrostatic interaction of the $\pi$-electrons with the central ion (and the remaining ligands) is taken into account[77] by using the perturbation theory. The ligand molecule is studied then in the electric field of the central ion and the other ligands.

Also for complexes with ligands such as $SCN^-$ similar absorption bands arise, besides charge–transfer bands. From what has been said it appears of importance for the analysis of complex spectra that the spectrum of the complex ion be compared with that of the free ligand as well as with the spectra of complexes not having a transition metal central ion. In this manner the bands that may possibly be present and which correspond to transitions in the term system of the ligand can be localized in the spectrum. These bands change their positions only slightly when the central ion or other ligands in the complex are changed. For the free atom or ion the Rydberg Series are obtained when, e.g. a $3d$ electron through excitation undergoes transitions to the $4p$, $4f$, . . . , $5p$, $5f$, . . . states, that is in general an $nl$ electron is excited to the $(n+1)(l\pm1)$, $(n+2)(l\pm1)$, . . . . $\infty(l\pm1)$ states. Transitions with $l = \pm 2$ are also known, $3d \to 4s$.

*Rydberg series* can also be observed in the spectra of unsaturated molecules, e.g. ethylene. For this molecule the longest wavelength transition of the series lies at 1745 Å. This is followed by other transitions up to

77. H. L. Schläfer: *Z. phys. Chem.* (Frankfurt), **19**, 265 (1959). R. L. Belford, A. E. Martell and M. Calvin: *J. inorg. nucl. Chem.*, **2**, 11 (1956).

ca. 240 kcal where the ionization limit (continuum) is reached and the electron is separated from the then positive ethylene. Such a series of bands which tend to a convergence limit is also designated as a Rydberg series.

The situation for Rydberg transitions in inorganic complexes is complicated because overlapping of states belonging to various values of the principle quantum number $n$ takes place. As an example, ligand orbitals with $n = 2$ overlap with orbitals of the central ion with $n = 3$ and 4 (formation of $\sigma$ bonds) for octahedral complexes of the first transition series having $H_2O$ as the ligand. For every central ion bound into a complex a series of energy states exists which can lead to the dissociation of an electron. However, it is not possible to distinguish rigorously between the long wavelength transitions within the energetically lowest-lying states of such a series and charge–transfer transitions which take place between states essentially localized on the central ion and those having predominantly ligand character. It is to be expected that the transitions at shorter wavelengths, for which the excited electron is farther removed from the central ion than the distance metal ion–ligand, are almost 'pure' Rydberg transitions. They lie, however, in the vacuum ultraviolet and have not yet been observed. For this purpose studies would have to be undertaken in the gas phase, limiting the possibilities to a very few

Table A.11. Schematic (simplified) tabulation of the electronic transitions possible for transition metal complexes

| Excited electron originally localized on | Nature of the transition | | Number of centers participating |
|---|---|---|---|
| central ion | $nd \rightarrow nd$ | ligand field bands (central ion bands) | 1 (central ion) |
| | $nd \rightarrow (n+1)(l \pm 1)$<br>$\rightarrow (n+2)(l \pm 1)$<br>$\rightarrow \ldots\ldots\ldots\ldots$<br>$\rightarrow \infty (l+1)$ | Rydberg spectra | 1 (central ion) |
| | $d \rightarrow \pi^*$ | Metal oxidation bands (charge–transfer bands) | 2 (central ion $\rightarrow$ ligand) |
| ligands | $\pi \rightarrow \pi^*$<br>$(n \rightarrow \pi^*)$ | (intra-ligand bands) | 1 (ligand) |
| | $\pi \rightarrow d$<br>$\sigma \rightarrow d$ | metal reduction bands (charge–transfer bands) | 2 (ligand $\rightarrow$ central ion) |

complexes. $d \rightarrow p$ spectra have been discussed for planar complexes, e.g. $[PtCl_4]^{2-}$.

Table A.11 systematically summarizes the possible types of absorption bands. The classification was carried out on the basis of the centre where the electron which is excited is originally localized. This classification is of course only a rough approximation. With the help of the extended electrostatic theory presented in sections 4–10 (without consideration of the MO method) the long wavelength parity forbidden $d \rightarrow d$ bands, given in the first row of Table A.11, can be understood in good agreement with experiment. The treatment of charge–transfer bands must be undertaken on the basis of the MO theory and is still in its very early stages[76,78].

78. Cf. e.g. C. K. Jørgensen: *Molec. Phys.*, **2**, 309 (1959), as well as [73, 72].

# 2. Magnetic properties of transition metal complexes

Although besides the ground state the higher (excited) states of the molecules in question are of interest for a discussion of the optical properties of metal complexes, a knowledge of the ground state alone is generally adequate to allow an understanding of the magnetic behaviour. The magnetic moment is obtained to the first approximation from the multiplicity (and thereby the spin) of the ground term. However, to explain finer details, such as the size of possible orbital moment contributions or the temperature dependence of the susceptibility, more information is necessary concerning the '*structure*' of the *ground state**. It is not sufficient to consider only orbital angular momentum. Rather spin–orbit coupling must be taken into account as well as possible interaction with energetically higher-lying states of the same irreducible representation and multiplicity.

In the following it will be shown what the ligand field theory can accomplish in this field.

## 2.1. General introduction. Survey of experimental findings

For transition metal complexes of the first series with the configuration $3d^1$–$3d^9$ of the central ion ($Ti^{3+}$–$Cu^{2+}$) two groups of complexes can be distinguished using the measured values of the paramagnetic susceptibility:

(a) Complexes for which the value of the effective magnetic moment (measured in *Bohr magnetons*, (B.M.))† found is almost equal to that

* A *ground state* will here be taken to be in the broader sense all those energy levels which are appreciably occupied at normal temperatures. That is, they lie of the order of $kT$ higher than the lowest level (about 200 $cm^{-1}$ at room temperature). In the most general case every level of the ground state is degenerate, i.e. it is described by more than one eigenfunction. Each of these eigenfunctions corresponds to a definite spin and orbital moment. The total moment of the system considered is obtained by averaging the moments of the individual states, taking into consideration the probability of each of being occupied.

† The effective magnetic moment is calculated from

$$\mu_{\text{eff}} = \left(\frac{N_{\text{L}}\beta_{\text{B}}^2}{3k}\right)^{-\frac{1}{2}} (\chi'_{\text{mol}}T)^{\frac{1}{2}} = 2{,}84\,(\chi'_{\text{mol}}T)^{\frac{1}{2}} \qquad [\text{B.M.}],$$

obtained from the formula for pure spin magnetism

$$\mu_{\text{eff.}} = \{4S(S+1)\}^{\frac{1}{2}} = \{n(n+2)\}^{\frac{1}{2}} \text{ [B.M.]}$$

$S$ is the spin quantum number and $n$ the number of unpaired electrons*. Table A.12 contains the values of moments obtained from the relation for complexes having from one to nine $d$ electrons. Complexes of this type are called high-spin (spin-free) complexes.

Table A.12. Effective spin moments in Bohr magnetons

| Number of unpaired electrons ($n$) | 1 | 2 | 3 | 4 | 5 |
|---|---|---|---|---|---|
| Multiplicity of the state $M = 2S+1$ | Doublet | Triplet | Quartet | Quintet | Sextet |
| $\mu_{\text{eff.}}$ [B.M.] | 1·73 | 2·83 | 3·87 | 4·90 | 5·92 |

(b) Certain complexes show lower moments than would be expected from the relation for pure spin magnetism given above, or are even diamagnetic.

For the $[Fe(CN)_6]^{3-}$ ion instead of the value 5·92 B.M. expected for five unpaired electrons, a value corresponding to approximately one unpaired electron is found. For $[Co(NH_3)]^{3+}$ diamagnetism is found rather than 4·90 B.M. Complexes of this type are denoted as low-spin (spin-pared) complexes.

Table A.13 gives a survey of values of magnetic moments measured for complexes of the first series of transition metals. It is seen that the experimental moments in the rule deviate rather strongly from values calculated

---

where $N_L$ is the Loschmidt number, $k$ the Boltzmann constant, $T$ the absolute temperature and $\chi'_{\text{mol}}$ the susceptibility per mole corrected for the diamagnetic contribution. $\beta_B$ is the Bohr magneton

$$\beta_B = \frac{e \cdot h}{4\pi m_e c}$$

($e$ = charge, $m_e$ = mass of the electron, $h$ = Planck's constant, $c$ = velocity of light).

Its value is $0{\cdot}927 \times 10^{-20}$ [erg/gauss$^{-1}$].

*In determining the number of unpaired electrons Hund's principle of maximum multiplicity is used. The occupation of the five $d$ orbitals, each of which according to the Pauli principle can be occupied by at the most two electrons of antiparallel spin, is such that first each state in the series $d^1$–$d^5$ is occupied by one electron, with the spins oriented parallel to one another. From $d^6$–$d^{10}$ each additional electron enters a state already occupied by one electron by *pairing its spin*.

Table A.13. Tabulation of the magnetic moments of complexes with central ions of the first transition metal series

| Ion | Number of *d* electrons | High-spin: Number of unpaired electrons | High-spin: Spin moment (theor.) [B.M.] | High-spin: Experimental values [B.M.] | Low-spin: Number of unpaired electrons | Low-spin: Spin moment (theor.) [B.M.] | Low-spin: Experimental values [B.M.] |
|---|---|---|---|---|---|---|---|
| $Ti^{3+}$ | 1 | 1 | 1·73 | 1·65—1·79 | | | |
| $V^{4+}$ | | | | 1·68—1·78 | | | |
| $V^{3+}$ | 2 | 2 | 2·83 | 2·75—2·85 | | | |
| $V^{2+}$ | 3 | 3 | 3·87 | 3·80—3·90 | | | |
| $Cr^{3+}$ | | | | 3·70—3·90 | | | |
| $Mn^{4+}$ | | | | 3·8 —4·0 | | | |
| $Cr^{2+}$ | 4 | 4 | 4·90 | 4·75—4·90 | 2 | 2·83 | 3·20—3·30 |
| $Mn^{3+}$ | | | | 4·90—5·00 | 2 | | 3·18 |
| $Mn^{2+}$ | 5 | 5 | 5·92 | 5·65—6·10 | 1 | 1·73 | 1·80—2·10 |
| $Fe^{3+}$ | | | | 5·70—6·0 | 1 | | 2·0 —2·5 |
| $Fe^{2+}$ | 6 | 4 | 4·90 | 5·10—5·70 | 0 | 0 | diamagn. |
| $Co^{3+}$ | | | | 4·3 | 0 | | diamagn. |
| $Co^{2+}$ | 7 | 3 | 3·87 | 4·30—5·20 | 1 | 1·73 | 1·8 |
| $Ni^{3+}$ | | | | | 1 | | 1·8 —2·0 |
| $Ni^{2+}$ | 8 | 2 | 2·83 | 2·80—3·50 | 0 | 0 | diamagn. |
| $Cu^{2+}$ | 9 | 1 | 1·73 | 1·70—2·20 | | | |

according to the formula for pure spin magnetism. This deviation is the strongest for the second half of the series, especially for $Co^{2+}$ and $Ni^{2+}$.

Low-spin complexes exist for $Cr^{2+}$, $Mn^{3+}(d^4)$; $Mn^{2+}$, $Fe^{3+}(d^5)$; $Fe^{2+}$, $Co^{3+}(d^6)$; $Co^{2+}$, $Ni^{3+}(d^7)$ and $Ni^{2+}(d^8)$.

## 2.2. Interpretation of low-spin complexes according to Pauling

The *ionic theory* of Kossel (1916)[1] and Magnus (1922)[2], (see section I.3a) which depicts a complex ion as being composed of rigid spheres, having point charges or dipoles at their centres and considers the electrostatic

1. W. Kossel: *Ann Phys.*, **49**, 229 (1916); *Naturwiss.*, **7**, 339, 360 (1919); **11**, 589 (1923).
2. A. Magnus: *Z. anorg. allg. Chem.*, **124**, 289 (1922); cf. also Garrik: *Philos. Mag.*, [7] **9**, 131 (1930); [7] **10**, 76 (1930); [7] **11**, 741 (1931).

interaction of the spheres, could give no explanation for the appearance of low-spin complexes.

Using the valence bond theory (see section 1.3b) Pauling (1931)[4] investigated further the basic assumption of Sidgwick (1923)[3], that the bond between the ligand and the central ion in complexes results from an electron transferred from a lone pair on the bond-forming ligand to the central ion. The interaction of this electron with the electron remaining on the ligand then takes place.

If bonds, that is $\sigma$ bonds, are to be formed between the ligands and the ion, then the central ion must have energetically low-lying orbitals available to accept the electrons given up by the ligands. If this is the case, then the functions characterizing these orbitals must be capable of mixing so that valence orbitals having certain directional properties arise (hybridization). These considerations are based upon the formation of electron pair bonds, as are found for example in the hydrogen molecule. That is, covalent bonding is assumed.

For an octahedral complex with a central ion of the first transition series two $3d$ states, a $4s$ and three $4p$ states are necessary for the formation of equivalent bonds to the vertices of an octahedron (see Appendix, pp. 484 ff.).

In Table A.14 the corresponding linear combinations are given. According to this it is the $d\gamma$ ($e_g$) states ($d_{x^2-y^2}$, $d_{z^2}$) with which ligands can combine to form $\sigma$ bonds. Table A.15 gives the states of the central ion required to form equivalent bonds for the coordination numbers 4 (tetrahedron, planar quadratic orientation), 6 (octahedron) and 8 (cube, antiprism of dodecahedron).

Table A.14. The six equivalent orbitals for octahedral ($O_h$) symmetry

$$\begin{aligned}
\psi_1 &= 1/\sqrt{6}s + 1/\sqrt{2}p_z + 1/\sqrt{3}d_{z^2} \\
\psi_2 &= 1/\sqrt{6}s - 1/\sqrt{2}p_z + 1/\sqrt{3}d_{z^2} \\
\psi_3 &= 1/\sqrt{6}s + 1/\sqrt{2}p_x + 1/\sqrt{12}d_{z^2} + 1/2d_{x^2-y^2} \\
\psi_4 &= 1/\sqrt{6}s - 1/\sqrt{2}p_x + 1/\sqrt{12}d_{z^2} + 1/2d_{x^2-y^2} \\
\psi_5 &= 1/\sqrt{6}s + 1/\sqrt{2}p_y + 1/\sqrt{12}d_{z^2} - 1/2d_{x^2-y^2} \\
\psi_6 &= 1/\sqrt{6}s - 1/\sqrt{2}p_y + 1/\sqrt{12}d_{z^2} - 1/2d_{x^2-y^2}
\end{aligned}$$

3. N. V. Sidgwick: *J. chem. Soc.*, **123**, 725 (1923); *Trans Faraday Soc.*, **19**, 469 (1923); *Chemy Ind.*, **42**, 316 (1923); cf. also *The Chemical Elements and their Compounds*, Oxford University Press 1950.
4. L. Pauling: *The Nature of the Chemical Bond*, 3rd edn, pp. 145 ff., Ithaca, New York, Cornell University Press 1960; *J. Am. chem. Soc.*, **53**, 1367 (1931); **54**, 988 (1932); *Proc. natn. Acad. Sci. U.S.A.*, **14**, 359 (1928).

Table A.15. Orbitals available for $\sigma$-bonds for the case of complexes with coordination numbers 4, 6 and 8

| Coordination number | Stereochemistry | Symmetry | Orbitals available for $\sigma$-bonds | Abbreviated notation |
|---|---|---|---|---|
| 4 | Tetrahedron | $T_d$ | $s, p_x, p_y, p_z$ | $sp^3$ |
| | Square | $D_{4h}$ | $d_{x^2-y^2}, s, p_x, p_y$ | $dsp^2$ |
| 6 | Octahedron | $O_h$ | $d_{x^2-y^2}, d_{z^2}, s, p_x, p_y, p_z$ | $d^2sp^3$ |
| | Cube | $O_h$ | $f_{xyz}, d_{xy}, d_{xz}, d_{yz}, s, p_x, p_y, p_z$ | $d^3sp^3f$ |
| 8 | Archimedian Antiprism | $D_{4d}$ | $d_{x^2-y^2}, d_{z^2}, d_{xz}, d_{yz}, s, p_x, p_y, p_z$ | $d^4sp^3$ |
| | Dodecahedron | $D_{2d}$ | | |

Kimball (1940)[5] studied the situation for other coordination numbers and geometrical orientations. Each of these states is to be occupied by a pair of ligand electrons having antiparallel spin. If every orbital is represented by a cell (as indicated in Figure A.70) the low-spin behaviour of octahedral $Fe^{3+}$, $Co^{3+}$, $Fe^{2+}$ and $Co^{2+}$ complexes can be understood.

Figure A.71 shows the interpretation of the magnetic behaviour of tetrahedral and square planar $Ni^{2+}$ and $Co^{2+}$ complexes using Pauling's theory.

| | 3d | | | | | 4s | 4p | | | 5s? | | Number of unpaired electrons |
|---|---|---|---|---|---|---|---|---|---|---|---|---|
| $Fe^{3+}$ | ↑ | ↑ | ↑ | ↑ | ↑ | | | | | | Free ions or high-spin complexes | 5 |
| $Co^{3+}$, $Fe^{2+}$ | ↑↓ | ↑ | ↑ | ↑ | ↑ | | | | | | | 4 |
| $Co^{2+}$ | ↑↓ | ↑↓ | ↑ | ↑ | ↑ | | | | | | | 3 |
| $[Fe(CN)_6]^{3-}$ | ↑↓ | ↑↓ | ↑ | xx | xx | xx | xx | xx | xx | | Low-spin complexes | 1 |
| $[Fe(CN)_6]^{4-}$, $[Co(NH_3)_6]^{3+}$ | ↑↓ | ↑↓ | ↑↓ | xx | xx | xx | xx | xx | xx | | | 0 |
| $[Co(CN)_6]^{4-}$ | ↑↓ | ↑↓ | ↑↓ | xx | xx | xx | xx | xx | xx | ↑ | | 1 |
| | | | | $3d^2$ | | $4s$ | $p^3$ | | | 5s? | | |

Figure A.70. Interpretation of the low spin octahedral $Fe^{3+}$, $Co^{3+}$, $Fe^{2+}$ and $Co^{2+}$ complexes according to Pauling.

The distribution of the electrons of the free ions $Fe^{3+}$, $Co^{3+}$ or $Fe^{2+}$ and $Co^{2+}$ over the five *d* states leads to 5, 4, or 3 unpaired electrons when the principle of maximum multiplicity is taken into account. High-spin complexes of these central ions, e.g. $[Fe(H_2O)_6]^{3+}$, $[CoF_6]^{3-}$, $[Fe(NH_3)_6]^{2+}$, $[Co(H_2O)_6]^{2+}$ show paramagnetism, as is to be expected for the corresponding numbers of unpaired electrons. For low-spin complexes the occupation $3d^24s4p^3$ with 12 electrons from the ligands (one pair per ligand) and 6 equivalent bonds to the central ion leads to magnetic moments corresponding to one unpaired electron or to diamagnetism. In Figures A.70 and A.71 the electrons of the central ion are represented by arrows (↑) and those contributed by the ligands by crosses ( × ).

The appearance of diamagnetic quadratic planar Ni(II) complexes as well as Co(II) complexes with reduced moments follows from the $3d4s4p^2$ hybridization, while tetrahedral $Ni^{2+}$ and $Co^{2+}$ complex ions with $sp^3$ occupation should be high-spin (Figure A.71).

In this manner the appearance of low-spin complexes can be understood after Pauling. A different type of bond is expected for high-spin and for

5. G. E. Kimball: *J. chem. Phys.*, **8**, 188 (1940).

| | | 3d | | | | | 4s | 4p | | | | Number of unpaired electrons |
|---|---|---|---|---|---|---|---|---|---|---|---|---|
| $Co^{2+}$ | | ↑↓ | ↑↓ | ↑ | ↑ | ↑ | | | | | Free ions | 3 |
| $Ni^{2+}$ | | ↑↓ | ↑↓ | ↑↓ | ↑ | ↑ | | | | | Free ions | 2 |
| $[CoX_4]^{2-}$ | $T_d$ | ↑↓ | ↑↓ | ↑ | ↑ | ↑ | XX | XX | XX | XX | high-spin | 3 |
| | $D_{4h}$ | ↑↓ | ↑↓ | ↑↓ | ↑ | XX | XX | XX | XX | | low-spin | 1 |
| $[NiX_4]^{2-}$ | $T_d$ | ↑↓ | ↑↓ | ↑↓ | ↑ | ↑ | XX | XX | XX | XX | high-spin | 2 |
| | $D_{4h}$ | ↑↓ | ↑↓ | ↑↓ | ↑↓ | XX | XX | XX | XX | | low-spin | 0 |

Figure A.71. Interpretation of the appearance of diamagnetic quadratic planar $Ni^{2+}$ and $Co^{2+}$ complexes after Pauling.

low-spin complexes. In the former no 3*d* states of the central ion are occupied by electrons contributed by the ligands. Rather a more ionic bond should be present, so that the magnetic moment corresponds to that of the free central ion, i.e. the number of unpaired electrons is given by *Hund's principle of maximum multiplicity*. For low-spin complexes *covalent* electron pair bonds should be present. 3*d* states of the central ion are occupied. Therefore it appears logical to distinguish between the bond states with the help of measured magnetic moments. This *magnetic bonding criterion* is still used to some extent today in texts on inorganic chemistry, although, as we shall see later, it is by no means valid.

The original ideas of Pauling were later extended with the assumption that for high-spin compounds hybridization takes place between *s*, *p* and *d* states having the *same* principal quantum numbers. For octahedral complexes of the first transition metal series, e.g. the high-spin compounds, would be designated by $4s4p^3 4d^2 (sp^3d^2)$ and low-spin by $3d^2 4s4p^3 (d^2sp^3)$ as is shown for $Fe^{3+}$ complexes in Figure A.72.

Taube[6] uses the notation 'outer' and 'inner-orbital' complexes for the two types, while Burstall and Nyholm[7] speak of 'higher and lower-level' complexes.

According to these concepts the formation of electron pair bonds between the central ion and the ligands must be assumed for high-spin as well as for low-spin complexes. In the one case the energetically lower-lying 3*d* and in the other the higher-lying 4*d* states are used. Thereby the originally rigorous classification of complexes on the basis of measured values of magnetic moments into those with ionic bonds and those with covalent bonds is relaxed.

---

6. H. Taube: *Chem. Rev.*, **50**, 69 (1952).
7. H. D. Burstall and R. S. Nyholm: *J. chem. Soc.*, **1952**, 3570.

| | 3d | | | | | 4s | 4p | | | 4d | | | | | | Number of unpaired electrons |
|---|---|---|---|---|---|---|---|---|---|---|---|---|---|---|---|---|
| $Fe^{3+}$ | ↑ | ↑ | ↑ | ↑ | ↑ | | | | | | | | | | | 5 |
| $[Fe(H_2O)_6]^{3+}$ | ↑ | ↑ | ↑ | ↑ | ↑ | xx | xx | xx | xx | xx | xx | | | | $sp^3d^2$ | 5 |
| $[Fe(CN)_6]^{3-}$ | ↑↓ | ↑↓ | ↑ | xx | xx | xx | xx | xx | xx | | | | | | $d^2sp^3$ | 1 |

Figure A.72. Outer and inner orbitals for the example of high and low-spin complexes of trivalent iron.

For a critique of the valence bond method as applied to complexes see, e.g. Jørgensen[8], Ballhausen[9] and Liehr[10]. According to Pauling's scheme $[Co(H_2O)_6]^{3+}$ would have to be formulated as

$$\left[\begin{array}{ccc} {}^{+}H_2O \searrow & & \nearrow OH_2^{+} \\ {}^{+}H_2O - & \overset{3-}{Co} & - OH_2^{+} \\ {}^{+}H_2O \nearrow & & \searrow OH_2^{+} \end{array}\right]$$

where the central ion has the charge $-3$. The existence of such strongly negatively-charged central ions appears very improbable.

Pauling's valence bond theory proves to be inadequate for the discussion of, e.g. $d^9$ complexes. According to this theory a planar $Cu^{2+}$ complex should show $dsp^2$ hybridization. That would mean, however, that one $d$ electron must be promoted into a higher-lying orbital, e.g. $4p_z$. This is not very probable from a chemical standpoint because $Cu^{2+}$ is strongly resistant to further oxidation. Also, such a postulation is not compatible with the results of paramagnetic resonance measurements which show unambiguously that $Cu^{2+}$ ion has nine $d$ electrons.

The difference between the valence bond theory in the Pauling form on the one hand and the ligand field theory (or MO theory), developed in the following section with the idea of interpreting magnetic properties on the other is seen at once, when it is considered that the antibonding $d\gamma^*$ ($e^*$) states do not appear in the valence bond theory. Therefore the use of the valence bond theory is always problematical when considering complexes for which such antibonding states are occupied. That is the case for high-spin complexes having more than three and for low-spin complexes with more than seven $d$ electrons.

---

8. C. K. Jørgensen; *Absorption Spectra and Chemical Bonding*, pp. 224 ff., Pergamon Press, Oxford–London–Paris 1962.
9. C. J. Ballhausen: Theories of Bonding in Coordination Compounds, *Proceedings of the 6th ICCC*, pp. 3 ff., MacMillan, New York, 1961.
10. A. D. Liehr: *J. chem. Educ.*, **39**, 135 (1962); and the reply by L. Pauling: *J. chem. Educ.*, **39**, 463 (1962).

## 2.3. Interpretation of the appearance of low-spin complexes on the basis of the ligand field theory

The extended ionic model, which at first glance appears to be incompatible with Pauling's covalent models can also be used to explain the appearance of low-spin complexes. This was shown by van Vleck (1935)[11,12]. However, his arguments were not further developed for some time, and the chemist always employed Pauling's concept when discussing magnetic properties. After about 1950 the problem was attacked once more from various sides on the basis of the ligand field theory.

### (a) *The one-electron scheme*

We shall discuss first the question of the appearance of high- and low-spin complexes by considering the splitting in the ligand field of a fivefold orbitally degenerate one-electron $d$ state. In order to obtain the ground state which determines the magnetic behaviour of the complex ion the individual states are to be filled corresponding to the number of $d$ electrons of the central ion in accordance with the Pauli principle.

For octahedral complexes we must consider (see section 1.7a) the splitting of the five $d$ states ($d_{xy}$, $d_{xz}$, $d_{yz}$, $d_{x^2-y^2}$, $d_{z^2}$), which in the free ion belong to the same family, into two families: The energetically lower-lying $d^x{}_y$, $d_{xz}$, $d_{yz}$ states, belonging to the $T_{2g}$ representation and forming the threefold orbitally degenerate $d\varepsilon$ state and the states $d_{x^2-y^2}$, $d_{z^2}$ which belong to the representation $E_g$ and form the two fold $d\gamma$ state. The energy states of the complex ion with a central ion having $N$ $d$ electrons are obtained in the zero*th* approximation from the possibilities of distribution of the $N$ electrons over the two families of the states corresponding to $d\varepsilon^n\, d\gamma^{N-n}$ ($0 \leqq n \leqq 6$, $0 \leqq N-n \leqq 4$). From the configuration $d\varepsilon^n\, d\gamma^{N-n}$, from which the ground state arises, the number of unpaired electrons and thereby the magnetic moment to be expected can be obtained.

As an example consider the case of $Co^{3+}$ (Figure A.73). For a weak ligand field the splitting $\Delta \equiv 10\,Dq$ is small. The distance between the $d\varepsilon$ and the $d\gamma$ state is so small that the six electrons are distributed over the two levels according to $d\varepsilon^4\, d\gamma^2$. This electron distribution leads to four unpaired electrons, as is the case for the free $Co^{3+}$ or $Fe^{2+}$ ion, so that corresponding complexes, as e.g. $[CoF_6]^{3+}$ or $[Fe(NH_3)_6]^{2+}$ show normal magnetic behaviour.

For a strong ligand field the splitting $\Delta$ is so large that now the six electrons occupy the three $d\varepsilon$ states by pairing spins. The number of

11. J. H. Van Vleck: *J. chem. Phys.*, **3**, 803, 807 (1935); *Phys. Rev.*, **41**, 202 (1932).
12. J. H. Van Vleck and A. Sherman: *Rev. mod. Phys.*, **7**, 167 (1935).

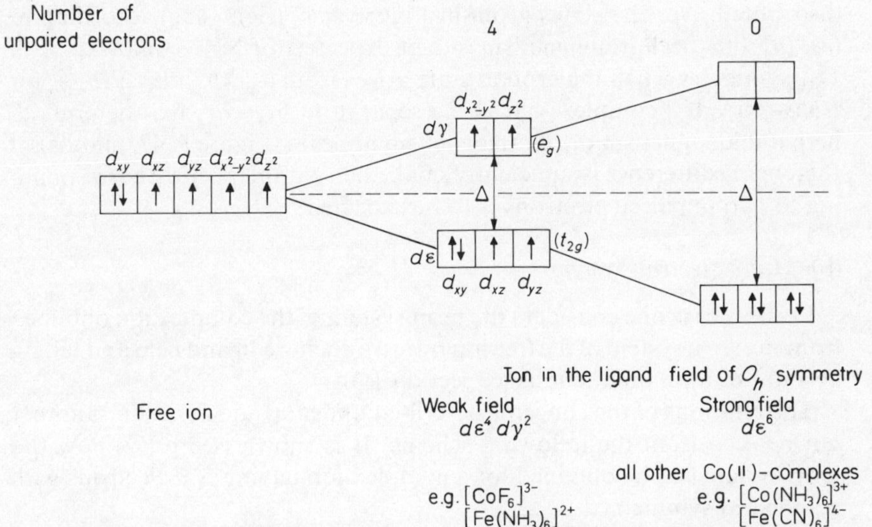


Figure A.73. Magnetic behaviour of octahedral $d^6$ complexes ($Co^{3+}$, $Fe^{2+}$) in the one-electron scheme.

unpaired electrons for the configuration $d\varepsilon^6$ is zero; complexes of this type are *diamagnetic*, as e.g. $[Co(NH_3)_6]^{3+}$ or $[Fe(CN)_6]^{4-}$.

Using analogous reasoning the appearance of low-spin complexes for octahedral $d^4$, $d^5$, and $d^7$ complexes can be understood. The ground states arise from the configurations

| | high-spin | low-spin |
|---|---|---|
| $d^4$ | $d\varepsilon^3\, d\gamma$ | $d\varepsilon^4$ |
| $d^5$ | $d\varepsilon^3\, d\gamma^2$ | $d\varepsilon^5$ |
| $d^7$ | $d\varepsilon^5\, d\gamma^2$ | $d\varepsilon^6\, d\gamma$ |

Two $d^5$ complexes are compared below as an example of the relation between splitting magnitude and magnetic properties.

$$[FeF_6]^{3-} \quad \Delta \approx 13\,900\ \mathrm{cm^{-1}}:\ \text{high–spin}$$
$$[Fe(CN_6)]^{3-} \quad \Delta \approx 30\,000\ \mathrm{cm^{-1}}:\ \text{low–spin}$$

Similar arguments may be carried out for complexes of lower symmetry. In this case one begins with the splitting diagram of the one-electron $d$ states in a ligand field of the symmetry in question and considers the distribution of electrons in those states for weak and strong fields.

Consider the appearance of diamagnetic planar quadratic Ni(II) complexes. The eight $d$ electrons of the central ion $Ni^{2+}$ are to be distributed over the states given in Figure A.47 (right side) according to $(e_g)^4(a_{1g})^2(b_{2g})^2$. Paramagnetism is to be expected for $Ni^{2+}$ complexes with $D_{4h}$ symmetry when the ground state arises from $(e_g)^4(b_{2g})^2(a_{1g})^1(b_{1g})^1$ for *trans*-$[NiA_4B_2]$ complexes. It is the separation $b_{1g}-a_{1g}$ that determines here if diamagnetism or paramagnetism appears (Figure A.47, middle). If this energy difference is sufficiently small, then paramagnetism corresponding to two unpaired electrons is to be expected.

(b) *Many-electron system*

For this case one considers the term system of the complex ion obtained from the term system of the free ion by introducing a ligand field and taking account of term interaction (see section 1.6).

The relations of the concepts described under (a) and (b) to one another can be seen from the following scheme. It is shown (see p. 52) how the term system can be obtained for a complex ion having $N$ $d$ electrons with octahedral symmetry.

term System of the complex ion (b) — one-electron term system (a)

$$d^N \xrightarrow[\text{electron interaction}]{\text{Step 1}} {}^{2S+1}L \xrightarrow[\text{ligand field + term interaction}]{\text{Step 2}} {}^{2S+1}\Gamma \xleftarrow[\text{electron interaction + configuration interaction}]{\text{Step 2}} d\varepsilon^n\, d\gamma^{N-n} \xleftarrow[\text{ligand field}]{\text{Step 1}} d^N$$

Under (a) we have discussed magnetic properties by carrying out only the first step from right to left in the above scheme. In this section the same question will be treated by using the term system of the complex ion with the states ${}^{2S+1}\Gamma$, which can be obtained, e.g. by using the two steps from left to right in the scheme as starting point.

The magnetic behaviour of the complex ion is determined by the multiplicity of its ground term. Low spin is expressed in the term diagram (i.e. the plot of the term energies as a function of ligand field strength (see section 1.8)) in that a term arising from an energetically higher-lying term of the free ion, and having a different multiplicity than the ground term of the free ion becomes the ground term of the complex ion. Term crossing is therefore found (see section 1.6f). We consider the example of $d^6$ complexes that has already been treated under (a) in the one-electron scheme.

Terms having the multiplicities 5, 3 and 1 arise from the configuration $d^6$. According to Hund's rules the ground term of the free ion is a ${}^5D$ term. The calculation shows that a term arising from the energetically higher-lying singlet state of the free ion crosses the ${}^5T_{2g}$ term, which arises from the ${}^5D$ term, at a certain critical value of the ligand field strength. For field strengths less than the critical field at the crossover point $P$ (see Figure A.37, p. 50) the $^5T_{2g}(d\varepsilon^4\, d\gamma^2)$ term is the ground term so that high spin complexes such as $[CoF_6]^{3-}$ or $[Fe(NH_3)_6]^{2+}$ arise. For field strengths greater than the critical field a $^1A_{1g}(d\varepsilon^6)$ term is the ground term and one finds diamagnetic complexes as e.g. $[Co(NH_3)_6]^{3+}$ or $[Fe(CN)_6]^{4-}$.

Also the type of absorption spectrum changes in a characteristic manner in going from high-spin to low-spin complexes (cf. p. 50).

While Figure A.37 depicts the situation schematically, the exact term diagrams for the configurations $d^2$ to $d^8$ for octahedral symmetry can be found in the papers of Orgel[13] and Tanabe and Sugano[14]. They are the results of quantitative calculations. A Tanabe–Sugano diagram for $d^6$ ions for $O_h$ symmetry was seen in Figure A.52, p. 74. This should be compared with the simplified schematic diagram of Figure A.37. The crossover point lies at $Dq/B = 2$.

Term crossings with an accompanying change of multiplicity and consequently low-spin behaviour can occur for octahedral $d^4$, $d^5$, $d^6$ and $d^7$ complexes. In Table A.16 the theoretically possible ground states for octahedral complex ions are summarized. For $d^4$ and $d^7$ the multiplicity is

Table A.16. Ground states of octahedral complex ions

| Number of $d$ electrons | 1 | 2 | 3 | 4 | 5 | 6 | 7 | 8 | 9 |
|---|---|---|---|---|---|---|---|---|---|
| high-spin | $^2T_{2g}$ $d\varepsilon$ | $^3T_{1g}$ $d\varepsilon^2$ | $^4A_{2g}$ $d\varepsilon^3$ | $^5E_g$ $d\varepsilon^3\, d\gamma$ | $^6A_{1g}$ $d\varepsilon^3\, d\gamma^2$ | $^5T_{2g}$ $d\varepsilon^4\, d\gamma^2$ | $^4T_{1g}$ $d\varepsilon^5\, d\gamma^2$ | $^3A_{2g}$ $d\varepsilon^6\, d\gamma^2$ | $^2E_g$ $d\varepsilon^6\, d\gamma^3$ |
| low-spin | | | | $^3T_{1g}$ $d\varepsilon^4$ | $^2T_{2g}$ $d\varepsilon^5$ | $^1A_{1g}$ $d\varepsilon^6$ | $^2E_g$ $d\varepsilon^6\, d\gamma$ | | |

For every state besides the irreducible representation $^{2S+1}\Gamma$ the configuration $d\varepsilon^n\, d\gamma^{N-n}$ from which it arises in the strong field case when electron interaction is considered is given.

reduced by two, for $d^5$ and $d^6$ however by four. For $d^7$ ground states with both conceivable multiplicities ($2S+1 = 4$ or 2) occur. For $d^5$ and $d^6$ the middle and for $d^4$ the smallest value of the possible multiplicities are missing. If one compares these results of the theory with the known magnetochemical data for octahedral complexes of corresponding metal ions, it is found that representatives of all thirteen complex types called

13. L. E. Orgel: *J. chem. Phys.*, **23**, 1004 (1955).
14. Y. Tanabe and S. Sugano: *J. phys. Soc. Japan*, **9**, 753, 766 (1954).

for by the theory are known. No examples of those types forbidden by the theory are found[15].

The ligand field theory enables us to understand the appearance of low-spin complexes without, as is the case for Pauling's theory, having to assume a covalent electron pair bond between the central ion and the ligand.

Low spin means within the framework of the ligand field theory simply that the strength of the ligand field is greater than a certain critical value. It is important to emphasize that the cross-over point can lie at quite different field strengths for different complexes. It therefore follows that even if one wanted to identify increasing strength of the ligand field with an increasing degree of 'covalence', an assumption which is not justified (see section 1.9), then the results of magnetic measurements still do not allow an unambiguous conclusion concerning the bonding to be drawn. Pauling himself[16] later (1948) partially revised his magnetic bonding criterion, which apparently has not been taken note of by many chemists.

In particular it follows from these considerations that the magnetic criterion by no means agrees in general with the stability criterion. The strength of the ligand field depends of course upon the bond strength. However, when the cross-over point for two types of complex ions lies at different field strengths, it is not necessary that a low-spin complex be a stable one and a high-spin complex be labile (see chapter 6, p. 189).

The magnetic behaviour can also be understood on the basis of the *molecular orbital theory* (see section 1.12)[17]. The extended ionic theory and Pauling's theory are, as Van Vleck[11] has shown, special cases of the MO theory: The electronic structure of the central ion and the ligands are brought into consideration. The one-electron term system obtained in this manner is filled with the electrons considered, beginning at the bottom. Each orbital can contain a maximum of two electrons having antiparallel spin, according to the Pauli exclusion principle.

Figure A.65, p. 105, shows a schematic representation of the MOs of an octahedral complex ion, where only $\sigma$ bonds are considered.*

The magnetic behaviour depends upon the occupation of the non-bonding $t_{2g}(d\varepsilon)$ states and the antibonding $e_g^*(d\gamma^*)$ states. Corresponding to the consideration (a) the separation $\Delta = 10\ Dq$ of both states determines

* With consideration of the $\pi$ bonds the position of the $t_{2g}$ state compared with the free ion is changed in that the central ion $t_{2g}$ state and the ligand $t_{2g}$ state 'repel' one another (cf. Figure A.67).

15. H. Hartmann: *Z. Naturf.*, **11a**, 884 (1956).
16. L. Pauling: *J. chem. Soc.*, **1948**, 1461.
17. F. Hund: *Z. Phys.*, **73**, 1, 565 (1931); **74**, 429 (1932). R. S. Mulliken: *Phys. Rev.*, **40**, 55 (1932); **41**, 49, 751 (1932); **43**, 279 (1933).

whether spin-pairing and thereby low-spin behaviour appears or not. For example the ground state of the high-spin $d^5$ complexes has the occupation $[(a_{1g})^2(t_{1u})^6(e_g)^2](t_{2g})^3(e_g^*)^2$ with $S = \frac{5}{2}$ (sextet); for low-spin $d^5$ complexes the occupation is $[(a_{1g})^2(t_{1u})^6(e_g)^2](t_{2g})^5$ with $S = \frac{1}{2}$ (doublet).

## 2.4. Orbital moment contribution and stereochemistry

It was Van Vleck[18] who first explained the fact that the magnetic moments of complexes of transition metals can be approximately obtained from the formula for pure spin magnetism on the basis of the influence of *electric fields of low symmetry* produced by the ligands.

*Two* magnetic moments are to be ascribed to every electron in an atom or molecule. In the *Bohr* model the orbiting of the electron about the nucleus is the cause of *orbital magnetism*; the electron spin accounts for *spin magnetism.* An atom with $N$ electrons is described magnetically by a system of $2N$ coupled moment vectors.

The magnetic moment of the nuclei (nuclear spin) is roughly three orders of magnitude smaller than the electron spin magnetism and can therefore be neglected in our discussion.

The spin magnetism as well as the orbital magnetism are coupled with the mechanical angular momenta. The spin magnetism arises from the mechanical angular momentum of the electrons about their own axes, the spin angular momentum. The orbital magnetism comes from the motion of the electrons about the nucleus, from the orbital angular momentum. Both angular momenta are quantized.

An atomic system can be considered to be a magnetic dipole with the moment

$$\vec{\mu} = (\vec{L} + 2\vec{S})\,\beta_B$$

where $\vec{L}$ and $\vec{S}$ are the total angular momentum and the total spin angular momentum (measured in units of $h/2\pi$). The arrows denote vector quantities. $\beta_B$ is the *Bohr* magneton.

For the case that the spin–orbit coupling can be essentially neglected, that is the *multiplet splitting* of the ground term is *small* compared with $kT$ ($\lambda \ll kT$, $\lambda$ = spin–orbit coupling constant for the ground term of the ion in question (cf. p. 93)), the effective magnetic moment is then given by:

$$\mu_{\text{eff.}} = \{L(L+1) + 4S(S+1)\}^{\frac{1}{2}} \text{ [B.M.]}.$$

One obtains in the other extreme case ($\lambda \gg kT$), when the multiplicity splitting is so great that virtually all ions are in the ground component

18. J. H. Van Vleck: *Theory of Magnetic and Electric Susceptibilities*, Oxford University Press 1932.

of the multiplet for the usual temperatures of measurement:

$$\mu_{\text{eff.}} = g\{J(J+1)\}^{\frac{1}{2}} \qquad \text{[B.M.]}$$

where $g$ is the *Landé factor*

$$g = 1 + \frac{J(J+1) + S(S+1) - L(L+1)}{2J(J+1)}$$

$J$, $S$ and $L$ are the quantum numbers which characterize the total angular momentum, the total spin angular momentum and the total orbital angular momentum. While this relation for $\mu_{\text{eff.}}$ to a good approximation describes the situation for complexes of the rare earth, it has been often attempted to apply the equation

$$\mu_{\text{eff.}} = \{L(L+1) + 4S(S+1)\}^{\frac{1}{2}} \quad \text{[B.M.]}$$

to complexes of the first series of transition metals. It is, however, not valid except for $Co^{2+}$ and $Ni^{2+}$ compounds. Rather the values of the moments are obtained to a good approximation when the formula for the pure spin magnetism is used

$$\mu_{\text{eff.}} = \{4S(S+1)\}^{\frac{1}{2}} \qquad \text{[B.M.]}$$

i.e., when the orbital angular momentum is neglected*.

Although the spin magnetic moment is insensitive to the influences of the environment of the metal ion (the spin moment of an unpaired electron cannot be reduced when the metal ion enters a complex and forms bonds to the ligands; it is possible that spin pairing with another electron takes place) the case is different for the orbital moment.

The situation can be understood qualitatively by considering the symmetry properties of the five $d$ functions which are shown in Figures A.38 and A.39, pp. 53 and 54.

In order for an electron to have an orbital angular momentum and thereby an orbital magnetic moment with reference to a given axis it must be possible to transform the orbital into a fully equivalent orbital by a rotation about that axis.

We shall start with the free ion and choose the $z$ axis. The $d_{x^2-y^2}$ orbital is brought into the fully equivalent $d_{xy}$ orbital by a rotation of 45° about the $z$ axis, and the $d_{xy}$ into the $d_{yz}$ by a rotation of 90° (see Figure A.38).

---

* $Co^{2+}$ and $Ni^{2+}$ complexes show higher moment values than those obtained using $\mu_{\text{eff.}} = \{4S(S+1)\}^{1/2}$ (cf. p. 120 and Table A.13). The effective magnetic moments lie in the direction of those given by $\mu_{\text{eff.}} = \{L(L+1)+4S(S+1)\}^{1/2}$.

This leads to an orbital momentum of $\pm 2$ or $\pm 1$ (measured in units of $h/2\pi$). The $d_{z^2}$ function gives no contribution with reference to the $z$ axis*.

If the ion is brought into a ligand field, the conditions are changed so that the degeneracy of the five $d$ states is to a certain degree removed. For octahedral field symmetry the states $d_{xy}$, $d_{xz}$, $d_{yz}$ are stabilized compared with the $d_{x^2-y^2}$, $d_{z^2}$ states. Thereby the equivalence of $d_{x^2-y^2}$ and $d_{xy}$ is no longer present because the functions correspond to states having different energies. Nevertheless $d_{xz}$ and $d_{yz}$ can still be transformed into one another by a rotation of 90° and contribute one unit to the orbital angular momentum.

In this manner it can be qualitatively understood why, when the symmetry is low enough so that $d_{xz}$ and $d_{yz}$ likewise have different energies, a ligand field partially or completely suppresses the orbital moment (quenching of orbital momentum). Likewise it is understandable that for certain complexes moments are found which much more closely approach those values calculated from the formula for pure spin magnetism or for other complexes values which are somewhat higher but still beneath the values given by $\mu_{\text{eff.}} = \{L(L+1)+4S(S+1)\}^{1/2}$†.

On the basis of the one-electron scheme (this chapter, section 3a) predictions can be made for which octahedral complexes of transition metals of the configuration $d^N$ ($N = 1, 2, \ldots, 9$) of the central ion *orbital angular momentum contributions* are to be expected and for which not.

Electrons in the $d\gamma$ states cannot contribute to the orbital moment because the $d_{x^2-y^2}$ and $d_{z^2}$ orbitals cannot be transformed into one another by a rotation about any axis.

---

* While the function $d_{z^2}$ represents a state for which the $z$ component of the orbital angular momentum is exactly $m_l\hbar = 0$, the functions $d_{xy}$, $d_{yz}$, $d_{xz}$, $d_{x^2-y^2}$ describe states having no definite values of the orbital angular momentum in the $z$ direction (cf. the explicit expressions for the five $d$ functions on p. 55). The functions which can be brought into one another by a rotation about the $z$ axis can, however, always be superimposed so that the resulting functions describe states having definite values of the $z$ components of the orbital angular momentum.

† If and to what degree the orbital momentum is quenched to $\mu_{\text{eff.}}$ through a ligand field is seen from a perturbation calculation. The orbital momentum for a state $\Psi(^{2S+1}\Gamma)$ is zero when the integral

$$(\Psi\,(^{2S+1}\Gamma),\ L\ \Psi\,(^{2S+1}\Gamma))$$

vanishes. $L$ is here the angular momentum operator. If $L$ transforms according to the irreducible representation $\Gamma'$, then this integral vanishes whenever the identity representation is not contained in the direct product of the representations (cf. part B, chapter 2, section 5)

$$\Gamma \dot{\times} \Gamma' \dot{\times} \Gamma$$

For $O_h$ symmetry $L$ belongs to $T_{1g}$. As an explicit consideration shows, the orbital momentum contribution must be completely quenched for $\Gamma = A_{1g}$, $A_{2g}$ or $E_g$, but not necessarily for $\Gamma = T_{1g}$ or $T_{2g}$.

In the case that all three $d\varepsilon$ states are each occupied by one electron, where the electrons have parallel spin, $d_{xz}$ and $d_{yz}$ are no longer equivalent. No equivalent state is obtained by a rotation of 90° about the $z$ axis because $d_{yz}$ already contains one electron. Thereby the configurations $d\varepsilon^3$ and $d\varepsilon^6$ do not contribute to the orbital moment.

*No orbital moment contributions* are to be expected for octahedral high-spin complexes of the configurations

$$d^3(d\varepsilon^3);\ d^4(d\varepsilon^3 d\gamma);\ d^5(d\varepsilon^3 d\gamma^2);$$
$$d^8(d\varepsilon^6 d\gamma^2) \text{ and } d^9(d\varepsilon^6 d\gamma^3)$$

while for

$$d^1(d\varepsilon);\ d^2(d\varepsilon^2);\ d^6(d\varepsilon^4 d\gamma^2) \text{ and } d^7(d\varepsilon^5 d\gamma^2)$$

a *nonzero orbital moment contribution* is possible.

For low-spin complexes of octahedral symmetry an *orbital moment contribution* should be seen for $d^4(d\varepsilon^4)$ and $d^5(d\varepsilon^5)$ and *none* for $d^6(d\varepsilon^6)$ and $d^7$ $(d\varepsilon^6\ d\gamma)$.

For tetrahedral complexes the $d\gamma$ states lie energetically lower than the $d\varepsilon$ states. The occupation with $N = 1$ to 9 electrons follows according to $d\gamma^n\ d\varepsilon^{N-n}$. From the term inversion compared with the tetrahedral structure it follows that in the cases for which one expects orbital moment contributions for $O_h$ symmetry for $T_d$ symmetry no contribution should appear. On the other hand orbital moment contributions should appear for tetrahedral complexes in those cases where they are not present for octahedral complexes. For high-spin tetrahedral complexes *no orbital moment contribution* should appear for

$d\gamma^1, d\gamma^2, d\gamma^2\ d\varepsilon^3, d\gamma^3\ d\varepsilon^3, d\gamma^4\ d\varepsilon^4$ while for

$d\gamma^2\ d\varepsilon, d\gamma^2\ d\varepsilon^2, d\gamma^4\ d\varepsilon^4, d\gamma^4\ d\varepsilon^5$ *an orbital moment* is expected.

A survey of the octahedral and tetrahedral transition metal complexes of the series $d^1$ to $d^9$ for which orbital moment contributions are to be expected from the considerations treated thus far is given in Table A.17.

Low-spin tetrahedral complexes are not known, at least not for central ions of the first transition metal series. This is understandable because the magnitude of the splitting for the tetrahedral field is much smaller than for the octahedral case, so that high-spin behaviour should be the rule $(-\Delta_{T_d} = \frac{4}{9}\Delta_{O_h}$ for equal central ion-ligands separations).

One comes to exactly analogous qualitative predictions about the expected moment contributions when starting from the many-electron system (cf. this chapter, section 3b). The splitting of the terms of the free ion in the ligand field leads to a term system of the complex ion having a

Table A.17. Expected orbital moment contributions to the magnetic moment for octahedral and tetrahedral complexes with the central ions of the configurations $d^1$ to $d^9$ inclusive

(shaded = orbital moment contribution to be expected)

| symmetry | $d^1$ | $d^2$ | $d^3$ | $d^4$ | $d^5$ | $d^6$ | $d^7$ | $d^8$ | $d^9$ |
|---|---|---|---|---|---|---|---|---|---|
| $O_h$ high-spin | $d\varepsilon^1$ $^2T_{2g}$ | $d\varepsilon^2$ $^3T_{1g}$ | $d\varepsilon^3$ $^4A_{2g}$ | $d\varepsilon^3\,d\gamma$ $^5E_g$ | $d\varepsilon^3\,d\gamma^2$ $^6A_{1g}$ | $d\varepsilon^4\,d\gamma^2$ $^5T_{2g}$ | $d\varepsilon^5\,d\gamma^2$ $^4T_{1g}$ | $d\varepsilon^6\,d\gamma^2$ $^3A_{2g}$ | $d\varepsilon^6\,d\gamma^3$ $^2E_g$ |
| low-spin | | | | $d\varepsilon^4$ $^3T_{1g}$ | $d\varepsilon^5$ $^2T_{2g}$ | $d\varepsilon^6$ $^1A_{1g}$ | $d\varepsilon^6\,d\gamma^1$ $^2E_g$ | | |
| $T_d$ high-spin | $d\gamma^1$ $^2E$ | $d\gamma^2$ $^3A_2$ | $d\gamma^2\,d\varepsilon$ $^4T_1$ | $d\gamma^2\,d\varepsilon^2$ $^5T_2$ | $d\gamma^2\,d\varepsilon^3$ $^6A_1$ | $d\gamma^3\,d\varepsilon^3$ $^5E$ | $d\gamma^4\,d\varepsilon^3$ $^4A_2$ | $d\gamma^4\,d\varepsilon^4$ $^3T_1$ | $d\gamma^4\,d\varepsilon^5$ $^2T_2$ |
| low-spin | | | $d\gamma^3$ $^2E$ | $d\gamma^4$ $^1A_1$ | $d\gamma^4\,d\varepsilon^1$ $^2T_2$ | $d\gamma^4\,d\varepsilon^2$ $^3T_1$ | | | |

ground term $^{2S+1}\Gamma$. For octahedral and tetrahedral complexes either threefold orbitally degenerate $T_1$ or $T_2$, twofold $E$, or onefold $A_1$ or $A_2$ appear as the ground terms (cf. Tables A.16 and A.17).

An orbital moment contribution can arise only when a threefold orbitally degenerate term is the ground state. Such a state is always slightly split (slight distortion of the coordination polyhedron, following from the *Jahn–Teller* theorem). This splitting is of the order of $kT$, so that a *Boltzmann* distribution of the ions over the splitting states arises.

A *onefold* state should give *no* orbital moment contribution. A *twofold state* is according to Bethe[19] a '*nonmagnetic*' state with reference to the orbital moment.

Figure A.74 gives the splitting diagrams for high-spin complexes with $D$ and $F$ ground states for $O_h$ and $T_d$ symmetry. It is immediately seen where orbital moment contributions are to be expected.

According to this, e.g., for octahedral Co(II) and tetrahedral Ni(II) complexes an orbital moment contribution to the magnetic moment should appear. Octahedral Ni(II) and tetrahedral Co(II) complexes should have moments, to the first approximation, as obtained from the formula for pure spin magnetism.

Table A.18 gives the moments determined experimentally for several octahedral and tetrahedral high-spin $Co^{2+}$ complexes. It is seen that the octahedral complexes have values of the effective magnetic moment from 4·8 to 5·1 B.M., while for tetrahedral complexes lower values of 4·3 to 4·7 B.M. are observed[20].

19. H. Bethe: *Ann. Phys.*, [5] **3**, 135 (1929).
20. R. S. Nyholm: *Q. Rev. chem. Soc.*, **7**, 377 (1953).

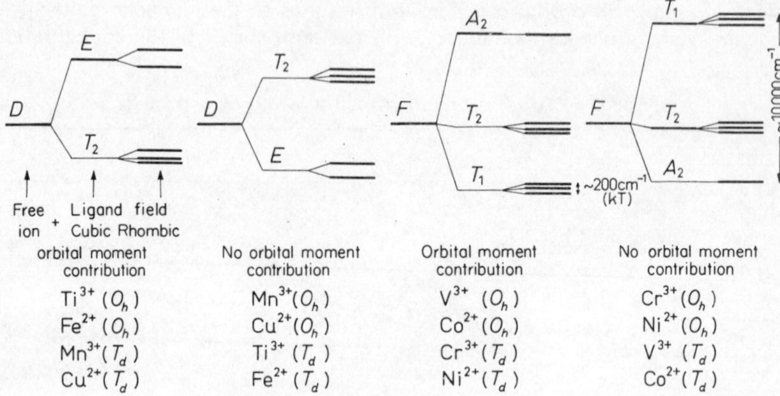


Figure A.74. The contribution of the orbital moment to the magnetic moment. Splitting diagrams for *D* and *P* states in cubic fields having small rhombic field components superimposed (schematic).

For octahedral complexes one finds therefore an appreciable orbital moment contribution. The $\mu_{\text{eff.}}$ values lie close to those expected from the relation $\{L(L+1)+4S(S+1)\}^{1/2}$ of 5·2 B.M. In agreement with the preceding considerations, the values found for the tetrahedral complexes lie lower, but are nevertheless greater than one would expect according to the formula for pure spin magnetism (3·87 B.M.).

Because of spin–orbit coupling, even when no splitting of the ground state appears in cases having no orbital moment contribution, an interaction (intermixing) with the higher states can appear, giving an orbital moment contribution. The more exact theory shows[21] that in the first

Table A.18. Magnetic moments of octahedral and tetrahedral high-spin Co(II) complexes after Nyholm[20]

| Octahedral complex Formula | $\mu_{\text{eff.}}$ [B.M.] | Tetrahedral complex Formula | $\mu_{\text{eff.}}$ [B.M.] |
|---|---|---|---|
| $[Co(H_2O)_6]Cl_2$ | 4·94 | $(pyH)_2[CoCl_4]$ | 4·74 |
| $[Co(H_2O)_6](ClO_4)_2 \cdot 6H_2O$ | 4·93 | $(pyH)_2[CoBr_4]$ | 4·67 |
| $[Co(NH_3)_6](ClO_4)_2$ | 5·04 | $Hg[Co(SCN)_4]$ | 4·33 |
| $[Co(py)_6](ClO_4)_2$ | 4·87 | $[CoCl_2, 2(C_2H_5)_3P]$ | 4·48 |
| $[Co(dipy)_3](ClO_4)_2$ | 4·86 | $[CoCl_2(py)_2]$ | 4·62 |

21. (a) Cf. B. N. Figgis and J. Lewis in *Modern Coordination Chemistry*, Chap. 6, pp. 427 ff., Interscience, New York–London 1960. (b) W. G. Penney and R. Schlapp: *Phys. Rev.*, **42**, 666 (1932).

half of the transition metal series for the complexes in question a decrease of the moment compared with the pure spin moment is to be expected, but in the second half of the series—as for $Co^{2+}$—an increase should be seen*.

For the tetrahedral Ni(II) complex $(Ph_3MeAs)_2$ $[NiCl_4]$ one finds $\mu_{eff.} = 4{\cdot}1$ B.M., while octahedral complexes have lower values between 2·9 and 3·2 B.M.

Thereby one obtains from a study of the magnetic moments for $Ni^{2+}$ and $Co^{2+}$ complexes of cubic symmetry a means of determining if tetrahedral or octahedral coordination is present, as was noted by Nyholm[20,22].

## 2.5. The refined theory of the paramagnetic susceptibility of transition metal complexes; consideration of spin–orbit coupling

The temperature dependence of the paramagnetic susceptibility is given in the ideal case by *Curie's Law*

$$\chi'_{mol} = \frac{C}{T}$$

where $C$ is the *Curie* constant and $T$ the absolute temperature. $\chi'_{mol}$ is the molar susceptibility corrected for the diamagnetism.

For complexes where this law is rigorously valid, the effective magnetic moment should not depend upon the temperature. In general, however, smaller or larger deviations from the Curie Law are found; the $\mu_{eff.}$ values depend upon the temperature. In many cases, at least within a certain temperature range, the dependence of $\chi'_{mol}$ upon $T$ is represented approximately by the *Curie–Weiss* Law

$$\chi'_{mol} = \frac{C}{T-\Theta}$$

$\Theta$ is the Weiss constant, an empirical quantity giving the deviation of the system in question from ideal behaviour, which is assumed when deriving Curie's Law.

For the majority of sufficiently 'magnetically dilute' complexes the values of the Weiss constant lie between 20 and 40°K. In certain cases

* The magnetic moment is obtained from $\mu = \mu_0[1-\alpha(\lambda/\Delta)]$, where $\mu_0 = \{4S(S+1)\}^{1/2}$ is the pure spin moment, $\Delta$ is the separation between the interacting states, $\alpha$ is a constant and $\lambda$ the spin–orbit coupling constant for the ground term of the ion in question ($\zeta_{nd} = \pm 2S\lambda$). $\zeta_{nd}$ is the (positive) spin–orbit coupling constant for one $d$ electron. The minus sign for $\lambda$ applies for the second half of the transition metal series (cf. p. 93).

22. R.S. Nyholm: *J. inorg. nucl. Chem.*, **8**, 401 (1958). B. N. Figgis and R. S. Nyholm: *J. chem. Soc.*, **1954**, 4463. R. S. Nyholm: *J. Proc. R. Soc. N.S.W.*, **89**, 8 (1955); *Rec. Chem. Progr.*, **19**, 45 (1958).

the appearance and the magnitude of $\Theta$ can be coordinated with properties of the systems studied (antiferromagnetic interactions).

If the exact 'structure' of the ground state of a complex, i.e., the position of the energetically low-lying states on the energy scale as well as the corresponding degrees of degeneracy is known, one can give explicit expressions for the *temperature dependence of the susceptibility*. Calculations of this sort were already carried out in 1932 by Penney and Schlapp[21b].

The terms which are thermally accessible can be determined from the terms of the free ion, when the electric field of the surrounding ligands is treated as a perturbation. In practice it is sufficient to consider only the nearest neighbours. For such calculations the spin–orbit coupling must be taken into account.

The Hamiltonian of the system has the form

$$H = H_0 + V_{\text{cryst.}} + \sum_i \xi(r_i)\, l_i s_i + \beta_B H (L + 2S).$$

$H_0$ is the Hamiltonian of the free metal ion without spin–orbit coupling, $\sum_i \xi(r_i) l_i s_i$ takes spin–orbit coupling into account, $V_{\text{cryst.}}$ is the electrostatic energy in the field of the neighbouring ions. $V_{\text{cryst.}}$ has the symmetry of the complex ion and is generally made up additively of a cubically (symmetric) major portion $V_{\text{cub.}}$ and a component of lower symmetry $V'$ (tetragonal or rhombic). The latter term corresponds to the influence of an external magnetic field of the strength $H$.

The total magnetic moment or susceptibility is obtained, under the assumption that the exchange interaction between the magnetic ions can be neglected, by taking a *statistical average value* of all occupied stationary states, using the *Boltzmann* distribution (cf. part B, chapter 5, section 5).

The expression

$$\chi = N_L \frac{\sum\limits_{n,m}\left[\dfrac{(E^{(1)}_{n,m})^2}{kT} - 2E^{(2)}_{n,m}\right] e^{-\frac{E^{(0)}_n}{kT}}}{\sum\limits_n e^{-\frac{E^{(0)}_n}{kT}}}$$

is applicable, where $N_L$ is the *Loschmidt number*, $E^{(0)}_n$ the energy without a magnetic field and $E^{(1)}_{n,m}$ as well as $E^{(2)}_{n,m}$ the coefficients of the first and second order *Zeeman energies*. They appear when the state energy $E_{n,m}$ of the system is expanded as a series in increasing powers of the magnetic field strength $H$:

$$E_{n,m} = E^{(0)}_n + E^{(1)}_{n,m} \cdot H + E^{(2)}_{n,m} \cdot H^2 + \cdots \tag{18}$$

In general magnetic fields remove the $g_n$ fold degeneracy of a state $n$. The states arising in this manner are denoted by the magnetic quantum number $m$.

Kotani[23,24] has carried out calculations of this sort for complex ions of configuration $d\varepsilon^1$ to $d\varepsilon^5$ for the case of octahedral symmetry of the

23. M. Kotani: *J. phys. Soc. Japan*, **4**, 293 (1949).
24. Cf. also B. N. Figgis: *Nature*, **182**, 1568 (1958).

crystal field $V_{cryst.} = V_{O_h}$. The results can be shown in diagrams giving the effective magnetic moment as a function of $kT/\zeta_d$ or $kT/|\lambda|$*.

The (solid) curves calculated by Kotani are depicted in Figure A.75. According to these $d\varepsilon^3$ compounds should obey Curie's Law exactly, i.e. $\mu_{eff.}$ should be independent of temperature. The dashed lines parallel to the abscissa give the $\mu_{eff.}$ values expected from the formula for pure spin magnetism for one and for two unpaired electrons.

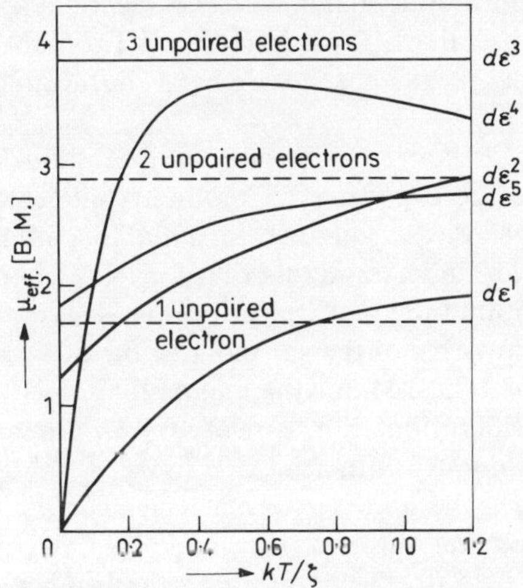


Figure A.75. Effective magnetic moment as a function of $kT/\xi$ after Kotani[23].

$kT$ is for room temperature $\sim 200\ \text{cm}^{-1}$. Octahedral complexes of the first transition metal series should accordingly have $kT/\zeta_{3d}$ values of from about 0·4 to 1 ($Ti^{3+}$ : $\zeta_{3d} = 154\ \text{cm}^{-1}$, $Co^{3+}$ : $\zeta_{3d} = 500\ \text{cm}^{-1}$). The corresponding values of the magnetic moment decrease in the right-hand portion of the diagram, i.e., they lie in the neighbourhood of the values expected from the formula $\{4S(S+1)\}^{1/2}$.

For compounds with central ions of the second and third transition metal series, however, the values for spin–orbit coupling constants $\zeta_{nd}$ are much higher (e.g., $Os^{4+}$ : $\zeta_{5d} = 6400\ \text{cm}^{-1}$). $kT/\zeta_{nd}$ decreases then of the order of from 0·02 to 0·1; i.e., the $\mu_{eff.}$ values lie in the left-hand region of the diagram. Thereby anomalously low values of the moments for such complexes are obtained as well as an appreciable temperature dependence.

---

* A summary of the one-electron spin–orbit coupling constants $\zeta_{nd}$ of the first and second transition metal series was given by T. M. Dunn (*Trans. Faraday Soc.*, **57**, 1441 (1961)).

For example $K_2[OsCl_6]$ ($d\varepsilon^4$) with two unpaired electrons has a moment of 1·40 B.M. at room temperature. $\mu_{eff.}$ is proportional to the square root of the absolute temperature (cf. the corresponding curve in Figure A.75).

For low-spin complexes, especially for the higher series of transition metals, the deviation of the observed moments from those obtained from the formula for pure spin magnetism on the basis of the number of unpaired electrons can be very large.

For $K_3[Fe(CN)_6]$ ($d\varepsilon^5$) a moment of 1·73 B.M. is to be expected from the formula for pure spin magnetism, corresponding to one unpaired electron. However, a value of 2·4 B.M. is found. From the *Kontani* diagram, using $\zeta_{3d} = 440\ cm^{-1}$ (value for the free $Fe^{3+}$ ion) a value $\mu_{eff.} = 2{\cdot}46$ B.M. is obtained.

With the help of *Kotani's theory* one can qualitatively understand why for $d\varepsilon^3$ complexes the experimental moments agree comparatively well with the value 3·87 B.M. from the formula for pure spin magnetism. Likewise it becomes understandable that for low-spin complexes with $d^4$ and $d^5$ configuration of the first transition metal series higher and for the corresponding complexes of the second and third transition series lower values than 2·83 or 1·73 B.M. for the magnetic moment are found.

Analogous considerations can be made for all other complexes having cubic ($O_h$ or $T_d$) symmetry with $d\varepsilon^n\, d\gamma^{N-n}$ or $d\gamma^n\, d\varepsilon^{N-n}$ configuration[25].

Figure A.76 shows the splitting for one $d$ electron ($d\varepsilon^1$) starting from the free ion and consecutively applying the ligand field, the spin–orbit coupling and the magnetic field (to the first approximation consideration of only the term with the first power in $H$).

Only a crude approximation of the temperature dependence of the susceptibility can be made using the simple *Kotani* theory. For a more accurate description it is in the rule necessary to use $V_{cryst.} = V_{O_h} + V'$. That is, a *field component* $V'$ *of lower symmetry* (rhombic or tetragonal) usually of the order of the spin–orbit coupling energy must be added. An *anisotropy* of the susceptibility[26] is obtained in this manner. When studying single crystals different values are found depending upon the orientation of the crystal in the magnetic field. If the sample measured is a crystalline powder, as is usually the case, an average value of the susceptibility is obtained.

For example Figgis[27] has recently calculated $\mu_{eff.}$ as a function of $kT/\zeta_d$ for cubic complexes with $T_2$ ground terms by taking spin–orbit coupling, the cubic field and an axially symmetric field component into

25. Cf. [21a] p. 431 J. S. Griffith: *Trans. Faraday Soc.*, **54**, 1109 (1958).
26. Cf. e.g. H. Kamimura: *J. phys. Soc. Japan*, **11**, 1171 (1956).
27. B. N. Figgis: *Trans. Faraday Soc.*, **57**, 198, 204 (1961); **56**, 1553 (1960).

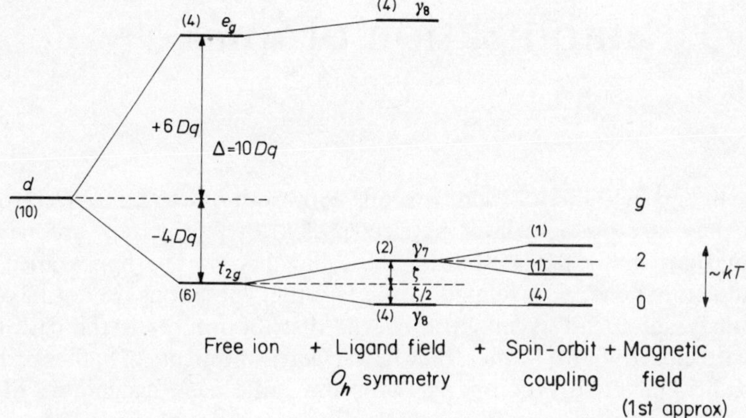


Figure A.76. Splitting of a *d* state in an octahedral field with consideration of various approximations (schematic). (Numbers in parentheses are the degrees of degeneracy. $\gamma_8$ and $\gamma_7$ are the corresponding double-valued representations; cf. [19].)

account. In this calculation two additional parameters appear, the splitting correction $\Delta'$ determined by $V''$ and the electron delocalization parameter $k$ introduced by Stevens. From a comparison of the experimental curves $\mu_{\text{eff.}} = f(T)$ with the curves calculated for various values of the parameters, the magnitude of the axially symmetric field component can be estimated*.

* For complex ions a *decrease* of the spin–orbit coupling constants by about 10–40 per cent compared with the free central ions is observed, as follows from the results of paramagnetic resonance measurements. For copper(II)-hexaquo ion $\zeta_{3d}$ is 620 cm$^{-1}$, while for free $Cu^{2+}/\zeta_{3d}$ = 820 cm$^{-1}$ (J. Owen: *Proc. Roy. Soc.*, **A 227**, 183 (1955); T. M. Dunn: *J. chem. Soc.* (London) 623 (1959); cf. also section 1.11, relativistic nephelauxetic effect.)

# 3. Stabilization of ions

The ligand field theory yields for ions with from one to nine $d$ electrons when these are brought into the electric field of the ligands, an energy greater than that calculated from the ionic theory. The appearance of this additional energy is related to the fact that these ions do not have a rigorously spherically symmetric charge distribution, as is the case for ions with closed shells. Rather, they have electric moments of higher order. An additional energy compared with the ionic case (calculated after Kossel and Magnus assuming a spherically symmetric charge distribution) arises because of the interaction of these moments with the ligands. This additional energy, the *ligand* (or *crystal field stabilization energy* (LFSE)) can be given for every ion of the series $d^1$–$d^9$ in a ligand field of given symmetry.

In analogy to the discussion of magnetic behaviour (section 2.3) a complete calculation of the crystal field stabilization energy could be carried out in order to determine as exactly as possible the energy difference between the ground term of the complex ion and the free ion. Either the weak-field or the strong-field method can be employed, but all possible term or configuration interactions must be considered (cf. p. 51). On the other hand it is also possible to carry out the calculation approximately within the one-electron scheme. This is advantageous because of its simplicity. Here the $d$ electrons are distributed over the one-electron $d$ states. The magnitude of the stabilization energy can be obtained from a simple summation, as will be shown in the next section for complexes having cubic symmetry.

### 3.1. Ligand field stabilization of cubic complexes

We shall now consider the case of octahedral complex ions in the *weak-field approximation.* We begin with the term system of the free ion in which each term is characterized by $^{2S+1}L$.

The total energy of stabilization $E_s$(tot.) is equal to the energy by which the ground term of the free ion $^{2S+1}L_{(0)}$ sinks when brought into the ligand field. It is therefore equal to the energy difference $E(^{2S+1}L_{(0)}) - E(^{2S+1}\Gamma_{(0)})$ in Figure A.77. To obtain the ligand field stabilization energy $E_s$(LFSE) a quantity equal to the energy shift in the presence of the ligand potential $E_s$(class.) must be subtracted. This is to the first approximation equal for all of the terms of the free ion and arises according to Kossel and Magnus from the classical *Coulomb* interaction. Figure A.77 shows in detail the various energy levels. To the right and to the left the terms of the free ion denoted by $^{2S+1}L$ are given. The ground term is designated by $^{2S+1}L_{(0)}$.

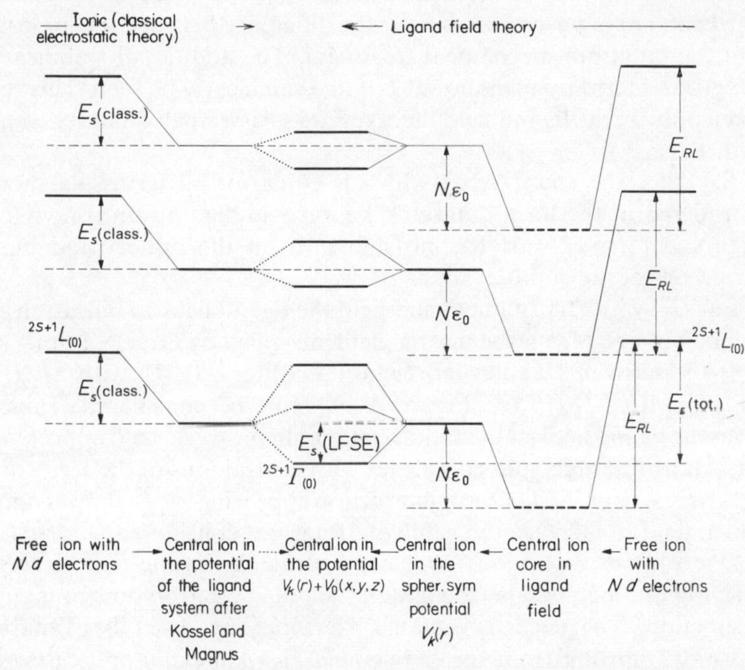


Figure A.77. The definition of the ligand field stabilization energy (LFSE).

The energy of the interaction between the central ion and the ligand field can be divided into two parts. The interaction between the central ion core (that is, the central ion without the $N$ outer $d$ electrons) leads to a stabilization of all terms by the energy $E_{RL}$ (the first step from right to left in Figure A.77). The repulsion between the $d$ electrons of the central ion and the ligands must also be considered. This gives a further contribution composed of the labilizing interaction $N\varepsilon_0$ ($d$ electrons in the spherically symmetric part $V_k(r)$ of the ligand potential; second step from right to left) and the splitting energy ($d$ electrons in non-spherically symmetric potential $V_0(x, y, z)$, third step from right to left). The ground term of the complex ion $^{2S+1}\Gamma_{(0)}$ is stabilized by the ligand field stabilization energy (LFSE) compared to the ground term without consideration of $V_0(x, y, z)$.

Going from left to right, in the first step the situation is depicted when the influence of the potential of the ligand system on the energy of the ground term $^{2S+1}L_{(0)}$ (and the higher terms) of the central ion is considered in the sense of the ionic theory. All terms sink in the first approximation by an amount $E_s$(class.). A comparison of the second with the third column of terms makes clear the difference between the ionic picture and the quantum mechanical treatment. The additional stabilization of the ground term by the amount $E_s$(LFSE) can clearly be seen. This result is given only by the ligand field theory. $E_s$(LFSE) is a small quantity compared with $E_s$(class.).

$E_s$(class.), the energy shift which is equal for all terms, has not been considered in the term diagrams we have studied up until now (cf. e.g., Figure A.15, p. 24). It has no influence on the optical and magnetic properties of the complex ions.

For $O_h$ symmetry of the ligand field the ligand field stabilization energy can be obtained from the matrix elements given by Orgel[1]. For $d^2$ and $d^7$ ions the term interaction (cf. section 1.6d) for $^3T_{1g}(F)$ with $^3T_{1g}(P)$ and $^4T_{1g}(F)$ with $^4T_{1g}(P)$ (cf. Figure A.29) must be considered. Table A.19 gives the ligand field stabilization energies in the weak-field approximation for octahedral high-spin complexes with central ions of the series $d^1$ to $d^9$ (cf. Table A.4, p. 75). The term interaction appearing for $d^2$ and $d^7$ complexes which tend to increase the value of $6Dq$ is not considered.

One obtains analogous results by using the *strong-field method* and carrying out a complete calculation including electron and configuration interaction. The necessary matrix elements are given by Tanabe and Sugano[2]. Starting from the *strong-field approximation* and carrying out only the *first step* (inclusion of the ligand field), we obtain a result quite easily. For octahedral symmetry, e.g., we consider the occupation of the $d\varepsilon$ and $d\gamma$ states which are stabilized by $-4Dq$ or $+6Dq$ compared to the spherically symmetric charge distribution (Figure A.43). In this approximation (without electron and configuration interaction) we obtain the LFSE of the ground term arising from the configuration $d\varepsilon^n\, d\gamma^{N-n}$, using

$$\text{LFSE(Oct.)} = -[4n - 6(N-n)]\, Dq.$$

where $N$ is the total number of $d$ electrons, $n$ is the number of $d\varepsilon$ electrons occupying the $t_{2g}$ state and $(N-n)$ is the number of $d\gamma$ electrons occupying the $e_g$ state.

1. L. E. Orgel: *J. chem. Soc.*, **1952**, 4756; *J. chem. Phys.*, **23**, 1004 (1955).
2. Y. Tanabe and S. Sugano: *J. phys. Soc. Japan*, **9**, 753 (1954).

Table A.19. Ligand field stabilization energies in the weak field approximation for $O_h$ symmetry (high-spin complexes) (cf. also Table A.20)

| Configuration of the central ion | LSFE in $Dq$ | Ground term of the complex ion |
|---|---|---|
| $d^1$ | $-4$ | $^2T_{2g}$ |
| $d^2$ | $-6$* | $^3T_{1g}$ |
| $d^3$ | $-12$ | $^4A_{2g}$ |
| $d^4$ | $-6$ | $^5E_g$ |
| $d^5$ | $0$ | $^6A_{1g}$ |
| $d^6$ | $-4$ | $^5T_{2g}$ |
| $d^7$ | $-6$* | $^4T_{1g}$ |
| $d^8$ | $-12$ | $^3A_{2g}$ |
| $d^9$ | $-6$ | $^2E_g$ |

* These values are increased when term interaction is considered.

Table A.20 contains stabilization energies obtained in this manner for high-spin and low-spin complexes of the series $d^1$ to $d^9$. In the last column the additional stabilization energy of the low-spin as compared to the high-spin complexes is given. The numbers written in brackets indicate the number of additional electron pairs with antiparallel spin which appear for low-spin complexes as compared with high-spin complexes*.

Table A.21 gives the corresponding values for tetrahedral complexes, where it should be considered that $-Dq(\text{tetr.}) = \frac{4}{9}Dq(\text{oct.})$ for equal separations and charges. As has already been mentioned (p. 134) low-spin tetrahedral complexes do not appear for the first series of transition metals because of the small $Dq$ values. The ligand field stabilization energy belonging to the ground term arising from the configuration $d\gamma^n\, d\varepsilon^{N-n}$ is given by

$$\text{LFSE(Tetr.)} = -[6n - 4(N-n)]\, Dq,$$

where $n$ is the number of $d\gamma$ and $N-n$ the number of $d\varepsilon$ electrons.

For the case of *low-spin* complexes, the *spin pairing energy P* must be considered as an additional term, so that the relations for the ligand field stabilization energy then become:

$$\text{LFSE (Oct.)} = -[4n - 6(N-n)]\, Dq + P,$$
$$\text{LFSE (Tetr.)} = -[6n - 4(N-n)]\, Dq + P.$$

* The high-spin complexes have the same number of electron pairs as the free ions.

Table A.20. LFSE for octahedral complexes in the strong field approximation

| | High-spin complexes | | | Low-spin complexes | | | Additional stabilization |
|---|---|---|---|---|---|---|---|
| $N$ | Configuration | Ground term | LFSE in $Dq$ | Configuration | Ground term | LFSE in $Dq$ | in $Dq$ |
| 1 | $d\varepsilon^1$ | $^2T_{2g}$ | $-4$ | | | | |
| 2 | $d\varepsilon^2$ * | $^3T_{1g}$ | $-8$ | | | | |
| 3 | $d\varepsilon^3$ | $^4A_{2g}$ | $-12$ | | | | |
| 4 | $d\varepsilon^3\,d\gamma$ | $^5E_g$ | $-6$ | $d\varepsilon^4$ | $^3T_{1g}$ | $-16$ | $-10$ (1) |
| 5 | $d\varepsilon^3\,d\gamma^2$ | $^6A_{1g}$ | 0 | $d\varepsilon^5$ | $^2T_{2g}$ | $-20$ | $-20$ (2) |
| 6 | $d\varepsilon^4\,d\gamma^2$ | $^5T_{2g}$ | $-4$ | $d\varepsilon^6$ | $^1A_{1g}$ | $-24$ | $-20$ (2) |
| 7 | $d\varepsilon^5\,d\gamma^2$ * | $^4T_{1g}$ | $-8$ | $d\varepsilon^6\,d\gamma$ | $^2E_g$ | $-18$ | $-10$ (1) |
| 8 | $d\varepsilon^6\,d\gamma^2$ | $^3A_{2g}$ | $-12$ | | | | |
| 9 | $d\varepsilon^6\,d\gamma^3$ | $^2E_g$ | $-6$ | | | | |

* The electron- as well as configuration-interaction, which in these cases lead to a change of the LFSE, produces LFSE values which lie between $-6Dq$ and $-8Dq$. McClure[6] uses $-7Dq$ as an average value. $^3T_{1g}$ and $^4T_{1g}$ are not pure $d\varepsilon^2$ or $d\varepsilon^5\,d\gamma^2$ states, but rather contain contributions from $d\varepsilon\,d\gamma$ or $d\varepsilon^4\,d\gamma^3$.

Table A.21. LFSE for tetrahedral complexes for the strong field case

| | High-spin complexes | | | Low-spin complexes | | | Additional stabilization |
|---|---|---|---|---|---|---|---|
| $N$ | Configuration | Ground term | LFSE in $Dq$ | Configuration | Ground term | LFSE in $Dq$ | in $Dq$ |
| 1 | $d\gamma^1$ | $^2E$ | $-6$ | | | | |
| 2 | $d\gamma^2$ | $^3A_2$ | $-12$ | | | | |
| 3 | $d\gamma^2\,d\varepsilon$* | $^4T_1$ | $-8$ | $d\gamma^3$ | $^2E$ | $-18$ | $-10$ (1) |
| 4 | $d\gamma^2\,d\varepsilon^2$ | $^5T_2$ | $-4$ | $d\gamma^4$ | $^1A_1$ | $-24$ | $-20$ (2) |
| 5 | $d\gamma^2\,d\varepsilon^3$ | $^6A_1$ | 0 | $d\gamma^4\,d\varepsilon^1$ | $^2T_2$ | $-20$ | $-20$ (2) |
| 6 | $d\gamma^3\,d\varepsilon^3$ | $^5E$ | $-6$ | $d\gamma^4\,d\varepsilon^2$ | $^3T_1$ | $-16$ | $-10$ (1) |
| 7 | $d\gamma^4\,d\varepsilon^3$ | $^4A_2$ | $-12$ | | | | |
| 8 | $d\gamma^4\,d\varepsilon^4$ * | $^3T_1$ | $-8$ | | | | |
| 9 | $d\gamma^4\,d\varepsilon^5$ | $^2T_2$ | $-4$ | | | | |

* Cf. footnote to Table A.20 $^4T_1$ and $^3T_1$ are not pure $d\gamma^2\,d\varepsilon$ or $d\gamma^4\,d\varepsilon^4$ states.

The ground term of the free ion is used as the standard for measuring the LFSE. The spin-pairing energy for octahedral symmetry correspond to one additional electron pair for $d^4$ and $d^7$ complexes and for $d^5$ and $d^6$ complexes to two additional electron pairs compared to the free ion.

For tetrahedral complexes one finds for $d^3$ and $d^6$ one and for $d^4$ and $d^5$ two additional electron pairs. The expressions given in Table A.22, column 3 are obtained for the spin pairing energy $P$ in the strong field approximation.

Table A.22. Total spin pairing energy $P$ as well as average spin-pairing energy $\Pi$ as a function of the electron interaction parameters $B$ and $C$*

| Number of *d*-electrons Oct.($O_h$) | Tetr.($T_d$) | Total spin-pairing energy $P$ | Average pairing energy $\Pi$ per $10Dq$ Ligand field stabilization |
|---|---|---|---|
| 4 | 6 | $6B+5C$ | $6B+5C$ |
| 5 | 5 | $15B+10C$ | $7{\cdot}5B+5C$ |
| 6 | 4 | $5B+8C$ | $2{\cdot}5B+4C$ |
| 7 | 3 | $4B+4C$ | $4B+4C$ |

* Determined from the matrix elements for the strong field case (without consideration of configuration interaction).

## 3.2. Stabilization and tendency of formation of low-spin complexes

Using the concepts developed in the preceding section we are in the position to study more completely the low-spin complexes which appear for the case of octahedral configurations, as was mentioned in Chapter 2. In low-spin complexes the electrons tend to occupy the energetically lower lying $t_{2g}$ states under spin pairing. In high-spin complexes the $t_{2g}$ and $e_g$ states are so occupied that the electrons have parallel spins as far as is possible. The number of unpaired spins expected in both cases for octahedral symmetry differs only for complexes with 4, 5, 6 and 7 electrons.

The question why some transition metal ions form low-spin complexes more easily than others can be discussed qualitatively after Griffith and Orgel[3] as follows: The gain in LFSE for octahedral low-spin $d^4$ and $d^7$ complexes is $+10Dq$ and for $d^5$ and $d^6$ complexes $+20Dq$ (cf. Table A.20, last column). The gain in stabilization energy is compensated for to a certain degree by the increasing energy of interaction between the *d*

3. J. S. Griffith and L. E. Orgel: *Q. Rev. chem. Soc.*, **11**, 381 (1957); cf. also L. E. Orgel: *An Introduction to Transition Metal Chemistry*, pp. 47 ff., Methuen, London; John Wiley, New York, 1960.

electrons. This interaction energy arises on the one hand from the fact that the electrons, when they are all in the $t_{2g}$ state lie closer to one another than when they are distributed over the $t_{2g}$ and the $e_g$ states. Thereby the classical *Coulomb* repulsion between them is greater. On the other hand, the purely quantum mechanical *exchange energy** must also be taken into consideration. This is zero for each electron pair with antiparallel spin, but leads to an additional stabilizing exchange contribution for each pair of electrons with parallel spins.

If the increase in repulsion energy in the cases $d^4$, $d^7$ or $d^5$, $d^6$ is denoted by $\Pi$ and $2\Pi$, then $\Pi$ is an *average pairing energy* per unit of ligand field stabilization $+10Dq\Pi$ can be considered approximately to consist of the sum of a *Coulomb* and an exchange term:

$$\Pi \cong \Pi_c + \Pi_e$$

where it can be assumed that the Coulomb contribution $\Pi_c$ varies only slightly within the series of $d^n$ complexes which come into question. The contribution of the exchange energy $\Pi_e$ can be estimated by determining for each individual case the number $Z$ of pairs of electrons with parallel spin and setting $\Pi_e \sim ZK$, where $K$ is the average energy of one pair.

Table A.23 gives the values of $Z$ for octahedral high-spin and low-spin complexes in the series $d^1$ to $d^9$.

Table A.23. Number of pairs of electrons with parallel spin for octahedral complexes of configurations $d^1$ to $d^9$ inclusive

| Number of | Z | |
|---|---|---|
| electrons | high-spin | low-spin |
| 1 | 0 | |
| 2 | 1 | |
| 3 | 3 | |
| 4 | 6 | 3 |
| 5 | 10 | 4 |
| 6 | 10 | 6 |
| 7 | 11 | 9 |
| 8 | 13 | |
| 9 | 16 | |

* The exchange energy is a purely quantum mechanical phenomenon and has no classical analogue. Its magnitude depends upon the orientation of the spins of the lone (unpaired) electrons. In general the exchange interaction leads to a stabilization of an atom which increases with the number of unpaired electrons with parallel spin.

For example, the ground states of the high-spin complexes with $d^5$ and $d^6$ configurations arise from $d\varepsilon^2\, d\gamma^2$ or $d\varepsilon^4\, d\gamma^2$ corresponding to the occupations:

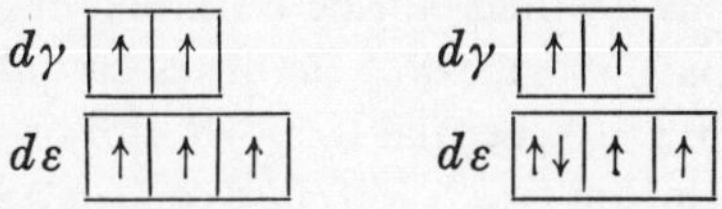

$$d\gamma \; \boxed{\uparrow \,|\, \uparrow} \qquad\qquad d\gamma \; \boxed{\uparrow \,|\, \uparrow}$$

$$d\varepsilon \; \boxed{\uparrow \,|\, \uparrow \,|\, \uparrow} \qquad\qquad d\varepsilon \; \boxed{\uparrow\downarrow \,|\, \uparrow \,|\, \uparrow}$$

The number of electron pairs with parallel spin is in both cases

$$\binom{5}{2} = \frac{5 \cdot 4}{1 \cdot 2} = 10.$$

For the corresponding low-spin complexes the ground states of which arise from the configurations $d^5$ and $d^6$ one finds

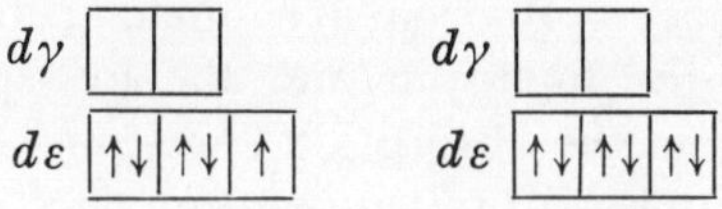

$$d\gamma \; \boxed{\phantom{\uparrow} \,|\, \phantom{\uparrow}} \qquad\qquad d\gamma \; \boxed{\phantom{\uparrow} \,|\, \phantom{\uparrow}}$$

$$d\varepsilon \; \boxed{\uparrow\downarrow \,|\, \uparrow\downarrow \,|\, \uparrow} \qquad\qquad d\varepsilon \; \boxed{\uparrow\downarrow \,|\, \uparrow\downarrow \,|\, \uparrow\downarrow}$$

so that:

$$Z = \binom{3}{2} + \binom{2}{2} = 4 \quad \text{or} \quad Z = \binom{3}{2} + \binom{3}{2} = 6.$$

The decrease of the exchange energy accompanying the transition from the high- and the low-spin case is greater for $d^5$ than for $d^6$. One obtains $10K - 4K = 6K$ compared with $10K - 6K = 4K$, i.e., as the decrease for $d^5$ $3K$ and for $d^6$ $2K$ per $10Dq$ stabilization energy. Using the $Z$ values from Table A.23 $6K - 3K = 3K$ is found for $d^4$. From this it can be concluded that $d^6$ compounds should show spin pairing and thereby be low-spin for lower ligand field strengths than $d^4$ and $d^5$ complexes*.

Besides $[CoF_6]^{3-}$ all other complexes of $Co^{3+}$, such as $[Co(H_2O)_6]^{3+}$, $[Co(NH_3)_6]^{3+}$, $[Coen_3]^{3+}$, are of the low-spin type, while $[Mn(H_2O)_6]^{2+}$, $[Mn(NH_3)_6]^{2+}$ or $[Fe(H_2O)_6]^{3+}$, $[Fe(NH_3)_6]^{3+}$ ($d^5$) and $[Cr(H_2O)_6]^{2+}$, $[Cr(H_2O)_6]^{2+}$ as well as $[Cr(NH_3)_6]^{2+}$ ($d^4$) are high-spin.

The criterion for the appearance of *low-spin* complexes is $10Dq > \Pi$.

The cross-over points described earlier (p. 50, p. 129) are defined within the framework developed here as $10Dq = \Pi$.

---

* For this qualitative argument the Jahn–Teller distortions which are particularly strong for low-spin complexes and produce deviations from $O_h$ symmetry are not considered.

The $Dq$ values can be determined from the absorption spectra of the appropriate complexes (p. 77). The average spin pairing energy can be expressed by means of the electron interaction parameters $B$ and $C$ (Table A.22, last column), which can be obtained approximately from the spectra of the free ions[4]. More exact $B$ values can also be obtained from the absorption spectra. $C$ is set equal to $\sim 4B$. If $B$ or $C$ had the same value for all ions, then the relation $\Pi(d^6) < \Pi(d^7) < \Pi(d^4) < \Pi(d^5)$ should hold. $\Pi$ is more strongly dependent upon the nature of the ligands than is $Dq$. For a given metal ion the ligands can be arranged in a series according to increasing $Dq$ values (spectrochemical series, cf. p. 78). One observes then, concerning the magnetic behaviour, that from a certain ligand on, namely when $\Delta \equiv 10Dq$ becomes greater than $\Pi$, then spin pairing occurs. It follows that all complexes with ligands lying to the right of this point in the spectrochemical series are of the low-spin type.

A special situation exists in the immediate neighbourhood of the cross-over point. Then it is conceivable that at room temperature an equilibrium could exist between the high-spin and the low-spin complexes. The magnetic behaviour and the absorption spectrum should then show an unusual behaviour with regard to temperature dependence.

### 3.3. Influence of ligand field stabilization on the thermodynamic properties of transition metal complexes

The stabilization of ions produced by a deviation from the spherically symmetric charge distribution as discussed in section 3.1 leads among other things to an understanding of the *heats of hydration* $\Delta H_H$ of transition metal ions.

The heat of hydration is defined as the heat liberated when the gaseous metal ion is dissolved in water. It can be assumed that the corresponding hexaquo ions arise as follows:

$$M^{n+}_{\text{gasf.}} \xrightarrow{H_2O} [M(H_2O)_6]^{n+}_{\text{aq.}} + \Delta H_H$$

In Figure A.78 the heats of hydration $-\Delta H_H$ in kcal/mol* are plotted as a function of the atomic number and connected by a solid line. The

* The $\Delta H_H$ values (correspondingly also the lattice energies, p. 152) are obtained from suitable thermodynamic cyclical processes[6].

4. A calculation of the critical field strength, which includes overlap of terms, can be made within the framework of a more quantitative theory. Cf. J. S. Griffith: *J. inorg. nucl. Chem.*, **2**, 1, 229 (1956). L. E. Orgel: *J. chem. Phys.*, **23**, 1819 (1955). J. S. Griffith: *The Theory of Transition Metal Ions*, pp. 237, Cambridge University Press 1961.

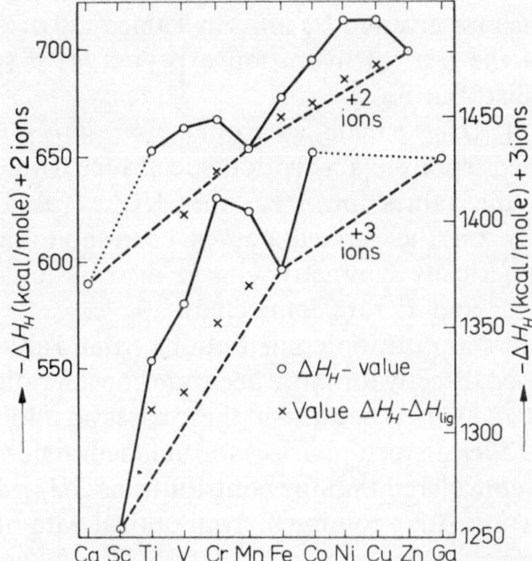


Figure A.78. Heats of hydration of transitions metal ions as a function of the atomic number. (O. G. Holmes and D. S. McClure: *J. Chem. Physics*, **26**, 1686 (1957).). Values of $\Delta H_{\text{lig}}$ determined from spectroscopic *Dq* values have been subtracted from all $\Delta H_H$ values. They lie quite close to the dashed curve for $\Delta H_H$, which is obtained by connecting the values for spherically symmetric $d^0$, $d^5$ and $d^{10}$ metal ions.

upper portion of Figure A.78 shows the situation for bivalent and the lower portion for trivalent metal ions of the first transition series.

Certain maxima and minima can be seen in the curves. If the corresponding aquo-complexes could be treated by an ionic model, it would be expected that with increasing atomic number, e.g., from $Ca^{2+}$ to $Zn^{2+}$, the heat of hydration $\Delta H_H$ should also increase monotonically*.

That this is not the case depends upon the fact that the charge distribution of the central ions in the aquo-complexes is not spherically symmetric. Only for $d^0$, $d^5$ and $d^{10}$ is there a spherically symmetric charge distribution; in the other cases the ligand field stabilization energy, Table A.19 or Table A.20, column 4, must be considered. $\Delta H_H$ can be assumed on this basis to be made up additively of two parts:

$$\Delta H_H = \Delta H_{\text{corr.}} + \Delta H_{\text{lig.}},$$

* According to the ionic theory the ionic radii should decrease monotonically with increasing atomic number (cf. p. 160). For the $\Delta H_H$ values an inverse relation is to be expected.

where $\Delta H_{\text{lig.}}$ is the LFSE given in $Dq$ units in Table A.19 or A.20 and $\Delta H_{\text{corr.}}$ is the portion of the heat of hydration expected for a spherically symmetrical charge distribution[5,6].

If the values of $\Delta H_{\text{lig.}}$ calculated from $-[4n-6(N-n)]Dq$ using the $Dq$ values obtained from the absorption spectra (cf. Table A.6, p. 78) are subtracted from the values for $\Delta H_H$, the $\Delta H_{\text{corr.}}$ values are obtained. They lie virtually on the dashed curves corresponding to the curve expected for a spherically symmetric charge distribution (obtained when the points for $d^0$, $d^5$ and $d^{10}$ are connected).

The ligand field stabilization contributions from Table A.19 or Table A.20 cannot be used directly for more accurate considerations of $d^4$ and $d^9$ complexes ($Cr^{2+}$ and $Cu^{2+}$) because in these cases an additional stabilization from *Jahn–Teller* distortion arises for octahedral coordination.

It is also to be considered that the contributions $\Delta H_{\text{lig.}}$ calculated using the values from Table A.20, column 4, from optical data of corresponding aquo-complexes (Tables of $Dq$ values p. 78) are already only approximate values because the influence of the ligand field stabilization on the ionic radius has been neglected (cf. section 3.5, pp. 160 ff.). A certain (small) part of the ligand field energy is needed to compensate for the normal repulsive forces. In this manner a small decrease of the ionic radius is produced[7].

One finds therefore that with the absence of *quantum mechanical* ligand field effects the heat of hydration (as well as other similar thermodynamic quantities) of the transition metal complexes should monotonically increase with increasing atomic number (dashed $\Delta H_{\text{corr.}}$ curves in Figure A.78). It should be mentioned once more that the differences found between the ions, as follows from the preceding considerations, can not be completely attributed to ligand field effects. The continuous rise of the $\Delta H_{\text{corr.}}$ curves from $Ti^{2+}$ to $Zn^{2+}$ is related to the decrease of the ionic radii and thereby to the parallel increase of the electronegativities. Only the deviations from monotonic behaviour of the curves result from typically ligand field effects.

Analogous considerations as for the heats of hydration apply to the heats of formation and the lattice energies. These are all visibly influenced by the ligand field effects.

The *lattice energy* is the heat of formation of a solid compound from its components in the gaseous state. For the case of singly charged anions

---

5. P. George, D. S. McClure, J. S. Griffith and L. E. Orgel: *J. chem. Phys.*, **24**, 1269 (1956).
6. P. George and D. S. McClure: *Prog. inorg. Chem.*, **1**, 381 (1959).
7. N. S. Hush and H. M. L. Pryce: *J. chem. Phys.*, **26**, 143 (1957) as well as **28**, 244 (1958). N. S. Hush: *Discuss. Faraday Soc.*, **26**, 145 (1958).

$X^-$ and $n$-fold positively charged metal ions $M^{n+}$ we have the reaction*

$$N^{n+}_{\text{gas.}} + nX^-_{\text{gas.}} \rightarrow \text{MX}_{n\,\text{sol.}} + \Delta H_G.$$

If the halogenides $MX_2$ (X = F, Cl, Br, I) are considered under the assumption of octahedral microsymmetry, one obtains the curves in Figure A.79 when $-\Delta H_G$ is plotted against the atomic number†. If straight lines are drawn between the points for $Ca^{2+}$ ($d^0$) and $Zn^{2+}$ ($d^{10}$), they pass close by the point for $Mn^{2+}$ ($d^5$). In these three cases the charge distribution is spherically symmetric and no ligand field stabilization is to be expected. The shape of the curves in Figure A.79 reflects the expected

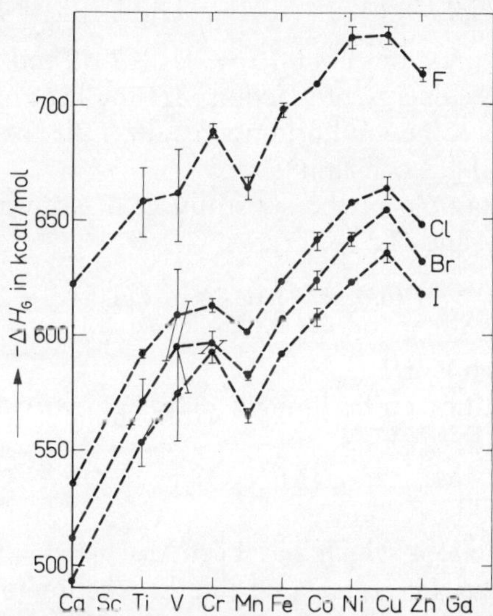


Figure A.79. Lattice energies of metallic halogenides $MX_2$ as a function of the atomic number[6].

ligand field stabilization, which is given by the values in column 4 of Table A.20 in multiples of $Dq$. It is completely similar to the curve observed for the heats of hydration (Figure A.78)

$$\Delta H_G = \Delta H_{\text{corr.}} + \Delta H_{\text{lig.}}.$$

---

* The lattice energy is often measured using the free neutral atoms as the reference point. In the following considerations the reference point used is the ions in their classical valence states[6].

† The given values for the lattice energies are enthalpy values and contain the (smaller) $P\Delta V$ terms.

The difference between the experimental values $\Delta H_G$ and those lying on the lines connecting $Zn^{2+}$ and $Ca^{2+}$ $\Delta H_{corr.}$ is $\Delta H_{lig.}$.

Ligand field stabilization effects are also seen for *heats of formation* as well as for *heats of reaction* of dissolved complex ions with transition metal central ions. They influence the free energies of reaction and thereby the corresponding equilibrium constants, the values of which are a measure of thermodynamic stability.

The stability of a transition metal complex ion in aqueous solution can be described quantitatively by the free energy of reaction $\Delta G$ or by the equilibrium constant $K$ of the reaction

$$[M(H_2O)_6]_{aq.} + qL_{aq.} \rightleftharpoons [ML_q(H_2O)_{6-q}]_{aq.} + qH_2O_{lig.}$$

The following expressions hold $\Delta G = \Delta H - T\Delta S$ and $-\Delta G = RT \ln K$, where $\Delta G$ is the free energy of reaction, $\Delta H$ the heat of reaction, $\Delta S$ the entropy of reaction, $K$ the equilibrium constant, $T$ the absolute temperature and $R$ the universal gas constant[8].

Consider the example of the substitution of six water molecules by six monodentate ligands L:

$$[M(H_2O)_6]_{aq.} + 6L_{aq.} \rightleftharpoons [ML_{6aq.}] + 6H_2O_{lig.} + \Delta H_{compl.}{}^*$$

The heat of reaction is $\Delta H_{compl.}$.

The above reaction can be thought of as being divided into two subreactions:

$$M_{gas.} \xrightarrow{H_2O} [M(H_2O)_6]_{aq.} + \Delta H_H,$$

the hydration reaction, which has been discussed already and where the heat of hydration $\Delta H_H$ is released and the reaction for the formation of a complex ion $[ML_6]$ in aqueous solution from a gaseous metal ion M and the ligands found in aqueous solution

$$M_{gas.} + 6L_{aq.} \rightleftharpoons [ML_6] + \Delta H_L.$$

The heat of reaction here is called $\Delta H_L$. It is readily seen that the expression

$$\Delta H_{compl.} = \Delta H_L - \Delta H_H$$

holds. $\Delta H_L$ and $\Delta H_H$ contain (besides a $\Delta H_{corr.}$ term) for transition metal ions, when the ions $d^0$, $d^{10}$ and $d^5$ are not considered, a ligand field

---

* Charges have been omitted to simplify the notation[8].

8. Cf. e.g. H. L. Schläfer: *Komplexbildung in Lösung—Methoden zur Bestimmung Zusammensetzung und der stabilitätskonstanten gelöster Komplexverbindungen*, Springer-Verlag, Berlin–Göttingen–Heidelberg, 1961.

stabilization term. Thus

$$\Delta H_L = \Delta H^{(L)}_{corr.} + \Delta H^{(L)}_{lig.}$$

$$\Delta H_H = \Delta H^{(H_2O)}_{corr.} + \Delta H^{(H_2O)}_{lig.}.$$

The ligand field stabilization portion of $\Delta H_{compl.}$ is equal to the difference between the $\Delta H_{lig.}$ values for the $[ML_6]$ and the $[M(H_2O)_6]$ complex:

$$\Delta H^{(compl.)}_{lig.} = \Delta H^{(L)}_{lig.} - \Delta H^{(H_2O)}_{lig.}$$

$$\Delta H^{(compl.)}_{lig.} = -[4n - 6(N - n)]\{Dq^{(L)} - Dq^{(H_2O)}\},$$

where $Dq^{(L)}$ and $Dq^{(H_2O)}$ are the values of the field strength parameters for the $[ML_6]$ or the hexaquo complex which can be obtained from pertinent absorption spectra. Accordingly, a characteristic movement of the $\Delta H_{compl.}$ values within a series of $[ML_6]$ complexes of the same charge as a function of the atomic number of the metal ion should become apparent and should be analogous to the situation found for the heats of hydration and lattice energies.

Unfortunately insufficient experimental values of $\Delta H_{compl.}$ for suitable $[ML_6]$ complexes are available at the present time, so that a test of the concepts sketched here is not possible.

However, values of the equilibrium constants $K$ for the formation or dissociation of numerous complexes are known and consequently the corresponding values of the free energies of the reactions. In order to utilize these values for a comparison with the predictions of the ligand field theory it is necessary to make assumptions concerning the change of the entropy of reaction $\Delta S$ for a series of formation reactions for complexes with different metal ions having the same charge. As the first approximation it can be assumed that this change can be neglected. Then the equilibrium constants $K$ for complex formation of given ligands with a series of equally charged central ions should give at least roughly the change of the heat of reaction $\Delta H_{compl.}$ in accordance with $-\Delta H_{compl.} \approx RT \ln K$.

In certain cases the values of the individual constants $k_1, k_2, \ldots, k_6$ for the formation or dissociation of a $[ML_{6-n}(H_2O)_n]$ complex are known*.

---

* These correspond to the equilibria

$$[M(H_2O)_6] + L \rightleftharpoons [M(H_2O)_5L] + H_2O \qquad k_1$$

$$[M(H_2O)_5L] + L \rightleftharpoons [M(H_2O)_4L_2] + H_2O \qquad k_2$$

$$\cdots\cdots\cdots\cdots\cdots\cdots\cdots\cdots\cdots\cdots$$

$$[M(H_2O)L_5] + L \rightleftharpoons [ML_6] + H_2O \qquad k_6$$

The overall constants $K_i$ are obtained from $K_i = \Pi^{i}_{j=1} k_j$. Thus $K_6$ for the equilibrium

$$[M(H_2O)_6] + 6L \rightleftharpoons [ML_6] + 6H_2O \quad \text{is} \quad K_6 = k_1 \cdot k_2 \cdot k_3 \cdot k_4 \cdot k_5 \cdot k_6.$$

(cf. reference 8).

As an example the successive reaction constants for the complex formation $k_1$, $k_2$ and $k_3$ are known for the ethylenediamine complexes of the bivalent metals $Mn^{2+}$, $Fe^{2+}$, $Co^{2+}$, $Ni^{2+}$, $Cu^{2+}$ and $Zn^{2+}$ of the type $[Men(H_2O)_4]^{2+}$, $[Men_2(H_2O)_2]^{2+}$, $[Men_3]^{2+}$. They are given in Table A.24. Every ethylenediamine molecule has two coordination positions so that when we make a comparison with the simple ammine complexes $[M(NH_3)_n(H_2O)_{6-n}]^{2+}$ and denote their successive constants of formation with $k'_1, k'_2, k'_3, \ldots, k'_6$, then $k_1$ corresponds to the product $k'_1 \cdot k'_2$.

Table A.24. Stability constants for the formation of $Men_3^{3+}$ complexes in 1 molar KCl 30°C

| | Mn | Fe | Co | Ni | Cu | Zn |
|---|---|---|---|---|---|---|
| $\log k_1$ | 2·73 | 4·28 | 5·89 | 7·52 | 10·55 | 5·71 |
| $\log k_2$ | 2·06 | 3·25 | 4·83 | 6·28 | 9·05 | 4·66 |
| $\log k_3$ | 0·88 | 1·99 | 3·10 | 4·26 | −1·0 | 1·72 |
| $\log K_3$ | 5·67 | 9·52 | 13·82 | 18·06 | 18·60 | 12·09 |
| $(\log K_3)_{corr.}$ | 5·67 | 6·95 | 8·24 | 9·52 | 10·80 | 12·09 |
| $\Delta \log K_3$ | 0 | 2·57 | 5·58 | 8·54 | 7·80 | 0 |
| $\Delta \log k_1$ | 0 | 0·85 | 1·95 | 3·00 | 5·44 | 0 |
| $\Delta \log k_2$ | 0 | 0·62 | 1·72 | 2·66 | 4·51 | 0 |
| $\Delta \log k_3$ | 0 | 0·94 | 1·88 | 2·88 | −1·11 | 0 |

The sum of the logarithms of the three successive constants gives the logarithm of the total constants $K_3$ for the formation of the trisethylenediamine complexes. These values are found in the fourth row of Table A.24. The corrected values for $Fe^{2+}$, $Co^{2+}$, $Ni^{2+}$ and $Cu^{2+}$, obtained by linear interpolation between the $\log K_3$ values for $Mn^{2+}$ ($d^5$) and $Zn^{2+}$ ($d^{10}$), are given in line 5. In line 6 the differences $\Delta \log K_3 = \log K_3 - (\log K_3)_{corr.}$, corresponding to the ligand field stabilization energy, can be found. Finally in the last three lines the values for $\Delta \log k_i$ for the individual formation constants, obtained in an analogous manner, are given.

One sees from Table A.24 by considering the $\Delta \log K_3$ values, that the additional stabilization in the series $Fe^{2+}$, $Co^{2+}$, $Ni^{2+}$ increases approximately in the proportion 1:2:3, as is to be expected from the theory (stabilization $\Delta H_{lig.}^{(compl.)} = -x\{Dq^{(en)} - Dq^{(H_2O)}\}$; $x = 4, 8, 12$ for $d^6, d^7, d^8$). Analogous results hold for the $\Delta \log k_i$ values of the successive formation reactions of the mono-, bis- and trisethylenediamine complexes.

For $Cu^{2+}$ ($d^9$) a stabilization energy of $\Delta H_{lig.}^{(compl.)} = -6\{Dq^{(3n)} - Dq^{(H_2O)}\}$ should be expected. Here, however, special effects appear, as copper complexes display strong *Jahn–Teller* distortions. Tetragonal symmetry is present for the $[Cuen_2(H_2O)_2]^{2+}$ ion. The two ethylenediamine molecules have four coordination positions in a plane. The $H_2O$ molecules in the trans positions are farther from the $Cu^{2+}$ ion. With the transition to $[Cuen_3]^{2+}$ a more octahedral symmetry is attained, accompanied by a decrease of the stabilization energy. The trisethylenediamine complex is unstable relative to the bisethylenediamine complex.

For the $[Nien_3]^{2+}$ ions a $Dq$ value of 1160 $cm^{-1}$ is obtained from the spectrum and for the hexaquo complex 880 $cm^{-1}$. $\Delta H_{\text{lig.}}^{(\text{compl.})}$ is then $-12(1160–880) = 3360\ cm^{-1}$ or $\approx 10$ kcal. This value is to be compared with $\approx 12{\cdot}3$ kcal obtained from $\Delta \log K_3 = \log K_3 - (\log K_3)_{\text{corr.}} = 8{\cdot}54$. Considering that the entropy term has been neglected and other approximations made, the agreement is satisfactory.

Figure A.80 shows the relative $\Delta H_L$ values obtained from $K_3$ values for $[Men_3]^{2+}$ complexes plotted against the atomic number of the metal ion (solid curve). The dashed curve calculated using $Dq$ values obtained when the $\Delta H_{\text{lig.}}^{(\text{L})}$ values are subtracted from the $\Delta H_L$ values. The $Dq$ values can be taken from the absorption spectra of the appropriate ethylenediamine complexes. As can be seen, the $\Delta H_L$ values obtained in this manner lie very near to the line connecting $Mn^{2+}$ and $Zn^{2+}$.

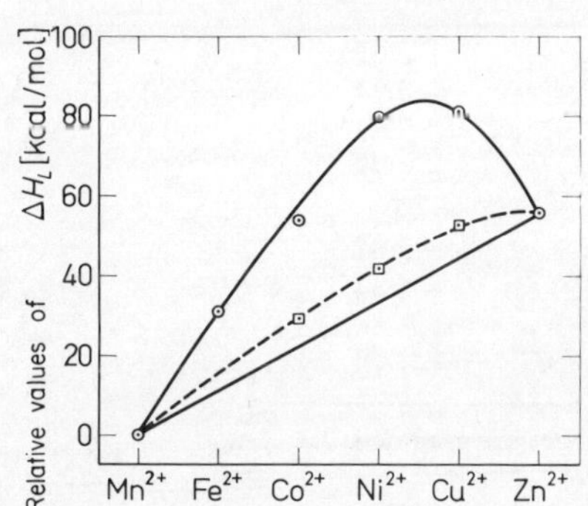


Figure A.80. Relative values of $\Delta H_L$ for $[Men_3]^{2+}$ complexes with M = $Mn^{2+}$, $Fe^{2+}$, $Co^{2+}$, $Ni^{2+}$ and $Zn^{2+}$. The interpolated curve (---) is obtained by subtracting the stabilization energies obtained from the $Dq$ values of the complexes.

The variation of the stabilization energies to be expected from the ligand field theory is in agreement with the rule concerning the stability of complexes given by Irving and Williams[9]:

$$Mn^{2+} < Fe^{2+} < Co^{2+} < Ni^{2+} < Cu^{2+} > Zn^{2+},$$

when complexes of these ions with the same ligands are compared. Deviations occur when low-spin complexes are formed with certain ligands[9].

9. H. Irving and J. P. Williams: *Nature*, **162**, 746 (1948); *J. chem. Soc.*, **1953**, 3192.

## 3.4. The relative stability of tetrahedral and octahedral high-spin complexes

With the help of data given in Table A.20 and A.21 several qualitative predictions can be made after Orgel[10] and McClure[11] concerning the relative stability of tetrahedral and octahedral complexes.

It must be considered thereby that for the tetrahedral case the splitting $\Delta \equiv 10Dq$ is smaller than for the octahedral case ($\frac{4}{9}$ of the splitting for the octahedron, assuming equal charges and equal nuclear separations (cf. p. 63 and Figure A.45). The terms are inverted, so that if one wishes to estimate LFSE values for tetrahedral coordination from known octahedral $Dq$ values, the data given in column 3 of Table A.25 are the determining factor.

Table A.25. Ligand field stabilization for $O_h$ and $T_d$ symmetry, referred to the field strength parameter for the case of octahedral coordination $Dq(O_h)$ for high-spin complexes $[Dq(T_d) = -\frac{4}{9}Dq(O_h)^*]$

| Number of *d*-electrons | LFSE in $Dq(O_h)$ Octahedral | Tetrahedral |
|---|---|---|
| 1 | −4 | −2·67 |
| 2 | −8 | −5·34 |
| 3 | −12 | −3·56 |
| 4 | −6 | −1·78 |
| 5 | 0 | 0 |
| 6 | −4 | −2·67 |
| 7 | −8 | −5·34 |
| 8 | −12 | −3·56 |
| 9 | −6 | −1·78 |

* Equal separations and charges are assumed.

It is of interest to ask to what degree it is possible to predict with the theory which complexes with central ions having the electronic configurations $d^N$ ($N = 0$ to 10) show preferentially octahedral or tetrahedral coordination.

One would expect that in the absence of ligand field effects, the tendency to form besides octahedral, tetrahedral complexes as well, should increase monotonically with increasing atomic number of the respective metal ion. The deviations from a spherically symmetric charge distribution of the

10. L. E. Orgel: *Report to the 10th Solvay Conference*, Brussels, May 1956, p. 289; cf. also
11. D. S. McClure: cf. [6] p. 410, as well as *J. phys. Chem. Solids*, **3**, 311 (1957).

central ion caused by the ligand field and the resulting stabilization effects should be expressed in terms of deviations from such monotonic behaviour.

Tetrahedral coordination should to a certain degree then appear when the difference between the stabilization energies for tetrahedral and for octahedral ligand orientation, the so-called *site preference energy*,

$$\Delta E_{\text{sp.}} = aDq(O_h) - bDq(T_d)$$

is small. *a* and *b* are numerical values to be taken from Tables A.20 and A.21. For the above condition to hold, the numerical values of the second term should show as small a difference as possible from the first term.

If the compounds of the divalent ions $Co^{2+}$, $Ni^{2+}$, $Cu^{2+}$ and $Zn^{2+}$ are considered, it is seen that $Zn^{2+}$ shows the greatest tendency to the coordination number four. As examples, $[Zn(NH_3)_4]^{2+}$ as well as a tetrahedral oxides are known. For $Cu^{2+}$ the tendency to form tetrahedral complexes is not appreciable. The tetrachloro compounds $[CuCl_4]^{2-}$ is known. $Co^{2+}$ has a great tendency to form tetrahedral complexes, which are in the rule blue. On the other hand, only very few examples for $Ni^{2+}$ are known (e.g., tetraphenylarsonium nickel(II)-tetrachloride.

These findings can be qualitatively understood when it is considered that the ligand field stabilization energy for octahedral complexes is maximal for $d^3$ and $d^8$ ($12Dq(O_h)$), while for tetrahedral coordination these maximal values appear for $d^2$ and $d^7$ ($-12Dq(T_d)$).

In this manner $\Delta E_{sp} = -12Dq(T_d) + 8Dq(O_h)$ is obtained for the $d^2$ and $d^7$ complexes and for the $d^3$ and $d^8$ complexes $\Delta E_{sp} = -8Dq(T_d) + 12Dq(O_h)$. The $Dq$ value for octahedral configuration is greater than for tetrahedral (Table A.25), although the stabilization energy for tetrahedral configuration is for $d^7$ $Co^{2+}$ in some cases so large that tetrahedral complexes become energetically favorable.

With such qualitative arguments one can understand then that $Co^{2+}$ (and $Ti^{2+}$) tend more strongly to tetrahedral coordination than does $Ni^{2+}$ (or $V^{2+}$).

No ligand field stabilization energy appears for $d^5$ and $d^{10}$ complexes because of the spherically symmetric charge distribution. Therefore a certain tendency to form tetrahedral complexes should be observed, as in fact is the case. For $Fe^{3+}$ the $[FeCl_4]^-$ ion is known, although $[FeCl_6]^{3-}$ appears to exist as well. Besides octahedral Mn(II) complexes $[MnX_4]^{2-}$ complexes also have been prepared.

These qualitative considerations introduced by Orgel[10] and McClure[11] have recently been criticized by Katzin[12]. The site preference energy gives

12. L. J. Katzin: *J. chem. Phys.*, **35**, 467 (1961).

only a very small contribution which in the rule is negligible compared with other energy magnitudes. Therefore arguments considering the appearance of tetrahedral complexes based only on $\Delta E_{sp}$ are questionable.

## 3.5. Ionic radii of transition metals of the iron group

The electron configuration of ions of the first series of transition metals is $(1s^2 2s^2 2p^6 3s^2 3p^6)3d^N$, $N = 1$ to 9. If a $3d$ electron is added to such an ion and the nuclear charge is simultaneously increased by one unit, then on the one hand the electron cloud contracts under the influence of a greater nuclear charge. On the other hand the mutual repulsion of the electrons increases. In total the first effect dominates; that is, the electron cloud should be concentrated closer to the nucleus for an increasing number of $d$ electrons. It should be expected on the basis of such ionic-theory arguments that the ionic radius decreases monotonically with increasing atomic number, as is schematically shown in Figure A.81.

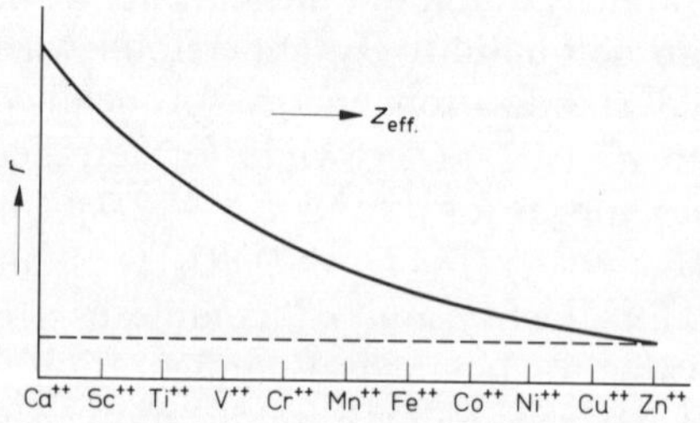


Figure A.81. Expected dependency of the ionic radius on the atomic number for the case of a rigorously spherically symmetrical charge distribution.

However, it is found that the radii of the complexed bi- and trivalent ions of the first transition metal series do not decrease monotonically with increasing atomic number, as is for example observed for the radii of the rare earth elements when going from lanthanum to lutetium (*lanthanide contraction*).

For the octahedral radii obtained from investigations of oxides or hydrates one finds (Figure A.82) an increase of the ionic radius for $d^4$ ($Mn^{3+}$, $Cr^{2+}$) and $d^5$ ($Fe^{3+}$, $Mn^{2+}$) as well as for $d^9$ ($Cu^{2+}$) and $d^{10}$ ($Zn^{2+}$).

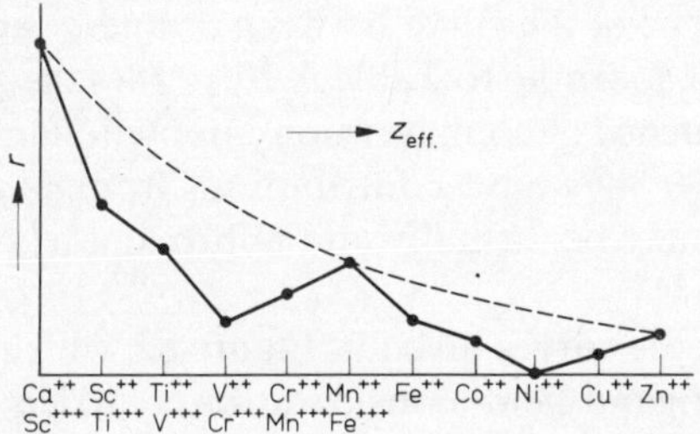


Figure A.82. Ionic radius for octahedral coordination as a function of the atomic number.

As has been shown by Van Santen and Van Wieringen[13,14], the deviations of the observed octahedral radii from those expected on the basis of ionic considerations for a spherically symmetric charge distribution can be qualitatively explained with the help of the ligand field theory using the one-electron model.

In an octahedral field the occupation of the $d\varepsilon$ and $d\gamma$ states with $N = 1$ to 9 $d$ electrons must be considered. The electron configurations of the ground states $d\varepsilon^n \, d\gamma^{N-n}$ are given in Table A.20, column 2. If the angular functions for $d\varepsilon$ and $d\gamma$ are considered (Figures A.38 and A.39, pp. 53, 54), one sees already from a rough consideration that a $d\gamma$ electron leads to a greater separation between the central ion and the six ligands than does a $d\varepsilon$ electron.

The occupation of the weakly antibonding $d_{xy}$, $d_{xz}$, $d_{yz}$ orbitals leads to a contraction of the electron cloud and thereby of the radius. However, the occupation of the strongly antibonding $d_{z^2}$, $d_{x^2-y^2}$ states produces an expansion of the electron cloud and an increase in the radius.

The results of these considerations are presented schematically in Figure A.82. The dashed curve gives the behaviour which one expects for spherically symmetric charge distributions and connects the points for the radii for $d^0$, $d^5$ and $d^{10}$. A decrease of the radius should follow in going from $d^0$ to $d^1$, $d^2$ and $d^3$ because the $d$ electrons fill the $t_{2g}$ state one after the other. The radii for $d^4$ and $d^5$ increase because here the $e_g$ state is being filled. The next three electrons occupy the $d\varepsilon$ state, leading to a decrease of the radii up to $d^8$. The radius for $d^9$ is, however, greater than that for $d^8$ because the additional electron occupies the $d\gamma$ orbital.

---

13. J. H. Van Santen and J. S. Van Wieringen: *Recl. Trav. chim. Pays-Bas Belg.*, **71**, 420 (1952).
14. N. S. Hush and M. H. L. Pryce: *J. chem. Phys.*, **28**, 244 (1958); **26**, 143 (1957).

The change of slope of the curve between $d^1$ and $d^2$ and between $d^6$ and $d^7$ arises because (cf. footnote to Table A.20, p. 146) the ground states of $d^2$ and $d^7$ as a consequence of configuration interaction are no longer $d\varepsilon^2$ or $d\varepsilon^5\, d\gamma^2$ states. Rather they have contributions from $d\varepsilon\, d\gamma$ or from $d\varepsilon^4\, d\gamma^3$. The ground states may be formally and approximately described as $d\varepsilon^{9/5}\, d\gamma^{1/5}$ or as $d\varepsilon^{24/5}\, d\gamma^{11/5}$.

A comparison of the curves given in Figure 82, which were obtained on the basis of the ligand-field considerations carried out above, with experimental findings shows very great deviations for $d^4(Mn^{3+})$ and $d^9(Cu^{2+})$. The explanation for this is that these ions do not form regular octahedral but strongly tetragonally distorted complexes. The stereochemistry of $d^4$ and $d^9$ complexes can be understood as a consequence of the *Jahn–Teller* theorem[15] (cf. chapter 4, p. 172).

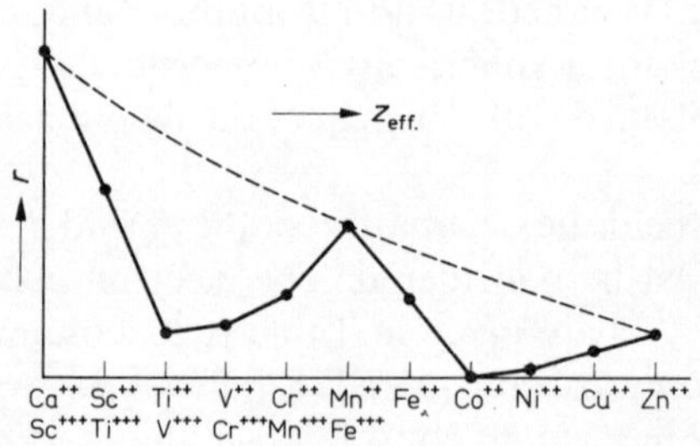


Figure A.83. Ionic radius for tetrahedral coordination as a function of the atomic number.

The tetrahedral radii depicted in Figure A.83 can be obtained analogously. The corresponding $d\gamma^n\, d\varepsilon^{N-n}$ configuration of Table A.21, column 2 are considered together with the fact that a $d\varepsilon$ electron leads to a greater separation between the central ion and the four ligands than does a $d\gamma$ electron. For tetrahedral coordination the $d\gamma(e)$ states are weakly $\pi$ antibonding or $\delta$ nonbonding and the $d\varepsilon(t_2)$ states are strongly antibonding.

Considerations of this sort may also be carried out for the octahedral radii of low-spin complexes (Table A.20, column 5). Because low-spin complexes have fewer $d\gamma$ electrons compared to the corresponding high-spin complexes, the ionic radii of the former should be smaller than for the latter. The decrease should be especially large in going from the

15. Cf. e.g. J. H. Van Vleck: *J. chem. Phys.*, **7**, 61, 72 (1939); *Discuss. Faraday Soc.*, **26**, 96 (1958).

high-spin to the low-spin type for $d^5$ and $d^6$ complexes. This is seen from the electron configuration of the respective ground states [$d^5$ (high-spin: $d\varepsilon^3\, d\gamma^2$, low-spin: $d\varepsilon^5$ (high-spin: $d\varepsilon^4\, d\gamma^2$, low-spin: $d\varepsilon^6$)].

# 4. Consequences of the Jahn–Teller theorem for transition metal complexes

Jahn and Teller[1] arrived at the conclusion, based on very general arguments, that the totally symmetric nuclear configuration is unstable for an orbitally degenerate electron state of a polyatomic molecule (linear molecules being excepted). The Jahn–Teller theorem is valid not only for the ground state, but for the excited electronic states as well. It follows from the theorem in particular that complexes with orbitally degenerate ground states should not exist. That is, if one determines that the electronic ground state for an assumed coordination polyhedron is orbitally degenerate, then the assumption is false in so far as the system should in reality have a structure which arises from the original coordination polyhedron by distortion in a manner to remove the orbital degeneracy. As a result of the descent in symmetry an entity must arise having a simple electronic ground state (i.e., non-orbitally-degenerate). In other words: There always exists a non-totally symmetric configuration, compared to which the totally symmetric configuration is unstable, so that the molecule distorts to attain a new form.

The theorem makes no *a priori* predictions concerning the magnitude and the direction of the distortion. Likewise nothing is said as to if the descent in symmetry appears as a *static deformation* of the coordination polyhedron and thereby influences the crystallographic separations or if the reduction of the field symmetry arises because the entity oscillates between several equivalent limiting structures having simple (non-degenerate) states. The latter case represents an exceedingly complicated dynamical problem (*dynamic Jahn–Teller effect*).

Considering for example octahedral transition metal complexes which are stable according to Jahn and Teller and which must show a distortion resulting from a descent in symmetry because the ground state is

1. H. A. Jahn and E. Teller: *Proc. R. Soc.*, **A 161,** 220 (1937). H. A. Jahn: *Proc. R. Soc.*, **A 164**, 117 (1938).

degenerate, one finds (cf. Tables A.16 and A.17):

| | | | |
|---|---|---|---|
| $d^3: {}^4A_{2g}(d\varepsilon^3)$ | $d^5: {}^6A_{1g}(d\varepsilon^3 d\gamma^2)$ | $d^8: {}^3A_{2g}(d\varepsilon^6 d\gamma^2)$ | high-spin |
| | $d^6: {}^1A_{1g}(d\varepsilon^6)$ | | low-spin |

as stable octahedral complexes with nondegenerate ground states.

Tetrahedral complexes with stable *Jahn–Teller ground states* are (cf. Table A.17):

| | | | |
|---|---|---|---|
| $d^2: {}^3A_2(d\gamma^2)$ | $d^5: {}^6A_1(d\gamma^2 d\varepsilon^3)$ | $d^7: {}^4A_2(d\gamma^4 d\varepsilon^3)$ | high-spin |
| | $d^4: {}^1A_1(d\gamma^4)$ | | low-spin. |

If a static effect is present identical ligands vibrate about equilibrium positions having varying distances from the central ion. When making structure determinations of crystals of such complexes one should observe the ligands in their various positions. Static deformations can also appear for complex ions in solution.

For the dynamic *Jahn–Teller* effect the height of the potential barrier hindering the transitions from one direction of distortion to another equivalent direction is comparable to $kT$ for a given absolute temperature $T$. Thus, an oscillation between the equivalent limiting structures appears. A study of the complex in question shows a statistical distribution of the complex ions over the various possible distortion states. In the time average a higher macroscopic symmetry can be observed in the crystal. This is illustrated with a simple example:

In Figure 84 the potential energy $V$ for a totally symmetric nuclear configuration having a doubly orbitally degenerate electronic state (e.g., $E_g$ for octahedral symmetry) is schematically drawn. The dashed curve gives $V$ as a function of the vibrational coordinate (normal coordinate) $\zeta$, of the totally symmetric vibration. Because no descent in symmetry appears for a totally symmetric vibration, the orbital degeneracy remains and no splitting is seen. The solid curve shows the potential energy as a function of a displacement coordinate $\xi$* corresponding to a non-totally symmetric vibration. As a result of the descent in symmetry associated with the non-totally symmetric displacement, the electronic state which is a doublet state for total symmetry splits into two nondegenerate states. The lower-lying state (lower branch of the solid line) is stabilized with regard to the totally symmetric configuration and has minima at $-\xi'$ and $+\xi'$. If the height of the potential barrier is $\Delta E > kt$, then a static effect is present; if $E \lesssim kT$, dynamic effect. In the latter case one can no longer say that

* In the figure it is assumed that the potential energy is symmetric with reference to $\xi$; $V(-\xi) = V(+\xi)$.

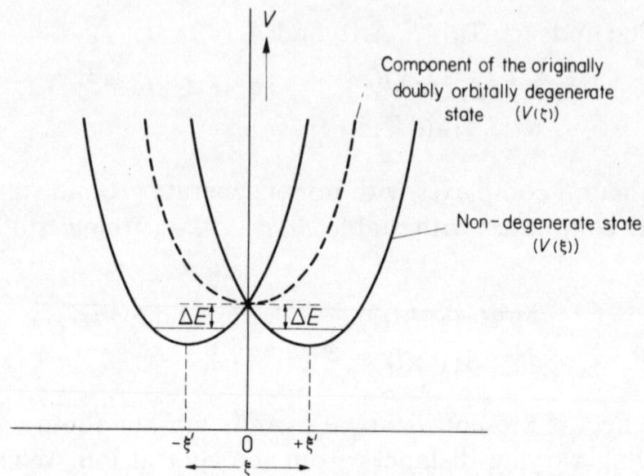


Figure A.84. Potential curve scheme to explain the Jahn-Teller theorem. (A cross-section of the potential surface.) Splitting of a doubly orbitally degenerate electron state into two non-degenerate components by a non-totally symmetrical vibration. $\xi = 0$ totally symmetrical configuration. Minima at $-\xi'$ and $+\xi'$.

the molecule exists in one or the other distortion state. Rather a resonance exists between the equivalent distortional states corresponding to the potential minima at $+\xi'$ and $-\xi'$.

There are two sorts of vibrations of an octahedral complex [$MA_6$], $\varepsilon_g$ and $\tau_{2g}$ (i.e., those which transform according to the representations $E_g$ and $T_{2g}$ of the group $O_h$) and those which distort the regular octahedral totally symmetric nuclear configuration having an orbitally degenerate ground state so that the degeneracy is removed in the sense of Jahn and Teller for the resulting deformed octahedron. In Figure A.85 these vibrations together with the other possible normal vibrations are depicted.

For example the $\varepsilon_g$ distortion leads to an octahedron either elongated or compressed in the direction of a fourfold axis (Figure A.86). There are therefore in each case three equivalent possibilities of tetragonal distortion because three fourfold axes pass through the vertices of the octahedron and the six ligands are identical. Thereby a complicated potential surface arises having three energy minima separated from one another by three potential barriers.

One can explain the distortion of the coordination tetrahedron of certain transition metal ions on the basis of a simple electrostatic consideration as is shown for $d^9$ ($Cu^{2+}$).

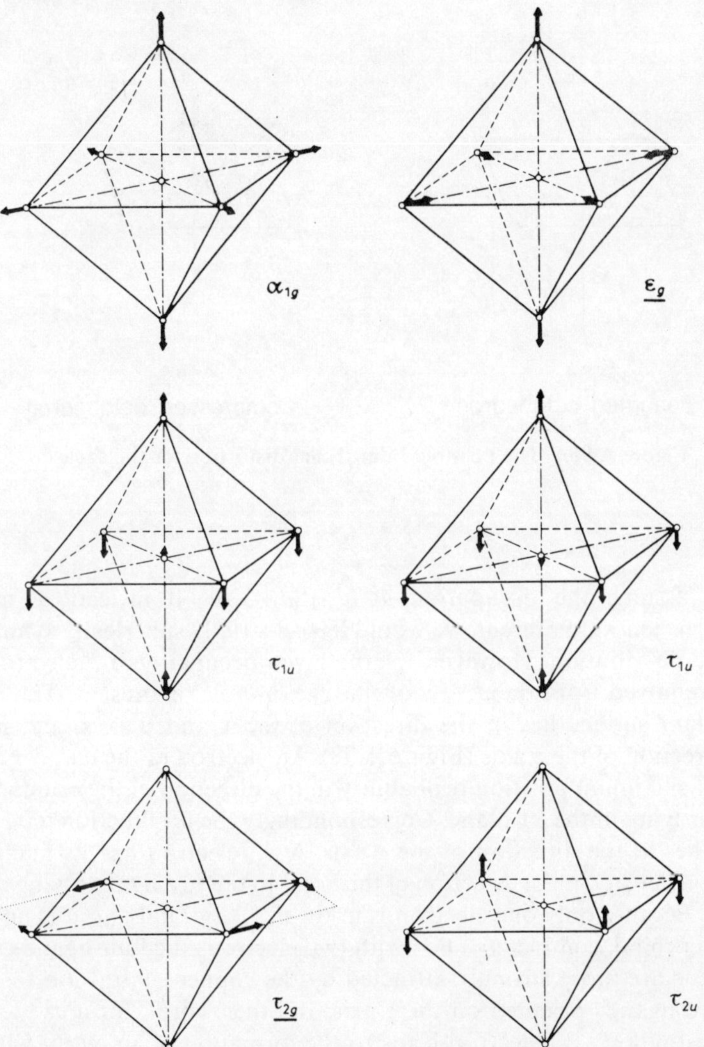


Figure A.85. Normal vibrations of a molecule of the type $XY_6$ (point group $O_h$). The underscored vibrations $\varepsilon_g$ and $\tau_{2g}$ are those which lead to a Jahn–Teller distortion of the octahedron.

A $[\overset{\text{II}}{\text{Cu}}\,\text{A}_6]$ complex has a $^2E_g$ ground state which arises from the configuration $d\varepsilon^6\, d\gamma^3$. Thereby the three $d\gamma$ electrons are distributed over the two $d_{z^2}$ and $d_{x^2-y^2}$ states. There are two possibilities

$$(d_{z^2})^2(d_{x^2-y^2})^1 \quad \text{and} \quad (d_{z^2})^1(d_{x^2-y^2})^2.$$

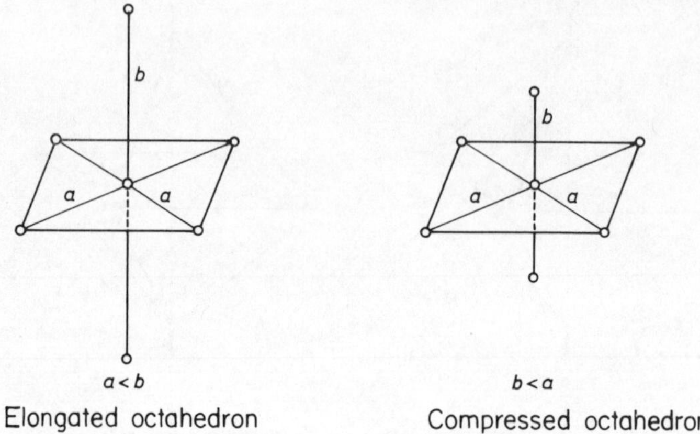


Figure A.86. The possible Jahn–Teller distortion for the $\varepsilon_g$ case.

We assume the occupation $d\varepsilon^6(d_{z^2})^2(d_{x^2-y^2})^1$. If instead of nine $d$ electrons ten were present, we would have a strictly spherically symmetric charge distribution. However, for the given occupation a $d$ electron has been removed from the $d_{x^2-y^2}$ orbital, i.e. a 'hole' is present. The $d_{x^2-y^2}$ boundary surface lies in the direction of the $x$ and $y$ axes, and not in the direction of the $z$ axis (Figure A.38). An electron in the $d_{x^2-y^2}$ orbital, has its maximum position probability in the direction of the bonds to the ligands lying in the $xy$ plane. Correspondingly the $d_{z^2}$ function (cf. Figure A.38) lies in the direction of the $z$ axis. A $d_{z^2}$ electron has its maximum electron density in the direction of the bond to the two ligands lying on the $z$ axis. Because only one electron is in the $d_{x^2-y^2}$ orbital and it is not, like the $d_{z^2}$ orbital, completely filled with two electrons the four ligands in the $xy$ plane are more strongly attracted by the copper central ion than are the two ligands oriented on the $z$ axis. In other words, for a $d^{10}$ ion the removal of a $d_{x^2-y^2}$ electron leads to the formation of an entity which is no longer spherically symmetric. This deviation from the spherically symmetric charge distribution causes the central ion to be less effectively shielded in the $x$ and $y$ directions than in the $z$ direction. The ligands lying in the $xy$ plane are attracted to the central ion by a force corresponding to a greater nuclear charge than that which is effective upon the ligands along the $z$ axis. The consequence is that a tetragonal distortion of the octahedron results so that the four ligands located in the $xy$ plane are closer to the central ion than the other two ligands lying along the $z$ axis (an octahedron elongated in the $z$ direction).

If the occupation $d\varepsilon^6(d_{x^2-y^2})^2(d_{z^2})^1$ is assumed, an electron is missing in the $d_{z^2}$ orbital and one would expect an exactly opposite distortion of the octahedron. The distance to the two ligands on the $z$ axis should be smaller than the distance to those in the $xy$ plane (an octahedron compressed in the $z$ direction). It cannot, however, be said on the basis of these elementary considerations which of the two distortions is energetically the more favourable.

In order to make statements concerning the nature and the magnitude of these effects it is necessary to carry out detailed calculations concerning the energy of the total complex as a function of all possible distortions. The configuration having the lowest total energy can be considered to be the equilibrium configuration.[2]

Figure A.87 gives in a schematically simplified manner the splitting of the $d\varepsilon$ and $d\gamma$ one-electron states when the octahedron is slightly elongated or compressed along the $z$ axis. The energetic order of the states is inverted when one goes from an elongated to a compressed octahedron. Both splittings $\delta_1$ and $\delta_2$ are small compared to the distances $d\varepsilon - d\gamma$, $\Delta \equiv 10Dq$, so that we are concerned with only comparatively small distortions of the coordination tetrahedron. The simple $d\gamma$ (progeny) terms lie, to the first approximation symmetrically about the original $d\gamma$ state ($+\frac{1}{2}\delta_2$ and $-\frac{1}{2}\delta_2$). The doubly degenerate state lies to the first approximation $\frac{1}{3}\delta_1$ lower (or higher) than the $d\varepsilon$ state from which it arises; the nondegenerate state $\frac{2}{3}\delta_1$ higher (or lower).

Let us consider an octahedral $d^9$ complex with $Cu^{2+}$ as the central ion. The ground state arises from the configuration $d\varepsilon^6\, d\gamma^3$. Compared to the tetrahedron the $d\varepsilon$ electrons of the trigonally distorted octahedron have the same energy: $4(\mp\frac{1}{3}\delta_1)+2(\pm\frac{2}{3}\delta_1) = 0$. The $d\gamma$ electrons are stabilized by $2(-\frac{1}{2}\delta_2)+1(+\frac{1}{2}\delta_2) = -\frac{1}{2}\delta_2$ compared to the octahedral case. This stabilization is the cause of tetragonal distortion for $d^9$ complexes.

Analogously (cf. Table A.26) such tetragonal distortions should show a stabilization of $-\frac{1}{2}\delta_2$ compared with the regular octahedron for all transition metal complexes; that is for

$d^4$ ${}^5E_g(d\varepsilon^3 d\gamma^1)$ $Cr^{2+}$, $Mn^{3+}$ high-spin
$d^7$ ${}^2E_g(d\varepsilon^6 d\gamma^1)$ $Co^{2+}$, $Ni^{3+}$ low-spin
$d^9$ ${}^2E_g(d\varepsilon^6 d\gamma^3)$ $Cu^{2+}$

One sees immediately from Figure A.87 that a distortion of the octahedron should appear for complexes with $n = 1, 2, 4$ and $5$ $d\varepsilon$ electrons. The values given in Table A.26 for the stabilization in each case compared

2. Cf. e.g. U. Öpik and M. H. L. Pryce: *Proc. R. Soc.*, **A 238**, 425 (1957).

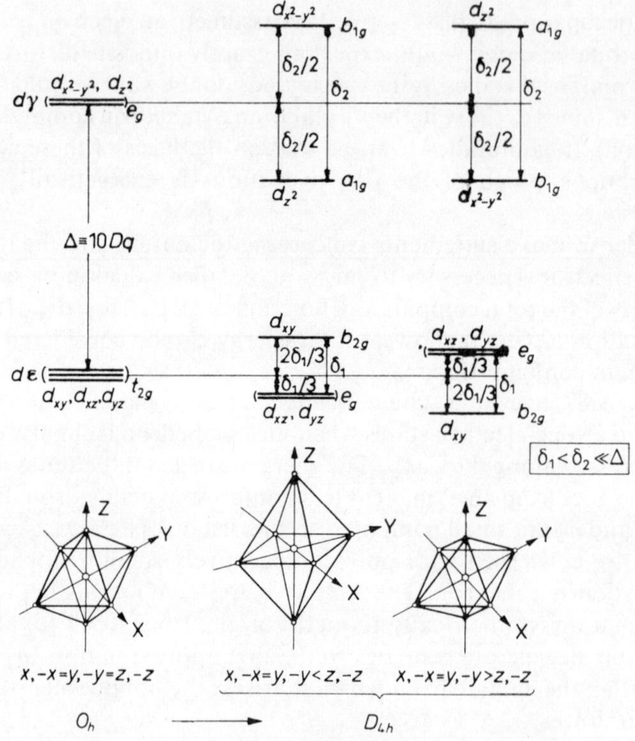


Figure A.87. A qualitative explanation of the Jahn–Teller theorem for $[MA_6]$-complexes.

with the undistorted octahedron are obtained from reasoning analogous to that used for the case of $d\gamma$ electrons.

*Jahn–Teller* distortions are expected for the following octahedral complex ions having 1, 2, 4 or 5 $d\varepsilon$ electrons:

| | | | |
|---|---|---|---|
| $d^1$ | $^2T_{2g}(d\varepsilon^1)$ | $Ti^{3+}$ | |
| $d^2$ | $^3T_{2g}(d\varepsilon^2)$ | $Ti^{2+}$, $V^{3+}$ | |
| $d^4$ | $^3T_{1g}(d\varepsilon^4)$ | $Cr^{2+}$, $Mn^{3+}$ | low-spin |
| $d^5$ | $^2T_{2g}(d\varepsilon^5)$ | $Mn^{2+}$, $Fe^{3+}$ | low-spin |
| $d^6$ | $^5T_{2g}(d\varepsilon^4 d\gamma^2)$ | $Fe^{2+}$, $Co^{3+}$ | high-spin |
| $d^7$ | $^4T_{1g}(d\varepsilon^5 d\gamma^2)$ | $Co^{2+}$, $Ni^{3+}$ | high-spin |

According to these qualitative considerations, the elongated octahedral orientation for $d\varepsilon^2$, $d\varepsilon^5$ and $d\varepsilon^5\, d\gamma^2$ complexes should be energetically more favorable than the compressed structure (Table A.26).

Table A.26. Jahn–Teller stabilization for tetragonally distorted octahedral ligand orientation estimated with the aid of Figure A.87 (expressed by the parameters $\delta_1$ and $\delta_2$, $\delta_1 < \delta_2 \ll \Delta$)

| | Number of $d\gamma$-electrons | Elongated octahedron: Configuration | Elongated octahedron: Stabilization | Compressed octahedron: Configuration | Compressed octahedron: Stabilization |
|---|---|---|---|---|---|
| $d\varepsilon^3$ or $d\varepsilon^6$ | 1 | $a_{1g}^1$ | $-\frac{\delta_2}{2}$ | $b_{1g}^1$ | $-\frac{\delta_2}{2}$ |
| | 3 | $a_{1g}^2 b_{1g}$ | $-\frac{\delta_2}{2}$ | $b_{1g}^2 a_{1g}$ | $-\frac{\delta_2}{2}$ |

| | Number of $d\varepsilon$-electrons | Elongated octahedron: Configuration | Elongated octahedron: Stabilization | Compressed octahedron: Configuration | Compressed octahedron: Stabilization |
|---|---|---|---|---|---|
| $d\gamma^0$ or $d\gamma^2$ | 1 | $e_g^1$ | $-\frac{\delta_1}{3}$ | $b_{2g}^1$ | $-\frac{2\delta_1}{3}$ |
| | 2 | $e_g^2$ | $-\frac{2\delta_1}{3}$ | $b_{2g}^1 e_g^1$ | $-\frac{\delta_1}{3}$ |
| | 4 | $e_g^3 b_{2g}^1$ | $-\frac{\delta_1}{3}$ | $b_{2g}^2 e_g^2$ | $-\frac{2\delta_1}{3}$ |
| | 5 | $e_g^4 b_{2g}^1$ | $-\frac{2\delta_1}{3}$ | $b_{2g}^2 e_g^3$ | $-\frac{\delta_1}{3}$ |

$\delta_1$ is much smaller than the spin-pairing energy, so that no pairing takes place.

According to Figure A.87 the splitting $\delta_1$ for the $d\varepsilon$ state is smaller than $\delta_2$ for the $d\gamma$ state. This can be qualitatively rationalized when one considers that within the framework of the MO theory (p. 106) the ($t_{2g}$) $d\varepsilon$ state is a nonbonding (or with consideration of $\pi$ bond contributions a weakly antibonding) state, the ($e_g$) $d\gamma$ state on the other hand is antibonding (or strongly antibonding)*. From this it follows that a distortion of the coordination octahedron should have a stronger effect on the latter than on the former. The term splittings associated with such *Jahn–Teller* distortions will therefore be small for complexes with $T_{2g}$ ground states compared with those of complexes with $E_g$ ground states. Experimental

* Cf. The spatial extension of the $t_{2g}$ and $e_g$ boundary surfaces (Figure A.38).

evidence should be most readily expected for complexes with $E_g$ ground states unless one studies complexes with pronounced $\pi$ bonds.

As has already been mentioned the theorem is also valid for excited states. In these cases a complicated dynamic problem must be considered. The comparatively short lifetime of an excited state does not allow the formation of a stable equilibrium configuration of the complex. If a complex is subjected to a dynamic *Jahn–Teller* effect in its ground state, then, as has been noted, resonance between various equivalent resonance structures of the nuclear configuration takes place. The *variation* of every complex molecule (caused by suitable vibrations) between these equivalent structures proceeds assuredly very rapidly. It is, however, slow compared to the mechanism of optical excitation, so that the absorption spectrum should show the momentarily reduced symmetry.

In conclusion we refer to experimental evidence for the appearance of the Jahn–Teller effect for transition metal complexes.

Although for transition metal complexes having orbitally degenerate ground states and octahedral symmetry a tetragonal or rhombic distortion is generally found, this is not necessarily a consequence solely of the *Jahn–Teller* theorem. Such orientations of complex components can result from the most diverse causes, i.e., packing effects in the crystalline lattice, repulsion between neighbouring ligands, etc. Structural anomalies can be determined experimentally by x-ray techniques and by electron paramagnetic resonance and nuclear magnetic resonance, as well as with electron and neutron diffraction methods.

A strong distortion is to be expected for octahedral complexes of $Cu^{2+}$ because the ground term is a $^2E_g$ term. Orgél and Dunitz[3] give a summary of structural data which shows that nearly all octahedrally coordinated $Cu^{2+}$ systems have $D_{4h}$ structure, in agreement with the demands of the Jahn–Teller theorem. For the majority of the complexes one finds the form of an elongated octahedron or as a limiting case the quadratic planar orientation of the ligands. The compressed octahedron is only rarely met, i.e., for the compounds $K_2CuF_4$ and $NaCuF_3$.

$Mn^{3+}$ has a $^5E_g$ ground state for the case of $O_h$ symmetry. Also, one finds here $D_{4h}$ structures, e.g., for $MnF_3$ as well as for MnF[1] ion, for which likewise the elongated octahedral orientation can be demonstrated.

$Cr^{2+}$ also has a $^5E_g$ ground term. One finds for example for $CrF_2$ the same structure as for $CuF_2$ ; four fluoride ions lie in the *xy* plane and have

3. L. E. Orgel: *An Introduction to Transition Metal Chemistry, Ligand Field Theory*, p. 60, Methuen, London; John Wiley, New York 1960. L. E. Orgel and J. D. Dunitz: *Nature*, **179**, 462 (1957). J. D. Dunitz and L. E. Orgel: *Adv. inorg. Chem. Radiochem.*, **2**, 1 (1960).

the same distances from $Cr^{2+}$. Two others lie on the $z$ axis and have identically greater separations.

For $Ni^{3+}$ the compound $NaNiO_2$ is known, which shows reduced paramagnetism and also has an elongated octahedral structure. Finally low-spin $Co^{2+}$ complexes, for which square planar structure is probable, are known.

If the fluorides $MF_2$ in the series M = $Cr^{2+}$, $Fe^{2+}$, $Co^{2+}$, $Ni^{2+}$. $Cu^{2+}$, $Zn^{2+}$, all of which show rutile structure, are considered, the $D_{4h}$ distortion described above is found only for $CrF_2$ and $CuF_2$. This result could hardly be attributable to packing effects because these should be equally evident for all compounds of the series.

Likewise in the series $MF_3$, M = $Cr^{3+}$, $Mn^{3+}$, $Fe^{3+}$ only the $Mn^{3+}$ compound is distorted.

From this it follows that the $d^4$ (high-spin), $d^7$ (low-spin) and $d^9$ ions differ in a characteristic manner from the other ions given in their stereochemical behaviour.

For complex ions with threefold orbitally degenerate $T_{1g}$ or $T_{2g}$ ground states the distortion of the coordination octahedron, as already mentioned, should be smaller than that for complexes having $E_g$ ground terms. Larger stereochemical effects are therefore not to be seen here. There exists some experimental evidence for the appearance of smaller distortions.

Besides the crystallographic evidence already discussed, which is applicable to the ground state, there are also spectroscopic measurements which, with suitable caution, can be interpreted as resulting from the Jahn–Teller theorem. These measurements are concerned with the splitting of excited $E_g$ states.

For $MA_6$ compounds with M = $Ti^{3+}$ or $Co^{3+}$ and $Fe^{2+}$ (high-spin) one finds as was shown in section 1.6b for the case of $Ti^{3+}$, a splitting of the excited $E_g$ states into two nondegenerate states. This takes place through tetragonal or rhombohedral distortion, so that an absorption band with a double structure or two separated bands is to be expected. (For the $T_{2g}$ ground state and also for excited $T_{2g}$ and $T_{1g}$ states, which are present for some complexes, only a small splitting is expected.)

The asymmetric broad band for some octahedrally coordinated $Ti^{3+}$ complexes such as $[Ti(H_2O)_6]^{3+}$ or caesium titanium alum or a shoulder indicating a double structure, as for hexureal titanium(III) ion[4] can be understood in this manner. Also Cotton and Meyers[5] could identify two bands for $[CoF_6]^{3+}$ and for $[Fe(H_2O)_6]^{2+}$.

---

4. H. Hartmann, H. L. Schläfer and K. H. Hansen: *Z. anorg. allg. Chem.*, **284**, 153 (1956).
5. F. A. Cotton and M. D. Meyers: *J. Am. chem. Soc.*, **82**, 5023 (1960).

The analysis of the envelope of absorption bands of complexes by reduction into gaussian curves is a virtually impossible task. All bands correspond to transitions between states of which at least one is subject to *Jahn–Teller distortion.* It must also be considered that for molecules with centres of symmetry because it is parity-forbidden the electronic transition is possible only in connection with vibrations of suitable symmetry species.

*Jahn-Teller* effects for excited states should be identifiable from an analysis of the vibrational structure of suitable absorption bands. Such a vibrational structure is observed for the broad spin-allowed bands only in exceptional cases, e.g., for $V^{3+}$ and $Cr^{3+}$ in corundum[6,7]. For $Cr^{3+}$ the ${}^4A_{2g} \rightarrow {}^4T_{2g}$ band shows a simple progression of a doubly degenerate $\varepsilon_g$ vibration. It can be concluded from this that the excited ${}^4T_{2g}$ state is unstable in the sense of the Jahn–Teller theorem. The spin-forbidden intercombination bands, e.g., ${}^4A_{2g} \rightarrow {}^2E_g$ for $d^3$ complexes having a small halfwidth offer a greater possibility to observe vibrational structure. A pronounced and easily resolvable vibrational structure is found for these bands for $[Cren_3]^{3+}$ and $[Crtn_3]^{3+}$ (en = $NH_2CH_2CH_2NH_2$, tn = $NH_2CH_2CH_2CH_2NH_2$).

Another case should be mentioned where the existence of a *Jahn–Teller* effect can be shown with certainty by using paramagnetic resonance measurements. The $g$ factor of the complex $[Cu(H_2O)_6]SiF_6$ is isotropic at room temperature but at low temperature an anisotropy is observed[8]. This behaviour can be explained when it is considered that the ${}^2E_g$ ground state is unstable according to the Jahn–Teller theorem, so that a *tetragonal* distortion of the coordination octahedron appears for stabilization. This leads (in the hyperspace of the vibrational coordinates) to a potential surface with three equivalent energy minima, located on a circle with the radius $r_0$ at $\varphi = \pi/3, \pi$ and $5\pi/3$. Saddle points lie between these minima at 0, $2\pi/3$, and $4\pi/3$. The height of the potential barrier $\Delta E$ between these minima is of the order of $kT$ at room temperature ($\approx 200\ \text{cm}^{-1}$). At low temperature ($\Delta E > kT$) the molecule is located in one of the three potential minima. That is, it has the form of a tetragonally distorted (elongated) octahedron. This is evidenced by the anisotropy of the $g$ factor ($g_x = g_y = g_\perp = 2{\cdot}11$; $g_z = g_\parallel = 2{\cdot}46$).

If the temperature is increased ($\Delta E \sim kT$) the molecule receives sufficient energy to surmount the energy barrier.

---

6. M. H. L. Pryce: *Discuss. Faraday Soc.*, **26**, 34 (1958).
7. R. A. Ford and O. F. Mill: *Spectrochim. Acta.*, **16**, 493 (1960); cf. also B. N. Grechusnikov and P. P. Feofilov: *J. exp. theor. Physics* (*USSR*), **29**, 384 (1955).
8. B. Bleany and D. J. E. Ingram: *Proc. phys. Soc. Lond.*, **A 63**, 408 (1950). B. Bleany and K. D. Bowers: *Proc. phys. Soc. Lond.*, **A 65**, 667 (1950); A. Abragam and M. H. L. Pryce: *Proc. phys. Soc. Lond.*, **A 63**, 409 (1950).

That is, the complex shows a sort of pseudo-rotation. The ligands carry out a coupled rotation about the octahedral bond directions and an isotropic $g$ factor ($g_x = g_y = g_z = g$) is obtained.

The Jahn–Teller theorem is valid not only for orbital degeneracy, but also (with the exception of *Kramers* degeneracy, related to the so-called time reversal*) for spin degeneracy. The energetic consequences of the theorem, which one should expect for spin degeneracy, are however exceedingly small, so that we shall not treat them here.

* Cf. V. Heine: *Group Theory in Quantum Mechanics*, pp. 164 ff. Pergamon Press, Oxford, 1960).

# 5. The kinetic stability of transition metal complexes

During the past ten years numerous systematic reaction kinetic studies of inorganic complex ions have been carried out. The vast majority of these studies are concerned with processes whereby one ligand in the coordination sphere is exchanged for another ligand, that is substitution reactions.

The capability of a complex ion to undergo *ligand exchange* is described by the concept of *kinetic stability*. Complexes for which the ligand exchange reactions proceed very rapidly are known as *labile*, and those for which only a slower or virtually no exchange is observed are termed *inert* complexes. No sharp boundary exists between these two classes. Taube has suggested calling those complexes labile for which at room temperature the complete reaction of about 0·1 M solution can be observed after one minute.

The kinetic stability is not to be confused with the thermodynamic stability discussed in section 3.3, which is based upon the formation or dissociation equilibria of complex ions, i.e., upon the equilibrium state. A quantitative measure of the thermodynamic stability is the value of the equilibrium constants $K$ or of the free energy of reaction $\Delta G$ for the equilibrium in question. Correspondingly one may take as a measure of the kinetic stability the value of reaction rate constants $k$ or the half life $\tau$.

An inert complex does not need to be thermodynamically stable (and vice versa), although one frequently finds that thermodynamically stable substances show slow ligand exchange reactions and that thermodynamically unstable substances react rapidly.

Two examples of cases for which the thermodynamic and the kinetic stability go in opposite directions are $[Co(NH_3)_6]^{3+}$ and $[Ni(CN)_4]^{2-}$. The $[Co(NH_3)_6]^{3+}$ ion is in acidic aqueous solution at room temperature completely inert to the exchange of $NH_3$ by $H_2O$ for weeks. Nevertheless the cobalt(III) hexammine ion is thermodynamically unstable because the equilibrium constant for the reaction

$$[Co(NH_3)_6]^{3+} + 6H_3O^+ \rightleftharpoons [Co(H_2O)_6]^{3+} + 6NH_4^+$$

is of the order of $10^{25}$. If the exchange of $CN^-$ in nickel(II) tetracyano complex is studied using radiocarbon $CN^-$ tracer in the solution, one sees that the complex is kinetically labile. However, it is thermodynamically stable because the overall dissociation constant for the reaction

$$[Ni(CN)_4]^{2-} + 6H_2O \rightleftharpoons [Ni(H_2O)_6]^{2+} + 4CN^-$$

is $10^{-22}$.

Of the octahedrally coordinated complexes of the first transition metal series especially those of $Cr^{3+}$ and $Co^{3+}$ are notable for their enormous stability. However, complexes of most of the other metal ions are in general more or less labile. For these ligands exchange reactions frequently proceed so rapidly that they cannot at all or only very inaccurately be followed by conventional kinetic experimental techniques. Because of the very slow reactions of the $Cr^{3+}$ (and $Co^{3+}$) complexes (half lives of hours or days) it is precisely these compounds on which the majority of the reaction kinetic studies have been carried out. Also, a large number of complexes of these ions can be isolated relatively easily.

Up to the present there exists no completely satisfactory interpretation of the observed relative kinetic stabilities of the complexes of the first transition metal series. However, one can make several statements which explain at least the exceptional position of chromium(III) and Co(III) complexes using the qualitative VB approach of Taube[1] and the semi-quantitative ligand field method according to Orgel[2] and C. K. Jørgensen[3].

Taube developed concepts concerning the kinetic behaviour of transition metal complexes with regard to substitution reactions on the basis of their electron configurations in the *valence bond theory* (cf. pp. 121 ff. and pp. 484 ff.). For octahedral complexes each of the six ligands contributes one electron pair to the bond. Six of these twelve ligand electrons occupy the six equivalent orbitals of the central ion. Two *d*, an *s* and three *p* orbitals of the central ion are available for the construction of these hybrids. For '*outer orbital*' complexes the six electrons are distributed over $sp^3d^2$, i.e., *d* orbitals having the same principal quantum number as the *s* and *p* orbitals which are used. For '*inner orbital*' complexes the electrons are distributed over $d^2sp^3$, i.e., one uses *d* orbitals having a principal quantum number one lower than the *s* and the *p* orbitals (Figure A.88) (cf. section 2.2, p. 124).

According to Taube labile complexes are those having an outer-orbital

1. H. Taube: *Chem. Rev.*, **50**, 69 (1952).
2. L. E. Orgel: *J. chem. Soc.*, **1952**, 4756; cf. also *An Introduction to Transition Metal Chemistry, Ligand Field Theory*, pp. 103 ff., Methuen, London; J. Wiley, New York, 1960.
3. C. K. Jørgensen: *Acta chem. scand.*, **9**, 605 (1955).

hybridization or those belonging to the class of inner orbital complexes having at least one empty (unoccupied) $d$ state. Inner-orbital complexes having all five $d$ states filled are kinetically stable. Accordingly it is immediately seen that $d^3$ as well as low-spin $d^4$, $d^5$ and $d^6$ complexes should be *inert* to ligand exchange (Figure A.88). $d^1$, $d^2$, $d^7$, $d^8$, $d^9$ as well as high-spin $d^4$, $d^5$ and $d^6$ complexes should be *labile*.

| | $(n-1)d$ | | | | | $ns$ | $np$ | | | $nd$ | | |
|---|---|---|---|---|---|---|---|---|---|---|---|---|
| $d^1$ | ↑ | | | | | XX | XX | XX | XX | XX | XX | Labile complexes $sp^3d^2$ |
| $d^2$ | ↑ | ↑ | | | | XX | XX | XX | XX | XX | XX | |
| $d^3$ | ↑ | ↑ | ↑ | XX | XX | XX | XX | XX | XX | | | Inert complexes $d^2sp^3$ |
| $d^4$ | ↑↓ | ↑ | ↑ | XX | XX | XX | XX | XX | XX | | | |
| $d^5$ | ↑↓ | ↑↓ | ↑ | XX | XX | XX | XX | XX | XX | | | |
| $d^6$ | ↑↓ | ↑↓ | ↑↓ | XX | XX | XX | XX | XX | XX | | | |

Figure A.88. Octahedral $sp^3d^2$ and $d^2sp^3$ complexes.

This is in qualitative agreement with experience: $d^3$ complexes i.e., all chromium(III) and vanadium(II) complexes are *inert* complexes. This is also true of complexes with higher transition metal ions Mo(III), W(III) and Re(IV). Diamagnetic $d^6$ complexes, i.e., all cobalt(III) complexes with the exception of the paramagnetic $[CoF_6]^{3-}$ ion, show great stability. Likewise, complexes with Fe(II), e.g., $[Fe(CN)_6]^{4-}$, $[Fe(phen)_3]^{2+}$, $[Fe(dipy)_3]^{2+}$ as well as complexes of the higher transition metal ions are stable: Ru(II), Os(II), Rh(III), Ir(III), Pd(IV) and Pt(IV).

$d^4$ complexes of the low-spin type such as $[Cr(CN)_6]^{4-}$, $[Cr(dipy)_3]^{2+}$, $[Mn(CN)_6]^{3-}$ as well as complexes with Re(III), Ru(IV) and Os(V) and Ir(IV) are likewise inert. Low-spin $d^5$ complexes such as $[Cr(dipy)_3]^+$, $[Mn(CN)_6]^{4-}$, $[Fe(phen)_3]^{3+}$, $[Fe(dipy)_3]^{3+}$, $[Fe(CN)_6]^{3-}$ as well as complexes with Ru(III), Os(III) and Ir(IV) are likewise inert.

However, complexes with Ti(III) and V(III) and high-spin manganese(II), iron(III), cobalt(II) and nickel(II) complexes, as well as those of Cu(II) are labile coordination complexes.

Differences in the kinetic stability which exist between the (labile) $d^1$ and $d^2$ complexes or between the (inert) $d^3$ and low-spin $d^4$ complexes cannot be understood with the help of the qualitative concepts of Taube.

The *ligand field theory* allows the question of the behaviour of transition metal complexes in substitution reactions to be more precisely treated than does the valence bond theory.

First of all something must be said concerning the nature of reaction mechanisms involved.

Using the terminology developed by Hughes and Ingold[4] the reactions of complex ions are divided into two types, $S_{N1}$ and $S_{N2}$.

An $S_{N1}$ reaction proceeds according to a two-step mechanism, where the first slow step is a *unimolecular dissociation process*

$$\text{M–X} = \text{M} + \text{X}$$

which is followed by the more rapid second step, the reaction of M or X with a second molecule or ion

$$\text{M} + \text{Y} = \text{M–Y}$$
$$\text{X} + \text{M}' = \text{M}'\text{–X}.$$

In the first dissociation step the coordination number is decreased by one so that an intermediary with a reduced coordination number arises. For octahedral complexes an intermediary with a coordination number 5 would be expected for an $S_{N1}$ mechanism.

For an $S_{N2}$ process the rate-determining step is a *bimolecular step* where one nucleophilic entity is replaced by another

$$\text{Y} + \text{M–X} = \text{Y}\cdots\text{M}\cdots\text{X} = \text{Y–M} + \text{X}.$$

The intermediary complex has a coordination number increased by one, i.e., for octahedral complexes one would have an intermediary with a coordination number 7 for an $S_{N2}$ reaction.

Between these limiting cases of the $S_{N1}$ and the $S_{N2}$ mechanism all other processes are conceivable. In practice it is often not possible to make an unambiguous classification because the reactions proceed according to an intermediary mechanism between the two limiting cases. This does not mean that the reactions in question follow a mechanism representing a mixture of $S_{N1}$ and $S_{N2}$ steps. This is to be sure conceivable. It is, however, in principle a question of how great the separation is in the transition state between Y and M, where Y is, for example, the solvent.

If the separation is so great that Y interacts only very weakly with M so that the influence on the energetics of the system is comparatively small, then the boundary case of the $S_{N1}$ dissociation mechanism is approached. For smaller separations between M and Y one comes to the boundary case of the $S_{N2}$ mechanism. For this reason the classification according to $S_{N1}$ and $S_{N2}$ reactions is problematical.

---

4. Cf. C. K. Ingold: *Structure and Mechanism in Organic Chemistry*, Chaps. 3 and 5, Cornell University Press, Ithaca 1953.

The substitution reactions, that is, the exchange of a ligand X of an octahedral complex $MX_6$ for a ligand Y begins in that Y approaches an empty region in the space surrounding the metal ion M. The exact path of this approach is not known. We can of course assume that Y approaches along a threefold axis of the octahedron or approximately in the direction of the centres of the edges of the octahedron. To bring about a chemical reaction Y must attain a degree of bonding to the central ion which is at least equivalent to the degree of bonding of the ligand X, which is to be replaced. X will then be expelled from the coordination sphere and replaced by Y. These primitive considerations are valid for an $S_{N1}$ as well as for an $S_{N2}$ process. The difference in particular is only that for an $S_{N1}$ mechanism X is already dissociated when Y is still relatively far away from M. Then Y is bound. For an $S_{N2}$ mechanism Y approaches closely enough for the intermediary

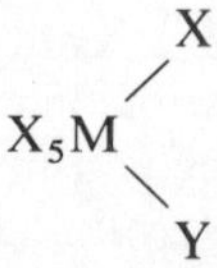


to form and then X dissociates.

Qualitatively the following predictions can be made on the basis of the ligand field theory: We consider first a substitution reaction for a high-spin octahedral $d^5$ complex. This has a strictly spherically symmetric charge distribution because the configuration from which the ground state arises is $d\varepsilon^3\ d\gamma^2$. In order to have approximately the same bonding between the metal ion and the old as well as the new ligands it is necessary to bring the latter up to the same separation from the central ion as the ligands which are already in the coordination sphere. For a spherically symmetric charge distribution the approach of the ligand is not hindered by special effects, as might be expected for a spatially oriented charge distribution. The situation for $Mn^{2+}$ or $Fe^{3+}$ is formally analogous to that observed for ions with a closed $d$ shell ($d^{10}$ configuration), that is, for other spherically symmetric ions in general, as well as for those which are not transition metal ions, e.g., $Mg^{2+}$ or $Al^{3+}$. Such complexes are therefore labile.

For diamagnetic $d^6$ complexes having a ground state arising from the $d\varepsilon^6$ configuration the deviation from a spherically symmetric charge distribution is greater than for the other transition metal complexes.

The electron density is almost zero in the bond direction and has maxima in the directions of the midpoints of the edges of the octahedron. For an approaching ion to attain an equivalent bond with the central ion as compared with the original ligands an electron must be transferred (promoted) from one of the $d\varepsilon$ states to a $d\gamma$ state. This, however, requires energy, so that it is qualitatively understandable that the approach of the ligand up to the requisite bond distance is greatly impeded.

A similar situation exists for $d^3$ complexes with a ground state arising from $d\varepsilon^3$. For a reaction to occur an electron must either be transferred to $d\gamma$, or an empty $d\varepsilon$ state must be created by the pairing of two $d\varepsilon$ electrons, a process which likewise requires energy.

Low-spin $d^6$ and $d^3$ complexes should then be kinetically inert. On the other hand $d^1$ and $d^2$ complexes have respectively two and one empty $d\varepsilon$ states available for bonding with the ligand and are labile.

$d^8$ complexes with a ground state arising from the configuration $d\varepsilon^6\ d\gamma^2$ should likewise be inert. An electron transfer is necessary if they are to react. According to experience octahedrally coordinated $Ni^{2+}$ complexes are by no means so stable as, e.g., $Cr^{3+}$ complexes. Orgel assumes that this is possibly evidence that electrons in $d\gamma$ states produce a weakening of the bond between the central ion and the ligand.

More quantitative information concerning the contribution of ligand field stabilization to the energy of activation for substitution reactions of octahedral complexes can be obtained when the ligand field stabilization energy (LFSE, chapter 3) for an octahedral configuration is compared with the stabilization of an assumed *transition state* of a particular structure. A tetragonal *pyramid** ($KZ = 5$) is used as the transition state for an $S_{N1}$ dissociation mechanism and a *pentagonal bipyramid* ($KZ = 7$) is assumed for an $S_{N2}$ mechanism. The stabilization energies can be obtained approximately for example by using the strong field method in a one-electron term system (cf. section 3.1). The one-electron $d$ states illustrated in Figure A.89 are filled with $d$ electrons ($d = 1$ to 9) from the bottom with consideration of the Pauli principle. For $d^4$, $d^5$, $d^6$ and $d^7$ for high-spin and low-spin complexes there are in each case two possible occupancies (section 2.3a).

The positions of the one-electron $d$ states on the energy scale are given in Figure A.89 in units of the field strength parameter $Dq$, which measures the strength of the ligand field for the case of octahedral symmetry. For lower field symmetry, that is for the quadratic pyramid or the pentagonal bipyramid, additional field parameters arise. In Figure A.89 the energies are given for these cases after Basolo and Pearson[5] in $Dq$ units. For the determination of the energies as a function of a single parameter $Dq$ for the two orientations of lower symmetry as well, a certain relation of the

* The tetragonal pyramid is favoured with regard to ligand field stabilization effects over the trigonal bipyramid, which is also a possible form of the transition state.

5. F. Basolo and R. G. Pearson: *Mechanisms of Inorganic Reactions. A Study of Metal Complexes in Solution*, p. 55, 108 ff., J. Wiley, New York–London 1958.

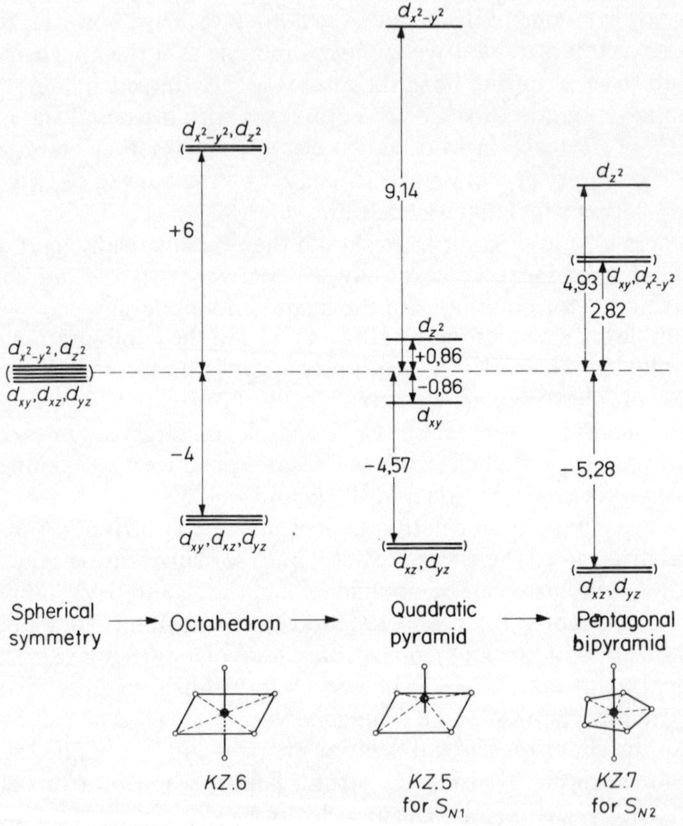


Figure A.89. Splitting of a one-electron *d* state for an octahedron, a quadratic bipyramid and a pentagonal bipyramid after Basolo and Pearson[5].

integrals over the radial portions of the wave functions which appear in the calculation ($B_2 = 2B_4$)* is assumed.

For octahedral symmetry $\bar{r}^4/R^6$ appears in the expression for *Dq*, where *R* is the ligand-metal ion separation and $\bar{r}^4$ is the quantum mechanical average value (expectation value) of the fourth power of the distance of the *d* electron from the nucleus. For lower symmetry additional terms in $\bar{r}^2/R^4$ appear. Basolo and Pearson[5] assume when estimating the

* For the definition of $B_2$ and $B_4$ see C. J. Ballhausen: *Kgl. Danske Vidensk Selsk. mat. fysiske Medd.*, **29**, No. 4 (1954); C. J. Ballhausen and C. K. Jørgensen: *Ebenda*, **29**, No. 14 (1955).

stabilization energies that both expression can be inter-converted with the help of a factor of about 3.

The values given in Tables A.27 and A.28 for high-spin and low-spin complexes are found from Figure A.89 by using appropriate occupancy

Table A.27. Ligand field stabilization energies (LFSE) for octahedral and quadratic-pyrimidal coordination as well as contributions to the energy of activation ($\Delta E_a$) in units of $Dq$ for substitution reactions showing octahedral coordination with the assumption of an $S_{N1}$ mechanism. (Strong field case, without including configuration interaction)[5].

| | High-spin | | | Low-spin | | |
|---|---|---|---|---|---|---|
| | LFSE in $Dq$ | | | LFSE in $Dq$ | | |
| | Octahedron | Quad. pyramid | $\Delta E_a$ in $Dq$ | Octahedron | Quad. pyramid | $\Delta E_a$ in $Dq$ |
| $d^0$ | 0 | 0 | 0 | | | |
| $d^1$ | −4 | −4·57 | −0·57 | | | |
| $d^2$ | −8 | −9·14 | −1·14 | | | |
| $d^3$ | −12 | −10·00 | +2·00 | | | |
| $d^4$ | −6 | −9·14 | −3·14 | −16 | −14·57 | +1·43 |
| $d^5$ | 0 | 0 | 0 | −20 | −19·14 | +0·86 |
| $d^6$ | −4 | −4·57 | −0·57 | −24 | −20·00 | +4·00 |
| $d^7$ | −8 | −9·14 | −1·14 | −18 | −19·14 | −1·14 |
| $d^8$ | −12 | −10·00 | +2·00 | | | |
| $d^9$ | −6 | −9·14 | −3·14 | | | |
| $d^{10}$ | 0 | 0 | 0 | | | |

* For the definition of $B_2$ and $B_4$ see, e.g., C. J. Ballhausen: *Kgl. Danske Vidensk. Selsk. mat. fysiske Medd.*, **29**, No. 4; C. J. Ballhausen and C. K. Jørgensen: *ibid*, **29**, No. 24 (1955).

of the states. For example the stabilization energy for $d^6$ complexes $E_S$ is given in units of $Dq$ by:

$$\left.\begin{array}{ll} d^6 \text{ high-spin } S_{N1}(d_{xz},d_{yz})^3(d_{xy})^1(d_{z^2})^1(d_{x^2-y^2})^1 & \\ \quad E_s = 3\cdot(-4{,}57) - 0{,}86 + 0{,}86 + 9{,}14 = & -\ 4{,}57 \\ \text{low-spin } S_{N1}(d_{xz},d_{yz})^4(d_{xy})^2 & \\ \quad E_s = 4\cdot(-4{,}57) + 2\cdot(-0{,}86) & = -20{,}00 \end{array}\right\} \text{quad. pyramid}$$

$$\left.\begin{array}{ll} \text{high-spin } S_{N2}(d_{xz},d_{yz})^3(d_{xy},d_{x^2-y^2})^2(d_{z^2})^1 & \\ \quad E_s = 3\cdot(-5{,}28) + 2\cdot(2{,}82) + 4{,}93 & = -\ 5{,}27 \\ \text{low-spin } S_{N2}(d_{xz},d_{yz})^4(d_{xy},d_{x^2-y^2})^2 & \\ \quad E_s = 4\cdot(-5{,}28) + 2\cdot(2{,}82) & = -15{,}48 \end{array}\right\} \text{pent. bipyramid}$$

Table A.28. Ligand field stabilization energies (LFSE) for octahedral and pentagonal-bipyrimidal coordination as well as contributions to the energy of activation ($\Delta E_a$) in $Dq$ units for substitution reactions showing octahedral coordination with the assumption of $S_{N2}$ mechanism. (The strong field case without consideration of electron and configuration interaction.)[5]

| | High-spin | | | Low-spin | | |
|---|---|---|---|---|---|---|
| | LFSE in $Dq$ | | | LFSE in $Dq$ | | |
| | Octahedron | Pentag. bipyramid | $\Delta E_a$ in $Dq$ | Octahedron | Pentag. bipyramid | $\Delta E_a$ in $Dq$ |
| $d^0$ | 0 | 0 | 0 | | | |
| $d^1$ | −4 | −5·28 | −1·28 | | | |
| $d^2$ | −8 | −10·56 | −2·56 | | | |
| $d^3$ | −12 | −7·74 | +4·26 | | | |
| $d^4$ | −6 | −4·93 | +1·07 | −16 | −13·02 | +2·98 |
| $d^5$ | 0 | 0 | 0 | −20 | −18·30 | +1·70 |
| $d^6$ | −4 | −5·28 | −1·28 | −24 | −15·48 | +8·52 |
| $d^7$ | −8 | −10·56 | −2·56 | −18 | −12·66 | +5·34 |
| $d^8$ | −12 | −7·74 | +4·26 | | | |
| $d^9$ | −6 | −4·93 | +1·07 | | | |
| $d^{10}$ | 0 | 0 | 0 | | | |

In columns 2 and 5 of Tables A.27 and A.28, the LFSE's for octahedral high- and low-spin complexes are given in multiples of the parameter $Dq$. Columns 3 and 6 give the stabilization energy for the case of a quadratic pyramidal orientation (coordination number 5) and for the pentagonal bipyramidal structure (coordination number 7).

Negative values of the quantity $\Delta E_a$ in columns 4 and 7 mean that the octahedral configuration is unstable with respect to the configuration of lower symmetry. $\Delta E_a$ gives the contribution to the activation energy and is calculated as the difference between the stabilization energy for the geometrical configuration corresponding to the transition state (columns 3 and 6) and the octahedral stabilization energy (columns 2 and 5). This indicates, however, that the contribution to the activation energy is virtually zero. Substitution reactions of corresponding complex ions should proceed comparatively rapidly, as is the case for spherically symmetric ions. Positive $\Delta E_a$ values indicate a contribution to the activation energy. The larger $\Delta E_a$, the slower the substitution should proceed.

It is seen from Table A.27 and A.28 that for $d^3$ and low-spin $d^4$, $d^5$ and $d^6$ compounds, independent of if the mechanism is $S_{N1}$ or $S_{N2}$, the reaction velocity for the substitution reactions for octahedral complexes should decrease as

$$d^5 > d^4 > d^3 > d^6.$$

Besides these, $d^8$ complexes should likewise be inert. The contribution to the activation energy is equal to that for $d^3$ complexes. This result differs from the predictions of the VB theory, according to which $d^8$ complexes should be kinetically labile*.

For $d^9$ compounds ($Cu^{2+}$) a contribution to the activation energy appears (Table A.28) when the substitution reaction proceeds according to an $S_{N2}$ mechanism ($+1{\cdot}07\ Dq$). $Cu^{2+}$ complexes are in the rule kinetically labile. According to the Jahn–Teller theorem (cf. chapter 4) $[\overset{\mathrm{II}}{\mathrm{Cu}}\ \mathrm{Ag}]$ complexes are always more or less tetragonally distorted so that the ligand field stabilization energy of $-6Dq$ is increased to about $-9Dq$. $\Delta E_a \approx +4Dq$ becomes correspondingly even larger, as can be seen from Table A.28. Because $Cu^{2+}$ compounds show rapid ligand exchange reactions, it can be concluded that an $S_{N2}$ mechanism is *not* likely to be involved.

The LFSE for the octahedral complexes as well as for the transition states could also be calculated by the weak-field method using term interaction (cf. section 3.1) (or using the strong-field method with electron and configuration interaction). Thereby more exact values can be obtained than those given in Tables A.27 and A.28 calculated by the strong-field method (without electron and configuration interaction). That is, however, not very meaningful when it is considered that although thereby $\Delta E_a$ is more exactly known, all such calculations yield only one fact, or a contribution to the activation energy. According to the *Arrhenius* relation $k = PZe^{-Ea/kT}$ the reaction rate constant $k$ depends not only upon the activation energy $E_a$, but also upon the pre-exponential factor $PZ$ as well. Using the treatment described above one obtains only the ligand field contribution to the activation energy $\Delta E_a$. In such considerations of kinetic behaviour $PZ$ is taken to be approximately constant, which is a very poor assumption.

Similar arguments can be applied to complexes with other structures, i.e., to tetrahedral or to square planar complexes.

---

* The high-spin $Ni^{2+}$ complexes are considered to be labile (for example in comparison to $Cr^{3+}$ complexes). However, compared to $Mn^{2+}$, $Co^{2+}$ and $Cu^{2+}$ as well as $Zn^{2+}$ complexes, they show slower ligand exchange.

# 6. Bonding

The bond in an NaCl crystal is characterized by the chemist as being essentially electrovalent (ionic), while in a compound such as chromium hexacarbonyl $[Cr(CO)_6]$ the limiting case of covalent bonding is very nearly realized.

At the outset it must be stated that '*covalence*' and '*electrovalence*' are not absolute concepts: they are meaningful only at a certain level of approximation. This condition must be kept in mind as we discuss the question of bonding in the transition metal complexes. First of all it must be established what is understood under 'electrovalence' and 'covalence' in the chemical sense.

In the limiting case of *electrovalence* the force holding the electrically charged components of a compound, the ions, together is of an electrostatic nature. For the given example of the sodium chloride crystal positive $Na^+$ and negative $Cl^-$ ions exist in the lattice.

The chemist usually associates the concept 'covalence' with the bonding type found in the ground state of the hydrogen molecule. Here the bond is produced by an electron pair shared by both protons.

The situation for the $H_2$ molecule was first studied by Heitler and London (1927). Their model corresponds well to the concepts which had been deduced by the chemists from the mass of empirical data on the nature and the properties of chemical bonding forces. The so-called *valence bond theory*, from which the concepts of covalent and electrovalent limiting cases*, as well as the concept of mesomerism stem, arose from

---

* The electrovalent and the covalent forms, e.g., of a $[Cr(NH_3)_6]^{3+}$ complex are formulated as

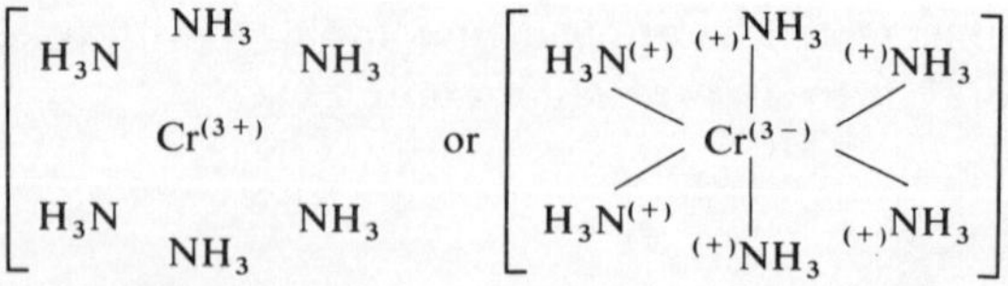

$$\underset{\text{electrovalent}}{\left[\begin{array}{ccc} & NH_3 & \\ H_3N & & NH_3 \\ & Cr^{(3+)} & \\ H_3N & & NH_3 \\ & NH_3 & \end{array}\right]} \quad \text{or} \quad \underset{\text{covalent}}{\left[\begin{array}{ccc} & {}^{(+)}NH_3 & \\ H_3N^{(+)} & | & {}^{(+)}NH_3 \\ \diagdown & Cr^{(3-)} & \diagup \\ \diagup & | & \diagdown \\ H_3N^{(+)} & {}^{(+)}NH_3 & {}^{(+)}NH_3 \end{array}\right]}$$

In the first case the triply positively charged chromium ion is surrounded by six $NH_3$ molecules. The bond arises from ion–dipole interaction. In the second case each bond is produced by an electron pair (symbolized by a valence dash). One electron from the lone pair on each $NH_3$ molecule is transferred to the central ion, which assumes the charge $-3$. This electron forms a bonding electron pair with the electron remaining on the $NH_3^{(+)}$ molecule ion.

these investigations carried out by Heitler and London on the $H_2$ molecule.

It is an overly optimistic generalization to apply these concepts developed from the treatment of the hydrogen molecule to all other cases without restriction, i.e., to consider every covalent bond as arising from an electron pair shared by two atoms. By merely replacing the classical valence 'dash' with an electron pair in the sense of the VB theory, one is not in the rule in the position to depict the energy states of molecules in agreement with experience.

For diatomic and polyatomic molecules the possibility of an MO treatment exist, as was shown in section 1.12. Accordingly the electrons move in MO's which in principle can extend over the entire molecule (i.e., over several nuclei). The MO theory gives a useful classification of electronic states. It appears to be reasonable to equate the concept of covalence with the *degree of electron delocalization* over several nuclei. Chromium hexacarbonyl is in this treatment a covalent compound in the sense that its properties can be described by assuming a considerable degree of electron delocalization.

The division into electrovalent and covalent complexes (*Normal* and '*Durchdringungskomplexe*' in the nomenclature of Biltz) is based historically on certain experimental criteria, e.g. *stereochemical properties, magnetic measurements*, results of *Raman* and i.r. *spectroscopic investigations, kinetic behaviour* in substitution reactions as well as *thermodynamic stability*.

First of all it should be considered what meaningful criteria for a division of the complexes into classes can be derived from these experimental data, and further if the classes arrived at from a consideration of these various criteria are identical. In that latter case the relation of the empirical classification with a possible theoretical classification according to the nature of the bonding is to be discussed.

Which of the given experimental criteria are to be used as a basis for classification is a question of expediency. One could utilize, e.g., the stability constants especially because their magnitudes determine if the complex formation is observable in preparative and in analytical work. On this basis complexes can be classified as *unstable* (*weak*) and as *stable* (*strong*). As one is interested not only in thermodynamic stability of complexes, but in some cases also in the other properties mentioned above it would naturally be convenient when all of the classification schemes would lead to the same division into classes. This is, happily, to a very large degree the case, although there is no complete coincidence of the classification schemes.

Weak complexes are often kinetically labile. The isolation of optical or stereo-isomers is in general not possible and the speed of exchange reactions is usually comparatively great. The complexes are frequently high-spin. In Raman and i.r. spectra no characteristic frequencies for the bond between the central ion and the ligands are observed. The central ion-ligand separations have values approximately equal to the sum of the corresponding ionic radii.

Strong complexes are to the contrary often kinetically stable. Optical and stereoisomers can be separated and exchange reactions usually proceed slowly. Reduced paramagnetism or diamagnetism is often found (low-spin complexes). X-ray investigations frequently show smaller separations of the components than one would expect on the basis of the corresponding ionic radii. In many cases characteristic Raman or i.r. frequencies are observed.

There are, however, a large number of complexes which cannot be fitted into these patterns. For example, as has been mentioned earlier, the diamagnetic $[Ni(CN)_4]^{2-}$ complex is exceedingly labile with regard to exchange of the $CN^-$ ligands. The diamagnetic $[Co(H_2O)_6]^{3+}$ ion exchanges very rapidly with water. Iron (III) haemoglobin and iron(III) haemin complexes show no exchange with radioactive $Fe^{3+}$ after two months and are therefore strong complexes. One observes, however, high-spin behaviour. In contrast to this the tris-2,2′-dipyridyl iron(II) complexes although markedly kinetically labile belong to the low-spin type.

That the magnetic criterion does not always agree with the kinetic criterion can be understood easily on the basis of the ligand field theory. We have seen in the preceding chapters that the magnitude which essentially determines the behaviour of the complexes, for cubic complexes (and we can to the first approximation treat the majority of complexes from the standpoint of cubic microsymmetry) is the field strength parameter $\Delta \equiv 10Dq$. Low-spin compounds arise because $Dq$ exceeds a certain critical value. This critical value lies at different $Dq$ values for complex ions with different central ions. For this reason a low-spin complex does not necessarily show a greater kinetic stability than another, high-spin complex, having a greater $Dq$ value than the former, although no spin pairing appears.

The spectrochemical series of the ligands (cf. p. 78f) shows that increasing $Dq$ values cannot be identified with an increasing degree of covalénce. (As an example $F^-$ has a greater $Dq$ value than $I^-$ but produces a stronger field.) Spin pairing and thereby a reduced magnetic moment or diamagnetism follow from every mechanism calling for a sufficiently large magnitude of $Dq$. The kinetic stability follows from this. However, one can predict nothing concerning the degree of covalence of the metal-ligand bond from such properties.

The at first sight astonishing fact that the magnetic moments can be explained by an extended electrostatic (ionic) model as well as by *Pauling*'s covalent model ceases to be remarkable when one considers the quantum mechanical *variation principle*, as has been treated especially by Hartmann[1]. This principle is related to the fact that the long-wavelength weak bands can be explained with the aid of the extended ionic theory as well as with the LCAO–MO theory. The explicit consideration of the electron interaction in the latter case brings of course great calculation difficulties.

All bonding cases can be roughly divided into three categories: the essentially pure electrovalent, the essentially pure covalent and the mixed cases (Figure A.90).

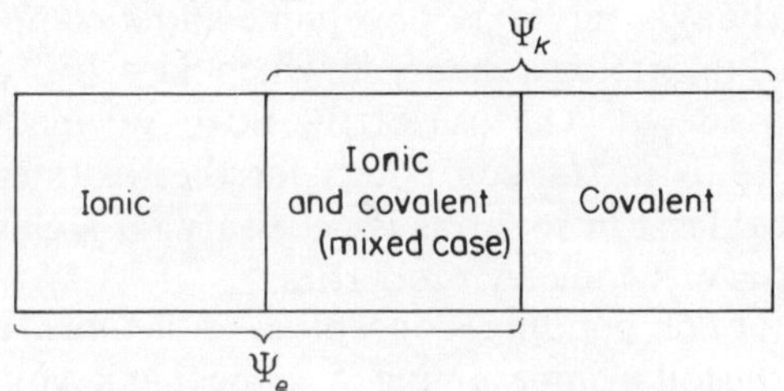


Figure A.90. Bonding cases.

Within the framework of the VB approximation of the quantum mechanical bonding theory the covalent and the electrovalent limiting forms correspond to wave functions $\psi_k$ and $\psi_e$, which we shall call accordingly the covalent and the electrovalent type. To the first approximation a fundamental principle of quantum mechanics, the variation principle, holds. According to this principle the energy associated with a wave function is already determined to the first order when the wave function itself is only known to the zeroth approximation. Because, however, the chemical and a large portion of the physical properties of molecules are essentially determined by energy magnitudes, a theory which works with relatively poor wave functions yields nevertheless results which are useful to the first order.

According to the scheme described in Figure A.90, one can expect that a theory utilizing $\psi_e$ functions should treat not only the ionic limiting cases, but also the mixed cases qualitatively correctly. It is correspondingly true that the mixed cases can be treated equally well using a theory which

1. H. Hartmann: *Z. phys. Chem.* (Frankfurt) **4**, 376 (1955); cf. also H. Hartmann and H. L. Schläfer: *Angew. Chem.*, **70**, 155 (1958).

employs $\psi_k$ functions. This means however, that in many mixed cases the question as to the bonding type can not be unambiguously answered, especially in so far as most chemical compounds cannot in the rule be rigorously classified alternatively as covalent or ionic. This applies also to complexes, so that from this point of view one must consider that the properties of many complexes can be described by two apparently contradictory models.

On the basis of the experience of the last few years for the vast majority of the mixed cases arising in the realm of complex chemistry it appears to be reasonable purely because of economy to use the analytically comparatively primitive extended ionic theory (electrostatic ligand field theory) as the starting point for a discussion. Only in those cases where this fails, e.g., for carbonyls, nitrosyls, phosphine and $\pi$ complexes, etc. is one in practice forced to employ a theory in which the rôle of covalent valence structures is considered. The use of the LCAO–MO method in its semiempirical form leads to starting points for the treatment of such compounds as well, at least in so far as is necessary for a classification of the terms on the basis of symmetry properties.

In a discussion of bonding conceptual difficulties arise which are illustrated with the following example. Assume that it would be possible to determine the electron density distribution in an $[MA_6]$ complex with great accuracy and that we would establish that this is identical with a distribution which is obtained by superposition of the electron density distribution of the free central ion and the free ligands when the components are arranged as in the complex ion. This result is in complete agreement with the predictions which follow from an ionic theory without polarization of the ligands A. Nevertheless covalent binding is by no means excluded: The bonding orbitals of the central ion M, as is well known, overlap to a certain degree with the orbitals belonging to lone pairs of electrons of the ligands A (e.g., for $A{=}NH_3$ they are orbitals of the lone pair electrons of nitrogen.) One must assume from this, consequently, that the electrons in these orbitals contribute somewhat to the electron density on the central ion M. This overlap charge can in principle be considered as belonging to each of the two partners which participate in the bond. How far we consider the bond to be covalent or ionic is partially dependent upon how we distribute the overlap charge over the two participating atoms.

The experimental quantity which is of interest to us in a discussion of bonding is thus essentially the *electron density distribution.* Were we able to determine this quantity with sufficient accuracy, which up until now has not been possible for complexes containing such heavy elements as the transition metals, then we would still have the problem of finding a more quantitative description of the bonding, that is, of finding the MO's corresponding to the electron density distribution.

In general several $\psi$ functions* can be assigned to a certain electron density distribution so that an unambiguous determination of $\psi$ from measurements of the electron density distribution is not possible. A more quantitative description of the bonding would then consist in discussing the degree of delocalization of certain electrons on the basis of the associated wave functions. Further one could consider which factors promote and which tend to hinder a delocalization of certain electrons.

In the following we shall treat briefly the question as to *which experimental data can contribute to a discussion of bonding*. From what has been said above it is clear that methods which give information concerning the *electron density distribution* are of importance. Today the usual refraction methods do not yet allow the electron density distribution of coordination complexes with transition metal ions to be determined with sufficient accuracy to permit the degree of electron delocalization to be established in this manner.

*Neutron diffraction* seems to be the most promising[2]. By means of this method in principle the density of unpaired electrons can be measured directly. The neutron has a magnetic moment, but no charge. It is scattered by the electrons, so that the effect depends critically upon the orientation of the electron spin. Such investigations are likely to become more important in the future.

There are other methods which exploit the magnetic properties of unpaired electrons. With their help one can study the $d$ electrons in paramagnetic compounds. Diamagnetic complexes and especially the paired ligand electrons, which are of great importance for the bond can, however, not be treated.

The methods which contributed the most to an understanding of the bonding problem are *electron paramagnetic resonance* (EPR) and *nuclear magnetic resonance* (NMR). The latter, however, has been of much less importance. For details the reader is referred to the comprehensive literature[3,4].

---

* As is well known, for $N$ electrons $\psi$ is a function of $3N$ space coordinates, while the electron density depends upon three coordinates.

2. C. G. Shull and E. O. Wollan: *Solid State Physics*, Vol. II, p. 137, Academic Press, New York–London 1956.
3. EPR: (a) W. Low: *Paramagnetic Resonance in Solids*; *Solid State Physics*, Supplement 2, Academic Press, New York–London 1960. (b) S. A. Altshuler and B. M. Kosyrew: *Paramagnetische Elektronenresonanz*. Verlg. Harry Deutsch, Frankfurt a. Main. Translation from Russian, 1964. (c) J. Owen: *J. inorg. nucl. Chem.*, **8**, 430 (Review article 1958) (*Suppl. to Scientific Research*,

In *electron paramagnetic resonance* a strong homogeneous field is applied to the sample. If a free paramagnetic ion with a ground state described by the quantum numbers $L$, $S$ and $J$ is brought into a magnetic field of field strength $H$, a splitting into $2J+1$ *Zeeman* levels takes place. These correspond to the possible orientations of the magnetic moment in the field. The energy of the individual states is given by $E = -g\beta_B H \cdot M_J$, where $g$ is the *Landé* factor

$$g = 1 + \frac{J(J+1) + S(S+1) - L(L+1)}{2J(J+1)},$$

$$\beta_B = \frac{eh}{4\pi m_e c}$$

is the *Bohr* magneton, $L$, $S$ and $J$ are the quantum numbers for the orbital angular momentum, the spin angular momentum, and the total angular momentum, and $M_J$ is the total magnetic angular momentum quantum number which determines the component of the total angular momentum in the direction of the external field.

If an alternating field of frequency $\nu$ (in the microwave region) is applied, its magnetic vector is polarized perpendicular to the direction of the magnetic field $H$, then magnetic dipole transitions between neighbouring *Zeeman* states ($\Delta M = \pm 1$) are induced. The resonance condition is then

$$\Delta E = h\nu = g \cdot \beta_B \cdot H.$$

---

**28**, Rome, 430 (1958)). (d) A. Carrington and H. C. Longuet-Higgins: *Q. Rev. chem. Soc.*, **14**, 427 (1960) (Electron resonance in crystalline transition metal compounds). (e) D. J. E. Ingram: *Spectroscopy at Radio and Microwave Frequencies*, Butterworth, London 1955. (f) Microwave and Radiofrequency Spectroscopy. *Discuss. Faraday Soc.*, **19** (1955); see also [4] (a) 4 and [4] (e). (g) G. Weber: *Bull. Ampère 10e année*, pp. 102 ff., fasc. spécial 1961. (h) K. H. Hausser: *Angew. Chem.*, **68**, 729 (1956).

4. NMR: (a) 1. E. R. Andrew: *Nuclear Magnetic Resonance*, Cambridge University Press. 2. H. S. Gutowsky: *Physical Methods in Chemical Analysis*, Vol. III, Academic Press, New York–London 1956. 3. G. E. Pake: *Am. J. Phys.*, **18**, 483, 473 (1950). 4. J. E. Wertz: *Chem. Rev.*, **55**, 829 (1955). (b) J. W. A. Pople, W. G. Schneider and H. J. Bernstein: *High Resolution Nuclear Magnetic Resonance*, McGraw-Hill, New York–San Francisco–Toronto–London 1959. (c) R. E. Richards: *J. inorg. nucl. Chem.*, **8**, 423 (Review Article) (1958). (d) E. Fluck: *Die kernmagnetische Resonanz und ihre Anwendung in der anorganischen Chemie*, Springer, Berlin–Göttingen–Heidelberg 1963. (e) G. E. Pake: *Solid State Physics*, Vol. II, p. 1, Academic Press, New York–London 1956. (f) J. D. Roberts: *Nuclear Magnetic Resonance*, McGraw-Hill, New York–San Francisco–Toronto–London 1959. (g) E. L. Muetterties and W. D. Phillips: *Adv. inorg. Chem. Radiochem.*, **4**, 231 (1962). (h) H. Sillescu: Magnetische Kernresonanz in der Komplexchemie, *Fortschr. chem. Forsch.*, **5**, 569 (1966).

In the simplest case, where only a single electron is present and $l = 0$, it follows that $g = 2$ (allowing for the relativistic effect more exactly 2·00229).

If the wavelength ($\lambda = c/\nu$) of the microwaves is expressed in centimeters and the strength of the applied field is measured in kilogauss, then one may write for the above resonance condition

$$H = \frac{21{\cdot}4178}{g \cdot \lambda}.$$

We shall discuss the fundamentals of the paramagnetic resonance method in terms of the behaviour of a single, isolated unpaired electron. In this, the most simple case, an orientation of the spin magnetic moment parallel and antiparallel to the field follows, so that two energy states result. The magnetic energy of the electron is then $-\beta_B H$ and $+\beta_B H$, depending upon the orientation of its moment (Figure A.91b). Electrons having both orientations appear. However, a Boltzmann distribution over the two states arises such that the energetically lower-lying state is the more densely populated. Absorption of microwaves of the frequency $\nu$ or the wavelength $\lambda$ produces a transition from the energetically lower to the higher state. The absorbed energy serves to invert the spin orientation of the electrons relative to the direction of the static field. In practice one maintains a constant frequency $\nu$ and varies the field strength $H$ until resonance absorption appears. This frequency lies in the centimeter wavelength region (e.g., for $g = 2$, $\lambda = 3$ cm, the $H \approx 3{\cdot}6$ kilogauss).

For several unpaired electrons the situation is more complicated. If paramagnetic ions are brought into a crystalline lattice, two types of interactions are seen: (a) magnetic dipole interactions and (b) interactions of the paramagnetic ions with the diamagnetic neighbours.

Magnetic dipole interactions can to a large degree be suppressed by studying crystals which have been diluted by an isomorphous diamagnetic salt. In the rule dilutions of above 1:1000 are sufficient. The interaction of the paramagnetic ions with the diamagnetic ligands (ions or dipoles) can be treated on the basis of the ligand field theory when the spin–orbit coupling is considered.

Under the influence of the ligand field the ground state of the paramagnetic ion splits into a number of progeny terms, depending upon the nature of the symmetry. One can measure the resonance transitions between the corresponding *Zeeman* levels by applying an external magnetic field. The $g$ factor in a crystal is no longer given by

$$g = 1 + \frac{J(J+1) + S(S+1) - L(L+1)}{2J(J+1)}$$

but is defined experimentally as the *spectroscopic splitting factor* $g = 21{\cdot}418/\lambda H$, i.e., by the measurement of $\lambda$ and $H$. $g$ depends upon the orientation of the external magnetic field with reference to the symmetry axes of the electric field generated by the ligands. It is seen that $g$ is a tensor which gives the anisotropy of the ligand field. The symmetry of the ligand field can be determined, therefore, from measurements of the paramagnetic resonance of oriented single crystals.

With reference to the question of the bonding state, especially with regard to the magnitude of the electron delocalization, the *hyperfine structure of the paramagnetic resonance* is of particular interest because it yields information concerning the distribution of an unpaired electron over the entire complex ion. One can observe the hyperfine structure caused by the surrounding ligands when one or more of the diamagnetic neighbours of the central ion have a *nuclear magnetic moment*. When an unpaired electron from an ion $M$ (central ion) is partially transferred to an atomic ion $X$ (ligand), then the magnitude of a possible hyperfine structure splitting is equal to the product of the hyperfine structure splitting of the free atom $X$ and the probability of finding the unpaired electron at $X$.

A nucleus with a nuclear moment $\mu_I$ and, e.g., a nuclear spin $I = \frac{1}{2}$ can likewise be oriented with its magnetic moment in the direction of the external magnetic field or opposed to it. The magnetic moment $\beta_B$ of a single unpaired electron interacts with that of the nucleus $\mu_I$. Depending upon if the spins are parallel or antiparallel one obtains a magnetic contribution to the energy which is proportional to $\pm\beta_B\mu_I$.

The nuclear spin in general does not follow when a transition in a magnetic field accompanied by the inversion of the electron spin takes place. When, therefore, the nuclear spin and the electron spin are parallel before the transition, they are antiparallel afterwards, and vice versa. The right side of Figure A.91c shows the corresponding energy states. One finds instead of a transition with the absence of nuclear spin, two transitions, the frequencies of which lie symmetrically about the frequency of the original transition. Their separation from one another depends upon the magnitude of the interaction between the nuclear spin and the electron spin and is denoted as *nuclear hyperfine structure splitting*.

A particularly instructive example of the application of hyperfine structure splitting of the paramagnetic resonance is the investigation of $[IrCl_6]^{2-}$ carried out by Griffith and Owen[5]. Here the one unpaired electron is a $4d$ electron of $Ir^{4+}$. Each Cl nucleus ($I = \frac{3}{2}$ for $^{35,37}$Cl) produces a hyperfine structure splitting where each line in the paramagnetic

---

5. J. H. E. Griffith and J. Owen: *Proc. R. Soc.*, **A 226**, 96 (1954).

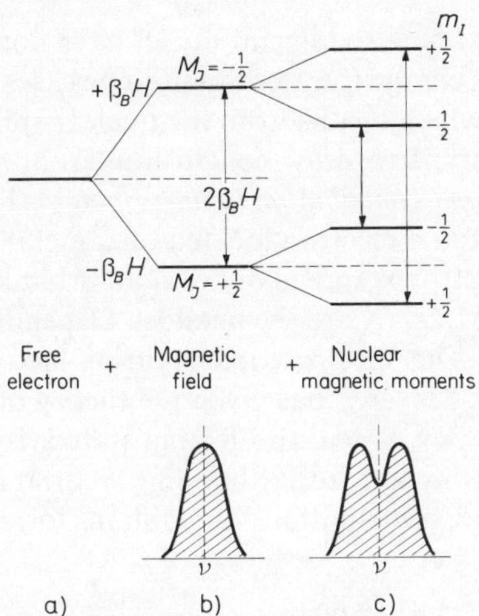


Figure A.91. Energy states of an electron in a magnetic field with consideration of the nuclear spin-electron spin interaction (↔ possible transitions: $\Delta M_p = \pm 1, \Delta m_J = 0$).

resonance spectrum splits into $2I+1 = 4$ components. Figure A.92 shows the spectrum of $Na_2[IrCl_6]$. From the distance between neighbouring components one can determine that the probability of finding the unpaired electron of the central ion in a $p\pi$ orbital of chlorine is 5 per cent. From this it follows that the position probability density of the unpaired electron on the $Ir^{4+}$ central ion is approximately 70 per cent and on the six chloride ligands about 30 per cent.

Similar investigations have been conducted by Tinkham[6] on $[MnF_6]^{4-}$ and $[FeF_6]^{4-}$ ($I = \frac{1}{2}$ for F), for which the probability of finding the unpaired electron of the central ion in a $p\sigma$ orbital of the $F^-$ ligand is 2 per cent. (The systems $Mn^{2+}$ and $Fe^{2+}$ in $ZnF_2$ were studied.)

The hyperfine splitting has also been studied for $[Mo(CN)_8]^{3-}$ ion where cyanide enriched with $^{13}C$ ($I = \frac{1}{2}$) was used[7].

Such investigations[8] yield, therefore, unambiguous evidence concerning an electron delocalization.

6. M. Tinkham: *Proc. R. Soc.*, **A 236**, 535, 549 (1956).
7. S. I. Weissman and M. Cohn: *J. chem. Phys.*, **27**, 1440 (1957).
8. J. Owen: *Discuss. Faraday Soc.*, **19**, 127 (1955).

In the cases treated here the ligand nuclei have nonzero nuclear spins. The magnitude of the charge transfer can in some cases be derived from the hyperfine structure, which results from the nuclear spin of the central ion.

A further possibility is to draw conclusions from the *reduction* of the *orbital magnetic moment contribution to the g facter*[8]. The orbital magnetic moment of an unpaired *d* electron located on a metal ion M is in general reduced when the electron is spread out over an orbital which extends over the entire complex $[MX_6]$ (X are the ligands). Depending upon the extent of this delocalization the spectroscopic splitting-factor is to a greater or lesser extent reduced. Stevens[9] has given the theory of this effect.

For $[Ni(H_2O)_6]^{2+}$ $3d^8$, $S = 1$, the ground state arises from $d\varepsilon^6\, d\gamma^2$ and both unpaired electrons are in antibonding $\sigma$ orbitals ($d_{x^2-y^2}$ and $d_{z^2}$). Neglecting the $\pi$ bond contributions one obtains for $g$ the expression*

$$g = 2{\cdot}0023 - \alpha^2 \cdot \frac{8\lambda_{(\text{fr. Ion})}}{\Delta},$$

where $\gamma_{(\text{fr. ion})}$ is the spin–orbit coupling constant for the free ion and $\Delta \equiv 10Dq$ the energy difference between the $d\varepsilon$ and the $d\gamma$ orbitals. $\alpha^2$ is the coefficient (cf. p. 105) giving the mixing of the orbitals $d_{x^2-y^2}$ and $d_{z^2}$ of the metal ion with the ligand. $\alpha^2$ is 1 for a purely ionic bond and 0·5 for a purely covalent bond. For the compound mentioned above one finds $\alpha^2 = 0{\cdot}83$ from the experimental values for $\lambda_{(\text{fr. ion})}$, $\Delta$ and $g$.

For $[IrCl_6]^{2-}$ the electron delocalization can be derived not only as already mentioned, from the paramagnetic resonance, but also from the spectroscopic splitting factor $g$. The ground state arises from the electron configuration $d\varepsilon^5$, $S = \frac{1}{2}$. From the measurement of $g$ the mixing coefficient $\beta^2$ can be determined, which gives the probability that the unpaired $d\varepsilon$ electron is located on the Ir central ion or on the Cl ligand in an antibonding $\pi$ orbital. From the $g$ factor one obtains $\beta^2 = 0{\cdot}66$, while the analysis of the hyperfine structure splitting yields 0·74.

Another means of studying electron delocalization is *nuclear magnetic resonance*. Atomic nuclei with an uneven number of protons or of neutrons

---

* A purely electrostatic theory[9] yields for a $d^8$ complex ion

$$g = 2{\cdot}0023 - 8\,\frac{\lambda_{(\text{fr. Ion})}}{\Delta}.$$

The experimental $g$ value can, however, be obtained from this expression only by using $\lambda_{(\text{compl. ion})}$ values ($\lambda_{(\text{compl. ion})} < \lambda_{(\text{fr. ion})}$) $g = 2{\cdot}0023 - 8\lambda_{(\text{compl. ion})}/\Delta$. On the basis of the MO theory $\lambda_{(\text{compl. ion})} \approx d^2\lambda_{(\text{fr. ion})}$.

9. K. W. H. Stevens: *Proc. R. Soc.*, **A 219**, 542 (1953); cf. also J. Owen: *Proc. R. Soc.*, **A 227**, 183 (1955).

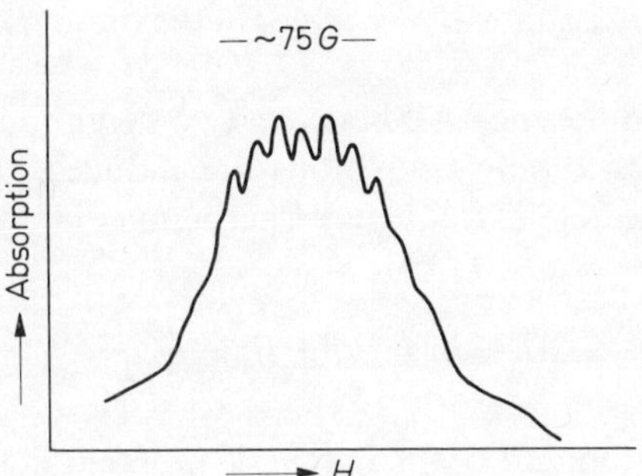


Figure A.92. Hyperfine structure splitting of paramagnetic resonance spectrum of $[IrCl_6]^{2-}$ after Griffith and Owen[5]. Single crystal of diamagnetic $Na[PtCl_6] \cdot 6H_2O$ in which about 0·5 per cent of the $Pt^{4+}$ has been replaced by $Ir^{4+}$.

or those in which protons as well as neutrons are present in uneven numbers have a nuclear spin which is characterized by the nuclear spin quantum number *I*. *I* can assume half integral and integral values. A nucleus with $I > 0$ has a magnetic moment. The nuclear magnetic moments are of the order of magnitude of the magnetic moment of the proton ($I = \frac{1}{2}$), the so-called nuclear magneton:

$$\beta_K = \frac{e \cdot h}{4 m_p \cdot c} = 5{\cdot}0493 \cdot 10^{-24}\,[\text{erg} \cdot \text{Gauß}^{-1}],$$

where $m_p$ is the mass of the proton (1836 times the mass of the electron). The relation corresponds to that for the magnetic moment of the electron. $\beta_K$ is obtained from the Bohr magneton $\beta_B$ by replacing $m_e$ by $m_p$ ($\beta_B = 1836\beta_K$).

If a nucleus with a spin quantum number *I* is brought into a homogeneous magnetic field of field strength $H_0$, then according to the quantum theory $2I+1$ energy states exist, each of which corresponds to a particular orientation of the nuclear spin in the magnetic field. The magnetic energy of such a state is given by

$$E = -g_I \beta_K \cdot H_0 \cdot m_I$$

where $g_I$ is the *nuclear g factor*, $m_I$ is the magnetic quantum number, which determines the component of the nuclear magnetic moment in the external field. It can assume $2I+1$ values in the order, $I, I-1, \ldots, -(I-1), -I$.

For the energy difference $\Delta E$ between two neighbouring states ($\Delta m_I = +1$) between which dipole transitions are induced by means of a high frequency magnetic field of frequency $\nu$ perpendicular to $H_0$, the resonance condition follows:

$$\Delta E = h\nu = g_I \beta_K \cdot H_0 = \frac{\mu_z \cdot H_0}{I},$$

where $\mu_z$ is the component of the magnetic moment in the direction of the external field ($\mu_z = g_I \cdot I \cdot \beta_K$).

For field strengths of 10,000 gauss the resonance frequencies fall in the region between 1 and 50 MHz.

Consider as an example an isolated nucleus with $I = \frac{1}{2}$, for which in the magnetic field $2I-1 = 2$ energy states $E(m_I = +\frac{1}{2}) = -\frac{1}{2}g_I \cdot \beta_K \cdot H_0$ and $E(m_I = -\frac{1}{2}) = +\frac{1}{2}g_I\beta_K H_0$ exist (Figure A.93). The resonance condition leads to $h\nu = 2\mu_z H_0$. One resonance line should therefore be observed. If one is concerned with protons the $H_0$ is 10,000 gauss and the resonance frequency $42{\cdot}58 \times 10^6$ Hz.

The application of nuclear magnetic resonance spectroscopy in chemistry is dependent upon the interaction arising between nuclear moments when the atoms combine to form compounds. Such couplings cause the NMR spectrum to assume a more complicated form than one single line. In general it may be said that a much smaller interaction with the environment takes place for NMR compared to EPR. A mechanism corresponding to spin-orbit coupling is completely absent and the magnetic dipole-dipole coupling of the nuclei is a factor of $\sim 10^6$ smaller than that for electrons. The nuclei are shielded from the environment by the electron shells. The nuclear spins are so little coupled with the molecular framework that they precess about the axis of the external magnetic field largely uninfluenced by thermal agitation. (Nuclei with $I \geqq 1$ behave somewhat differently.)

A nuclear magnet generates in its environment a weak additional magnetic field with a component in the direction of the externally applied field $H_0$, which we denote as $H_{\text{loc.}}$. $H_{\text{loc.}}$ increases or diminishes $H_0$ to a small extent, depending upon the momentary orientation of the nuclear spins producing $H_{\text{loc.}}$. In a crystal one obtains a resulting magnetic field of the strength $H_0 + H_{\text{loc.}}$, where $H_{\text{loc.}}$ originates from all the neighbouring nuclei. In this manner a broadening of the resonance lines appears. The closer the nuclei are to one another, the greater is the value of $H_{\text{loc.}}$, and the broader are the resonance lines. Frequently certain arrangements of nuclei give resonance lines of characteristic form. One can often in this

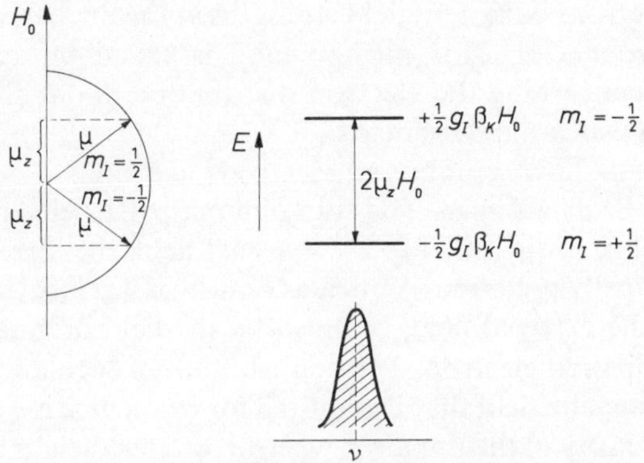


Figure A.93. Nucleus with $I = \frac{1}{2}$ in a magnetic field of strength $H_0$. (Possible transitions $\Delta m_I = \pm 1$.)

manner determine the arrangements and the nuclear separations of the atoms in crystals.

Nuclei with $I > \frac{1}{2}$ have in addition to the magnetic dipole moment an electric quadrupole moment as well, which arises from the non-spherically symmetric nuclear charge distribution. The interaction of the quadrupole moment with the electric field gradient of the electrons of a molecule can lead to greater splitting and to broadening of the NMR lines.

The local magnetic fields and the quadrupole interactions, as was explained, broaden the resonance lines of solids. When the molecules studied are relatively free to rotate, i.e., when liquids or solutions are investigated, such line-broadening vanishes to a large degree and much narrower resonance lines are observed.

The resonance frequency of a nucleus depends upon the field strength at the nucleus. This is, however, not completely identical with the field strength $H_0$ of the external magnetic field. Besides the influence of the neighbouring nuclei already mentioned, the field strength $H_k$ at the nucleus also depends upon the shielding of the surrounding electron clouds:

$$H_K = H_0 - H_0\sigma = H_0(1-\sigma).$$

$\sigma$ is the dimensionless magnetic shielding constant.

$\sigma$ may be divided into two parts, the first depends upon the diamagnetism and the second upon the second order temperature-dependent paramagnetism.

The applications of NMR to chemical problems centre around the field strength at the nucleus $H_K$, which is influenced by the movement of the paired electrons. Different field strengths at the nucleus cause different resonance frequencies. This phenomenon is called the *chemical shift**. Conclusions concerning the electron distribution about the nucleus can be drawn from such measurements.

If compounds having unpaired electrons are studied, analogous, but much larger effects are seen. The spin of an unpaired electron is oriented either parallel or antiparallel to the external field, the former orientation being energetically preferred. A nucleus on which it is located is influenced not only by the external field, but also by the field originating from the spin of the unpaired electron. The contribution of such an unpaired electron to the magnetic field that is required for resonance is proportional to the electron density of the unpaired electron on the nucleus in question. In this manner the electron density in the neighbourhood of the nucleus is accessible to external determination.

Proton resonance measurements have been carried out, e.g., on cyclopentadienyl complexes[10,11] as well as on certain acetyl acetonate complexes[12]. For both classes of compounds a large degree of electron delocalization could be shown. For the example of acetyl acetonates $[\overset{III}{M}(CH_3COCHCOCH_3)_3]$ one finds that unpaired electrons of the central ion have a nonzero probability of being located on the $CH_3$ and CH groups.

As an example, the resonance frequencies of the ring protons of the CH groups for a complex with $V^{3+}$ as central ion are greatly shifted compared to the frequency of the corresponding paramagnetic complex, with $Al^{3+}$

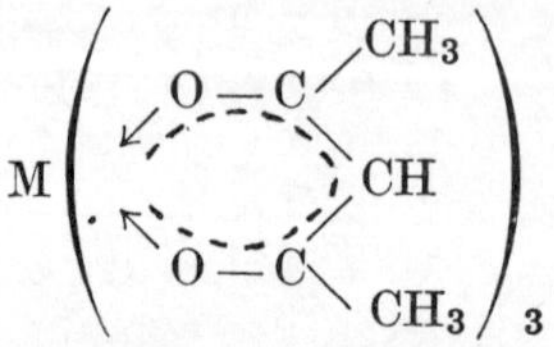

M( O—C(CH₃) ⋯ CH ⋯ O—C(CH₃) )₃

---

* The resonance field strength $H_v$ of a nucleus in a certain compound $V$ is compared with the resonance field strength $H_s$ of the same nucleus in a standard substance $S$. The chemical shift is then defined as

$$\delta = (H_v - H_s)/H_s$$

where $\delta$ is a dimensionless quantity.

10. H. M. McConnel and C. M. Holm: *J. chem. Phys.*, **28**, 749 (1958).
11. Cf. also E. O. Fischer and H. Werner: *Chem. Ber.*, **95**, 695 (1962). H. P. Fritz, H. Keiler and E. O. Fischer: *Naturwiss.*, **48**, 518 (1961); E. O. Fischer and K. Ulm: *Z. Naturf.*, **156**, 59 (Structure of sandwich complexes) (1960).
12. A Forman, J. N. Murrell and L. E. Orgel: *J. chem. Phys.*, **31**, 1129 (1959).

The magnitude of this shift becomes understandable because the $d$ electrons in the $t_{2g}$ states (the ground state of $V^{3+}$ for $O_h$ symmetry arises approximately from $t_{2g}^2$) are delocalized to a noticeable degree more than the $\pi$ electron system of the ligands and actually have a finite position probability in the 1$s$ states of the $H$ atoms.

The information which one can obtain with NMR is in principle the same as is yielded by EPR. For technical reasons the two methods are, however, often complimentary. Frequently the first method leads to difficulties when the electron density originating from an unpaired electron is so great that the signal becomes too broad. Then often a study of the hyperfine structure of the paramagnetic resonance leads to the desired result. On the other hand one frequently has good success with NMR when the nuclear moments are too small to allow a resolvable hyperfine structure in the EPR to be found.

A further possibility in certain cases to obtain information concerning the electron distribution consists in studies carried out with the aid of the *Mössbauer* effect[13,14]. Here one is concerned with the resonance fluorescence of recoil-free $\gamma$ radiation of the atomic nuclei. The frequency of the resonance fluorescence depends upon the bonding state of the atoms. This influences the energy levels of the nuclei so that the energy necessary to excite the resonance fluorescence is somewhat smaller or greater than the energy of the $\gamma$ quanta emitted from the radiation source. To obtain resonance the radiation source and the absorber must be in motion relative to one another. If the intensity of the resonance fluorescence is plotted against this relative velocity the corresponding *Mössbauer* spectra are obtained. $^{57}Fe$ proves to be especially suited for such investigations, and therefore a series of studies has been made on iron complexes. $^{61}Ni$ can also be used.

We have discussed those methods which offer the possibility of studying the electron delocalization directly. There are also other physical measurements with the aid of which indirect semiquantitative predictions of the degree of delocalization can be obtained. Of special importance are investigations of optical spectra.

We have already become acquainted with the reduction of the electron interaction parameter $B$ in complexes compared with the $B$ values of the

---

13. E. Fluck, W. Kerler and W. Neuwirth: *Angew. Chem.*, **75**, 461 (1963); as well as various articles in *Proceedings 8 ICCC*., Vienna, Sept. 7–11, 1964, pp. 1–14, Springer-Verlag, Vienna–New York.
14. V. I. Goldansky: The Mössbauer Effect and its Application in Chemistry, (Translation from the Russian) *Atomic Energy Review* (*I.A.E.A.*) *Vienna*, Vol. 1, No. 4 (1963); see also E. Fluck: The Mössbauer Effect and its Application in Chemistry, *Adv. inorg. Chem. Radiochem.*, **6**, 433–89 (1964).

free ions, which can be obtained from the corresponding optical spectra (nephelauxetic effect, cf. p. 82), as well as the analogous phenomenon for the values of the spin–orbit coupling constants $\zeta_{nd}$ (relativistic nephelauxetic effect, cf. p. 97). Both effects give a measure of the electron delocalization, at least when one compares complexes with the same central ion.

The diminution of the $B$ values can be expanded upon using the example of $Mn^{2+}$ complexes of octahedral symmetry. The sharp bands of small half-width for $[Mn^{II}A_6]$ complexes (in Figure A.58 denoted by a star), corresponding to the transition $(d\varepsilon^3\, d\gamma^2)^6A_{1g} \rightarrow (d\varepsilon^3\, d\gamma^2)^4A_{1g}$, $^4E_g$ lie according to the theory at $10B+5C$ or, expressed by the appropriate Slater radial integrals at $10F_2+125F_4$. Its position is to be compared with the term difference $^6S-^4G$ of the free $Mn^{2+}$ ion, which is also $10B+5C$.

For the free ion one finds from an analysis of the emission spectrum for $\Delta E(^6S-^4G) = 26{,}846\ \text{cm}^{-1}$. For $[Mn(H_2O)_6]^{2+}$ for example ca. 25,000 $\text{cm}^{-1}$ is obtained from the position of the sharp band for $\Delta E(^6A_{1g} \rightarrow {}^4A_{1g}, {}^4E_g)$. With $C = 4B$ it follows that $B_{(\text{compl. ion})} = 25{,}000:30 = 833$ $\text{cm}^{-1}$, while one obtains $B_{(\text{fr. ion})} = 26{,}846:30 = 895\ \text{cm}^{-1}$. The $\beta_{35}$ value is then $833:895 = 0{\cdot}93$. Systematic investigations show that for all Mn(II) complexes which are obtained in this manner $B_{(\text{compl. ion})}$ values are smaller than $895\ \text{cm}^{-1}$ (e.g. for $[MnF_6]^{4-}$ $\beta_{35} = 0{\cdot}94$ and for $[Mn(NCS)_6]^{4-}$ $\beta_{35} = 0{\cdot}91$, etc.*

Analogous results hold for all transition metal complex ions. The values of $B_{(\text{compl. ion})}$ obtained from the absorption spectra and for chromium(III) complexes also from phosphorescence spectra[15] on the basis of ligand field theory are always smaller than the values for the corresponding free metal ions.

The interpretation of this effect is (cf. p. 83), that the electron clouds of the $d$ electrons are expanded as a result of covalent interaction and thereby the repulsion between the various $d$ electrons is diminished. The radial distribution of the $d$ electrons is differently influenced by different environments. One can then, on the basis of values for $B_{(\text{compl. ion})}$ obtain a qualitative measure of the electron delocalization within the series of transition metal complexes. The ligands ordered according to decreasing $\beta$ values in a nephelauxetic series (cf. p. 93) are further ordered according to increasing tendency to form covalent bonds with the central ion. This is contrasted

---

* One differentiates between $B_{55}$ and $B_{35}$ and the parameters $\beta_{55}$ and $\beta_{35}$. The $B_{55}$ or $\beta_{55}$ values are taken from transitions where the states which combine with one another arise from the configuration $d\varepsilon^n$ ($t_2^n$ or $\gamma_5^n$), while the $B_{35}$ or $\beta_{35}$ are obtained from transitions arising from $d\varepsilon^n\, d\gamma^m(t_2^n e^m, \gamma_5^n\gamma_3^m)$, (cf. C. K. Jørgensen: *Advances in Chemical Physics*, Vol. 5, pp. 33 ff., Interscience, New York–London, 1963).

15. Cf. G. B. Porter and H. R. Schläfer: *Ber. Bunsenges. phys. Chem.*, **68**, 316 (1964); *Z. phys. Chem.* (Frankfurt), **40**, 280 (1964).

with the spectrochemical series (cf. p. 78), i.e., ordering of the ligands according to increasing values of the field strength parameter *Dq*.

A completely analogous effect is observed for the spin-orbit coupling constant $\zeta_{nd}$, which one can obtain for complexes from paramagnetic resonance measurements. Table A.29 gives several of these data. Not the values of the one-electron coupling constants, $\zeta_{nd}$ but of $\lambda$, the constant which is referred to the ground term, are given. (cf. p. 93). Only in exceptional cases is it possible to draw conclusions concerning the magnitude of the spin–orbit coupling constant from the optical absorption spectra. This is, e.g., for octahedral 5*d* complexes such as $[ReF_6]^{2-}$, $[OsF_6]^{2-}$, $[IrF_6]^{2-}$ and $[PtF_6]^{2-}$ the case, where structure on the spin-forbidden absorption bands is observed resulting from the spin–orbit coupling. It is also true under certain circumstances for tetrahedral $Ni^{2+}$ and $Co^{2+}$ complexes.

Table A.29. Comparison of spin–orbit coupling constants $\lambda$ for free ions with those in crystals

| Metal ion | Compound | $\lambda_{\text{free ion}}$ $\text{cm}^{-1}$ | $\lambda_{\text{compl. ion}}$ $\text{cm}^{-1}$ | $\beta = \frac{\lambda_{\text{compl. ion}}}{\lambda_{\text{free ion}}}$ |
|---|---|---|---|---|
| $V^{3+}$ | Alum† | 104 | 64 | 0·62 |
| $Cr^{3+}$ | in MgO§ | 91 | 63 | 0·69 |
| | | | or 46 | 0·51 |
| | Alum† | | 57 | 0·63 |
| $V^{2+}$ | in MgO§ | 55 | 34 | 0·62 |
| | Tutton-salts* | | 44 | 0·80 |
| $Co^{2+}$ | in MgO§ | −178 | −151 | 0·85 |
| $Ni^{2+}$ | in MgO§ | −324 | −250 | 0·77 |
| | Tutton-salts* | | −270 | 0·83 |
| $Cu^{2+}$ | Tutton-salts* | −829 | −695 | 0·84 |

* Tutton-salts are compounds of the form $M'_2M''(SO_4)_2 \cdot 6H_2O$ with $M' = K^+, Rb^+, NH_4^+$; $M''$ = a bivalent metal ion such as $V^{2+}$, $Ni^{2+}$, $Cr^{2+}$,.... Each unit cell of the monoclinic crystal contains two complexes $M''(H_2O)_6$. Four water molecules are located 1·9 Å from M″ and almost form a square. The other two $H_2O$ molecules are 2·15 Å from M″. The symmetry is therefore almost tetragonal.

† Alums have the formula $M'M'''(SO_4)_2 \cdot 12H_2O$ with $M' = K^+, Na^+, Rb^+, Ca^+, NH_4^+$, and $M'''$ = a trivalent metal such as $Ti^{3+}$, $V^{3+}$, $Cr^{3+}$,.... The crystals have cubic symmetry. Each unit cell contains four complexes $M'''(H_2O)_6$ located so as to give various allotropic modifications. The appearance of the $\alpha$, $\beta$, or $\gamma$ modifications apparently depends upon the varying magnitudes of the monovalent ions. For the $\alpha$ modification the octahedral $M'''(H_2O)_6$ complexes are somewhat deformed along the trigonal crystal axis. All four cubic axes of the octahedron have been rotated about the [111] crystal axis.

§ Bi- or trivalent metal ions imbedded in MgO with approximately octahedral micro-symmetry.

With this we have completed our discussion of the essential methods for the determination of electron delocalization. Indirect evidence of delocalization and thereby of covalent bonding can be obtained also from the deviation of the lattice energies calculated from the simple ionic theory from the actual values. If differences are observed between the nuclear separations determined X-ray crystallographically and those obtained from the ionic radii, this is generally an indication of covalent binding.

Also a decrease of the CO valence vibrational frequency in the metal carbonyls can be taken as evidence of covalence; it is associated with the $\pi$ bond.

In complexes where the formal charge of the metal is zero, the bond must be covalent (carbonyls, acetylene complexes). The same should hold also for complex ions with anomalously low oxidation states of the metal ion, e.g., certain $\alpha, \alpha'$-dipyridyl complexes.

There is also other evidence from absorption spectra that in certain cases orbitals of metal ions overlap with those of the ligands to a certain degree so that the *d* orbitals of the central ions are not pure metal ion orbitals. This is seen from the intensity of the parity forbidden $d \rightarrow d$ bands (cf. p. 84 ff).

When a purely electrostatic ligand field model is assumed as an approximation, the nonzero intensity of these transitions could be understood only on the basis of interaction of the electron motion with vibrations of suitable symmetry species (irreducible representation) or in the absence of a symmetry centre with static hemihedral fields. Both effects make possible a mixing of *d* orbitals with other orbitals of the metal ions, e.g., the *p* orbitals. There are, however, cases in which these two reasons are insufficient to explain the observed intensities. Then one must assume that the intensity results from an overlapping and mixing of the *d* orbitals with various other orbitals of the ligands.

These considerations show that the question of chemical bonding presents an exceedingly complicated problem when one desires more exact information. An unambiguous, quantitative answer cannot yet be given. This is partially because, as we have seen, only comparatively few experimental methods, the application of which is also very limited, can yield information. Also, principal underlying difficulties arise when one attempts to delimit chemical concepts more exactly in a physical sense. This is because chemical concepts originally derived from the realm of experience, e.g., covalence and electrovalence, can only be inadequately translated into the language of physics. The possibilities which exist for obtaining a more exact definition of these concepts for the coordination complexes have been discussed.

The advantage of the ligand field theory, which offers the chemist a useful tool, is that its predictions are based to a large extent upon energy magnitudes. These can be understood on the basis of the variation principle of quantum mechanics. The group-theoretical statements which are essential to the theory are likewise not dependent upon the particular sort of chemical bond being studied.

Therefore, the chemist can on the one hand be satisfied in having a useful theory which embraces a broad group of phenomena. On the other hand he will attempt further to make theoretical contributions to solving the fascinating central problem of bonding and, in so far as is possible, attempt to test them experimentally.

PART B

# 1. The theory of free atoms and ions

The starting point of the ligand field theory is the theory of free atoms or of free atomic ions. This will be sketched briefly in this chapter in as much detail as is necessary for the treatment of the ligand field theory. The theory of spin–orbit coupling of atomic systems was not included here, but a separate section of chapter 5 will be devoted to it.

In order to keep the size of the following chapter within bounds it was necessary to assume that the reader is acquainted with the elementary concepts of quantum theory as well as with such mathematical terms as operator, normalization, orthogonality, hermiticity, eigenfunctions, eigenvalues etc. Readers who are not yet familiar with such concepts are referred to Appendix I of this book as well as to several introductory texts which can be used to obtain the necessary background:

L. Pauling and E. B. Wilson, chapters I, II, III[1]

H. Eyring, J. Walter and G. E. Kimball, chapters I, II, III[2]

W. Kauzmann, *Mathematical Principles*, Part 1, pp. 13–108[3].

The theory of free atoms and ions is treated in numerous books. Of these *The Theory of Atomic Spectra*[4] by E. U. Condon and G. H. Shortley, already a classic text and reference work, assumes an exceptional place. Condon and Shortley is especially recommended to those who wish to study the theory of free atoms beyond the treatment of this book. One finds not only a virtually complete treatment of the theory of free atoms but also a highly developed mathematical formalism. Other references which are also concerned with the theory of free atoms are listed in the bibliography on pp. 498 ff.

---

1. L. Pauling and E. B. Wilson, Jr.: *Introduction to Quantum Mechanics*, McGraw-Hill, New York–London 1935.
2. H. Eyring, J. Walter and G. E. Kimball: *Quantum Chemistry*, 9th Edn, John Wiley, New York–London 1960.
3. W. Kauzmann: *Quantum Chemistry*, Academic Press, New York 1957.
4. E. U. Condon and G. H. Shortley: *The Theory of Atomic Spectra*, University Press, Cambridge, 1935.

## 1.1. One-electron systems

### 1.1.1. *The energy*

Consider the motion of an electron (mass $m$, charge $-e$) in the field of a potential $V$. The size of the potential depends only upon the distance from the centre of the potential source. The field arising from this potential will be called a *central field*. For the description of electron motion one chooses a spherical polar coordinate system* $(r, \theta, \phi)$, the origin of which coincides with the centre of the potential source. If $V = V(r)$, then the *Schrödinger* equation for the motion of the electron is†:

$$\left[-\frac{\hbar^2}{2m}\Delta - eV(r)\right]\psi(r,\theta,\phi) = E\psi(r,\theta,\phi). \tag{1–1}$$

where $\hbar = h/2\pi$, with $h$ = Planck's constant,

$$\Delta \equiv \frac{1}{r^2}\frac{\partial}{\partial r}\left(r^2\frac{\partial}{\partial r}\right) + \frac{1}{r^2 \sin\theta}\frac{\partial}{\partial\theta}\left(\sin\theta\frac{\partial}{\partial\theta}\right) + \frac{1}{r^2\sin^2\theta}\frac{\partial^2}{\partial\phi^2}$$

the *Laplacian* operator,

$\psi(r, \theta, \phi)$ the wave function of the electron;

$r, \theta, \phi$ are the spatial coordinates of the electron.

$E$ is the total energy of the electron.

If one assumes that $\psi(r, \theta, \phi)$ can be separated as follows‡

$$\psi(r,\theta,\phi) = R(r)\cdot Y(\theta,\phi), \tag{1–2}$$

then equation (1–1) can be reduced to two differential equations, one in $R(r)$ and one in $Y(\theta, \phi)$:

$$\frac{1}{\sin\theta}\frac{\partial}{\partial\theta}\left(\sin\theta\frac{\partial Y}{\partial\theta}\right) + \frac{1}{\sin^2\theta}\frac{\partial^2 Y}{\partial\phi^2} + \beta Y = 0, \tag{1–3}$$

---

* The following relations exist between the spherical polar coordinates and the coordinates of the original Cartesian system (cf. Figure B.1)

$$x = r\cdot\sin\theta\cdot\cos\phi$$
$$y = r\cdot\sin\theta\cdot\sin\phi$$
$$z = r\cdot\cos\theta.$$

The positive $z$ axis is defined by $\theta = 0$, the positive $x$ axis by $\theta = \pi/2$, $\phi = 0 \times r$, $\theta$, and $\phi$ have the range of values

$$r: 0 \ldots \infty$$
$$\theta: 0 \ldots \pi$$
$$\phi: 0 \ldots 2\pi$$

The volume element is $d\tau = r^2\, dr \,.\, \sin\theta \,.\, d\theta \,.\, d\phi$.

† Here the influence of the electron spin and relativistic effects are neglected.

‡ The solution, only briefly sketched here, is found in detail, e.g., in L. Pauling and E. B. Wilson, Jr., chapter V[1].

$$\frac{1}{r^2}\frac{d}{dr}\left(r^2\frac{dR}{dr}\right)+\left[-\frac{\beta}{r^2}+\frac{8\pi^2 m}{h^2}(E+eV(r))\right]R=0. \tag{1–4}$$

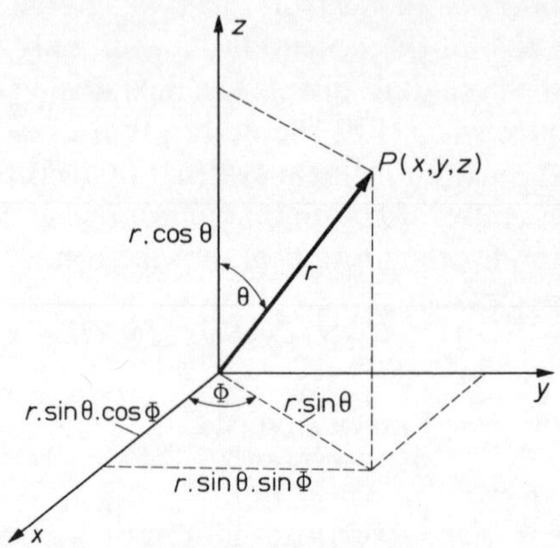


Figure B.1. Cartesian coordinates and spherical polar coordinates.

(a) *The solutions* $Y(\theta, \phi)$. Equation (1–3) has physically meaningful solutions only when the separation constant $\beta$ assumes the values

$$\beta=l(l+1), \quad l=0, 1, 2, 3, \ldots$$

In this case the *spherical harmonics** $Y_{l,m_l}(\theta, \phi)$ are solution of Equation (1–3):

$$Y_{l,m_l}(\theta,\phi)=(-1)^l\sqrt{\frac{2l+1}{4\pi}\cdot\frac{(l+m_l)!}{(l-m_l)!}}\cdot\frac{1}{2^l l!}\cdot\frac{1}{\sin^{m_l}\theta}\times$$
$$\times\frac{d^{l-m_l}}{(d\cos\theta)^{l-m_l}}\sin^{2l}\theta\, e^{i m_l\phi} \tag{1–5}$$
$$l=0, 1, 2, 3, \ldots$$
$$m_l=-l, \ -l+1, \ldots, \ -1, 0, +1, \ldots, l-1, l.$$

The $Y_{l,m_l}(\theta, \phi)$ are orthogonal and normalized to one, i.e.†,

$$\int_0^{\pi}\int_0^{2\pi} Y^*_{l,m_l}(\theta,\phi)\, Y_{l',m_l'}(\theta,\phi)\sin\theta\cdot d\theta\cdot d\phi=\delta_{l,l'}\cdot\delta_{m_l,m_l'}, \tag{1–6}$$

* We have chosen the same 'phase factor' as used by E. U. Condon and G. H. Shortley[4], p. 52. It is equal to $-1$ for odd positive $m_l$ and otherwise always $+1$. *Warning*: The choice of phase factor is not uniform in the literature.

† The integral in (1–6) is formed with the angular dependent part of the volume element $d\tau$: $\sin\theta \,.\, d\theta \,.\, d\phi$.

where $Y^*$ is the complex conjugate of $Y$. $\delta_{ij}$ is the *Kronecker symbol*:

$$\delta_{ij} = \begin{matrix} 1 \\ 0 \end{matrix} \quad \text{for} \quad \begin{matrix} i = j \\ i \neq j. \end{matrix} \tag{1–7}$$

The integral in (1–6) vanishes when $l = l'$ and $m_l = m_l'$ does not hold (*orthogonality*). For the case $l = l'$ and $m_l = m_l'$ the integral has the value 1 (*normalization to one*).

The spherical harmonics $Y_{l,m_l}(\theta, \phi)$ for $l = 0, 1, 2, 3, 4$ are given specifically in Table B.1.

Frequently one writes for $Y_{l,m_l}(\theta, \phi)$ the product

$$Y_{l,m_l}(\theta, \phi) = \Theta_{l,m_l}(\theta) \cdot \Phi_{m_l}(\phi) \tag{1–8}$$

where

$$\Phi_{m_l}(\phi) = \frac{1}{\sqrt{2\pi}} e^{i m_l \phi}. \tag{1–9}$$

The $\theta_{l,m_l}(\theta)$ are called associated Legendre polynomials*:

$$\Theta_{l,m_l}(\theta) = (-1)^l \sqrt{\frac{2l+1}{2} \cdot \frac{(l+m_l)!}{(l-m_l)!}} \cdot \frac{1}{2^l l!} \cdot \frac{1}{\sin^{m_l}\theta} \cdot \frac{d^{l-m_l}}{(d\cos\theta)^{l-m_l}} \sin^{2l}\theta.$$

(b) *The solutions R(r).* We shall restrict our discussion in the following to such potentials $V(r)$ which attract the electron to the potential source and lead to stationary states.

For a given $\beta = l(l+1)$ equation (1–4) has physically meaningful solutions only for certain values of the energy $E$. The eigenfunction $R_l(r)$ and energy eigenvalues $E_l$ belonging to a particular value of $l$ are distinguished by an index $n$: $R_{n,l}(r)$ or $E_{n,l}$. Energy eigenvalues with different pairs of indices $n, l$ have in general different values when the potential is of the form $V(r)$.

An important special case is the *Kepler problem*†. Here $V(r)$ is a Coulomb potential that is produced by a point charge $+Ze$:

$$V(r) = Ze/r. \tag{1–10}$$

---

* See e.g., E. U. Condon and G. H. Shortley[4], pp. 50 ff.

† The atomic *Kepler* problem is occasionally referred to as the problem of a *hydrogen-like atom.* If equation (1–10) is substituted into equation (1–1) one obtains the *Schrödinger* equation for an electron moving in the field of an (infinitely massive) atomic nucleus of charge $Z$.

Table B.1. Spherical harmonics for $l = 0, 1, 2, 3, 4$

| $l$ | $m_l$ | $Y_{l,m_l}(\theta, \phi)$ | $Y_{l,m_l}(x, y, z)$ |
|---|---|---|---|
| 0 | 0 | $\sqrt{\frac{1}{2\pi}}\sqrt{\frac{1}{2}}$ | $\sqrt{\frac{1}{4\pi}}$ |
| 1 | 0 | $\sqrt{\frac{1}{2\pi}}\sqrt{\frac{3}{2}}\cos\theta$ | $\sqrt{\frac{3}{4\pi}}\frac{z}{r}$ |
| | $\pm 1$ | $\mp\sqrt{\frac{1}{2\pi}}\sqrt{\frac{3}{4}}\sin\theta\, e^{\pm i\phi}$ | $\mp\sqrt{\frac{3}{8\pi}}\frac{x \pm iy}{r}$ |
| 2 | 0 | $\sqrt{\frac{1}{2\pi}}\sqrt{\frac{5}{8}}(2\cos^2\theta - \sin^2\theta)$ | $\sqrt{\frac{5}{4\pi}}\sqrt{\frac{1}{4}}\frac{3z^2 - r^2}{r^2}$ |
| | $\pm 1$ | $\mp\sqrt{\frac{1}{2\pi}}\sqrt{\frac{15}{4}}\cos\theta\sin\theta\, e^{\pm i\phi}$ | $\mp\sqrt{\frac{5}{4\pi}}\sqrt{\frac{3}{2}}\frac{z(x \pm iy)}{r^2}$ |
| | $\pm 2$ | $\sqrt{\frac{1}{2\pi}}\sqrt{\frac{15}{16}}\sin^2\theta\, e^{\pm i2\phi}$ | $\sqrt{\frac{5}{4\pi}}\sqrt{\frac{3}{8}}\frac{(x \pm iy)^2}{r^2}$ |

| $l$ | $m_l$ | $Y_{l,m_l}(\theta, \phi)$ | $Y_{l,m_l}(x, y, z)$ |
|---|---|---|---|
| 3 | 0 | $\sqrt{\frac{1}{2\pi}}\sqrt{\frac{7}{8}}(2\cos^3\theta - 3\cos\theta\sin^2\theta)$ | $\sqrt{\frac{7}{4\pi}}\sqrt{\frac{1}{4}}\frac{z(5z^2-3r^2)}{r^3}$ |
| | $\pm 1$ | $\mp\sqrt{\frac{1}{2\pi}}\sqrt{\frac{21}{32}}(4\cos^2\theta\sin\theta - \sin^3\theta)e^{\pm i\phi}$ | $\mp\sqrt{\frac{7}{4\pi}}\sqrt{\frac{3}{16}}(x\pm iy)\frac{(5z^2-r^2)}{r^3}$ |
| | $\pm 2$ | $\sqrt{\frac{1}{2\pi}}\sqrt{\frac{105}{16}}\cos\theta\sin^2\theta\, e^{\pm i2\phi}$ | $\sqrt{\frac{7}{4\pi}}\sqrt{\frac{15}{8}}\frac{z(x\pm iy)^2}{r^3}$ |
| | $\pm 3$ | $\mp\sqrt{\frac{1}{2\pi}}\sqrt{\frac{35}{32}}\sin^3\theta\, e^{\pm i3\phi}$ | $\mp\sqrt{\frac{7}{4\pi}}\sqrt{\frac{5}{16}}\frac{(x\pm iy)^3}{r^3}$ |
| 4 | 0 | $\sqrt{\frac{1}{2\pi}}\sqrt{\frac{9}{128}}(35\cos^4\theta - 30\cos^2\theta + 3)$ | $\sqrt{\frac{9}{4\pi}}\sqrt{\frac{1}{64}}\frac{(35z^4-30z^2r^2+3r^4)}{r^4}$ |
| | $\pm 1$ | $\mp\sqrt{\frac{1}{2\pi}}\sqrt{\frac{45}{32}}\sin\theta\,(7\cos^3\theta - 3\cos\theta)\, e^{\pm i\phi}$ | $\mp\sqrt{\frac{9}{4\pi}}\sqrt{\frac{5}{16}}(x\pm iy)\frac{(7z^3-3zr^2)}{r^4}$ |
| | $\pm 2$ | $\sqrt{\frac{1}{2\pi}}\sqrt{\frac{45}{64}}\sin^2\theta\,(7\cos^2\theta - 1)\, e^{\pm i2\phi}$ | $\sqrt{\frac{9}{4\pi}}\sqrt{\frac{5}{32}}(x\pm iy)^2\frac{(7z^2-r^2)}{r^4}$ |
| | $\pm 3$ | $\mp\sqrt{\frac{1}{2\pi}}\sqrt{\frac{315}{32}}\sin^3\theta\cos\theta\, e^{\pm i3\phi}$ | $\mp\sqrt{\frac{9}{4\pi}}\sqrt{\frac{35}{16}}\frac{z(x\pm iy)^3}{r^4}$ |
| | $\pm 4$ | $\sqrt{\frac{1}{2\pi}}\sqrt{\frac{315}{256}}\sin^4\theta\, e^{\pm i4\phi}$ | $\sqrt{\frac{9}{4\pi}}\sqrt{\frac{35}{128}}\frac{(x\pm iy)^4}{r^4}$ |

The $R(r)$ equation (1–4) has in this case for given $\beta = l(l+1)$ physically meaningful solutions only when $E$ assumes one of the values

$$E_n = -\frac{Z^2}{n^2} W_H, \quad n = l+1,\ l+2,\ l+3, \ldots \tag{1–11}$$

where $W_H = 2\pi^2 me^4/h^2$ is the ionization energy of the hydrogen atom. The associated eigenfunctions are*

$$R_{n,l}(r) = -\left[\left(\frac{2Z}{na_0}\right)^3 \frac{(n-l-1)!}{2n[(n+l)!]^3}\right]^{1/2} e^{-\varrho/2} \cdot L_{n+l}^{2l+1}(\varrho) \cdot \varrho^l \tag{1–12}$$

where the *associated Laguerre* polynomials are given by

$$L_{n+l}^{2l+1}(\varrho) = \sum_{k=0}^{n-l-1} (-1)^{k+1} \frac{[(n+l)!]^2}{(n-l-1-k)!(2l+1+k)!k!} \varrho^k \tag{1–13a}$$

with

$$\varrho = \frac{2Z}{na_0} r, \quad a_0 = h^2/4\pi^2 me^2. \tag{1–13b}$$

The radial functions $R_{n,l}(r)$ are orthogonal and normalized to one†

$$\int_0^\infty R_{n,l}(r) R_{n',l}(r) r^2 dr = \delta_{n,n'}. \tag{1–14}$$

When the potential is of the form (1–10) the eigenfunctions of the *Schrödinger* equation (1–1) are the orthonormalized functions

$$\psi_{n,l,m_l}(r,\theta,\phi) = R_{n,l}(r)\, Y_{l,m_l}(\theta,\phi) \tag{1–15}$$

with the energy eigenvalues:

$$E_n = -\frac{Z^2}{n^2} W_H. \tag{1–11}$$

Every state is thus characterized by three integers: the total or *principal* quantum number $n$, the *orbital* angular momentum (azimuthal) quantum number $l$ and the *magnetic* orbital angular momentum quantum number $m_l$. The quantum numbers have the ranges of values

$$n = 1, 2, 3, \ldots$$
$$l = 0, 1, 2, \ldots, n-1,$$
$$m_l = -l,\ -l+1,\ \ldots,\ +l-1,\ +l.$$

* The explicit forms of the radial functions $R_{n,l}(r)$ for $n = 1$ to 6 can be found in Pauling and Wilson[1], pp. 135 f.

† The integral in equation (1–14) is formed with the $r$ dependent part of the volume element $d\tau$.

For a given $l$ there exist thus $2l+1$ different $m_l$ values, and for a given $n$ there are $n$ different $l$ values. Therefore $\sum_{l=0}^{n-1}(2l+1) = n^2$ different pairs of values $l$, $m_l$ belong to a given principal quantum number $n$.
Example:

$$\begin{aligned} n=3, \quad & l=2, \quad m_l = -2, -1, 0, +1, +2 \\ & l=1, \quad m_l = -1, 0, +1 \\ & l=0, \quad m_l = 0. \end{aligned}$$

Because in the case of the Kepler problem the energy $E_n$ of a state according to equation (1–11) depends solely upon the principal quantum number $n$, there are $n^2$ states of the same energy which differ from one another by the quantum numbers $l$ and $m_l$. One says the $n^2$ states of the same energy are $n^2$-fold (energetically) degenerate.

The totality of states belonging to a given principal quantum number form a *shell* (or principal shell). Therefore the K, L, M, N... shells correspond to $n = 1, 2, 3, 4, \ldots$. The term system of the states for $n = 1, 2, 3, 4$ is schematically given in Figure B.2a.

As already mentioned the $l$ degeneracy is removed in going from a Coulomb to a non-Coulomb potential $V(r)$; states with the same principal quantum number $n$ but different orbital angular momentum quantum number $l$ have different energies. The $m_l$ degeneracy, however, remains (cf. Figure B.2b).

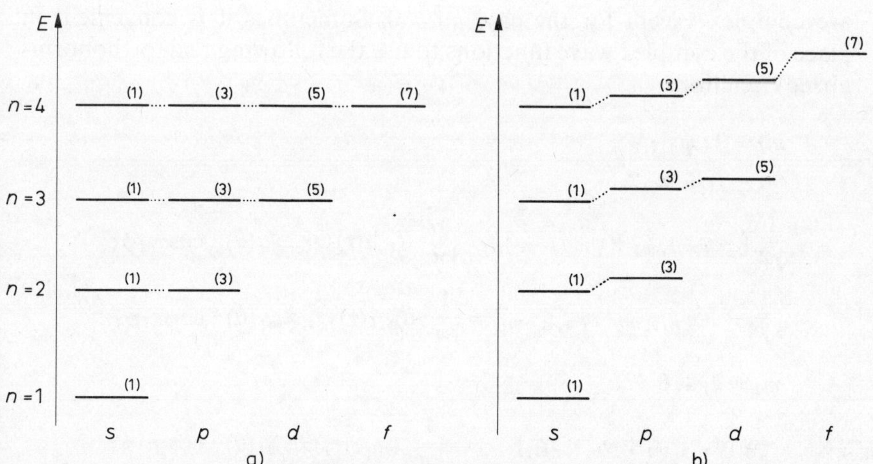


Figure B.2. Term scheme of an atomic one-electron system (a) for a Coulomb potential; (b) for a non-Coulomb potential. The numbers in brackets give the degree of $m_l$ degeneracy, $2l+1$, for the terms shown.

*One-electron states* with $l = 0, 1, 2, 3, 4, \ldots$ are denoted with lower case Latin letters $s, p, d, f, g, \ldots$. An electron in an $l = 0, 1, 2, 3, 4, \ldots$ state is called an $s, p, d, f, g, \ldots$ electron. If besides the orbital angular momentum quantum number one wishes to give the principal quantum number of an electron state as well, the value of the principal quantum number is written before the character symbol for $l$. For example '3*d* electron' indicates that an electron is in a state with $n = 3$ (M shell) and $l = 2$.

The splitting into states of different $l$ values for the case of non-Coulomb potentials is characterized by saying that the shells defined above split into subshells. These subshells are named after the corresponding character symbols for the electrons:

K shell → 1s subshell

L shell → 2s subshell
2p subshell

M shell → 3s subshell
3p subshell
3d subshell

etc.

The wave functions (1-15)

$$\psi_{n,l,m_l}(r,\theta,\phi) = R_{n,l}(r)\, Y_{l,m_l}(\theta,\phi) = R_{n,l}(r)\, \Theta_{l,m_l}(\theta) \cdot \frac{1}{\sqrt{2\pi}} e^{i m_l \phi}$$

are complex except for the case $m_l = 0$. Sometimes it is convenient in place of the complex wave functions to use the following real (orthonormalized) functions

$$m_l = 0 : \psi_{n,l,0},$$

$$m_l = 1, 3, 5, \ldots :$$

$$\begin{aligned} \frac{1}{\sqrt{2}}[-\psi_{n,l,m_l} + \psi_{n,l,-m_l}] &= \frac{1}{\sqrt{\pi}} R_{n,l}(r)\,\Theta_{l,-m_l}(\theta) \cdot \cos m_l\phi, \\ \frac{1}{i\sqrt{2}}[-\psi_{n,l,m_l} - \psi_{n,l,-m_l}] &= \frac{1}{\sqrt{\pi}} R_{n,l}(r)\,\Theta_{l,-m_l}(\theta) \cdot \sin m_l\phi, \end{aligned} \tag{1–16}$$

$$m_l = 2, 4, 6, \ldots :$$

$$\begin{aligned} \frac{1}{\sqrt{2}}[\psi_{n,l,m_l} + \psi_{n,l,-m_l}] &= \frac{1}{\sqrt{\pi}} R_{n,l}(r)\,\Theta_{l,m_l}(\theta) \cdot \cos m_l\phi, \\ \frac{1}{i\sqrt{2}}[\psi_{n,l,m_l} - \psi_{n,l,-m_l}] &= \frac{1}{\sqrt{\pi}} R_{n,l}(r)\,\Theta_{l,m_l}(\theta) \cdot \sin m_l\phi. \end{aligned}$$

These real functions are all (with the exception of $m_l = 0$) linear combinations of the form

$$\psi_{n,l,m_l} \pm \psi_{n,l,-m_l}. \tag{1–17}$$

Because $\psi_{n,l,m_l}$ and $\psi_{n,l,-m_l}$ are eigenfunctions of the same energy eigenvalue $E_{n,l}$ for spherically symmetric potentials, their linear combinations, equation (1–17), must also be eigenfunctions of the same energy eigenvalue $E_{n,l}$*.

In Table B.2 the real orthonormalized linear combinations of the spherical harmonics are summarized for $l = 0, 1, 2$. The conventional abbreviations for these real linear combinations are given in the last column of the table.

In Figure B.3 schematic representations of the real functions $\psi(1s)$; $\psi(2p_x)$, $\psi(2p_y)$, $\psi(2p_z)$, $\psi(3d_{z^2})$, $\psi(3d_{x^2-y^2})$, $\psi(3d_{xy})$, $\psi(3d_{xz})$, $\psi(3d_{yz})$ are given. The solid lines indicate planes of constant absolute value of these functions.

### 1.1.2. *The orbital angular momentum*†

A particle moving with the translational momentum $\vec{p}$ has in classical mechanics with reference to the origin 0, the angular momentum

$$\vec{l} = \vec{r} \times \vec{p}, \tag{1–18}$$

where $\vec{r}$ is the vector from the origin 0 to the particle (cf. Figure B.4). The

---

* One can easily convince himself of this by substituting $\Psi_{n,l,m_l} \pm \Psi_{n,l,-m_l}$ into the *Schrödinger* equation (1–1) and observing that because of

$$\left[-\frac{\hbar^2}{2m}\Delta - eV(r)\right]\psi_{n,l,m_l} = E_{n,l}\,\psi_{n,l,m_l}$$

and

$$\left[-\frac{\hbar^2}{2m}\Delta - eV(r)\right]\psi_{n,l,-m_l} = E_{n,l}\,\psi_{n,l,-m_l}$$

it holds that:

$$\begin{aligned}&\left[-\frac{\hbar^2}{2m}\Delta - eV(r)\right](\psi_{n,l,m_l} \pm \psi_{n,l,-m_l})\\ &= \left[-\frac{\hbar^2}{2m}\Delta - eV(r)\right]\psi_{n,l,m_l} \pm \left[-\frac{\hbar^2}{2m}\Delta - eV(r)\right]\psi_{n,l,-m_l}\\ &= E_{n,l}\,\psi_{n,l,m_l} \pm E_{n,l}\,\psi_{n,l,-m_l} = E_{n,l}(\psi_{n,l,m_l} \pm \psi_{n,l,-m_l}).\end{aligned}$$

(Cf. also Appendix, pp. 469 ff.)

† The theory of angular momentum is treated in detail by E. U. Condon and G. H. Shortley[4], chapter 3.

Table B.2. Real orthonormalized linear combinations of the spherical harmonics $Y_{l,m_l}(\theta, \phi)$ for $l = 0, 1, 2$

| $l$ | Linear combination | Symbol |
|---|---|---|
| 0 | $\frac{1}{\sqrt{2\pi}}\sqrt{\frac{1}{2}}$ | $s$ |
| 1 | $\frac{1}{\sqrt{2\pi}}\sqrt{\frac{3}{2}}\cos\theta = \frac{1}{\sqrt{2\pi}}\sqrt{\frac{3}{2}}\cdot\frac{1}{r}\cdot z$ | $p_z$ |
| | $\frac{1}{\sqrt{\pi}}\sqrt{\frac{3}{4}}\sin\theta\cos\phi = \frac{1}{\sqrt{\pi}}\sqrt{\frac{3}{4}}\cdot\frac{1}{r}\cdot x$ | $p_x$ |
| | $\frac{1}{\sqrt{\pi}}\sqrt{\frac{3}{4}}\sin\theta\sin\phi = \frac{1}{\sqrt{\pi}}\sqrt{\frac{3}{4}}\cdot\frac{1}{r}\cdot y$ | $p_y$ |
| 2 | $\frac{1}{\sqrt{2\pi}}\sqrt{\frac{5}{8}}(2\cos^2\theta - \sin^2\theta) = \frac{1}{\sqrt{2\pi}}\sqrt{\frac{5}{8}}\frac{1}{r^2}(3z^2 - r^2)$ | $d_{z^2}$ |
| | $\frac{1}{\sqrt{\pi}}\sqrt{\frac{15}{4}}\cos\theta\sin\theta\cos\phi = \frac{1}{\sqrt{\pi}}\sqrt{\frac{15}{4}}\frac{1}{r^2}\cdot xz$ | $d_{xz}$ |
| | $\frac{1}{\sqrt{\pi}}\sqrt{\frac{15}{4}}\cos\theta\sin\theta\sin\phi = \frac{1}{\sqrt{\pi}}\sqrt{\frac{15}{4}}\cdot\frac{1}{r^2}\cdot yz$ | $d_{yz}$ |
| | $\frac{1}{\sqrt{\pi}}\sqrt{\frac{15}{16}}\sin^2\theta\cos 2\phi = \frac{1}{\sqrt{\pi}}\sqrt{\frac{15}{16}}\frac{1}{r^2}(x^2 - y^2)$ | $d_{x^2-y^2}$ |
| | $\frac{1}{\sqrt{\pi}}\sqrt{\frac{15}{16}}\sin^2\theta\sin 2\phi = \frac{1}{\sqrt{\pi}}\sqrt{\frac{15}{4}}\frac{1}{r^2}\cdot xy$ | $d_{xy}$ |

classical angular momentum has as its quantum mechanical analogue the operator

$$\vec{\boldsymbol{l}} = \vec{i}\,\boldsymbol{l}_x + \vec{j}\,\boldsymbol{l}_y + \vec{k}\,\boldsymbol{l}_z \tag{1–19}$$

$\vec{i}, \vec{j}, \vec{k}$ are unit vectors in a cartesian coordinate system. The square of the

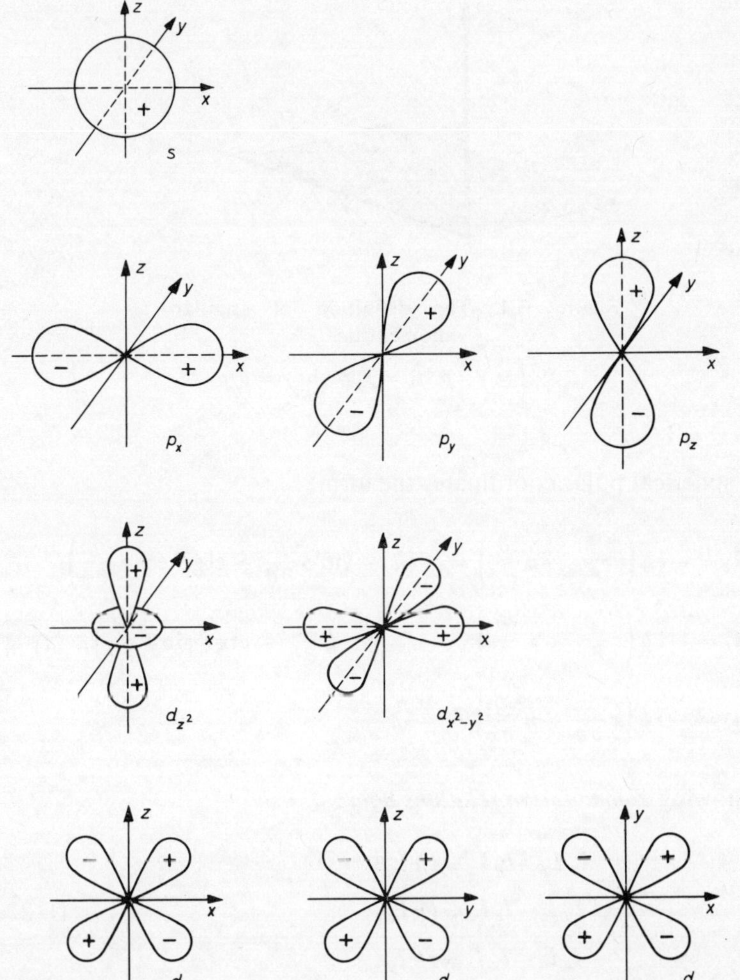


Figure B.3. Angular dependent portions of the wave functions for the real linear combinations given by equation (1–16).

classical angular momentum $|\vec{l}|^2$ corresponds to the operator

$$\boldsymbol{l}^2 = \boldsymbol{l}_x^2 + \boldsymbol{l}_y^2 + \boldsymbol{l}_z^2. \tag{1–20}$$

The components $\boldsymbol{l}_x$, $\boldsymbol{l}_y$, $\boldsymbol{l}_z$ of the operator $\vec{\boldsymbol{l}}$ have in cartesian coordinates

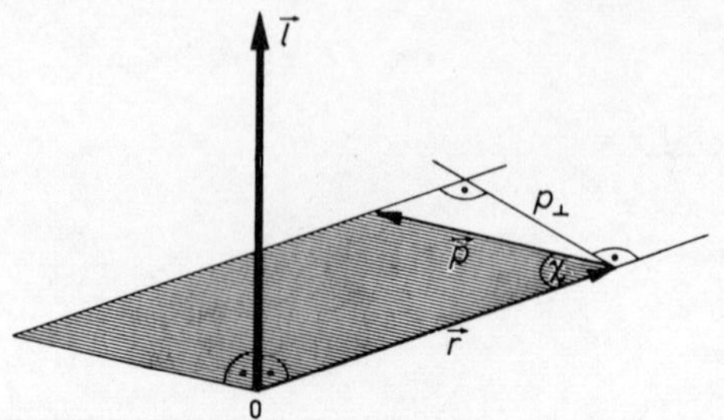


Figure B.4. The definition of angular momentum

$$\vec{l} = \vec{r} \times \vec{p}, |\vec{l}| = |\vec{r}||\vec{p}| \sin \chi = |\vec{r}| p_\perp .$$

and in spherical polar coordinates the form

$$\begin{aligned} \boldsymbol{l}_x &= -i\hbar\left(y\frac{\partial}{\partial z} - z\frac{\partial}{\partial y}\right) = -i\hbar\left(-\sin\phi\frac{\partial}{\partial\theta} - \operatorname{ctg}\theta\cos\phi\frac{\partial}{\partial\phi}\right), \\ \boldsymbol{l}_y &= -i\hbar\left(z\frac{\partial}{\partial x} - x\frac{\partial}{\partial z}\right) = -i\hbar\left(\cos\phi\frac{\partial}{\partial\theta} - \operatorname{ctg}\theta\sin\phi\frac{\partial}{\partial\phi}\right), \\ \boldsymbol{l}_z &= -i\hbar\left(x\frac{\partial}{\partial y} - y\frac{\partial}{\partial x}\right) = -i\hbar\frac{\partial}{\partial\phi}. \end{aligned} \qquad (1\text{–}21)$$

The following *commutation relations hold*:

$$\begin{aligned} \boldsymbol{l}_x\boldsymbol{l}_y - \boldsymbol{l}_y\boldsymbol{l}_x &= i\hbar\,\boldsymbol{l}_z, \\ \boldsymbol{l}_y\boldsymbol{l}_z - \boldsymbol{l}_z\boldsymbol{l}_y &= i\hbar\,\boldsymbol{l}_x, \\ \boldsymbol{l}_z\boldsymbol{l}_x - \boldsymbol{l}_x\boldsymbol{l}_z &= i\hbar\,\boldsymbol{l}_y, \end{aligned} \qquad (1\text{–}22\text{a})$$

and

$$\boldsymbol{l}^2\boldsymbol{l}_k - \boldsymbol{l}_k\boldsymbol{l}^2 = 0, \qquad k = x,\ y,\ z. \qquad (1\text{–}22\text{b})$$

The spherical harmonics $Y_{l,m_l}$ are eigenfunctions of the operators $\boldsymbol{l}^2$ and $\boldsymbol{l}_z$:

$$\boldsymbol{l}^2\,Y_{l,m_l}(\theta,\phi) = l(l+1)\hbar^2\,Y_{l,m_l}(\theta,\phi), \quad l = 0, 1, 2, \ldots \qquad (1\text{–}23)$$

$$\boldsymbol{l}_z\,Y_{l,m_l}(\theta,\phi) = m_l\hbar\,Y_{l,m_l}(\theta,\phi), \qquad m_l = -l, -l+1, \ldots, l-1, l$$

For example*:

$$\boldsymbol{l}^2\, Y_{2,1}(\theta,\phi) = 2(2+1)\,\hbar^2\, Y_{2,1}(\theta,\phi),$$
$$\boldsymbol{l}_z\, Y_{2,1}(\theta,\phi) = 1\cdot\hbar\, Y_{2,1}(\theta,\phi).$$

However, the $Y_{l,m_l}$ are *not* eigenfunctions of the operators $\boldsymbol{l}_x$ and $\boldsymbol{l}_y$.

Because the wave functions $\psi_{n,l,m_l}$ are products of the radial functions, $R_{n,l}(r)$ and the spherical harmonics $Y_{l,m_l}(\theta, \phi)$ but the operators $\boldsymbol{l}^2, \boldsymbol{l}_x, \boldsymbol{l}_y, \boldsymbol{l}_z$ operate on only the angular parts $Y_{l,m_l}$ it follows that

$$\boldsymbol{l}^2\,\psi_{n,l,m_l} = l(l+1)\,\hbar^2\,\psi_{n,l,m_l}, \quad l = 0, 1, 2, \ldots \tag{1–25}$$

$$\boldsymbol{l}_z\,\psi_{n,l,m_l} = m_l\,\hbar\,\psi_{n,l,m_l}, \quad m_l = -l, -l+1, \ldots, -1, 0, 1, \ldots, l-1, l. \tag{1–26}$$

As is the case for $Y_{l,m_l}$, the $\psi_{n,l,m_l}$ are *not* eigenfunctions of the component operators $\boldsymbol{l}_x$ and $\boldsymbol{l}_y$.

On the basis of the eigenvalue equations (1–25) and (1–26) we can give the orbital angular momentum quantum number $l$ and the magnetic orbital angular momentum quantum number $m_l$ a physical interpretation: An electron described by the wave function $\psi_{n,l,m_l}$ has an orbital angular momentum with the absolute value according to equation (1–25) of $\sqrt{l(l+1)}\hbar$ and the $z$ component whose value according to equation (1–26) is $m_l\hbar$. From the fact that $\psi_{n,l,m_l}$ is not an eigenfunction of $\boldsymbol{l}_x$ and $\boldsymbol{l}_y$ we conclude that the $x$ and the $y$ components of the orbital angular momentum do not have sharply defined values.

The appearance of energetic degeneracies of the states discussed in section 1.1.1 can be reformulated within the theory of the orbital angular momentum properties of the electron. For the case that the electron moves in a non-Coulomb potential $V(r)$ we find: Electrons in states having the same principal quantum number but different angular momenta have different energies. The orientation of the orbital angular momentum with reference to the $z$ axis of the coordinate system is irrelevant for the energy.

With the aid of the commutation relations (1–22) and the eigenvalue equations (1–23, 24) the following important relations for the so-called *shifting* operators (*raising* and *lowering* operators) $\boldsymbol{l}_+ \equiv \boldsymbol{l}_x + i\boldsymbol{l}_y$ and

---

* For example explicitly because $\boldsymbol{l}_z = i\hbar(\partial/\partial\phi)$ and

$$Y_{2,1} = -\frac{1}{\sqrt{2\pi}}\sqrt{\frac{15}{4}}\cos\theta\,\sin\theta\cdot e^{i\phi},\ \boldsymbol{l}_z\,Y_{2,1} = -i\hbar\left(-\frac{1}{\sqrt{2\pi}}\right)\sqrt{\frac{15}{4}}\cos\theta\,\sin\theta\frac{\partial}{\partial\phi}e^{i\phi}$$
$$= -i\hbar\left(-\frac{1}{\sqrt{2\pi}}\right)\sqrt{\frac{15}{4}}\cos\theta\,\sin\theta\cdot i\cdot e^{i\phi} = \hbar\left(-\frac{1}{\sqrt{2\pi}}\right)\sqrt{\frac{15}{4}}\cos\theta\,\sin\theta\, e^{i\phi} = \hbar\,Y_{2,1}.$$

$l_- \equiv l_x - il_y$ can be derived:

$$(l_x + il_y)\, Y_{l,m_l} = \hbar \sqrt{(l+m_l+1)(l-m_l)}\; Y_{l,m_l+1},$$
$$(l_x - il_y)\, Y_{l,m_l} = \hbar \sqrt{(l+m_l)(l-m_l+1)}\; Y_{l,m_l-1}. \tag{1–27}$$

If a function $Y_{l,m_l}$ is known, all other $Y_{l,m_l'}$ with $m_l' \neq m_l$ can be calculated by the (perhaps repeated) application of the shifting operators.

It must be emphasized that although the spherical harmonics $Y_{l,m_l}$ are eigenfunctions of the operators $l^2$ and $l_z$, the real linear combinations of $Y_{l,m_l}$ for $m_l \neq 0$ given in equation (1–16) are eigenfunctions only of the operator $l^2$ and not of $l_z$ (and $l_x$ and $l_y$).

### 1.1.3. *The electron spin*

Besides its mass and charge the electron also has an intrinsic angular momentum, called the spin, and a magnetic moment proportional to the spin. According to experiment the electron spin has the value $\sqrt{\frac{1}{2}(\frac{1}{2}+1)}\hbar$; its component in a chosen direction has the value either $+\frac{1}{2}\hbar$ or $-\frac{1}{2}\hbar$. (We shall select for this purpose the direction of the $z$ axis in a cartesian coordinate system.) In order to completely describe the state of an electron it is therefore necessary to use a wave function which besides the spatial coordinates $r, \theta, \phi$ also contains a fourth coordinate, the 'spin coordinate' $\sigma$, having the possible values $+\frac{1}{2}$ and $-\frac{1}{2}$. The complete wave function of the electron may then be written as $\psi(r, \theta, \phi; \sigma)$.

If no forces are operative on the magnetic moment proportional to the electron spin, then the wave function $\psi(r, \theta, \phi; \sigma)$ may be written as a product:

$$\psi(r, \theta, \phi; \sigma) = \psi(r, \theta, \phi) \cdot \chi(\sigma), \tag{1–28}$$

where $\psi(r, \theta, \phi)$ is the spatial part of the wave function which has already been introduced and $\chi(\sigma)$ is the spin function.

Similarly as for the orbital angular momentum an operator $s$ may be associated with the spin whose cartesian components $s_x, s_y, s_z$ and whose square $s^2$ obey the following commutation relations:

$$s_x s_y - s_y s_x = i\hbar s_z \qquad \text{[cyclical in } x, y, z] \tag{1–29}$$

and

$$s^2 s_k - s_k s^2 = 0 \quad k = x, y, z \tag{1–30}$$

For the operators $s_z$ and $s^2$ two linearly independent eigenfunctions, $\alpha(\sigma)$

and $\beta(\sigma)$, may be given:

$$\begin{aligned} s_z\,\alpha(\sigma) &= +\frac{1}{2}\hbar\,\alpha(\sigma), \\ s_z\,\beta(\sigma) &= -\frac{1}{2}\hbar\,\beta(\sigma), \end{aligned} \tag{1–31}$$

$$\begin{aligned} s^2\,\alpha(\sigma) &= \frac{1}{2}\left(\frac{1}{2}+1\right)\hbar^2\alpha(\sigma), \\ s^2\,\beta(\sigma) &= \frac{1}{2}\left(\frac{1}{2}+1\right)\hbar^2\beta(\sigma). \end{aligned} \tag{1–32}$$

It follows from the theory*, that $\alpha(\sigma)$ and $\beta(\sigma)$ for $\sigma = +\frac{1}{2}$ and $\sigma = -\frac{1}{2}$ have respectively the values

$$\begin{aligned} \alpha(1/2) &= 1, \quad \beta(1/2) = 0, \\ \alpha(-1/2) &= 0, \quad \beta(-1/2) = 1 \end{aligned} \tag{1–33}$$

The spin eigenfunctions are orthonormalized†:

$$\begin{aligned} &\sum_{\sigma=-\frac{1}{2},\frac{1}{2}} \alpha^*(\sigma)\,\alpha(\sigma) = 1, \\ &\sum_{\sigma=-\frac{1}{2},\frac{1}{2}} \beta^*(\sigma)\,\beta(\sigma) = 1, \\ &\sum_{\sigma=-\frac{1}{2},\frac{1}{2}} \alpha^*(\sigma)\,\beta(\sigma) = 0. \end{aligned} \tag{1–34}$$

For the so-called *spin shifting* operators $s_+ \equiv s_x + is_y$ and $s_- \equiv s_x - is_y$ the following relations hold:

$$\begin{aligned} (s_x - is_y)\alpha &= \hbar\beta \\ (s_x + is_y)\alpha &= 0 \\ (s_x - is_y)\beta &= 0 \\ (s_x + is_y)\beta &= \hbar\alpha. \end{aligned} \tag{1–35}$$

The *Schrödinger* equation (1–1) given on p. 207 describes a system in which the forces operative on the spin magnetic moment of the electron have been neglected. The Hamiltonian in (1–1) does not operate on the spin coordinate. Therefore $\psi_{n,l,m_l}\cdot\alpha$ and $\psi_{n,l,m_l}\cdot\beta$ are solutions of (1–1) having the same energy eigenvalue: The orientation of the spin with reference to the $z$ axis of the coordinate system has no effect on the energy.

---

* Cf., e.g., H. Hartmann[5], pp. 73 ff.

† Because the range of values of the spin variables $\sigma$ contains only the two values $+\frac{1}{2}$ and $-\frac{1}{2}$, we substitute for integration summation over the two values $\sigma = +\frac{1}{2}$ and $\sigma = -\frac{1}{2}$.

5. H. Hartmann: *Theorie der chemischen Bindung*, Springer-Verlag, Berlin–Göttingen–Heidelberg, 1954.

It is conventional to use in addition to the quantum numbers $n$, $l$, $m_l$ the eigenvalue of $s_z$ as a fourth index for the complete wave function: $\psi_{n,l,m_l,m_s}$.

$$\begin{aligned} \psi_{n,l,m_l} \cdot \alpha &= \psi_{n,l,m_l,+\frac{1}{2}}, \\ \psi_{n,l,m_l} \cdot \beta &= \psi_{n,l,m_l,-\frac{1}{2}}. \end{aligned} \tag{1–36}$$

$m_s$ is called the *magnetic spin quantum number*.

## 1.2. Many-electron systems

### 1.2.1. *The Schrödinger equation*

Consider a system consisting of an atomic nucleus with the charge $+Ze$ and $N$ electrons (mass $m$, charge $-e$). If the nucleus is held fixed, the *Schrödinger* equation of the system* is given by:

$$\left[-\frac{\hbar^2}{2m}\sum_{i=1}^{N}\Delta_i - \sum_{i=1}^{N}\frac{Ze^2}{r_i} + \frac{1}{2}\sum_{\substack{i=1 \\ i\neq j}}^{N}\sum_{j=1}^{N}\frac{e^2}{r_{ij}}\right]\Psi = W\Psi. \tag{1–37}$$

The following definitions hold:

$\Delta_i$ — the Laplacian which operates on the spatial coordinates $(r_i, \theta_i, \phi_i)$ of the $i$th electron,

$-Ze^2/r_i$ — the Coulomb energy of the interaction between the atomic nucleus and the $i$th electron; $r_i$ is the distance between the $i$th electron and the atomic nucleus,

$e^2/r_{ij}$ — the Coulomb energy of the interaction between the $i$th and the $j$th electron,

$\Psi$ — the wave function of the $N$ electrons $\Psi$ is dependent upon the spatial and the spin-coordinates of all $N$ electrons:

$$\Psi = \Psi(\vec{r}_1, \sigma_1; \vec{r}_2, \sigma_2; \ldots; \vec{r}_N, \sigma_N).$$

$\vec{r}_i$ is an abbreviation for $r_i, \theta_i, \phi_i$,

$W$ — the total energy of the $N$ electrons.

The *Schrödinger* equation (1–37) cannot be solved exactly for $N > 1$. If one wishes to study the wave functions $\Psi$ and the energies $W$ approximation methods must be used in attempting to solve equation (1–37). We shall use the so-called *central-field approximation* in conjunction with a

---

* The effect of the electron spin as well as relativistic effects have been neglected.

perturbation calculation. This procedure is known in the literature as the *Slater Theory of atoms and ions**.

### 1.2.2. *The central-field approximation*†

We shall use an artifice to arrive at an approximate solution of the *Schrödinger* equation (1–37). The quantity $+\sum_{i=1}^{N} eV(r_i) - \sum_{i=1}^{N} eV(r_i)$ is added to the *Hamiltonian*

$$\boldsymbol{H} = -\frac{\hbar^2}{2m}\sum_{i=1}^{N}\Delta_i - \sum_{i=1}^{N}\frac{Ze^2}{r_i} + \frac{1}{2}\sum_{\substack{i=1 \\ i\neq j}}^{N}\sum_{j=1}^{N}\frac{e^2}{r_{ij}} + \sum_{i=1}^{N} e\,V(r_i) - \sum_{i=1}^{N} e\,V(r_i) \qquad (1\text{–}38)$$

and $\boldsymbol{H}$ is subsequently divided into two parts $\boldsymbol{H}_0$ and $\boldsymbol{H}_1$ :

$$\boldsymbol{H} = \boldsymbol{H}_0 + \boldsymbol{H}_1 \qquad (1\text{–}39)$$

with

$$\boldsymbol{H}_0 = -\frac{\hbar^2}{2m}\sum_{i=1}^{N}\Delta_i - \sum_{i=1}^{N} e\,V(r_i) \qquad (1\text{–}40)$$

and

$$\boldsymbol{H}_1 = -\sum_{i=1}^{N}\left[\frac{Ze^2}{r_i} - e\,V(r_i)\right] + \frac{1}{2}\sum_{\substack{i=1 \\ i\neq j}}^{N}\sum_{j=1}^{N}\frac{e^2}{r_{ij}}. \qquad (1\text{–}41)$$

If for $V(r)$ a function is chosen such that $\boldsymbol{H}_1$ (everywhere in coordinate space) is as small as possible, then $\boldsymbol{H}_1$ may be treated as a perturbation operator of an unperturbed system that is described by the *Schrödinger* equation

$$\boldsymbol{H}_0\,\Psi_0 = E\,\Psi_0 \qquad (1\text{–}42)$$

---

* The procedure mentioned here is treated in detail by E. U. Condon and G. H. Shortley[4], chapter 6. We shall limit ourselves to a presentation of those parts of the procedure which are essential for the ligand field theory.

† In the German text the expression 'Abschirmfeldnäherung', shielding-field approximation, is used as a specialization of the more general term 'Zentralfeldnäherung', central-field approximation, used by Condon and Shortley[4] and others.

If $\boldsymbol{H}_{\mathrm{i}}$ is neglected in comparison to $\boldsymbol{H}_0$, one speaks of the so-called central-field approximation. $\sum_{i=1}^{N} V(r_i)$ is called the *shielding potential**.

The differential equation (1–42) can be solved with a product of one-electron functions $\psi^{(i)}(\vec{r}_i, \sigma_i)$. One finds that

$$\Psi_0(\vec{r}_1, \sigma_1; \vec{r}_2, \sigma_2; \ldots; \vec{r}_N, \sigma_N) = \psi^{(1)}(\vec{r}_1, \sigma_1) \cdot \psi^{(2)}(\vec{r}_2, \sigma_2) \ldots \psi^{(N)}(\vec{r}_N, \sigma_N) \tag{1–43}$$

is an eigenfunction of (1–42) having the eigenvalue

$$E = \varepsilon^{(1)} + \varepsilon^{(2)} + \cdots + \varepsilon^{(N)} \tag{1–44}$$

The $\psi^{(i)}(\vec{r}, \sigma)$ and $\varepsilon^{(i)}$ are respectively eigenfunctions and eigenvalues of the *one-electron equation*

$$\left[-\frac{\hbar^2}{2m}\Delta - eV(r)\right]\psi(\vec{r}, \sigma) = \varepsilon\psi(\vec{r}, \sigma). \tag{1–45}$$

This is the *Schrödinger* equation of an electron moving under the influence of a shielding potential $V(r)$. The solutions of this equation have already been given in section 1.1†:

$$\psi_{n_i, l_i, m_{l_i}, m_{s_i}}(\vec{r}, \sigma) = R_{n_i, l_i}(r)\, Y_{l_i, m_{l_i}}(\theta, \phi) \begin{Bmatrix} \alpha(\sigma) \\ \beta(\sigma) \end{Bmatrix}, \tag{1–46}$$

with the eigenvalues $\varepsilon_{n_i, l_i}$. With this and considering equations (1–43, 44) the eigenfunctions and eigenvalues of equation (1–42) are known. If we make use of the abbreviations $a_i$ for $n_i, l_i, m_{l_i}, m_{s_i}$ and $(i)$ for $(\vec{r}_i, \sigma_i)$, then they become

$$\Psi_0 = \psi_{a_1}(1) \cdot \psi_{a_2}(2) \ldots \psi_{a_N}(N), \tag{1–47}$$

or

$$E = \varepsilon_{n_1, l_1} + \varepsilon_{n_2, l_2} + \cdots + \varepsilon_{n_N, l_N}. \tag{1–48}$$

$\Psi_0$ in equation (1–47) is not the only eigenfunction of (1–42) having the eigenvalue $E = \varepsilon_{n_1, l_1} + \ldots + \varepsilon_{n_N, l_N}$. For example the function

$$\psi_{a_1}(1) \cdot \psi_{a_2}(2) \ldots \psi_{a_r}(s)\psi_{a_s}(r) \ldots \psi_{a_N}(N) \tag{1–49}$$

* This notation becomes clear when one uses the special form $-\sum_i Z_{\text{eff.}} e^2/r_i$ for $-\sum_i eV(r_i)$, that is, in the transition from $\boldsymbol{H}$ to $\boldsymbol{H}_0$ the substitution $-\sum_i Ze^2/r_i + \frac{1}{2}\sum_{i \neq j} e^2/r_{ij} \to -\sum_i Z_{\text{eff.}} e^2/r_i$ is used. Such a substitution is in general optimal when $Z_{\text{eff.}} \leqq Z$ is chosen, that is, when the electron–electron interaction is formally approximately compensated for by a diminution of the nuclear charge: The electrons 'shield' the nucleus.

† Equation (1–45) is identical with the *Schrödinger* equation (1–1). Accordingly the solutions (1–15, 36) derived in section 1.1. from (1–1) are also solutions of (1–45).

belongs to the same eigenvalue. This function is obtained by exchanging the coordinates of the $r$th and the $s$th electron from equation (1–47). There are in total $N!$ different *product functions* which can be obtained from equation (1–47) by performing all possible permutations* of the electron coordinates. All have the same eigenvalue (1–48). Likewise every linear combination of these $N!$ functions is an eigenfunction of (1–42) with the eigenvalue (1–48). From experience and from the general theory it can be seen that only linear combinations of the form

$$\Phi = \frac{1}{\sqrt{N!}} \sum_P (-1)^P P \Psi_0 \tag{1–50}$$

are physically meaningful. $1/\sqrt{N!}$ is a normalization factor. The summation in equation (1–50) is taken over all permutations $P$ of the electron coordinates. The factor $(-1)^P$ has the result that all functions which arise from $\Psi_0$ by even permutations of the electron coordinates have the coefficient $+1$ in the linear combination (1–50). Functions which arise from an odd permutation of the electron coordinates have the coefficient $-1$ in (1–50). If the coordinates of two electrons are exchanged in $\Phi$, then the sign of $\Phi$ is also changed. One says that $\Phi$ is an *antimetric* function (with reference to exchange of the coordinates). $\Phi$ is often called the *Heisenberg–Slater function* (HS function). It is conventional to use the abbreviated form for $\Phi$

$$\Phi(a_1; a_2; \ldots; a_N) \tag{1–51}$$

In

$$\Phi(n_1, l_1, m_{l_1}, m_{s_1};\ n_2, l_2, m_{l_2}, m_{s_2};\ \ldots;\ n_N, l_N, m_{l_N}, m_{s_N}). \tag{1–52}$$

the set of quantum numbers $n_i, l_i, m_{l_i}, m_{j_i}$ abbreviated in equation (1–51) by $a_i$ is given explicitly. If all one-electron states have the same quantum number $n$ and the same angular momentum quantum number $l$, that is, belong to the same subshell, then the $n_i$ and $l_i$ in (1–52) are usually omitted:

$$\Phi(m_{l_1}, m_{s_1};\ m_{l_2}, m_{s_2};\ \ldots;\ m_{l_N}, m_{s_N}). \tag{1–53}$$

Also instead of $m_s = +\frac{1}{2}$ we shall write $+$ and $-$ as superscripts of the appropriate $m_l$, e.g. $\Phi(\overset{+}{m_{l_1}}; \overset{-}{m_{l_2}}; \ldots; \overset{+}{m_{l_N}})$.

---

* We shall take all possible permutations of the electron coordinates to mean all possible exchanges of the electron coordinates. In the *identity* permutation no exchanges are performed, the functions remain unchanged. If an even number of pairs of electron coordinates are exchanged, one speaks of an even permutation. The identity permutation is the simplest even permutation. Likewise a permutation is odd when an odd number of exchanges of pairs are made. For example equation (1–49) arises from (1–47) through an odd permutation: *one* pair of electrons is exchanged.

The functions $\Phi$ defined by equation (1–50) can also be written in determinantal form:

$$\Phi = \frac{1}{\sqrt{N!}} \begin{vmatrix} \psi_{a_1}(\vec{r}_1, \sigma_1) & \psi_{a_2}(\vec{r}_1, \sigma_1) & \dots & \psi_{a_N}(\vec{r}_1, \sigma_1) \\ \psi_{a_1}(\vec{r}_2, \sigma_2) & \psi_{a_2}(\vec{r}_2, \sigma_2) & \dots & \psi_{a_N}(\vec{r}_2, \sigma_2) \\ \vdots & \vdots & & \vdots \\ \psi_{a_1}(\vec{r}_N, \sigma_N) & \psi_{a_2}(\vec{r}_N, \sigma_N) & \dots & \psi_{a_N}(\vec{r}_N, \sigma_N) \end{vmatrix}. \tag{1–54}$$

A function $\Phi$ vanishes identically when two or more sets of quantum numbers $a_i = (n_i, l_i, m_{l_i}, m_{s_i})$ are the same*. In other words: every electron state characterized by a certain set $n_i$, $l_i$, $m_{l_i}$, $m_{s_i}$ may be 'occupied' at most by one electron (Pauli exclusion principle).

If the one-electron functions $\psi_{a_i}$ which appear in equation (1–50) or (1–54) are orthonormalized, i.e. †,

$$\sum_{\sigma=\frac{1}{2},-\frac{1}{2}} \iiint_{-\infty}^{+\infty} d\tau \psi^*_{a_i}(\vec{r}, \sigma) \psi_{a_j}(\vec{r}, \sigma) = \delta_{n_i, n_j} \delta_{l_i, l_j} \delta_{m_{l_i}, m_{l_j}} \delta_{m_{s_i}, m_{s_j}}, \tag{1–55}$$

then $\Phi$ is also normalized

$$\sum_{\sigma_1=\frac{1}{2},-\frac{1}{2}} \cdots \sum_{\sigma_N=\frac{1}{2},-\frac{1}{2}} \iiint_{-\infty}^{+\infty} d\tau_1 \cdots \iiint_{-\infty}^{+\infty} d\tau_N \Phi^* \Phi = 1, \tag{1–56}$$

and two functions $\Phi(a_1, a_2, \dots, a_N)$ and $\Phi(a'_1, a'_2, \dots, a'_N)$ are orthogonal when all $a_i$ which appear in $\Phi$ do not also appear in $\Phi'$:

$$\sum_{\sigma_1=\frac{1}{2},-\frac{1}{2}} \cdots \sum_{\sigma_N=\frac{1}{2},-\frac{1}{2}} \iiint_{-\infty}^{+\infty} d\tau_1 \dots \iiint_{-\infty}^{+\infty} d\tau_N \Phi^* \Phi' = 0$$
$$\{a_1, a_2, \dots, a_N\} \neq \{a'_1, a'_2, \dots, a'_N\}. \tag{1–56′}$$

### 1.2.3. *Electron configurations*

In the following we wish to investigate which energy values $E$ an atomic $N$-electron system can have in the central-field approximation when the *Pauli* principle is taken into account.

In section 1.1. we saw that for the case of a non-Coulomb potential $V(r)$, $2l+1$ orbitally degenerate states are associated with an energy value

* In this case the $\Phi$ determinant has two or more equal columns and is therefore identically zero.

† Because the complete one-electron functions $\psi(\vec{r}, \sigma)$ depend upon the spatial and spin coordinates we must not only integrate over the spatial coordinate, but sum over the range of values of the spin coordinate $\sigma = +\frac{1}{2}, -\frac{1}{2}$ as well. $\delta_{ij}$ is the Kronecker symbol defined by equation (1–7).

$\varepsilon_{n,l}$ of the one-electron system. If we also consider that because of the two possible spin states ($m_s = +\frac{1}{2}$ or $m_s = -\frac{1}{2}$) two different (although energetically equal) states arise from each of these $2l+1$ orbitally degenerate states, then there are in total $2(2l+1)$ different one-electron states for each energy eigenvalue $\varepsilon_{n,l}$. According to the Pauli exclusion principle each of these states may be occupied by at most one electron. That is, in the expression for the energy $E = \varepsilon_{n_1,l_1} + \varepsilon_{n_2,l_2} + \ldots + \varepsilon_{n_N,l_N}$ for an $N$-electron system a given term in the sum $\varepsilon_{n,l}$ may appear at most $2(2l+1)$ times. For example the one-electron states with the energy $\varepsilon_{2p}$ can be occupied at most by $2\times(2\times1+1) = 6$ electrons and the one-electron states with the energy $\varepsilon_{3d}$ at most by $2\times(2\times2+1) = 10$ electrons.

To indicate that the one-electron states of $\varepsilon_{n,l}$ are occupied by $a$ electrons and those of $\varepsilon_{n',l'}$ with $b$ electrons etc., one writes symbolically $(nl)^a(n'l')^b\ldots$. In this manner the occupation of the one-electron states of an atom or ion, the '*electron configuration*,' can be represented in compact form.

If all of the one-electron states of a shell or subshell are occupied by electrons, one speaks of a *closed* shell or subshell. The electrons in closed shells or subshells are referred to in general as *inner shell electrons* and those in open shells or subshells as *valence electrons.* Together with the atomic nucleus the inner shell electrons form the *inner shell* of the atom or ion.

In the following we shall consider those atomic systems which have only $d$ electrons outside a closed shell or subshell. In particular we shall be interested in the electron configurations of the transition metal ions $[R](nd)^N$. $[R]$ is here an abbreviation for the electron configuration of the closed shells and subshells. In particular

(a) for ions of the first transition metal series
$[R] \equiv (1s)^2(2s)^2(2p)^6(3s)^2(3p)^6 = [\mathrm{Ar}]$ (Argon configuration)
$n = 3$,

(b) for ions of the second transition metal series
$[R] = (1s)^2(2s)^2(2p)^6(3s)^2(3p)^6(3d)^{10}(4s)^2(4p)^6 = [\mathrm{Kr}]$
(Krypton configuration)
$n = 4$,

(c) for ions of the third transition metal series
$[R] = (1s)^2(2s)^2(2p)^6(3s)^2(3p)^6(3d)^{10}(4s)^2(4p)^6(4d)^{10}(4f)^{14}(5s)^2(5p)^6 = [\mathrm{Xe}]$
(Xenon configuration)
$n = 5$

$N$, the number of $(nd)$ electrons, can be $0, 1, 2, 3, \ldots, 10$.

The energy of a system of electrons with the configuration $[R](nd)^N$ is within the central-field approximation according to equation (1–48) equal to the sum of the one-electron energies:

$$E = E_R + N \cdot \varepsilon_{nd},$$

where $E_R$ stands for the sum of the one-electron energies of the closed shell electrons; e.g., $E_R$ for the ions of the first transition metal series is

$$E_R = 2\varepsilon_{1s} + 2\varepsilon_{2s} + 6\varepsilon_{2p} + 2\varepsilon_{3s} + 6\varepsilon_{3p}.$$

The $\Phi$ functions of a particular electron configuration are obtained when within this electron configuration all possible occupations allowed by the Pauli exclusion principle are taken into account.

One must then determine which $a_i = (n_i, l_i, m_{l_i}, m_{s_i})$ are to be substituted into (1–52). Because all the one-electron states of the closed shells and subshells are according to definition filled with electrons, the corresponding $a_i$ are already determined. We abbreviate the totality of these $a_i$ with $A_R$. It is further known that the remaining undetermined $a_i$ must characterize the one-electron states of the nd subshell. The determination of these remaining $a_i$ will be explained in the following with several examples.

1. *Configuration* $[R](nd)^1$. The one $(nd)$ valence electron is in one of the ten states of the $(nd)$ subshell, which differ from one another only by the $m_l, m_s$ value pair. There exist then ten different $a_i$ and thus ten functions $\Phi(A_R; m_l, m_s)$:

$$\Phi(A_R; 2^+), \quad \Phi(A_R; 1^+), \quad \Phi(A_R; 0^+), \quad \Phi(A_R; -1^+), \quad \Phi(A_R; -2^+),$$
$$\Phi(A_R; 2^-), \quad \Phi(A_R; 1^-), \quad \Phi(A_R; 0^-), \quad \Phi(A_R; -1^-), \quad \Phi(A_R; -2^-).$$

It is conventional to omit the symbol $A_R$ in the brackets to simplify the notation:

$$\Phi(2^+), \Phi(1^+), \Phi(0^+), \Phi(-1^+), \Phi(-2^+), \Phi(2^-), \Phi(1^-), \Phi(0^-), \Phi(-1^-), \Phi(-2^-).$$

All ten $\Phi$ functions describe states of the same energy in the central-field approximation. The $d^1$ system is then tenfold energetically degenerate.

2. *Configuration* $[R](nd)^2$. In this case we have two $d$ electrons. If we bring one of these electrons into a particular one of the ten degenerate one-electron states of the *nd* subshell, then the second electron because of the Pauli exclusion principle can only occupy one of the remaining nine states. If the first electron is in the state $m_l, m_s = 2^+$, then the other can occupy *one* of the states $m_l', m_s' = 2^-, 1^+, 1^-, 0^+, 0^-, -1^+, -1^-, -2^+, -2^-$. Therefore the following functions $\Phi(m_l, m_s; m_l', m_s')$ are possible:

$$\Phi(2^+; 2^-), \quad \Phi(2^+; 1^+), \quad \Phi(2^+; 1^-), \quad \Phi(2^+; 0^+)$$
$$\Phi(2^+; 0^-), \quad \Phi(2^+; -1^+), \Phi(2^+; -1^-), \Phi(2^+; -2^+), \Phi(2^+; -2^-).$$

However, we have not yet exhausted all possibilities. If one electron is for example located in the state $2^-$, then the other can occupy the states $2^+, 1^+, 1^-, 0^+, 0^-, -1^+, -1^-, -2^+, -2^-$. In Table B.3 all possible $\Phi$

Table B.3. $\Phi$ functions of the configuration $d^2$

| $M_L$ \ $M_S$ | 1 | 0 | −1 |
|---|---|---|---|
| 4 | | $\Phi(2^+;2^-)$ | |
| 3 | $\Phi(2^+;1^+)$ | $\Phi(2^+;1^-), \Phi(2^-;1^+)$ | $\Phi(2^-;1^-)$ |
| 2 | $\Phi(2^+;0^+)$ | $\Phi(2^+;0^-), \Phi(2^-;0^+), \Phi(1^+;1^-)$ | $\Phi(2^-;0^-)$ |
| 1 | $\Phi(2^+;-1^+), \Phi(1^+;0^+)$ | $\Phi(2^+;-1^-), \Phi(2^-;-1^+)$ $\Phi(1^+;0^-), \Phi(1^-;0^+)$ | $\Phi(2^-;-1^-), \Phi(1^-;0^-)$ |
| 0 | $\Phi(2^+;-2^+), \Phi(1^+;-1^+)$ | $\Phi(2^+;-2^-), \Phi(2^-;-2^+), \Phi(1^+;-1^-), \Phi(1^-;-1^+), \Phi(0^+;0^-)$ | $\Phi(2^-;-2^-), \Phi(1^-;-1^-)$ |
| −1 | $\Phi(1^+;-2^+), \Phi(0^+;-1^+)$ | $\Phi(1^+;-2^-), \Phi(1^-;-2^+), \Phi(0^+;-1^-), \Phi(0^-;-1^+)$ | $\Phi(1^-;-2^-), \Phi(0^-;-1^-)$ |
| −2 | $\Phi(0^+;-2^+)$ | $\Phi(0^+;-2^-), \Phi(0^-;-2^+), \Phi(-1^+;-1^-)$ | $\Phi(0^-;-2^-)$ |
| −3 | $\Phi(-1^+;-2^+)$ | $\Phi(-1^+;-2^-), \Phi(-1^-;-2^+)$ | $\Phi(-1^-;-2^-)$ |
| −4 | | $\Phi(-2^+;-2^-)$ | |

functions of the configuration $(nd)^2$ are summarized. They are ordered according to the quantities $M_L = m_l + m_l'$ and $M_S = m_s + m_s'$. For functions differing only in the order of $m_l$, $m_s$ and $m_l'$, $m_s'$, e.g., $\Phi(2^+;1^-)$ and $\Phi(1^-;2^+)$ in each case only a single function is included in the table. Because of the antisymmetry of the $\Phi$ functions, a change of the order of $m_l$, $m_s$ and $m_l'$, $m_s'$ means only a change of the sign of the $\Phi$ functions. $\Phi$ functions which differ only in sign are physically equivalent. Because all $\Phi$ functions of the Table B.3 belong to the same energy in the central-field approximation, the $(nd)^2$ system in the approximation is 45-fold degenerate.

3. *Configuration* $[R](nd)^3$. The three valence electrons are to be distributed over the ten degenerate $d$ states, again taking the Pauli principle into consideration. If, for example, two electrons have occupied the states $m_l, m_s = 2^+$ and $m_l', m_s' = 2^-$ then the third electron can enter only one of

the states $m_l'', m_s'' = 1^+, 1^-, 0^+, 0^-, -1^+, -1^-, -2^+, -2^-$. From this we derive the following functions $\Phi(m_l, m_s; m_l', m_s', m_l'', m_s'')$:

$$\Phi(2^+;2^-;1^+),\quad \Phi(2^+;2^-;1^-),\quad \Phi(2^+;2^-;0^+),\quad \Phi(2^+;2^-;0^-),$$
$$\Phi(2^+;2^-;-1^+),\ \Phi(2^+;2^-;-1^-),\ \Phi(2^+;2^-;-2^+),\ \Phi(2^+;2^-;-2^-).$$

The remaining $\Phi$ functions are found by considering all other possible occupations of the $d$ states. As in the case of $d^2$ one can summarize all physically different* $\Phi$ functions, again ordered according to the quantities $M_L = m_l + m_l' + m_l''$ and $M_S = m_s + m_s' + m_s''$, in a table (cf. Figure B.5). In Table B.4 the upper right hand portion (heavy borders) of the scheme is given explicity. The scheme can be easily completed with the aid of Table B.4. One obtains the functions for a given $M_L$ and $M_S$ value from the functions

for $M_L, -M_S$, by changing the signs of all $m_s$ values,
for $-M_L, M_S$, by changing the signs of all $m_l$ values,
for $-M_L, -M_S$, by changing the signs of all $m_l$ and of all $m_s$ values.

| $M_L$ \ $M_S$ | $\frac{3}{2}$ | $\frac{1}{2}$ | $-\frac{1}{2}$ | $-\frac{3}{2}$ |
|---|---|---|---|---|
| 5 | | | | |
| 4 | | | | |
| 3 | | | | |
| 2 | | | | |
| 1 | | | | |
| 0 | | | | |
| −1 | | | | |
| −2 | | | | |
| −3 | | | | |
| −4 | | | | |
| −5 | | | | |

Figure B.5. Scheme of the $M_L, M_S$ table for $d^3$ systems.

According to this prescription we find, e.g., from the functions with $M_L = 1$ and $M_S = \frac{3}{2}$ the following functions for $M_L = 1$ and $M_S = -\frac{3}{2}$: $\Phi(2^-;1^-;-2^-)$, $\Phi(2^-;0^-;-1^-)$.

---

* Of the physically equivalent $\Phi$ functions which differ only in the order of the $m_l$, $m_s$ values, only a single function is considered.

On the basis of the scheme in Figure B.5 and Table B.4 one sees that 120 physically different $\Phi$ functions belong to the configuration $d^3$. The $d^3$ configuration is *120-fold degenerate* in the central-field approximation.

Table B.4. A section of the $M_L, M_S$ Table for $d^3$ systems

| $M_L \backslash M_S$ | $\frac{3}{2}$ | $\frac{1}{2}$ |
|---|---|---|
| 5 | | $\Phi(2^+;2^-;1^+)$ |
| 4 | | $\Phi(2^+;2^-;0^+), \Phi(2^+;1^+;1^-)$ |
| 3 | $\Phi(2^+;1^+;0^+)$ | $\Phi(2^+;1^+;0^-), \Phi(2^+;1^-;0^+)$<br>$\Phi(2^-;1^+;0^+), \Phi(2^+;2^-;-1^+)$ |
| 2 | $\Phi(2^+;1^+;-1^+)$ | $\Phi(2^+;2^-;-2^+), \Phi(2^+;1^+;-1^-),$<br>$\Phi(2^+;1^-;-1\ ), \Phi(2^+;0^+;0^-),$<br>$\Phi(2^-;1^+;-1^+), \Phi(1^+;1^-\ 0^+)$ |
| 1 | $\Phi(2^+;1^+;-2^+)$<br>$\Phi(2^+;0^+;-1^+)$ | $\Phi(2^+;1^+;-2^-), \Phi(2^+;1^-;-2^+),$<br>$\Phi(2^+;0^+;-1^-), \Phi(2^+;0^-;-1^+),$<br>$\Phi(2^-;1^+;-2^+), \Phi(2^-;0^+;-1^+),$<br>$\Phi(1^+;1^-;-1^+), \Phi(1^+;0^+;0^-)$ |
| 0 | $\Phi(2^+;0^+;-2^+)$<br>$\Phi(1^+;0^+:-1^+)$ | $\Phi(2^+;0^+;-2^-), \Phi(2^+;0^-;-2^+),$<br>$\Phi(2^+;-1^+;-1^-), \Phi(2^-;0^+;-2^+),$<br>$\Phi(1^+;1^-;-2^+), \Phi(1^+;0^+;-1^-),$<br>$\Phi(1^+;0^-;-1^+), \Phi(1^-;0^+;-1^+)$ |

The $\Phi$ functions for the configurations $[R](nd)^4$, $[R](nd)^5, \ldots, [R](nd)^9$ can be obtained according to the same procedure we have applied to $d^1$, $d^2$, and $d^3$ ions. The number of $\Phi$ functions increases greatly up to the configuration $[R](nd)^5$ but decreases to the same degree with a decreasing number of $d$ electrons until $d^9$ is reached. As is apparent, it becomes quite laborious to list all of the $\Phi$ functions in the notation which has been employed up to this point. Even for the case of $d^9$ one has, as we shall see, only ten different $\Phi$ functions. Each of these contains, however, nine $m_l$, $m_s$ values. In the case of $d^8$ eight notations concerning $m_l$, $m_s$ values must be made for each $\Phi$-function, etc. Nature does, however, offer some help in simplifying the system of notation. This will be shown with several examples.

4. *Configuration* $[R](nd)^9$. We wish to distribute nine $d$ electrons over ten one-electron states. For each of the possible occupations one one-electron state remains empty. That is, one electron is missing in the

corresponding closed $d$ subshell, or the $d$ subshell has a hole. Instead of giving the possible occupations of the ten one-electron states by the nine electrons we can state in which of the ten one-electron states the hole is located, which one of the states is empty. In the brackets of the $\Phi$ functions we write the $m_l, m_s$ value of the hole and not the nine $m_l, m_s$ values of the electrons. The one hole can be in any of the ten one-electron states, as can the one electron in the $d^1$ case. One finds therefore ten $\Phi$ functions:

$$\Phi(2^+),\ \Phi(2^-),\ \Phi(1^+),\ \Phi(1^-),\ \Phi(0^+),\ \Phi(0^-),\ \Phi(-1^+),\ \Phi(-1^-),\ \Phi(-2^+),\ \Phi(-2^-).$$

This set of $\Phi$ functions is formally identical with the set of $\Phi$ functions for the $d^1$ case. Of course one must be exceedingly careful to note that the brackets of the $\Phi$ function for this $d^9$ case contain the $m_l, m_s$ *value of the hole.* Because ten physically different $\Phi$ functions belong to the $d^9$ configuration, a tenfold degeneracy is present in the central-field approximation.

5. *Configuration* $[R](nd)^8$. The eight electrons leave two holes empty in the $d$ subshell. Similar to the $d^9$ case we can distribute the two holes instead of the eight electrons over the ten one-electron states. The distribution possibilities are of course equivalent to those of the $d^2$ case. If one writes in the brackets of the $\Phi$ function the $m_l, m_s$ *values of the holes* one obtains exactly the functions given in Table B.3 for the $d^2$ case. Again it must be emphasized that the $\Phi$ functions of the $d^8$ and $d^2$ systems are only formally identical. If Table B.3 is used for the $d^8$ case, it must be considered that the values $M_L = m_l + m_l'$ and $M_S = m_s + m_s'$ refer to the holes. The $d^8$ ion and the $d^2$ ion are in the central-field approximation *45-fold degenerate.* As for the $d^1$ and the $d^9$ case, and the $d^2$ and the $d^8$ case, one finds a *formal equivalence of electrons and holes* for the $d^3$ and the $d^7$ case and the $d^4$ and the $d^6$ case. For practical calculations this proves to be a great aid. We shall not give explicitly the $\Phi$ functions for the configurations $[R](nd)^4$, $[R](nd)^5$, $[R](nd)^6$, $[R](nd)^7$. They can be obtained without difficulty when one applies the procedure given here. In the central-field approximation the $d^4$ system is 210-fold, the $d^5$ system 252-fold, the $d^6$ system 210-fold and the $d^7$ system 120-fold degenerate.

### 1.2.4. *Total angular momentum and total spin*

The operator $\boldsymbol{L}$ of the orbital angular momentum of an $N$-electron system is defined by

$$\boldsymbol{L}^2 = \boldsymbol{L}_x^2 + \boldsymbol{L}_y^2 + \boldsymbol{L}_z^2 \tag{1–57}$$

$$\boldsymbol{L}_k = \boldsymbol{l}_{k_1} + \boldsymbol{l}_{k_2} + \cdots + \boldsymbol{l}_{k_N}, \quad k = x, y, z. \tag{1–58}$$

Here, e.g., $l_{x_i}$ is the operator which represents the $x$ component of the orbital angular momentum of the $i$th electron. The commutation relations hold:

$$\boldsymbol{L}_x \boldsymbol{L}_y - \boldsymbol{L}_y \boldsymbol{L}_x = i\hbar \boldsymbol{L}_z \qquad \text{cyclic in } x, y, z \tag{1–59}$$

$$\boldsymbol{L}^2 \boldsymbol{L}_k - \boldsymbol{L}_k \boldsymbol{L}^2 = 0 \qquad k = x, y, z. \tag{1–60}$$

Correspondingly the operator $\boldsymbol{S}$ of the spin of an $N$ electron system is defined by

$$\boldsymbol{S}^2 = \boldsymbol{S}_x^2 + \boldsymbol{S}_y^2 + \boldsymbol{S}_z^2 \tag{1–61}$$

$$\boldsymbol{S}_k = \boldsymbol{s}_{k1} + \boldsymbol{s}_{k2} + \cdots + \boldsymbol{s}_{kN}, \qquad k = x, y, z, \tag{1–62}$$

and has the commutation relations

$$\boldsymbol{S}_x \boldsymbol{S}_y - \boldsymbol{S}_y \boldsymbol{S}_x = i\hbar \boldsymbol{S}_z \qquad \text{cyclic in } x, y, z \tag{1–63}$$

and

$$\boldsymbol{S}^2 \boldsymbol{S}_k - \boldsymbol{S}_k \boldsymbol{S}^2 = 0, \qquad k = x, y, z. \tag{1–64}$$

The operators $\boldsymbol{S}^2$, $\boldsymbol{S}_x$, $\boldsymbol{S}_y$, $\boldsymbol{S}_z$ commute with the operators $\boldsymbol{L}^2$, $\boldsymbol{L}_x$, $\boldsymbol{L}_y$, $\boldsymbol{L}_z$, c.g. $\boldsymbol{S}^2\boldsymbol{L}_y - \boldsymbol{L}_y\boldsymbol{S}^2 = 0$.

Because $\boldsymbol{L}^2, \boldsymbol{L}_z, \boldsymbol{S}^2, \boldsymbol{S}_z$ commute with one another, functions must exist which are simultaneously eigenfunctions of all the operators*. These functions are denoted by $\Psi(L, M_L, S, M_S)$:

$$\boldsymbol{L}^2 \Psi(L, M_L, S, M_S) = L(L+1)\hbar^2 \Psi(L, M_L, S, M_S) \qquad L = 0, 1, 2, 3, \ldots^{\dagger} \tag{1–65}$$

$$\boldsymbol{L}_z \Psi(L, M_L, S, M_S) = M_L \hbar \Psi(L, M_L, S, M_S) \qquad M_L = -L, -L+1, \ldots, L-1, L \tag{1–66}$$

$$\boldsymbol{S}^2 \Psi(L, M_L, S, M_S) = S(S+1)\hbar^2 \Psi(L, M_L, S, M_S) \qquad S = 0, \frac{1}{2}, 1, \frac{3}{2}, 2, \frac{5}{2}, 3, \ldots \tag{1–67}$$

$$\boldsymbol{S}_z \Psi(L, M_L, S, M_S) = M_S \hbar \Psi(L, M_L, S, M_S) \qquad M_S = -S, -S+1, \ldots, S-1, S. \tag{1–68}$$

If an $N$ electron system is in a state which can be described by $\Psi(L, M_L, S, M_S)$, then the total angular momentum and the total spin as well as their

---

* The theorem holds: if two operators $\boldsymbol{A}$ and $\boldsymbol{B}$ commute, then there exists a set of functions which are simultaneously eigenfunctions of $\boldsymbol{A}$ and $\boldsymbol{B}$. Cf., e.g., Eyring, Walter and Kimball[2], p.

† The eigenvalue equation for $\boldsymbol{L}^2$ can also have solutions with $L = \frac{1}{2}, \frac{3}{2}, \frac{5}{2}, \ldots$. For our problems only integral $L$ values are of interest.

$z$ components have the sharp values $\sqrt{L(L+1)}\hbar$, $\sqrt{S(S+1)}\hbar$, $M_L\hbar$ or $M_S\hbar$.*

The following relations hold for the shifting operators $\boldsymbol{L}_\pm \equiv \boldsymbol{L}_x \pm i\boldsymbol{L}_y$ and $\boldsymbol{S}_\pm \equiv \boldsymbol{S}_x \pm i\boldsymbol{S}_y$ for the total system

$$(\boldsymbol{L}_x \pm i\,\boldsymbol{L}_y)\Psi(L, M_L, S, M_S) = \hbar\sqrt{(L \pm M_L + 1)(L \mp M_L)}\;\Psi(L, M_L \pm 1, S, M_S), \tag{1–69}$$

$$(\boldsymbol{S}_x \pm i\,\boldsymbol{S}_y)\Psi(L, M_L, S, M_S) = \hbar\sqrt{(S \pm M_S + 1)(S \mp M_S)}\;\Psi(L, M_L, S, M_S \pm 1). \tag{1–70}$$

That is, if, e.g., the operator $\boldsymbol{L}_x + i\boldsymbol{L}_y$ operates on a $\Psi$ function of the eigenvalues $M_L\hbar$ of the operator $\boldsymbol{L}_z$, then one obtains, except for the factor $\hbar\sqrt{(L+M_L+1)(L-M_L)}$, the $\Psi$ function of the eigenvalue $(M_L+1)\hbar$ of the operator $L_z$, $L$, $S$, and $M_S$ are not changed by this process. If a function $\Psi(L, M_L, S, M_S)$ is known, then all other functions $\Psi(L, M'_L, S, M'_S)$ with $M'_L = -L, \ldots, L$ and $M'_S = -S, \ldots, S$ can be determined by repeated application of the shifting operators.

The antisymmetrized product functions $\Phi$ which we have introduced satisfy the relations:

$$\begin{aligned}\boldsymbol{L}^2\Phi(m_{l_1}, m_{s_1}; m_{l_2}, m_{s_2}; \ldots; m_{l_N}, m_{s_N}) &= \hbar^2\left(\sum_{i=1}^{N} l_i(l_i+1) + 2\sum_{i<j}^{N}\sum^{N} m_{l_i} m_{l_j}\right)\times\\ &\times\Phi(m_{l_1}, m_{s_1}; m_{l_2}, m_{s_2}; \ldots; m_{l_N}, m_{s_N})\\ &+\hbar^2\sum_{i\neq j}^{N}\sum^{N}\sqrt{(l_i+m_{l_i}+1)(l_i-m_{l_i})}\sqrt{(l_j-m_{l_j}+1)(l_j+m_{l_j})}\times\\ &\times\Phi(m_{l_1}, m_{s_1}; \ldots; m_{l_i}+1, m_{s_i}; \ldots; m_{l_j}-1, m_{s_j}; \ldots; m_{l_N}, m_{s_N})\end{aligned} \tag{1–71}$$

$$\begin{aligned}\boldsymbol{S}^2\Phi(m_{l_1}, m_{s_1}; m_{l_2}, m_{s_2}; \ldots; m_{l_N}, m_{s_N}) &= \hbar^2\left(\sum_{i=1}^{N}|m_{s_i}|(|m_{s_i}|+1) + 2\sum_{i<j}^{N}\sum^{N} m_{s_i} m_{s_j}\right)\times\\ &\times\Phi(m_{l_1}, m_{s_1}; m_{l_2}, m_{s_2}; \ldots; m_{l_N}, m_{s_N})\\ &+\hbar^2\sum_{i\neq j}^{N}\sum^{N}\sqrt{(3/2+m_{s_i})(1/2-m_{s_i})}\sqrt{(3/2-m_{s_j})(1/2+m_{s_j})}\times\\ &\times\Phi(m_{l_1}, m_{s_2}; \ldots; m_{l_i}, m_{s_i}+1; \ldots; m_{l_j}, m_{s_j}-1; \ldots; m_{l_N}, m_{s_N})\end{aligned} \tag{1–72}$$

$$\begin{aligned}\boldsymbol{L}_z\Phi(m_{l_1}, m_{s_1}; m_{l_2}, m_{s_2}; \ldots; m_{l_N}, m_{s_N}) &= \hbar M_L \Phi(m_{l_1}, m_{s_1}; m_{l_2}, m_{s_2}; \ldots; m_{l_N}, m_{s_N})\\ M_L &= m_{l_1} + m_{l_2} + \cdots + m_{l_N}\end{aligned} \tag{1–73}$$

* $L$, $S$, $M_L$, $M_S$ are called the total angular momentum quantum number, total spin quantum number, magnetic total angular momentum quantum number and magnetic total spin quantum number.

$$S_z\,\Phi(m_{l_1},m_{s_1};m_{l_2},m_{s_2};\ldots;m_{l_N},m_{s_N})=\hbar M_S\Phi(m_{l_1},m_{s_1};m_{l_2},m_{s_2};\ldots;m_{l_N},m_{s_N})$$
$$M_S=m_{s_1}+m_{s_2}+\cdots+m_{s_N} \tag{1–74}$$

$$(L_x\pm i\,L_y)\,\Phi\,(m_{l_1},m_{s_1};m_{l_2},m_{s_2};\ldots;m_{l_N},m_{s_N}) \tag{1–75}$$
$$=\hbar\sum_{i=1}^{N}\sqrt{(l_i\pm m_{l_i}+1)(l_i\mp m_{l_i})}\;\Phi(m_{l_1},m_{s_1};\ldots;m_{l_i}\pm 1,m_{s_i};\ldots;m_{l_N},m_{s_N})$$

$$(S_x\pm i\,S_y)\,\Phi(m_{l_1},m_{s_1};m_{l_2},m_{s_2};\ldots;m_{l_N},m_{s_N}) \tag{1–76}$$
$$=\hbar\sum_{l=1}^{N}\sqrt{(3/2\pm m_{s_i})(1/2\mp m_{s_i})}\;\Phi\,(m_{l_1},m_{s_1};\ldots;m_{l_i},m_{s_i}\pm 1;\ldots;m_{l_N},m_{s_N})$$

From these relations it follows that:

1. Because of equations (1–71, 72) the $\Phi$ are in general not eigenfunctions of $\boldsymbol{L}^2$ and $\boldsymbol{S}^2$, i.e., the $\Phi$ in general describe states in which the squares of the absolute values of the total angular momentum and the total spin do not have defined values. The $\Phi$ are therefore in general not identical with the $\Psi(L, M_L, S, M_S)$.

2. The $\Phi$ functions are according to equations (1–73, 74) eigenfunctions of $\boldsymbol{L}_z$ and $\boldsymbol{S}_z$; they represent states in which the $z$ components of the total angular momentum and total spin have the sharply defined values

$$\hbar M_L=\hbar\,(m_{l_1}+m_{l_2}+\cdots+m_{l_N})\ \text{or}$$
$$\hbar M_S=\hbar\,(m_{s_1}+m_{s_2}+\cdots+m_{s_N})$$

One sees then the physical reason for the arrangement of the $\Phi$ functions in the $M_L$, $M_S$ tables: $\Phi$ functions in the same *row* represent states with the same $z$ component of the total angular momentum; those in the same *column* stand for states with the same $z$ component of the total spin*.

It is easily seen that every linear combination of $\Phi$ functions from the same box of an $M_L$, $M_S$ table is an eigenfunction of $\boldsymbol{L}_z$ and $\boldsymbol{S}_z$ having the eigenvalues (likewise belonging in the same box) $M_L\hbar$ or $M_S\hbar$.

### 1.2.5 *The states* $^{2S+1}L$

What total angular momentum and what total spin can an $N$-electron system have which is described by a given electron configuration in the central-field approximation?

* We make use here and in the following of the fact that electrons of the closed shells and subshells do not contribute to the total angular momentum and total spin or to their $z$ components.

In order to answer this question one begins by collecting all the $\Phi$ functions of a particular electron configuration and arranging these in an $M_L, M_S$ table (cf. the example for $(nd)^2$, Table B.3). The number of states corresponding to a given $M_L, M_S$ pair is equal to the number of the $\Phi$ functions in the appropriate box of the table. For example, we see from Table B.3 that an $(nd)^2$ system with the quantum numbers $M_L = 1$, $M_S = 0$ has four states. From this one can derive the possible values of the total angular momentum and the total spin when it is considered that according to equations (1–65, 66, 67, 68) the quantum number $L$ has the $M_L$ values $-L, -L+1, \ldots, L-1, L$ and the quantum number $S$ the $M_S$ values $-S, -S+1, \ldots, S-1, S$; that, therefore, $L \geqq |M_L|$ and $S \geqq |M_S|$ hold. We shall illustrate the procedure of finding the states corresponding to the quantum numbers $L$, $M_L$, $S$, $M_S$ using the example of the $(nd)^2$ configuration.

According to Table B.3 no state with $M_S = \pm 1$ exists for the highest $M_L$ value $M_L = 4$; however, a state with $M_S = 0$ does exist. This state must, therefore, belong to $L \geqq 4$ and $S \geqq 0$. Because 4 is the highest $M_L$ value appearing and because no state with $M_S > 0$ exists for $M_L = 4$, the state with $M_L = 4$, $M_S = 0$ can only belong to $L = 4$ and $S = 0$. For $L = 4$, $S = 0$ there are in total nine states with $M_L = 4, 3, 2, 1, 0, -1, -2, -3, -4$, $M_S = 0$. Therefore we cross out in each case one state* in the appropriate box of Table B.3. The highest $M_L$ and $M_S$ values of the remaining states are $M_L = 3$ and $M_S = 1$. They belong apparently to $L = 3$, $S = 1$. For $L = 3$, $S = 1$ there are 21 states $M_L = 3, 2, 1, 0, -1, -2, -3$, $M_S = 1, 0, -1$. Again in each case we cross out one state in the appropriate box. The state with $M_L = 2$, $M_S = 0$ has the highest $M_L$ and $M_S$ value of the remaining states. $L = 2$, $S = 0$ belong to it. The four states with $M_L = 1, 0, -1, -2$, $M_S = 0$ are likewise states of $L = 2$, $S = 0$. If the procedure described here is applied further one also finds the states with $L = 1$, $S = 1$ and $L = 0$, $S = 0$.

The totality of states belonging to a given $L$ and $S$ value (and to a given electron configuration) is called a *term*. In analogy to the one-electron system, for which the states corresponding to $l = 0, 1, 2, 3, \ldots$ were denoted as $s, p, d, f \ldots$ states, one writes for terms of many-electron systems with $L = 0, 1, 2, 3, \ldots$ the symbols $S, P, D, F, \ldots$. In order to characterize terms with regard to their $S$ values one writes the value of

* We are speaking specifically of the states and not of the $\Phi$ functions in the table, because—as we have seen earlier—the $\Phi$ functions in general are not equal to the wave functions $\Psi(L, M_L, S, M_S)$. The $\Psi$ are in general linear combinations of the $\Phi$ in a given box. It can be shown that for each box there are exactly as many $\Psi$ functions as there are $\Phi$ functions in the box. When *counting* the possible states one may cross out in the table in the place of the 'used' states in each case one $\Phi$.

$2S+1$ to the upper left of the symbol for $L$, e.g., $^1S$ or $^4F$. The value of $2S+1$ is the *multiplicity* of the corresponding term. Terms with $S=0$ (i.e., multiplicity $2S+1=1$) are singlets, with $S=\frac{1}{2}$ (i.e., $2S+1=2$) doublets, with $S=1$ (i.e., $2S+1=3$) triplets, etc. Because the $2L+1$ $M_L$ values $-L, -L+1, \ldots L-1, L$ and the $2S+1$ $M_S$ values $-S, -S+1, \ldots, S-1, S$ correspond to given $L$ and $S$, a term is a totality of $(2L+1)(2S+1)$ states. For example the term $^4F$ consists of $(2\times 3+1)\times 4 = 28$ states.

| $L$ | $S$ | Term | Number of States |
|---|---|---|---|
| 4 | 0 | $^1G$ | 9 |
| 3 | 1 | $^3F$ | 21 |
| 2 | 0 | $^1D$ | 5 |
| 1 | 1 | $^3P$ | 9 |
| 0 | 0 | $^1S$ | 1 |
| | | Total | 45 |

The terms of the transition metal ions for $d^1$ to $d^9$ are collected in Table B.5. When considering this table it is seen that a $d^N$ ion has the same terms as a $d^{10-N}$ ion. This situation has its origin in the formal *equivalence of electrons and holes* and the fact that closed shells and subshells make no contribution to the total angular momentum and to the total spin. It is, therefore, irrelevant for the momentum properties if a single $d$ electron is added to the closed inner shell of the ion or if a $d$ electron is removed from an ion with a closed $d$ subshell. If two ions have

Table B.5. The terms belonging to the configurations $d^N$ *

| Configuration | Terms |
|---|---|
| $d^1, d^9$ | $^2D$ |
| $d^2, d^8$ | $^3F\ ^3P$<br>$^1G\ ^1D\ ^1S$ |
| $d^3, d^7$ | $^4F\ ^4P$<br>$^2H\ ^2G\ ^2F\ a\,^2D\ b\,^2D\ ^2P$ |
| $d^4, d^6$ | $^5D$<br>$^3H\ ^3G\ a\,^3F\ b\,^3F\ ^3D\ a\,^3P\ b\,^3P$<br>$^1I\ a\,^1G\ b\,^1G\ ^1F\ a\,^1D\ b\,^1D\ a\,^1S\ b\,^1S$ |
| $d^5$ | $^6S$<br>$^4G\ ^4F\ ^4D\ ^4P$<br>$^2I\ ^2H\ a\,^2G\ b\,^2G\ a\,^2F\ b\,^2F\ a\,^2D\ b\,^2D\ c\,^2D\ ^2P\ ^2S$ |

* If in a given configuration several terms with the same $L$ and the same $S$ appear, we shall distinguish between them by using the additional symbols $a, b, c, \ldots$.

the same angular momentum properties, then they must also have the same terms because the terms of an ion are determined by its angular momentum properties.

1.2.6. *The functions* $\Psi(L, M_L, S, M_S)$

In determining the functions $\Psi(L, M_L, S, M_S)$ a procedure originating from Gray and Wills is very useful*. It is based on the following:

1. The $\Psi(L, M_L, S, M_S)$ may be represented by a linear combination of $\Phi$ functions of the appropriate $M_L, M_S$ box.
2. If a function $\Psi(L, M_L, S, M_S)$ is known, all other functions $\Psi(L, M'_L, S, M'_S)$ of the same term are obtained by (repeated) application of the shifting operators $\boldsymbol{L}_x \pm i\boldsymbol{L}_y$ and $\boldsymbol{S}_x \pm i\boldsymbol{S}_y$.
3. The functions $\Psi(L, M_L, S, M_S)$ are orthogonal and normalized†:

$$(\Psi(L, M_L, S, M_S), \Psi(L', M'_L, S', M'_S)) = \delta_{L,L'} \cdot \delta_{M_L, M'_L} \cdot \delta_{S,S'} \cdot \delta_{M_S, M'_S}.$$

The use of the shifting operators is characteristic of the *Gray–Wills* procedure.

We shall explain the procedure using the example of the $(nd)^2$ configuration. The state with $M_L = 4$, $M_S = 0$ belongs to $L = 4$, $S = 0$ ($^1G$ term) as was seen when counting the possible states. In Table B.3 the function $\Phi(2^+; 2^-)$ is in the appropriate box. Because this box contains only a single $\Phi$ function, this must be identical with the corresponding function $\Psi(4, 4, 0, 0)$‡:

$$\Psi(4, 4, 0, 0) = \Phi(2^+; 2^-). \tag{1–77}$$

A wave function of the $^1G$ term is thus known. The other $\Psi(4, M_L, 0, M_S)$ with $M_L = 3, 2, 1, 0, -1, -2, -3, -4$, $M_S = 0$ which likewise belong to the $^1G$ term, are obtained with the aid of the lowering operator $\boldsymbol{L}_x - i\boldsymbol{L}_y$. If this operator is applied to the left-hand side of equation (1–77) one obtains according to equation (1–69)

$$\begin{aligned}(\boldsymbol{L}_x - i\boldsymbol{L}_y)\Psi(4, 4, 0, 0) &= \hbar\sqrt{(4-4+1)(4+4)}\,\Psi(4, 3, 0, 0) \\ &= 2\sqrt{2}\,\hbar\,\Psi(4, 3, 0, 0).\end{aligned} \tag{1–78}$$

* Gray and Wills, *Phys. Rev.*, **38**, 248 (1931). Cf. also Condon and Shortley[4], pp. 226 ff.

† We use in the following an abbreviation for integrals:

$$(f, g) \equiv \sum_{\sigma_1, \cdots, \sigma_N} \iiint d\tau_1 \ldots d\tau_N f^*(\vec{r}_1, \sigma_1; \ldots; \vec{r}_N, \sigma_N)\, g(\vec{r}_1, \sigma_1; \ldots; \vec{r}_N, \sigma_N)$$

Note: The complex conjugate of the term to the left of the comma is to be taken.

‡ That $(2^+; 2^-)$ are really eigenfunction of $\boldsymbol{L}^2$ and $\boldsymbol{S}^2$ with the eigenvalues $4(4+1)\hbar^2$ or 0 can be easily confirmed through insertion into equation (1–71) or (1–72).

Application of the operator $L_x - iL_y$ to the right-hand side of equation (1–77) gives because of equation (1–75)

$$(L_x - iL_y)\Phi(2^+; 2^-)$$
$$= \hbar\sqrt{(2-2+1)(2+2)}\Phi(1^+; 2^-) + \hbar\sqrt{(2-2+1)(2+2)}\Phi(2^+; 1^-).$$
$$= 2\hbar[\Phi(2^+; 1^-) - \Phi(2^-; 1^+)]. \tag{1–79}$$

Because of equation (1–77) the right-hand sides of (1–78) and (1–79) must be equal. We have

$$2\sqrt{2}\,\hbar\Psi(4, 3, 0, 0) = 2\hbar[\Phi(2^+; 1^-) - \Phi(2^-; 1^+)]$$

or

$$\Psi(4, 3, 0, 0) = \frac{1}{\sqrt{2}}[\Phi(2^+; 1^-) - \Phi(2^-; 1^+)]. \tag{1–80}$$

With this we have obtained another function of the $^1G$ term. $\Psi(4, 3, 0, 0)$ is, as was to be expected, a linear combination of the $\Phi$ functions of the box with $M_L = 3$, $M_S = 0$. By repeated application of the lowering operator $L_x - iL_y$ to equation (1–80) all remaining functions of the $^1G$ term are obtained.

When calculating the functions $\Psi(3, M_L, 1, M_S)$ of the $^3F$ term we begin with the box $M_L = 3$, $M_S = 1$ of Table B.3. Here only the single function $\Phi(2^+; 1^+)$ is present. It must apparently be equal to $\Psi(3, 3, 1, 1)$:

$$\Psi(3, 3, 1, 1) = \Phi(2^+; 1^+) \tag{1–81}$$

With the help of the lowering operator $L_x - iL_y$ we then calculate the functions $\Psi(3, M_L, 1, 1)$, $M_L = 2, 1, 0, -1, -2, -1$. We move so to speak downwards in the $(M_S = 1)$ column of the Table B.3. The $\Psi$ functions of the $^3F$ term with $M_S = 0, -1$ are obtained from the functions $(3, M_L, 1, 1)$ by applying the spin lowering operator $S_x - iS_y$ and taking equations (1–70) and (1–76) into account. Here we move to the right in Table B.3. While counting the possible states of the $(nd)^2$ system we had determined that there is one term $^1D$. One of the states of the $^1D$ term must be describable by a linear combination of the $\Phi$ functions of the box with $M_L = 2$, $M_S = 0$:

$$\Psi(2, 2, 0, 0) = c_1\Phi(2^+; 0^-) + c_2\Phi(1^+; 1^-) + c_3\Phi(2^-; 0^+).$$

$c_1$, $c_2$, $c_3$ are constant coefficients of the linear combination. From the calculations for the terms $^1G$ and $^3F$ given above the functions for the box with $(M_L = 2, M_S = 0)$ are:

$$^1G\text{ Term:}\quad \Psi(4,2,0,0) = \frac{1}{\sqrt{14}}[\sqrt{3}\,\Phi(2^+;0^-) + 2\sqrt{2}\,\Phi(1^+;1^-) - \sqrt{3}\,\Phi(2^-;0^+)],$$

$$^3F\text{ Term:}\quad \Psi(3,2,1,0) = \frac{1}{\sqrt{2}}[\Phi(2^+;0^-) + \Phi(2^-;0^+)].$$

Because $\Psi(2,2,0,0)$, $\Psi(3,2,1,0)$ and $\Psi(4,2,0,0)$ are eigenfunctions of $\boldsymbol{L}^2$ having different eigenvalues ($L = 2, 3$ or $4$), they must be orthogonal. Considering (1–56′) it must, therefore, hold that:

$$\big(\Psi(4,2,0,0), \Psi(2,2,0,0)\big) = \frac{1}{\sqrt{14}}[\sqrt{3}c_1 + 2\sqrt{2}c_2 - \sqrt{3}c_3] = 0,$$

$$\big(\Psi(3,2,1,0), \Psi(2,2,0,0)\big) = \frac{1}{\sqrt{2}}[c_1 + c_3] = 0,$$

or

$$c_2 = -\sqrt{3/2}\,c_1 \quad \text{und} \quad c_3 = -c_1.$$

From this it also follows that

$$\Psi(2,2,0,0) = c_1[\Phi(2^+;0^-) - \sqrt{3/2}\,\Phi(1^+;1^-) - \Phi(2^-;0^+)].$$

The remaining undetermined constant $c_1$ is determined from the normalization condition:

$$\big(\Psi(2,2,0,0), \Psi(2,2,0,0)\big) = 1$$

which yields

$$c_1 = \pm\sqrt{2/7}.$$

The sign of the $\Psi$ function is physically unimportant. We may choose for example the plus sign: $c_1 = \sqrt{\tfrac{2}{7}}$

$$\Psi(2,2,0,0) = \frac{1}{\sqrt{7}}[\sqrt{2}\Phi(2^+;0^-) - \sqrt{3}\Phi(1^+;1^-) - \sqrt{2}\Phi(2^-;0^+)].$$

We have then determined one $\Psi$ function of the $^1D$ term. The remaining functions of this term are found by applying the lowering operator $\boldsymbol{L}_x - i\boldsymbol{L}_y$.

Using this procedure all $\Psi$ functions of the remaining terms may be determined.

The Gray–Wills procedure can be applied to any configuration we choose.

In Tables B.6 and B.7 the $\Psi$ functions for the configurations $d^2$ and $d^3$ are collected.

Table B.6. $\Psi$ functions for the configuration $d^2$

| Term | $\Psi(L, M_L, S, M_S)$ | $\Sigma c_i \Phi_i(m_l, m_s; m_l', m_s')$ |
|---|---|---|
| $^1G$ | $\Psi(4,4,0,0)$ | $\Phi(2^+;2^-)$ |
| | $\Psi(4,3,0,0)$ | $\sqrt{1/2}[\Phi(2^+;1^-)-\Phi(2^-;1^+)]$ |
| | $\Psi(4,2,0,0)$ | $\sqrt{3/14}\Phi(2^+;0^-)-\sqrt{3/14}\Phi(2^-;0^+)+\sqrt{8/14}\Phi(1^+;1^-)$ |
| | $\Psi(4,1,0,0)$ | $\sqrt{1/14}\Phi(2^+;-1^-)-\sqrt{1/14}\Phi(2^-;-1^+)$ $+\sqrt{6/14}\Phi(1^+;0^-)-\sqrt{6/14}\Phi(1^-;0^+)$ |
| | $\Psi(4,0,0,0)$ | $\sqrt{1/70}\Phi(2^+;-2^-)-\sqrt{1/70}\Phi(2^-;-2^+)+\sqrt{16/70}$ $\Phi(1^+;-1^-)-\sqrt{16/70}\Phi(1^-;-1^+)+\sqrt{36/70}\Phi(0^+;0^-)$ |
| | $\Psi(4,-1,0,0)$ | $\sqrt{1/14}\Phi(1^+;-2^-)-\sqrt{1/14}\Phi(1^-;-2^+)$ $+\sqrt{6/14}\Phi(0^+;-1^-)-\sqrt{6/14}\Phi(0^-;-1^+)$ |
| | $\Psi(4,-2,0,0)$ | $\sqrt{3/14}\Phi(0^+;-2^-)-\sqrt{3/14}\Phi(0^-;-2^+)$ $+\sqrt{8/14}\Phi(-1^+;-1^-)$ |
| | $\Psi(4,-3,0,0)$ | $\sqrt{1/2}[\Phi(-1^+;-2^-)-\Phi(-1^-;-2^+)]$ |
| | $\Psi(4,-4,0,0)$ | $\Phi(-2^+;-2^-)$ |
| $^3F$ | $\Psi(3,3,1,1)$ | $\Phi(2^+;1^+)$ |
| | $\Psi(3,2,1,1)$ | $\Phi(2^+;0^+)$ |
| | $\Psi(3,1,1,1)$ | $\sqrt{6/10}\Phi(2^+;-1^+)+\sqrt{4/10}\Phi(1^+;0^+)$ |
| | $\Psi(3,0,1,1)$ | $\sqrt{1/5}\Phi(2^+;-2^+)+\sqrt{4/5}\Phi(1^+;-1^+)$ |
| | $\Psi(3,-1,1,1)$ | $\sqrt{6/10}\Phi(1^+;-2^+)+\sqrt{4/10}\Phi(0^+;-1^+)$ |
| | $\Psi(3,-2,1,1)$ | $\Phi(0^+;-2^+)$ |
| | $\Psi(3,-3,1,1)$ | $\Phi(-1^+;-2^+)$ |
| | The functions for $M_S=0$ and $M_S=-1$ are obtained from the functions given here by applying the lowering operator $S_x - iS_y$ once or twice, respectively. | |
| $^1D$ | $\Psi(2,2,0,0)$ | $\sqrt{2/7}\Phi(2^+;0^-)-\sqrt{2/7}\Phi(2^-;0^+)-\sqrt{3/7}\Phi(1^+;1^-)$ |
| | $\Psi(2,1,0,0)$ | $\sqrt{6/14}\Phi(2^+;-1^-)-\sqrt{6/14}\Phi(2^-;-1^+)$ $-\sqrt{1/14}\Phi(1^+;0^-)+\sqrt{1/14}\Phi(1^-;0^+)$ |
| | $\Psi(2,0,0,0)$ | $\sqrt{4/14}\Phi(^{2+};-2^-)-\sqrt{4/14}\Phi(2^-;-2^+)$ $+\sqrt{1/14}\Phi(1^+;-1^-)-\sqrt{1/14}\Phi(1^-;-1^+)$ $-\sqrt{4/14}\Phi(0^+;0^-)$ |

Table B.6.—continued

| Term | $\Psi(L, M_L, S, M_S)$ | $\Sigma c_i\Phi_i(m_l, m_s; m_l', m_s')$ |
|---|---|---|
| | $\Psi(2,-1,0,0)$ | $\sqrt{6/14}\Phi(1^+;-2^-)-\sqrt{6/14}\Phi(1^-;-2^+)$ $-\sqrt{1/14}\Phi(0^+;-1^-)+\sqrt{1/14}\Phi(0^-;-1^+)$ |
| | $\Psi(2,-2,0,0)$ | $\sqrt{2/7}\Phi(0^+;-2^-)-\sqrt{2/7}\Phi(0^-;-2^+)$ $-\sqrt{3/7}\Phi(-1^+;-1^-)$ |
| $^3P$ | $\Psi(1,1,1,1)$ | $-\sqrt{2/5}\Phi(2^+;-1^+)+\sqrt{3/5}\Phi(1^+;0^+)$ |
| | $\Psi(1,0,1,1)$ | $-\sqrt{4/5}\Phi(2^+;-2^+)+\sqrt{1/5}\Phi(1^+;-1^+)$ |
| | $\Psi(1,-1,1,1)$ | $-\sqrt{2/5}\Phi(1^+;-2^+)+\sqrt{3/5}\Phi(0^+;-1^+)$ |
| The functions for $M_S = 0$ and $M_S = -1$ are obtained from the functions given here by applying the lowering operator $\boldsymbol{S}_x - i\boldsymbol{S}_y$ once or twice, respectively. | | |
| $^1S$ | $\Psi(0,0,0,0)$ | $\sqrt{1/5}[\Phi(2^+;-2^-)-\Phi(2^-;-2^+)-\Phi(1^+;-1^-)$ $+\Phi(1^-;-1^+)+\Phi(0^+;0^-)]$ |

It must be emphasized that the $\Psi$ functions (as linear combinations of the $\Phi$) in the central-field approximation of equation (1–42) are eigenfunctions having the same eigenvalues as the $\Phi$. One says the $\Phi$ basis is

Table B.7. $\Psi$ functions for the $^4F$ and $^4P$ term of the $d^3$ configuration

| Term | $\Psi(L, M_L, S, M_S)$ | $\Sigma c_i\Phi_i(m_l, m_s; m_l', m_s'; m_l'', m_s'')$ |
|---|---|---|
| $^4F$ | $\Psi(3,3,3/2,3/2)$ | $\Phi(2^+;1^+;0^+)$ |
| | $\Psi(3,2,3/2,3/2)$ | $\Phi(2^+;1^+;-1^+)$ |
| | $\Psi(3,1,3/2,3/2)$ | $\sqrt{2/5}\Phi(2^+;1^+;-2^+)+\sqrt{3/5}\Phi(2^+;0^+;-1^+)$ |
| | $\Psi(3,0,3/2,3/2)$ | $\sqrt{4/5}\Phi(2^+;0^+;-2^+)+\sqrt{1/5}\Phi(1^+;0^+;-1^+)$ |
| | $\Psi(3,-1,3/2,3/2)$ | $\sqrt{2/5}\Phi(2^+;-1^+;-2^+)+\sqrt{3/5}\Phi(1^+;0^+;-2^+)$ |
| | $\Psi(3,-2,3/2,3/2)$ | $\Phi(1^+;-1^+;-2^+)$ |
| | $\Psi(3,-3,3/2,3/2)$ | $\Phi(0^+;-1^+;-2^+)$ |
| $^4P$ | $\Psi(1,1,3/2,3/2)$ | $-\sqrt{3/5}\Phi(2^+;1^+;-2^+)+\sqrt{2/5}\Phi(2^+;0^+;-1^+)$ |
| | $\Psi(1,0,3/2,3/2)$ | $-\sqrt{1/5}\Phi(2^+;0^+;-2^+)+\sqrt{4/5}\Phi(1^+;0^+;-1^+)$ |
| | $\Psi(1,-1,3/2,3/2)$ | $-\sqrt{3/5}\Phi(2^+;-1^+;-2^+)+\sqrt{2/5}\Phi(1^+;0^+;-2^+)$ |

The functions for $M_S = \frac{1}{2}$, $M_S = -\frac{1}{2}$ and $M_S = -\frac{3}{2}$ are obtained from the functions given here by applying the lowering operator $\boldsymbol{S}_x - i\boldsymbol{S}_y$ once, twice or three times, respectively.

*equivalent* to the $\Psi$ basis. When, for example, a perturbation calculation is carried out in connection with the central-field approximation, it must have no influence on the result if one starts from the $\Phi$ basis or from the $\Psi$ basis.

### 1.2.7. *Electron interaction*

We arrived at the central-field approximation of equation (1–42) because we neglected the operator $H_1$ (1–41), a part of the complete *Hamiltonian*, equation (1–38). In the central-field approximation all states of a configuration have the same energy, they are degenerate. In describing the states in the central-field approximation one can equally well use the $\Phi$ basis or the $\Psi$ basis. We wish to approximately correct the error made by neglecting the operator $H_1$. The *perturbation theory for degenerate systems* gives us the necessary procedure*. In the following we limit ourselves to states arising from the same configuration.

In the central-field approximation (in an 'unperturbed' system) let there be $r$ states, e.g., for the configuration $d^2$, $r = 45$, all of which have the energy $E_0$. The orthonormalized functions corresponding to these states are called

$$\Phi_1, \Phi_2, \ldots, \Phi_r$$

or

$$\Psi_1, \Psi_2, \ldots, \Psi_r.$$

The operator $H_1$ which was neglected in the central-field approximation is now treated as a *perturbation* of the 'unperturbed' system. When the perturbation is applied the energies of the single states of the system are in general changed. We denote these energy changes with $\Delta E$. According to the perturbation theory for degenerate systems the $\Delta E$ are given to the first approximation by the roots of the following *secular determinant*†:

$$\begin{vmatrix} H_{11}-\Delta E & H_{12} \ldots\ldots & H_{1r} \\ H_{21} & H_{22}-\Delta E \ldots & H_{2r} \\ \vdots & \vdots \quad \ddots & \vdots \\ H_{r1} & H_{r2} \ldots & H_{rr}-\Delta E \end{vmatrix} = 0. \tag{1–82}$$

If one begins with the $\Phi$ basis, the matrix elements $H_{ij}$ are defined by the integrals‡ $H_{ij} = (\Phi_i, H_1\Phi_j)$. If the $\Psi$ basis is used, then $H_{ij} = (\Psi_i, H_1\Psi_j)$

---

* Cf. i.e. Pauling and Wilson, *op. cit.*, chapter 6.

† Cf. Appendix, p. 474.

‡ We use the notation given in the footnote on p. 238. The explicit form of the integral $H_{ij}$ is obtained by setting f = $\Phi_i$ and g = $H_1\Phi_j$.

holds. The determinant (1–82) has as many columns and rows as there are degenerate $\Phi$ or $\Psi$ functions. For the $(nd)^2$ configuration, for example, one has a $45 \times 45$ determinant.

In principle one determines the $r$ roots $\Delta E_1, \Delta E_2, \ldots, \Delta E_r$ of the secular determinant, equation (1–82), by calculating the integrals $H_{ij}$ and resolving the determinant according to powers of $\Delta E$. The zeros of the polynomial

$$(\Delta E)^r - (H_{11} + H_{22} + \cdots + H_{rr})(\Delta E)^{r-1} + \cdots + (H_{11} H_{22} \cdots H_{rr}) = 0 \tag{1–82a}$$

are the roots of the secular determinant. How extensive this problem can become is seen from the example of the $(nd)^2$ configuration. Here there are $45 \times 45 = 2025$ matrix elements $H_{ij}$ to calculate.

The solution of the problem is greatly simplified when two theorems concerning eigenvalue problems are used*:

*Theorem 1:*

If two functions $f_i$ and $f_j$ are eigenfunctions of a Hermitian operator $\boldsymbol{a}$ belonging to *different* eigenvalues $F_i$ and $F_j$, i.e.,

$$\begin{aligned} \boldsymbol{a} f_i &= F_i f_i, \\ \boldsymbol{a} f_j &= F_j f_j, \end{aligned} \qquad F_i \neq F_j,$$

and if $\boldsymbol{a}$ commutes with the Hermitian operator $\boldsymbol{b}$, i.e.,

$$\boldsymbol{a}\boldsymbol{b} - \boldsymbol{b}\boldsymbol{a} = 0,$$

then the integral

$$(f_i, \boldsymbol{b} f_j) = 0$$

is valid. (*Non-combination theorem.*)

*Theorem 2:*

If a Hermitian operator $\boldsymbol{b}$ commutes with $\boldsymbol{L}^2$, $\boldsymbol{S}^2$, $\boldsymbol{L}_z$, $\boldsymbol{S}_z$ then it holds for two angular momentum eigenfunctions $\Psi_i(L, M_L, S, M_S)$ and $\Psi_j(L, M_L, S, M_S)$ that:

$$(\Psi_i(L, M_L, S, M_S), \boldsymbol{b}\Psi_j(L, M_L, S, M_S)) = (\Psi_i(L, M'_L, S, M'_S), \boldsymbol{b}\Psi_j(L, M'_L, S, M'_S)),$$

i.e., the integrals are independent of the $M_L$ and $M_S$ values. The case $\Psi_i = \Psi_j$ is included†.

* Cf., e.g. Condon and Shortley[4], p. 16 and p. 49.

† The case $i \neq j$ arises only if two or more equal terms arise from one configuration, for example $d^3$: two $^2D$ terms.

We make several comments concerning the application of the theorems:

The operators $\boldsymbol{L}^2$, $\boldsymbol{S}^2$, $\boldsymbol{L}_z$, $\boldsymbol{S}_z$ and $\boldsymbol{H}_1$ are Hermitian. $\boldsymbol{L}^2$, $\boldsymbol{S}^2$, $\boldsymbol{L}_z$, $\boldsymbol{S}_z$ commute with $\boldsymbol{H}_1$. In theorem 1 we may set $\boldsymbol{a} = \boldsymbol{L}^2$ or $\boldsymbol{S}^2$, $\boldsymbol{L}_z$ or $\boldsymbol{S}_z$ and in both theorems $\boldsymbol{b} = \boldsymbol{H}_1$.

If the $\Phi$ basis is chosen, theorem 2 cannot be applied because the $\Phi$ are not in general eigenfunctions of $\boldsymbol{L}^2$ and $\boldsymbol{S}^2$. The $\Phi$ are, however, eigenfunctions of $\boldsymbol{L}_z$ and $\boldsymbol{S}_z$. According to theorem 1, a matrix element $H_{ij} = (\Phi_i, \boldsymbol{H}_1\Phi_j)$ vanishes when $\Phi_i$ and $\Phi_j$ do not have the same eigenvalues $M_L\hbar$ or $M_S\hbar$, that is, when $\Phi_i$ and $\Phi_j$ are not in the same $M_L, M_S$ box. If the secular determinant (1–82) is formulated from the beginning so that the matrix elements $H_{ij}$ which are formed with $\Phi$ functions from the same $M_L, M_S$ box are as close as possible to one another, then the secular determinant assumes the form of a *block determinant**. Because the subdeterminants along the principal diagonal may each individually be set equal to zero, the secular determinant, which is usually of high order, is reduced to a set of subdeterminants of lower order. The order of each subdeterminant is equal to the number of $\Phi$ functions in the corresponding $M_L, M_S$ box.

A further aid for practical calculations with the $\Phi$ basis is the so-called *diagonal sum rule*. This is based on the fact that the sum of the roots of a secular determinant is equal to the sum of the diagonal elements $H_{ii}$ (cf. (1–82a)):

$$\Delta E_1 + \Delta E_2 + \cdots + \Delta E_r = H_{11} + H_{22} + \cdots + H_{rr}.$$

Calculations with the $\Phi$ basis is advantageous when only the *energy* of an atomic many-electron system is desired. If besides the energy the functions $\Psi(L, M_L, S, M_S)$ are also to be determined, it is then more convenient to calculate the energy using the $\Psi$ basis. We wish to follow the latter case further and will not treat the diagonal sum rule†.

If one chooses as the basis for the calculation of $\Delta E$ the $\Psi(L, M_L, S, M_S)$, then the secular determinant (1–82) assumes a particularly simple form. The $\Psi$ belonging to the various terms are distinguished by the $L, S$ value pair. According to theorem 1 all off-diagonal elements $H_{ij} = (\Psi_i, \boldsymbol{H}_1\Psi_j)$ which are formed with $\Psi$ functions of different terms vanish.

If each term appears only once (example: $(nd)^2$ configuration), then all matrix elements with $\Psi$ functions of the same term belonging to different pairs of values $M_L, M_S$ vanish as well. In this case all off-diagonal elements are zero and only the diagonal elements $H_{ii}$ are in general nonzero. The

---

* Cf. Appendix, p. 476.

† Detailed examples for the application of the diagonal sum rule are found, e.g., in Condon and Shortley[4], pp. 191 ff.

secular determinant is then a step determinant in which subdeterminants of the first order of the form $H_{ii}-\Delta E$ lie along the principal diagonal:

$$\begin{vmatrix} H_{11}-\Delta E & 0 & 0 & \cdots\cdots & 0 \\ 0 & H_{22}-\Delta E & 0 & \cdots\cdots & 0 \\ 0 & 0 & H_{33}-\Delta E & \cdots\cdots & 0 \\ \vdots & \vdots & \vdots & \ddots & \vdots \\ 0 & 0 & 0 & \cdots\cdots & H_{rr}-\Delta E \end{vmatrix} = 0 .$$

Because the principal determinant must vanish, each of the subdeterminant blocks is then individually set equal to zero:

$$H_{ii}-\Delta E=0, \quad i=1, 2, \ldots, r.$$

All functions $\Psi(L, M_L, S, M_S)$ of a term are eigenfunctions of $\boldsymbol{L}^2$ and $\boldsymbol{S}^2$ with the same eigenvalues $L(L+1)\hbar^2$ or $S(S+1)\hbar^2$. According to theorem 2 the diagonal elements $H_{ii}$ which are formed with the $\Psi$ functions of the same term are all the same: The energies of the states of the same term are all changed by the same amount $\Delta E$ under the perturbation $\boldsymbol{H}_1$. States belonging to different terms, however, undergo in general different changes in energy under perturbation. In other words: The states of a given electron configuration which are all degenerate in the central-field approximation energetically split into terms under the influence of a perturbation $\boldsymbol{H}_1$. States of the same term remain degenerate. Because the $\Delta E = H_{ii}$ of the states of the same term are equal, it is sufficient to calculate only one matrix element $H_{ii}$ for each term. This value is equal to the change in energy undergone by every state of the term.

Example: $(nd)^2$ configuration (cf. Figure B.6)

The terms $^1S$, $^3P$, $^1D$, $^3F$, $^1G$ appear only once. The secular determinant is thus completely diagonal. Because the changes in energy of the states of a given term are all the same, we need to give only one diagonal element for each term:

$$\begin{aligned}
\Delta E(^1S) &= (\Psi(0,0,0,0), \boldsymbol{H}_1\Psi(0,0,0,0)), \\
\Delta E(^3P) &= (\Psi(1,1,1,1), \boldsymbol{H}_1\Psi(1,1,1,1)), \\
\Delta E(^1D) &= (\Psi(2,2,0,0), \boldsymbol{H}_1\Psi(2,2,0,0)), \\
\Delta E(^3F) &= (\Psi(3,3,1,1), \boldsymbol{H}_1\Psi(3,3,1,1)), \\
\Delta E(^1G) &= (\Psi(4,4,0,0), \boldsymbol{H}_1\Psi(4,4,0,0)).
\end{aligned}$$

If a term appears several times (e.g., $(nd)^3$ configuration: the $^2D$ term is double), say $k$ times, then for each possible set of values $L$, $M_L$, $S$, $M_S$

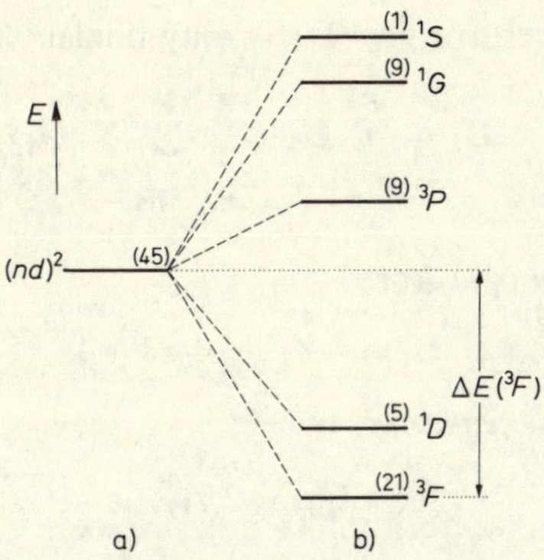


Figure B.6. Term diagram for the $(nd)^2$ configuration: (a) in the central-field approximation; (b) with consideration of electron interaction. The numbers in brackets give the degree of degeneracy of the individual terms.

there exist exactly $k$ linearly independent $\Psi$ functions $\Psi_a(L, M_L, S, M_S)$, $\Psi_b(L, M_L, S, M_S), \ldots, \Psi_k(L, M_L, S, M_S)$. From the theorems 1 and 2 it follows for this case: the part of the secular determinant which corresponds to the term appearing $k$ times is a step-determinant having $(2L+1)(2S+1)$ subdeterminants of order $k$ along the diagonal. Every subdeterminant contains matrix elements which are formed with $\Psi$ functions having the same set of values $L$, $M_L$, $S$, $M_S$. The subdeterminants have the same roots $\Delta E$. It is, therefore, sufficient when the roots of a *single* subdeterminant are calculated. Because the subdeterminants of a term appearing $k$ times are of degree $k$, one finds in general $k$ different roots $\Delta E$: A term appearing $k$ times splits under the influence of the perturbation $\boldsymbol{H}_1$ in general into $k$ terms. Each of these terms is $(2L+1)(2S+1)$-fold degenerate.

1.2.8. *The matrix elements* $(\Psi, \boldsymbol{H}_1 \Psi)$

Because the $\Psi$ functions are linear combinations of the $\Phi$

$$\Psi_\varkappa = c_{\varkappa_1}\Phi_1 + c_{\varkappa_2}\Phi_2 + \cdots + c_{\varkappa_r}\Phi_r + \cdots,$$
$$\Psi_\lambda = c_{\lambda_1}\Phi_1 + c_{\lambda_2}\Phi_2 + \cdots + c_{\lambda_s}\Phi_s + \cdots,$$

an integral $(\Psi_\kappa, \boldsymbol{H}_1 \Psi_\lambda)$ is made up additively of subintegrals of the form

$$h_{rs} = (\Phi_r, \boldsymbol{H}_1 \Phi_s) \tag{1–83}$$

The perturbation operator $\boldsymbol{H}_1$ (1–41) may be decomposed into two parts

$$\boldsymbol{H}_1 = \sum_{i=1}^{N} \boldsymbol{U}_i + \frac{1}{2} \sum_{i=1}^{N} \sum_{\substack{j=1 \\ i \neq j}}^{N} \boldsymbol{Q}_{ij}, \tag{1–84}$$

with the *one-electron operators*

$$\boldsymbol{U}_i = -Ze^2/r_i + eV(r_i) \tag{1–85}$$

and the *two-electron operators*

$$\boldsymbol{Q}_{ij} = e^2/r_{ij}. \tag{1–86}$$

If in the integral $h_{rs}$ we express the perturbation operator $\boldsymbol{H}_1$ as in (1–84) in terms of the $\boldsymbol{U}_i$ and $\boldsymbol{Q}_{ij}$, then we obtain

$$h_{rs} = \sum_{i=1}^{N} (\Phi_r, \boldsymbol{U}_i \Phi_s) + \frac{1}{2} \sum_{i=1}^{N} \sum_{\substack{j=1 \\ i \neq j}}^{N} (\Phi_r, \boldsymbol{Q}_{ij} \Phi_s). \tag{1–87}$$

As we have seen earlier (cf. p. 225) the $\Phi$ functions are antisymmetrized products of one-electron functions $\psi_{a_i}$. In particular let the two functions $\Phi_r$ and $\Phi_s$ have the following form:

$$\Phi_r = \Phi_r(a_1^{(r)}, a_2^{(r)}, \ldots, a_N^{(r)}),$$
$$\Phi_s = \Phi_s(a_1^{(s)}, a_2^{(s)}, \ldots, a_N^{(s)}).$$

The $a_i$ are the symbols for the four quantum numbers $n_i$, $l_i$, $m_{l_i}$ $m_{s_i}$ defined on p. 225. In order to distinguish the $a_i$ in $\Phi_r$ from the $a_i$ in $\Phi_s$ the superscripts $r$ or $s$ were added to the $a_i$. The calculation of the $h_{rs}$ begins with a comparison of $\Phi_r$ with $\Phi_s$. First it is seen if one of the $a_1^{(s)}, a_2^{(s)}, \ldots, a_N^{(s)}$ is equal to $a_1^{(r)}$, if one is equal to $a_2^{(r)}$, etc. If equal $a$ exist, one permutes the $a_i^{(s)}$ until all $a^{(s)}$ which are equal to an $a^{(r)}$ occupy the same place in the function $\Phi_s$ as the corresponding $a^{(r)}$ in the function $\Phi_r$. It must, however, be kept in mind that $\Phi_s$ changes its sign under odd permutation of the $a^{(s)}$ and keeps the original sign under even permutation. Assume that the $a^{(s)}$ have been ordered according to this procedure. The function thus arising from $\Phi_s$ is called $\Phi_s'$. $\Phi_s' = \pm\Phi_s$ where the minus sign is taken when $\Phi_s'$ arises from $\Phi_s$ under an odd permutation of the $a^{(s)}$.

Example: Let

$$\Phi_r = \Phi_r(2^+; 2^-; 0^+; 0^-),$$
$$\Phi_s = \Phi_s(2^-; 1^+; 1^-; 0^+).$$

In both $\Phi$ functions $2^-$ and $0^+$ appear. In order to bring $2^-$ and $0^+$ in $\Phi_s$ to the same positions which they have in $\Phi_r$ we exchange in $\Phi_s$ $2^-$ with $1^+$

and $1^-$ with $0^+$:

$$\Phi'_s = \Phi'_s(1^+; 2^-; 0^+; 1^-).$$

We have thus *exchanged two pairs* and the permutation is even. The $\Phi_s$ function resulting from the permutation has the same sign as $\Phi_s$: $\Phi'_s = \Phi_s$. The $a_i$ in the function $\Phi'_s$ are denoted with a bar: $\Phi'_s = \Phi'_s(\bar{a}_1^{(s)}, \bar{a}_2^{(s)}, \ldots, \bar{a}_N^{(s)})$. The special notation is necessary because the series $\bar{a}_1^{(s)}, \bar{a}_2^{(s)}, \ldots, \bar{a}_N^{(s)}$ in general arises from a permutation of the series $a_1^{(s)}, a_2^{(s)}, \ldots, a_N^{(s)}$ and is therefore not identical with it. We write then the integral $h_{rs}$ with $\Phi'_s$:

$$h_{rs} = \pm\{\sum_{i=1}^{N} (\Phi_r, U_i\Phi'_s) + \frac{1}{2}\sum_{\substack{i=1\\i\neq j}}^{N}\sum_{j=1}^{N} (\Phi_r, Q_{ij}\Phi'_s)\}. \tag{1–88}$$

It should be remembered that the one-electron functions $\psi_a$, which compose the $\Phi$ functions are orthogonal and normalized to one. The theory states that under this condition four cases are to be differentiated when calculating the $h_{rs}$ with reference to the $\Phi_r$ and $\Phi_s$ and that the following two types of integrals involving one-electron functions $\psi_a$ appear:

$$U(a_i^{(r)}; \bar{a}_i^{(s)}) \equiv (\psi_{a_i^{(r)}}(i), U_i\psi_{\bar{a}_i^{(s)}}(i)), \tag{1–89a}$$

$$V(a_i^{(r)}; a_j^{(r)}; \bar{a}_p^{(s)}; \bar{a}_q^{(s)}) \equiv (\psi_{a_i^{(r)}}(i)\psi_{a_j^{(r)}}(j), Q_{ij}\psi_{\bar{a}_p^{(s)}}(i)\psi_{\bar{a}_q^{(s)}}(j)). \tag{1–89b}$$

The four cases and the corresponding $h_{rs}$ values are:

(a) $\Phi_r = \Phi'_s$. Both functions are the same. For every $a_i^{(r)}$ there is an equal $\bar{a}_i^{(s)}$: $a_i^{(r)} = \bar{a}_i^{(s)} = a_i$.

Example:

$$\bar{a}_i^{(s)}: a_i^{(r)} = \bar{a}_i^{(s)} = a_i.$$

$$\Phi_r = \Phi_r(2^+; 2^-; 0^+), \quad \Phi'_s = \Phi'_s(2^+; 2^-; 0^+)$$

$$h_{rs} = \pm\{\sum_{i=1}^{N} U(a_i; a_i) + \frac{1}{2}\sum_{\substack{i=1\\i\neq j}}^{N}\sum_{j=1}^{N} [V(a_i; a_j; a_i; a_j) - V(a_i; a_j; a_j; a_i)]\}$$

(b) $\Phi_r \neq \Phi'_s$. Except for $a_p^{(r)}$ there exists for every $a_i^{(r)}$, $i \neq p$, an equal $\bar{a}_i^{(s)}$: $a_i^{(r)} = \bar{a}_i^{(s)} = a_i$ for $i \neq p$ and $a_p^{(r)} \neq \bar{a}_p^{(s)}$.

Example: $\Phi_r = \Phi_r(2^+; 2^-; 0^+)$, $\Phi'_s = \Phi'_s(2^+; 1^-; 0^+)$. $a_p^{(r)}$ is here $2^-$ and $\bar{a}_p^{(s)} = 1^-$.

$$h_{rs} = \pm\{U(a_p^{(r)}; \bar{a}_p^{(s)}) + \sum_{\substack{i=1\\(i\neq p)}}^{N} [V(a_p^{(r)}; a_i; \bar{a}_p^{(s)}; a_i) - V(a_p^{(r)}; a_i; a_i; \bar{a}_p^{(s)})]\}$$

(c) $\Phi_r \neq \Phi'_s$. Except for $a_p^{(r)}$ and $a_q^{(r)}$ there exists for every $a_i^{(r)}$ an equal $\bar{a}_i^{(s)}$: $a_i^{(r)} = \bar{a}_i^{(s)} = a_i$ for $i \neq p, q$, and $a_p^{(r)} \neq \bar{a}_p^{(s)}$, $a_q^{(r)} \neq \bar{a}_q^{(s)}$.

Example: $\Phi_r = \Phi_r(2^+; 2^-; 0^+)$, $\Phi'_s = \Phi'_s(1^+; 1^-; 0^+)$

Here we have $a_p^{(r)} = 2^+; \bar{a}_p^{(s)} = 1^+$ and $a_q^{(r)} = 2^-; \bar{a}_q^{(s)} = 1^-$.

$$\Phi_r = \Phi_r(2^+; 2^-; 0^+),\ \Phi'_s = \Phi'_s(1^+; 1^-; 0^+)$$
$$a_p^{(r)} = 2^+; \bar{a}_p^{(s)} = 1^+ \qquad a_q^{(r)} = 2^-; \bar{a}_q^{(s)} = 1^-.$$
$$h_{rs} = \pm\{V(a_p^{(r)}; a_q^{(r)}; \bar{a}_p^{(s)}; \bar{a}_q^{(s)}) - V(a_p^{(r)}; a_q^{(r)}; \bar{a}_q^{(s)}; \bar{a}_p^{(s)})\}$$

(d) $\Phi_r \neq \Phi'_s$. There exist more than two $a^{(r)}$ for which there are no equal $\bar{a}^{(s)}$.

Example:

$$\Phi_r = \Phi_r(2^+; 2^-; 0^+),\ \Phi'_s = \Phi'_s(1^+; 1^-; -2^+)$$
$$h_{rs} = 0.$$

After having expressed the $h_{rs}$ in terms of the integrals $U$ and $V$ we turn to the problem of evaluating these integrals specifically. We state without proof:

$$U(a_i; a_j) = I(n_i, l_i; n_j, l_j)\,\delta_{l_i, l_j}\,\delta_{m_{l_i}, m_{l_j}}\,\delta_{m_{s_i}, m_{s_j}} \tag{1–90a}$$

with

$$I(n_i, l_i; n_j, l_j) = \int_0^\infty R^*_{n_i, l_i}(r)\,\boldsymbol{U}\,R_{n_j, l_j}(r)\,r^2\,dr, \tag{1–90b}$$

where the $R_{n,l}(r)$ are the radial parts of the one-electron functions $\psi_{n,l,m_l,m_s}$. Further we have:

$$V(a_i; a_j; a_p; a_q) = \delta_{m_{s_i}, m_{s_p}}\,\delta_{m_{s_j}, m_{s_q}}\,\delta_{m_{l_i}+m_{l_j},\, m_{l_p}+m_{l_q}} \times$$
$$\times \sum_{k=0}^{\infty} c^k(l_i, m_{l_i}; l_p, m_{l_p})\,c^k(l_q, m_{l_q}; l_j, m_{l_j})\,R^k(n_i, l_i; n_j, l_j; n_p, l_p; n_q, l_q) \tag{1–91}$$

with

$$c^k(l_i, m_{l_i}; l_j, m_{l_j}) = \left(\frac{4\pi}{2k+1}\right)^{\frac{1}{2}} \int_0^\pi \int_0^{2\pi} Y^*_{l_i, m_{l_i}}(\theta, \phi)\,Y_{k, m_{l_i}-m_{l_j}}(\theta, \phi) \times$$
$$\times Y_{l_j, m_{l_j}}(\theta, \phi) \sin\theta\,d\theta\,d\phi \tag{1–92a}$$

and

$$R^k(n_i, l_i; n_j, l_j; n_p, l_p; n_q, l_q) =$$
$$= e^2 \int_0^\infty \int_0^\infty R^*_{n_i, l_i}(r_i)\,R^*_{n_j, l_j}(r_j)\,\frac{r_<^k}{r_>^{k+1}}\,R_{n_p, l_p}(r_i)\,R_{n_q, l_q}(r_j)\,r_i^2 r_j^2\,dr_i\,dr_j. \tag{1–92b}$$

Here $r_<$ and $r_>$ are respectively the smaller and the larger of $r_i$ and $r_j$*. In particular we have

$$R^k(n_i, l_i; n_j, l_j; n_i, l_i; n_j, l_j) = F^k(n_i, l_i; n_j, l_j), \quad (1\text{–}93)$$

$$R^k(n_i, l_i; n_j, l_j; n_j, l_j; n_i, l_i) = G^k(n_i, l_i; n_j, l_j). \quad (1\text{–}94)$$

If all values $n$ and $l$ are equal, then $R^k = F^k$ holds. Often the functions with superscripts $F^k$ are replaced by functions with subscripts $F_k$. In particular for $l_i = l_j = 2$ the relations

$$\begin{aligned} F_0 &= F^0, \\ F_2 &= \frac{1}{49} F^2, \\ F_4 &= \frac{1}{441} F^4. \end{aligned} \quad (1\text{–}95a)$$

follow by definition. In the ligand field theory it has become conventional to express the $F_k$ functions with $l_i = l_j = 2$ in terms of the so-called *Racah* parameters $A$, $B$, $C$. These are defined as follows

$$\begin{aligned} A &= F_0 - 49F_4, \\ B &= F_2 - 5F_4, \\ C &= 35F_4. \end{aligned} \quad (1\text{–}95b)$$

The relation

$$c^k(l_i, m_{l_i}; l_j, m_{l_j}) = (-1)^{m_{l_j} - m_{l_i}} c^k(l_j, m_{l_j}; l_i, m_{l_i}). \quad (1\text{–}96)$$

holds for the coefficients $c^k$ defined in equation (1–92). The nonzero values of $c^k$ are tabulated for $l = 0, 1, 2$ in Table B.8.

We shall carry out the calculation of perturbation energies for a simple example.

The term $^3F$ belongs to the $(nd)^2$ configuration. On p. 246 it was established that

$$\Delta E(^3F) = \big(\Psi(3, 3, 1, 1),\ H_1 \Psi(3, 3, 1, 1)\big)$$

holds. From equation (1–81) we see that

$$\Psi(3, 3, 1, 1) = \Phi(2^+; 1^+).$$

* It follows that

$$\int_0^\infty dr_1 \int_0^\infty dr_2\, f(r_1, r_2) \frac{r_<^k}{r_>^{k+1}} =$$

$$= \int_0^\infty dr_1 \int_0^{r_1} dr_2\, f(r_1, r_2) \frac{r_2^k}{r_1^{k+1}} + \int_0^\infty dr_1 \int_{r_1}^\infty dr_2\, f(r_1, r_2) \frac{r_1^k}{r_2^{k+1}}.$$

Table B.8a. $c^k(l, m_l, l', m_l'$ for $l+l'$ odd. We write $c^k = \pm\sqrt{x/D_k}$. Only the sign of the root and $x$ are entered in the table. $D_k$ depends only upon $l$ and $l'$ and is given at the head of each column. After Condon and Shortley[4], pp. 178–9.

| $ll'$ | $m_l$ | $m_l'$ | $k = 1$ | 3 |
|---|---|---|---|---|
| $sp$ | 0 | ±1 | $-\sqrt{1/3}$ | |
| | 0 | 0 | +1 | |
| $pd$ | ±1 | ±2 | $-\sqrt{6/15}$ | $+\sqrt{3/245}$ |
| | ±1 | ±1 | +3 | −9 |
| | ±1 | 0 | −1 | +18 |
| | 0 | ±2 | 0 | +15 |
| | 0 | ±1 | −3 | −24 |
| | 0 | 0 | +4 | +27 |
| | ±1 | ∓2 | 0 | +45 |
| | ±1 | ∓1 | 0 | −30 |

Table B.8b. $c^k(l, m_l\,; l', m_l')$ for $l+l'$ even

| $ll'$ | $m_l$ | $m_l'$ | $k = 0$ | 2 | 4 |
|---|---|---|---|---|---|
| $ss$ | 0 | 0 | +1 | | |
| $sd$ | 0 | ±2 | | $+\sqrt{1/5}$ | |
| | 0 | ±1 | | −1 | |
| | 0 | 0 | | +1 | |
| $pp$ | ±1 | ±1 | +1 | $-\sqrt{1/25}$ | |
| | ±1 | 0 | 0 | +3 | |
| | 0 | 0 | +1 | +4 | |
| | ±1 | ∓1 | 0 | −6 | |
| $dd$ | ±2 | ±2 | +1 | $-\sqrt{4/49}$ | $+\sqrt{1/441}$ |
| | ±2 | ±1 | 0 | +6 | −5 |
| | ±2 | 0 | 0 | −4 | +15 |
| | ±1 | ±1 | +1 | +1 | −16 |
| | ±1 | 0 | 0 | +1 | +30 |
| | 0 | 0 | +1 | +4 | +36 |
| | ±2 | ∓2 | 0 | 0 | +70 |
| | ±2 | ∓1 | 0 | 0 | −35 |
| | ±1 | ∓1 | 0 | −6 | −40 |

This leads to

$$\Delta E(^3F) = \left(\Phi(2^+; 1^+),\ \boldsymbol{H}_1\Phi(2^+; 1^+)\right).$$

$\Delta E$ consists therefore of only one $h_{rs}$ integral, and we have the case (a) explained on p. 249. It therefore holds that

$$\begin{aligned}\Delta E(^3F) = U(2^+; 2^+) + U(1^+; 1^+) \\ + \frac{1}{2}\{V(2^+; 1^+; 2^+; 1^+) - V(2^+; 1^+; 1^+; 2^+) \\ + V(1^+; 2^+; 1^+; 2^+) - V(1^+; 2^+; 2^+; 1^+)\},\end{aligned}$$

With consideration of equations (1–90, 91, 93) this becomes

$$\begin{aligned}\Delta E(^3F) = 2I(nd; nd) \\ + \frac{1}{2}\sum_{k=0}^{\infty}\{c^k(2, 2; 2, 2)c^k(2, 1; 2, 1) \\ - c^k(2, 2; 2, 1)c^k(2, 2; 2, 1) \\ + c^k(2, 1; 2, 1)c^k(2, 2; 2, 2) \\ - c^k(2, 1; 2, 2)c^k(2, 1; 2, 2)\}F^k(nd; nd)\end{aligned}$$

and because of equation (1–96)

$$\begin{aligned}\Delta E(^3F) = 2I(nd; nd) + \sum_{k=0}^{\infty}\{c^k(2, 2; 2, 2)c^k(2, 1; 2, 1) \\ - c^k(2, 2; 2, 1)c^k(2, 2; 2, 1)\}F^k(nd; nd).\end{aligned}$$

The values of the coefficients $c^k$ can be taken from Table B.8b

$$\begin{aligned}\Delta E(^3F) &= 2I(nd; nd) + \{1 \cdot 1 - 0 \cdot 0\}F^0(nd; nd) \\ &\quad + \left\{\left(-\frac{\sqrt{4}}{7}\right)\frac{1}{7} - \frac{\sqrt{6}}{7}\,\frac{\sqrt{6}}{7}\right\}F^2(nd; nd) \\ &\quad + \left\{\frac{1}{21}\left(-\frac{\sqrt{16}}{21}\right) - \left(-\frac{\sqrt{5}}{21}\right)\left(-\frac{\sqrt{5}}{21}\right)\right\}F^4(nd; nd) \\ &= 2I(nd; nd) \\ &\quad + F^0(nd; nd) - \frac{8}{49}F^2(nd; nd) - \frac{9}{441}F^4(nd; nd)\end{aligned}$$

or with consideration of equation (1–95)

$$\Delta E(^3F) = 2I(nd; nd) + A - 8B.$$

In Table B.9 the energies $\Delta E(^{2S+1}L) - NI$ are given for the configurations $(nd)^N$, $N = 2, 3, 4, 5$. As a result of the equivalence of electrons and holes the term separations in a $d^N$ system are equal to those of a $d^{10-N}$ system.

Therefore the term separations for configurations $(nd)^N$ with $N = 6, 7, 8$ can also be obtained from Table **B**.9.

The term diagram of the free $Cr^{3+}$ ion in Figure A.14 of Part A gives us an idea as to the accuracy with which term systems can be calculated using the Slater theory of free atoms and atomic ions as explained here. In the left-hand portion of the figure the empirically determined term positions are shown, with reference to the energy of the ground state term, $^4F$. The theoretical term energies are functions of the Racah parameter $A, B, C$, cf. Table B.9 for $d^3$. The term separations depend solely upon $B$ and $C$. If $B$ is chosen so that the calculated energies of the term $^4P$ referred to the energy of the $^4F$ ground state term agree with the empirical $^4P$ energy $E(^4P)-E(^4F) = 15B$, then one finds $B \approx 920\,\text{cm}^{-1}$. If further we choose $C = 4B$, cf. pp. 76 ff., then one obtains $C = 3680\,\text{cm}^{-1}$. The term positions given in the right-hand side of Figure A.14 are obtained with the help of these parameter values.

Table B.9. The energies $\Delta E(^{2S+1}L)-N\cdot I$ for the configurations $(nd)^N$, $N = 2, 3, 4, 5$, expressed in terms of the *Racah* parameters $A, B, C$

| $d^2$ | $d^3$ |
|---|---|
| $^3F = A-8B$ | $^4F = 3A-15B$ |
| $^3P = A+7B$ | $^4P = 3A$ |
| $^1G = A+4B+2C$ | $^2H = {}^2P = 3A-6B+3C$ |
| $^1D = A-3B+2C$ | $^2G = 3A-11B+3C$ |
| $^1S = A+14B+7C$ | $^2F = 3A+9B+3C$ |
| | $a, b\,^2D = 3A+5B+5C\pm(193B^2+$ |
| | $+8BC+4C^2)^{\frac{1}{2}}$ |
| $d^4$ | $d^5$ |
| $^5D = 6A-21B$ | $^6S = 10A-35B$ |
| $^3H = 6A-17B+4C$ | $^4G = 10A-25B+5C$ |
| $^3G = 6A-12B+4C$ | $^4F = 10A-13B+7C$ |
| $a,b\,^3F = 6A-5B+\frac{11}{2}C$ | $^4D = 10A-18B+5C$ |
| $\pm\frac{3}{2}(68B^2+4BC+C^2)^{\frac{1}{2}}$ | $^4P = 10A-28B+7C$ |
| $^3D = 6A-5B+4C$ | $^2I = 10A-24B+8C$ |
| $a,b\,^3P = 6A-5B+\frac{11}{2}C$ | $^2H = 10A-22B+10C$ |
| | $a\,^2G = 10A-13B+8C$ |
| $\pm\frac{1}{2}(912B^2-24BC+9C^2)^{\frac{1}{2}}$ | $b\,^2G = 10A+3B+10C$ |
| $^1I = 6A-15B+6C$ | $a\,^2F = 10A-9B+8C$ |
| $a,b\,^1G = 6A-5B+\frac{15}{2}C$ | $b\,^2F = 10A-25B+10C$ |
| $\pm\frac{1}{2}(708B^2-12BC+9C^2)^{\frac{1}{2}}$ | $c\,^2D = 10A-4B+10C$ |
| $^1F = 6A+6C$ | $a, b\,^2D = 10A-3B+11C$ |
| $a,b\,^1D = 6A+9B+\frac{15}{2}C$ | $\pm 3(57B^2+2BC+C^2)^{\frac{1}{2}}$ |
| $\pm\frac{3}{2}(144B^2+8BC+C^2)^{\frac{1}{2}}$ | $^2P = 10A+20B+10C$ |
| $a,b\,^1S = 6A+10B+10C$ | $^2S = 10A-3B+8C$ |
| $\pm 2(193B^2+8BC+4C^2)^{\frac{1}{2}}$ | |

# 2. Group theory

Complex ions consist of a central ion, in the case of the metal complexes considered here a transition metal ion, about which the ligands (molecules or ions) are grouped. For polynuclear complexes several central ions are present which are bound to one another by bridge ligands. The ligands surround the central ion in a certain arrangement in space, thus giving the complex ion a specific geometrical structure. These structures have a definite *symmetry.* Many physical properties of such complexes can be qualitatively understood purely from a knowledge of the symmetry of the complex ions. The relationship between symmetry and physical properties can be seen by applying the *theory of groups and representations* to the quantum mechanics of these compounds.

We shall begin this chapter by defining the terms symmetry and symmetry operations and explaining them on the basis of specific examples. Next those elements of group theory which are of importance for the ligand field theory will be introduced, again with reference to specific examples. Finally the connection between group theory and quantum mechanics will be treated.

Also, in this chapter several general theorems, in particular from the theory of groups and representations, will be introduced without proof and applied.

In the bibliography on p. 499 ff. several volumes are cited which treat group theory and the theory of representations. These contain proofs of the theorems used here. The quite readable introduction by Cotton is especially to be recommended.

## 2.1. Symmetry operations and symmetry groups

### 2.1.1. *Definition of a symmetry operation and of a group*

First we must establish what is to be understood under 'symmetry of a geometrical structure' and how this symmetry is to be described. Consider as an example an equilateral triangle having the vertices $a$, $b$ and $c$. (cf. Figure B.7). All vertices are identical. The triangle lies in the plane of the paper and the midpoint of the triangle coincides with the origin of the $xy$ coordinate system. We shall now consider what possibilities exist to move the triangle so that afterwards vertices of the triangle lie at those

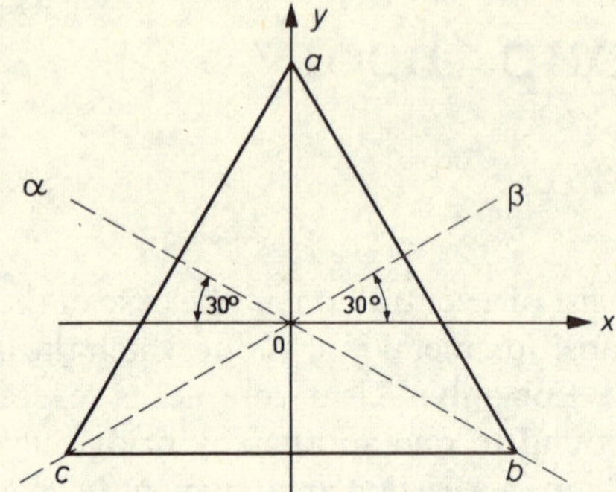


Figure B.7. The $z$ axis passes through the point 0 perpendicular to the plane of the paper.

points in the $xy$ plane (plane of the paper) where vertices of the triangle were originally located. The labelling of the vertices is here irrelevant. We, therefore, study movements such that after their completion the labelling of the vertices appears to be changed compared to the original situation. It is simply required that the triangle be transformed into itself. Such movements are called *symmetry operations or covering operations.* The symmetry of an entity is determined by the totality of its symmetry operations.

Six different covering operations may be given for the equilateral triangle and these are denoted by the letters $E, \sigma_1, \sigma_2, \sigma_3, C_3, C_3^2$.

$E$: The *identity* operation. The triangle remains in its original position; it is left at rest. Symbolically we write:

$\sigma_1$: *Reflexion* of the triangle in the $yz$ plane (the $z$ axis is perpendicular to the plane of the paper.) Vertex $a$ is not moved, $b$ and $c$ are interchanged:

$\sigma_2$: *Reflexion* of the triangle in the $\alpha z$ plane ($\alpha$ is a straight line inclined by an angle of $-30°$ to the $x$ axis) (cf. Figure B.7).

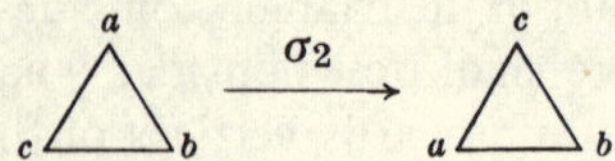

$\sigma_3$ : *Reflexion* of the triangle in the $\beta z$ plane ($\beta$ is a straight line inclined by an angle of $+30°$ to the $x$ axis):

$C_3$ : *Rotation* of the triangle (in the plane of the paper) by an angle of $2\pi/3$ in a clockwise direction about the midpoint of the triangle

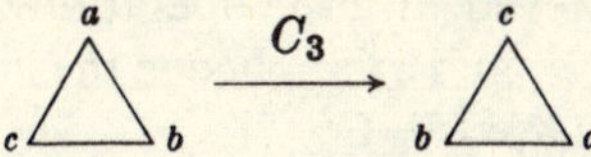

$C_3^2$ : *Rotation* of the triangle (in the plane of the paper) by an angle $4\pi/3$ in a clockwise direction about the midpoint of the triangle

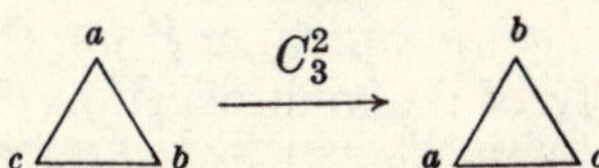

Other symmetry operations may be given for the equilateral triangle. They are, however, always equivalent to one of the operations given above. For example, the result of rotating the triangle (in the plane of the paper) by the angle $-2\pi/3$, which of course is also a covering operation, is identical with the result of our rotation operation $C_3^2$.

Because the given symmetry operations $E, \sigma_1, \sigma_2, \sigma_3, C_3, C_3^2$ leave the position of the triangle unchanged, except for the labelling of the vertices, the result of two operations carried out consecutively must have the same property as a *single* symmetry operation. Further, because $E, \sigma_1, \sigma_2, \sigma_3, C_3, C_3^2$ are all possible symmetry operations of the equilateral triangle, two of the symmetry operations, applied one after another in a given order, should have the same effect as one of the operations from $E, \sigma_1, \sigma_2, \sigma_3, C_3, C_3^2$.

As an example let us apply consecutively the symmetry operations $\sigma_1$ and $\sigma_2$ in the order first $\sigma_1$ and then $\sigma_2$. It is conventional to write the operation to be carried out first to the right of the operation which is to follow; our combined operation is then symbolized by $\sigma_3\sigma_1$.

If $\sigma_1$ is applied to the triangle, then we obtain

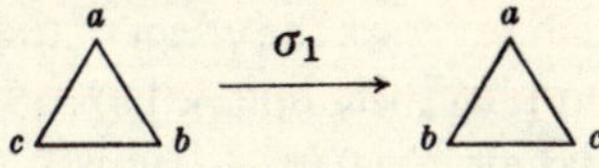

The operation $\sigma_3$ converts the triangle 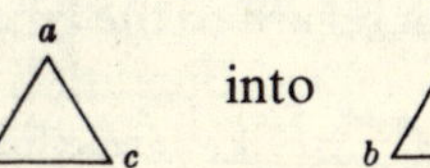 into :

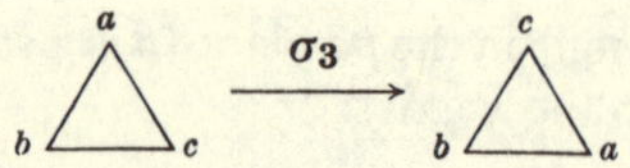

Summarizing we may write:

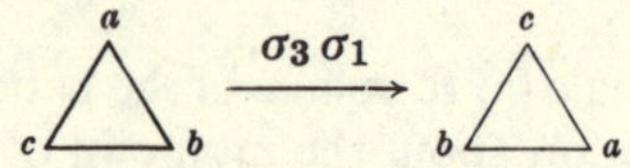

The same effect is achieved with the operation $C_3 : \sigma_3\sigma_1 = C_3$. Correspondingly one finds that $\sigma_1$ and $\sigma_3$ applied in the inverse order are equivalent to the operation $C_3^2 : \sigma_1\sigma_3 = C_3^2$.

An operation formed by combining two operations $P$ and $Q$, $PQ$ or $QP$, is called formally the '*product*' of the '*factors*' $P$ and $Q$. Combining $P$ and $Q$ to $PQ$ or $QP$, in our case the consecutive application of symmetry operations, is called formally '*multiplication*' of $P$ and $Q$.

An important property of the multiplication of symmetry operations can be seen from the examples introduced above $\sigma_3\sigma_1 = C_3$ and $\sigma_1\sigma_3 = C_3^2$. Because $C_3 \neq C_3^2$, then $\sigma_3\sigma_1 \neq \sigma_1\sigma_3$, i.e., a different order of the factors leads in our example to different results. One says that the multiplication of symmetry operations is in general *noncommutative*

$$PQ \neq QP.$$

Analogously to the product formation $\sigma_3\sigma_1 = C_3$ and $\sigma_1\sigma_3 = C_3^2$ all possible products of two operations from $E, \sigma_1, \sigma_2, \sigma_3, C_3, C_3^2$ can be formed. The results are presented in Table B.10, a so-called multiplication table. The top row contains the factors which are to the right in the product (which are to be applied first) and the left-hand column the factors which stand to the left in the products (which are to be applied second). The result of the multiplication of two factors is to be found at the intersection of the appropriate row and column. If first $C_3^2$ and then $\sigma_2$ are to be applied, then we go down in the column headed by $C_3^2$ and to the right in the row which begins with $\sigma_2$. At the intersection of row and column we find $\sigma_1$. Therefore, $\sigma_2 C_3^2 = \sigma_1$. According to Table B.10, to every ordered product of two operations one operation is uniquely associated which produces the same effect as that product.

Also, we see from the multiplication table that for each operation there is exactly one operation which reverses the effect of the first operation. If, e.g., first $C_3^2$ and then $C_3$ are applied, the final effect is the same as that of the identity operation $E : C_3 C_3^2 = E$, i.e., the triangle remains in its original position with regard to the labelling of the vertices. One says that $C_3$ is the *reciprocal* or *inverse* operation of $C_3^2$ and writes symbolically $C_3 = (C_3^2)^{-1}$. Of course $C_3^2$ is likewise the reciprocal of $C_3 : C_3^2 = C_3^{-1}$. If for two

Table B.10. Multiplication table for the symmetry group $C_{3v}$

right-hand factor (first operation); left-hand factor (second operation)

| | $E$ | $\sigma_1$ | $\sigma_2$ | $\sigma_3$ | $C_3$ | $C_3^2$ |
|---|---|---|---|---|---|---|
| $E$ | $E$ | $\sigma_1$ | $\sigma_2$ | $\sigma_3$ | $C_3$ | $C_3^2$ |
| $\sigma_1$ | $\sigma_1$ | $E$ | $C_3$ | $C_3^2$ | $\sigma_2$ | $\sigma_3$ |
| $\sigma_2$ | $\sigma_2$ | $C_3^2$ | $E$ | $C_3$ | $\sigma_3$ | $\sigma_1$ |
| $\sigma_3$ | $\sigma_3$ | $C_3$ | $C_3^2$ | $E$ | $\sigma_1$ | $\sigma_2$ |
| $C_3$ | $C_3$ | $\sigma_3$ | $\sigma_1$ | $\sigma_2$ | $C_3^2$ | $E$ |
| $C_3^2$ | $C_3^2$ | $\sigma_2$ | $\sigma_3$ | $\sigma_1$ | $E$ | $C_3$ |

symmetry operations $P$ and $Q$ the relation $PQ = E$ holds, where $E$ is the identity operation, then in general $P = Q^{-1}$ is called the reciprocal operation of $Q$ and $Q = P^{-1}$ is the reciprocal operation to $P$. Each of the symmetry operations of the equilateral triangle $E, \sigma_1, \sigma_2, \sigma_3, C_3, C_3^2$ has a reciprocal operation, in particular $E^{-1} = E, \sigma_1^{-1} = \sigma_1, \sigma_2^{-1} = \sigma_2, \sigma_3^{-1} = \sigma_3$, $C_3^{-1} = C_3^2$, $(C_3^2)^{-1} = C_3$. The operations $E, \sigma_1, \sigma_2, \sigma_3$ are their own reciprocals.

A further important property of the symmetry operations of the equilateral triangle can be obtained from the multiplication table. If one wishes to calculate the effect of three consecutively applied operations $PQR$ with the help of the multiplication table, there are two ways of proceeding. On the one hand the product $QR$ can be found first, after which the result is multiplied by $P$ from the left: $P(QR)$, or one can determine first the result $PQ$ and then multiply this product from the right with $R$: $(PQ)R$. Independent of which operations from the set $E, \sigma_1, \sigma_2, \sigma_3, C_3, C_3^2$ one uses for $P$, $Q$ and $R$, with the aid of the multiplication table (Table B.10) the result

$$P(QR) = (PQ)R$$

is always obtained. In such a case multiplication is said to be *associative*.

The results which have just been obtained from a study of the symmetry operations of the equilateral triangle and their connecting properties show that the set of operations $E, \sigma_1, \sigma_2, \sigma_3, C_3, C_3^2$ forms a *group*. In general a group is defined as a set of elements $P, Q, R, \ldots$ which satisfies the following conditions:

(1) There exists a relation ('multiplication') such that to every 'product' of two elements from the set one element of the set is uniquely associated.

(2) The set contains the unit element $E$, which satisfies the relation

$$EP = PE = P,$$

where $P$ can be every element of the set.

(3) The associative law of multiplication is valid for all elements

$$P(QR) = (PQ)R,$$

whereby $P$ can be every element of the set.

(4) For every element $P$ of the set there exists an element $Q$ of the set such that

$$QP = PQ = E.$$

$Q = P^{-1}$ is the element which is reciprocal to $P$.

Because the symmetry operations of the equilateral triangle satisfy all of these conditions, when the consecutive execution of operations is taken as their product, the set $E, \sigma_1, \sigma_2, \sigma_3, C_3, C_3^2$ does form a group. In particular because its elements are *symmetry operations* it is called the *symmetry group* of the equilateral triangle. Conventionally this special group is denoted with the symbol* $C_{3v}$.

The number of elements which constitute the group is called the *order of the group*. The symmetry group $C_{3v}$ has then the order 6.

### 2.1.2. *Classes and subgroups*

Using the example of the concrete symmetry group $C_{3v}$, the terms *class* and *subgroup* will be introduced. It is apparent that the six symmetry operations of $C_{3v}$ can be divided into three types: the identity operation $E$, the reflections $\sigma_1, \sigma_2, \sigma_3$ and the rotations $C_3, C_3^2$. One says, $[E]$, $[\sigma_1, \sigma_2, \sigma_3]$, $[C_3, C_3^2]$ form in each case a *class* of the group $C_{3v}$. As was the case for the group of the triangle, the division into classes for other symmetry groups can usually be carried out immediately on the basis of geometric considerations. This is true because a class contains only similar operations†. The rigorous definition of a class is: Two elements $Q$ and $R$ belong to the same class when they satisfy the relations $P^{-1}QP = R$, where $P$ is an

* We shall use the nomenclature of A. M. Schönflies, cf. *Theorie der Kristallstruktur*, Borntraeger, Berlin 1919.

† The term 'class of operations' has the following geometric meaning. If two operations belong to the same class it is always possible to substitute a new coordinate system for the original such that one operation is replaced by the other. If in the case of the group of the triangle one chooses the $y$ axis as the perpendicular bisector of the side $\overline{ac}$ (Figure B.7), then the operation $\sigma_1$ leads to the same result as was given by $\sigma_2$ in the old coordinate system.

element of the group and $P^{-1}$ is its reciprocal. A group usually has several classes; an element may belong to only one class.

If for example in the case $C_{3v}$ the element $E$ is taken as $Q$ and $P$ is allowed to become all of the elements of the group, then one obtains the first of the following columns. Independent of which element of the group is taken as $P$, one always obtains on the right-hand side of the relation the result $E$. That is, $E$ forms a class by itself. If we choose for $Q$, $\sigma_1$, $\sigma_2$ or $\sigma_3$, then $\sigma_1$, $\sigma_2$ and $\sigma_3$ and no other element of the group appears on the right-hand side of the corresponding columns. Therefore $\sigma_1, \sigma_2, \sigma_3$ form a class. From the last two columns it follows that $C_3$ and $C_3^2$ constitute a class, exactly as is to be expected on the basis of geometry.

| | | | | | |
|---|---|---|---|---|---|
| $EEE = E$ | $E\sigma_1E = \sigma_1$ | $E\sigma_2E = \sigma_2$ | $E\sigma_3E = \sigma_3$ | $EC_3E = C_3$ | $EC_3^2E = C_3^2$ |
| $\sigma_1E\sigma_1 = E$ | $\sigma_1\sigma_1\sigma_1 = \sigma_1$ | $\sigma_1\sigma_2\sigma_1 = \sigma_3$ | $\sigma_1\sigma_3\sigma_1 = \sigma_2$ | $\sigma_1C_3\sigma_1 = C_3^2$ | $\sigma_1C_3^2\sigma_1 = C_3$ |
| $\sigma_2E\sigma_2 = E$ | $\sigma_2\sigma_1\sigma_2 = \sigma_3$ | $\sigma_2\sigma_2\sigma_2 = \sigma_2$ | $\sigma_2\sigma_3\sigma_2 = \sigma_1$ | $\sigma_2C_3\sigma_2 = C_3^2$ | $\sigma_2C_3^2\sigma_2 = C_3$ |
| $\sigma_3E\sigma_3 = E$ | $\sigma_3\sigma_1\sigma_3 = \sigma_2$ | $\sigma_3\sigma_2\sigma_3 = \sigma_1$ | $\sigma_3\sigma_3\sigma_3 = \sigma_3$ | $\sigma_3C_3\sigma_3 = C_3^2$ | $\sigma_3C_3^2\sigma_3 = C_3$ |
| $C_3^2EC_3 = E$ | $C_3^2\sigma_1C_3 = \sigma_3$ | $C_3^2\sigma_2C_3 = \sigma_1$ | $C_3^2\sigma_3C_3 = \sigma_2$ | $C_3^2C_3C_3 = C_3$ | $C_3^2C_3^2C_3 = C_3^2$ |
| $C_3EC_3^2 = E$ | $C_3\sigma_1C_3^2 = \sigma_2$ | $C_3\sigma_2C_3^2 = \sigma_3$ | $C_3\sigma_3C_3^2 = \sigma_1$ | $C_3C_3C_3^2 = C_3$ | $C_3C_3^2C_3^2 = C_3^2$ |

Let us now consider the term *subgroup.* It frequently occurs that taken by itself a portion of the set of elements constituting a group fulfils all of the conditions imposed on a group. This portion of the total group is called a subgroup of the group (or as is sometimes said for the sake of clarity of the 'larger group'). From the multiplication table, Table B.10, for the group $C_{3v}$ for example, one can see that $C_{3v}$ has the following four subgroups, for which the appropriate multiplication tables are given:

$E, \sigma_1$:

| | $E$ | $\sigma_1$ |
|---|---|---|
| $E$ | $E$ | $\sigma_1$ |
| $\sigma_1$ | $\sigma_1$ | $E$ |

$E, \sigma_2$:

| | $E$ | $\sigma_2$ |
|---|---|---|
| $E$ | $E$ | $\sigma_2$ |
| $\sigma_2$ | $\sigma_2$ | $E$ |

$E, \sigma_3$:

| | $E$ | $\sigma_3$ |
|---|---|---|
| $E$ | $E$ | $\sigma_3$ |
| $\sigma_3$ | $\sigma_3$ | $E$ |

$E, C_3, C_3^2$:

| | $E$ | $C_3$ | $C_3^2$ |
|---|---|---|---|
| $E$ | $E$ | $C_3$ | $C_3^2$ |
| $C_3$ | $C_3$ | $C_3^2$ | $E$ |
| $C_3^2$ | $C_3^2$ | $E$ | $C_3$ |

One can easily convince himself that each of these four sets of elements fulfils the group rules. As is seen from the example, a group may have several subgroups. Different from the division of the elements of the group into classes, it is possible for an element of the group to belong to several subgroups. For example, the unit element $E$ by definition of a group must be contained in every subgroup.

### 2.1.3. *Special symmetry groups*

We shall now list those symmetry groups which are especially important for the ligand field theory. It is useful to define first the symbols conventionally used:

$E$: The identity operation which leaves the figure in its original position.

$C_n$: Rotation about the symmetry axes through an angle $2\pi/n$. Of all possible rotations $C_n$ about a symmetry axis, that rotation with the greatest $n$ determines the *multiplicity* of the symmetry axis. The multiplicity may be $n_{max} = 1$ (no symmetry axis), 2, 3, 4, 5, . . . .

$\sigma$: Reflection in a plane of symmetry. Three cases are distinguished.

$\sigma_h$: The plane of symmetry is perpendicular to the principal symmetry axis (to the axis about which the rotation $C_n$ with the greatest $n$ is possible).

$\sigma_v$: The plane of symmetry contains the principal symmetry axis.

$\sigma_d$: The plane of symmetry contains the principal symmetry axis and bisects the angle between two symmetry axes with $n = 2$, which are perpendicular to the axis of rotation.

$S_n$: Improper rotation, (Rotary reflection) i.e., rotation about the axis by the angle $2\pi/n$ followed by a reflection in a plane which is perpendicular to the axis of rotation.

$i$: Inversion through a centre of symmetry.

It should be noted that such reflections $\sigma$ and improper rotations $S_n$ can always be represented by a combination of the inversion $i$ with one or more rotations.

Equipped with the necessary nomenclature, let us now consider those symmetry groups which are of most importance for the ligand field theory.

2.1.3.1. *The rotation–reflection group of the sphere, Symbol* $R_{3i}$. This group is the symmetry group of a sphere. A sphere is brought into itself when the sphere (cf. Figure B.8)

(a) is rotated by an arbitrary angle about an arbitrary axis through its centre.
(b) is reflected in a centre of symmetry (centre of the sphere), i.e., inverted.
(c) is rotated as in (a), and then reflected.

It is apparent that the order of this group is infinite. One of the numerous subgroups of the rotation-reflection group of the sphere is the *rotation group of the sphere*. It is defined as the set of rotations described by (a) and contains therefore only half as many elements as the rotation-reflection group of the sphere. The rotation through an angle of zero degrees is to be compared with the inversion given in (b).

The rotation-reflection group of the sphere is of great importance for us because the symmetry groups which appear in the ligand field theory are all subgroups of this larger group.

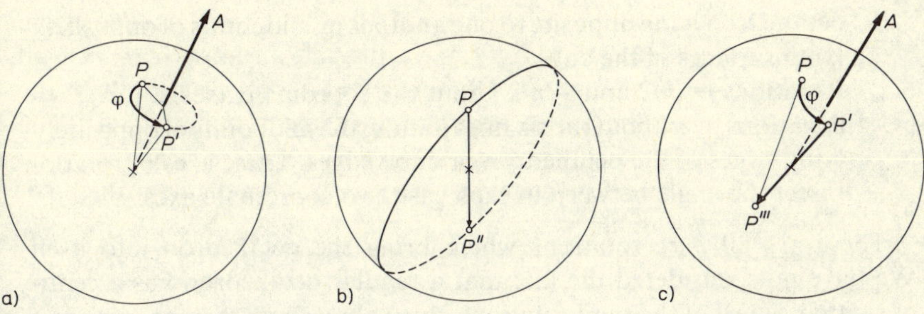


Figure B.8. The rotation-reflection group of the sphere. (a) Rotation through an angle $\varphi$ about an axis $A$: $P \to P'$. (b) Inversion (through the centre) $P \to P''$. (c) Rotation through an angle $\varphi$ about an axis $A(P \to P')$, followed by inversion $(P' \to P'''): P \to P'''$.

2.1.3.2. *The point group of the regular octahedron, symbol* $O_h$. This is the most important symmetry group for the ligand field theory and has as elements the following symmetry operations (cf. Figure B.9)*:

$E$: Identity operation.

$8C_3$: Rotations by $2\pi/3 = 120°$ and $-2\pi/3$ about the four diagonals of the cube. The diagonals of the cube are for the octahedron the lines joining the midpoints of oppositely-lying (triangular) surfaces.

$3C_2$: Rotations by $\pi$ about the three coordinate axes $x$, $y$, $z$ sketched in Figure B.9. The axes $x$, $y$, $z$ connect in each case vertices of the

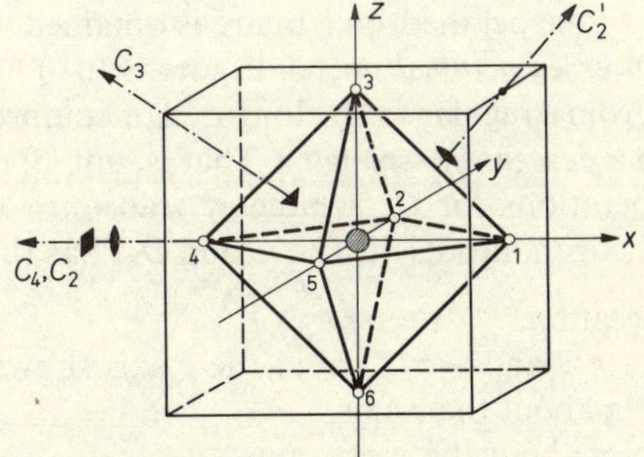


Figure B.9. Regular octahedron and cube, symmetry $O_h$. ⬮, ▲, ■ are symbols for two, three or fourfold rotational axes.

* Operations which belong to the same class will be written together and the number of these operations given. $3C_2$ means for example that there are three rotations through the angle $2\pi/2 = \pi$ and that these form a class.

octahedron lying opposite to one another or midpoints of oppositely-lying surfaces of the cube.

$6C_4$ : Rotations by $\pi/2$ and $-\pi/2$ about the coordinate axes $x$, $y$, $z$.

$6C_2'$ : Rotations by $\pi$ about the six lines joining the midpoints of oppositely-lying edges of the octahedron or cube edges. These axes of rotation bisect the angle between in each case two coordinate axes.

These are all *pure* rotations which bring the octahedron into itself. We have not considered the fact that a regular octahedron has a centre of symmetry. All of the pure rotations given above may then be combined with the inversion $i$. The complete group $O_h$ is obtained when we add the 24 elements $iE = i$, $8iC_3$, $3iC_2$, $6iC_4$, $6iC_2'$, to the 24 elements of the pure rotations*. The (holohedral) octahedral group of the rotation-reflection group of the sphere $O_h$ is therefore of order 48. It is apparent that this is a subgroup of the rotation-reflection group of the sphere because all the elements of $O_h$ are also elements of the rotation-reflection group of the sphere.

Examples of complex ions with the symmetry $O_h$ are:

$[Ti(H_2O)_6]^{3+}$, $[Cr(NH_3)_6]^{3+}$, $[CoF_6]^{3-}$, $[MoCl_6]^{3-}$, $[Fe(CN)_6]^{3-}$, $[Cr(CO(NH_2)_2)_6]^{3+}$.

2.1.3.3. *The point group of the elongated or compressed octahedron, symbol $D_{4h}$.* Let us assume that two pairs of opposite vertices of a regular octahedron are held fixed. Then by changing to the same degree the distance of the third pair of vertices from the centre of the octahedron along the principal axis of the octahedron a figure is obtained which is called an *elongated* or *compressed octahedron.* (cf. Figure B.10†.)

The transition from a regular to an elongated or compressed octahedron is connected with a *descent in symmetry.* That is, not all operations which are symmetry operations for $O_h$ symmetry transform the elongated or compressed octahedron into itself. The group $D_{4h}$ has the elements

$E$: Identity operation.

$2C_4$: Rotations of $\pi/2$ and $-\pi/2$ about the $z$ axis shown in Figure B.10.

$C_2$: Rotation of $\pi$ about the $z$ axis.

$2C_2'$: Rotations of $\pi$ about the $x$ or $y$ axis.

---

* The 24 elements of pure rotations compose the so-called hemihedral group of the octahedron.

† We include the case where the two opposite vertices are infinitely far from the centre, so that the original octahedron degenerates into a *square.* A figure which arises from a regular octahedron when two opposite vertices of the octahedron are considered as being equivalent to one another but not to the other four vertices also has $D_{4h}$ symmetry.

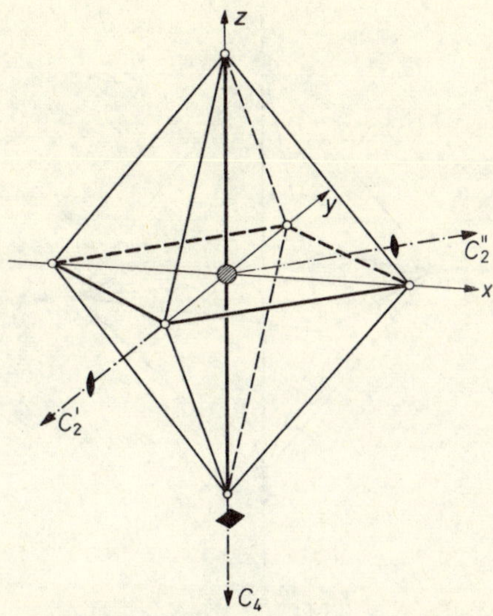


Figure B.10. An octahedron elongated uniformly along the $z$ axis, symmetry $D_{4h}$. ⬮ and ■ are symbols for two or fourfold rotational axes.

$2C_2''$: Rotations of $\pi$ about the two axes which bisect the angle between the $x$ and $y$ axes.

Because the elongated or compressed octahedron has an inversion centre, in order to obtain the complete group $D_{4h}$ we must add the 8 improper rotations, $iE = i$, $2iC_4 = 2S_4$, $iC_2 = \sigma_h$, $2iC_2 = 2\sigma_v$, $2iC_2'' = 2\sigma_d$ to the 8 proper rotations. The group $D_{4h}$ has the order 16. Because all of its elements are also elements of $O_h$, the group $D_{4h}$ is a subgroup of the octahedral group $O_h$ and consequently is also a subgroup of the rotation-reflection group of the sphere. Examples of complex ions with the symmetry $D_{4h}$ are:

$$\text{trans–}[Co(NH_3)_4Cl_2]^+, \text{ trans–}[Co(NH_3)_4Br_2]^+$$
$$[PtCl_4]^{2-}, [Pt(CN)_4]^{2-}, [Ni(CN)_4]^{2-}.$$

2.1.3.4. *The point group of the nonuniformly elongated or compressed octahedron. Symbol* $C_{4v}$. Starting from a figure with symmetry $D_{4h}$ one can reduce the symmetry even further by elongating or compressing the octahedron nonuniformly, that is, the two octahedral vertices effected are moved to different degrees along the principal axis (Figure B.11).

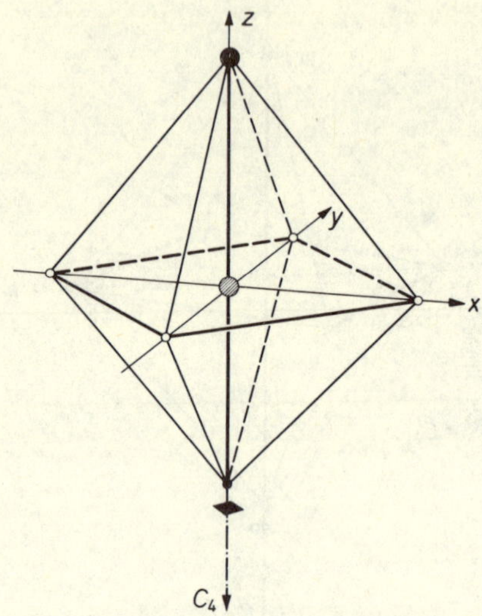


Figure B.11. An octahedron elongated non-uniformly along the $z$ axis. Symmetry $C_{4v}$. ■ is the symbol for a fourfold rotational axis.

An *unequally elongated* or *compressed octahedron* is thus obtained having the symmetry $C_{4v}$. It is readily seen that this figure does not have a centre of inversion. Now only the following symmetry operations are possible:

$E$: Identity operation.
$2C_4$: Rotations of $\pi/2$ and $-\pi/2$ about the $z$ axis.
$C_2$: Rotations of $\pi$ about the $z$ axis.
$2\sigma_v$: Reflections in the $xz$ plane or the $yz$ plane.
$2\sigma_d$: Reflections in the planes containing the $z$ axis and the bisector of the angle between the $x$ and $y$ axes.

The group $C_{4v}$ has the order 8 and is a subgroup of the rotation-reflection group of the sphere, the group $O_h$ and the group $D_{4h}$.

Examples of complexes having the symmetry $C_{4v}$ are:

$$[Cr(NH_3)_5Cl]^{2+},\ [Co(NH_3)_5Br]^{2+},\ [Cu(NH_3)_5H_2O]^{2+}.$$

2.1.3.5. *The point group of the regular tetrahedron, symbol $T_d$*. In Figure B.12 a regular tetrahedron is inscribed in a cube. The tetrahedron does not have a centre of symmetry. The point group $T_d$ of the tetrahedron is composed of the following symmetry operations:

| | |
|---|---|
| $E$: | Identity operation. |
| $8C_3$: | Rotations of $2\pi/3 = 120°$ and $-2\pi/3$ about the four diagonals of the cube. |
| $3C_2$: | Rotations by $\pi$ about the three coordinate axes $x$, $y$, $z$ sketched in Figure B.12. |
| $6S_4 := 6\sigma_h C_4$: | Rotations by $\pi/2$ and $-\pi/2$ about the coordinate axes $x$, $y$, $z$, followed by reflection in the plane perpendicular to these axes which contains the midpoint of the tetrahedron. |
| $6\sigma_d = 6iC_2'$: | Rotations by $\pi$ about the axes which connect the midpoints of two oppositely-lying cube edges, followed by reflection in the midpoint of the tetrahedron. |

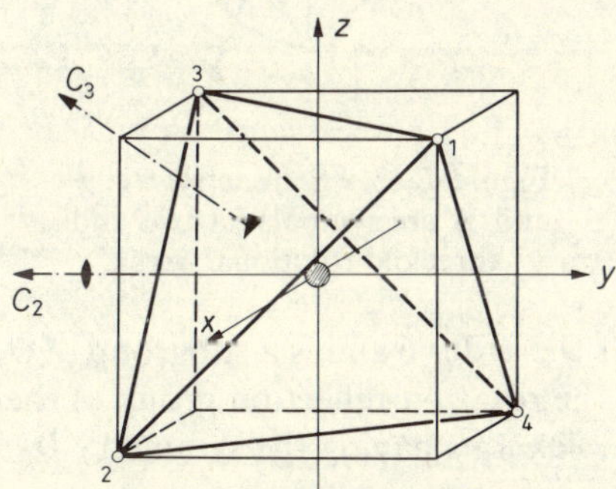


Figure B.12. Regular tetrahedron, symmetry $T_d$. ⬮ and ▲ are symbols for two and threefold rotational axes.

The group $T_d$ has the order 24. It is a subgroup of $O_h$ and consequently of the rotation-reflection group of the sphere. The essential difference between the groups $T_d$ and $O_h$ is that only $O_h$ has a centre of inversion.

Examples of complex ions with symmetry $T_d$ are:

$$[CoCl_4]^{2-},\ [Co(NCS)_4]^{2-},\ [Cu(CN)_4]^{2-},\ [CuCl_4]^{2-}.$$

2.1.3.6. *The point group $D_3$*. This point group is realized for complex ions having three ligands, each one of which occupies two coordination positions of the central ion. In Figure B.13 such a figure is schematically shown. The heavy curved lines represent ligands. The group $D_3$ is made up of the following symmetry operations:

$E$: Identity operation.
$2C_3$: Rotations of $2\pi/3 = 120°$ and $-2\pi/3$ about the axis sketched in Figure B.13 and denoted with $C_3$.
$3C_2$: Rotations by $\pi$ about the axes which in each case connect the centre of a curve with the midpoint of the figure.

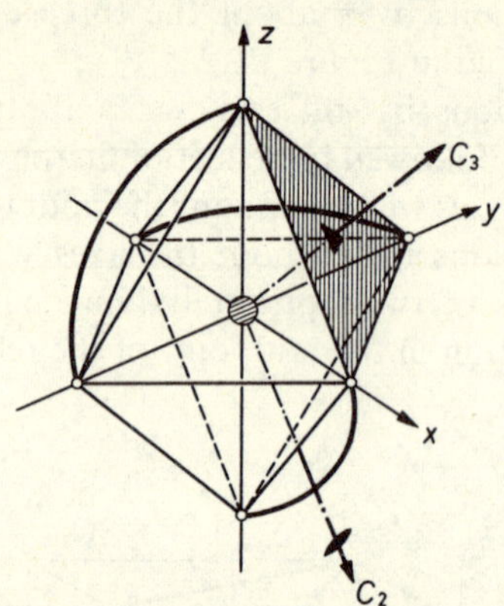


Figure B.13. Symmetry $D_3$. ⬧ and ▲ are symbols for two and threefold rotational axes.

The group $D_3$ has the order 6 and is a subgroup of $O_h$ and consequently also a subgroup of the rotation-reflection group of the sphere.

Examples of complex ions having the symmetry $D_3$ are:

$$[\mathrm{Cr\ ox_3}]^{3-},\ [\mathrm{Co\ en_3}]^{3+},\ [\mathrm{Cr\ tn_3}]^{3+}.$$

ox = oxalate ion,
en = ethylenediamine,
tn = trimethylenediamine

2.1.3.7. *The point group* $C_{2v}$. The figures shown in Figure B.14 belong to the point group $C_{2v}$. They are invariant under four symmetry operations, the identity operation, a twofold rotation, and reflection in two planes of symmetry. The figure denoted by *cis*-$[MA_4B_2]$ has the symmetry operations:

$E$: The identity operation.
$C_2$: Rotation by $\pi$ about the bisector of the angle between the $x$ and $y$ axes (corresponding to the symmetry operation $C_2''$ in $D_{4h}$, cf. Figure B.10).
$\sigma_v$: Reflection in the $xy$ plane (corresponding to $\sigma_h$ in $D_{4h}$).
$\sigma_v'$: Reflection in the plane containing the $z$ axis and the bisector of the angle between the $x$ and $y$ axes (corresponding to $\sigma_d$ in $D_{4h}$).

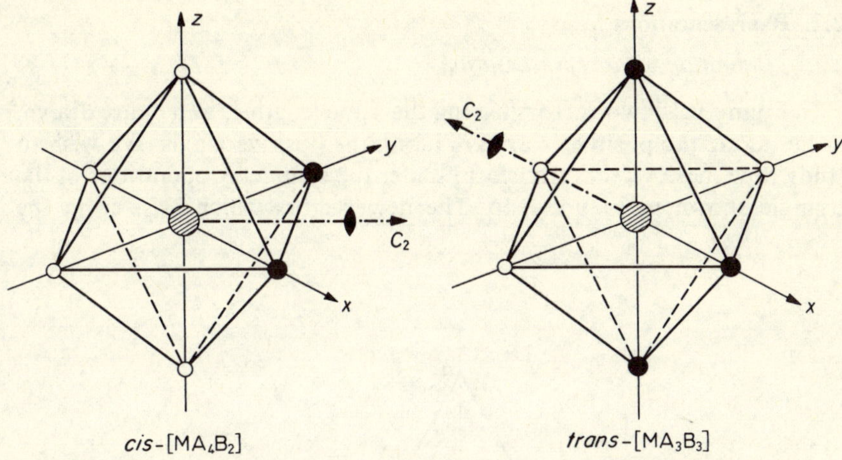


Figure B.14. Symmetry $C_{2v}$. ⬮ is the symbol for twofold rotational axes.

The symmetry operations for the *trans*-[$MA_3B_3$]-figure are

$E$: The identity operation.
$C_2$: Rotation by $\pi$ about the $x$ axis (corresponding to $C_2'$ in $D_{4h}$).
$\sigma_v$: Reflection in the $xy$ plane (corresponding to $\sigma_h$ in $D_{4h}$).
$\sigma_v'$: Reflection in the $xz$ plane (corresponding to $\sigma_v$ in $D_{4h}$).

The symmetry operations for both figures are also symmetry operations of the group $D_{4h}$. $C_{2v}$ is, therefore, a subgroup of $D_{4h}$ and consequently also a subgroup of $O_h$ and $R_{3i}$.

Examples of complex ions with the symmetry $C_{2v}$ are

$$\text{cis-}[Co(NH_3)_4Cl_2]^+, \quad \text{trans-}[Co(NH_3)_3(H_2O)_3]^{3+}.$$

Summarizing we present a scheme for the symmetry groups described, $R_{3i}$, $O_h$, $C_{4v}$, $T_d$, $D_3$, $C_{2v}$ from which it can be seen which groups are subgroups of the larger groups*:

$$\begin{array}{l} \quad\quad\quad\quad\ \nearrow D_{4h} \rightarrow C_{4v} \rightarrow C_{2v} \\ R_{3i} \rightarrow O_h \rightarrow T_d \\ \quad\quad\quad\quad\ \searrow D_3 \end{array}$$

---

* This scheme is naturally in so far incomplete as the groups $R_{3i}$, $O_h$ and $D_{4h}$ have many more subgroups than the scheme indicates. A complete description of the subgroups belonging to the larger groups can be found for example in E. B. Wilson, Jr., J. C. Decius and P. C. Cross: *Molecular Vibrations*, pp. 333 ff., McGraw-Hill Book Company, Inc., New York—Toronto—London, 1955.

## 2.2. Representations

### 2.2.1. *Definition of a representation*

In Figure B.15 two vectors having the same length $\vec{r}_1$ and $\vec{r}_2$ are drawn. $\vec{r}_1$ lies along the positive $x$ axis, $\vec{r}_2$ along the positive $y$ axis. We wish to study how these vectors transform under the symmetry operations of the triangle shown in Figure B.15. The new vectors which arise from the original vectors under the influence of a symmetry operation $R$ can always be described as a linear combination of $\vec{r}_1$ and $\vec{r}_2$.

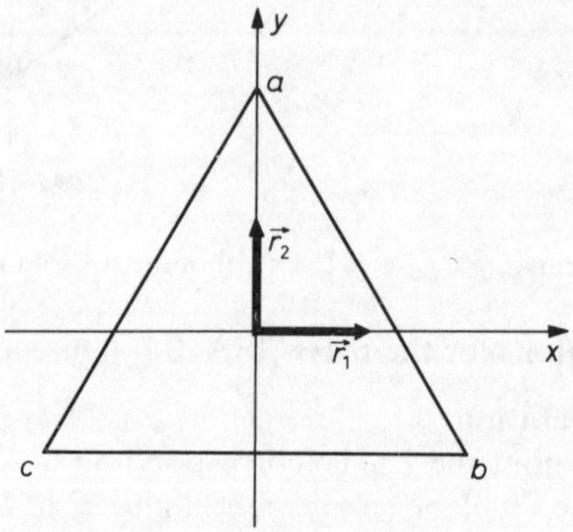


Figure B.15.

$$R\vec{r}_1 = \Gamma(R)_{1,1}\vec{r}_1 + \Gamma(R)_{2,1}\vec{r}_2,$$
$$R\vec{r}_2 = \Gamma(R)_{1,2}\vec{r}_1 + \Gamma(R)_{2,2}\vec{r}_2,$$

or more concisely

$$R\vec{r}_i = \sum_{j=1}^{2} \Gamma(R)_{j,i}\vec{r}_j, \quad i = 1,2.$$

The $\Gamma(R)_{j,i}$ are numbers which depend upon the symmetry operations $R$ which are executed. For each operation we summarize the appropriate $\Gamma(R)_{j,i}$ in a matrix scheme:

$$\Gamma(R) = \begin{pmatrix} \Gamma(R)_{1,1} & \Gamma(R)_{1,2} \\ \Gamma(R)_{2,1} & \Gamma(R)_{2,2} \end{pmatrix}.$$

For our example one finds without difficulty the following transformation equations and $\boldsymbol{\Gamma}$ matrices:

$$\left.\begin{aligned} E\vec{r}_1 &= \vec{r}_1 \\ E\vec{r}_2 &= \vec{r}_2 \end{aligned}\right\} \qquad \Gamma(E) = \begin{pmatrix} 1 & 0 \\ 0 & 1 \end{pmatrix},$$

$$\left.\begin{aligned} \sigma_1\vec{r}_1 &= -\vec{r}_1 \\ \sigma_1\vec{r}_2 &= \vec{r}_2 \end{aligned}\right\} \qquad \Gamma(\sigma_1) = \begin{pmatrix} -1 & 0 \\ 0 & 1 \end{pmatrix},$$

$$\left.\begin{aligned} \sigma_2\vec{r}_1 &= \frac{1}{2}\vec{r}_1 - \frac{\sqrt{3}}{2}\vec{r}_2 \\ \sigma_2\vec{r}_2 &= -\frac{\sqrt{3}}{2}\vec{r}_1 - \frac{1}{2}\vec{r}_2 \end{aligned}\right\} \Gamma(\sigma_2) = \begin{pmatrix} \frac{1}{2} & -\frac{\sqrt{3}}{2} \\ -\frac{\sqrt{3}}{2} & -\frac{1}{2} \end{pmatrix},$$

$$\left.\begin{aligned} \sigma_3\vec{r}_1 &= \frac{1}{2}\vec{r}_1 + \frac{\sqrt{3}}{2}\vec{r}_2 \\ \sigma_3\vec{r}_2 &= \frac{\sqrt{3}}{2}\vec{r}_1 - \frac{1}{2}\vec{r}_2 \end{aligned}\right\} \Gamma(\sigma_3) = \begin{pmatrix} \frac{1}{2} & \frac{\sqrt{3}}{2} \\ \frac{\sqrt{3}}{2} & -\frac{1}{2} \end{pmatrix},$$

$$\left.\begin{aligned} C_3\vec{r}_1 &= -\frac{1}{2}\vec{r}_1 - \frac{\sqrt{3}}{2}\vec{r}_2 \\ C_3\vec{r}_2 &= \frac{\sqrt{3}}{2}\vec{r}_1 - \frac{1}{2}\vec{r}_2 \end{aligned}\right\} \Gamma(C_3) = \begin{pmatrix} -\frac{1}{2} & \frac{\sqrt{3}}{2} \\ -\frac{\sqrt{3}}{2} & -\frac{1}{2} \end{pmatrix},$$

$$\left.\begin{aligned} C_3^2\vec{r}_1 &= -\frac{1}{2}\vec{r}_1 + \frac{\sqrt{3}}{2}\vec{r}_2 \\ C_3^2\vec{r}_2 &= -\frac{\sqrt{3}}{2}\vec{r}_1 - \frac{1}{2}\vec{r}_2 \end{aligned}\right\} \Gamma(C_3^2) = \begin{pmatrix} -\frac{1}{2} & -\frac{\sqrt{3}}{2} \\ \frac{\sqrt{3}}{2} & -\frac{1}{2} \end{pmatrix}.$$

As a second example a vector pair $\vec{r}'_1$ and $\vec{r}'_2$ is drawn in Figure B.16. $\vec{r}'_2$ lies along the $y$ axis, $\vec{r}'_1$ is inclined to the $x$ axis by an angle of $-30°$. Completely in analogy with our first example we wish to subject our

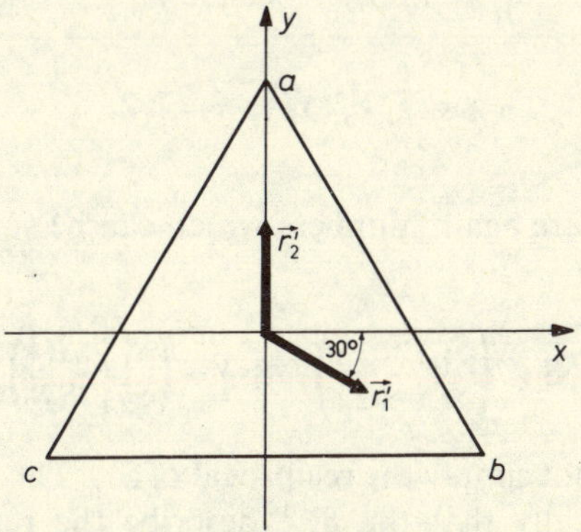


Figure B.16.

vectors $\vec{r}'_1, \vec{r}'_2$ to the symmetry operations of the equilateral triangle and present the appropriate transformation equations

$$R\vec{r}'_i = \sum_{j=1}^{2} \Gamma'(R)_{j,i}\, \vec{r}'_j, \quad i = 1,2$$

and matrices

$$\boldsymbol{\Gamma}'(R) = \begin{pmatrix} \Gamma'(R)_{1,1} & \Gamma'(R)_{1,2} \\ \Gamma'(R)_{2,1} & \Gamma'(R)_{2,2} \end{pmatrix}$$

The coefficients $\Gamma'(R)_{j,i}$ and the matrices $\boldsymbol{\Gamma}'(R)$ are primed to distinguish them from the first example. For the symmetry operations of the equilateral triangle one obtains:

$$\left.\begin{aligned} E\vec{r}'_1 &= \vec{r}'_1 \\ E\vec{r}'_2 &= \vec{r}'_2 \end{aligned}\right\} \quad \boldsymbol{\Gamma}'(E) = \begin{pmatrix} 1 & 0 \\ 0 & 1 \end{pmatrix}, \quad \left.\begin{aligned} \sigma_3\vec{r}'_1 &= \vec{r}'_2 \\ \sigma_3\vec{r}'_2 &= \vec{r}'_1 \end{aligned}\right\} \quad \boldsymbol{\Gamma}'(\sigma_3) = \begin{pmatrix} 0 & 1 \\ 1 & 0 \end{pmatrix},$$

$$\left.\begin{aligned} \sigma_1\vec{r}'_1 &= -\vec{r}'_1 - \vec{r}'_2 \\ \sigma_1\vec{r}'_2 &= \vec{r}'_2 \end{aligned}\right\} \quad \boldsymbol{\Gamma}'(\sigma_1) = \begin{pmatrix} -1 & 0 \\ -1 & 1 \end{pmatrix}, \quad \left.\begin{aligned} C_3\vec{r}'_1 &= -\vec{r}'_1 - \vec{r}'_2 \\ C_3\vec{r}'_2 &= \vec{r}'_1 \end{aligned}\right\} \quad \boldsymbol{\Gamma}'(C_3) = \begin{pmatrix} -1 & 1 \\ -1 & 0 \end{pmatrix},$$

$$\left.\begin{aligned} \sigma_2\vec{r}'_1 &= \vec{r}'_1 \\ \sigma_2\vec{r}'_2 &= -\vec{r}'_1 - \vec{r}'_2 \end{aligned}\right\} \quad \boldsymbol{\Gamma}'(\sigma_2) = \begin{pmatrix} 1 & -1 \\ 0 & -1 \end{pmatrix}, \quad \left.\begin{aligned} C_3^2\vec{r}'_1 &= \vec{r}'_2 \\ C_3^2\vec{r}'_2 &= -\vec{r}'_1 - \vec{r}'_2 \end{aligned}\right\} \quad \boldsymbol{\Gamma}'(C_3^2) = \begin{pmatrix} 0 & -1 \\ 1 & -1 \end{pmatrix}.$$

The vector pairs $\vec{r}'_1, \vec{r}'_2$ and $\vec{r}_1, \vec{r}_2$ used in the two examples are not independent of one another; the vectors $\vec{r}_1, \vec{r}_2$ can be expressed in terms of the vectors $\vec{r}'_1, \vec{r}'_2$ and vice versa. This is accomplished through transformation equations of the form:

$$\vec{r}'_i = \sum_{j=1}^{2} S_{j,i}\, \vec{r}_j, \quad i = 1,2,$$

$$\vec{r}_i = \sum_{j=1}^{2} S^{-1}_{j,i}\, \vec{r}'_j, \quad i = 1,2.$$

The $S_{j,i}$ and $S^{-1}_{j,i}$ are again numbers which can be summarized in matrix form:

$$\mathbf{S} = \begin{pmatrix} S_{1,1} & S_{1,2} \\ S_{2,1} & S_{2,2} \end{pmatrix}, \quad \mathbf{S}^{-1} = \begin{pmatrix} S^{-1}_{1,1} & S^{-1}_{1,2} \\ S^{-1}_{2,1} & S^{-1}_{2,2} \end{pmatrix}.$$

$S^{-1}_{j,i}$ is in general not simply the reciprocal of $S_{j,i}$. The $-1$ in $S^{-1}_{j,i}$ is only to indicate symbolically that the $S^{-1}_{j,i}$ describe the reversal of the transformations characterized by $S_{j,i}$. A comparison of the two vector pairs

$\vec{r}_1, \vec{r}_2$ and $\vec{r}'_1, \vec{r}'_2$ shows that

$$\left.\begin{aligned}\vec{r}'_1 &= \frac{\sqrt{3}}{2}\vec{r}_1 - \frac{1}{2}\vec{r}_2\\ \vec{r}'_2 &= \vec{r}_2\end{aligned}\right\} \quad \mathbf{S} = \begin{pmatrix} \frac{\sqrt{3}}{2} & 0 \\ -\frac{1}{2} & 1 \end{pmatrix}$$

and

$$\left.\begin{aligned}\vec{r}_1 &= \frac{2}{\sqrt{3}}\vec{r}'_1 + \frac{1}{\sqrt{3}}\vec{r}'_2\\ \vec{r}_2 &= \vec{r}'_2\end{aligned}\right\} \quad \mathbf{S}^{-1} = \begin{pmatrix} \frac{2}{\sqrt{3}} & 0 \\ \frac{1}{\sqrt{3}} & 1 \end{pmatrix}$$

hold. It can be easily seen that

$$\mathbf{S}\,\mathbf{S}^{-1} = \mathbf{S}^{-1}\,\mathbf{S} = \begin{pmatrix} 1 & 0 \\ 0 & 1 \end{pmatrix}$$

and thereby is equal to the *unit matrix* $\mathbf{E}$*.

The two examples lead to an important relation. For every symmetry operation $R$ the matrices $\boldsymbol{\Gamma}(R)$ and $\boldsymbol{\Gamma}'(R)$ are related by

$$\boldsymbol{\Gamma}'(R) = \mathbf{S}^{-1}\,\boldsymbol{\Gamma}(R)\,\mathbf{S}, \tag{2–1}$$

where $\boldsymbol{S}$ and $\boldsymbol{S}^{-1}$ are the transformation matrices defined above.

Example:

$$\mathbf{S}^{-1}\,\boldsymbol{\Gamma}(C_3)\,\mathbf{S} = \begin{pmatrix} \frac{2}{\sqrt{3}} & 0 \\ \frac{1}{\sqrt{3}} & 1 \end{pmatrix} \begin{pmatrix} -\frac{1}{2} & \frac{\sqrt{3}}{2} \\ -\frac{\sqrt{3}}{2} & -\frac{1}{2} \end{pmatrix} \begin{pmatrix} \frac{\sqrt{3}}{2} & 0 \\ -\frac{1}{2} & 1 \end{pmatrix}.$$

If first of all we take the matrix product $\mathbf{S}^{-1}\boldsymbol{\Gamma}(C_3)$, we obtain

$$\mathbf{S}^{-1}\,\boldsymbol{\Gamma}(C_3) = \begin{pmatrix} -\frac{1}{\sqrt{3}} & 1 \\ -\frac{2}{\sqrt{3}} & 0 \end{pmatrix}.$$

Finally we obtain

$$\mathbf{S}^{-1}\,\boldsymbol{\Gamma}(C_3)\,\mathbf{S} = \begin{pmatrix} -\frac{1}{\sqrt{3}} & 1 \\ -\frac{2}{\sqrt{3}} & 0 \end{pmatrix} \begin{pmatrix} \frac{\sqrt{3}}{2} & 0 \\ -\frac{1}{2} & 1 \end{pmatrix} = \begin{pmatrix} -1 & 1 \\ -1 & 0 \end{pmatrix}.$$

---

* The elements of matrix algebra are summarized in the Appendix, pp. 47 ff.

This is precisely equal to $\boldsymbol{\Gamma}'(C_3)$. The reader can easily verify that the relation (2–1) is satisfied for all other symmetry operations of the equilateral triangle.

At the beginning of this chapter we studied the symmetry operations $E, \sigma_1, \sigma_2, \sigma_3, C_3, C_3^2$ of the group of the triangle. The consecutive application of symmetry operations is the definition of a combination of the elements. The results of combining all pairs of elements are summarized in the multiplication Table B.10. It is then logical to prepare a multiplication table for the transformation matrices $\boldsymbol{\Gamma}(E), \boldsymbol{\Gamma}(\sigma_1), \boldsymbol{\Gamma}(\sigma_2), \boldsymbol{\Gamma}(\sigma_3), \boldsymbol{\Gamma}(C_3), \boldsymbol{\Gamma}(C_3^2)$ introduced at the beginning of this section as well. We select matrix multiplication as the combining operation for the matrices

$$[\boldsymbol{\Gamma}(P)\boldsymbol{\Gamma}(Q)]_{kl} = \sum_i \Gamma(P)_{ki}\Gamma(Q)_{il}.$$

By this means one arrives at the multiplication table of the transformation matrices, Table B.11. A comparison with the multiplication table for the symmetry operations of the equilateral triangle, Table B.10

Table B.11. Multiplication table for the transformation matrices for the vector pair $\vec{r}_1, \vec{r}_2$

| left-hand factor \ right-hand factor | $\boldsymbol{\Gamma}(E)$ | $\boldsymbol{\Gamma}(\sigma_1)$ | $\boldsymbol{\Gamma}(\sigma_2)$ | $\boldsymbol{\Gamma}(\sigma_3)$ | $\boldsymbol{\Gamma}(C_3)$ | $\boldsymbol{\Gamma}(C_3^2)$ |
|---|---|---|---|---|---|---|
| $\boldsymbol{\Gamma}(E)$ | $\boldsymbol{\Gamma}(E)$ | $\boldsymbol{\Gamma}(\sigma_1)$ | $\boldsymbol{\Gamma}(\sigma_2)$ | $\boldsymbol{\Gamma}(\sigma_3)$ | $\boldsymbol{\Gamma}(C_3)$ | $\boldsymbol{\Gamma}(C_3^2)$ |
| $\boldsymbol{\Gamma}(\sigma_1)$ | $\boldsymbol{\Gamma}(\sigma_1)$ | $\boldsymbol{\Gamma}(E)$ | $\boldsymbol{\Gamma}(C_3)$ | $\boldsymbol{\Gamma}(C_3^2)$ | $\boldsymbol{\Gamma}(\sigma_2)$ | $\boldsymbol{\Gamma}(\sigma_3)$ |
| $\boldsymbol{\Gamma}(\sigma_2)$ | $\boldsymbol{\Gamma}(\sigma_2)$ | $\boldsymbol{\Gamma}(C_3^2)$ | $\boldsymbol{\Gamma}(E)$ | $\boldsymbol{\Gamma}(C_3)$ | $\boldsymbol{\Gamma}(\sigma_3)$ | $\boldsymbol{\Gamma}(\sigma_1)$ |
| $\boldsymbol{\Gamma}(\sigma_3)$ | $\boldsymbol{\Gamma}(\sigma_3)$ | $\boldsymbol{\Gamma}(C_3)$ | $\boldsymbol{\Gamma}(C_3^2)$ | $\boldsymbol{\Gamma}(E)$ | $\boldsymbol{\Gamma}(\sigma_1)$ | $\boldsymbol{\Gamma}(\sigma_2)$ |
| $\boldsymbol{\Gamma}(C_3)$ | $\boldsymbol{\Gamma}(C_3)$ | $\boldsymbol{\Gamma}(\sigma_3)$ | $\boldsymbol{\Gamma}(\sigma_1)$ | $\boldsymbol{\Gamma}(\sigma_2)$ | $\boldsymbol{\Gamma}(C_3^2)$ | $\boldsymbol{\Gamma}(E)$ |
| $\boldsymbol{\Gamma}(C_3^2)$ | $\boldsymbol{\Gamma}(C_3^2)$ | $\boldsymbol{\Gamma}(\sigma_2)$ | $\boldsymbol{\Gamma}(\sigma_3)$ | $\boldsymbol{\Gamma}(\sigma_1)$ | $\boldsymbol{\Gamma}(E)$ | $\boldsymbol{\Gamma}(C_3)$ |

shows that both multiplication tables are identical when the transformation matrices are assigned to the symmetry operations in the following manner:

$$\begin{aligned} E &\to \boldsymbol{\Gamma}(E), & \sigma_3 &\to \boldsymbol{\Gamma}(\sigma_3), \\ \sigma_1 &\to \boldsymbol{\Gamma}(\sigma_1), & C_3 &\to \boldsymbol{\Gamma}(C_3), \\ \sigma_2 &\to \boldsymbol{\Gamma}(\sigma_2), & C_3^2 &\to \boldsymbol{\Gamma}(C_3^2). \end{aligned}$$

The statement that for this assignment the multiplication tables are equal means that to every product $PQ = R$ of symmetry operations, the product of the corresponding transformation matrices $\boldsymbol{\Gamma}(P)\boldsymbol{\Gamma}(Q) = \boldsymbol{\Gamma}(R)$ can be uniquely associated:

$$PQ = R \longrightarrow \boldsymbol{\Gamma}(P)\boldsymbol{\Gamma}(Q) = \boldsymbol{\Gamma}(R)$$

(Association is consecutive execution of operations) (Association is matrix multiplication)

Example: From the assignment $\sigma_1 \rightarrow \boldsymbol{\Gamma}(\sigma_1), \sigma_2 \rightarrow \boldsymbol{\Gamma}(\sigma_2)$ because of

$$\sigma_1\sigma_2 = C_3$$

and

$$\boldsymbol{\Gamma}(\sigma_1)\boldsymbol{\Gamma}(\sigma_2) = \begin{pmatrix} -1 & 0 \\ 0 & 1 \end{pmatrix} \begin{pmatrix} \frac{1}{2} & -\frac{\sqrt{3}}{2} \\ -\frac{\sqrt{3}}{2} & -\frac{1}{2} \end{pmatrix} = \begin{pmatrix} -\frac{1}{2} & \frac{\sqrt{3}}{2} \\ -\frac{\sqrt{3}}{2} & -\frac{1}{2} \end{pmatrix} = \boldsymbol{\Gamma}(C_3)$$

the assignment

$$\sigma_1\sigma_2 = C_3 \rightarrow \boldsymbol{\Gamma}(\sigma_1)\boldsymbol{\Gamma}(\sigma_2) = \boldsymbol{\Gamma}(C_3)$$

actually follows.

In general one calls the set of square matrices $\boldsymbol{\Gamma}(P), \boldsymbol{\Gamma}(Q), \boldsymbol{\Gamma}(R), \ldots$, which can be assigned to the elements $P, Q, R, \ldots$ of a group so that the product of the associated matrices corresponds to the product of the two group elements

$$PQ = R \rightarrow \boldsymbol{\Gamma}(P)\boldsymbol{\Gamma}(Q) = \boldsymbol{\Gamma}(R),$$

a *representation* $\Gamma$ of the group*.

If the matrices of a representation consist of $n$ rows (and $n$ columns), then one says the representation has the *dimensionality* $n$. All matrices belonging to the same representation must of course have the same dimensionality. Otherwise matrix multiplication between them would not be possible. One-dimensional representations having matrices all of which

* This assignment need not be reversibly unique. More than one element of the group can correspond to the same matrix of the representation, e.g.

$$\begin{array}{ccc} \boldsymbol{\Gamma}(P) & & \boldsymbol{\Gamma}(Q) \\ \swarrow \; \downarrow \; \searrow & , & \swarrow \; \searrow \quad , \ldots \\ P \; P' \; P'' & & Q \quad Q' \end{array}$$

One speaks of a homomorphism. If the correspondence is reversibly unique (a one-to-one correspondence), then the converse is also true and every representation matrix corresponds to one and only one group element. We have an isomorphism.

are equal to one (1) are called *identical representations* (or totally symmetrical representations).

On the basis of the definition of the representation it can be seen that the matrix transformations treated above $\boldsymbol{\Gamma}(E)$, $\boldsymbol{\Gamma}(\sigma_1)$, $\boldsymbol{\Gamma}(\sigma_2)$, $\boldsymbol{\Gamma}(\sigma_3)$, $\boldsymbol{\Gamma}(C_3)$, $\boldsymbol{\Gamma}(C_3^2)$ form a representation $\Gamma$ of the group of the triangle $C_{3v}$—a representation of dimensionality two.

The primed transformation matrices described on p. 272 $\boldsymbol{\Gamma}'(E)$, $\boldsymbol{\Gamma}'(\sigma_1)$, $\boldsymbol{\Gamma}'(\sigma_2)$, $\boldsymbol{\Gamma}'(\sigma_3)$, $\boldsymbol{\Gamma}'(C_3)$, $\boldsymbol{\Gamma}'(C_3^2)$ are likewise a two dimensional representation of the group $C_{3v}$, because they also have the same multiplication table as the symmetry operations of $C_{3v}$. We shall call this representation $\Gamma'$.

### 2.2.2. *Equivalent representations: similarity transformations*

We remember that the matrices of the representation $\Gamma$ describe the transformations of the vector pair $\vec{r}_1, \vec{r}_2$ and the matrices of the representation $\Gamma'$ the transformation of the vector pair $\vec{r}'_1, \vec{r}'_2$ under the symmetry operations of the triangle. Thus the vector pair $\vec{r}_1, \vec{r}_2$ is called the *basis* which *induces* the representation $\Gamma$ and the vector pair $\vec{r}'_1, \vec{r}'_2$ the basis which *induces* $\Gamma'$.

Two representations $\Gamma$ and $\Gamma'$ between whose matrices $\boldsymbol{\Gamma}(R)$ and $\boldsymbol{\Gamma}'(R)$ the relationship

$$\boldsymbol{\Gamma}'(R) = \mathbf{S}^{-1}\boldsymbol{\Gamma}(R)\mathbf{S} \tag{2–2}$$

holds for all $R$ are called *equivalent representations.* It is required that $\mathbf{S}$ be a nonsingular matrix, i.e., the determinant of $\mathbf{S}$ cannot assume the value zero. A relation of the form (2–2) is called a *similarity transformation.*

The representations $\Gamma$ and $\Gamma'$ of the group $C_{3v}$ fulfil the relation (2–2), as we have seen on p. 273, and are therefore equivalent representations. $\mathbf{S}$ is in this example the matrix of the transformation which brings the basis $\vec{r}_1, \vec{r}_2$ into $\vec{r}'_1, \vec{r}'_2$.

For the symmetry group $C_{3v}$ other representations besides $\Gamma$ and $\Gamma'$ may be given. As examples consider the two one-dimensional representations $\Gamma''$ and $\Gamma'''$:

$$\boldsymbol{\Gamma}''(E) = (1), \quad \boldsymbol{\Gamma}''(\sigma_1) = (1), \quad \boldsymbol{\Gamma}''(\sigma_2) = (1), \quad \boldsymbol{\Gamma}''(\sigma_3) = (1), \quad \boldsymbol{\Gamma}''(C_3) = (1), \quad \boldsymbol{\Gamma}''(C_3^2) = (1),$$

$$\boldsymbol{\Gamma}'''(E) = (1), \quad \boldsymbol{\Gamma}'''(\sigma_1) = (-1), \quad \boldsymbol{\Gamma}'''(\sigma_2) = (-1), \quad \boldsymbol{\Gamma}'''(\sigma_3) = (-1), \quad \boldsymbol{\Gamma}'''(C_3) = (1), \quad \boldsymbol{\Gamma}'''(C_3^2) = (1).$$

$\Gamma''$ is the identical representation of the group $C_{3v}$.

2.2.3. *Characters*

The sum of the diagonal elements of an ($l$-dimensional) representation matrix

$$\boldsymbol{\Gamma}(R) = \begin{pmatrix} \Gamma(R)_{1,1} & \Gamma(R)_{1,2} & \dots & \Gamma(R)_{1,l} \\ \Gamma(R)_{2,1} & \Gamma(R)_{2,2} & \dots & \Gamma(R)_{2,l} \\ \vdots & \vdots & \ddots & \vdots \\ \Gamma(R)_{l,1} & \Gamma(R)_{l,2} & \dots & \Gamma(R)_{l,l} \end{pmatrix}$$

is called the *character*

$$\chi_\Gamma(R) = \Gamma(R)_{1,1} + \Gamma(R)_{2,2} + \cdots + \Gamma(R)_{l,l}$$

of the representation matrix $\boldsymbol{\Gamma}(R)$. The totality of the characters $\chi_\Gamma(P), \chi_\Gamma(Q), \chi_\Gamma(R), \dots$ belonging to a representation $\Gamma$ is the *character system* of this representation:

Example: The representations $\Gamma$ and $\Gamma'$ of the group $C_{3v}$ (cf. pp. 271 and 272) have the following character systems:

$$\chi_\Gamma(E) = 1 + 1 = 2, \quad \chi_\Gamma(\sigma_1) = -1 + 1 = 0, \quad \chi_\Gamma(\sigma_2) = 1/2 - 1/2 = 0,$$
$$\chi_\Gamma(\sigma_3) = 1/2 - 1/2 = 0, \; \chi_\Gamma(C_3) = -1/2 - 1/2 = -1, \; \chi_\Gamma(C_3^2) = -1/2 - 1/2 = -1,$$

or

$$\chi_{\Gamma'}(E) = 1 + 1 = 2, \quad \chi_{\Gamma'}(\sigma_1) = -1 + 1 = 0, \quad \chi_{\Gamma'}(\sigma_2) = 1 - 1 = 0,$$
$$\chi_{\Gamma'}(\sigma_3) = 0 + 0 = 0, \quad \chi_{\Gamma'}(C_3) = -1 + 0 = -1, \quad \chi_{\Gamma'}(C_3^2) = 0 - 1 = -1.$$

From this example we formulate the general theorem: *equivalent representations have equal character systems.* The converse is also valid: *If two representations have equal character systems, they are equivalent,* that is, they differ at most by a similarity transformation.

From a comparison of the characters $\chi_\Gamma$ as well as the characters $\chi_{\Gamma'}$ of the example above one obtains further (the generally valid theorem): *Characters of elements of the same class are equal; the character is a function of the class.*

Example: Representation $\Gamma$:

$$\chi_\Gamma(E) = 2, \quad \chi_\Gamma(\sigma_1) = \chi_\Gamma(\sigma_2) = \chi_\Gamma(\sigma_3) = 0, \quad \chi_\Gamma(C_3) = \chi_\Gamma(C_3^2) = -1,$$

Representation $\Gamma'$:

$$\chi_{\Gamma'}(E) = 2, \quad \chi_{\Gamma'}(\sigma_1) = \chi_{\Gamma'}(\sigma_2) = \chi_{\Gamma'}(\sigma_3) = 0, \quad \chi_{\Gamma'}(C_3) = \chi_{\Gamma'}(C_3^2) = -1.$$

### 2.2.4. *Reduction of representations*

A representation is *reducible* when a similarity transformation exists which brings *all* matrices of the representation into the (*same*) form*

$$\boldsymbol{\Gamma}(P)=\begin{pmatrix} \boxed{\boldsymbol{\Gamma}_1(P)} & & 0 \\ & \boxed{\boldsymbol{\Gamma}_2(P)} & \\ 0 & & \boxed{\boldsymbol{\Gamma}_3(P)} \ddots \end{pmatrix}, \boldsymbol{\Gamma}(Q)=\begin{pmatrix} \boxed{\boldsymbol{\Gamma}_1(Q)} & & 0 \\ & \boxed{\boldsymbol{\Gamma}_2(Q)} & \\ 0 & & \boxed{\boldsymbol{\Gamma}_3(Q)} \ddots \end{pmatrix}, \ldots$$

One says that the representation $\Gamma$ decomposes into the representations $\Gamma_1, \Gamma_2, \ldots$, and writes then formally†

$$\Gamma=\Gamma_1 \dot{+} \Gamma_2 \dot{+} \Gamma_3 \dot{+} \cdots$$

or

$$\Gamma=\dot{\sum_i}\Gamma_i.$$

Example: Consider a three-dimensional representation $\Gamma'$ of the group of the triangle $C_{3v}$:

$$\boldsymbol{\Gamma}'(E)=\begin{pmatrix} 1 & 0 & 0 \\ 0 & 1 & 0 \\ 0 & 0 & 1 \end{pmatrix}, \boldsymbol{\Gamma}'(\sigma_1)=\begin{pmatrix} -1 & 0 & 0 \\ -1 & 1 & 0 \\ -7 & 0 & 1 \end{pmatrix}, \boldsymbol{\Gamma}'(\sigma_2)=\begin{pmatrix} 7 & 1 & -2 \\ 12 & 3 & -4 \\ 30 & 5 & -9 \end{pmatrix}$$

$$\boldsymbol{\Gamma}'(\sigma_3)=\begin{pmatrix} -6 & -1 & 2 \\ 7 & 2 & -2 \\ -14 & -2 & 5 \end{pmatrix}, \boldsymbol{\Gamma}'(C_3)=\begin{pmatrix} -7 & -1 & 2 \\ 5 & 2 & -2 \\ -19 & -2 & 5 \end{pmatrix}, \boldsymbol{\Gamma}'(C_3^2)=\begin{pmatrix} 6 & 1 & -2 \\ 13 & 3 & -4 \\ 28 & 5 & -9 \end{pmatrix}$$

---

* The matrices $\Gamma(R)$ consist of blocks $\Gamma_1(R), \Gamma_2(R), \ldots$, arranged along the diagonals of the matrices. The matrix elements lying outside these boxes are all zero.

Example:

$$\mathbf{A}=\begin{pmatrix} 1 & 0 & 0 & 0 & 0 \\ 2 & 0 & 0 & 0 & 0 \\ 0 & 0 & 7 & 1/2 & 0 \\ 0 & 0 & 3 & 0 & 2 \\ 0 & 0 & 4 & 1 & 6 \end{pmatrix}=\begin{pmatrix} \mathbf{A}_1 & 0 \\ 0 & \mathbf{A}_2 \end{pmatrix}.$$

† The dots over the $+$ sign and the summation sign $\sum$ indicate that the 'addition' is only to be carried out formally in the sense described above.

If the $\boldsymbol{\Gamma}'$ matrices are subjected to a similarity transformation

$$\boldsymbol{\Gamma}(R) = \mathbf{S}^{-1}\boldsymbol{\Gamma}'(R)\,\mathbf{S}$$

where

$$\mathbf{S} = \begin{pmatrix} 0 & 1 & 0 \\ 2 & 0 & 1 \\ 1 & 3 & 1 \end{pmatrix}, \quad \mathbf{S}^{-1} = \begin{pmatrix} 3 & 1 & -1 \\ 1 & 0 & 0 \\ -6 & -1 & 2 \end{pmatrix},$$

we find that

$$\boldsymbol{\Gamma}(E) = \begin{pmatrix} 1 & 0 & 0 \\ 0 & 1 & 0 \\ 0 & 0 & 1 \end{pmatrix}, \boldsymbol{\Gamma}(\sigma_1) = \begin{pmatrix} 1 & 0 & 0 \\ 0 & -1 & 0 \\ 0 & -1 & 1 \end{pmatrix}, \boldsymbol{\Gamma}(\sigma_2) = \begin{pmatrix} 1 & 0 & 0 \\ 0 & 1 & -1 \\ 0 & 0 & -1 \end{pmatrix},$$

$$\boldsymbol{\Gamma}(\sigma_3) = \begin{pmatrix} 1 & 0 & 0 \\ 0 & 0 & 1 \\ 0 & 1 & 0 \end{pmatrix}, \boldsymbol{\Gamma}(C_3) = \begin{pmatrix} 1 & 0 & 0 \\ 0 & -1 & 1 \\ 0 & -1 & 0 \end{pmatrix}, \boldsymbol{\Gamma}(C_3^2) = \begin{pmatrix} 1 & 0 & 0 \\ 0 & 0 & -1 \\ 0 & 1 & -1 \end{pmatrix}.$$

The similarity transformation described decomposes the representation $\Gamma'$:

$$\Gamma \doteq \Gamma_1 \dot{+} \Gamma_2,$$

where $\Gamma_1$ is the (one-dimensional) box in the upper left-hand corners and $\Gamma_2$ is the (two-dimensional) box in the lower right-hand corners of the $\Gamma$ matrices.

A representation $\Gamma$ is called *irreducible* when no similarity transformation exists which causes the representation to decompose into representations of lower dimensionalities. *A reducible representation $\Gamma$ can be decomposed into its irreducible components $\Gamma_1, \Gamma_2, \ldots \Gamma_k$ in only one way.*

$$\Gamma = n_1\Gamma_1 \dot{+} n_2\Gamma_2 \dot{+} \cdots \dot{+} n_k\Gamma_k \dot{+} \cdots = \dot{\sum_i} n_i\Gamma_i.$$

In this expression $n_i$ indicates how many times the reducible representation $\Gamma_i$ is contained in $\Gamma$.

Two theorems concerning irreducible representations remain to be given:

*For every group there exist exactly as many nonequivalent irreducible representations as there are classes in the group.*

For example, the group $C_{3v}$ has three and the group $O_h$ ten irreducible representations.

*The sum of the squares of the dimensionalities $l_1, l_2, \ldots, l_r$ of the (non-equivalent) irreducible representations $\Gamma_1, \Gamma_2, \ldots, \Gamma_r$ of a group is equal to the order h of the group:*

$$\sum_{i=1}^{r} l_i^2 = h. \tag{2–3}$$

The group of the triangle $C_{3v}$, e.g., has three nonequivalent irreducible representations and has the order 6. The dimensionalities $l_1$, $l_2$, $l_3$ of these irreducible representations are found from the equation

$$l_1^2 + l_2^2 + l_3^2 = 6.$$

The number 6 can be decomposed in one and only one way into the sum of the squares of three integers,

$$1^2 + 1^2 + 2^2 = 6,$$

i.e., the group of the triangle $C_{3v}$ has two oné-dimensional and one two-dimensional irreducible representations.

Analogously one finds for the octahedral group $O_h$ (order 48, ten irreducible representations):

$$1^2 + 1^2 + 1^2 + 1^2 + 2^2 + 2^2 + 3^2 + 3^2 + 3^2 + 3^2 = 48.$$

The group $O_h$ has then four one-dimensional, two two-dimensional and four three-dimensional irreducible representations.

### 2.2.5. *Character tables*

The characters of the irreducible representations of a group are generally given in the form of a square system, a *character table.* If we are concerned with symmetry groups, symmetry operations characteristic of each class, in each case preceded by the number of members of the class, are written one after another in the first line of the table. The left-hand column contains the symbols of the irreducible representations. The values of the appropriate characters are given at the points of intersection of the columns and rows. We see for example from the following character table (Table B.12) that for $C_{3v}$, the group of the triangle, the character of a matrix $\mathbf{A}_2(\sigma_v)$ of the irreducible representation $A_2$ and corresponding to the class of reflections $[\sigma_1, \sigma_2, \sigma_3] = 3\sigma_v$ has the value $\chi_{A_2(\sigma_v)} = -1$.

In theoretical chemistry and molecular physics it is conventional (after Mulliken) to denote one-dimensional irreducible representations with the letters $A$ and $B$, two-dimensional representations with $E$, three-dimensional with $T$ or $F$. To distinguish between various irreducible representations of the same dimensionality the subscripts 1 or 2 are used. If a system has a centre of inversion, each representation symbol appears twice. The symbol for the irreducible representation which is symmetric with respect to inversion $i$ is indicated by a g (*gerade*) in addition to the numerical subscript, e.g., $A_{2g}$; the irreducible representation which is antimetric with regard to inversion is denoted by the additional index $u$ (*ungerade*), e.g., $A_{2u}$.

Table B.12. Character table for the irreducible representations of the group $C_{3v}$

| $C_{3v}$ | $E$ | $2C_3$ | $3\sigma_v$ |
|---|---|---|---|
| $A_1$ | 1 | 1 | 1 |
| $A_2$ | 1 | 1 | −1 |
| $E$ | 2 | −1 | 0 |

On pp. 479 ff. the character tables of the point groups discussed on pp. 263 ff. are given. An additional column gives the symbols introduced by Bethe for the irreducible representations, as these are also frequently found in the literature.

### 2.2.6. *The reduction formula*

It often occurs in the ligand field theory that given a representation one wishes to determine into which irreducible representations it can be decomposed. Generally only the characters are known for the given representation and for the irreducible representations of the group. Fortunately a relation is known in which, besides the order of the group under consideration, only the characters of the representation $\Gamma$ to be reduced and the irreducible representations $\Gamma_1, \Gamma_2, \Gamma_3, \ldots$ appear. This relation indicates how often a, say the $i$th irreducible representation $\Gamma_i$, is contained in $\Gamma$. It is formulated:

$$n_{\Gamma_i} = \frac{1}{h} \sum_R \chi^*_{\Gamma_i}(R)\chi_\Gamma(R) \qquad (2\text{–}4a),$$

where

$n_{\Gamma_i}$ a number indicating how many times the irreducible representation $\Gamma_i$ is contained in $\Gamma$,

$h$ the order of the group,

$\chi_{\Gamma_i}(R)$ the character of the $i$th irreducible representation $\Gamma_i$ for the group element $P$,

$\chi_\Gamma(R)$ the character of the representation which is to be reduced, $\Gamma$, for the group element $R$,

$\sum_R$ a summation to be taken over *all group elements.*

The star in (2–4a) indicates that the complex conjugate of $\chi_{\Gamma_i}(R)$ is to be taken. For the cases which we shall consider the characters of the irreducible representations are real, so that $\chi^*_{\Gamma_i}(R) = \chi_{\Gamma_i}(R)$.

Occasionally one finds relation (2–4a) in the literature in a somewhat different form. Because the characters of a representation for all elements of the same class are identical, one may also write for equation (2–4a):

$$n_{\Gamma_i} = \frac{1}{h} \sum_k \lambda_k \chi^*_{\Gamma_i}(R_k) \chi_\Gamma(R_k) \qquad (2\text{–}4b),$$

where

- $n_{\Gamma_i}$ is the number which indicates how many times the irreducible representation $\Gamma_i$ is contained in $\Gamma$
- $h$ is the order of the group,
- $\lambda_k$ is the number of elements in the $k$th class,
- $\chi_{\Gamma_i}(R_k)$ is the character of the $i$th irreducible representation $\Gamma_i$ for any element $R_k$ from the $k$th class,
- $\chi_\Gamma(R_k)$ is the character of the representation which is to be reduced, $\Gamma$, for any element $R_k$ from the $k$th class,
- $\sum_k$ is a summation to be taken over all classes.

Example: Given is a seven-dimensional representation $\Gamma$ of the tetrahedral group $T_d$ having the following characters:

| | $E$ | $8C_3$ | $3C_2$ | $6S_4$ | $6\sigma_d$ |
|---|---|---|---|---|---|
| $\chi_\Gamma(R_k) =$ | 7 | 1 | −1 | −1 | −1 |

With the aid of the character table of the irreducible representations of $T_d$ (see p. 480) we obtain using equation (2–4b):

$$n_{A_1} = \frac{1}{24}\{1\cdot1\cdot7+8\cdot1\cdot1\ +3\cdot1\cdot(-1)+6\cdot1\cdot(-1)+6\cdot1\cdot(-1)\}=0,$$

$$n_{A_2} = \frac{1}{24}\{1\cdot1\cdot7+8\cdot1\cdot1\ +3\cdot1\cdot(-1)+6\cdot(-1)(-1)+6\cdot(-1)(-1)\}=1,$$

$$n_E = \frac{1}{24}\{1\cdot2\cdot7+8\cdot(-1)\cdot1+3\cdot2\cdot(-1)+6\cdot0\cdot(-1)+6\cdot0\cdot(-1)\}=0,$$

$$n_{T_1} = \frac{1}{24}\{1\cdot3\cdot7+8\cdot0\cdot1\ +3\cdot(-1)(-1)+6\cdot1\cdot(-1)+6\cdot(-1)(-1)\}=1,$$

$$n_{T_2} = \frac{1}{24}\{1\cdot3\cdot7+8\cdot0\cdot1\ +3\cdot(-1)(-1)+6\cdot(-1)(-1)+6\cdot1\cdot(-1)\}=1.$$

The given seven-dimensional representation $\Gamma$ decomposes accordingly into the one-dimensional irreducible representation $A_2$ and the three-dimensional irreducible representations $T_1$ and $T_2$:

$$\Gamma = A_2 \dot{+} T_1 \dot{+} T_2.$$

We now wish to treat several examples which are of special interest in the ligand field theory. It often occurs that one goes from regular octahedral structures (group $O_h$) to figures of lower symmetry, having symmetry groups which are subgroups of $O_h$. As will be explained later, when treating the term splitting the question arises which irreducible representations of the subgroup are contained in the irreducible representations of the group $O_h$. The answer is obtained easily by using the reduction formulae (2–4a) and (2–4b).

As an example let us consider a descent in symmetry $O_h \to D_{4h}$. In order to be able to apply the reduction formula it must first of all be established which symmetry elements $O_h$ and $D_{4h}$ have in common and from which classes of the octahedral group the classes of $D_{4h}$ arise. From section 2.1.3.2 and section 2.1.3.3 we obtain the following assignment of the classes (cf. also Figures B.9 and B.10):

| $O_h$ | | $D_{4h}$ | $O_h$ | | $D_{4h}$ |
|---|---|---|---|---|---|
| $E$ | $\to$ | $E$ | $i$ | $\to$ | $i$ |
| $3C_2$ | $\nearrow$ | $C_2$ | $3iC_2$ | $\nearrow$ | $iC_2 = \sigma_h$ |
| | $\searrow$ | $2C_2'$ | | $\searrow$ | $2iC_2' = 2\sigma_v$ |
| $6C_4$ | $\to$ | $2C_4$ | $6iC_4$ | $\to$ | $2iC_4 = 2S_4$ |
| $6C_2'$ | $\to$ | $2C_2''$ | $6iC_2'$ | $\to$ | $2iC_2'' = 2\sigma_d$ |

Using these relations we can now obtain from the character table for $O_h$ the characters of the symmetry operations of $D_{4h}$ for each irreducible $O_h$ representation. One sees for example from the $T_{2g}$ line of the $O_h$ character table that (cf. p. 479)

| | $E$ | $2C_4$ | $C_2$ | $2C_2'$ | $2C_2''$ | $i$ | $2S_4$ | $\sigma_h$ | $2\sigma_v$ | $2\sigma_d$ |
|---|---|---|---|---|---|---|---|---|---|---|
| $\chi =$ | 3 | −1 | −1 | −1 | 1 | 3 | −1 | −1 | −1 | 1 |

This is the character system of a representation of $D_{4h}$. Whether this character system is reducible or not is determined by applying the reduction formula (2–4b) in connection with the character table for $D_{4h}$. In this manner the numbers $n_{\Gamma_i}$ which indicate how often the irreducible representation $\Gamma_i$ of $D_{4h}$ is contained in the given representation are determined:

$$n_{A_{1g}} = \frac{1}{16}\{1\cdot 1\cdot 3 + 2\cdot 1\cdot(-1) + 1\cdot 1\cdot(-1) + 2\cdot 1\cdot(-1) + 2\cdot 1\cdot 1 + 1\cdot 1\cdot 3 + 2\cdot 1\cdot(-1) + 1\cdot 1\cdot(-1) + 2\cdot 1\cdot(-1) + 2\cdot 1\cdot 1\} = 0,$$

$$n_{A_{1u}}=\frac{1}{16}\{1\cdot1\cdot3+2\cdot1\cdot(-1)+1\cdot1\cdot(-1)+2\cdot1\cdot(-1)+2\cdot1\cdot1$$
$$+1\cdot(-1)\cdot3+2\cdot(-1)\cdot(-1)+1\cdot(-1)\cdot(-1)+2\cdot(-1)\cdot(-1)$$
$$+2\cdot(-1)\cdot1\}=0,$$

$$n_{A_{2g}}=\frac{1}{16}\{1\cdot1\cdot3+2\cdot1\cdot(-1)+1\cdot1\cdot(-1)+2\cdot(-1)\cdot(-1)+2\cdot(-1)\cdot1$$
$$+1\cdot1\cdot3+2\cdot1\cdot(-1)+1\cdot1\cdot(-1)+2\cdot(-1)\cdot(-1)+2\cdot(-1)\cdot1\}$$
$$=0,$$

$$n_{A_{2u}}=\frac{1}{16}\{1\cdot1\cdot3+2\cdot1\cdot(-1)+1\cdot1\cdot(-1)+2\cdot(-1)\cdot(-1)$$
$$+2\cdot(-1)\cdot1+1\cdot(-1)\cdot3+2\cdot(-1)\cdot(-1)+1\cdot(-1)\cdot(-1)+2\cdot1\cdot(-1)$$
$$+2\cdot1\cdot1\}=0,$$

$$n_{B_{1g}}=\frac{1}{16}\{1\cdot1\cdot3+2\cdot(-1)\cdot(-1)+1\cdot1\cdot(-1)+2\cdot1\cdot(-1)$$
$$+2\cdot(-1)\cdot1+1\cdot1\cdot3+2\cdot(-1)\cdot(-1)+1\cdot1\cdot(-1)$$
$$+2\cdot1\cdot(-1)+2\cdot(-1)\cdot1\}=0,$$

$$n_{B_{1u}}=\frac{1}{16}\{1\cdot1\cdot3+2\cdot(-1)\cdot(-1)+1\cdot1\cdot(-1)+2\cdot1\cdot(-1)$$
$$+2\cdot(-1)\cdot1+1\cdot(-1)\cdot3+2\cdot1\cdot(-1)+1\cdot(-1)\cdot(-1)$$
$$+2\cdot(-1)\cdot(-1)+2\cdot1\cdot1\}=0,$$

$$n_{B_{2g}}=\frac{1}{16}\{1\cdot1\cdot3+2\cdot(-1)\cdot(-1)+1\cdot1\cdot(-1)+2\cdot(-1)\cdot(-1)$$
$$+2\cdot1\cdot1+1\cdot1\cdot3+2\cdot(-1)\cdot(-1)+1\cdot1\cdot(-1)+2\cdot(-1)\cdot(-1)$$
$$+2\cdot1\cdot1\}=1,$$

$$n_{B_{2u}}=\frac{1}{16}\{1\cdot1\cdot3+2\cdot(-1)\cdot(-1)+1\cdot1\cdot(-1)+2\cdot(-1)\cdot(-1)$$
$$+2\cdot1\cdot1+1\cdot(-1)\cdot3+2\cdot1\cdot(-1)+1\cdot(-1)\cdot(-1)+2\cdot1\cdot(-1)$$
$$+2\cdot(-1)\cdot1\}=0,$$

$$n_{E_g}=\frac{1}{16}\{1\cdot2\cdot3+2\cdot0\cdot(-1)+1\cdot(-2)\cdot(-1)+2\cdot0\cdot(-1)+2\cdot0\cdot1$$
$$+1\cdot2\cdot3+2\cdot0\cdot(-1)+1\cdot(-2)\cdot(-1)+2\cdot0\cdot(-1)+2\cdot0\cdot1\}=1,$$

$$n_{E_u}=\frac{1}{16}\{1\cdot2\cdot3+2\cdot0\cdot(-1)+1\cdot(-2)\cdot(-1)+2\cdot0\cdot(-1)+2\cdot0\cdot1$$
$$+1\cdot(-2)\cdot3+2\cdot0\cdot(-1)+1\cdot2\cdot(-1)+2\cdot0\cdot(-1)+2\cdot0\cdot1\}=0,$$

that is, the irreducible representations $B_{2g}$ and $E_g$ of the group $D_{4h}$ are each contained once in the $O_h$ representation $T_{2g}$:

$$T_{2g}(O_h)=B_{2g}(D_{4h})\dot{+}E_g(D_{4h}).$$

Following the above method the irreducible $D_{4h}$ components can be determined for all of the other octahedral representations. One proceeds in exactly the same manner when going from $O_h$ to another subgroup of $O_h$ or when going from a subgroup $U$ of $O_h$ to a group which is likewise a subgroup of $U$, e.g., $D_{4h} \rightarrow C_{4v}$.

The results of such investigations are generally tabulated in so-called *correlation schemes.* From these the decomposition of the irreducible representations of a larger group into the irreducible representations of a subgroup can be read directly. The Table B.13a and b are such correlation schemes. Table B.13a shows which irreducible representations of the groups

Table B.13. Correlation Schemes

(a) for subgroups of $O_h$

| $O_h$ | $T_d$ | $D_{4h}$ | $D_3$ |
|---|---|---|---|
| $A_{1g}$ | $A_1$ | $A_{1g}$ | $A_1$ |
| $A_{1u}$ | $A_2$ | $A_{1u}$ | $A_1$ |
| $A_{2g}$ | $A_2$ | $B_{1g}$ | $A_2$ |
| $A_{2u}$ | $A_1$ | $B_{1u}$ | $A_2$ |
| $E_g$ | $E$ | $A_{1g}+B_{1g}$ | $E$ |
| $E_u$ | $E$ | $A_{1u}+B_{1u}$ | $E$ |
| $T_{1g}$ | $T_1$ | $A_{2g}+E_g$ | $A_2+E$ |
| $T_{1u}$ | $T_2$ | $A_{2u}+E_u$ | $A_2+E$ |
| $T_{2g}$ | $T_2$ | $B_{2g}+E_g$ | $A_1+E$ |
| $T_{2u}$ | $T_1$ | $B_{2u}+E_u$ | $A_1+E$ |

(b) for subgroups of $D_{4h}$

| $D_{4h}$ | $C_{4v}$ | $C_{2v}$ |
|---|---|---|
| $A_{1g}$ | $A_1$ | $A_1$ |
| $A_{1u}$ | $A_2$ | $A_2$ |
| $A_{2g}$ | $A_2$ | $B_1$ |
| $A_{2u}$ | $A_1$ | $B_2$ |
| $B_{1g}$ | $B_1$ | $A_1$ |
| $B_{1u}$ | $B_2$ | $A_2$ |
| $B_{2g}$ | $B_2$ | $B_1$ |
| $B_{2u}$ | $B_1$ | $B_2$ |
| $E_g$ | $E$ | $A_2+B_2$ |
| $E_u$ | $E$ | $A_1+B_1$ |

$T_d$, $D_{4h}$, $D_3$ are contained in the irreducible representations of $O_h$. For example, the result given above $T_{2g}(O_h) = B_{2g}(D_{4h})+E_g(D_{4h})$ is obtained from Table B.13a by taking in the column '$O_h$' the row containing $T_{2g}$ and going to the right until the column '$D_{4h}$' is reached and reading $B_{2g}+E_g$. Table B.13b is a correlation scheme for the subgroups $C_{4v}$ and $C_{2v}$ of $D_{4h}$. The Tables B.13a and B.13b can be combined. Let us assume we wish to determine which irreducible representations of $C_{2v}$ are contained in an irreducible representation of $O_h$. First of all we find in Table B.13a those particular $D_{4h}$ representations which belong to the $O_h$ representation and determine subsequently using Table B.13b the $C_{2v}$ representation of a particular $D_{4h}$ representation. For example:

$$T_{2g}(O_h) \begin{cases} \nearrow B_{2g}(D_{4h}) \rightarrow B_1(C_{2v}) \\ \searrow E_g(D_{4h}) \begin{cases} \nearrow A_2(C_{2v}) \\ \searrow B_2(C_{2v}) \end{cases} \end{cases}$$

that is,

$$T_{2g}(O_h) = B_{2g}(D_{4h}) \dot{+} E_g(D_{4h}) = B_1(C_{2v}) \dot{+} A_2(C_{2v}) \dot{+} B_2(C_{2v}).$$

## 2.3. Coordinate transformations

Consider an orthogonal cartesian $xyz$ coordinate system (cf. Figure B.17). A given point $P$ *fixed in space* is described by the coordinates $x, y, z$. If the $xyz$ coordinate system is rotated in any manner about the

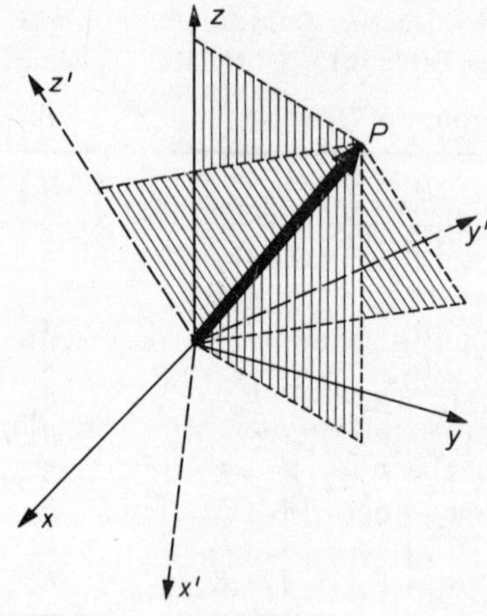


Figure B.17.

origin or is reflected in the origin, a new (primed) $x', y', z'$ coordinate system is obtained. In this new system the coordinates of the fixed point $P$ are $x', y', z'$. The new coordinates $x', y', z'$ of $P$ may be expressed in the old unprimed coordinates $x, y, z$ by using the following relations:

$$\begin{aligned} x' &= a_{11}x + a_{12}y + a_{13}z, \\ y' &= a_{21}x + a_{22}y + a_{23}z, \\ z' &= a_{31}x + a_{32}y + a_{33}z, \end{aligned} \tag{2–5}$$

where the $a_{ij}$ are numbers dependent upon the nature of the transformation. To simplify the notation one often writes $x_1, x_2, x_3$ for $x, y, z$ and $x'_1, x'_2, x'_3$ for $x', y', z'$, etc. The three relations given above can be concisely formulated as

$$x'_i = \sum_{j=1}^{3} a_{ij}x_j, \qquad i = 1, 2, 3. \tag{2–6}$$

Frequently matrix notation is used

$$\mathbf{x}' = \mathbf{A}\,\mathbf{x}, \tag{2–7}$$

where $\mathbf{x}'$ and $\mathbf{x}$ are column matrices

$$\mathbf{x}' \equiv \begin{pmatrix} x_1' \\ x_2' \\ x_3' \end{pmatrix} \equiv \begin{pmatrix} x' \\ y' \\ z' \end{pmatrix}, \qquad \mathbf{x} \equiv \begin{pmatrix} x_1 \\ x_2 \\ x_3 \end{pmatrix} \equiv \begin{pmatrix} x \\ y \\ z \end{pmatrix}$$

and $\mathbf{A}$ is the so-called *transformation matrix*, the following system of coefficients $a_{ij}$:

$$\mathbf{A} \equiv \begin{pmatrix} a_{11} & a_{12} & a_{13} \\ a_{21} & a_{22} & a_{23} \\ a_{31} & a_{32} & a_{33} \end{pmatrix}.$$

The multiplication of the column matrix $\mathbf{x}$ with the quadratic transformation matrix $\mathbf{A}$ is defined by equation (2–6).

If the transformation is a rotation, then the $a_{ij}$ depend upon the nature and the magnitude of the rotation which brings the $xyz$ system into the $x'y'z'$ system. Rotations of coordinate systems are conventionally decomposed into three partial rotations through angles $\phi$, $\theta$, $\psi$ about three well-defined axes, the so-called Eulerian angles (cf. Figure B.18). The first rotation through the angle $\phi$ (in a counter-clockwise direction) about the $z$ axis brings the $xyz$ system into a $\xi\eta\zeta$ intermediary system. Subsequently the $\xi\eta\zeta$ system is rotated by the angle $\theta$ in.a counter-clockwise direction about the $\xi$ axis. This leads to a new intermediary system $\xi'\eta'\zeta'$. Finally a rotation through the angle $\psi$ in a counter-clockwise direction about the $\zeta'$ axis is carried out and the desired $x'y'z'$ system is obtained*. Expressed in terms of the Eulerian angles $\phi$, $\theta$, $\psi$ the transformation matrix for a rotation is:

$$\mathbf{A} = \begin{pmatrix} \cos\psi\cos\phi - \cos\theta\sin\phi\sin\psi & \cos\psi\sin\phi + \cos\theta\cos\phi\sin\psi & \sin\psi\sin\theta \\ -\sin\psi\cos\phi - \cos\theta\sin\phi\cos\psi & -\sin\psi\sin\phi + \cos\theta\cos\phi\cos\psi & \cos\psi\sin\theta \\ \sin\theta\sin\phi & -\sin\theta\cos\phi & \cos\theta \end{pmatrix} \tag{2–8}$$

* A detailed explanation of the Eulerian angles can be found in H. Goldstein: *Classical Mechanics*, pp. 107 ff., Addison-Wesley, Reading, Mass. U.S.A.

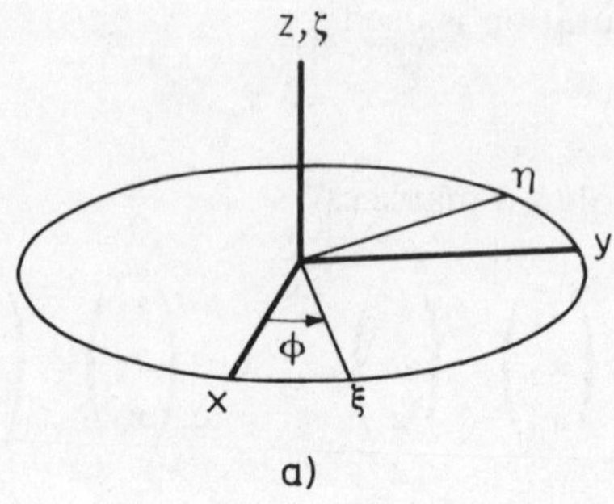


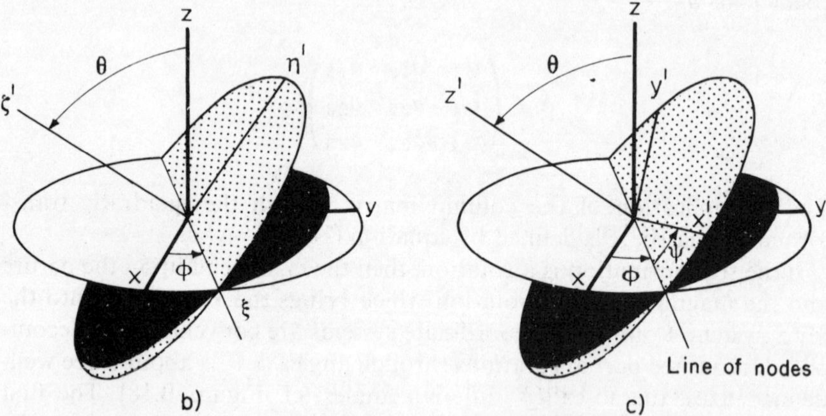


Figure B.18. The rotations defining the Eulerian angles (from H. Goldstein: *Classical Mechanics*, Addison-Wesley Publ., Reading, Mass.–London, 1959).

Example: The $x'y'z'$ system is to be obtained by rotating the $xyz$ system by an angle of $\pi/6$ about the $z$ axis. For the *Eulerian* angles, then $\phi = \pi/6, \theta = 0, \psi = 0$ and

$$\mathbf{A}(\phi=\pi/6, \theta=0, \psi=0) = \begin{pmatrix} \frac{\sqrt{3}}{2} & \frac{1}{2} & 0 \\ -\frac{1}{2} & \frac{\sqrt{3}}{2} & 0 \\ 0 & 0 & 1 \end{pmatrix}.$$

Taking the matrix product $\mathbf{Ax}$ yields the column matrix

$$\mathbf{A\,x} = \begin{pmatrix} \frac{\sqrt{3}}{2} \cdot x_1 + \frac{1}{2} \cdot x_2 + 0 \cdot x_3 \\ -\frac{1}{2} \cdot x_1 + \frac{\sqrt{3}}{2} \cdot x_2 + 0 \cdot x_3 \\ 0 \cdot x_1 + 0 \cdot x_2 + 1 \cdot x_3 \end{pmatrix}.$$

Because $\mathbf{Ax}$ is equal to $\mathbf{x}'$ we obtain finally the transformation matrix written as components:

$$x_1' = \frac{\sqrt{3}}{2}\,x_1 + \frac{1}{2}\,x_2\,,$$
$$x_2' = -\frac{1}{2}\,x_1 + \frac{\sqrt{3}}{2}\,x_2\,,$$
$$x_3' = x_3\,.$$

Besides the pure rotations, the inversion, a reflection through the origin of the coordinate system, is an especially important transformation:

$$x_1' = -x_1\,,\quad x_2' = -x_2\,,\quad x_3' = -x_3\,. \tag{2–9}$$

The *inversion matrix* $\mathbf{S}$ is then:

$$\mathbf{S} = \begin{pmatrix} -1 & 0 & 0 \\ 0 & -1 & 0 \\ 0 & 0 & -1 \end{pmatrix} \tag{2–10}$$

We turn now to the important case in which after the transformation from an $xyz$ system to an $x'y'z'$ system has been carried out a further transformation to a third system $x''y''z''$ follows. We are then interested in the relation between the coordinates of the fixed point $P$ in the unprimed system and its coordinates in the doubly primed system. The transformation from the unprimed system to the primed system is again described by a matrix $\mathbf{A}$: $\mathbf{x}' = \mathbf{Ax}$. The transition from a primed system to a doubly primed system is given by the matrix $\boldsymbol{B}$ having the elements $b_{kl}$:

$$\mathbf{x}'' = \mathbf{B}\mathbf{x}'\,, \tag{2–11a}$$

or written as components:

$$x_k'' = \sum_{l=1}^{3} b_{kl}x_l'\,,\quad k = 1,2,3\,. \tag{2–11b}$$

The relationship between $\mathbf{x}''$ and $\mathbf{x}$ is obtained when the substitutions $\mathbf{x}' = \mathbf{Ax}$ into equation (2–11a) or $x_l' = \sum_{j=1}^{3} a_{lj}x_j$ into equation (2–11b) (according to equation (2–6)) are made:

$$\mathbf{x}'' = \mathbf{B}\mathbf{A}\mathbf{x}\,, \tag{2–12a}$$

$$x_k'' = \sum_{l=1}^{3}\sum_{j=1}^{3} b_{kl}a_{lj}x_j = \sum_{j=1}^{3}\left[\sum_{l=1}^{3} b_{kl}a_{lj}\right]x_j\,,\quad k = 1,2,3\,. \tag{2–12b}$$

It must of course also be possible to go directly from the unprimed $xyz$ system to the doubly-primed $x''y''z''$ system. The transformation is given by the matrix $\mathbf{C}$ with the elements $c_{kj}$

$$\mathbf{x}'' = \mathbf{C}\mathbf{x}, \tag{2–13a}$$

$$x''_k = \sum_{j=1}^{3} c_{kj} x_j, \quad k = 1,2,3. \tag{2–13b}$$

A comparison of the equations (2–12) with the equations (2–13) leads to

$$\mathbf{C} = \mathbf{B}\mathbf{A}, \tag{2–14a}$$

$$c_{kj} = \sum_{l=1}^{3} b_{kl} a_{lj}. \tag{2–14b}$$

Equation (2–14b) is precisely the definition of matrix multiplication and shows how the element $c_{nm}$ of the product matrix $\mathbf{C}$ are to be calculated from the elements $a_{nm}$ and $b_{nm}$ of the matrices to be multiplied together. One obtains the matrix $\mathbf{C}$, which gives the direct transformation from the $xyz$ system to the $x''y''z''$ system, by forming the product of the matrix $\mathbf{A}$ (transition from $xyz$ system to $x'y'z'$ system) with the matrix $\mathbf{B}$ (transition from $x'y'z'$ system to the $x''y''z''$ system) according to the rules of matrix multiplication:

$$\mathbf{x} \xrightarrow{\mathbf{A}} \mathbf{x}' \xrightarrow{\mathbf{B}} \mathbf{x}'',$$

$$\mathbf{x} \xrightarrow{\mathbf{C}=\mathbf{B}\mathbf{A}} \mathbf{x}''.$$

For every coordinate transformation $\mathbf{A}$ there must of course exist a transformation which reverses the effect of the coordinate transformation $\mathbf{A}$. Such a transformation is called the *reciprocal* or *inverse* of $\mathbf{A}$ and is denoted by the symbol $\mathbf{A}^{-1}$. If

$$\mathbf{x}' = \mathbf{A}\mathbf{x}, \tag{2–7}$$

holds, then $\mathbf{A}^{-1}$ is defined by the relation

$$\mathbf{x} = \mathbf{A}^{-1}\mathbf{x}'. \tag{2–15}$$

By substituting equation (2–15) into (2–7) and (2–7) into (2–15) one obtains

$$\mathbf{x}' = \mathbf{A}\mathbf{A}^{-1}\mathbf{x}', \quad \mathbf{x} = \mathbf{A}^{-1}\mathbf{A}\mathbf{x},$$

i.e.,

$$\mathbf{A}\mathbf{A}^{-1} = \mathbf{A}^{-1}\mathbf{A} = \mathbf{E},$$

where **E** is the unit matrix

$$\mathbf{E}=\begin{pmatrix}1 & 0 & 0\\ 0 & 1 & 0\\ 0 & 0 & 1\end{pmatrix}$$

We shall now consider a function which assigns a certain functional value to every fixed point in space. Such 'spatially fixed' functional values must naturally be independent of the coordinate system used to describe them. In an unprimed cartesian $xyz$ system let the functional value at the point $P$ (with the coordinates $\{x_1, x_2, x_3\} \equiv \boldsymbol{x}$) be given by $f(x_1, x_2, x_3) = f(\boldsymbol{x})$. The same point $P$ has in a primed $x'y'z'$ system the coordinates $\{x'_1, x'_2, x'_3) \equiv \boldsymbol{x}'$, and the functional value at the point $P$ should be described by $f'(x'_1, x'_2, x'_3) \equiv f'(\boldsymbol{x})'$*. Because we have agreed that $f(\boldsymbol{x})$ and $f'(\boldsymbol{x})$ give the value of the same function at the same point $P$, they must be identical

$$f'(\mathbf{x}')=f(\mathbf{x}). \qquad (2\text{–}16)$$

If we assume that the primed coordinate system arises from the unprimed system through a transformation **A**

$$\mathbf{x}'=\mathbf{A}\mathbf{x},$$

then we may write for equation (2–16)

$$f'(\mathbf{x}')=f'(\mathbf{A}\mathbf{x})=f(\mathbf{x}). \qquad (2\text{–}17)$$

In order to indicate that the transition from $f$ to $f'$ takes place in connection with a coordinate transformation **A**, $f'$ is written symbolically as $\boldsymbol{A}f$:

$$f'(\mathbf{x}')\equiv\boldsymbol{A}f(\mathbf{x}')=\boldsymbol{A}f(\mathbf{A}\mathbf{x})=f(\mathbf{x}). \qquad (2\text{–}18)$$

$\boldsymbol{A}$ can then be considered to be an operator which changes the function $f$ so that the value of the new function $\boldsymbol{A}f$ for $\boldsymbol{x}'$ is equal to the value of the function $f$ for $\boldsymbol{x}$ where $\boldsymbol{x}'$ and $\boldsymbol{x}$ describe the same point in space (in different coordinate systems).

Example: We consider in two-dimensional space the function $f(x_1, x_2) = ax_1^2+bx_2^2$. The curves of constant value of the function are ellipses (cf. Figure B.19). We choose as the transformation $\boldsymbol{A}$ which leads to a primed coordinate system $x'_1, x'_2$, a rotation by 90° in a clockwise direction:

$$x'_1=-x_2, \quad x'_2=x_1.$$

Expressed in the new coordinates $x'_1, x'_2$ the function $f(x_1, x_2)$ is:

$$f(x_1,x_2)=ax_1^2+bx_2^2=bx_1'^2+ax_2'^2=\boldsymbol{A}f(x'_1,x'_2).$$

* The prime on the function symbol $f'$ is only to distinguish between $f$ and $f'$; differentiation is not indicated thereby.

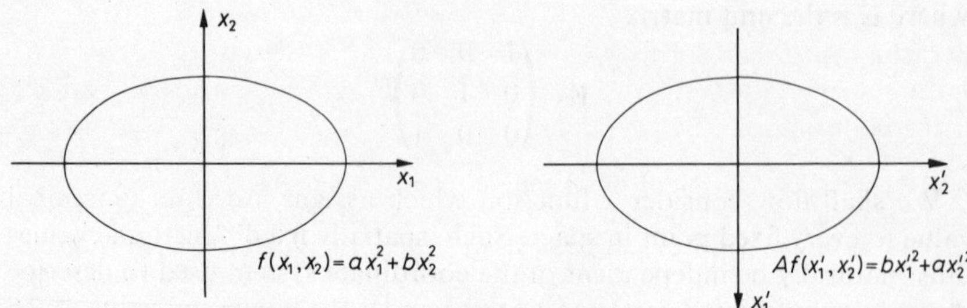


Figure B.19.

The function $Af$ is obtained from $f$ by interchanging the coefficients $a$ and $b$. $\boldsymbol{A}$ is then here an operation which interchanges $a$ and $b$.

If two coordinate transformations $\mathbf{x}' = \mathbf{Ax}$ and $\mathbf{x}'' = \mathbf{Bx}'$ are carried out consecutively, then because of $\mathbf{x}'' = \mathbf{BAx}$, completely analogously to the case of a single transformation the relation

$$\boldsymbol{BA}f(\mathbf{x}'') = \boldsymbol{BA}f(\mathbf{BAx}) = f(\mathbf{x}) \tag{2–19}$$

holds.

These present considerations can also be applied to functions of more than three variables. The many-electron functions $\Psi$, which were introduced in chapter 1 are examples of such functions. The $\Psi$ are of course spatially fixed functions because the physical situations expressed by these $\Psi$ functions must naturally be independent of the coordinate system employed. Let us assume that $\Psi$ is the wavefunction of a system of $N$ electrons. The space coordinates of the $i$th electron in the unprimed $xyz$ coordinate system is denoted by $\{x_{1i}, x_{2i}, x_{3i}\} \equiv \boldsymbol{x}_{ii}$. For the spatially dependent part of the wavefunction we may write

$$\Psi = \Psi(\mathbf{x}_1, \mathbf{x}_2, \ldots, \mathbf{x}_i, \ldots, \mathbf{x}_N).$$

We go from the unprimed $xyz$ coordinate system by means of a transformation $\mathbf{x}' = \mathbf{Ax}$ to a primed $x'y'z'$ system. The spatial coordinates of the $N$ electrons transform in the same manner:

$$\mathbf{x}_1' = \mathbf{Ax}_1, \mathbf{x}_2' = \mathbf{Ax}_2, \ldots, \mathbf{x}_i' = \mathbf{Ax}_i, \ldots, \mathbf{x}_N' = \mathbf{Ax}_N. \tag{2–20}$$

We shall call the wave function which describes in the $x'y'z'$ system the same situation as the function $\Psi(x_1, \ldots, x_N)$ in the $xyz$ system $\boldsymbol{A}\Psi(x_1', \ldots, x_N')$. Considering the relation which exists between the unprimed and the primed coordinates, the relation

$$\boldsymbol{A}\Psi(\mathbf{x}_1', \mathbf{x}_2', \ldots, \mathbf{x}_N') = \boldsymbol{A}\Psi(\mathbf{Ax}_1, \mathbf{Ax}_2, \ldots, \mathbf{Ax}_N) = \Psi(\mathbf{x}_1, \mathbf{x}_2, \ldots, \mathbf{x}_N) \tag{2–21}$$

must hold. As in equation (2–18) $A$ can also here be considered to be an operator which transforms $\Psi$ so that the value of the new function $A\Psi$ for the coordinates $x'_1, x'_2, \ldots, x'_N$ is equal to the value of the old function $\Psi$ for the coordinates $x_1, x_2, \ldots, x_N$.

The behaviour of functions of several, let us say $n$, variables under coordinate transformations can be formulated in a very general manner. One considers the $n$ variables as components of an $n$-dimensional vector $\{x_1, x_2, \ldots, x_n\} \equiv \mathbf{x}$ in an $n$-dimensional space. A real orthogonal transformation brings the unprimed coordinate system $x_1, x_2, \ldots, x_n$ into a primed coordinate system

$$\begin{aligned} x'_1 &= R_{11}x_1 + R_{12}x_2 + \cdots + R_{1n}x_n, \\ x'_2 &= R_{21}x_1 + R_{22}x_2 + \cdots + R_{2n}x_n, \\ &\vdots \\ x'_n &= R_{n1}x_1 + R_{n2}x_2 + \cdots + R_{nn}x_n, \end{aligned}$$

or

$$\mathbf{x}' = \mathbf{R}\mathbf{x}.$$

$\mathbf{R}$ is here an $n$-dimensional matrix of the coefficients $R_{ij}$. If one wishes to express a function $f(x)$ in the primed coordinate system, then it must be replaced by a new function $Rf$ in the primed coordinates:

$$Rf(\mathbf{x}') = Rf(\mathbf{R}\mathbf{x}) = f(\mathbf{x}).$$

The coordinate transformations which are of importance for electron wave functions are especially rotations and reflections in a three-dimensional physical space. The three spatial coordinates of each electron are transformed into one another, but not into the spatial coordinates of the other electrons. The generalized formulation of the transformation problem (in an $n$-dimensional space) which has just been described offers us thus no advantages. We will, therefore, use the formulation given earlier (equation 2–20, 21) in which the transformations of the spatial coordinates are expressed separately for the individual electrons.

## 2.4. Group theory and quantum mechanics

### 2.4.1. *The symmetry group of the Hamiltonian*

The methods of group theory and the theory of representations can often be applied advantageously to the solution of quantum mechanical problems. In order to demonstrate the relationships we begin with the results obtained in the preceding section. It was shown there that every (real linear orthogonal) coordinate transformation $\mathbf{A}$ can be associated with a (linear unitary) operator $\mathbf{A}$, so that for every function $f(\mathbf{x})$

$$f(\mathbf{x}) = Af(\mathbf{A}\mathbf{x}) \tag{2–22}$$

holds. If equation (2–22) is valid for every function $f(x)$ then it must in particular be valid also for a function $\varphi(\mathbf{x})$ which arises from the application of the operator, e.g., the Hamiltonian* $\boldsymbol{H}(\mathbf{x})$ to an arbitrary function $\psi(\mathbf{x})$

$$\varphi(\mathbf{x}) = \boldsymbol{H}(\mathbf{x})\psi(\mathbf{x}). \tag{2–23}$$

Likewise

$$\begin{aligned}\varphi(\mathbf{x}) = \boldsymbol{H}(\mathbf{x})\,\psi(\mathbf{x}) = \boldsymbol{A}\varphi(\mathbf{A}\mathbf{x}) = \boldsymbol{A}\,[\boldsymbol{H}(\mathbf{A}\mathbf{x})\,\psi(\mathbf{A}\mathbf{x})]\\ = \boldsymbol{A}\,\boldsymbol{H}(\mathbf{A}\mathbf{x})\,\boldsymbol{A}^{-1}\psi(\mathbf{x})\end{aligned} \tag{2–24}$$

hold.

From this it follows that the operator $\boldsymbol{A}\boldsymbol{H}(\mathbf{A}\mathbf{x})\boldsymbol{A}^{-1}$ produces the same effect as the Hamiltonian $\boldsymbol{H}(\mathbf{x})$:

$$\boldsymbol{H}(\mathbf{x}) = \boldsymbol{A}\,\boldsymbol{H}(\mathbf{A}\mathbf{x})\,\boldsymbol{A}^{-1}. \tag{2–25}$$

In the following we shall restrict ourselves to the particular coordinate transformations **A**, which leave the Hamiltonian *invariant*. In order to distinguish these special transformations from the general transformations **A** we shall denote them with **R**. By way of definition it holds for **R** that

$$\boldsymbol{H}(\mathbf{R}\mathbf{x}) = \boldsymbol{H}(\mathbf{x}). \tag{2–26}$$

For a transformation **R** equation (2–25) becomes

$$\boldsymbol{H}(\mathbf{x}) = \boldsymbol{R}\boldsymbol{H}(\mathbf{x})\,\boldsymbol{R}^{-1} \tag{2–27}$$

or

$$\boldsymbol{R}\boldsymbol{H}(\mathbf{x}) = \boldsymbol{H}(\mathbf{x})\,\boldsymbol{R}, \tag{2–28}$$

i.e., the operator **R** associated with the transformation ***R*** *commutes* with the Hamiltonian.

The following theorems are valid for the transformations **R** and the operators ***R***:

(1) The set of all coordinate transformations **R** which leave the Hamiltonian invariant

$$\boldsymbol{H}(\mathbf{R}\mathbf{x}) = \boldsymbol{H}(\mathbf{x}),$$

form a *group.*

(2) The set of all operators **R** (associated with the coordinate transformations ***R***) which commute with the Hamiltonian

$$\boldsymbol{R}\boldsymbol{H}(\mathbf{x}) = \boldsymbol{H}(\mathbf{x})\,\boldsymbol{R},$$

form a group, the *group of the Hamiltonian.*

---

* The symbol **x** in the brackets after the ***H*** stands in general for the coordinates as well as for the derivatives with respect to the coordinates.

Of the coordinate transformations **R** the *symmetry transformations** are of special interest in the ligand field theory. The symmetry transformations and the symmetry operations associated with them each form a group; the group of the symmetry operators is called the *symmetry group of the Hamiltonian.*

In those cases of interest, the Hamiltonian $\boldsymbol{H}(\mathbf{x})$ always appears as a sum of a term $\boldsymbol{T}(\mathbf{x})$ representing the kinetic energy of the particles and the potential energy term $\boldsymbol{V}(\mathbf{x})$:

$$\boldsymbol{H}(\mathrm{x}) = \boldsymbol{T}(\mathrm{x}) + \boldsymbol{V}(\mathrm{x}). \tag{2–29}$$

If one wishes to represent the symmetry group of the Hamiltonian, all of the symmetry operations which leave both the kinetic energy operator and the potential energy operator invariant must be found. Finding the symmetry group of each Hamiltonian of the form (2–29) is simplified by the following:

The kinetic energy operator, $\boldsymbol{T}(\mathbf{x})$, is invariant under all symmetry transformations of the rotation-reflection group of the sphere and therefore commutes with all of the associated operators

$$\begin{aligned} \boldsymbol{T}(\mathbf{R}\mathrm{x}) &= \boldsymbol{T}(\mathrm{x}), \\ \boldsymbol{R}\boldsymbol{T}(\mathrm{x}) &= \boldsymbol{T}(\mathrm{x})\boldsymbol{R}. \end{aligned} \tag{2–30}$$

The symmetry group of the Hamiltonian of the form (2–29) is therefore determined by the symmetry operations which leave the potential energy operator invariant:

$$\begin{aligned} \boldsymbol{V}(\mathbf{R}\mathrm{x}) &= \boldsymbol{V}(\mathrm{x}), \\ \boldsymbol{R}\boldsymbol{V}(\mathrm{x}) &= \boldsymbol{V}(\mathrm{x})\boldsymbol{R}. \end{aligned} \tag{2–31}$$

That is the reason why the symmetry group of the Hamiltonian $\boldsymbol{H}(\mathbf{x})$ can usually be easily found from the *geometry* of the system, which is described by $\boldsymbol{H}$

Example: If an electron moves in the potential of six equal octahedrally oriented point charges, then the potential energy $\boldsymbol{V}(\mathbf{x})$ is obviously invariant under all symmetry operations of the octahedral group $O_h$, because each of these operations brings the potential generating octahedron

* If $\boldsymbol{H}(\mathbf{x})$ describes the motion of a particle (e.g., an electron) then all coordination transformations **R** are symmetry transformations. If $\boldsymbol{H}(\mathbf{x})$ on the other hand is a many-particle operator, then the set of all coordinate transformations is made up of symmetry transformations and such coordinate transformations which take account of the interchangeability of the particles.

into itself (except for a physically irrelevant numbering of the equal point charges).

If both sides of the *Schrödinger* Equation

$$\boldsymbol{H}(\mathbf{x})\psi(\mathbf{x})=E\psi(\mathbf{x}) \tag{2–32}$$

are subjected to a symmetry operator $\boldsymbol{R}$

$$\boldsymbol{R}\boldsymbol{H}(\mathbf{x})\,\psi(\mathbf{x})=\boldsymbol{R}E\psi(\mathbf{x}),$$

one finds, because $\boldsymbol{R}$ and $\boldsymbol{H}(\mathbf{x})$ commute and $E$ is constant, that

$$\boldsymbol{H}(\mathbf{x})\,\boldsymbol{R}\psi(\mathbf{x})=E\boldsymbol{R}\psi(\mathbf{x}). \tag{2–33}$$

It can be seen from equations (2–32) and (2–33) that if $\boldsymbol{R}$ is a symmetry operator of the group of the Hamiltonian $\boldsymbol{H}(\mathbf{x})$, then $\psi(\mathbf{x})$ and $\boldsymbol{R}\psi(\mathbf{x})$ are both eigenfunctions of $\boldsymbol{H}(\mathbf{x})$. $\psi(\mathbf{x})$ and $\boldsymbol{R}\psi(\mathbf{x})$ are functions having the same eigenvalue $E$.

For the case that an eigenvalue $E_i$ of the Schrödinger equation is nondegenerate, that is, that only *one* eigenfunction $\psi_i$ exists having the given eigenvalue, then obviously it must hold that

$$\boldsymbol{R}\psi_i=c\psi_i,$$

where $c$ is a constant. If the function $\Psi_i$ is normalized (to one) and if $\boldsymbol{R}\Psi_i$ is also supposed to be normalized, then it must be true that

$$\boldsymbol{R}\psi_i=\pm 1\cdot\psi_i. \tag{2–34}$$

We thus see that under the influence of the symmetry operators $\boldsymbol{R}$ the eigenfunction of a nondegenerate eigenvalue induces a representation that is one-dimensional and therefore, irreducible.

The case of nondegenerate eigenvalues is of secondary importance in the ligand field theory. More important are the cases of degenerate eigenvalues, and these will be discussed in the next section.

### 2.4.2. *Degenerate systems*

2.4.2.1. *General treatment.* Let the *Schrödinger* equation have a g-fold degenerate eigenvalue:

$$\boldsymbol{H}\psi_i=E\psi_i,\quad i=1,2,\ldots,g. \tag{2–35}$$

The $\psi_1,\psi_2,\ldots,\psi_g$ form a set of g linearly independent functions. Application of a symmetry operator $\boldsymbol{R}$ of the symmetry group of $\boldsymbol{H}$ yields when the commutability of $\boldsymbol{R}$ and $\boldsymbol{H}$ is considered:

$$\boldsymbol{R}\boldsymbol{H}\psi_i=\boldsymbol{H}\boldsymbol{R}\psi_i=E\boldsymbol{R}\psi_i,$$

or

$$H(R\psi_i) = E(R\psi_i), \tag{2-36}$$

i.e., if an arbitrary symmetry operator $R$ (of the symmetry group of $H$) is allowed to act upon an arbitrary function $\psi_i$ of the degenerate set, then a new function $R\psi_i$ is obtained, which is likewise an eigenfunction of $H$ and in particular has the same energy eigenvalue $E$ as does $\psi_i$. If, however, the $R\psi_i$ are eigenfunctions having the eigenvalue $E$, it follows that they can be expressed as linear combinations of the $\psi_i$*:

$$R\psi_i = \sum_{j=1}^{g} \Gamma_{ji}(R)\psi_j, \qquad i = 1, 2, \ldots, g. \tag{2-37}$$

Correspondingly for another symmetry operator $S$ of the symmetry group of $H$ it holds that

$$S\psi_j = \sum_{l=1}^{g} \Gamma_{lj}(S)\psi_l, \qquad j = 1, 2, \ldots, g, \tag{2-38}$$

and the consecutive application of the operators $R$ and $S$ to a function $\psi_i$ yields

$$\left.\begin{aligned} SR\psi_i &= \sum_{j=1}^{g} \Gamma_{ji}(R)\, S\psi_j \\ &= \sum_{j=1}^{g} \sum_{l=1}^{g} \Gamma_{ji}(R)\Gamma_{lj}(S)\psi_l \end{aligned}\right\} \quad i = 1, 2, \ldots, g. \tag{2-39}$$

On the other hand by definition

$$SR\psi_i = \sum_{l=1}^{g} \Gamma_{li}(SR)\psi_l. \tag{2-40}$$

The $g \times g$ coefficients $\Gamma_{ji}(R)$ in (2–37) can be considered as elements of a matrix†

$$\Gamma(R) = \begin{pmatrix} \Gamma_{11}(R) & \Gamma_{12}(R) & \ldots & \Gamma_{1g}(R) \\ \Gamma_{21}(R) & \Gamma_{22}(R) & \ldots & \Gamma_{2g}(R) \\ \vdots & \vdots & & \vdots \\ \Gamma_{g1}(R) & \Gamma_{g2}(R) & \ldots & \Gamma_{gg}(R) \end{pmatrix}$$

---

* If that were not the case, then the eigenvalue $E$ would be by definition more than $g$-fold degenerate, in contradiction to the original assumption.

† It should be noted that the transformation coefficients $\Gamma_{ji}, j = 1, 2, 3, \ldots, g$, for the function $\psi_i$ stand in the $i$th column of the matrix $\Gamma$. One then says that the function $\psi_i$ transforms as the $i$th column of the representation matrix.

and correspondingly the $g \times g$ coefficients $\Gamma_{lj}(S)$ in (2–38) as elements of a matrix $\boldsymbol{\Gamma}(S)$ and likewise the $g \times g$ coefficients $\Gamma_{li}(SR)$ in (2–40) as elements of a matrix $\boldsymbol{\Gamma}(SR)$. A comparison of the right-hand sides of equations (2–39) and (2–40) shows that the matrix $\boldsymbol{\Gamma}(SR)$ is equal to the product of the matrices $\boldsymbol{\Gamma}(S)$ and $\boldsymbol{\Gamma}(R)$:

$$\boldsymbol{\Gamma}(\boldsymbol{SR}) = \boldsymbol{\Gamma}(\boldsymbol{S})\,\boldsymbol{\Gamma}(\boldsymbol{R}). \tag{2–41}$$

Taking the definitions and results for representations of section 2.2 into account the conclusions stated above may be summarized as follows:

*The matrices which describe the symmetry transformations of a set of degenerate eigenfunctions form a representation of the symmetry group of the Hamiltonian**.

We shall now show that every set of linearly independent linear combinations of the functions $\psi_i$ introduced above induces a representation of the symmetry group of the Hamiltonian. This representation differs from that induced by $\psi_i$ only by a similarity transformation.

The new set of functions consists of a linear combination of the old functions $\psi_i$:

$$\phi_j = \sum_{i=1}^{g} S_{ij}\psi_i, \qquad j = 1,2,\ldots,g. \tag{2–42}$$

The $g \times g$ coefficients $S_{ij}$ may be considered to be elements of a (non-singular) matrix $\boldsymbol{S}$, the reciprocal of which is denoted by $\boldsymbol{S}^{-1}$.

The effect of a symmetry operator $\boldsymbol{R}$ (of the symmetry group of $\boldsymbol{H}$) on the function $\phi_j$ may be easily seen when equation (2–37) is considered

$$\begin{aligned}\boldsymbol{R}\phi_j &= \sum_{i=1}^{g} S_{ij}\boldsymbol{R}\psi_i = \sum_{i=1}^{g}\sum_{k=1}^{g} S_{ij}\Gamma_{ki}(\boldsymbol{R})\psi_k \\ &= \sum_{i=1}^{g}\sum_{k=1}^{g}\sum_{l=1}^{g} S_{ij}\Gamma_{ki}(\boldsymbol{R})S^{-1}_{lk}\phi_l.\end{aligned} \tag{2–43}$$

If on the other hand

$$\boldsymbol{R}\phi_j = \sum_{l=1}^{g} \Gamma'_{lj}(\boldsymbol{R})\phi_l, \tag{2–44}$$

is taken, then a comparison of the right-hand sides of equations (2–43) and (2–44) shows that

$$\boldsymbol{\Gamma}'(\boldsymbol{R}) = \mathbf{S}^{-1}\boldsymbol{\Gamma}(\boldsymbol{R})\mathbf{S}, \tag{2–45}$$

* If the basis of functions which induces the representation is orthonormalized, then the matrices $\boldsymbol{\Gamma}$ are *unitary*: $\boldsymbol{\Gamma}(\boldsymbol{R})^{\dagger}\boldsymbol{\Gamma}(\boldsymbol{R}) = 1$, where the $ij$th element of the matrix $\boldsymbol{\Gamma}(\boldsymbol{R})^{\dagger}$ is defined by $\Gamma_{ij}(\boldsymbol{R})^{\dagger} = \Gamma^*_{ji}(\boldsymbol{R})$.

i.e., *the representations of two degenerate sets of functions which are connected with one another by a nonsingular linear transformation differ by a similarity transformation.*

The representation induced by a degenerate set of functions will be either irreducible or reducible. If the representation is irreducible, then all functions of the set have the same *symmetry species.* They describe a *term* having this symmetry species. If the representation on the other hand is reducible, that is, contains two or more irreducible components, then the set of functions can be subjected to a transformation (2–42), so that a new set of functions with the following properties arises: The new set of functions decomposes into families of functions so that the individual families each induce one of the irreducible representations contained in the reducible representation. Each of the families of functions describes a term which may be characterized by the symmetry species of the family. Because in the case of a reducible representation all of the progeny terms are energetically the same, one speaks of *accidental degeneracy.*

We shall now use an example of importance in the ligand field theory, the spherical harmonics, to clarify the methods and considerations described formally in this section.

2.4.2.2. *Spherical harmonics.* The *spherical harmonics* $Y_{l,m_l}(\theta, \phi)$ were introduced in chapter 1 as solutions of the angular dependent differential equation of the atomic one-electron problem. We obtained as the spatially dependent parts of the one-electron wavefunctions (cf. (1–15)):

$$\begin{aligned} \psi_{n,l,m_l}(r,\theta,\phi) &= R_{n,l}(r)\, Y_{l,m_l}(\theta,\phi) \\ &= R_{n,l}(r)\Theta_{l,m_l}(\theta)\, \frac{e^{i m_l \phi}}{\sqrt{2\pi}}\,. \end{aligned} \tag{2–46}$$

According to section 1.1.1, p. 207, the functions

$$\psi_{n,l,m_l}, \quad m_l = -l, -l+1, \ldots, l-1, l, \tag{2–47}$$

form a degenerate set of functions when a general spherically symmetric potential $V(r)$ is present; they describe states of the same energy. A set of functions (2–47) consists of $2l+1$ functions. In the case of $d$ functions, for example, we have fivefold (orbital) degeneracy.

We wish to study here the transformation behaviour of these functions under the influence of symmetry operations of the rotation-reflection group of the sphere (cf. p. 262). These include rotations about axes through the origin of the coordinate system, inversion through the origin and operations made up of consecutively executed rotations and inversion (improper rotations). The representation matrices or their characters may be obtained immediately from the transformation behaviour of the wave functions.

Let us consider first a *rotation* through an angle $\varphi$. If we are interested simply in the character of the associated representation matrix, it suffices to study the rotation about an arbitrary axis, because as was shown on p. 262, the character is a function of the class. But all rotations through the angle $\varphi$, independent of about which axis of the rotation-reflection group of the sphere, belong to the same class. We derive the character associated with a rotation through an angle $\varphi$ from the transformation behaviour of a rotation by the angle $\varphi$ about the $z$ axis. The $z$ axis is chosen as the rotation axis for the sake of simplicity in obtaining the desired results. For a rotation of $\phi' = \phi + \varphi$ one obtains:

$$\begin{aligned}\boldsymbol{R}_{\varphi}\psi_{n,l,m_l}(r,\theta,\phi) &= \psi_{n,l,m_l}(r,\theta,\phi') = \psi_{n,l,m_l}(r,\theta,\phi+\varphi)\\ &= R_{n,l}(r)\Theta_{l,m_l}(\theta)\,\frac{e^{i m_l(\phi+\varphi)}}{\sqrt{2\pi}}\\ &= R_{n,l}(r)\Theta_{l,m_l}(\theta)\frac{e^{i m_l \phi}}{\sqrt{2\pi}}\cdot e^{i m_l \varphi},\end{aligned}$$

i.e., the function $\psi_{n,l,m_l}$ is brought into itself except for the factor $e^{-im_l\varphi}$ by a rotation through the angle $\varphi$ about the $z$ axis:

$$\boldsymbol{R}_{\varphi}\psi_{n,l,m_l} = e^{i m_l \varphi}\,\psi_{n,l,m_l}.$$

The general relation (2–37) assumes here this special form. Only the coefficients $\Gamma(\boldsymbol{R}_{\varphi})$ having two equal indices are different from zero; the remaining elements vanish. The corresponding matrix $\Gamma(\boldsymbol{R}_{\varphi})$ for a basis (2–47) can then be written immediately*:

$$\Gamma(\boldsymbol{R}_{\varphi}) \equiv \mathbf{D}_l(\varphi) = \begin{pmatrix} e^{-il\varphi} & 0 \ldots\ldots & 0\\ 0 & e^{i(-l+1)\varphi} \ldots & 0\\ \vdots & \vdots \quad \ddots & \vdots\\ 0 & 0 \ldots\ldots & e^{il\varphi}\end{pmatrix} \tag{2–48}$$

The character $\chi_l(\varphi)$ (= the sum of the diagonal elements) of a rotation

* We shall also employ here the convention that representation matrices based on the rotation-reflection group of the sphere are denoted by $\mathbf{D}$ (*Darstellung*) with $l$ as a subscript when, as here, the induced basis functions are $\psi_{n,l,m_l}$ or spherical harmonics $Y_{l,m_l}$.

through the angle $\varphi$ is therefore equal to

$$\begin{aligned}\chi_l(\varphi) &= e^{-il\varphi} + e^{i(-l+1)\varphi} + \cdots + e^{il\varphi} \\ &= e^{-il\varphi}(1 + e^{i\varphi} + \cdots + e^{i2l\varphi}) \\ &= \frac{\sin\left(l + \frac{1}{2}\right)\varphi}{\sin\frac{\varphi}{2}} .\end{aligned} \tag{2–49}$$

We then obtain the special values

$$\begin{aligned}\chi(\pi) &= (-1)^l, \\ \chi(\pi/2) &= \begin{matrix} 1 \\ -1 \end{matrix} \quad \text{for} \quad \begin{matrix} l = 0, 1, 4, 5, \ldots \\ l = 2, 3, 6, 7, \ldots \end{matrix}\end{aligned} \tag{2–50}$$

Similar considerations show that an *inversion* through the origin, $x' = -x$, $y' = -y$, $z' = -y$, leads to the result

$$\boldsymbol{R}_i \psi_{n,l,m_l} = (-1)^l \psi_{n,l,m_l}, \tag{2–51}$$

i.e., inversion brings a function $\psi_{n,l,m_l}$ into itself except for the factor $(-1)^l$. $\psi$ functions with odd $l$ change their sign; one speaks therefore of *odd* (*ungerade*) functions or says that these functions have *odd* parity. Accordingly the $p, f, h, \ldots$ functions have odd parity. $\psi$ functions with even $l$, i.e., $s, d, g, \ldots$ functions transform into themselves without a change of sign under (2–51) and are called therefore *even* (*gerade*) functions; they have *even parity*.

Which representation matrix is induced by a basis (2–47) under the inversion? Because all functions of such a basis have the same $l$ value, one finds using (2–51):

$$\boldsymbol{\Gamma}(\boldsymbol{R}_i) \equiv \mathbf{D}_l(i) = \begin{pmatrix} (-1)^l & 0 \ldots & 0 \\ 0 & (-1)^l \ldots & 0 \\ \vdots & \vdots \ddots & \vdots \\ 0 & 0 \ldots & (-1)^l \end{pmatrix}, \tag{2–52}$$

with the character

$$\chi_l(i) = (2l + 1)(-1)^l. \tag{2–53}$$

For a basis of $d$ functions for example ($l = 2$) we obtain

$$\chi_2(i) = (2 \cdot 2 + 1)(-1)^2 = +5. \tag{2–54}$$

The character of a representation matrix which is induced by a basis (2–47) and describes a rotation through the angle $\varphi$ followed by an inversion is:

$$\chi_l(i\varphi) = (-1)^l \frac{\sin\left(l+\frac{1}{2}\right)\varphi}{\sin\frac{\varphi}{2}} \tag{2–55}$$

We now have the necessary formalism to be able to determine which representation a basis (2–47) induces when it is subjected to a group of symmetry operations consisting of rotations, inversion and combined operations. One can choose as an example all point groups which are subgroups of the rotation-reflection group of the sphere. We shall treat the symmetry group $O_h$. The symmetry operations which make up this group are given on p. 263 ff.

The character corresponding to the identity operation $E$ is $\chi_l(E) = 2l+1$, because in each case the function is brought into itself and the basis consists in total of $2l+1$ functions. The characters for the remaining operations are calculated with the aid of the relations (2–49, 53, 55). The results for the $l$ values, $l = 0, 1, 2, 3, 4, 5$ are given in Table B.14.

We ask if a representation of a subgroup of the rotation-reflection group of the sphere which is induced by a basis (2–47) is reducible. Using the character relation (2–4b) together with Table B.14 and the character table for $O_h$ one obtains for the example $O_h$, e.g., for $l = 2$ ($d$ functions):

$$n_{A_{1g}} = n_{A_{2g}} = n_{A_{1u}} = n_{A_{2u}} = n_{E_u} = n_{T_{1g}} = n_{T_{1u}} = n_{T_{2u}} = 0,$$
$$n_{E_g} = 1,$$
$$n_{T_{2g}} = 1,$$

i.e., the representation of the symmetry group $O_h$ induced by a basis of $d$ functions $\psi_{n,2,m_l}$ is reducible. It contains as irreducible components the irreducible representations $E_g$ and $T_{2g}$ of the group $O_h$.

All point groups may be analyzed according to the above procedure, for each basis $\psi_{n,l,m_l}$. That is, one determines in this manner which irreducible components of a given point group are contained in a representation induced by a basis $\psi_{n,l,m_l}$, $m_l = -1, \ldots, +1$. It is essential for this procedure that one go *directly* from the rotation-reflection group of the sphere to the given point group.

Another path which is of special practical value in the ligand field theory can also be chosen. The complex ions are generally regular octahedral figures (Group $O_h$) or belong to symmetry groups which are subgroups of $O_h$. Because deviations from octahedral symmetry have often only a very small influence on the properties of complex ions, it is frequently sufficient to consider nonoctahedral structures in a rough approximation as octahedral and to calculate the deviations from octahedral symmetry

Table B.14. Characters of representation matrices which are induced by spherical harmonics $Y_{l,m_l}$ or from atomic functions $\psi_{n,l,m_l}$

| | Angle of rotation $\varphi$ | $l = 0$ $s$ | $l = 1$ $p$ | $l = 2$ $d$ | $l = 3$ $f$ | $l = 4$ $g$ | $l = 5$ $h$ |
|---|---|---|---|---|---|---|---|
| $\chi_l(E)$ | $0$ | 1 | 3 | 5 | 7 | 9 | 11 |
| $\chi_l(C_2)$ | $\pi$ | 1 | $-1$ | 1 | $-1$ | 1 | $-1$ |
| $\chi_l(C_3)$ | $\frac{2\pi}{3}$ | 1 | 0 | $-1$ | 1 | 0 | $-1$ |
| $\chi_l(C_4)$ | $\frac{\pi}{2}$ | 1 | 1 | $-1$ | $-1$ | 1 | 1 |
| $\chi_l(i)$ | $0$ | 1 | $-3$ | 5 | $-7$ | 9 | $-11$ |
| $\chi_l(iC_2)$ | $\pi$ | 1 | 1 | 1 | 1 | 1 | 1 |
| $\chi_l(iC_3)$ | $\frac{2\pi}{3}$ | 1 | 0 | $-1$ | $-1$ | 0 | 1 |
| $\chi_l(iC_4)$ | $\frac{\pi}{2}$ | 1 | $-1$ | $-1$ | 1 | 1 | $-1$ |

as a small perturbation. In this connection, as we shall see later, it is useful to resolve the question as to which irreducible representations of the point group with which we are concerned (in so far as it is a subgroup of $O_h$) are contained in the *individual* representations of $O_h$. The answer is found not by going *directly* from the rotation-reflection group of the sphere to the subgroup of $O_h$, but rather by proceeding *step-wise* first to an irreducible $O_h$ representation having the basis $\psi_{n,l,m_l}$. The $O_h$ representations are then individually studied to find the irreducible components of the subgroup of $O_h$ in question. We have treated this second step already in section 2.6 for those subgroups of the $O_h$ group which most frequently occur. The correlation schemes, Table B.13, give the desired information. The results for the bases $\psi_{n,l,m_l}$, $l = 0, 1, 2, 3, 4, 5$ obtained with the aid of the correlation scheme B.13 are collected in Table B.15 for the groups $O_h$, $T_d$, $D_3$, $D_{4h}$, $C_{4v}$, $C_{2v}$. From these one determines e.g., that a $d$ electron basis $\psi_{n,2,m_l}$, $m_l = -2, -1, 0, +1, +2$ induces a representation of $O_h$ containing the irreducible $O_h$ components $E_g$ and $T_{2g}$. With a descent in symmetry $O_h \rightarrow D_3$ the octahedral representation $T_{2g}$ decomposes into the irreducible $D_3$ representations $A_1$ and $E$.

Table B.15 shows that the representations induced by the bases (2–47) are in general reducible. If the representation matrices for the bases (2–47) are set up explicitly, it is found that these matrices are not in general in reduced form, that is, they do not in general appear as block matrices (cf. p. 278). If the reducibility of a representation has been determined,

Table B.15. Correlation schemes of the irreducible representations of the group $O_h$, $T_d$, $D_3$, $D_{4h}$, $C_{4v}$, $C_{2v}$ induced by the bases $\psi_{n,l,m_l}$, $l = 0, 1, 2, 3, 4, 5$.

| | $O_h$ | $T_d$ | $D_3$ | $D_{4h}$ | $C_{4v}$ | $C_{2v}$ |
|---|---|---|---|---|---|---|
| $s(l = 0)$ | $A_{1g}$ | $A_1$ | $A_1$ | $A_{1g}$ | $A_1$ | $A_1$ |
| $p(l = 1)$ | $T_{1u}$ | $T_2$ | $A_2 \dotplus E$ | $A_{2u} \dotplus E_u$ | $A_1 \dotplus E$ | $A_1 \dotplus B_1 \dotplus B_2$ |
| $d(l = 2)$ | $E_g$ | $E$ | $E$ | $A_{1g} \dotplus B_1$ | $A_1 \dotplus B_1$ | $2A_1$ |
| | $T_{2g}$ | $T_2$ | $A_1 \dotplus E$ | $B_{2g} \dotplus E_g$ | $B_2 \dotplus E$ | $A_2 \dotplus B_1 \dotplus B_2$ |
| $f(l = 3)$ | $A_{2u}$ | $A_1$ | $A_2$ | $B_{1u}$ | $B_2$ | $A_2$ |
| | $T_{1u}$ | $T_2$ | $A_2 \dotplus E$ | $A_{2u} \dotplus E_u$ | $A_1 \dotplus E$ | $A_1 \dotplus B_1 \dotplus B_2$ |
| | $T_{2u}$ | $T_1$ | $A_1 \dotplus E$ | $B_{2u} \dotplus E_u$ | $B_1 \dotplus E$ | $A_1 \dotplus B_1 \dotplus B^2$ |
| $g(l = 4)$ | $A_{1g}$ | $A_1$ | $A_1$ | $A_{1g}$ | $A_1$ | $A_1$ |
| | $E_g$ | $E$ | $E$ | $A_{1g} \dotplus B_{1g}$ | $A_1 \dotplus B_1$ | $2A_1$ |
| | $T_{1g}$ | $T_1$ | $A_2 \dotplus E$ | $A_{2g} \dotplus E_g$ | $A_2 \dotplus E$ | $A_2 \dotplus B_1 \dotplus B_2$ |
| | $T_{2g}$ | $T_2$ | $A_1 \dotplus E$ | $B_{2g} \dotplus E_g$ | $B_2 \dotplus E$ | $A_2 \dotplus B_1 \dotplus B_2$ |
| $h(l = 5)$ | $E_u$ | $E$ | $E$ | $A_{1u} \dotplus B_{1u}$ | $A_2 \dotplus B_2$ | $2A_2$ |
| | $T_{1u}$ | $T_2$ | $A_2 \dotplus E$ | $A_{2u} \dotplus E_u$ | $A_1 \dotplus E$ | $A_1 \dotplus B_1 \dotplus B_2$ |
| | $T_{1u}$ | $T_2$ | $A_2 \dotplus E$ | $A_{2u} \dotplus E_u$ | $A_1 \dotplus E$ | $A_1 \dotplus B_1 \dotplus B_2$ |
| | $T_{2u}$ | $T_1$ | $A_1 \dotplus E$ | $B_{2u} \dotplus E_u$ | $B_1 \dotplus E$ | $A_1 \dotplus B_1 \dotplus B_2$ |

then it must be possible to generate new functions through a suitable linear combination of the basis functions (2–47) such that they decompose into families so that in each case one of these families induces an irreducible representation contained in the reducible representation. In other words, the new functions must be so constituted and their separation into families must proceed so that a new function goes into only one linear combination of the new functions of the same family under the influence of *all* symmetry operations of the group*. The number of new functions belonging to a family must of course be equal to the dimensionality of the irreducible representation which they induce.

How are the 'new functions' found? There are at least two ways. The method which usually leads most rapidly to results is that of planned trial: One considers how the old functions behave under symmetry operations and, taking this into consideration, the 'correct' linear combination is 'invented'. (Of course the problem of demonstrating that these results are the correct ones still remains.) With practice this procedure (which is not at all esoteric) usually brings one more rapidly to the correct result than does the second procedure. This allows the calculation of the correct linear combinations directly without relying upon intuition and is based

* When we speak of a linear combination, we include the special case where the new function goes into itself or into a single function of the same family.

upon the use of *projection operators*. The application of such operators to an 'old function' tells if these functions contain certain components of the 'new functions'. Slater* treats the method of projection operators in considerable detail.

We shall be satisfied here in giving the symmetry-adapted linear combinations of the $\psi_{n,l,m_l}$ functions for the symmetries $O$, $D_{4h}$, $D_3$ in tabular form without derivation. In the Tables B.16, 17, 18 only the linear combinations of the angular dependent portions $Y_{l,m_l}(\theta, \phi)$ of the $\psi_{n,l,m_l}$ functions are given and not of the $\psi_{n,l,m_l}$ functions themselves. The reason for this is readily apparent. The members of a basis (2–47) differ only in the $m_l$ values $m_l = -l, -l+1, \ldots, l-1, l$; they have the same radial dependence $R_{n,l}(r)$ but different angular dependencies $Y_{l,m_l}(\theta, \phi)$. Now the symmetry operators making up the point groups can operate only on the angular dependent portion $Y_{l,m_l}(\theta, \phi)$. The radially dependent portions $R_{n,l}(r)$ are invariant because the distance $r$ of the reference point from the centre of the coordinates remains constant, i.e., $R_{n,l}(r') = R_{n,l}(r)$. The symmetry-adapted linear combinations of the $\psi_{n,l,m_l}$ functions can be obtained from Tables B.16, 17, and 18 simply by multiplying the linear combinations in these tables by the radially dependent portions $R_{n,l}(r)$.

The symmetry-adapted functions for the group $O_h$ are obtained in the following manner from Table B.16 for $O$: If the basis functions have even parity, then the index $g$ is simply added to the representation symbol in the second column of the table, for odd basis functions the index $u$ is used.

Table B.16. Symmetry adapted linear combinations of spherical harmonics $Y_{l,m_l}$ for the symmetry group $O$. The transition according to $O_h$ is obtained by adding the index $g$ or $u$ to the representation symbol depending upon if the atomic functions have even or odd parity

| | | | |
|---|---|---|---|
| $l = 0$ | $A_1$ | | $Y_{0,0}$ |
| $l = 1$ | $T_1$ | $a$ | $Y_{1,1}$ |
| | | $b$ | $Y_{1,0}$ |
| | | $c$ | $Y_{1,-1}$ |
| $l = 2$ | $E$ | $a$ | $Y_{2,0}$ |
| | | $b$ | $\sqrt{1/2}(Y_{2,2} + Y_{2,-2})$ |
| | $T_2$ | $a$ | $Y_{2,-1}$ |
| | | $b$ | $\sqrt{1/2}(Y_{2,2} - Y_{2,-2})$ |
| | | $c$ | $-Y_{2,1}$ |

* J. S. Slater: *Quantum Theory of Molecules and Solids*, Vol. 1, pp. 172 ff. and 323 ff. McGraw-Hill Book Company, Inc., New York–San Francisco–Toronto–London 1963.

Table B.16.—continued

| $l = 0$ | $A_1$ | | $Y_{0,0}$ |
|---|---|---|---|
| $l = 3$ | $A_2$ | | $\sqrt{1/2}(Y_{3,2} - Y_{3,-2})$ |
| | $T_1$ | $a$ | $-\sqrt{1/8}(\sqrt{5}Y_{3,-3} + \sqrt{3}T_{3,1})$ |
| | | $b$ | $Y_{3,0}$ |
| | | $c$ | $-\sqrt{1/8}(\sqrt{5}Y_{3,3} + \sqrt{3}Y_{3,-1})$ |
| | $T_2$ | $a$ | $-\sqrt{1/8}(\sqrt{3}Y_{3,3} - \sqrt{5}Y_{3,-1})$ |
| | | $b$ | $\sqrt{1/2}(Y_{3,2} + Y_{3,-2})$ |
| | | $c$ | $-\sqrt{1/8}(\sqrt{3}Y_{3,-3} - \sqrt{5}Y_{3,1})$ |
| $l = 4$ | $A_1$ | | $\sqrt{1/12}\{\sqrt{7}Y_{4,0} + \sqrt{5/2}(Y_{4,4} + Y_{4,-4})\}$ |
| | $E$ | $a$ | $-\sqrt{1/12}\{\sqrt{5}Y_{4,0} - \sqrt{7/2}(Y_{4,4} + Y_{4,-4})\}$ |
| | | $b$ | $\sqrt{1/2}(Y_{4,2} + Y_{4,-2})$ |
| | $T_1$ | $a$ | $-\sqrt{1/8}(Y_{4,-3} + \sqrt{7}Y_{4,1})$ |
| | | $b$ | $\sqrt{1/2}(Y_{4,4} - Y_{4,-4})$ |
| | | $c$ | $\sqrt{1/8}(Y_{4,3} + \sqrt{7}Y_{4,-1})$ |
| | $T_2$ | $a$ | $\sqrt{1/8}(\sqrt{7}Y_{4,3} - Y_{4,-1})$ |
| | | $b$ | $\sqrt{1/2}(Y_{4,2} - Y_{4,-2})$ |
| | | $c$ | $\sqrt{1/8}(Y_{4,1} - \sqrt{7}Y_{4,-3})$ |
| $j = 1/2$ | $E'$ | $a$ | $Y_{1/2,1/2}$ |
| | | $b$ | $Y_{1/2,-1/2}$ |
| $j = 3/2$ | $G'$ | $a$ | $Y_{3/2,3/2}$ |
| | | $b$ | $Y_{3/2,1/2}$ |
| | | $c$ | $Y_{3/2,-1/2}$ |
| | | $d$ | $Y_{3/2,-3/2}$ |
| $j = 5/2$ | $E''$ | $a$ | $\sqrt{1/6}(Y_{5/2,5/2} - \sqrt{5}Y_{5/2,-3/2})$ |
| | | $b$ | $\sqrt{1/6}(Y_{5/2,-5/2} - \sqrt{5}Y_{5/2,3/2})$ |
| $j = 5/2$ | $G'$ | $a$ | $-\sqrt{1/6}(Y_{5/2,3/2} + \sqrt{5}Y_{5/2,-5/2})$ |
| | | $b$ | $Y_{5/2,1/2}$ |
| | | $c$ | $-Y_{5/2,-1/2}$ |
| | | $d$ | $\sqrt{1/6}(\sqrt{5}Y_{5/2,5/2} + Y_{5/2,-3/2})$ |

Table B.17. Symmetry adapted linear combinations of spherical harmonics $Y_{l,m_l}$ for the symmetry group $D_{4h}$

| | | | |
|---|---|---|---|
| $l=0$ | $A_{1g}$ | | $Y_{0,0}$ |
| $l=1$ | $A_{2u}$ | | $-iY_{1,0}$ |
| | $E_u$ | $a$ | $i\sqrt{1/2}(Y_{1,1}-Y_{1,-1})$ |
| | | $b$ | $\sqrt{1/2}(Y_{1,1}+Y_{1,-1})$ |
| $l=2$ | $A_{1g}$ | | $Y_{2,0}$ |
| | $B_{1g}$ | | $\sqrt{1/2}(Y_{2,2}+Y_{2,-2})$ |
| | $B_{2g}$ | | $-i\sqrt{1/2}(Y_{2,2}-Y_{2,-2})$ |
| | $E_g$ | $a$ | $i\sqrt{1/2}(Y_{2,-1}+Y_{2,1})$ |
| | | $b$ | $-\sqrt{1/2}(Y_{2,-1}-Y_{2,1})$ |
| | $A_{2u}$ | | $-iY_{3,0}$ |
| | $B_{1u}$ | | $\sqrt{1/2}(Y_{3,2}-Y_{3,-2})$ |
| | $B_{2u}$ | | $-i\sqrt{1/2}(Y_{3,2}+Y_{3,-2})$ |
| $l=3$ | $E_u$ | $a$ | $i\sqrt{1/16}\{\sqrt{5}(Y_{3,3}-Y_{3,-3})-\sqrt{3}(Y_{3,1}-Y_{3,-1})\}$ |
| | | $b$ | $-\sqrt{1/16}\{\sqrt{5}(Y_{3,3}+Y_{3,-3})+\sqrt{3}(Y_{3,1}+Y_{3,-1})\}$ |
| | $E_u$ | $a$ | $-i\sqrt{1/16}\{\sqrt{3}(Y_{3,3}-Y_{3,-3})+\sqrt{5}(Y_{3,1}-Y_{3,-1})\}$ |
| | | $b$ | $\sqrt{1/16}\{\sqrt{3}(Y_{3,3}+Y_{3,-3})-\sqrt{5}(Y_{3,1}+Y_{3,-1})\}$ |
| | $A_{1g}$ | | $\sqrt{1/12}\{\sqrt{7}Y_{4,0}+\sqrt{5/2}(Y_{4,4}+Y_{4,-4})\}$ |
| | $A_{1g}$ | | $-\sqrt{1/12}\{\sqrt{5}Y_{4,0}-\sqrt{7/2}(Y_{4,4}+Y_{4,-4})\}$ |
| | $A_{2g}$ | | $-i\sqrt{1/2}(Y_{4,4}-Y_{4,-4})$ |
| | $B_{1g}$ | | $\sqrt{1/2}(Y_{4,2}+Y_{4,-2})$ |
| $l=4$ | $B_{2g}$ | | $-i\sqrt{1/2}(Y_{4,4}-Y_{4,-2})$ |
| | $E_g$ | $a$ | $-i\sqrt{1/16}\{Y_{4,3}+Y_{4,-3}+\sqrt{7}(Y_{4,1}+Y_{4,-1})\}$ |
| | | $b$ | $\sqrt{1/16}\{Y_{4,3}-Y_{4,-3}-\sqrt{7}(Y_{4,1}-Y_{4,-1})\}$ |
| | $E_g$ | $a$ | $i\sqrt{1/16}\{\sqrt{7}(Y_{4,3}+Y_{4,-3})-(Y_{4,1}+Y_{4,-1})\}$ |
| | | $b$ | $-\sqrt{1/16}\{\sqrt{7}(Y_{4,3}-Y_{4,-3})+(Y_{4,1}-Y_{4,-1})\}$ |

Table B.18. Symmetry-adapted linear combinations of the spherical harmonics $Y_{l,m_l}$ for the symmetry group $D_3$

| | | | |
|---|---|---|---|
| $l=0$ | $A_1$ | | $Y_{0,0}$ |
| $l=1$ | $A_2$ | | $i\sqrt{1/3}\{\sqrt{1/2}[(Y_{1,1}-Y_{1,-1})-i(Y_{1,1}+Y_{1,-1})]-Y_{1,0}\}$ |
| | $E$ | $a$ | $-\sqrt{1/4}\{(Y_{1,1}+Y_{1,-1})-i(Y_{1,1}-Y_{1,-1})\}$ |
| | | $b$ | $\sqrt{1/12}\{(Y_{1,1}+Y_{1,-1})+i(Y_{1,1}-Y_{1,-1}+2\sqrt{2}Y_{1,0})\}$ |
| $l=2$ | $A_1$ | | $i\sqrt{1/6}\{(Y_{2,1}+Y_{2,-1}-Y_{2,2}+Y_{2,-2})+i(Y_{2,1}-Y_{2,-1})\}$ |
| | $E$ | $a$ | $Y_{2,0}$ |
| | | $b$ | $\sqrt{1/2}(Y_{2,2}+Y_{2,-2})$ |
| | $E$ | $a$ | $\sqrt{1/12}\{(Y_{2,1}-Y_{2,-1})-i(Y_{2,1}+Y_{2,-1}+2Y_{2,2}-2Y_{2,-2})\}$ |
| | | $b$ | $\sqrt{1/4}\{(Y_{2,1}-Y_{2,-1})+i(Y_{2,1}+Y_{2,-1})\}$ |
| $l=3$ | $A_1$ | | $i\sqrt{1/6}\{\sqrt{1/8}[(1-i)(-\sqrt{3}Y_{3,3}+\sqrt{5}Y_{3,-1})$ <br> $+(1+i)(\sqrt{3}Y_{3,-3}-\sqrt{5}Y_{3,1})]-(Y_{3,2}+Y_{3,-2})\}$ |
| | $A_2$ | | $\sqrt{1/2}(Y_{3,2}-Y_{3,-2})$ |
| | $A_2$ | | $i\sqrt{1/6}\{\sqrt{1/8}[(1-i)(-\sqrt{5}Y_{3,-3}-\sqrt{3}Y_{3,1})$ <br> $+(1+i)(\sqrt{5}Y_{3,3}+\sqrt{}Y_{3,-1})]-\sqrt{2}Y_{3,0}\}$ |
| | $E$ | $a$ | $\sqrt{1/32}\{(1-i)(\sqrt{5}Y_{3,-3}+\sqrt{3}Y_{3,1})$ <br> $+(1+i)(\sqrt{5}Y_{3,3}+\sqrt{3}Y_{3,-1})\}$ |
| | | $b$ | $-\sqrt{1/12}\{\sqrt{1/8}[(1+i)(\sqrt{5}Y_{3,-3}+\sqrt{3}Y_{3,1})$ <br> $+(1-i)(\sqrt{5}Y_{3,3}+\sqrt{3}Y_{3,-1})]-i2\sqrt{2}Y_{3,0}\}$ |
| | $E$ | $a$ | $\sqrt{1/12}\{\sqrt{1/8}[(1+i)(\sqrt{3}Y_{3,3}-\sqrt{5}Y_{3,-1})$ <br> $+(1-i)(\sqrt{3}Y_{3,-3}-\sqrt{5}Y_{3,1})]-i2(Y_{3,2}+Y_{3,-2})\}$ |
| | | $b$ | $-\sqrt{1/32}\{(1-i)(-\sqrt{3}Y_{3,3}+\sqrt{5}Y_{3,-1})$ <br> $+(1+i)(-\sqrt{3}Y_{3,-3}+\sqrt{5}Y_{3,-1})\}$ |
| $l=4$ | $A_1$ | | $\sqrt{1/12}\{\sqrt{7}Y_{4,0}+\sqrt{5/2}(Y_{4,4}+Y_{4,-4})\}$ |
| | $A_1$ | | $i\sqrt{1/6}\{\sqrt{1/8}[(1-i)(\sqrt{7}Y_{4,3}-Y_{4,-1})$ <br> $+(1+i)(-Y_{4,1}+\sqrt{7}Y_{4,-3})]-(Y_{4,2}-Y_{4,-2})\}$ |

Table B.18.—continued

| $l=0$ | $A_1$ | | $Y_{0,0}$ |
|---|---|---|---|
| | $A_2$ | | $-i\sqrt{1/6}\{\sqrt{1/8}[(1-i)(Y_{4,-3}+\sqrt{7}Y_{4,1})$ $+(1+i)(Y_{4,3}+\sqrt{7}Y_{4,-1})]+(Y_{4,4}-Y_{4,-4})\}$ |
| | $E$ | $a$ | $-\sqrt{1/12}\{\sqrt{5}Y_{4,0}-\sqrt{7/2}(Y_{4,4}+Y_{4,-4})\}$ |
| | | $b$ | $\sqrt{1/2}(Y_{4,2}+Y_{4,-2})$ |
| | $E$ | $a$ | $-\sqrt{1/32}\{(1-i)(-Y_{4,-3}-\sqrt{7}Y_{4,1})$ $+(1+i)(Y_{4,3}+\sqrt{7}Y_{4,-1})\}$ |
| | | $b$ | $\sqrt{1/12}\{\sqrt{1/8}[(1-i)(-Y_{4,-3}-\sqrt{7}Y_{4,1})$ $+(1-i)(Y_{4,3}+\sqrt{7}Y_{4,-1})]+i2(Y_{4,4}-Y_{4,-4})\}$ |
| | $E$ | $a$ | $-\sqrt{1/12}\{\sqrt{1/8}[(1+i)(\sqrt{7}Y_{4,3}-Y_{4,-1})$ $+(1-i)(Y_{4,1}-\sqrt{7}Y_{4,-3})]+i2(Y_{4,2}-Y_{4,-2})\}$ |
| | | $b$ | $-\sqrt{1/32}\{(1-i)(\sqrt{7}Y_{4,3}-Y_{4,-1})$ $+(1+i)(Y_{4,1}-\sqrt{7}Y_{4,-3})\}$ |

The symmetry-adapted functions for the group $T_d$ may also be taken from the Table B.16. If the basis functions are even, then the representation symbols given in the second column of the table apply, for odd basis functions the indices 1 and 2 must be interchanged. (See the correlation scheme in Table B.13a).

2.4.3. *Perturbation theory*

Consider a system whose Schrödinger equation has a $g$-fold degenerate eigenvalue:

$$\boldsymbol{H}_0\,\psi_i=E\psi_i,\quad i=1,2,\ldots,g. \tag{2–56}$$

If we subject this system to a *perturbation* described by an operator $\boldsymbol{H}_1$, then approximate statements concerning the eigenvalues and eigenfunctions of the perturbed system can be obtained using the *quantum mechanical perturbation theory*. The application of group theoretical principles is frequently advantageous. Two cases can be distinguished with respect to the symmetries of the unperturbed and the perturbed systems: The symmetry of $\boldsymbol{H}_1$ is the same as that of $\boldsymbol{H}_0$ (or higher) or $\boldsymbol{H}_1$ has a lower symmetry than $\boldsymbol{H}_0$.

2.4.3.1. *Symmetric perturbations.* One speaks of a *symmetric perturbation* when $\boldsymbol{H}_1$ is invariant under all symmetry operations of the symmetry group of $\boldsymbol{H}_0$. Two theorems hold for symmetric perturbations:

(a) If the representation of the symmetry group of $\boldsymbol{H}_0$ which is induced by a degenerate basis $\psi_i$, $i = 1, 2, \ldots, g$ is *irreducible* then a symmetric perturbation leads under no circumstances to a removal of the degeneracy.

(b) If the representation of the symmetry group of $\boldsymbol{H}_0$ which is induced by a degenerate basis $\psi_i$, $i = 1, 2, \ldots, g$ is *reducible* (cf. p. 278 ff.),

$$\Gamma = \sum_j n_j \Gamma_j,$$

then the degeneracy can be removed. However, the originally $g$-fold degenerate term can split into at most $\sum_j n_j$ (= number of irreducible components of $\Gamma$) energetically different terms.

An example of case (b) has already been given in chapter 1, pp. 209 ff. An *accidental* degeneracy is present for the atomic *Kepler* problem. There all states of the same principal quantum number $n$ have the same energy. For $n > 1$ the functions of the same principal quantum number $n$ induce a reducible representation of the symmetry group of the *Kepler* problem, the rotation-reflection group of the sphere. Functions of the same subshell that is, those having the same $l$, transform according to the irreducible representations of the rotation-reflection group of the sphere. When the *Kepler* system is subjected to a spherically symmetric perturbation (as for example is produced by a non-Coulombic shielding potential $V(r)$), the accidental degeneracy is, however, removed and a splitting of the energy levels of the subshell occurs.

2.4.3.2. *Asymmetric perturbation.* If the perturbation operation $\boldsymbol{H}_1$ is of lower symmetry than the Hamiltonian $\boldsymbol{H}_0$ of the unperturbed system, then one speaks of an *asymmetric perturbation.* Such a perturbation is typical for the ligand field theory where it often occurs that the symmetry group of the perturbation operator $\boldsymbol{H}_1$ is a subgroup of the symmetry group of $\boldsymbol{H}_0$. There $\boldsymbol{H}_0$ describes the motion of the electrons in a transition metal ion which is not complexed and its symmetry group is therefore the rotation-reflection group of the sphere. $\boldsymbol{H}_1$ represents the energy of the interaction between the electrons of the transition metal ion and the ligands of its complex ion. The symmetry group of $\boldsymbol{H}_1$ is determined by the geometry of the complex ion. For octahedral complex ions, for example, the group $O_h$ is the symmetry group of $\boldsymbol{H}_1$.

The symmetry group of the Hamiltonian of a perturbed system

$$\boldsymbol{H} = \boldsymbol{H}_0 + \boldsymbol{H}_1,$$

is of course the same as the symmetry group of $\boldsymbol{H}_1$ if $\boldsymbol{H}_1$ has lower symmetry than $\boldsymbol{H}_0$. The eigenvalues and eigenfunctions of the Schrödinger equation of $\boldsymbol{H}$ are classified according to the irreducible representation of the symmetry group of $\boldsymbol{H}_1$.

What group theoretical results can be obtained concerning the states of an asymmetrically perturbed system if the states of the unperturbed system are known? For a $g$-fold eigenvalue of the unperturbed system with the eigenfunctions $\psi_i$, $i = 1, 2, \ldots, g$ one proceeds as follows when answering this question:

1. For the symmetry operations which compose the symmetry group of the perturbation operator $\boldsymbol{H}_1$, the appropriate representation $\Gamma$ is determined from the transformation behaviour of the basis of the $\psi_i$. In practice it is adequate to determine the character of this representation $\Gamma$.

If, for example, the unperturbed system has spherical symmetry and the eigenfunctions are of the form $\psi_{n,l,m_l} = R_{n,l}(r) Y_{l,m_l}(\theta, \phi)$ and the perturbing operator has symmetry $O_h$, then the results given on pp. 302 ff. are obtained.

2. It is determined if the representation $\Gamma$ is reducible, i.e., if it contains irreducible components $\Gamma_j$ belonging to the symmetry group of the perturbation operator. The character relations (2–4a) or (2–4b) are of use here.

$$\Gamma = \sum_j n_j \Gamma_j.$$

The results may be found in Table B.15 for the example of a descent in symmetry from spherical to $O_h$ symmetry. We find from the table that, e.g., the representation $\Gamma$ which is induced by a basis of (five) $d$ functions contains as irreducible components the irreducible representations $E_g$ and $T_{2g}$ each once.

3. Use is then made of the following theorem:

Under the influence of an asymmetric perturbation $\boldsymbol{H}_1$ a term which is degenerate in the unperturbed system and whose functions transform as the representation $\Gamma$ of the group of the perturbation operation can split into $n_1$ terms of the species $\Gamma_1$, $n_2$ terms of the species $\Gamma_2$, etc. The degeneracy of a progeny term is equal to the dimensionality of the irreducible representation to which these progeny terms belong.

It follows from the example we have just chosen that a term which is fivefold orbitally degenerate in an unperturbed spherically symmetric system and is described by five $d$ functions can split under the influence of a perturbation of symmetry $O_h$ into a (twofold orbitally) degenerate $E_g$ term and a threefold orbitally degenerate $T_{2g}$ term.

In this section we shall consider no further explicit examples, because all problems of the ligand field theory are examples of the methods and theorems, sketched above. The subjects to be handled in chapter 3 on ligand field theory will give occasion enough to apply the material which has been discussed in this section.

It should be emphasized that the group theoretical methods allow predictions as to if and into how many progeny terms of what species a degenerate term splits under the influence of a perturbation. However, purely from group theoretical considerations nothing can be said about the *magnitude* of the splitting. Only an explicit perturbation calculation can give us this information.

## 2.5. Product representations

Consider a symmetry group $G$ with the elements $E, \ldots, R, \ldots$ where the two sets of functions $f_1^A, f_2^A, \ldots, f_{nA}^A$ and $g_1^B, g_2^B, \ldots, g_{nB}^B$ are bases of representations $\Gamma_A$ (dimensionality $n_A$) or $\Gamma_B$ (dimensionality $n_B$) of this group. One may then write for the symmetry operators $\boldsymbol{R}$ (cf. p. 296):

$$\boldsymbol{R} f_i^A = \sum_{j=1}^{n_A} a_{j,i}(\boldsymbol{R}) f_j^A ,$$

$$\boldsymbol{R} g_k^B = \sum_{l=1}^{n_B} b_{l,k}(\boldsymbol{R}) g_l^B .$$

From this it follows that

$$\boldsymbol{R}(f_i^A g_k^B) = (\boldsymbol{R} f_i^A)(\boldsymbol{R} g_k^B) = \sum_{j=1}^{n_A} \sum_{l=1}^{n_B} a_{j,i}(\boldsymbol{R}) b_{l,k}(\boldsymbol{R}) f_j^A g_l^B = \sum_{j=1}^{n_A} \sum_{l=1}^{n_B} c_{jl,ik}(\boldsymbol{R}) f_j^A g_l^B$$

where

$$c_{jl,ik}(\boldsymbol{R}) = a_{j,i}(\boldsymbol{R}) b_{l,k}(\boldsymbol{R}) .$$

That is, the set of all possible products $f_i^A g_k^B (i = 1, 2, \ldots, n_A ; k = 1, 2, \ldots, n_B)$ is likewise a basis of a representation of the group $G$. The corresponding *product representation* $\Gamma_C = \Gamma_A \dot{\times} \Gamma_B$ has the dimensionality $n_A \cdot n_B$. The elements of the representation matrices are $c_{jl,ik}$. The characters $\chi_C(\boldsymbol{R})$ of the representation matrices have an important property*:

---

* The proof of this theorem is quite simple

$$\chi_C(\boldsymbol{R}) = \sum_{j=1}^{n_A} \sum_{l=1}^{n_B} c_{jl,jl}(\boldsymbol{R}) = \sum_{j=1}^{n_A} \sum_{l=1}^{n_B} a_{j,j}(\boldsymbol{R}) b_{l,l}(\boldsymbol{R}) = \chi_A(\boldsymbol{R}) \chi_B(\boldsymbol{R}) .$$

*The characters $\chi_C(\boldsymbol{R})$ of the representation of a product basis $f_i^A g_k^B$ are equal to the products of the characters $\chi_A(\boldsymbol{R})$ and $\chi_B(\boldsymbol{R})$ of the representations induced by the bases $f_i^A$ or $g_k^B$.*

$$\chi_C(\boldsymbol{R}) = \chi_A(\boldsymbol{R}) \cdot \chi_B(\boldsymbol{R}). \tag{2–57}$$

Example: For the symmetry group $T_d$ with the aid of the character table p. 480 one obtains from (2–57) for example for the representations $A_1 \dot{\times} A_2$, $A_2 \dot{\times} E$, $E \dot{\times} T_2$, $A_2 \dot{\times} E \dot{\times} T_1$ the character system given in the following table:

| $T_d$ | $E$ | $8C_3$ | $3C_2$ | $6S_4$ | $6\sigma_d$ |
|---|---|---|---|---|---|
| $A_1$ | 1 | 1 | 1 | 1 | 1 |
| $A_2$ | 1 | 1 | 1 | $-1$ | $-1$ |
| $E$ | 2 | $-1$ | 2 | 0 | 0 |
| $T_1$ | 3 | 0 | $-1$ | 1 | $-1$ |
| $T_2$ | 3 | 0 | $-1$ | $-1$ | 1 |
| $A_1 \dot{\times} A_2$ | $1\cdot 1 = 1$ | $1\cdot 1 = 1$ | $1\cdot 1 = 1$ | $1(-1) = -1$ | $1(-1) = -1$ |
| $A_2 \dot{\times} E$ | $1\cdot 2 = 2$ | $1(-1) = -1$ | $1\cdot 2 = 2$ | $(-1)\cdot 0 = 0$ | $(-1)\cdot 0 = 0$ |
| $E \dot{\times} T_2$ | $2\cdot 3 = 6$ | $(-1)\cdot 0 = 0$ | $2(-1) = -2$ | $0\cdot(-1) = 0$ | $0\cdot 1 = 0$ |
| $A_2 \dot{\times} E \dot{\times} T_1$ | $1\cdot 2\cdot 3 = 6$ | $1(-1)0 = 0$ | $1\cdot 2(-1)$ $= 2$ | $(-1)0\cdot 1$ $= 0$ | $(-1)\cdot 0(-1)$ $- 0$ |

As we have just demonstrated with an example, one can of course form the product representation from more than two representations. The characters of the product representations are then simply the products of the characters of all of the representations which make up the product representation.

The representation of a product basis is in general reducible, also then when the factors of the product basis belong to irreducible representations. The character relations (2–4a) or (2–4b) give information concerning the irreducible components of a reducible product representation. We find for the product representation $A_2 \dot{\times} E \dot{\times} T_1$ given in the example above:

$$n_{A_1} = \frac{1}{24}[1\cdot 1\cdot 6 + 8\cdot 1\cdot 0 + 3\cdot 1\cdot(-2) + 6\cdot 1\cdot 0 + 6\cdot 1\cdot 0] = 0,$$

$$n_{A_2} = \frac{1}{24}[1\cdot 1\cdot 6 + 8\cdot 1\cdot 0 + 3\cdot 1\cdot(-2) + 6\cdot(-1)\cdot 0 + 6\cdot(-1)\cdot 0] = 0,$$

$$n_E = \frac{1}{24}[1\cdot 2\cdot 6 + 8\cdot(-1)\cdot 0 + 3\cdot 2\cdot(-2) + 6\cdot 0\cdot 0 + 6\cdot 0\cdot 0] = 0,$$

$$n_{T_1} = \frac{1}{24}[1\cdot 3\cdot 6 + 8\cdot 0\cdot 0 + 3\cdot(-1)(-2) + 6\cdot 1\cdot 0 + 6\cdot(-1)\cdot 0] = 1,$$

$$n_{T_2} = \frac{1}{24}[1 \cdot 3 \cdot 6 + 8 \cdot 0 \cdot 0 \quad + 3 \cdot (-1)(-2) + 6 \cdot (-1) \cdot 0 + 6 \cdot 1 \cdot 0] \quad = 1,$$

i.e., the six-dimensional product representation $A_2 \dot{\times} E \dot{\times} T_1$ contains the two three-dimensional irreducible representations $T_1$ and $T_2$:

$$A_2 \dot{\times} E \dot{\times} T_1 = T_1 \dot{+} T_2.$$

In Table B.19 the irreducible components of the representations of all products of two components are given for the symmetry group $O$.

The formation of product representations plays an important role in the calculation of quantum mechanical interaction integrals. Let us consider first a definite integral

$$I = \int_{Q_1}^{Q_2} (\text{Integrand}) \cdot \mathrm{d}Q,$$

where the integrand depends upon one or more variables for which we write summarily $Q$. $\mathrm{d}Q$ can then be the volume element $\mathrm{d}\tau$ and the integrand is a function of the spatial coordinates. The *value* of the integral $I$ must of course be independent of the specific, arbitrary choice of the variables $Q$. If the integrand is made up of operators and $\psi$ functions which belong to a molecular system having a particular symmetry, then the value $I$ of the integral can only be different from zero when the integrand or portions thereof are invariant under all symmetry transformations of the symmetry group in question. Translated into the parlance of representation theory: the integral can only then be nonzero when the integrand or portions thereof transform as the totally symmetric, i.e., the identity, representation.

The integrals found in the quantum theory of atomic and molecular systems of the form

$$(f, \boldsymbol{O}\, g),$$

where $\boldsymbol{O}$ is an operator*, and $f$ and $g$ are wave functions which are usually chosen so that they belong to bases of the irreducible representations of the symmetry group of the system under consideration. The representation to which the integrand belongs, is found by forming the product

* In particular the operator of course can be $I$, so that integrals of the type $(f, g)$ are also included.

Table B.19. Irreducible components of the representation products $\Gamma_i \dot{\times} \Gamma_j$ of the double group $O'$. The upper left-hand portion of the scheme bordered by the dotted line belongs to the simple group $O$. The scheme for $O_h$ follows from this when it is considered that $g$(erade) $\dot{\times}$ $g$(erade) = $g$(erade), $g \dot{\times} u = u \dot{\times} g = u$, $u \dot{\times} u = g$ holds.

| $O'$ | $A_1$ | $A_2$ | $E$ | $T_1$ | $T_2$ | $E'$ | $E''$ | $G'$ | Bethe's nomenclature |
|---|---|---|---|---|---|---|---|---|---|
| $A_1$ | $A_1$ | $A_2$ | $E$ | $T_1$ | $T_2$ | $E'$ | $E''$ | $G'$ | $\Gamma_1$ |
| $A_2$ | $A_2$ | $A_1$ | $E$ | $T_2$ | $T_1$ | $E''$ | $E'$ | $G'$ | $\Gamma_2$ |
| $E$ | $E$ | $E$ | $A_1 \dotplus A_2 \dotplus E$ | $T_1 \dotplus T_2$ | $T_1 \dotplus T_2$ | $G'$ | $G'$ | $E' \dotplus E'' \dotplus G'$ | $\Gamma_3$ |
| $T_1$ | $T_1$ | $T_2$ | $T_1 \dotplus T_2$ | $A_1 \dotplus E \dotplus T_1 \dotplus T_2$ | $A_2 \dotplus E \dotplus T_1 \dotplus T_2$ | $E' \dotplus G'$ | $E'' \dotplus G'$ | $E' \dotplus E'' \dotplus 2G'$ | $\Gamma_4$ |
| $T_2$ | $T_2$ | $T_1$ | $T_1 \dotplus T_2$ | $A_2 \dotplus E \dotplus T_1 \dotplus T_2$ | $A_1 \dotplus E \dotplus T_1 \dotplus T_2$ | $E'' \dotplus G'$ | $E' \dotplus G'$ | $E' \dotplus E'' \dotplus 2G'$ | $\Gamma_5$ |
| $E'$ | $E'$ | $E''$ | $G'$ | $E' \dotplus G'$ | $E'' \dotplus G'$ | $A_1 \dotplus T_1$ | $A_2 \dotplus T_2$ | $E \dotplus T_1 \dotplus T_2$ | $\Gamma_6$ |
| $E''$ | $E''$ | $E'$ | $G'$ | $E'' \dotplus G'$ | $E' \dotplus G'$ | $A_2 \dotplus T_2$ | $A_1 \dotplus T_1$ | $E \dotplus T_1 \dotplus T_2$ | $\Gamma_7$ |
| $G'$ | $G'$ | $G'$ | $E' \dotplus E'' \dotplus G'$ | $E' \dotplus E'' \dotplus 2G'$ | $E' \dotplus E'' \dotplus 2G'$ | $E \dotplus T_1 \dotplus T_2$ | $E \dotplus T_1 \dotplus T_2$ | $A_1 \dotplus A_2 \dotplus E \dotplus 2T_1 \dotplus 2T_2$ | $\Gamma_8$ |
| | $\Gamma_1$ | $\Gamma_2$ | $\Gamma_3$ | $\Gamma_4$ | $\Gamma_5$ | $\Gamma_6$ | $\Gamma_7$ | $\Gamma_8$ | |

Bethe's nomenclature.

representation

$$\Gamma_{\text{Integrand}} = \Gamma^*(f) \dot{\times} \Gamma(\boldsymbol{O}) \dot{\times} \Gamma(g).$$

The character relations (2–4a) or (2–4b) allow further the determination of the irreducible components $\Gamma_i$ of $\Gamma_{\text{integrand}}$.

$$\Gamma_{\text{Integrand}} = \dot{\sum_i} n_{\Gamma_i} \Gamma_i.$$

$\Gamma_{\text{integrand}}$ contains the identity representation $\Gamma_{id}$ of course if and only if the integrand or portions thereof are totally symmetric. The statement established above concerning when the integrals vanish can be formulated as follows:

*An integral $(f, \boldsymbol{O}g)$ can be nonzero only if the identity representation $\Gamma_{\text{id}}$ is contained in $\Gamma_{\text{integrand}}$,*

$$\Gamma_{\text{Integrand}} \supset \Gamma_{\text{id}}.$$

Because further the theorem:

*The product representation $\Gamma_{AB}$ of two irreducible representations $\Gamma_A$ and $\Gamma_B$, $\Gamma_{AB} = \Gamma_A \times \Gamma_B$ contains the identity representation $\Gamma_{\text{id}}$ if and only if $\Gamma_A = \Gamma_B$*

holds, we can specialize the first-mentioned theorem:

*If an operator $\boldsymbol{O}$ transforms as the identity representation $\Gamma_{\text{id}}$, then the integral $(f, \boldsymbol{O}g)$ can be different from zero only if $f$ and $g$ belong to the same irreducible representation $\Gamma(f) = \Gamma(g)$ (non-combination rule).*

Several other theorems concerning integrals can also be derived. These are summarized here for later use, where the symbols used have the meaning:

$f_1^A, f_2^A, \ldots, f_{nA}^A$ are functions which form a basis of the $A$th irreducible representation $\Gamma_A$ of the symmetry group of the Hamiltonian $\boldsymbol{H}$. The lower index on the $f$ gives the column of the representation matrices according to which the $f$ function under consideration transforms. $n_A$ is the dimensionality of $\Gamma_A$.

$g_1^B, g_2^B, \ldots, g_{nB}^B$ is a functional basis of the $B$th irreducible representation $\Gamma_B$ of the symmetry group of $\boldsymbol{H}$.

$\boldsymbol{H}$ is the Hamiltonian (which by definition transforms as the identity representation $\Gamma_{\text{id}}$ of the symmetry group of $\boldsymbol{H}$).

*Theorem 1a:*

$$(f_i^A, g_j^B) = 0, \qquad \text{except for } A = B, i = j$$

*Theorem 1b:*

$$(f_i^A, g_i^A) = (f_j^A, g_j^A)$$

*Theorem 2a* (non-combination rule):

$$(f_i^A, Hg_j^B) = 0, \qquad \text{except for } A = B, i = j$$

*Theorem 2b:*

$$(f_i^A, Hg_i^A) = (f_j^A, Hg_j^A)$$

## 2.6. Double groups

In section 2.4.2.2 the transformation behaviour of the spherical harmonics $Y_{l,m_l}$, $l = 0, 1, 2, \ldots$ was studied. We found for the character of the representation matrix induced from a basis $Y_{l,m_l}$ ($m_l = -l, -l+1, \ldots, l-1, l$) by a rotation through the angle $\phi$

$$\chi_l(\phi) = \frac{\sin\left(l + \frac{1}{2}\right)\phi}{\sin\frac{1}{2}\phi} \tag{2–49}$$

This character relation is assuredly applicable for the spatial portions of the atomic one-electron and many-electron functions* because these show the same transformation behaviour under rotation as the spherical harmonics with integral index $l$.

Under certain conditions, as will be discussed in chapter 5, it is useful to characterize atomic states by the *total angular momentum*, that is, the resultants of the orbital and spin angular momenta. The quantum numbers $J$ of the total angular momentum are obtained additively from the orbital quantum number $L$ and the spin quantum number $S$. $L$ is always integral but $S$ is integral only for systems having an even number of electrons. For systems with an odd number of electrons $S$ is half-integral†. The same holds for $J$.

If $J$ is integral then the character of the transformation behaviour of the appropriate wave functions under a rotation may be characterized

---

* For many-electron functions upper case italic Latin letters $L$ will be used for the orbital angular momentum quantum number.

† Cf. p. 233.

immediately with the help of (2–49):

$$\chi_J(\phi) = \frac{\sin\left(J+\frac{1}{2}\right)\phi}{\sin\frac{1}{2}\phi}. \tag{2–58}$$

In particular it is seen that a rotation by $2\pi$ which brings the system into itself is an identity operation*,

$$\chi_J(\phi+2\pi) = \chi_J(\phi), \qquad J \text{ integral.}$$

The situation becomes problematical for half-integral $J$. In this case

$$\chi_J(\phi+2\pi) = -\chi_J(\phi).$$

Here a rotation of $2\pi$ leads to a change in the sign of the characters. The representations for half-integral $J$ are therefore *double-valued* and thus not true representations. Besides this change of sign under a rotation of $2\pi$ one observes that

$$\chi_J(\phi+4\pi) = \chi_J(\phi)$$

and

$$\chi_J(\pi) = \chi_J(3\pi) = 0$$

hold, that is, a rotation of $4\pi$ is an identity operation and (only) for rotations of $\pi$ is the representation single-valued.

In order to determine the double-valued representations, we follow the method of Bethe†. He constructed the fiction that an atomic or molecular system is brought into itself not by a rotation of $2\pi$ but rather by a rotation of $4\pi$. The rotation of $4\pi$ is thus to be *equated* to the *identity operation E*. The rotation of $2\pi$ was treated by Bethe not as an identity operation but rather as a new group element $R$. With the aid of this new group element $R$ he constructed from the elements of the original simple group the corresponding double group by adding to the elements $E, A, B, C, \ldots$ of the simple group the elements $ER = R, AR, BR, CR, \ldots$. The *double group* formed in this manner consists of exactly twice as many elements as the simple group. The double group also contains, therefore, more classes than the simple group, although not necessarily twice as many.

When dividing the elements of a double group into classes, we can start with the fact‡ that the character is a class property. Because of the dissimilarity of characters in the double group a rotation through the angle $\phi$

*This result is obtained without difficulty when $\phi+2\pi$ is substituted into (2–58).

†H. A. Bethe: *Ann. Physik*, **3** (5), 135 (1929).

‡Cf. p. 277.

does not in general belong to the same class as the rotation (about the same axis) through the angle $\phi+2\pi$, the case $\phi = \pi$ being excepted. From this it may be derived that classes of the simple rotation group which contain rotations of $\pi$ always correspond to one class in the double group. All other classes of the simple rotation group correspond to *two* classes in the double group. In general one obtains the following types of classes of the double group:

$$\begin{aligned}&\{E\}\\&\{R\}\\&\{C_2, C_2 R\}\\&\{C_n, C_n^{n-1} R\}\\&\{C_n^m, C_n^{n-m} R\}\end{aligned} \tag{2–59}$$

It has already been shown (cf. p. 279) that the number of irreducible representations of a group (and thereby also of the double group) is equal to the number of classes of the group. The dimensionalities of the irreducible representations are found with the formula (2–3).

Concerning Nomenclature: Double groups are denoted with a prime on the group symbol, e.g., $O_h$ and $O'_h$ to distinguish them from the simple groups. For irreducible representations either *Mulliken's* system of notation is used and primed, e.g., $A'_1$, $E'$, etc., or the system of Bethe, according to which all irreducible representations are called $\Gamma$ and distinguished from one another by the use of indices. We shall use Bethe's system of notation in connection with double groups.

Example: The double group $O'$.
The simple group $O$ consists of 24 elements (cf. pp. 263 ff.)

$$E \quad 8C_3 \quad 3C_2 \quad 6C_4 \quad 6C'_2.$$

From these we obtain the 48 elements of the double group $O'$

$$E \quad R \quad 8C_3 \quad 8C_3R \quad 3C_2 \quad 3C_2R \quad 6C_4 \quad 6C_4R \quad 6C'_2 \quad 6C'_2R.$$

Under consideration of (2–59) we obtain the following division into classes

$$\{E\}\{R\}\begin{Bmatrix}4C_3\\4C_3^2R\end{Bmatrix}\begin{Bmatrix}4C_3^2\\4C_3R\end{Bmatrix}\begin{Bmatrix}3C_2\\3C_2R\end{Bmatrix}\begin{Bmatrix}3C_4\\3C_4^3R\end{Bmatrix}\begin{Bmatrix}3C_4^3\\3C_4R\end{Bmatrix}\begin{Bmatrix}6C'_2\\6C'_2R\end{Bmatrix}$$

There exist therefore 8 classes and thus 8 irreducible representations. The dimensionalities $l_i$, $i = 1, 2, \ldots, 8$ of the eight representations are determined according to (2–3):

$$48 = l_1^2+l_2^2+l_3^2+l_4^2+l_5^2+l_6^2+l_7^2+l_8^2.$$

Only the following decomposition of 48 into the sum of the squares of eight numbers is possible:

$$48 = 1^2 + 1^2 + 2^2 + 2^2 + 2^2 + 3^2 + 3^2 + 4^2.$$

Therefore the double group $O'$ has two one-dimensional, three two-dimensional, two three-dimensional and a four-dimensional representation. On p. 480 the character table of the double group $O'$ is given. The formation of product representations for double groups is subject to the same rules as hold for simple groups. In Table B.19, p. 315, for the double group $O'$ the irreducible components of the representations for all direct products of two representations are summarized.

# 3. Ligand field theory

## 3.1. The model

The model upon which the *ligand field theory* is based has already been explained in detail in Part A, section 1.4. In this theory we are interested chiefly in the electron orbitals which can be assigned to the unfilled shells of the central ion (3*d* or 4*d* subshell). In the ligand field theory in a restricted sense, the ionic or electrostatic ligand field theory (sometimes also called the crystal field theory*), the assumption is made that the electrons referred to above move in the potential field of the central ion core and in the electrostatic potential of the ligands. The latter part of this assumption implies that the typically quantum mechanical exchange forces between the electrons of the central ion and the electrons of the ligands are neglected. In other words one can then say that the ligand field theory is the theory of an *intra-complex Stark effect*, where the 'external' electrostatic field which produces the Stark effect arises from the charge distribution in the ligand system and is operative upon the electrons of the central ion.

The *Hamiltonian* for the motion of $N$ electrons in an unfilled shell of the central ion has the following form†, when one chooses the model for complexes as described above:

$$\boldsymbol{H} = \sum_{i=1}^{N}\left[-\frac{1}{2}\Delta_i - Z^*/r_i\right] + \frac{1}{2}\sum_{\substack{i=1\\i\neq j}}^{N}\sum_{j=1}^{N} 1/r_{ij} + \sum_{i=1}^{N}\boldsymbol{V}(\vec{r}_i) + \sum_{i=1}^{N}\boldsymbol{H}\,so\,(i). \tag{3–1}$$

The symbols have the meaning:

- $-\frac{1}{2}\Delta_i$ the kinetic energy of the $i$th electron;
- $-Z^*/r_i$ the potential energy of the $i$th electron in the field of the central ion core, where $Z^*$ is the effective charge of the core and $r_i$ is the distance between the central ion nucleus and the $i$th electron;
- $1/r_{ij}$ the energy of the Coulomb interaction between the $i$th and the $j$th electron ($r_{ij}$ is the distance between the $i$th and the $j$th electron);

---

* Cf. footnote p. 17.

† To simplify the notation we use atomic units in the following, cf. pp. 498 f.

$V(\vec{r}_i)$ the energy of the Coulomb interaction between the *i*th electron and *all* ligands.

$H_{SO}(i)$ the spin–orbit coupling energy of the *i*th electron;

The *Hamiltonian* of equation (3–1) differs from the *Hamiltonian* of the free atomic ion corresponding to the central ion (cf. equation (1–37))* in that the additional term $\sum_{i=1}^{N} V(\vec{r}_i)$ is present. As is the case for the *Schrödinger* equation of the free atoms or atomic ions with many electrons the *Schrödinger* equation with the *Hamiltonian* (3–1) *cannot* be solved *exactly*. We must limit ourselves to searching for approximate solutions. *Perturbation calculations* are here extremely helpful. As is the case for all perturbation calculations one begins with the problem of an 'unperturbed' system which is *exactly* soluble. It is natural to choose for the 'unperturbed' system the *Hamiltonian*

$$\boldsymbol{H}_{00} = \sum_{i=1}^{N} \left[ -\frac{1}{2} \Delta_i - Z^*/r_i \right] \tag{3–2}$$

$H_{00}$ represents the energy of a system of $N$ electrons which move *independently* of *one another* in the effective potential of the central ion core. The general solutions of the associated *Schrödinger* equation

$$\boldsymbol{H}_{00}\Phi = E_{00}\Phi \tag{3–3}$$

have already been introduced in chapter 1. Because here the $N$ electrons are supposed to be $d$ electrons of a transition metal ion, the solution functions $\Phi$ are antisymmetrized products of one-electron $d$ functions†. All $\Phi$ functions of the same electron configuration $(nd)^N$ are degenerate and describe states of the energy

$$E_{00} = N \cdot \varepsilon_{nd}, \tag{3–4}$$

where $\varepsilon_{nd}$ is the energy of a single $nd$ electron moving in the effective potential of the central ion core.

Starting from the set of $\Phi$ functions of an electron configuration $d^N$ we shall attempt to obtain approximate solutions of the *Schrödinger* equation with the *Hamiltonian* (3–1). In principle it is possible with the aid of such a set of degenerate $\Phi$ functions to carry out a perturbation

* In (1–37) the spin–orbit coupling operator $\Sigma H_{SO}(i)$ has been neglected because in chapter 1 a vanishingly small spin–orbit coupling was assumed.

† Concerning the definition of the $\Phi$ functions and their form for electron configurations $d^N$ see pp. 225 ff.

calculation for the complete perturbation operator

$$\boldsymbol{H}_1 = \frac{1}{2} \sum_{i=1}^{N} \sum_{\substack{j=1 \\ i \neq j}}^{N} 1/r_{ij} + \sum_{i=1}^{N} \boldsymbol{V}(\vec{r}_i) + \sum_{i=1}^{N} \boldsymbol{H}_{SO}(i) \tag{3–5}$$

For this purpose we use the *secular determinant*

$$\|(\Phi_r, \boldsymbol{H}_1 \Phi_s) - \Delta E \cdot \delta_{rs}\| = 0, \quad r, s = 1, 2, \ldots, \eta, \tag{3–6}$$

($\eta$ = degree of degeneracy)

which is characteristic of perturbation calculations for degenerate systems and which has as roots the perturbation energies $\Delta E_k (k = 1, 2, \ldots, \eta)$. This procedure proves to be exceedingly complicated, especially for many-electron systems, because the secular determinants which appear are in general of a very high order and their roots can be obtained only with great effort. Also, some characteristic results of the ligand field theory are lost in a maze of calculation. It must, however, be emphasized that the procedure outlined above can in principle be carried out, leads, and must lead, to the same results as the methods which will be described in the following section.

It is conventional to divide the perturbation calculation into well-defined separate steps. In order to do this the terms of the complete perturbation operator (3–5) are ordered according to their magnitudes. In the first part of the perturbation calculation one takes as the perturbation operator the term from (3–5) which gives the largest contribution to the perturbation energy. Next the perturbation calculation is carried out with the second largest term from (3–5), etc.

The comparison of the calculated results with experiment shows that three different cases with regard to the relative magnitudes of the energy terms of (3–5) can be distinguished:

(a) The *weak crystalline field case**

$$\frac{1}{2} \sum_{i \neq j} \sum 1/r_{ij} > \sum_i \boldsymbol{V}(\vec{r}_i) > \sum_i \boldsymbol{H}_{SO}(i)$$

(b) The *strong crystalline field case**

$$\sum_i \boldsymbol{V}(\vec{r}_i) > \frac{1}{2} \sum_{i \neq j} \sum 1/r_{ij} > \sum_i \boldsymbol{H}_{SO}(i)$$

(c) The *strong spin–orbit coupling case*

$$\sum_i \boldsymbol{H}_{SO}(i) > \sum_i \boldsymbol{V}(\vec{r}_i), \ \frac{1}{2} \sum_{i \neq j} \sum 1/r_{ij}$$

* In case (a) and case (b) the terms 'weak' and 'strong' refer to the comparison with the electron interaction.

For the complexes of the first series of transition metals the electron interaction energy is of the same order of magnitude as the ligand field energy. These complexes cannot be unambiguously assigned to case (a) or (b). This has no effect on the result of the perturbation calculation because in such a case at least the first two steps of the perturbation calculation would be carried out. Case (b) is found especially for complexes of the second and third transition metal series. The rare earth complexes (the $4f$ subshell is not complete) are characteristic of case (c).

### 3.2. Ligand fields

Let us assume that the system of all ligands of a given complex ion has a spatial charge distribution having the density $\rho(\vec{R}) \equiv \rho(R, \Theta, \Phi)$. Here $\vec{R}$ is the vector from the nucleus of the central ion to a point having the spherical coordinates $R, \Theta, \Phi$, and having the charge density $\rho(\vec{R})$ (cf. Figure B.20). According to classical electrostatics such a charge distribution

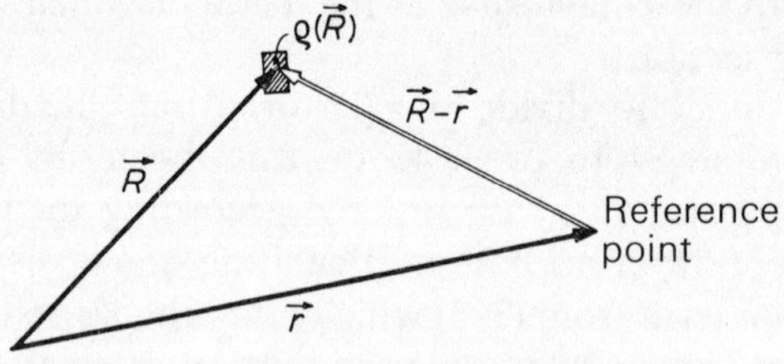


Figure B.20.

$\rho(\vec{R})$ produces at the reference point $\vec{r} \equiv \{r, \theta, \phi\}$ the potential

$$V(\vec{r}) = \int \frac{\varrho(\vec{R})}{|\vec{R}-\vec{r}|}\, d\tau_{\vec{R}}, \tag{3–7}$$

where the integration is to be taken over the entire space of the coordinates $R, \Theta, \Phi$. If in (3–7) $1/|\vec{R}-\vec{r}|$ is replaced by the well-known expansion in spherical harmonics*

$$1/|\vec{R}-\vec{r}| = \sum_{\lambda=0}^{\infty} \frac{4\pi}{2\lambda+1} \frac{r_<^{\lambda}}{r_>^{\lambda+1}} \sum_{\varkappa=-\lambda}^{+\lambda} Y_{\lambda,\varkappa}(\theta,\phi) \cdot Y_{\lambda,\varkappa}^{*}(\Theta,\Phi), \tag{3–8}$$

* Cf. e.g. H. Eyring, J. Walter and G. E. Kimball: *Quantum Chemistry*, pp. 369 ff. John Wiley, London 1938; see also chapter 1, p. 259.

then we obtain

$$V(\vec{r})=\sum_{\lambda=0}^{\infty}\sum_{\varkappa=-\lambda}^{+\lambda}\frac{4\pi}{2\lambda+1}\,Y_{\lambda,\varkappa}(\theta,\phi)\int\varrho(\vec{R})\,\frac{r_<^{\lambda}}{r_>^{\lambda+1}}\,Y^*_{\lambda,\varkappa}(\Theta,\Phi)\,\mathrm{d}\tau_{\vec{R}}\,. \tag{3–9}$$

For our problems $V(\vec{r})$ is of interest only for the space in which the electrons of the central ion move. But especially in these regions the charge density of the ligands should become vanishingly small for our chosen model. Therefore in the region of interest the expressions $r_< = r$ and $r_> = R$ always hold. Then $V(\vec{r})$ assumes the special form:

$$V(\vec{r})=V(r,\theta,\phi)=\sum_{\lambda=0}^{\infty}\sum_{\varkappa=-\lambda}^{+\lambda}A_{\lambda,\varkappa}\cdot r^{\lambda}\cdot Y_{\lambda,\varkappa}(\theta,\phi) \tag{3–10}$$

with

$$A_{\lambda,\varkappa}=\frac{4\pi}{2\lambda+1}\int\frac{\varrho(\vec{R})\,Y^*_{\lambda,\varkappa}(\Theta,\Phi)}{R^{\lambda+1}}\,\mathrm{d}\tau_{\vec{R}}\,. \tag{3–11}$$

The expression for $V(\vec{r})$ is an infinite series. Of course it is impossible to give $V(\vec{r})$ completely because that would mean the calculation of infinitely many integrals of the form $A_{\lambda,\kappa}$. Fortunately in the ligand field theory only a very few terms of the series expansion of $V(\vec{r})$ play a role. Which terms these are can be determined easily from the following facts:

1. If the charge distribution $\rho(\vec{R})$ of the ligand system has a particular symmetry, then $V(\vec{r})$ must also have this same symmetry, i.e., $V(\vec{r})$ must be *invariant under all symmetry operations* of the appropriate symmetry group. This is of course quite generally valid and is not restricted to application in thc ligand ficld theory.

2. Particularly in the ligand field theory the terms $A_{\lambda,\kappa}r^{\lambda}Y_{\lambda,\kappa}(\theta,\phi)$ of the series expansion of $V(\vec{r})$ appear as portions of the perturbation operator in the perturbation calculation. The subintegrals contained in the secular determinants (3–6)

$$\left(\Phi_r,\sum_{i=1}^{N}V(\vec{r}_i)\Phi_s\right)$$

may always be reduced to integrals with *one-electron functions*

$$\psi_{n,l,m_l}=R_{n,l}(r)\,Y_{l,m_l}(\theta,\phi)$$

because the terms $V(\vec{r}_i)$ are *one-electron operators* (cf. p. 249 ff.):

$$(\psi_{n,l,m_l},A_{\lambda,\varkappa}\,r^{\lambda}Y_{\lambda,\varkappa}\,\psi_{n,l,m_l'})=A_{\lambda,\varkappa}\,(R_{n,l},r^{\lambda}R_{n,l})\,(Y_{l,m_l},Y_{\lambda,\varkappa}\,Y_{l,m_l'})\,.$$

It can be shown generally* that an integral over three spherical harmonics $(Y_{l,m_l}, Y_{\lambda,\kappa} Y_{l,m_l})$ is nonzero only if†

(a) the relation $\lambda \leqq 2l$ is satisfied. If one considers the $d$ electron functions, i.e., $l = 2$, then it follows that $\lambda \leqq 4$. In this case one can neglect all terms in the series expansion of $V(\vec{r})$ with $\lambda > 4$ because these make no contribution in the calculation of perturbation energies.

(b) $\lambda$ is *even*. Terms in the series expansion of $V(\vec{r})$ with odd $\lambda$ in principle can make no contribution to the perturbation energy. They may be neglected in the expansion.

The use of only the theorems (2a) and (2b) leads to a great simplification of the expression for the ligand potential. When treating systems of $d$ electrons, for example, one may use the abbreviated form of $V(\vec{r})$

$$V(\vec{r}) = A_{0,0} r^0 Y_{0,0} + \sum_{\kappa=-2}^{+2} A_{2,\kappa} r^2 Y_{2,\kappa} + \sum_{\kappa=-4}^{+4} A_{4,\kappa} r^4 Y_{4,\kappa} \tag{3–12}$$

because $\lambda \leqq 4$ and $\lambda$ is even. It should be emphasized that this abbreviated form does not describe the complete potential but only that part which plays a role in the ligand field theory of $d$ electrons.

The expression (3–12) holds for all $d$ electron systems, independent of the symmetry of the ligand system which produces the potential. Symmetry considerations according to point 1 above lead in general to further simplifications of the expression for the potential. We shall exemplify this using the case of $O_h$ symmetry, which has such a dominant position in the ligand field theory of complexes.

Consider a ligand system of octahedral symmetry, say six point charges occupying the vertices of a regular octahedron. We let a principal axis of the octahedron coincide with the $z$ axis of the coordinate system, cf. Figure B.9. It is apparent that this system of point charges is invariant under all symmetry operations of the group $O_h$‡. The potential of this system of point charges likewise must be invariant.

---

* J. A. Gaunt: *Trans. Roy. Soc.*, **A 228**, 151 (1929).

† We consider here only the most important case for the ligand field theory, namely that all $N$ electrons are described by one-electron functions of the same subshell, i.e., that all $N$ electrons are $d$ electrons. There are modifications of the ligand field theory where the $N$ electrons are represented by one-electron functions which arise from a mixture of functions of different orbital angular momentum quantum numbers $l$ and $l'$, that is from $d$ and $p$ functions. One finds integrals of the type

$$(\psi_{n,l,m_l}, A_{\lambda,\kappa} r^\lambda Y_{\lambda,\kappa} \psi_{n,l',m_l'}) = A_{\lambda,\kappa}(R_{n,l}, r^\lambda R_{n,l'})(Y_{l,m_l}, Y_{\lambda,\kappa} Y_{l',m_l'}).$$

The integrals $(Y_{l,m_l}, Y_{\lambda,\kappa} Y_{l',m_l'})$ have in general a nonzero value when (a) $|l-l'| \leqq \lambda \leqq |l+l'|$ holds and (b) $l+l'+\lambda$ is even. The case considered above with $l = l'$ is included as a special case.

‡ Cf. pp. 263 f.

If the coordinate system is rotated about the $z$ axis through the angle $\pi/2$ (corresponding to the symmetry operation $C_4$), then a point which had the coordinates $r$, $\theta$, $\phi$ in the old coordinate system has in the new coordinate system the coordinates $r' = r$, $\theta' = \theta$, $\phi' = \phi - \pi/2$. Application of this operation $C_4$ to a spherical harmonic yields:

$$C_4 Y_{\lambda,\varkappa}(\theta,\phi) = C_4 \Theta_{\lambda,\varkappa}(\theta)\Phi_\varkappa(\phi) = C_4 \Theta_{\lambda,\varkappa}(\theta)\, e^{i\varkappa\phi}/\sqrt{2\pi} = \Theta_{\lambda,\varkappa}(\theta')e^{i\varkappa\phi'}/\sqrt{2\pi}$$
$$= \Theta_{\lambda,\varkappa}(\theta)\, e^{i\varkappa(\phi-\pi/2)}/\sqrt{2\pi} = e^{-i\varkappa\pi/2}\,\Theta_{\lambda,\varkappa}(\theta)\, e^{i\varkappa\phi}/\sqrt{2\pi} = e^{-i\varkappa\pi/2}Y_{\lambda,\varkappa}(\theta,\phi),$$

i.e., the spherical harmonics except for a factor $e^{-i\kappa\pi/2}$ are brought into themselves by $C_4$. Application of $C_4$ to the potential $V(\vec{r})$ (3–12) gives the result:

$$C_4V(\vec{r}) = A_{0,0}r^0e^{-i0\pi/2}Y_{0,0} + \sum_{\varkappa=-2}^{+2} A_{2,\varkappa}r^2e^{-i\varkappa\pi/2}Y_{2,\varkappa} + \sum_{\varkappa=-4}^{+4} A_{4,\varkappa}r^4e^{-i\varkappa\pi/2}Y_{4,\varkappa}.$$

Because of the invariance requirement this expression must be equal to $V(\vec{r})$:

$$C_4V(\vec{r}) = V(\vec{r}).$$

This is, however, only satisfied, if the factors $e^{-i\kappa\pi/2}$ are equal to 1; i.e., the exponent must be equal to an integral multiple of $2\pi\imath$:

$$-i\varkappa\pi/2 = i\,2\pi n, \quad n = 0, \pm 1, \pm 2, \pm 3, \ldots$$

Invariance is only achieved if

$$\varkappa = 0, \pm 4, \pm 8, \ldots$$

holds. This restriction reduces $V(\vec{r})$ to

$$V(\vec{r}) = A_{0,0}r^0Y_{0,0} + A_{2,0}r^2Y_{2,0} + A_{4,-4}r^4Y_{4,-4} + A_{4,0}r^4Y_{4,0} + A_{4,4}r^4Y_{4,4}. \tag{3–13}$$

If this coordinate system is reflected in the $xz$ plane (symmetry operation $\sigma_{xz}$), then the relation $x' = x$, $y' = -y$, $z' = z$ exists between the old and the new coordinates. If the spherical harmonics are expressed in terms of cartesian coordinates (see Table B.1) the following transformation properties are easily found:

$$\sigma_{xz}Y_{0,0} = Y_{0,0},$$
$$\sigma_{xz}Y_{2,0} = Y_{2,0},$$
$$\sigma_{xz}Y_{4,0} = Y_{4,0},$$
$$\sigma_{xz}Y_{4,4} = Y_{4,-4},$$
$$\sigma_{xz}Y_{4,-4} = Y_{4,4}.$$

From this it follows:

$$\sigma_{xz}V(\vec{r}) = A_{0,0}r^0 Y_{0,0} + A_{2,0}r^2 Y_{2,0} + A_{4,-4}r^4 Y_{4,4} + A_{4,0}r^4 Y_{4,0} + \\ + A_{4,4}r^4 Y_{4,-4}.$$

$V(\vec{r})$ is therefore invariant under the symmetry operation $\sigma_{xz}$ only if $A_{4,4} = A_{4,-4}$:

$$V(\vec{r}) = A_{0,0}r^0 Y_{0,0} + A_{2,0}r^2 Y_{2,0} + A_{4,0}r^4 Y_{4,0} + A_{4,4}r^4(Y_{4,4} + Y_{4,-4}).$$

A further simplification of $V(\vec{r})$ is achieved when we consider that the rotation of the coordinate system $C_3(x' = y, y' = z, z' = x)$ is also a symmetry operation;

$$\begin{aligned} C_3V(\vec{r}) = {} & A_{0,0}r^0 Y_{0,0} + A_{2,0}r^2 \sqrt{5/4\pi}\sqrt{1/4}\, r^{-2}(3z^2 - r^2) \\ & + A_{4,0}r^4\sqrt{9/4\pi}\sqrt{1/64}\, r^{-4}(35x^4 - 30x^2r^2 + 3r^4) \\ & + A_{4,4}r^4\sqrt{9/4\pi}\sqrt{35/128}\, r^{-4}[(y+iz)^4 + (y-iz)^4] \\ = {} & A_{0,0}r^0 Y_{0,0} + A_{2,0}r^2\sqrt{5/4\pi}\sqrt{1/4}\, r^{-2}(2x^2 - y^2 - z^2) \\ & + A_{4,0}r^4\sqrt{9/4\pi}\sqrt{1/64}\, r^{-4}(8x^4 + 3y^4 + 3z^4 - 24x^2y^2 - 24x^2z^2 + 6y^2z^2) \\ & + A_{4,4}r^4\sqrt{9/4\pi}\sqrt{35/128}\, r^{-4}(2y^4 - 12y^2z^2 + 2z^4). \end{aligned}$$

$V(\vec{r})$ itself expressed in cartesian coordinates is

$$\begin{aligned} V(\vec{r}) = {} & A_{0,0}r^0 Y_{0,0} + A_{2,0}r^2\sqrt{5/4\pi}\sqrt{1/4}\, r^{-2}(-x^2 - y^2 + 2z^2) \\ & + A_{4,0}r^4\sqrt{9/4\pi}\sqrt{1/64}\, r^{-4}(3x^4 + 3y^4 + 8z^4 + 6x^2y^2 - 24x^2z^2 - 24y^2z^2) \\ & + A_{4,4}r^4\sqrt{9/4\pi}\sqrt{35/128}\, r^{-4}(2x^4 - 12x^2y^2 + 2y^4). \end{aligned}$$

The invariance of $V(\vec{r})$ under $C_3$ is achieved when the coefficients of the individual powers of $x, y, z$ in $C_3V(\vec{r})$ and $V(\vec{r})$ are equal. Because the coefficients, e.g., from $x^2$ in $C_3V(\vec{r})$ and $V(\vec{r})$ are different, it must hold that $A_{2,0} = 0$. Further setting the coefficients of $z^4$ (when we neglect the common factor $\sqrt{9/4\pi}$) equal to one another yields

$$A_{4,0}\sqrt{1/64}\cdot 3 + A_{4,4}\sqrt{35/128}\cdot 2 = A_{4,0}\sqrt{1/64}\cdot 8$$

or

$$A_{4,4} = \sqrt{5/14}A_{4,0}.$$

Finally we obtain

$$V_{O_h}(r,\theta,\phi) = A_{0,0}r^0 Y_{0,0} + A_{4,0}r^4[Y_{4,0} + \sqrt{5/14}(Y_{4,4} + Y_{4,-4})]. \quad (3\text{–}14)$$

Because besides the regular octahedron, e.g., a cube also has $O_h$ symmetry, $V_{O_h}$ holds also for ligand systems with cubic structure (coordinate systems as in Figure B.9). The coefficients $A_{0,0}$ and $A_{4,0}$ of course must be determined independently for ligand systems with octahedral structure and for those with cubic structure.

If the expression for the potential of a given ligand system is known in the general symmetry-invariant form, then the coefficients $A_{\lambda,\kappa}$ must be calculated in order to obtain the explicit potential form.

In the ionic ligand field theory it often suffices to approximate the charge distribution of the *individual* ligands by point charges or point dipoles. Let us consider the effect of assuming point charges or point dipoles on the coefficients $A_{\lambda,\kappa}$ (3–11).

(a) The ligands act as *point charges.* The $k$th ligand has the charge $-q_k$ and the coordinates $R_k$, $\Theta_k$, $\Phi_k$. The integral in (3–11) becomes a summation over the ligands

$$A^{(L)}_{\lambda,\varkappa} = \frac{4\pi}{2\lambda+1}\sum_k \frac{-q_k\, Y^*_{\lambda,\varkappa}(\Theta_k,\Phi_k)}{R_k^{\lambda+1}}\,.$$

If in particular all ligands have the same charge $-q$ and the same distance $R$ from the nucleus of the central ion, then we have

$$A^{(L)}_{\lambda,\varkappa} = -\frac{4\pi}{2\lambda+1}\frac{q}{R^{\lambda+1}}\sum_k Y^*_{\lambda,\varkappa}(\Theta_k,\Phi_k)$$

or

$$A^{(L)}_{\lambda,\varkappa} = a^{(L)}_{\lambda}\sum_k Y^*_{\lambda,\varkappa}(\Theta_k,\Phi_k) \tag{3–15a}$$

where

$$a^{(L)}_{\lambda} = -\frac{4\pi}{2\lambda+1}\frac{q}{R^{\lambda+1}}\,. \tag{3–15b}$$

(b) The ligands act as *point dipoles* whose moment vectors coincide with the lines connecting ligand and central ion. The $k$th ligand has the moment $\mu_k = q_k \cdot d_k$ (negative end of the dipole faces the central ion) and the coordinates $R_k$, $\Theta_k$, $\Phi_k$. Also in this case the integral (3–11) becomes a summation over the ligands. Consider first the contribution made by the ligand $k$ to the sum. For this purpose we first assume the point dipole of $k$ to be a finite dipole consisting of two charges $-q_k$ and $+q_k$ having the separation $d_k$, cf. Figure B.21. The distance of the charge $-q_k$ from the central ion is $R_k - d_k/2$; that of the charge $+q_k$ is $R_k + \frac{1}{2}d_k$. Both charges give then the contribution to the sum

$$\frac{-q_k\, Y^*_{\lambda,\varkappa}(\Theta_k,\Phi_k)}{(R_k-d_k/2)^{\lambda+1}} + \frac{q_k\, Y^*_{\lambda,\varkappa}(\Theta_k,\Phi_k)}{(R_k+d_k/2)^{\lambda+1}}$$

$$= \frac{-q_k}{[R_k^2-(d_k/2)^2]^{\lambda+1}}\left[(R_k+d_k/2)^{\lambda+1} - (R_k-d_k/2)^{\lambda+1}\right] Y^*_{\lambda,\varkappa}(\Theta_k,\Phi_k)\,.$$

According to the binomial expansion theorem we may write

$$\frac{-q_k}{[R_k^2-(d_k/2)^2]^{\lambda+1}}\left[R_k^{\lambda+1} + (\lambda+1)R_k^{\lambda}d_k/2 + \cdots - R_k^{\lambda+1} + (\lambda+1)R_k^{\lambda}d_k/2 - \cdots\right]\times$$
$$\times\, Y^*_{\lambda,\varkappa}(\Theta_k,\Phi_k)\,.$$

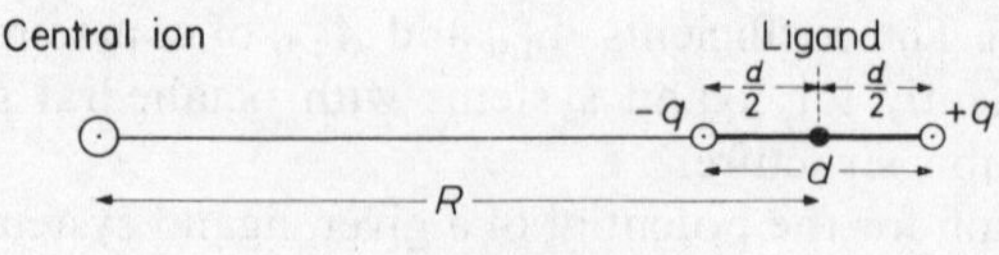


Figure B.21.

When point dipoles are assumed $d_k \ll R_k$ holds. With this the contribution of $k$th ligand becomes:

$$-(\lambda+1)\frac{q_k d_k}{R_k^{\lambda+2}} Y^*_{\lambda,\varkappa}(\Theta_k, \Phi_k) = -(\lambda+1)\frac{\mu_k}{R_k^{\lambda+2}} Y^*_{\lambda,\varkappa}(\Theta_k, \Phi_k).$$

$A_{\lambda,\kappa}$ is then for point dipoles

$$A^{(D)}_{\lambda,\varkappa} = -4\pi\frac{\lambda+1}{2\lambda+1}\sum_k \frac{\mu_k}{R_k^{\lambda+2}} Y^*_{\lambda,\varkappa}(\Theta_k, \Phi_k).$$

For equal point dipoles $\mu$ with equal distances $R$ from the central ion we have

$$A^{(D)}_{\lambda,\varkappa} = -4\pi\frac{\lambda+1}{2\lambda+1}\frac{\mu}{R^{\lambda+2}}\sum_k Y^*_{\lambda,\varkappa}(\Theta_k, \Phi_k)$$

or

$$A^{(D)}_{\lambda,\varkappa} = a^{(D)}_{\lambda}\sum_k Y^*_{\lambda,\varkappa}(\Theta_k, \Phi_k) \quad (3\text{–}16a)$$

with

$$a^{(D)}_{\lambda} = -4\pi\frac{\lambda+1}{2\lambda+1}\frac{\mu}{R^{\lambda+2}}. \quad (3\text{–}16b)$$

As an example we calculate the potential of a ligand system consisting of six point charges $-q$ on the vertices of a regular octahedron. We call the distance ligand–central ion $R$ and take the angular coordinate of the six ligands to be (cf. Figure B.9)

$$\Theta_1=\pi/2,\ \Phi_1=0;\quad \Theta_2=\pi/2,\ \Phi_2=\pi/2;\quad \Theta_3=0;$$
$$\Theta_4=\pi/2,\ \Phi_4=\pi;\quad \Theta_5=\pi/2,\ \Phi_5=3\pi/2;\quad \Theta_6=\pi.$$

The figure has the symmetry $O_h$ and the potential expression (3–14) holds. $A_{0,0}$ and $A_{4,0}$ are to be calculated. According to (3–15a) we have

$$A^{(L)}_{0,0} = a^{(L)}_0[Y^*_{0,0}(\Theta_1,\Phi_1)+Y^*_{0,0}(\Theta_2,\Phi_2)+Y^*_{0,0}(\Theta_3,\Phi_3)$$
$$+Y^*_{0,0}(\Theta_4,\Phi_4)+Y^*_{0,0}(\Theta_5,\Phi_5)+Y^*_{0,0}(\Theta_6,\Phi_6)].$$

Because $Y_{0,0}$ is a constant equal to $1/\sqrt{2}\times 1/\sqrt{2\pi}$ (cf. Table B.1), we may write immediately

$$A^{(L)}_{0,0} = a^{(L)}_0\cdot 6\cdot\frac{1}{\sqrt{2}}\frac{1}{\sqrt{2\pi}} = \frac{3}{\sqrt{\pi}}a^{(L)}_0.$$

Further we obtain:

$$A_{4,0}^{(L)}=a_4^{(L)}[Y_{4,0}^*(\Theta_1,\Phi_1)+Y_{4,0}^*(\Theta_2,\Phi_2)+Y_{4,0}^*(\Theta_3,\Phi_3)+Y_{4,0}^*(\Theta_4,\Phi_4)$$
$$+Y_{4,0}^*(\Theta_5,\Phi_5)+Y_{4,0}^*(\Theta_6,\Phi_6)].$$

With the aid of Table B.1 and by substituting the coordinate value of the ligands this becomes

$$A_{4,0}^{(L)}=a_4^{(L)}[4\cdot\frac{1}{\sqrt{2\pi}}\cdot\frac{3}{8\sqrt{2}}\cdot 3+2\cdot\frac{1}{\sqrt{2\pi}}\cdot\frac{3}{8\sqrt{2}}(35-30+3)]$$
$$=\frac{21}{4}\cdot\frac{1}{\sqrt{\pi}}\cdot a_4^{(L)}.$$

The abbreviated *potential* of the point charge octahedron is therefore

$$V_{\text{Oct.}}^{(L)}=\frac{3}{\sqrt{\pi}}\,a_0^{(L)}\cdot r^0 Y_{0,0}+\frac{21}{4\sqrt{\pi}}\,a_4^{(L)}r^4[Y_{4,0}+\sqrt{5/14}(Y_{4,4}+Y_{4,-4})].\quad(3\text{–}17)$$

Here according to (3–15b) we have

$$a_0^{(L)}=-4\pi\cdot\frac{q}{R}\quad\text{and}\quad a_4^{(L)}=-\frac{4\pi}{9}\cdot\frac{q}{R^5}.$$

When treating a system of six octahedrally oriented point dipoles the potential expression (3–17) can be used when $a_0^{(L)}$ and $a_4^{(L)}$ are replaced by $a_0^{(D)}$ and $a_4^{(D)}$ respectively, because $A_{\lambda\kappa}^{(L)}$ and $A_{\lambda,\chi}^{(D)}$ differ only in these terms

$$a_0^{(D)}=-4\pi\cdot\frac{\mu}{R^2},\quad a_4^{(D)}=-\frac{20\pi}{9}\cdot\frac{\mu}{R^6}.$$

The procedure described above may be applied without difficulty to other ligand systems. The results for the most important cases are given in Table B.20. In making this tabulation the potential expression (3–12) was assumed. That is, the values hold only for the study of *d* electron systems. Several remarks should be made concerning the results:

1. The *first* term of all potential expressions is independent of $\vec{r}$, the reference point from which the potential is measured. For point charge ligand systems it describes exactly the potential of the interior of a spherical shell on which the charges of all ligands are equally distributed. The radius of the sphere is equal to the ligand–central ion distance. (In the case of point dipole ligands a sphere of equal size has an electric double layer which is to be thought of as arising from an equal distribution of all ligand dipoles over its surface.) The contribution of the first term is greater than the contributions of all other terms of a potential expression because it is

Table B.20. Effective potentials of various ligand systems (for $d$ electron systems)

| Ligand system | Effective potential for $d$ electron systems |
| --- | --- |
| Six ligands on the vertices of a regular *octahedron*, cf. Figure B.9, $O_h$ | $V^{(i)}_{\text{oct.}} = \frac{3}{\sqrt{\pi}} a_0^{(i)} Y_{0,0} + \frac{21}{4\sqrt{\pi}} a_4^{(i)} r^4 \times$ $\times [Y_{4,0} + \sqrt{5/14}(Y_{4,4} + Y_{4,-4})]$ |
| eight ligands on the vertices of a *cube*, cf. Figure B.9, $O_h$ | $V^{(i)}_{\text{cube}} = \frac{4}{\sqrt{\pi}} a_0^{(i)} Y_{0,0} - \frac{14}{3\sqrt{\pi}} a_4^{(i)} r^4 \times$ $\times [Y_{4,0} + \sqrt{5/14}(Y_{4,4} + Y_{4,-4})]$ |
| four ligands on the vertices of a regular *tetrahedron*, cf. Figure B.12, $T_d$ | $V^{(i)}_{\text{tetra.}} = \frac{2}{\sqrt{\pi}} a_0^{(i)} Y_{0,0} - \frac{7}{3\sqrt{\pi}} a_4^{(i)} r^4 \times$ $\times [Y_{4,0} + \sqrt{5/14}(Y_{4,4} + Y_{4,-4})]$ |
| four ligands on the corners of a *square*, cf. Figure B.10, $D_{4h}$ | $V^{(i)}_{\text{square}} = \frac{2}{\sqrt{\pi}} a_0^{(i)} Y_{0,0} - \frac{\sqrt{5}}{\sqrt{\pi}} a_2^{(i)} r^2 Y_{2,0} +$ $+ \frac{9}{4\sqrt{\pi}} a_4^{(i)} r^4 [Y_{4,0} + \sqrt{35/18}(Y_{4,4} + Y_{4,-4})]$ |
| eight ligands on the vertices of a *archimedean antiprism*, $D_{4d}$ | $V^{(i)}_{\text{anti.}} = \frac{4}{\sqrt{\pi}} a_0^{(i)} Y_{0,0} - \frac{14}{3\sqrt{\pi}} a_4^{(i)} r^4 Y_{4,0}$ |
| $a_0^{(L)} = -4\pi \frac{q}{R}$, $a_2^{(L)} = -\frac{4\pi}{5} \frac{q}{R^3}$, | $a_4^{(L)} = -\frac{4\pi}{9} \frac{q}{R^5}$ |
| $a_0^{(D)} = -4\pi \frac{\mu}{R^2}$, $a_2^{(D)} = -\frac{12\pi}{5} \frac{\mu}{R^4}$, | $a_4^{(D)} = -\frac{20\pi}{9} \frac{\mu}{R^6}$ |
| general $D_{4h}$ potential | $V_{D_{4h}} = A_{0,0} Y_{0,0} + A_{2,0} r^2 Y_{2,0} + A_{4,0} r^4 Y_{4,0} +$ $+ A_{4,4} r^4 (Y_{4,4} + Y_{4,-4})$ |

proportional to $1/R$ (for dipole systems to $1/R^2$), while the remaining terms go as much higher powers of $1/R$*.

2. By comparing the potential expression for octahedron, cube, and tetrahedron it is seen that all show the same angular dependency. If the proportion of the corresponding terms is formed with

$$Y_{4,0}+\sqrt{5/14}\,(Y_{4,4}+Y_{4,-4})$$

and in the various structures the same central ion ligand distances $R$ and equal charges $-q$ or equal dipole moments $\mu$ are assumed, one finds the important numerical relation

$$\text{octahedron}:\text{cube}:\text{tetrahedron} = 1:-8/9:-4/9$$

### 3.3. $d^1$ systems

The general *Hamiltonian* (3–1) has the particular form for the case $N=1$†

$$\boldsymbol{H}=-\frac{1}{2}\Delta-Z^*/r-V(\vec{r})+\boldsymbol{H}_{SO}. \tag{3–18}$$

If we limit ourselves to the treatment of the transition metal complexes, in particular to those of the first series, then the contribution of the spin–orbit coupling to the total energy is according to experience small compared to the remaining terms in (3–18). It is then permissible, especially when interpreting optical properties, in good approximation, to work with the abbreviated Hamiltonian

$$\boldsymbol{H}=-\frac{1}{2}\Delta-Z^*/r-V(\vec{r}). \tag{3–19}$$

The Schrödinger equation having this Hamiltonian is in general not exactly soluble. We shall therefore attempt to find approximate solutions using perturbation calculations.

For the Hamiltonian of the *unperturbed* system we take

$$\boldsymbol{H}_{00}=-\frac{1}{2}\Delta-Z^*/r \tag{3–20}$$

and as perturbation operator

$$\boldsymbol{H}_1=-V(\vec{r}). \tag{3–21}$$

---

* On p. 58 the first, angular-independent, part of the potential was denoted with $V_k$ and the angular dependent part for the case of octahedral symmetry with $V_0$.

† For the energy operator $V(\vec{r})$ of the electrostatic interaction between $d$ electrons and ligands we have written explicitly the product of negative electric charges (in atomic units $-1$) with the potential $V(\vec{r})$ of the ligands.

We have thus made the following assumption for the unperturbed system: the one electron moves alone in the potential of the central ion core. In other words: $\boldsymbol{H}_{00}$ represents the energy of the one electron in a *free* transition metal ion—*not bound in a complex*. The solutions of the corresponding *Schrödinger* equation

$$\boldsymbol{H}_{00}\psi = E_{00}\psi \tag{3–22}$$

are known from the theory of free ions, cf. section 1.1.1. If the electron considered is an *nd* electron, then the eigenvalue $E_{00}$ is equal to the atomic one-electron energy $\varepsilon_{nd}$:

$$E_{00} = \varepsilon_{nd}. \tag{3–23}$$

The solutions of (3–22) belonging to the one-electron energy $\varepsilon_{nd}$ are the five different orbitals

$$\begin{gathered}\psi_{n,2,m_l} = R_{n,2}(r)\,Y_{2,m_l}(\theta,\phi),\\ m_l = -2,\ -1,\ 0,\ +1,\ +2\end{gathered} \tag{3–24}$$

or any arbitrary set of five (linearly independent) linear combinations of $\varphi_1, \varphi_2, \varphi_3, \varphi_4, \varphi_5$ of the functions $\psi_{n,2m_l}$. A five-fold orbital degeneracy is then present. Because each of the five orbitals can appear with $\alpha$ or $\beta$ spin the total degree of degeneracy is 10. The totality of 10 states described by

$$\varphi_i \begin{Bmatrix}\alpha\\ \beta\end{Bmatrix}, \quad i = 1,2,3,4,5, \tag{3–25}$$

gives a $^2D$ term. The group of the *Schrödinger* equation (3–22) is the rotation-reflection group of the sphere because of the spherical symmetry of $\boldsymbol{H}_{00}$.

Starting with a 10 fold degenerate basis (3–25) we shall use perturbation theory for degenerate systems to investigate the changes the $^2D$ term of the free $d^1$ ion undergoes when it is brought into a complex as a central ion. The appropriate perturbation is described by the operator $\boldsymbol{H}_1$. The energy changes are found to be the roots $\Delta\varepsilon$ of the $10\times 10$ secular determinant

$$\left|\begin{array}{ccc:ccc}
H_{1\alpha,1\alpha}-\Delta\varepsilon & H_{1\alpha,2\alpha}\cdots & H_{1\alpha,5\alpha} & H_{1\alpha,1\beta} & \cdots & H_{1\alpha,5\beta}\\
H_{2\alpha,1\alpha} & H_{2\alpha,2\alpha}-\Delta\varepsilon\,.. & H_{2\alpha,5\alpha} & \vdots & \ddots & \vdots\\
\vdots & \vdots\quad\ddots & \vdots & \vdots & & \vdots\\
H_{5\alpha,1\alpha} & H_{5\alpha,2\alpha}\cdots & H_{5\alpha,5\alpha}-\Delta\varepsilon & H_{5\alpha,1\beta} & \cdots & H_{5\alpha,5\beta}\\
\hdashline
H_{1\beta,1\alpha} & \cdots & H_{1\beta,5\alpha} & H_{1\beta,1\beta}-\Delta\varepsilon & H_{1\beta,2\beta}\cdots & H_{1\beta,5\beta}\\
\vdots & \ddots & \vdots & H_{2\beta,1\beta} & H_{2\beta,2\beta}-\Delta\varepsilon\,.. & H_{2\beta,5\beta}\\
\vdots & & \vdots & \vdots & \vdots\quad\ddots & \vdots\\
H_{5\beta,1\alpha} & \cdots & H_{5\beta,5\alpha} & H_{5\beta,1\beta} & H_{5\beta,2\beta}\cdots & H_{5\beta,5\beta}-\Delta\varepsilon
\end{array}\right| \tag{3–26}$$

with

$$H_{i\alpha,j\alpha}=(\varphi_i\alpha,\ \boldsymbol{H}_1\varphi_j\alpha),\quad H_{i\beta,j\beta}=(\varphi_i\beta,\ \boldsymbol{H}_1\varphi_j\beta),$$
$$H_{i\alpha,j\beta}=(\varphi_i\alpha,\ \boldsymbol{H}_1\varphi_j\beta),\quad H_{i\beta,j\alpha}=(\varphi_i\beta,\ \boldsymbol{H}_1\varphi_j\alpha).$$

Because the operator $\boldsymbol{H}_1$ is a ligand field operator and therefore has no effect on the spin coordinates and because the spin functions $\alpha$ and $\beta$ form an orthonormalized system, it follows immediately that*

$$\begin{aligned} H_{i\alpha,j\alpha}&=H_{i\beta,j\beta}=(\varphi_i,\boldsymbol{H}_1\varphi_j)\equiv H_{ij},\\ H_{i\alpha,j\beta}&=H_{i\beta,j\alpha}=0 \end{aligned} \tag{3–27}$$

The elements in the upper right-hand quarter and in the lower left-hand quarter of the determinant (3–26) are all equal to zero. The complete determinant is thus equal to the product of the upper left-hand and the lower right-hand determinants. Because of (3–27) these subdeterminants are equal element for element and have the same roots. The roots of the complete determinant (3–26) are obtained when the roots of *one* subdeterminant are found and it is considered that each of these roots is double because of $\alpha$–$\beta$ spin degeneracy.

In the following we restrict ourselves to the consideration of the subdeterminant

$$||H_{ij}-\delta_{ij}\Delta\varepsilon||=0,\quad i,j=1,2,\ldots,5 \tag{3–28a}$$

with

$$H_{ij}=(\varphi_i,\boldsymbol{H}_1\varphi_j). \tag{3–28b}$$

The symmetry group of the ligand field operator $\boldsymbol{H}_1 = -V(\bar{r})$ is always a subgroup of the rotation-reflection group of the sphere (which describes the transformation properties of the operator $\boldsymbol{H}_{00}$ of the unperturbed system). We then have the case of an unsymmetrical perturbation, treated in section 2.4.3.2 and the theorems and methods given there may be applied.

Using Table B.15 we can see which progeny terms can arise from the fivefold orbitally degenerate $D$ term considered here.

---

* Because $\boldsymbol{H}_1$ does not operate on the spin coordinates, the integral over the space coordinates can be brought in front of the integral over spin coordinates. Examples:

$$H_{i\alpha,j\alpha}=(\varphi_i\alpha,\boldsymbol{H}_1\varphi_j\alpha)=(\varphi_i,\boldsymbol{H}_1\varphi_j)\cdot(\alpha,\alpha).$$

$\alpha$ is normalized, that is $(\alpha,\alpha)=1$ and therefore we have

$$H_{i\alpha,j\alpha}=(\varphi_i,\boldsymbol{H}_1\varphi_j).$$

It is necessary to calculate all elements $H_{ij}$ of the secular determinant in order to determine the *magnitude* of the splitting. The functions $\psi_{n,2,m_l}$ (3–24) or any arbitrary set of (linearly-independent) linear combinations of these functions $\psi_{n,2,m_l}$ can be used as the $\varphi$ basis. It is, however, possible that all 25 $H_{ij}$ are nonzero and that the determination of the roots of (3–28a) will be tedious. It is preferable to choose as the $\varphi$ basis those linear combinations of $\psi_{n,2,m_l}$ which transform as *irreducible representations* of the symmetry group of $\boldsymbol{H}_1$ to which the resulting terms belong. This is determined by group theoretical methods. Then, for example, the non-combination rule 2a, p. 317 can be used to show that usually many of the $H_{ij}$ elements vanish. The 'correct', i.e., *symmetry adapted linear combinations* (of the spherical harmonics) are given in the Tables B.16, 17, 18 for the most important ligand field symmetries. For the explicit calculation of those elements which do not necessarily vanish (for group theoretical reasons) the potential expressions expanded in spherical harmonics are used (Table B.20).

### 3.3.1. *Ligand fields of $O_h$ symmetry*

From Table B.15 it can be seen that the fivefold orbitally degenerate $D$ term splits into a double $E_g$ term and a threefold $T_{2g}$ term under the influence of a ligand field of $O_h$ symmetry:

$$D \rightarrow E_g \dotplus T_{2g}.$$

The secular determinant (3–28a) must then have two roots, of which one is doubly and the other threefold degenerate. The 'correct' $\varphi$ basis is, according to Table B.16* :

$$\begin{aligned}
E_g &\begin{cases} \varphi_1 = \psi_{n,2,0} & = e_g^a \\ \varphi_2 = \dfrac{1}{\sqrt{2}}(\psi_{n,2,2} + \psi_{n,2,-2}) & = e_g^b \end{cases} \\
T_{2g} &\begin{cases} \varphi_3 = \dfrac{1}{\sqrt{2}}(\psi_{n,2,2} - \psi_{n,2,-2}) & = t_{2g}^0 \\ \varphi_4 = -\psi_{n,2,1} & = -t_{2g}^+ \\ \varphi_5 = \psi_{n,2,-1} & = t_{2g}^- \end{cases}
\end{aligned} \tag{3–29}$$

The first two functions $\varphi_1$, $\varphi_2$ transform as the irreducible representation $E_g$ of $O_h$, the last three $\varphi_3$, $\varphi_4$, $\varphi_5$ as $T_{2g}$ of $O_h$. To the extreme right the symbols are given which are sometimes used to denote the $\varphi$ functions for $O_h$ symmetry.

* The functions $\phi$ used in Part A, p. 55 ff. are likewise symmetry adapted. Except for the phase factor, which is unimportant for the calculation of the energies, the $\phi$ functions differ from the $\varphi$ functions given here only in that in Part A linear combinations of the functions $\varphi_4$ and $\varphi_5$ are used instead of these functions themselves.

When setting up the secular determinant (3–28a) we utilize the theorems 2a and 2b, p. 317. According to theorem 2a the integrals $H_{i,j} = (\varphi_i \boldsymbol{H}_1 \varphi_j)$ are zero when $\varphi_i$ and $\varphi_j$ belong to different irreducible representations of the symmetry group of $\boldsymbol{H}_1$. In the present case it holds that

$$H_{1,3} = H_{1,4} = H_{1,5} = H_{3,1} = H_{4,1} = H_{5,1} = H_{2,3} = H_{2,4} = H_{2,5} = H_{3,2} = H_{4,2} = H_{5,2} = 0.$$

Further according to theorem 2a all integrals $H_{i,j} = (\varphi_i \boldsymbol{H}_1 \varphi_j)$ vanish when $\varphi_i$ and $\varphi_j$ belong to the irreducible representation of the symmetry group of $\boldsymbol{H}_1$, but transform as different columns of the representation matrices. This means here that

$$H_{1,2} = H_{2,1} = 0$$

and

$$H_{3,4} = H_{4,3} = H_{3,5} = H_{5,3} = H_{4,5} = H_{5,4} = 0.$$

Accordingly all nondiagonal elements of the secular determinant are zero:

$$\begin{vmatrix} H_{1,1} - \Delta\varepsilon & 0 & 0 & 0 & 0 \\ 0 & H_{2,2} - \Delta\varepsilon & 0 & 0 & 0 \\ 0 & 0 & H_{3,3} - \Delta\varepsilon & 0 & 0 \\ 0 & 0 & 0 & H_{4,4} - \Delta\varepsilon & 0 \\ 0 & 0 & 0 & 0 & H_{5,5} - \Delta\varepsilon \end{vmatrix} = 0.$$

As roots of the secular determinant one then finds

$$\Delta\varepsilon_1 = H_{1,1}, \quad \Delta\varepsilon_2 = H_{2,2}, \quad \Delta\varepsilon_3 = H_{3,3}, \quad \Delta\varepsilon_4 = H_{4,4}, \quad \Delta\varepsilon_5 = H_{5,5}.$$

If the theorem 2b, p. 317 is also taken into consideration the following results are obtained: The functions $\varphi_1$ and $\varphi_2$ of the symmetry species $E_g$ describe a doubly orbitally degenerate term having the perturbation energy

$$\Delta\varepsilon(E_g) = H_{1,1} = H_{2,2}$$

and the functions $\varphi_3$, $\varphi_4$, $\varphi_5$ of the species $T_{2g}$ describe a triply orbitally degenerate term having the perturbation energy

$$\Delta\varepsilon(T_{2g}) = H_{3,3} = H_{4,4} = H_{5,5}.$$

We are left then with the task of evaluating the integrals $H_{ii}$ explicitly. With consideration of (3–29), (3–24) and (3–14) we find

$$\begin{aligned}
\Delta\varepsilon(E_g) &= H_{1,1} = -\int \psi^*_{n,2,0} V_{O_h} \psi_{n,2,0}\,\mathrm{d}\tau \\
&= -\int_0^\infty\int_0^\pi\int_0^{2\pi} R^*_{n,2}(r) Y^*_{2,0}(\theta,\phi)\, V_{O_h}(r,\theta,\phi) R_{n,2}(r) Y_{2,0}(\theta,\phi) r^2\,\mathrm{d}r \sin\theta\,\mathrm{d}\theta\,\mathrm{d}\phi \\
&= -A_{0,0}\int_0^\infty R^*_{n,2}(r) r^0 R_{n,2}(r) r^2\,\mathrm{d}r \int_0^\pi\int_0^{2\pi} Y^*_{2,0}(\theta,\phi) Y_{0,0} Y_{2,0}(\theta,\phi) \sin\theta\,\mathrm{d}\theta\,\mathrm{d}\phi \\
&\quad - A_{4,0}\int_0^\infty R^*_{n,2}(r) r^4 R_{n,2}(r) r^2\,\mathrm{d}r \times \\
&\quad \times \int_0^\pi\int_0^{2\pi} Y^*_{2,0}(\theta,\phi)[Y_{4,0} + \sqrt{5/14}(Y_{4,4} + Y_{4,-4})] Y_{2,0}(\theta,\phi) \sin\theta\,\mathrm{d}\theta\,\mathrm{d}\phi.
\end{aligned}$$

It is conventional to abbreviate the radial integrals as

$$\int_0^\infty R^*_{n,2}(r) r^k R_{n,2}(r) r^2\,\mathrm{d}r = \overline{r^k}. \tag{3–30}$$

$\Delta\varepsilon(E_g)$ is reduced in this manner to

$$\begin{aligned}
\Delta\varepsilon(E_g) &= -A_{0,0}\overline{r^0}\int_0^\pi\int_0^{2\pi} Y^*_{2,0}(\theta,\phi) Y_{0,0} Y_{2,0}(\theta,\phi) \sin\theta\,\mathrm{d}\theta\,\mathrm{d}\phi \\
&\quad - A_{4,0}\overline{r^4}\int_0^\pi\int_0^{2\pi} Y^*_{2,0}(\theta,\phi)[Y_{4,0} + \sqrt{5/14}(Y_{4,4} + Y_{4,-4})] Y_{2,0}(\theta,\phi) \sin\theta\,\mathrm{d}\theta\,\mathrm{d}\phi.
\end{aligned}$$

Using the values of the integrals over the angular coordinates as given in Table B.21 we obtain

$$\begin{aligned}
\Delta\varepsilon(E_g) &= -A_{0,0}\overline{r^0}\cdot\frac{1}{2\sqrt{\pi}} - A_{4,0}\overline{r^4}\left[\frac{6}{14\sqrt{\pi}} + \sqrt{\frac{5}{14}}\,(0+0)\right] \\
&= -\frac{A_{0,0}}{2\sqrt{\pi}}\cdot\overline{r^0} - 6\cdot\frac{A_{4,0}}{14\sqrt{\pi}}\cdot\overline{r^4}.
\end{aligned} \tag{3–31a}$$

In order to determine $\Delta\varepsilon(T_{2g})$ we evaluate the integral $H_{4,4}$:

$$\begin{aligned}
\Delta\varepsilon(T_{2g}) &= H_{4,4} = -\int \psi^*_{n,2,1} V_{O_h} \psi_{n,2,1}\,\mathrm{d}\tau \\
&= -A_{0,0}\overline{r^0}\int_0^\pi\int_0^{2\pi} Y^*_{2,1}(\theta,\phi) Y_{0,0} Y_{2,1}(\theta,\phi) \sin\theta\,\mathrm{d}\theta\,\mathrm{d}\phi \\
&\quad - A_{4,0}\overline{r^4}\int_0^\pi\int_0^{2\pi} Y^*_{2,1}(\theta,\phi)[Y_{4,0} + \sqrt{5/14}(Y_{4,4} + Y_{4,-4})] Y_{2,1}(\theta,\phi) \sin\theta\,\mathrm{d}\theta\,\mathrm{d}\phi.
\end{aligned}$$

With the aid of Table B.21 this becomes

$$\Delta\varepsilon(T_{2g}) = -\frac{A_{0,0}}{2\sqrt{\pi}}\cdot\overline{r^0} + 4\cdot\frac{A_{4,0}}{14\sqrt{\pi}}\cdot\overline{r^4}. \tag{3–31b}$$

Table B.21.

Values of the integrals $\int_0^{\pi}\int_0^{2\pi} Y^*_{2,ml}(\theta,\phi)Y_{\lambda,\kappa}(\theta,\phi)Y_{2,ml'}(\theta,\phi)\sin\theta\, d\theta\, d\phi$

| | $\lambda = 0$ | $\lambda = 2$ | $\lambda = 4$ |
|---|---|---|---|
| $2\sqrt{\pi}(Y_{2,0}, Y_{\lambda,0}Y_{2,0})$ | 1 | $2\sqrt{5/7}$ | 6/7 |
| $2\sqrt{\pi}(Y_{2,\pm 1}, Y_{\lambda,0}Y_{2,\pm 1})$ | 1 | $\sqrt{5/7}$ | $-4/7$ |
| $2\sqrt{\pi}(Y_{2,\pm 2}, Y_{\lambda,0}Y_{2,\pm 2})$ | 1 | $-2\sqrt{5/7}$ | 1/7 |
| $2\sqrt{\pi}(Y_{2,\pm 2}, Y_{\lambda,\pm 4}Y_{2,\mp 2})$ | 0 | 0 | $\sqrt{70/7}$ |

A comparison of the perturbation energies $\Delta\varepsilon(E_g)$ and $\Delta\varepsilon(T_{2g})$ shows the following:

1. Both contain the same (positive) term $-(A_{0,0}/2\sqrt{\pi})\cdot \overline{r^0}$ (which is often denoted by $\varepsilon_0$ alone). This common term which arises from the angular-independent portion of the potential expression $V_{O_h}$ causes a *shift* of the progeny terms $E_g$ and $T_{2g}$ to the same degree, that is, it does not lead to a splitting (cf. Figure B.22).

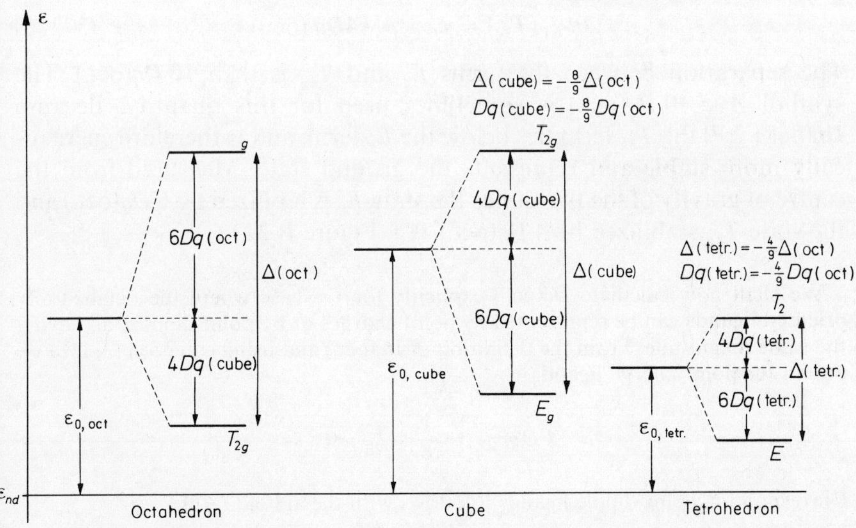


Figure B.22. Term schemes for $d^1$ systems with point charge ligands, $\Delta \equiv 10\,Dq$.

2. The *splitting* of the $D$ term into the terms $E_g$ and $T_{2g}$ is caused purely by the last two terms

$$-6 \cdot \frac{A_{4,0}}{14\sqrt{\pi}} \cdot \overline{r^4} \quad \text{or} \quad +4 \cdot \frac{A_{4,0}}{14\sqrt{\pi}} \cdot \overline{r^4}$$

3.3.1.1. *Octahedral structure.* If the system of ligands consists of six identical ligands which occupy the vertices of a *regular octahedron*, then according to Table B.20 the following relations hold

$$-A_{4,0}^{(\text{Oct.})} = -\frac{21}{4} \cdot \frac{1}{\sqrt{\pi}} \cdot a_4^{(i)}$$

or

$$-\frac{A_{4,0}^{(\text{Oct.})}}{14\sqrt{\pi}} \cdot \overline{r^4} = -\frac{3}{8\pi} \cdot a_4^{(i)} \overline{r^4}.$$

This quantity is set equal to 1 $Dq$(oct.):

$$1 Dq(\text{Oct.}) = -\frac{3}{8\pi} \cdot a_4^{(i)} \overline{r^4} = -\frac{A_{4,0}^{(\text{Oct.})}}{14\sqrt{\pi}} \cdot \overline{r^4}.$$

The so-called cubic *field strength parameter* $Dq$(oct.) is positive. For $d^1$ complex ions having octahedral structure we find (cf. equation 3–31a, b).

$$\Delta\varepsilon_{\text{Oct.}}(E_g) = \varepsilon_{0,\text{Oct.}} + 6Dq\,(\text{oct.}) \qquad (3\text{–}31c)$$

$$\Delta\varepsilon_{\text{Oct.}}(T_{2g}) = \varepsilon_{0,\text{Oct.}} - 4Dq\,(\text{oct.}) \qquad (3\text{–}31d)$$

The separation between the terms $E_g$ and $T_{2g}$ is then 10 $Dq$(oct.). The symbol $\Delta \equiv 10\,Dq(\text{oct.})$ is sometimes used for this quantity. Because $Dq(\text{oct.}) \geqq 0$ the $T_{2g}$ term lies below the $E_g$ term, and is therefore energetically more stable and represents the ground state. Measured from the centre of gravity of the two states the state $E_g$ is labilized by 6 $Dq$(oct.) and the state $T_{2g}$ stabilized by 4 $Dq$(oct.). (Cf. Figure B.22.)

We shall now calculate $Dq$(oct.) explicitly for the case where the octahedrally oriented ligands can be represented by point charges or by point dipoles, all having the same magnitude. From the definition of $Dq$(oct.) and using equation (3–15b) we obtain for point charge ligands:

$$1\,Dq_{(\text{Oct.})}^{(L)} = \frac{1}{6} \cdot \frac{q}{R^5} \cdot \overline{r^4}.$$

Correspondingly for dipole ligands it follows with the aid of (3–16b)

$$1\,Dq_{(\text{Oct.})}^{(D)} = \frac{5}{6} \cdot \frac{\mu}{R^6} \cdot \overline{r^4}.$$

The $Dq$(oct.) values depend upon the charge $q$ or upon the dipole moment $\mu$ of the ligands, upon the distance $R$ between the ligands and the central ion and finally upon the effective nuclear charge of the central ion. (When calculating $\overline{r^4}$ the radial functions $R_{n,2}(r)$ containing the effective nuclear charge as a parameter are used.)

That the results given by an explicit calculation of $Dq$(oct.) values in the manner indicated above are inadequate for a description of actual cases has already been explained in Part A, p. 77 ff. It is much more reasonable to adjust the $Dq$(oct.) values on the basis of observed spectra. The explicit calculation of $Dq$ values is, however, of importance in the history of the ligand field theory. Ilse and Hartmann* could show using such estimates of the $Dq$ values that the long wavelength portion of the spectra of transition metal complexes can be explained essentially on the basis of a perturbation of the $d$ electron systems of the central ions by the ligand fields.

3.3.1.2. *Cubic structure.* For a system of eight equal ligands on the vertices of a cube one obtains from Table B.20 the potential expression

$$A_{0,0}^{(\text{cube})} = \frac{4}{\sqrt{\pi}} \cdot a_0^{(i)},$$

$$A_{4,0}^{(\text{cube})} = -\frac{14}{3} \cdot \frac{1}{\sqrt{\pi}} \cdot a_4^{(i)}.$$

A comparison with the coefficients $A_{0,0}$ and $A_{4,0}$ for the octahedral system gives

$$A_{0,0}^{(\text{cube})} = \frac{4}{3} \cdot A_{0,0}^{(\text{Oct.})},$$

$$A_{4,0}^{(\text{cube})} = -\frac{8}{9} \cdot A_{4,0}^{(\text{Oct.})} \quad \text{or} \quad \frac{A_{4,0}^{(\text{cube})}}{14\sqrt{\pi}} \cdot \overline{r^4} = \frac{8}{9}\, Dq(\text{Oct.}).$$

The perturbation energies become according to (3–31a, b)

$$\begin{aligned} \Delta\varepsilon_{\text{cube}}(E_g) &= \varepsilon_{0,\text{cube}} - 6 \cdot \frac{8}{9}\, Dq(\text{Oct.}), \\ \Delta\varepsilon_{\text{cube}}(T_{2g}) &= \varepsilon_{0,\text{cube}} + 4 \cdot \frac{8}{9}\, Dq(\text{Oct.}). \end{aligned} \tag{3–33}$$

We conclude: The $E_g$ term is the ground state term and the $T_{2g}$ term is an excited term because $Dq(\text{oct.}) \geqq 0$. A *term inversion* is therefore associated with the transition from the octahedral system to the cubic system. Also the separation between the two terms is here a factor of $\frac{8}{9}$ smaller than in the octahedral system†. $\varepsilon_{0,\text{cube}}$ is however a factor of $\frac{4}{3}$ greater than $\varepsilon_{0,\text{oct.}}$. These relations are shown in Figure B.22. The expression $\Delta(\text{cube}) = 10\, Dq(\text{cube}) = -\frac{8}{9}\Delta(\text{oct.}) = -10 \times \frac{8}{9}\, Dq(\text{oct.})$ was used here.

---

* Ilse, F. E. and H. Hartmann: *Z. Physik. Chem.*, **197**, 239 (1951); *Z. Naturforsch.*, **6a**, 751 (1951). Dissertation F. E. Ilse, Frankfurt/Main 1946.

† Assuming identical ligands and identical ligand–central ion distances in both systems.

3.3.2. *Ligand fields of symmetry $T_d$*

According to Table B.15 the fivefold orbitally degenerate $D$ state splits into a doubly degenerate $E$ state and a threefold $T_2$ state:

$$D \rightarrow E + T_2 .$$

We obtain the 'correct' $\varphi$ basis from Table B.16:

$$\begin{aligned} E &\begin{cases} \varphi_1 = \psi_{n,2,0} \\ \varphi_2 = \frac{1}{\sqrt{2}} (\psi_{n,2,2} + \psi_{n,2,-2}) \end{cases} \\ T_2 &\begin{cases} \varphi_3 = \frac{1}{\sqrt{2}} (\psi_{n,2,2} - \psi_{n,2,-2}) \\ \varphi_4 = -\psi_{n,2,1} \\ \varphi_5 = \psi_{n,2,-1} \end{cases} \end{aligned} \tag{3–34}$$

These are the same $\varphi$ functions as for $O_h$ symmetry. The arguments which held in that case lead also here to the conclusion that the secular determinant derived from the $\varphi$ basis (3–34) can be completely diagonalized and has the following roots

$$\Delta\varepsilon_{\text{Tetr.}}(E) = H_{1,1} = H_{2,2},$$
$$\Delta\varepsilon_{\text{Tetr.}}(T_2) = H_{3,3} = H_{4,4} = H_{5,5},$$

where

$$H_{i,i} = -\int \varphi_i^* V_{T_d} \varphi_i \, d\tau .$$

When evaluating the integrals $H_{i,i}$ the considerations for the $O_h$ case can be followed to a large extent. If the ligand system consists of four identical ligands on the vertices of a regular tetrahedron, then the abbreviated potential expression is according to Table B.20

$$V_{\text{Tetr.}}^{(i)} = \frac{2}{\sqrt{\pi}} \cdot a_0^{(i)} Y_{0,0} - \frac{7}{3} \cdot \frac{1}{\sqrt{\pi}} \cdot a_4^{(i)} r^4 [Y_{4,0} + \sqrt{5/14}\, (Y_{4,4} + Y_{4,-4})],$$

i.e., $V_{\text{tetr.}}^{(i)}$ differs from the octahedral potential only in the coefficients of the spherical harmonics:

$$A_{0,0}^{(\text{Tetr.})} = 2/3 \cdot A_{0,0}^{(\text{Oct.})},$$
$$A_{4,0}^{(\text{Tetr.})} = -4/9 \cdot A_{4,0}^{(\text{Oct.})}.$$

It is therefore permissible to use the expressions (3–31a) for the perturbation energy $\Delta\varepsilon(E)$ and (3–31b) for $\Delta\varepsilon(T_2)$ directly and one obtains in this manner

$$\Delta \varepsilon_{\text{Tetr.}}(E) = \varepsilon_{0,\text{Tetr.}} - 6 \cdot \frac{4}{9}\, Dq(\text{Oct.}),$$
$$\Delta \varepsilon_{\text{Tetr.}}(T_2) = \varepsilon_{0,\text{Tetr.}} + 4 \cdot \frac{4}{9}\, Dq(\text{Oct.}). \qquad (3\text{–}35)$$

i.e., the ground term is the $E$ term and the excited term is the $T_2$ term. Therefore the energetic order of the terms is the inverse of that for octahedral systems, but the same as for systems having cubic structure. The $E$-$T_2$ separation for the tetrahedron is half as large as the $E_g$-$T_{2g}$ separation for the cube and equal to $\frac{4}{9}$ of the $E_g$-$T_{2g}$ separation for the octahedron:

$$\Delta(\text{octahedron}):\Delta(\text{cube}):\Delta(\text{tetrahedron}) = 1 : -\tfrac{8}{9} : -\tfrac{4}{9}. \qquad (3\text{–}36\text{a})$$

The following holds for the energy term $\varepsilon_{0,\text{tetr.}}$:

$$\varepsilon_{0,\text{tetr.}} = \tfrac{1}{2}\varepsilon_{0,\text{cube}} = \tfrac{2}{3}\varepsilon_{0,\text{oct.}} \qquad (3\text{–}36\text{b})$$

(Cf. also Figure B.22.)

### 3.3.3. *Ligand fields of symmetry $D_{4h}$*

According to Table B.15 the fivefold orbitally degenerate $D$ term splits into three simple terms $A_{1g}$, $B_{1g}$, $B_{2g}$ and a doubly degenerate term $E_g$:

$$D \rightarrow A_{1g} \dotplus B_{1g} \dotplus B_{2g} \dotplus E_g.$$

The symmetry adapted linear combinations of the $\psi_{n,l,m_l}$ functions are according to Table B.17:

$$\begin{aligned}
A_{1g}:&\quad \varphi_1 = \psi_{n,2,0}\\
B_{1g}:&\quad \varphi_2 = \frac{1}{\sqrt{2}}(\psi_{n,2,2} + \psi_{n,2,-2})\\
B_{2g}:&\quad \varphi_3 = -\frac{i}{\sqrt{2}}(\psi_{n,2,2} - \psi_{n,2,-2})\\
E_g:&\quad \begin{cases} \varphi_4 = \dfrac{i}{\sqrt{2}}(\psi_{n,2,-1} + \psi_{n,2,1})\\ \varphi_5 = -\dfrac{1}{\sqrt{2}}(\psi_{n,2,-1} - \psi_{n,2,1}) \end{cases}
\end{aligned} \qquad (3\text{–}37)$$

It is seen from theorem 2a, p. 317 that the secular determinant obtained using this $\varphi$ basis is completely diagonal. When theorem 2b is considered we have further $H_{4,4} = H_{5,5}$. The term energies are then

$$\Delta \varepsilon(A_{1g}) = H_{1,1},$$
$$\Delta \varepsilon(B_{1g}) = H_{2,2},$$

$$\Delta\varepsilon(B_{2g}) = H_{3,3},$$
$$\Delta\varepsilon(E_g) \;= H_{4,4} = H_{5,5}. \tag{3–38}$$

We shall now discuss an example of practical importance, namely when the ligand system is an *elongated* or *compressed* octahedron.

The appropriate potential (cf. Table B.20)

$$V_{D4h} = A_{0,0}(D_{4h})Y_{0,0} + A_{2,0}(D_{4h})r^2Y_{2,0} + A_{4,0}(D_{4h})r^4Y_{4,0} + A_{4,4}(D_{4h})r^4(Y_{4,4}+Y_{4,-4}),$$

is treated as the potential of a regular octahedron

$$V_{\text{Oct.}} = A_{0,0}(\text{Oct.})Y_{0,0} + A_{4,0}(\text{Oct.})r^4\{Y_{4,0} + \sqrt{5/14}\cdot(Y_{4,4}+Y_{4,-4})\}$$

having a superimposed tetragonal component $\Delta V_{D4h}$:

$$V_{D_{4h}} = V_{\text{Oct.}} + \Delta V_{D_{4h}},$$

where (cf. $V_{D4h}$ with $V_{\text{oct.}}$)

$$\Delta V_{D_{4h}} = [A_{0,0}(D_{4h}) - A_{0,0}(\text{Oct.})]Y_{0,0} + A_{2,0}(D_{4h})r^2Y_{2,0} + [A_{4,0}(D_{4h}) - A_{4,0}(\text{Oct.})]r^4Y_{4,0} + [A_{4,4}(D_{4h}) - \sqrt{5/14}\cdot A_{4,0}(\text{Oct.})]r^4(Y_{4,4}+Y_{4,-4}),$$

or

$$\Delta V_{D4h} = PY_{0,0} + Sr^2Y_{2,0} + Tr^4Y_{4,0} + Ur^4(Y_{4,4}+Y_{4,-4})$$

with

$$P = A_{0,0}(D_{4h}) - A_{0,0}(\text{Oct.}),$$
$$S = A_{2,0}(D_{4h}),$$
$$T = A_{4,0}(D_{4h}) - A_{4,0}(\text{Oct.}),$$
$$U = A_{4,4}(D_{4h}) - \sqrt{5/14}\cdot A_{4,0}(\text{Oct.}).$$

If the octahedron is elongated or compressed along the $z$ axis and if the distances between the two ligands lying on the $z$ axis and the central ion are denoted with $R_{|}$, and the remaining four central ion–ligand distances with $R_{\square}$, cf. Figure B.23, then using the definition of $A_{\lambda,\kappa}$ in (3–11) and assuming point charge ligands having the charge $-q$ (the procedure for calculating the $A_{\lambda,\kappa}$ was discussed on pp. 329 ff.) one finds:

$$P = 4\sqrt{\pi}\,q[1/R_{\square} - 1/R_{|}],$$

$$S = \frac{4\sqrt{\pi}}{\sqrt{5}}\,q[1/R_{\square}^3 - 1/R_{|}^3],$$

$$T = \frac{4\sqrt{\pi}}{3} q[1/R_{\square}^5 - 1/R_{|}^5],$$
$$U = 0. \tag{3–39a}$$

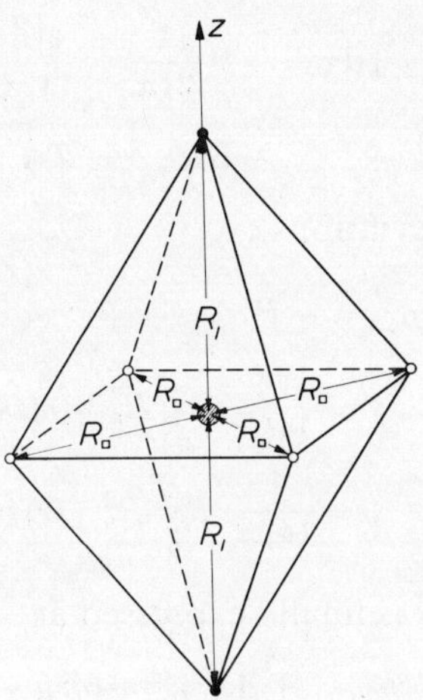


Figure B.23.

Consequently it holds for

| elongated octahedra | compressed octahedra | |
|---|---|---|
| $(R_| > R_\square)$ | $(R_| < R_\square)$ | |
| $P > 0$ | $P < 0$ | |
| $S > 0$ | $S < 0$ | (3–39b) |
| $T > 0$ | $T < 0$ | |
| $U = 0$ | $U = 0$ | |

The perturbation energies (3–38) produced by a ligand field potential $V_{D4h} = V_{\text{oct.}} + \Delta V_{D4h}$ are obtained using the perturbation operator

$$V = -V_{D_{4h}} = -V_{\text{Oct.}} - \Delta V_{D_{4h}}.$$

As an example it holds that

$$\Delta\varepsilon(A_{1g}) = \int \psi_{n,2,0}^* V \psi_{n,2,0}\, d\tau$$
$$= -\int \psi_{n,2,0}^* V_{\text{Oct.}} \psi_{n,2,0}\, d\tau - \int \psi_{n,2,0}^* \Delta V_{D_{4h}} \psi_{n,2,0}\, d\tau.$$

The value of the first integral has already been determined on pp. 338 ff.

$$-\int \psi_{n,2,0}^* V_{\text{Oct.}} \psi_{n,2,0}\, d\tau = \varepsilon_{0,\text{Oct.}} + 6Dq.$$

When

$$\psi_{n,2,0} = R_{n,2}(r) \cdot Y_{2,0}(\theta, \phi)$$

and the explicit form of $\Delta V_{D4h}$ given above are substituted into the second integral one finds using Table B.21:

$$-\int \psi^*_{n,2,0} \Delta V_{D4h} \psi_{n,2,0} \, d\tau = -P \frac{1}{2\sqrt{\pi}} - \frac{2\sqrt{5}}{14\sqrt{\pi}} S \int_0^\infty R^2_{n,2}(r) \, r^2 \, r^2 \, dr$$
$$-6 \cdot \frac{1}{14\sqrt{\pi}} T \int_0^\infty R^2_{n,2}(r) r^4 r^2 \, dr.$$

If we adopt the abbreviations

$$\Delta \varepsilon_{0,D4h} = -P \frac{1}{2\sqrt{\pi}},$$

$$\text{D}s = \frac{\sqrt{5}}{14\sqrt{\pi}} S \int_0^\infty R^2_{n,2}(r) r^2 r^2 \, dr,$$

$$\text{D}t = \frac{1}{14\sqrt{\pi}} T \int_0^\infty R^2_{n,2}(r) r^4 r^2 \, dr,$$

the perturbation energy is finally expressed as

$$\Delta\varepsilon(A_{1g}) = \varepsilon_{0,\text{Oct.}} + \Delta\varepsilon_{0,D_{4h}} + 6Dq - 2Ds - 6Dt.$$

Following the same procedure one obtains the remaining perturbation energies

$$\Delta\varepsilon(B_{1g}) = \varepsilon_{0,\text{Oct.}} + \Delta\varepsilon_{0,D_{4h}} + 6Dq + 2Ds - Dt,$$
$$\Delta\varepsilon(B_{2g}) = \varepsilon_{0,\text{Oct.}} + \Delta\varepsilon_{0,D_{4h}} - 4Dq + 2Ds - Dt,$$
$$\Delta\varepsilon(E_g) = \varepsilon_{0,\text{Oct.}} + \Delta\varepsilon_{0,D_{4h}} - 4Dq - Ds + 4Dt.$$

It is clearly seen from these energy expressions how the term splittings arise in the transition from the regular octahedron to the elongated or compressed octahedron (cf. Table B.13a) as a result of the tetragonal potential contribution $\Delta V_{D4h}$.

$$E_g(O_h) \rightarrow A_{1g}(D_{4h}) \dotplus B_{1g}(D_{4h}),$$
$$T_{2g}(O_h) \rightarrow B_{2g}(D_{4h}) \dotplus E_g(D_{4h}),$$

From the signs of the parameters $P$, $S$, $T$ in equation (3–39b) one obtains easily the signs of $\Delta\varepsilon_{0,D4h}$, $Ds$, $Dt$:

| elongated octahedron | compressed octahedron |
|---|---|
| $(R_{\mid} > R_{\square})$ | $(R_{\mid} < R_{\square})$ |
| $\Delta \varepsilon_{0,D4h} < 0$ | $\Delta \varepsilon_{0,D4h} > 0$ |
| $Ds > 0$ | $Ds < 0$ |
| $Dt > 0$ | $Dt < 0$ |

The splittings for elongated octahedra are shown schematically in Part A, Figure 47.

The case of *square planar* complex structures is contained in the example discussed here as the boundary case for $R_| \to \infty$. The appropriate parameters $\Delta\varepsilon_{0,D4h}$, $Ds$, $Dt$ are obtained by setting the quantities $1/R_|$, $1/R_|^3$ or $1/R_|^5$ equal to zero in equation (3–39a).

## 3.4. $d^N$ systems

If a transition metal ion bound in a complex contains $N$ $d$ electrons in its outermost shell, then according to the ligand field theory, the *Hamiltonian* (3–1) is applicable. We shall limit ourselves in this section to the treatment of systems in which the spin–orbit coupling makes only a very minor contribution to the total energy, so that the spin–orbit coupling operator

$$\sum_{i=1}^{N} \boldsymbol{H}_{SO}(i)$$

may be neglected in equation (3–1). This approximation is applicable principally to complexes of the first transition metal series. Our discussion is based on the *Hamiltonian*

$$\boldsymbol{H} = \sum_{i=1}^{N}\left[-\frac{1}{2}\Delta_i - Z^*/r_i\right] + \sum_{i<j}^{N}\sum^{N} 1/r_{ij} + \sum_{i=1}^{N} \boldsymbol{V}(\vec{r}_i). \tag{3–40}$$

Approximate solutions of the associated *Schrödinger* equation are determined using the method given in section 3.1. Here an 'unperturbed' system of $N$ $d$ electrons which move independently of one another in the core potential of the central ion is first considered:

$$\boldsymbol{H}_{00}\,\Phi = E_{00}\Phi \tag{3–41}$$

with

$$\boldsymbol{H}_{00} = \sum_{i=1}^{N}\left[-\frac{1}{2}\Delta_i - Z^*/r_i\right]. \tag{3–42}$$

If the $N$ electrons are *nd* electrons, then the *Schrödinger* equation (3–41) has as solutions

$$E_{00} = N \cdot \varepsilon_{nd}.$$

as has already been discussed in section 3.1. The solution functions $\Phi$ are *antisymmetrized products of one-electron d functions.*

If, for example, the number $N$ of the $d$ electrons is equal to 2, i.e., if we have a $d^2$ system, then

$$E_{00} = 2\varepsilon_{nd},$$

and the $\Phi$ functions are according to chapter 1, p. 229 the 45 functions $\Phi_{(m_l m_s; m'_l m'_s)}$ summarized in Table B.3. All of these 45 $\Phi$ functions describe states of the same energy $2\varepsilon_{nd}$; we have a 45-fold degeneracy.

The portion of the *Hamiltonian* (3–40) which has not yet been considered, the *perturbation operator*

$$\boldsymbol{H}_1 = \sum_{i<j}\sum 1/r_{ij} + \sum_i \boldsymbol{V}(\vec{r}_i) = \boldsymbol{H}_{\text{el.}} + \boldsymbol{H}_{\text{Lig.}}, \qquad (3\text{–}43),$$

is taken into account by a perturbation calculation starting with the $\Phi$ basis or, in general, with a set of (linearly independent) linear combinations of the $\Phi$. If the functions which make up such a set are denoted by $\Theta_k$, $k = 1, 2, \ldots, \eta$, whereby $\eta$ is the *degree of degeneracy*, then the *perturbation energies* generated by $H_1$ are obtained as the roots $\Delta E_k$, $k = 1, 2, \ldots, \eta$ of the *secular determinant*

$$||(\Theta_r, \boldsymbol{H}_1\Theta_s) - \delta_{r,s}\Delta E|| = 0\,, \qquad (3\text{–}44a)$$

$$r, s = 1, 2, \ldots, \eta$$

or

$$||(\Theta_r, \boldsymbol{H}_{\text{el.}}\Theta_s) + (\Theta_r, \boldsymbol{H}_{\text{Lig.}}\Theta_s) - \delta_{r,s}\Delta E|| = 0. \qquad (3\text{–}44b)$$

If it is to be expected that the contribution of the electron interaction energy to the perturbation energy $\Delta E$, represented by the operator $\boldsymbol{H}_{\text{el}}$, is of the same order of magnitude as the contribution of the energy of interaction between the *d* electron system and the ligand system, represented by the operator $\boldsymbol{H}_{\text{Lig.}}$, then of course the complete secular problem (3–44) must be solved. This is for example the case for complexes with transition metal ions of the first series (3*d* electron systems). Before treating this problem we wish to consider two limiting cases. The insights obtained in this manner simplify on the one hand the solution of the complete problem (3–44); on the other hand we become familiar with two greatly simplified models and their limitations. We shall consider the '*weak*' and '*strong*' (ligand) *field limiting cases.*

### 3.4.1. *The weak-field case*

This case arises when the contribution of the electron interaction to the perturbation energy is much larger than the contribution of the electron–ligand interaction. It is logical here to study as the *first step* the influence of the electron interaction on the unperturbed system. The complete degeneracy of the unperturbed system is in general partially or completely removed as a result of the electron interaction. The completely degenerate term of the unperturbed system is split. In the *second step* we study how the electron–ligand interaction affects those terms which appear as a result of the splitting induced by the electron interaction. The essential contribution of the ligand field theory thus comes into play in the second step of the weak-field limiting case.

*Step 1.* The degenerate basis of the $\Phi$ functions having the eigenvalue $E_{00} = N\varepsilon_{nd}$ for the unperturbed system (with the *Hamiltonian* (3–42)) is taken as the starting point for a perturbation calculation using the perturbation operator of the electron interaction

$$\boldsymbol{H}_{\text{el.}} = \sum_{i<j}\sum 1/r_{ij}.$$

The roots of the secular determinant

$$||(\Phi_r, \boldsymbol{H}_{\text{el.}}\Phi_s) - \delta_{r,s}\Delta E_{\text{el.}}|| = 0, \quad r,s = 1,2,\ldots,\eta \tag{3–45}$$

($\eta$ = degree of degeneracy)

are the perturbation energies resulting because of the electron interaction.

The methods and results of this first step have already been discussed in detail in chapter 1. Taken together with *Hamiltonian* $\boldsymbol{H}_{00}$ of the unperturbed system $\boldsymbol{H}_{\text{el.}}$ forms precisely the *Hamiltonian* of the *free atomic ion*:

$$\boldsymbol{H}_{\text{Ion}} = \boldsymbol{H}_{00} + \boldsymbol{H}_{\text{el.}}.$$

In chapter 1 it was shown that the degeneracy of the case without electron interaction is in general partially or completely removed when electron interaction is present: The states which are completely degenerate in the unperturbed system decompose energetically into families of states, that is into *terms.* The symbols $^{2S+1}L$ were used to denote these terms where $2S+1$ was the *multiplicity* characteristic of the total spin and $L$ was the *total angular momentum quantum number.* We saw that a term $^{2S+1}L$ can be described by the totality of the functions

$$\Psi(L, M_L, S, M_S) \qquad \begin{array}{l} M_L = -L, -L+1, \ldots, L-1, L \\ M_S = -S, -S+1, \ldots, S-1, S \end{array}$$

where the $\Psi$ are eigenfunctions of the operators of the total angular momentum and of the total spin. The $\Psi$ functions can be represented as linear combinations of the antisymmetrized products $\Phi$*. The first step in the weak-field case yields then the *term system of a free central ion,* one that is *not bound in a complex.*

---

* It should be kept in mind that the $\Phi$ functions themselves are in general not solutions of (3–45). That is, they are not functions such that when substituted into (3–45) only the diagonal elements of (3–45) are nonzero; they do not *diagonalize* the secular determinant. The $\Psi$ functions, however, diagonalize the secular determinant and are thereby solution functions of (3–45).

The terms $^{2S+1}L$ and their energies (expressed by the *Racah* parameters $A$, $B$, $C$) are collected in Tables B.5 and B.9 for atomic ions having the electron configuration $d^N$, $N = 1, 2, \ldots, 9$; the functions $\Psi(L, M_L, S, M_S)$ for all terms of the configuration $d^2$ and for the $^4F$ and $^4P$ term of the configuration $d^3$ are given in Tables B.6 and B.7. From Table B.9 it can for example be seen that the following terms and energies result from the configuration $d^2$ under the influence of electron interaction:

$$
\begin{aligned}
\boldsymbol{d^2}:\quad & {}^1S \;\; \Delta E_{\mathrm{el.}}(^1S) = A + 14B + 7C \\
& {}^1D \;\; \Delta E_{\mathrm{el.}}(^1D) = A - \;\; 3B + 2C \\
& {}^1G \;\; \Delta E_{\mathrm{el.}}(^1G) = A + \;\; 4B + 2C \\
& {}^3P \;\; \Delta E_{\mathrm{el.}}(^3P) = A + 7B \\
& {}^3F \;\; \Delta E_{\mathrm{el.}}(^3F) = A - 8B
\end{aligned}
$$

The following should be noted concerning the functions $\Psi(L, M_L, S, M_S)$. The $\Psi$ functions belonging to the same term, that is, which have the same $L$ and the same $S$ and differ only in the $M_L, M_S$ value pair, form a $(2S+1)(2L+1)$ fold degenerate basis. Each set of $(2S+1)(2L+1)$ (linearly independent) linear combinations of the $\Psi$ functions of a term is suited equally well for describing the term: The linear combinations are also eigenfunctions of the total orbital angular momentum and total spin operators and thus represent states of the same energy, the term energy.

*Step 2.* In the second step of the weak-field case we study the influence exerted by the interaction between the $d$ electrons of the central ion and the ligand system, represented by the operator

$$\boldsymbol{H}_{\mathrm{Lig.}} = \sum_i \boldsymbol{V}(\vec{r}),$$

on the single terms $^{2S+1}L$ of the (until now free) central ion.

In order to determine the perturbation produced by the ligand field on the single terms of the central ion, a *perturbation calculation with the operator* $\boldsymbol{H}_{\mathrm{Lig.}}$ is carried out—a separate perturbation calculation for each term. If the (free) central ion considered has a term $^{2S+1}L$, then in order to determine the perturbation of the energy of this term by the ligand field, it is necessary that the roots of the following secular determinant be found:

$$||(\Theta_r, \boldsymbol{H}_{\mathrm{Lig.}}\Theta_s) - \delta_{r,s}\Delta E_{\mathrm{Lig.}}|| = 0, \tag{3–46}$$

where the $\Theta_k$, $k = 1, 2, \ldots, (2S+1)(2L+1)$ are wave functions of the term $^{2S+1}L$. The $\Theta_k$ may be either the $\Psi(L, M_L, S, M_S)$ or a set of $(2S+1)(2L+1)$ (linear independent) linear combinations of these $\Psi$.

The solution of the secular problem can as a rule be greatly simplified through the use of group theoretical arguments. First one considers if a term $^{2S+1}L$ can split under the influence of the ligand field and which symmetry species these splitting terms have. The symmetry group $G$ of the perturbation operator $\boldsymbol{H}_{\text{Lig.}}$ is a subgroup of the rotation-reflection group of the sphere (The *Hamiltonian* of the free ion $\boldsymbol{H}_{\text{ion}} = \boldsymbol{H}_{00} + \boldsymbol{H}_{\text{el.}}$ is invariant under the operations of this group). The functions of the term considered $^{2S+1}L$ obtained in Step 1 $\Psi(L, M_L, S, M_S)$, $M_L = -L, \ldots, +L$, $M_S = -S, \ldots, +S$ induce, when subjected to the operations of the symmetry group $G$ of the perturbation operator $\boldsymbol{H}_{\text{Lig.}}$, a representation of $G$ which is in general reducible. When determining the transformation behaviour of the $\Psi(L, M_L, S, M_S)$ it must be considered that the $\Psi(L, M_L, S, M_S)$ are sums of products of one-electron $d$ functions and therefore behave as spherical harmonics $Y_{L,M_L}$ under *rotations.* On the other hand they are always invariant under *inversion* (because they are made up of even $d$ functions), i.e., they have even parity, cf. section 2.4.2.2. For systems having symmetry groups with even and odd irreducible representations the splitting terms always are to be assigned to the *even* irreducible representations, independent of if $L$ is even or odd.

The following splitting terms (designated by their representation symbols) arise from the terms $^{2S+1}L$ of the configuration $d^2$ for example under the influence of a perturbation $\boldsymbol{H}_{\text{Lig}}$ of symmetry $O_h$ (cf. Table B.15):

$$^1S \rightarrow {}^1A_{1g}$$
$$^1D \rightarrow {}^1E_g \dot{+} {}^1T_{2g}$$
$$^1G \rightarrow {}^1A_{1g} \dot{+} {}^1E_g \dot{+} {}^1T_{1g} \dot{+} {}^1T_{2g}$$
$$^3P \rightarrow {}^3T_{1g}$$
$$^3F \rightarrow {}^3A_{2g} \dot{+} {}^3T_{1g} \dot{+} {}^3T_{2g}$$

In order to calculate the *magnitude* of the splittings one could in principle substitute as $\Theta_k$ functions in the secular determinant for the term $^{2S+1}L$ the functions $\Psi(L, M_L, S, M_S)$ and determine the roots $\Delta E_{\text{Lig.}}$. It is, however, more advantageous when for the $\Theta_k$ linear combinations of the $\Psi$ are used which transform as the *irreducible representations of the symmetry group* of $\boldsymbol{H}_{\text{Lig.}}$, i.e., which are *symmetry adapted.* There exists then the possibility of using the theorems given in section 2.5, pp. 317 to greatly simplify the secular problem.

3.4.1.1. *Example of a $d^2$ ion in a ligand field of $O_h$ symmetry.* As a typical example we shall consider the influence of a ligand field of $O_h$ symmetry on the $^1S$, $^1D$, $^1G$, $^3P$, $^3F$ terms which arise from the $d^2$ central ion configuration.

*Term $^1S$.* As we have seen the $^1S$ has only the single splitting product $^1A_{1g}$. Because the functions $\Psi(L, M_L, S, M_S)$ have the same transformation properties as the spherical harmonics* we can take those symmetry adapted linear combinations of the $\Psi(L, M_L, S, M_S)$ which have the 'correct' linear combinations of the $Y_{L,M_L}$ functions directly from Table B.16. According to this the functions associated with the species $A_{1g}$ is equal to $Y_{00}$. Consequently it holds (for singlet functions $S = 0, M_S = 0$)†:

$$\Theta(^1A_{1g}|^1S) = \Psi(0,0,0,0).$$

The relations between the $\Psi(L, M_L, S, M_S)$ functions and the anti-symmetrized products $\Phi$ of the one-electron functions are given in Table B.6. Accordingly we find

$$\Theta(^1A_{1g}|^1S) = \frac{1}{\sqrt{5}}[\Phi(2^+;-2^-)-\Phi(2^-;-2^+)-\Phi(1^+;-1^-)+\Phi(1^-;-1^+) \\ +\Phi(0^+;0^-)]. \qquad (3\text{–}47)$$

Because a $^1S$ term and thereby the progeny term $^1A_{1g}$ as well are simple (nondegenerate) terms, that is only *one* function $\Theta(^1A_{1g}|^1S)$ is necessary for their description, it follows that the secular determinant (3–46) is of the first degree

$$|(\Theta(^1A_{1g}|^1S), \boldsymbol{H}_{\text{Lig.}}\Theta(^1A_{1g}|^1S)) - \Delta E_{\text{Lig.}}(^1A_{1g}|^1S)| = 0$$

and has the one solution

$$\Delta E_{\text{Lig.}}(^1A_{1g}|^1S) = (\Theta(^1A_{1g}|^1S), \boldsymbol{H}_{\text{Lig.}}\Theta(^1A_{1g}|^1S)).$$

If we substitute for $\Theta$ the linear combination (3–47) then we obtain

$$\Delta E_{\text{Lig.}}(^1A_{1g}|^1S) =$$
$$\frac{1}{5}\{(\Phi(2^+;-2^-), \boldsymbol{H}_{\text{Lig.}}\Phi(2^+;-2^+)) - (\Phi(2^+;-2^-), \boldsymbol{H}_{\text{Lig.}}\Phi(2^-;-2^+))$$
$$-(\Phi(2^+;-2^-), \boldsymbol{H}_{\text{Lig.}}\Phi(1^+;-1^-)) + (\Phi(2^+;-2^-), \boldsymbol{H}_{\text{Lig.}}\Phi(1^-;-1^+))$$
$$+(\Phi(2^+;-2^-), \boldsymbol{H}_{\text{Lig.}}\Phi(0^+;0^-))$$
$$-(\Phi(2^-;-2^+), \boldsymbol{H}_{\text{Lig.}}\Phi(2^+;-2^-)) + (\Phi(2^-;-2^+), \boldsymbol{H}_{\text{Lig.}}\Phi(2^-;-2^+))$$
$$+(\Phi(2^-;-2^+), \boldsymbol{H}_{\text{Lig.}}\Phi(1^+;-1^-)) - (\Phi(2^-;-2^+), \boldsymbol{H}_{\text{Lig.}}\Phi(1^-;-1^+))$$
$$-(\Phi(2^-;-2^+), \boldsymbol{H}_{\text{Lig.}}\Phi(0^+;0^-))$$
$$-(\Phi(1^+;-1^-), \boldsymbol{H}_{\text{Lig.}}\Phi(2^+;-2^-)) + (\Phi(1^+;-1^-), \boldsymbol{H}_{\text{Lig.}}\Phi(2^-;-2^+))$$

---

* Except for the fact that the $\Psi(L, M_L, S, M_S)$ functions belonging to $d^N$ systems always have even parity.

† The symmetry adapted linear combinations of the $\Psi(L, M_L, S, M_S)$ functions are denoted with $\Theta$. The symbols for the symmetry species and the original term are given in parentheses: $\Theta(^{2S+1}\Gamma_i{}^{2S+1}L)$. If the irreducible representation $\Gamma_i$ is poly-dimensional the specific $\Theta$ functions are distinguished from one another by means of indices: $\Theta_r(^{2S+1}\Gamma_i{}^{2S+1}L)$ with $r = 1, 2, \ldots, g_i$ ($g_i$ = dimensionality of $\Gamma_i$).

$$
\begin{aligned}
&+(\Phi(1^+;-1^-), \boldsymbol{H}_{\text{Lig.}}\Phi(1^+;-1^-)) - (\Phi(1^+;-1^-), \boldsymbol{H}_{\text{Lig.}}\Phi(1^-;-1^+))\\
&-(\Phi(1^+;-1^-), \boldsymbol{H}_{\text{Lig.}}\Phi(0^+;0^-))\\
&+(\Phi(1^-;-1^+), \boldsymbol{H}_{\text{Lig.}}\Phi(2^+;-2^-)) - (\Phi(1^-;-1^+), \boldsymbol{H}_{\text{Lig.}}\Phi(2^-;-2^+))\\
&-(\Phi(1^-;-1^+), \boldsymbol{H}_{\text{Lig.}}\Phi(1^+;-1^-)) + (\Phi(1^-;-1^+), \boldsymbol{H}_{\text{Lig.}}\Phi(1^-;-1^+))\\
&+(\Phi(1^-;-1^+), \boldsymbol{H}_{\text{Lig.}}\Phi(0^+;0^-))\\
&+(\Phi(0^+;0^-), \quad \boldsymbol{H}_{\text{Lig.}}\Phi(2^+;-2^-)) - (\Phi(0^+;0^-), \quad \boldsymbol{H}_{\text{Lig.}}\Phi(2^-;-2^+))\\
&-(\Phi(0^+;0^-), \quad \boldsymbol{H}_{\text{Lig.}}\Phi(1^+;-1^-)) + (\Phi(0^+;0^-), \quad \boldsymbol{H}_{\text{Lig.}}\Phi(1^-;-1^+))\\
&+(\Phi(0^+;0^-), \quad \boldsymbol{H}_{\text{Lig.}}\Phi(0^+;0^-))\}.
\end{aligned}
\tag{3–48}
$$

Because $\boldsymbol{H}_{\text{Lig.}}$ is a sum of *one-electron operators* $\boldsymbol{H}_{\text{Lig.}} = V(\vec{r}_1) + V(\vec{r}_2)$ the two-electron integrals which appear can be reduced to one-electron integrals. Integrals having two identical $\Phi$ functions are according to rule (a), on p. 249, equal to the sum of the corresponding one-electron integrals:

$$
\begin{aligned}
(\Phi(2^+;-2^-), \boldsymbol{H}_{\text{Lig.}}\Phi(2^+;-2^-)) &= (2^+|\boldsymbol{V}|2^+) + (-2^-|\boldsymbol{V}|-2^-)\\
(\Phi(2^-;-2^+), \boldsymbol{H}_{\text{Lig.}}\Phi(2^-;-2^+)) &= (2^-|\boldsymbol{V}|2^-) + (-2^+|\boldsymbol{V}|-2^+)\\
(\Phi(1^+;-1^-), \boldsymbol{H}_{\text{Lig.}}\Phi(1^+;-1^-)) &= (1^+|\boldsymbol{V}|1^+) + (-1^-|\boldsymbol{V}|-1^-)\\
(\Phi(1^-;-1^+), \boldsymbol{H}_{\text{Lig.}}\Phi(1^-;-1^+)) &= (1^-|\boldsymbol{V}|1^-) + (-1^+|\boldsymbol{V}|-1^+)\\
(\Phi(0^+;0^-), \quad \boldsymbol{H}_{\text{Lig.}}\Phi(0^+;0^-)) &= (0^+|\boldsymbol{V}|0^+) + (0^-|\boldsymbol{V}|0^-)
\end{aligned}
$$

with

$$(m_l, m_s|\boldsymbol{V}|m_l', m_s') = \int \psi^*_{n,l,m_l}\chi^*_{m_s}\boldsymbol{V}\psi_{n,l,m_l'}\chi_{m_s'}\,d\tau\,d\sigma$$

All remaining integrals in the energy expression (3–48) contain in each case two $\Phi$ functions which differ from one another in both $m_l$, $m_s$ values. According to rule (c), p. 249, they are all zero. Thus we have

$$
\begin{aligned}
\Delta E_{\text{Lig.}}(^1A_{1g}|^1S) = \frac{1}{5}\{&(2^+|\boldsymbol{V}|2^+) + (-2^-|\boldsymbol{V}|-2^-)\\
&+(2^-|\boldsymbol{V}|2^-) + (-2^+|\boldsymbol{V}|-2^+)\\
&+(1^+|\boldsymbol{V}|1^+) + (-1^-|\boldsymbol{V}|-1^-)\\
&+(1^-|\boldsymbol{V}|1^-) + (-1^+|\boldsymbol{V}|-1^+)\\
&+(0^+|\boldsymbol{V}|0^+) + (0^-|\boldsymbol{V}|0^-)\}.
\end{aligned}
$$

Integration over the spin coordinate in $(2^+|V|2^+)$ gives

$$(2^+|\boldsymbol{V}|2^+) = (\psi_{n,2,2}, \boldsymbol{V}\psi_{n,2,2})(\alpha,\alpha) = (2|\boldsymbol{V}|2),$$

because $(\alpha, \alpha) = 1$ where $(2|V|2)$ is the integral that is to be taken only over the spatial coordinates and which contains the spin-free spatial functions:

$$\Delta E_{\mathrm{Lig.}}(^1A_{1g}|\,^1S) = \frac{2}{5}\{(2|V|2) + (-2|V|-2) + (1|V|1) + (-1|V|-1) + (0|V|0)\}.$$

Analogous results are given by integration over the spin coordinates of the remaining one-electron integrals. The following expression is then obtained for the perturbation energy:

$$(2|V|2) \equiv \int \psi^*_{n,2,2}\, V\, \psi_{n,2,2}\, \mathrm{d}\tau.$$

A further simplification of the energy expression may be obtained when it is considered that $\psi^*_{n,2,m} = \psi_{n,2,-m}$ and that $V$ commutes with the functions $\psi_{n,2,m}$:

$$(m|V|m) \equiv \int \psi^*_{n,2,m} V \psi_{n,2,m} \mathrm{d}\tau = \int \psi_{n,2,-m} V \psi_{n,2,m} \mathrm{d}\tau = \int \psi_{n,2,m} V \psi_{n,2,-m} \mathrm{d}\tau = \int \psi^*_{n,2,-m} V \psi_{n,2,-m} \mathrm{d}\tau \equiv (-m|V|-m). \tag{3–49}$$

Bearing this in mind we find

$$\Delta E_{\mathrm{Lig.}}(^1A_{1g}|^1S) = \frac{2}{5}\{2(2|V|2) + 2(1|V|1) + (0|V|0)\}.$$

The value of the one-electron integral $(2|V|2)$ is calculated by substituting for $V$ the negative of the explicit potential expression (3–14) for the $O_h$ systems*:

$$\begin{aligned}(2|V|2) &= -\int \psi^*_{n,2,2} V_{O_h} \psi_{n,2,2} \mathrm{d}\tau \\ &= -\int_0^\infty\int_0^\pi\int_0^{2\pi} R^*_{n,2} Y^*_{2,2} V_{O_h} R_{n,2} Y_{2,2} r^2 \mathrm{d}r \sin\theta\, \mathrm{d}\theta\, \mathrm{d}\phi \\ &= -A_{0,0}\int_0^\infty R^*_{n,2} r^0 R_{n,2} r^2 \mathrm{d}r \int_0^\pi\int_0^{2\pi} Y^*_{2,2} Y_{0,0} Y_{2,2} \sin\theta \mathrm{d}\theta \mathrm{d}\phi \\ &\quad -A_{4,0}\int_0^\infty R^*_{n,2} r^4 R_{n,2} r^2 \mathrm{d}r \int_0^\pi\int_0^{2\pi} Y^*_{2,2}[Y_{4,0} + \sqrt{5/14}(Y_{4,4} + Y_{4,-4})] Y_{2,2} \sin\theta \mathrm{d}\theta \mathrm{d}\phi.\end{aligned}$$

Using Table B.21 and the abbreviation (3–30) this becomes

$$\begin{aligned}(2|V|2) &= -A_{0,0}\overline{r^0}\frac{1}{2\sqrt{\pi}} - A_{4,0}\overline{r^4}\left[\frac{1}{7\cdot 2\pi} + \sqrt{5/14}(0+0)\right] \\ &= -A_{0,0}\overline{r^0}\frac{1}{2\sqrt{\pi}} - A_{4,0}\overline{r^4}\frac{1}{14\sqrt{\pi}}.\end{aligned} \tag{3–50a}$$

---

* The operator $V$ of the electrostatic electron–ligand interaction is equal to the negative ligand potential $V_{O_h}$ because the charge of the $d$ electron is negative.

According to the definition of $\varepsilon_0$ and $Dq$ for octahedral systems, p. 340, this can finally be written as

$$(2\,|V|\,2) = \varepsilon_0 + 1\,Dq. \tag{3–50a}$$

The integrals $(1|\,V|1)$ and $(0|\,V|0)$ have been evaluated earlier (cf. pp. 338 ff.)

$$(1\,|V|\,1) = -A_{0,0}\overline{r^0}\,\frac{1}{2\sqrt{\pi}} + 4\cdot A_{4,0}\overline{r^4}\,\frac{1}{14\sqrt{\pi}} = \varepsilon_0 - 4\,Dq, \tag{3–50b}$$

$$(0\,|V|\,0) = -A_{0,0}\overline{r^0}\,\frac{1}{2\sqrt{\pi}} - 6\cdot A_{4,0}\overline{r^4}\,\frac{1}{14\sqrt{\pi}} = \varepsilon_0 + 6\,Dq. \tag{3–50c}$$

With this we obtain

$$\Delta E_{\text{Lig.}}(^1A_{1g}|^1S) = \frac{2}{5}\,\{2(\varepsilon_0 + 1\,Dq) + 2(\varepsilon_0 - 4\,Dq) + (\varepsilon_0 + 6\,Dq)\}$$

or

$$\Delta E_{\text{Lig.}}(^1A_{1g}|^1S) = 2\varepsilon_0. \tag{3–51}$$

$^1D$ *Term.* As has been shown above the $^1D$ term splits into a $^1E_g$ and a $^1T_{2g}$ progeny term. Using Tables B.16 and B.6 one finds the following two symmetry adapted $^1E_g$ functions

$$\begin{aligned}\Theta_1(^1E_g|^1D) &= \Psi(2,0,0,0)\\ &= \sqrt{4/14}\,\Phi(2^+;-2^-) - \sqrt{4/14}\,\Phi(2^-;-2^+) + \sqrt{1/14}\,\Phi(1^+;-1^-)\\ &\quad \sqrt{1/14}\,\Phi(1^-;-1^+) - \sqrt{4/14}\,\Phi(0^+;0^-),\\ \Theta_2(^1E_g|^1D) &= \frac{1}{\sqrt{2}}\,[\Psi(2,2,0,0) + \Psi(2,-2,0,0)]\\ &= \frac{1}{\sqrt{2}}[\sqrt{2/7}\,\Phi(2^+;0^-) - \sqrt{2/7}\,\Phi(2^-;0^+) - \sqrt{3/7}\,\Phi(1^+;1^-)\\ &\quad + \sqrt{2/7}\,\Phi(0^+;-2^-) - \sqrt{2/7}\,\Phi(0^-;-2^+) - \sqrt{3/7}\,\Phi(-1^+;-1^-)].\end{aligned}$$

The three $T_{2g}$ functions are

$$\begin{aligned}\Theta_1(^1T_{2g}|^1D) &= \Psi(2,-1,0,0)\\ &= \sqrt{6/14}\,\Phi(1^+;-2^-) - \sqrt{6/14}\,\Phi(1^-;-2^+) - \sqrt{1/14}\,\Phi(0^+;-1^-)\\ &\quad + \sqrt{1/14}\,\Phi(0^-;-1^+),\\ \Theta_2(^1T_{2g}|^1D) &= \frac{1}{\sqrt{2}}\,[\Psi(2,2,0,0) - \Psi(2,-2,0,0)]\\ &= \frac{1}{\sqrt{2}}\,[\sqrt{2/7}\,\Phi(2^+;0^-) - \sqrt{2/7}\,\Phi(2^-;0^+) - \sqrt{3/7}\,\Phi(1^+;1^-)\\ &\quad - \sqrt{2/7}\,\Phi(0^+;-2^-) + \sqrt{2/7}\,\Phi(0^-;-2^+) + \sqrt{3/7}\,\Phi(-1^+;-1^-)],\\ \Theta_3(^1T_{2g}|^1D) &= -\Psi(2,1,0,0)\\ &= -\sqrt{6/14}\,\Phi(2^+;-1^-) + \sqrt{6/14}\,\Phi(2^-;-1^+) + \sqrt{1/14}\,\Phi(1^+;0^-)\\ &\quad - \sqrt{1/14}\,\Phi(1^-;0^+).\end{aligned}$$

The energy perturbation which the $^1D$ term experiences under the influence of the ligand field is yielded by the secular problem with the five given $\Theta$ functions as a basis. The appropriate $5 \times 5$ determinant:

$$
\begin{array}{l}
\Theta_1({}^1E_g|{}^1D) \\
\Theta_2({}^1E_g|{}^1D) \\
\Theta_1({}^1T_{2g}|{}^1D) \\
\Theta_2({}^1T_{2g}|{}^1D) \\
\Theta_3({}^1T_{2g}|{}^1D)
\end{array}
\left|
\begin{array}{cc|ccc}
\multicolumn{2}{c|}{E_g} & & & \\
\times & \times & & 0 & \\
\times & \times & & & \\
\hline
 & & \times & \times & \times \\
 & 0 & \times & \times & \times \\
 & & \times & \times & \times \\
 & & \multicolumn{3}{c}{T_{2g}}
\end{array}
\right| = 0
$$

decomposes for group theoretical reasons into a $2 \times 2$ determinant with the basis $\Theta_1({}^1E_g|{}^1D)$, $\Theta_2({}^1E_g|{}^1D)$ and a $3 \times 3$ determinant with the basis $\Theta_1({}^1T_{2g}|{}^1D)$, $\Theta_2({}^1T_{2g}|{}^1D)$, $\Theta_3({}^1T_{2g}|{}^1D)$ because all nondiagonal elements $(\Theta_i({}^1E_g|{}^1D), \boldsymbol{H}_{\text{lig.}}\Theta_j({}^1T_{2g}|{}^1D))$ are zero (cf. theorem 2a, p. 317). Besides this the nondiagonal elements vanish not only in the $2 \times 2$ determinant but also in the $3 \times 3$ determinant according to theorem 2a. Finally theorem 2b gives information concerning the equality of diagonal elements, so that we may write for the roots of the secular determinant and therefore for the perturbation energies

$$
\begin{aligned}
\Delta E_{\text{Lig.}}({}^1E_g|{}^1D) &= (\Theta_1({}^1E_g|{}^1D), \boldsymbol{H}_{\text{Lig.}}\Theta_1({}^1E_g|{}^1D)) \\
&= (\Theta_2({}^1E_g|{}^1D), \boldsymbol{H}_{\text{Lig.}}\Theta_2({}^1E_g|{}^1D)), \\
\Delta E_{\text{Lig.}}({}^1T_{2g}|{}^1D) &= (\Theta_1({}^1T_{2g}|{}^1D), \boldsymbol{H}_{\text{Lig.}}\Theta_1({}^1T_{2g}|{}^1D)) \\
&= (\Theta_2({}^1T_{2g}|{}^1D), \boldsymbol{H}_{\text{Lig.}}\Theta_2({}^1T_{2g}|{}^1D)) \\
&= (\Theta_3({}^1T_{2g}|{}^1D), \boldsymbol{H}_{\text{Lig.}}\Theta_3({}^1T_{2g}|{}^1D)).
\end{aligned}
$$

The perturbation energy of the doubly degenerate $^1E_g$ progeny term can be determined by a calculation of the integral $(\Theta_1({}^1E_g|{}^1D), \boldsymbol{H}_{\text{Lig.}}\Theta_1({}^1E_g|{}^1D))$. Here one proceeds as in the calculation of the corresponding integral of the $^1S$ term; in particular

1. $\Theta_1({}^1E_g|{}^1D)$ is replaced by the $\Phi$ linear combination given above,
2. the integrals containing $\Phi$ functions which appear in this manner are reduced under consideration of the rules on p. 249 to one-electron integrals.
3. one integrates over the spin coordinates and uses
4. relation (3–49).

As a preliminary result we find

$$\Delta E_{\text{Lig.}}(^1E_g|^1D) = \frac{16}{14}(2|\boldsymbol{V}|2) + \frac{4}{14}(1|\boldsymbol{V}|1) + \frac{8}{14}(0|\boldsymbol{V}|0),$$

and finally with consideration of (3–50a, b, c)

$$\Delta E_{\text{Lig.}}(^1E_g|^1D) = 2\varepsilon_0 + \frac{24}{7}Dq. \tag{3–52a}$$

Following the same procedure one obtains

$$\Delta E_{\text{Lig.}}(^1T_{2g}|^1D) = 2\varepsilon_0 - \frac{16}{7}Dq. \tag{3–52b}$$

$^1G$ *Term.* The $^1G$ term has four progeny terms: $^1A_{1g}$, $^1E_g$, $^1T_{1g}$, $^1T_{2g}$. When symmetry adapted functions are used as a basis the $9 \times 9$ secular determinant of the $^1G$ term decomposes into a $1 \times 1$ determinant (belonging to $^1A_{1g}$), and a $2 \times 2$ determinant (to $^1E_g$), a $3 \times 3$ determinant (to $^1T_{1g}$) and a $3 \times 3$ determinant (to $^1T_{2g}$). According to Tables B.16 and B.6 the function of the progeny term $^1A_{1g}$ is:

$$\begin{aligned}\Theta(^1A_{1g}|^1G) &= \sqrt{7/12}\,\Psi(4,0,0,0) + \sqrt{5/24}\,\Psi(4,4,0,0) + \sqrt{5/24}\,\Psi(4,-4,0,0)\\ &= \sqrt{1/120}\,\{\Phi(2^+;-2^-) - \Phi(2^-;-2^+) + 4\Phi(1^+;-1^-)\\ &\quad -4\Phi(1^-;-1^+) + 6\Phi(0^+;0^-) + 5\Phi(2^+;2^-)\\ &\quad +5\Phi(-2^+;-2^-)\}\end{aligned}$$

The perturbation energy

$$\Delta E_{\text{Lig.}}(^1A_{1g}|^1G) = (\Theta(^1A_{1g}|^1G), \boldsymbol{H}_{\text{Lig.}}\,\Theta(^1A_{1g}|^1G))$$

is calculated in that first the steps above ($^1D$ term) denoted by 1, 2, 3, and 4 are carried out. One obtains then

$$\begin{aligned}\Delta E_{\text{Lig.}}(^1A_{1g}|^1G) &= \frac{13}{15}(2|\boldsymbol{V}|2) + \frac{8}{15}(1|\boldsymbol{V}|1) + \frac{9}{15}(0|\boldsymbol{V}|0)\\ &\quad + \frac{1}{6}[(2|\boldsymbol{V}|-2) + (-2|\boldsymbol{V}|2)].\end{aligned}$$

The one-electron integrals of the type $(2|V|-2)$ and $(-2|V|2)$ appear in a different manner than in examples treated earlier. Their calculation follows according to the same scheme as the calculation of the integrals $(2|V|2)$, $(1|V|1)$, $(0|V|0)$ with the result:

$$(2|\boldsymbol{V}|-2) = (-2|\boldsymbol{V}|2) = 5Dq. \tag{3–53}$$

With this and with consideration of (3–50) we obtain

$$\Delta E_{\text{Lig.}}(^1A_{1g}|^1G) = 2\varepsilon_0 + 4Dq.$$

Following this same procedure one obtains for the remaining progeny terms of the $^1G$ term the perturbation energies

$$\Delta E_{\text{Lig.}}(^1E_g|^1G) = 2\varepsilon_0 + \frac{4}{7}Dq,$$

$$\Delta E_{\text{Lig.}}(^1T_{1g}|^1G) = 2\varepsilon_0 + 2Dq,$$

$$\Delta E_{\text{Lig.}}(^1T_{2g}|^1G) = 2\varepsilon_0 - \frac{26}{7}Dq.$$

We shall now consider finally the behaviour of the triplet terms $^3P$ and $^3F$ under the influence of the ligand field. The symmetry adapted $\Theta$ functions of a triplet term $^3L$ can be subdivided into three sets of $\Theta$ functions, where the functions of the different sets are eigenfunctions of different eigenvalues $M_S\hbar$ of the operator $S_z$ ($z$ component of the total spin):

$$\Theta_i(^3\Gamma_j|^3L, M_S=1)$$
$$\Theta_i(^3\Gamma_j|^3L, M_S=0)$$
$$\Theta_i(^3\Gamma_j|^3L, M_S=-1)$$

Because the ligand field operator $\boldsymbol{H}_{\text{Lig.}}$ has no effect upon the spin coordinates, $\Theta$ functions which belong to different $M_S$ values with respect to $\boldsymbol{H}_{\text{Lig.}}$ cannot combine with one another. When the $\Theta$ functions are appropriately numbered the secular determinant formed by using all the $\Theta$ functions decomposes into three subdeterminants, one for $M_S = 1$, one for $M_S = 0$ and one for $M_S = -1$. According to theorem 2b, p. 317 these three subdeterminants must have the same roots. Accordingly it suffices for the determination of the perturbation energies to set up only one of these subdeterminants and to find its roots.

$^3P$ *Term.* The $^3P$ term has only one progeny term: $^3T_{1g}$. We use the set of $\Theta$ functions with $M_S = 1$ to calculate the perturbation energy. According to Tables B.16 and B.6 the appropriate $\Theta$ functions are

$$\Theta_1(^3T_{1g}|^3P, M_S=1) = \Psi(1,1,1,1)$$
$$= -\sqrt{2/5}\,\Phi(2^+; -1^+) + \sqrt{3/5}\,\Phi(1^+; 0^+),$$
$$\Theta_2(^3T_{1g}|^3P, M_S=1) = \Psi(1,0,1,1)$$
$$= -\sqrt{4/5}\,\Phi(2^+; -2^+) + \sqrt{1/5}\,\Phi(1^+; -1^+),$$
$$\Theta_3(^3T_{1g}|^3P, M_S=1) = \Psi(1,-1,1,1)$$
$$= -\sqrt{2/5}\,\Phi(1^+; -2^+) + \sqrt{3/5}\,\Phi(0^+; -1^+).$$

According to theorem 2, p. 317 the secular determinant formed using these three $\Theta$ functions is completely diagonal and all three diagonal elements are equal. The value of the threefold root is therefore:

$$\Delta E_{\text{Lig.}}(^3T_{1g}|^3P) = (\Theta_i(^3T_{1g}|^3P, M_S=1), \boldsymbol{H}_{\text{Lig.}}\Theta_i(^3T_{1g}|^3P, M_S=1)),$$
$$i = 1,2,3.$$

If the function $\Theta_2$ is chosen for the calculation of $\Delta E_{\text{Lig.}}$, then according to the procedure already used several times we obtain

$$\begin{aligned}\Delta E_{\text{Lig.}}(^3T_{1g}|^3P) &= \frac{4}{5}[(2^+|\boldsymbol{V}|2^+) + (-2^+|\boldsymbol{V}|-2^+)] \\ &\quad + \frac{1}{5}[(1^+|\boldsymbol{V}|1^+) + (-1^+|\boldsymbol{V}|-1^+)] \\ &= \frac{8}{5}(2^+|\boldsymbol{V}|2^+) + \frac{2}{5}(1^+|\boldsymbol{V}|1^+)\end{aligned}$$

or with consideration of (3–50)

$$\Delta E_{\text{Lig.}}(^3T_{1g}|^3P) = 2\varepsilon_0.$$

$^3F$ *Term.* We shall use the symmetry adapted $\Theta$ functions with $M_S = 1$ to calculate the $^3F$ progeny terms $^3A_{2g}$, $^3T_{1g}$ and $^3T_{2g}$. The perturbation energy of the $^3A_{2g}$ term is then

$$\Delta E_{\text{Lig.}}(^3A_{2g}|^3F) = (\Theta(^3A_{2g}|^3F, M_S=1), \boldsymbol{H}_{\text{Lig.}}\Theta(^3A_{2g}|^3F, M_S=1))$$

with (see Tables B.16 and B.6)

$$\begin{aligned}\Theta(^3A_{2g}|^3F, M_S=1) &= \frac{1}{\sqrt{2}}[\Psi(3,2,1,1) - \Psi(3,-2;1,1)] \\ &= \frac{1}{\sqrt{2}}[\Phi(2^+;0^+) - \Phi(0^+;-2^+)].\end{aligned}$$

Reduction of the expression for the perturbation energy to the energy quantities $\varepsilon_0$ and $Dq$ gives, when use is made of the procedure explained above,

$$\Delta E_{\text{Lig.}}(^3A_{2g}|^3F) = 2\varepsilon_0 + 12Dq.$$

Completely analogously one obtains the perturbation energies of the remaining progeny terms of $^3F$:

$$\Delta E_{\text{Lig.}}(^3T_{1g}|^3F) = 2\varepsilon_0 - 6Dq,$$
$$\Delta E_{\text{Lig.}}(^3T_{2g}|^3F) = 2\varepsilon_0 + 2Dq.$$

The energies of the terms which arise from the terms $^1S$, $^1D$, $^1G$, $^3P$, $^3F$ of the configuration $d^2$ under the influence of an octahedral ligand field are given graphically in Figure B.24 as a function of the quantity $Dq$ for the weak-field case. Here the term positions for the value of the abscissa

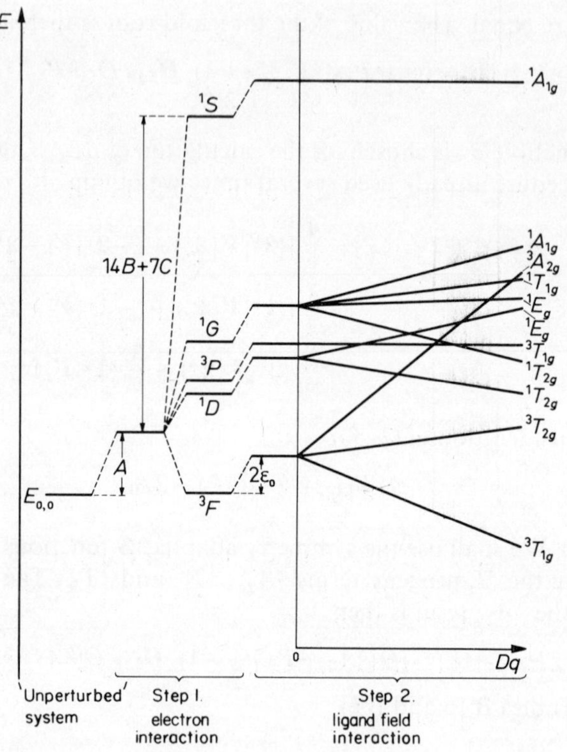


Figure B.24. Term splitting diagram for octahedrally bound $d^2$ systems for the weak-field case (without term interaction) with $C/B = 4$.

$Dq = 0$ were calculated in their energetic order with $C = 4B$, cf. Part A, p. 317 and Part B, p. 360.

We refer to the '*centre of gravity*' *theorem*, which is rigorously valid in the weak-field case. It states that the arithmetic mean of the *perturbation energies* of all the progeny terms which arise from the same $^{2S+1}L$ term, formed under consideration of the degree of degeneracy, is equal to the energy factor $N\varepsilon_0$, which is common for all progeny terms, where $N$ is the number of $d$ electrons:

$$\frac{\sum_i g_i \Delta E_{\text{Lig.}}\,(^{2S+1}\Gamma_i|^{2S+1}L)}{\sum_i g_i} = N\varepsilon_0 \qquad \text{(3–54)}$$

($g_i$ = total degree of degeneracy orbit × spin of the progeny term $^{2S+1}\Gamma_i$)

For example for the progeny terms of the $^1G$ term of the configuration $d^2$ ($N = 2$) one finds in agreement with the centre of gravity theorem that

$$\frac{1}{1+2+3+3}\left[1\cdot(2\varepsilon_0+4Dq)+2\cdot\left(2\varepsilon_0+\frac{4}{7}Dq\right)+3\cdot(2\varepsilon_0+2Dq)\right.$$
$$\left.+3\cdot\left(2\varepsilon_0-\frac{26}{7}Dq\right)\right]=2\varepsilon_0.$$

The centre of gravity theorem may be used in this sense as a control on the energies calculated for progeny terms.

3.4.1.2. *Example of a $d^3$ ion in a ligand field of $O_h$ symmetry.* We limit ourselves here to the study of the progeny terms of the $^4F$ term, which is the ground state term in the free $d^3$ ion. Under the influence of a regular octahedral ligand field the terms $^4A_{2g}$, $^4T_{1g}$ and $^4T_{2g}$ result from the $^4F$ term. With consideration of the fact that symmetry adapted $\Theta$ functions with different $M_S$ values cannot combine with one another with reference to $H_{\text{Lig.}}$ and with consideration of theorem 2b, p. 317, concerning the identity of energy integrals similarly as with the treatment of the triplet terms of the $d^2$ configuration one needs to evaluate only very few energy integrals. The perturbation energy of the $^4A_{2g}$ progeny term is obtained by using the function (cf. Tables B.16 and B.17)

$$\Theta(^4A_{2g}|^4F, M_S=3/2)=\frac{1}{\sqrt{2}}[\Psi(3,2,3/2,3/2)-\Psi(3,-2,3/2,3/2)]$$
$$=\frac{1}{\sqrt{2}}[\Phi(2^+;1^+;-1^+)-\Phi(1^+;-1^+;-2^+)]$$

for

$$\Delta E_{\text{Lig.}}(^4A_{2g}|^4F)=(2^+|V|2^+)+2(1^+|V|1^+)-(2^+|V|-2^+),$$

or because of (3–50) for

$$\Delta E_{\text{Lig.}}(^4A_{2g}|^4F)=3\varepsilon_0-12Dq.$$

The perturbation energy of the $^4T_{1g}$ progeny term

$$\Delta E_{\text{Lig.}}(^4T_{1g}|^4F)=3\varepsilon_0+6Dq$$

and of the $^4T_{2g}$ term

$$\Delta E_{\text{Lig.}}(^4T_{2g}|^4F)=3\varepsilon_0-2Dq.$$

are calculated analogously.

One sees that the centre of gravity theorem is satisfied by forming:

$$\frac{1}{4+12+12}[4(3\varepsilon_0-12Dq)+12(3\varepsilon_0+6Dq)+12(3\varepsilon_0-2Dq)]=3\varepsilon_0.$$

### 3.4.2. *Strong-field case*

We have this case when the *electron interaction energy is small in comparison to the interaction between the electrons and the ligand system.* In this sense the strong-field case is the supplement and opposite of the weak-field case. As with the latter method we proceed here in two steps although the order of the steps is exactly reversed, compared to the weak-field case. In the first step we consider which properties a $d$ electron system has when it is bound in a complex *but experiences no electron interaction.* The second step gives information about the changes the properties determined in step one undergo when in addition the electron interaction is taken into consideration.

*Step 1.* The $\Phi$ functions of the unperturbed system (with the *Hamiltonian* (3–42)) having the eigenvalue $E_{00} = N\varepsilon_{nd}$ or in general a set of (linearly independent) linear combinations $\Theta$ of these functions is taken as a starting basis* for a perturbation calculation using the ligand field operator

$$\boldsymbol{H}_{\text{Lig.}} = \sum_{i=1}^{N} \boldsymbol{V}(\vec{r}_i).$$

The roots $\Delta\tilde{E}_{\text{Lig.}}$ of the secular determinant†

$$\|(\Theta_r, \boldsymbol{H}_{\text{Lig.}}\Theta_s) - (\Theta_r,\Theta_s)\Delta\tilde{E}_{\text{Lig.}}\| = 0, \quad r,s = 1,2,\ldots,\eta \qquad (3\text{–}55)$$

$$(\eta = \text{degree of degeneracy})$$

are the values of the perturbation energies which the $d$ electron system without electron interaction experiences under the influence of the ligand field. It is evident that the roots $\Delta\tilde{E}_{\text{Lig.}}$ are most easily obtained when from the beginning the $\Theta$ functions are chosen so that the secular determinant is for the most part diagonal. As in the examples treated earlier group and representation theoretical arguments are helpful in selecting the $\Theta$ functions: We shall use for the perturbation calculation $\Theta$ basis functions which transform according to the irreducible representations of the symmetry group of the perturbation operator $\boldsymbol{H}_{\text{Lig.}}$. Thus we have the possibility of applying the theorems 1 and 2, p. 317, concerning when integrals in the secular determinant vanish or are equal.

When searching for symmetry adapted $\Theta$ functions the following method is recommended. One begins with the selection of suitable one-electron $d$ functions. It is to be kept in mind that antisymmetrized products

* It should once more be mentioned here that linear combinations of the $\Phi$'s are also eigenfunctions of the unperturbed system having the energy eigenvalue $E_{00} = N\varepsilon_{nd}$.

† We use here the symbols $\Delta\tilde{E}_{\text{Lig.}}$ to distinguish from the energy expressions $\Delta E_{\text{Lig.}}$ which appear in the second step of the weak field treatment.

of the type of the $\Phi$ functions are eigenfunctions of the unperturbed system having the eigenvalue $E_{00} = N\varepsilon_{nd}$, independent of from which set of one-electron $d$ functions they come. It is convenient for a spherically symmetric problem of free ions to use one-electron $d$ functions with the space parts

$$\psi_{n,2,m_l} = R_{n,2}(r)Y_{2,m_l}(\theta,\phi),\ m_l = -2, -1, 0, +1, +2$$

However, here it is useful to start from space functions which transform as the irreducible representations of the symmetry group of $\boldsymbol{H}_{\text{Lig.}}$. Such a symmetry adapted set of functions is obtained by a suitable linear combination of the $\psi_{n,2,m_l}$ according to the procedure explained earlier (cf. section 2.4.2.2 concerning spherical harmonics). The space parts which transform according to the irreducible representation $\Gamma_1$ of the symmetry group of $\boldsymbol{H}_{\text{Lig.}}$ are denoted by

$$\phi_1(\Gamma_1), \phi_2(\Gamma_1), \ldots, \phi_{g_1}(\Gamma_1),$$

where $g_1$ is the dimensionality of $\Gamma_1$. The space parts belonging to the other irreducible representations $\Gamma_2$ are denoted analogously*.

All possible antisymmetrized products of the symmetry-adapted one-electron space parts $\phi_i(\Gamma_j)$ are constructed, taking account of the *Pauli* principle. A $\phi_i(\Gamma_j)$ can be occupied by a maximum of two electrons which must differ in their spins. For a $d^2$ system, for example, the antisymmetrized products are of the form

$$\Phi(\phi_i^+(\Gamma_j); \phi_i^-(\Gamma_j)), \Phi(\phi_i^+(\Gamma_j); \phi_{i'}^-(\Gamma_{j'})), \Phi(\phi_i^-(\Gamma_j); \phi_{i'}^+(\Gamma_{j'})),$$
$$\Phi(\phi_i^+(\Gamma_j); \phi_{i'}^+(\Gamma_{j'})), \Phi(\phi_i^-(\Gamma_j); \phi_{i'}^-(\Gamma_{j'})),$$

where the signs + and − stand symbolically for the spin functions $\alpha$ and $\beta$. The antisymmetrized products of the $\phi$ functions induce in general a reducible representation of the symmetry group of $\boldsymbol{H}_{\text{Lig.}}$ and are not in the rule symmetry adapted many-electron functions. These may, however, be generated by a suitable linear combination of the antisymmetrized products.

Which (reducible) representation is induced by the antisymmetrized products is seen without difficulty when use is made of the results obtained in section 2.5 concerning product representations. For this purpose all possible antisymmetrized products of a problem are ordered in families which have as members those antisymmetrized products in which the

---

* Of course in general there will not be $\phi$ functions belonging to every irreducible representation of the symmetry group of $\boldsymbol{H}_{\text{Lig.}}$. In the case of $O_h$ symmetry two of the five $\phi$ functions belong to the symmetry species $E_g$ and three to the species $T_{2g}$. There are no $\phi$ functions belonging to the remaining irreducible representations of $O_h$.

same number of $\phi$ functions of the symmetry species (irreducible representation) $\Gamma_1$ and the same number of $\phi$ functions of the symmetry species $\Gamma_2$, etc. appear. If such a family consists of antisymmetrized products with $a$ $\phi$ functions of the species $\Gamma_1$ and $b$ $\phi$ functions of the species $\Gamma_2$, etc., then this family is denoted by the symbol $\gamma_1^a\gamma_2^b\ldots$. One says the family has the configuration* $\gamma_1^a\gamma_2^b\ldots$. The representation which is induced by the space parts of the members of this family is according to section 2.5 equal to the *product representation*

$$\underbrace{\Gamma_1 \dot{\times} \Gamma_1 \dot{\times} \cdots \dot{\times} \Gamma_1}_{a \text{ times}} \dot{\times} \underbrace{\Gamma_2 \dot{\times} \Gamma_2 \dot{\times} \cdots \dot{\times} \Gamma_2}_{b \text{ times}} \dot{\times} \cdots$$

The irreducible components contained in the product representation are obtained by reducing the representation. Knowledge of these components simplifies the determination of the symmetry adapted linear combinations of the antisymmetrized products.

We shall denote the symmetry adapted linear combinations with $\Xi$. They are characterized with regard to the spatial functions by the configuration $\gamma_1^a\gamma_2^b\ldots$ and its irreducible representation $\Gamma_i$: $\Xi(\Gamma_i\gamma_1^a\gamma_2^b\ldots)$. Up to this point only the spatial functions were considered in the classification but not the spin. Within a family of antisymmetrized products the total spin may be used as a further means of classification. For this purpose the antisymmetrized products are arranged according to their $M_S$ values. One can then use the procedure described in chapter 1, pp. 235 ff. to obtain the total spin states. As further classification symbols one has the multiplicity $2S+1$ and the $M_S$ value: $\Xi(^{2S+1}\Gamma_i|\gamma_1^a\gamma_2^a\ldots|M_S)$. If $\Gamma_i$ is many-dimensional there are several $\Xi$ functions for a $^{2S+1}\Gamma_i|\gamma_1^a\gamma_2^b\ldots|M_S$. We shall distinguish these functions from one another by indices on $\Xi$. On pp. 368 ff. it is shown using the example of the $d^2$ system for the case of octahedral symmetry how the $\Xi_r(^{2S+1}\Gamma_i|\gamma_1^a\gamma_2^b\ldots|M_S)$ are found.

If the symmetry adapted $\Xi$ functions are known the roots $\Delta\tilde{E}_{\text{Lig.}}$ of the secular determinant (3–55) can be easily determined. According to theorem 2a, p. 317, two functions $\Xi_r$ and $\Xi_s$ cannot combine with one another with reference to $\boldsymbol{H}_{\text{Lig.}}$ when both functions

1. belong to different irreducible representations,
2. transform according to different columns of the same irreducible representation,
3. have different multiplicity $2S+1$†,
4. have different $M_S$ values†.

---

* Because the $\phi$ functions are adapted to the symmetry of the ligand field, one speaks also of the '*ligand field configuration*'.

† $\boldsymbol{H}_{\text{Lig.}}$ does not operate on spin variables.

In addition it can be seen that $\Xi$ functions of different configurations $\gamma_1^a\gamma_2^b\ldots$ and $\gamma_1^{a'}\gamma_2^{b'}\ldots$ cannot combine with one another with respect to $\boldsymbol{H}_{\text{Lig.}}$. The explanation is the following: $\Xi$ functions are linear combinations of antisymmetrized products. The antisymmetrized products of a function $\Xi$ with the configuration $\gamma_1^a\gamma_2^b\ldots$ differ—when the antisymmetrized products are written in determinantal form—from the antisymmetrized products of a function $\Xi'$ having the configuration $\gamma_1^{a'}\gamma_2^{b'}\ldots$ by at least $|a-a'|+|b-b'|+\ldots$ columns. $\Xi$ functions of different configurations differ therefore by at least one column of their antisymmetrized products. Now the perturbation operator $\boldsymbol{H}_{\text{Lig.}}$ is equal to a sum of one-electron operators. The rules on p. 249 apply for the determination of integrals of the type $(\Xi, \boldsymbol{H}_{\text{Lig.}}\ \Xi')$. If the antisymmetrized products which compose the functions $\Xi$ and $\Xi'$ have two or more different columns, then the integral $(\Xi, \boldsymbol{H}_{\text{Lig.}}\ \Xi')$ is zero. If the antisymmetrized products differ in only *one* column, then the many-electron integral may be reduced to one-electron integrals of the type $(\phi_i(\Gamma_j), \boldsymbol{V}\phi_{i'}(\Gamma_{j'}))$, where $\phi_i(\Gamma_j)$ and $\phi_{i'}(\Gamma_{j'})$ are the one-electron functions by which the antisymmetrized products of the function $\Xi$ differ from those of the function $\Xi'$. Because $\phi_i(\Gamma_j)$ and $\phi_{i'}(\Gamma_{j'})$ according to our assumption belong to different irreducible representations of the symmetry group of $\boldsymbol{H}_{\text{Lig.}}$ and thereby also of $\boldsymbol{V}$, the remaining one-electron integrals, and with them also in this case the integral $(\Xi, \boldsymbol{H}_{\text{Lig.}}\ \Xi')$ must vanish.

The secular determinant (3–55) is therefore completely diagonal when each irreducible representation is contained not more than once in the product representations of the individual configurations.

The values of the diagonal elements

$$(\Xi_r({}^{2S+1}\Gamma_j|\gamma_1^a\gamma_2^b\ldots|M_S), \boldsymbol{H}_{\text{Lig.}}\ \Xi_r({}^{2S+1}\Gamma_j|\gamma_1^a\gamma_2^b\ldots|M_S))$$

can be easily calculated. $\Xi_r$ is a linear combination of antisymmetrized products $\Phi_u$:

$$\Xi_r = c_1\Phi_1 + c_2\Phi_2 + \cdots = \sum_u c_u\Phi_u,$$

with

$$\sum_u c_u^* c_u = 1 \qquad \text{(normalization condition)} \tag{3–56}$$

The diagonal element may then be written:

$$(\Xi_r, \boldsymbol{H}_{\text{Lig.}}\Xi_r) = \sum_u\sum_{u'} c_u^* c_{u'}(\Phi_u, \boldsymbol{H}_{\text{Lig.}}\Phi_{u'}). \tag{3–57}$$

The integrals with $\Phi_u \equiv \Phi_{u'}$ appearing in the double sum are according to rule (a), p. 249, equal to the sum of the one-electron integrals $(\phi_i(\Gamma_j),$

$V\phi_i(\Gamma_j)$) over all functions $\phi_i(\Gamma_j)$ which appear in $\phi_u$:

$$(\Phi_u, \boldsymbol{H}_{\text{Lig.}} \Phi_u) = \sum_{\phi \text{ in } \Phi_u} (\phi_i(\Gamma_j), \boldsymbol{V} \phi_i(\Gamma_j)). \tag{3–58}$$

This sum is, however, equal to the sum of the one-electron energies of the $d^1$ states represented by the $\phi$ functions which appear. Because $d^1$ states whose $\phi$ functions are of the same irreducible representation have the same energy, the values of the integrals $(\Phi_u, \boldsymbol{H}_{\text{Lig.}} \Phi_u)$ are determined unambiguously by the configuration $\gamma_1^a \gamma_2^b \ldots$ which is present.

Further it is apparent that

$$(\Phi_u, \boldsymbol{H}_{\text{Lig.}} \Phi_u) = (\Phi_{u'}, \boldsymbol{H}_{\text{Lig.}} \Phi_{u'}), \tag{3–59}$$

because $\phi$ functions of the same configuration are contained in $\Phi_u$ as well as in $\Phi_{u'}$.

The integrals $(\Phi_u, \boldsymbol{H}_{\text{Lig.}} \Phi_{u'})$ vanish for $\Phi_u \neq \Phi_{u'}$. $\Phi_u$ then differs from $\Phi_{u'}$ by at least one column. If the difference is more than one column, then it is apparent that the integral vanishes according to the rules (c) and (d) on p. 249. If $\Phi_u$ and $\Phi_{u'}$ differ only because in a column of $\Phi_u$ there is the function $\Phi_u$ and in the corresponding column of $\Phi_{u'}$ the function $\phi_{u'} \neq \phi_u$, then the many-electron integral $(\Phi_u, \boldsymbol{H}_{\text{Lig.}} \Phi_{u'})$ can be reduced to one-electron integrals $(\phi_u, V_{u'})$ using rule (b), p. 249. But because $\Phi_u$ and $\Phi_{u'}$ belong to the same configuration $\gamma_1^a \gamma_2^b \ldots$ the $\phi_u$ and $\phi_{u'}$ can differ only in that they represent different one-electron spin states (+ or −) or (and) belong to different columns of the same irreducible representation of the symmetry group of $\boldsymbol{H}_{\text{Lig.}}$ and thus of $\boldsymbol{V}$. In any case the remaining one-electron integral $(\phi_u, V\phi_{u'})$ and with it $(\Phi_u, \boldsymbol{H}_{\text{Lig.}} \Phi_{u'})$ must be zero. Taking these results into consideration along with equations (3–56, 57, 58, 59) it is seen that

$$\begin{aligned}
(\Xi_r, \boldsymbol{H}_{\text{Lig.}} \Xi_r) &= \sum_u \sum_{u'} c_u^* c_{u'} (\Phi_u, \boldsymbol{H}_{\text{Lig.}} \Phi_{u'}) \\
&= \sum_u \sum_{u'} c_u^* c_{u'} \delta_{u,u'} (\Phi_u, \boldsymbol{H}_{\text{Lig.}} \Phi_u) \\
&= \sum_u c_u^* c_u (\Phi_u, \boldsymbol{H}_{\text{Lig.}} \Phi_u) \\
&= [\sum_{\phi \text{ in } \Phi_u} (\phi_i(\Gamma_j), \boldsymbol{V} \phi_i(\Gamma_j))] \sum_u c_u^* c_u \\
&= \sum_{\phi \text{ in } \Phi_u} (\phi_i(\Gamma_j), \boldsymbol{V} \phi_i(\Gamma_j))
\end{aligned}$$

holds. If the perturbation energy of a one-electron state with the irreducible representation $\Gamma_j$ is denoted by

$$\Delta \varepsilon(\gamma_j) = (\phi_i(\Gamma_j), \boldsymbol{V} \phi_i(\Gamma_j)),$$

then the perturbation energy of a many-electron state described by $\Xi_r({}^{2S+1}\Gamma_j|\gamma_1^a\gamma_2^b\ldots|M_S)$ becomes

$$\Delta\bar{E}_{\text{Lig.}}(\gamma_1^a\gamma_2^b\ldots) = (\Xi_r({}^{2S+1}\Gamma_j|\gamma_1^a\gamma_2^b\ldots|M_S), \boldsymbol{H}_{\text{Lig.}}\Xi_r({}^{2S+1}\Gamma_j|\gamma_1^a\gamma_2^b\ldots|M_S))$$
$$= a\cdot\Delta\varepsilon(\gamma_1)+b\cdot\Delta\varepsilon(\gamma_2)+\cdots, \tag{3–60}$$

i.e., the value of the perturbation energy is equal to the sum of the one-electron perturbation energies of all of the one-electron states denoted by the configuration $\gamma_1^a\gamma_2^b\ldots$. It is therefore determined unambiguously and alone by the configuration. There exists, however, *degeneracy of the totality of states which arise from the same configuration.*

The totality of the many-electron states

arising from the same configuration $\gamma_1^a\gamma_2^b\ldots$,
with the same total spin quantum number $S$
and with the same irreducible representation $\Gamma_j$

are said to constitute a *term*. The functions of the same term are then

$$\Xi_r({}^{2S+1}\Gamma_j|\gamma_1^a\gamma_2^b\ldots|M_S)$$
$$(g_j = \text{dimensionality of } \Gamma_j)$$
$$M_S = -S, -S+1, \ldots, S-1, S.$$

According to the results obtained above terms arising from the same configuration have *the same energy*—insofar as the electron interaction is neglected, as has been assumed up to this point for the first step of our consideration of the strong-field case.

*Step 2.* Continuing from the first step where the influence of a ligand field on a $d^N$ electron system without electron interaction was considered, we shall now consider the consequences of the electron interaction in the second step. The electron interaction measured on the energetic effects of the ligand field interaction is assumed to be weak, in keeping with the definition of the strong field. Consideration of the electron interaction should therefore bring only a small correction to the results obtained in the first step. It will be adequate to a good approximation if the influence of the electron interaction is always studied *separately* for the totality of the degenerate states, that is, those which arise from the same configuration $\gamma_1^a\gamma_2^b\ldots$.

The corrections are determined with the aid of a *perturbation calculation* using the electron interaction operator

$$\boldsymbol{H}_{\text{el.}} = \sum_{i<j}^{N}\sum^{N} 1/r_{ij}$$

as perturbation operator. The perturbation energies produced by $H_{el.}$ are determined as roots $\Delta E_{el.}$ of secular determinants—one for each configuration $\gamma_1^a\gamma_2^b\ldots$. The functions $\Xi_r(2^{S+1}\Gamma_i|\gamma_1\gamma_2\ldots|M_S)$ belonging to the same configuration $\gamma_1^a\gamma_2^b\ldots$ are always to be used as basis functions for the individual secular determinants. The $\Xi$ functions in a determinant can differ only in the index $r$, in $S$, $\Gamma_i$ and $M_S$. The secular determinant of the configuration $\gamma_1^a\gamma_2^b\ldots$ is:

$$\|(\Xi_r({}^{2S+1}\Gamma_i|\gamma_1^a\gamma_2^b\ldots|M_S), \boldsymbol{H}_{el.}\Xi_{r'}({}^{2S'+1}\Gamma_{i'}|\gamma_1^a\gamma_2^b\ldots|M'_{S'})) \\ -(\Xi_r({}^{2S+1}\Gamma_i|\gamma_1^a\gamma_2^b\ldots|M_S),\Xi_{r'}({}^{2S'+1}\Gamma_{i'}|\gamma_1^a\gamma_2^b\ldots|M'_{S'}))\,\Delta\tilde{E}_{el.}\|=0.$$

Because $\boldsymbol{H}_{el.}$ does not operate upon the spin variables, those integrals containing two $\Xi$ functions with different $S$ and (or) $M_S$ values are zero. Further, integrals with $\Xi$ functions which transform according to different irreducible representations $\Gamma_i$ of the symmetry group of the ligand field or those which belong to the same irreducible representation but transform according to different columns of this irreducible representation (index $r$)* vanish. If the terms ${}^{2S+1}\Gamma_i$ arise at most once from a configuration $\gamma_1^a\gamma_2^b\ldots$, then the secular determinant is completely diagonal. Also, diagonal elements formed with $\Xi$ functions of the same term must be equal according to theorem 2b, p. 317.

The roots are then

$$\Delta\tilde{E}_{el.}({}^{2S+1}\Gamma_i|\gamma_1^a\gamma_2^b\ldots)=(\Xi_r({}^{2S+1}\Gamma_i|\gamma_1^a\gamma_2^b\ldots|M_S), \boldsymbol{H}_{el.}\Xi_r({}^{2S+1}\Gamma_i|\gamma_1^a\gamma_2^b\ldots|M_S)). \tag{3–61}$$

Because accordingly the perturbation energies are functions of the term ${}^{2S+1}\Gamma_i$, the degeneracy of different terms of the same configuration is in general removed under the influence of the electron interaction: *The states of a configuration split energetically into the terms* ${}^{2S+1}\Gamma_i$.

Because the $\Xi$ functions are sums of antisymmetrized products of (symmetry adapted) one-electron functions $\phi_u^{\pm}$, the calculation rules (cf. p. 249) for the two-electron operators $1/r_{ij}$ can be used for the evaluation of the integrals. The perturbation energies $\Delta E_{el.}$ can then be reduced to sums of integrals of the form

$$\left(\phi_u^{\pm}(1)\phi_{u'}^{\pm}(2), \frac{1}{r_{1,2}}\,\phi_{u''}^{\pm}(1)\phi_{u'''}^{\pm}(2)\right)$$

* $\boldsymbol{H}_{el.}$ belongs to the rotation-reflection group of the sphere, that is to a larger group of the point groups appearing in the ligand field theory. The $\Xi$ are classified according to the irreducible representations of this group. The theorems concerning non-combination and equality of integrals which were derived for the subgroup hold also for integrals with the operator $\boldsymbol{H}_{el.}$.

The $\phi_u$ are themselves linear combinations of the $\psi_{n,2,m_l}$. $\Delta\tilde{E}_{el.}$ can finally be reduced to a sum of integrals of the type

$$\left(\psi^{\pm}_{n,2,m_l}(1)\psi^{\pm}_{n,2,m_l'}(2), \frac{1}{r_{1,2}}\psi^{\pm}_{n,2,m_l''}(1)\psi^{\pm}_{n,2,m_l'''}(2)\right)$$

These integrals and with them $\Delta\tilde{E}_{el.}$ as well may be expressed by the well-known quantities $F_0, F_2, F_4$ introduced by Slater or by the *Racah* parameters $A, B, C$ (cf. chapter 1, pp. 251 ff.).

3.4.2.1. *Example of a $d^2$ ion in a ligand field of $O_h$ symmetry.* The functions $\varphi_1, \ldots \varphi_5$ could be chosen as symmetry adapted one-electron functions according to (3–29). Following convention, however (and also for the sake of variety), we shall work with a basis of *real* one-electron functions:

$$E_g \begin{cases} \phi_1 = \psi_{n,2,0} & = d_{z^2} \quad \sim 3z^2 - r^2 \\ \phi_2 = \frac{1}{\sqrt{2}}(\psi_{n,2,2} + \psi_{n,2,-2}) & = d_{x^2-y^2} \sim x^2 - y^2 \end{cases}$$

$$T_{2g} \begin{cases} \phi_3 = \frac{1}{i\sqrt{2}}(\psi_{n,2,2} - \psi_{n,2,-2}) & = d_{xy} \quad \sim xy \\ \phi_4 = -\frac{1}{\sqrt{2}}(\psi_{n,2,1} - \psi_{n,2,-1}) & = d_{xz} \quad \sim xz \\ \phi_5 = -\frac{1}{i\sqrt{2}}(\psi_{n,2,1} + \psi_{n,2,-1}) & = d_{yz} \quad \sim yz \end{cases}$$

That these real functions are symmetry adapted is readily apparent. Except for the sign and the $i$ factors $\phi_1, \phi_2, \phi_3$ are completely identical with the 'old' functions $\varphi_1, \varphi_2, \varphi_3$. $\phi_4$ and $\phi_5$ arise from a linear combination of the old $T_{2g}$ functions $\varphi_4$ and $\varphi_5$ and are therefore also $T_{2g}$ functions.

Both $d$ electrons can occupy the three configurations $(e_g)^2$, $(e_g)^1(t_{2g})^1$, $(t_{2g})^2$.

*Configuration* $(e_g)^2$. All antisymmetrized products belonging to the configuration $(e_g)^2$ are*:

$$\begin{array}{ll} \Phi(\phi_1^+; \phi_1^-) & M_s = 0 \\ \Phi(\phi_1^+; \phi_2^+) & M_s = 1 \\ \Phi(\phi_1^+; \phi_2^-), \Phi(\phi_1^-; \phi_2^+) & M_s = 0 \\ \Phi(\phi_1^-; \phi_2^-) & M_s = -1 \\ \Phi(\phi_2^+; \phi_2^-) & M_s = 0 \end{array}$$

---

* Only one of the antisymmetrized products which differ from one another in the sign need be considered; cf. p. 229.

Because there is one antisymmetrized product for $M_S = 1$, one for $M_S = -1$ and four antisymmetrized products for $M_S = 0$, it can be concluded that from $(e_g)^2$

three triplet states $(2S+1 = 3)$ and
three singlet states $(2S+1 = 1)$

result.

$\Phi(\phi_1^+; \phi_2^+)$ is certainly a triplet function. We shall call it

$$\chi_{1,2}(S=1, M_S=1) = \Phi(\phi_1^+; \phi_2^+).$$

The remaining triplet functions are found by applying the spin lowering operator $\boldsymbol{S}_-$ (cf. pp. 334 f.):*

$$\chi_{1,2}(1,0) = \frac{1}{\sqrt{2}}[\Phi(\phi_1^+; \phi_2^-) + \Phi(\phi_1^-; \phi_2^+)].$$

Setting the right-hand sides of the equations equal to one another gives

$$\boldsymbol{S}_-\chi_{1,2}(1,0) = [(1+0)(1-0+1)]^{1/2}\chi_{1,2}(1,-1) = \sqrt{2}\chi_{1,2}(1,-1),$$

$$\boldsymbol{S}_-\frac{1}{\sqrt{2}}[\Phi(\phi_1^+; \phi_2^-) + \Phi(\phi_1^-; \phi_2^+)] = \frac{1}{\sqrt{2}}[\Phi(\phi_1^-; \phi_2^-) + \Phi(\phi_1^-; \phi_2^-)]$$
$$= \sqrt{2}\Phi(\phi_1^-; \phi_2^-).$$
$$\chi_{1,2}(1,-1) = \Phi(\phi_1^-; \phi_2^-).$$

Applying $\boldsymbol{S}_-$ once again and setting the right-hand sides equal one obtains

$$\chi_{1,2}(1,1) = \Phi(\phi_1^+; \phi_2^+),$$
$$\chi_{1,2}(1,0) = \frac{1}{\sqrt{2}}[\Phi(\phi_1^+; \phi_2^-) + \Phi(\phi_1^-; \phi_2^+)],$$
$$\chi_{1,2}(1,-1) = \Phi(\phi_1^-; \phi_2^-).$$

There are thus three triplet functions

$$\boldsymbol{S}_-\chi_{1,2}(1,1) = [(1+1)(1-1+1)]^{1/2}\chi_{1,2}(1,0) = \sqrt{2}\chi_{1,2}(1,0),$$
$$\boldsymbol{S}_-\Phi(\phi_1^+; \phi_2^+) = \Phi(\phi_1^+; \phi_2^-) + \Phi(\phi_1^-; \phi_2^+).$$

All other functions which can be constructed from the antisymmetrized products and are orthogonal to the three triplet functions must be singlet

---

* In the future we shall write in the brackets following the $\chi$ symbol only the value of $S$ (first) and $M_S$ (second).

functions:

$$\chi_{1,1}(0,0)=\Phi(\phi_1^+;\phi_1^-),$$

$$\chi_{1,2}(0,0)=\frac{1}{\sqrt{2}}[\Phi(\phi_1^+;\phi_2^-)-\Phi(\phi_1^-;\phi_2^+)],$$

$$\chi_{2,2}(0,0)=\Phi(\phi_2^+;\phi_2^-).$$

After having arranged the functions according to the spin quantum numbers $S$ and $M_S$, linear combinations of the $\chi$ functions are to be formed, such that their space parts transform according to the irreducible representations of the group $O_h$. Which irreducible components are contained in the representation induced by the six $\chi$ functions is determined when the product representation $E_g \times E_g$ corresponding to the configuration $(e_g)^2$ is formed and reduced. According to Table B.19 we find

$$E_g \dot{\times} E_g = A_{1g} \dot{+} A_{2g} \dot{+} E_g .$$

In order to determine the linear combinations of the $\chi$ functions belonging to the irreducible representations the transformation behaviour of the space parts of the $\chi$ functions is studied. The transformation properties of the one-electron spatial parts $\phi$ are summarized in Table B.22. Using this table the transformation behaviour of the $\chi$ is determined without difficulty. It must, however, be kept in mind that*

$$\Phi(a\phi_a+b\phi_b;c\phi_c+d\phi_d)=$$
$$=ac\Phi(\phi_a;\phi_c)+ad\Phi(\phi_a;\phi_d)+bc\Phi(\phi_b;\phi_c)+bd\Phi(\phi_b;\phi_d)$$

holds. For example, the function $\chi_{1,2}(1,1) = \Phi(\phi_1^+;\phi_2^+)$ transforms as follows under the symmetry operations of the group $O_h$†:

$$\boldsymbol{E}\Phi(\phi_1^+;\phi_2^+)=\Phi(\phi_1^+;\phi_2^+),$$

$$\boldsymbol{C}_3\Phi(\phi_1^+;\phi_2^+)=\Phi(-\frac{1}{2}\phi_1^+ +\frac{\sqrt{3}}{2}\phi_2^+;-\frac{\sqrt{3}}{2}\phi_1^+ -\frac{1}{2}\phi_2^+)$$

$$=\frac{\sqrt{3}}{4}\Phi(\phi_1^+;\phi_1^+)+\frac{1}{4}\Phi(\phi_1^+;\phi_2^+)-\frac{3}{4}\Phi(\phi_2^+;\phi_1^+)$$

$$-\frac{\sqrt{3}}{4}\Phi(\phi_2^+;\phi_2^+).$$

Because their columns are equal $\Phi(\phi_1^+;\phi_1^+)$ and $\Phi(\phi_2^+;\phi_2^+)$ vanish identically. Because $\Phi(\phi_2^+;\phi_1^+) = -\Phi(\phi_1^+;\phi_2^+)$ we obtain finally

$$\boldsymbol{C}_3\Phi(\phi_1^+;\phi_2^+)=\Phi(\phi_1^+;\phi_2^+).$$

---

* This relation is easily obtained by expanding the determinants.

† The operations specified at the top of Table B.22 are used.

Table B.22. Transformation behaviour of the real $d$ functions (cf. p. 369) for the symmetry operations of the group $O$. From each class only one operation is considered

| | $E$ | $C_3\begin{pmatrix} x \to y \\ y \to z \\ z \to x \end{pmatrix}$ | $C_2\begin{pmatrix} x \to -x \\ y \to -x \\ z \to z \end{pmatrix}$ | $C_4\begin{pmatrix} x \to y \\ y \to -x \\ z \to z \end{pmatrix}$ | $C_2'\begin{pmatrix} x \to y \\ y \to x \\ z \to -z \end{pmatrix}$ | |
|---|---|---|---|---|---|---|
| $\phi_1$ | $\phi_1$ | $-\frac{1}{2}\phi_1+\frac{\sqrt{3}}{2}\phi_2$ | $\phi_1$ | $\phi_1$ | $\phi_1$ | $E_g$ |
| $\phi_2$ | $\phi_2$ | $-\frac{\sqrt{3}}{2}\phi_1-\frac{1}{2}\phi_2$ | $\phi_2$ | $-\phi_2$ | $-\phi_2$ | |
| $\phi_3$ | $\phi_3$ | $\phi_5$ | $\phi_3$ | $-\phi_3$ | $\phi_3$ | $T_{2g}$ |
| $\phi_4$ | $\phi_4$ | $\phi_3$ | $-\phi_4$ | $\phi_5$ | $-\phi_5$ | |
| $\phi_5$ | $\phi_5$ | $\phi_4$ | $-\phi_5$ | $-\phi_4$ | $-\phi_4$ | |

Table B.23. Transformation behaviour of the $\chi$ functions of the configuration $(e_g)^2$. Operations as in Table B.22

| | $E$ | $C_3$ | $C_2$ | $C_4$ | $C_2'$ | |
|---|---|---|---|---|---|---|
| $\chi_{1,2}(1, M_S)$ $M_S =$ $-1, 0, +1$ | $\chi_{1,2}(1, M_S)$ | $\chi_{1,2}(1, M_S)$ | $\chi_{1,2}(1, M_S)$ | $-\chi_{1,2}(1, M_S)$ | $-\chi_{1,2}(1, M_S)$ | ${}^3A_{2g}$ |
| $\chi_{1,1}(0, 0)$ | $\chi_{1,1}(0, 0)$ | $\sqrt{1/16}\chi_{1,1}(0, 0)$ $-\sqrt{6/16}\chi_{1,2}(0, 0)$ $+\sqrt{9/16}\chi_{2,2}(0, 0)$ | $\chi_{1,1}(0, 0)$ | $\chi_{1,1}(0, 0)$ | $\chi_{1,1}(0, 0)$ | ${}^1A_{1g} \dotplus {}^1E_g$ |
| $\chi_{1,2}(0, 0)$ | $\chi_{1,2}(0, 0)$ | $\sqrt{6/16}\chi_{1,1}(0, 0)$ $-\sqrt{4/16}\chi_{1,2}(0, 0)$ $-\sqrt{6/16}\chi_{2,2}(0, 0)$ | $\chi_{1,2}(0, 0)$ | $-\chi_{1,2}(0, 0)$ | $-\chi_{1,2}(0, 0)$ | |
| $\chi_{2,2}(0, 0)$ | $\chi_{2,2}(0, 0)$ | $\sqrt{9/16}\chi_{1,1}(0, 0)$ $+\sqrt{6/16}\chi_{1,2}(0, 0)$ $+\sqrt{1/16}\chi_{2,2}(0, 0)$ | $\chi_{2,2}(0, 0)$ | $\chi_{2,2}(0, 0)$ | $\chi_{2,2}(0, 0)$ | |

Further the relations

$$C_2\Phi(\phi_1^+;\phi_2^+)=\Phi(\phi_1^+;\phi_2^+),$$
$$C_4\Phi(\phi_1^+;\phi_2^+)=\Phi(\phi_1^+;-\phi_2^+)=-\Phi(\phi_1^+;\phi_2^+),$$
$$C_2'\Phi(\phi_1^+;\phi_2^+)=\Phi(\phi_1^+;-\phi_2^+)=-\Phi(\phi_1^+;\phi_2^+)$$

hold.

The transformation behaviour of the remaining $\chi$ functions can be determined completely analogously. These results are to be found in Table B.23. A comparison with the character table for $O_h$ shows that each of the functions $\chi_{1,2}(1,1)\chi_{1,2}(1,0)$, $\chi_{1,2}(1,-1)$ transforms as the irreducible representation $A_{2g}$ of $O_h$. Because these three functions belong to $S = 1$ and differ only in their $M_S$ values it may be concluded that a triplet term $^3A_{2g}$ is present and that its three states (with $M_S = 1, 0, -1$) can be described by the symmetry adapted functions

$$\Xi(^3A_{2g}|(e_g)^2|M_S=1) \quad =\chi_{1,2}(1,1)$$
$$\Xi(^3A_{2g}|(e_g)^2|M_S=0) \quad =\chi_{1,2}(1,0)$$
$$\Xi(^3A_{2g}|(e_g)^2|M_S=-1)=\chi_{1,2}(1,-1)$$

It was determined above that on the one hand a triplet term $^3A_{2g}$ and three singlet states arise from the configuration $(e_g)^2$. On the other hand the space parts of the six $\chi$ functions induce a representation with the irreducible components $A_{1g}\dot{+}A_{2g}\dot{+}E_g$. The irreducible component $A_{2g}$ is already 'used up' for the triplet term $^3A_{2g}$. The remaining irreducible components $A_{1g}$ and $E_g$ are to be assigned to singlet terms $^1A_{1g}$ and $^1E_g$. Only those functions which have not yet been used $\chi_{1,1}(0,0)$, $\chi_{1,2}(0,0)$, $\chi_{2,2}(0,0)$ may be employed to describe these singlet terms. It must be possible to construct from these functions a linear combination whose space part transforms as $A_{1g}$ and, therefore, is invariant under all symmetry transformations of $O_h$.

A linear combination with undetermined coefficients

$$\Xi(^1A_{1g}|(e_g)^2|M_S=0)=a\chi_{1,1}(0,0)+b\chi_{1,2}(0,0)+c\chi_{2,2}(0,0)$$

is assumed for the function sought and the coefficients $a, b, c$ are determined so that the desired invariance is obtained. If $C_4$ is used we obtain with the aid of Table B.23

$$C_4[a\chi_{1,1}(0,0)+b\chi_{1,2}(0,0)+c\chi_{2,2}(0,0)]$$
$$=a\chi_{1,1}(0,0)-b\chi_{1,2}(0,0)+c\chi_{2,2}(0,0).$$

The result is only equal to the original function if $b = 0$. In this case we have

$$\Xi(^1A_{1g}|(e_g)^2|M_S=0)=a\chi_{1,1}(0,0)+c\chi_{2,2}(0,0).$$

Application of $C_3$ gives

$$\begin{aligned} C_3[a\chi_{1,1}(0,0)+c\chi_{2,2}(0,0)] &= a[\sqrt{1/16}\chi_{1,1}(0,0)-\sqrt{6/16}\chi_{1,2}(0,0)+\sqrt{9/16}\chi_{2,2}(0,0)] \\ &+c[\sqrt{9/16}\chi_{1,1}(0,0)+\sqrt{6/16}\chi_{1,2}(0,0)+\sqrt{1/16}\chi_{2,2}(0,0)] \\ &=[\sqrt{1/16}a+\sqrt{9/16}c]\chi_{1,1}(0,0)+[-\sqrt{6/16}a+\sqrt{6/16}c]\chi_{1,2}(0,0) \\ &+[\sqrt{9/16}a+\sqrt{1/16}c]\chi_{2,2}(0,0). \end{aligned}$$

In order for the expression on the right-hand side to be equal to $a\chi_{1,1}(0,0)+c\chi_{2,2}(0,0)$, it must hold that $a = c$:

$$\Xi(^1A_{1g}|(e_g)^2|M_S=0)=a[\chi_{1,1}(0,0)+\chi_{2,2}(0,0)].$$

The as yet undetermined normalization constant has the value $a = 1/\sqrt{2}$ because $\chi_{1,1}(0,0)$ and $\chi_{2,2}(0,0)$ are orthonormalized:

$$\Xi(^1A_{1g}|(e_g)^2|M_S=0)=\frac{1}{\sqrt{2}}[\chi_{1,1}(0,0)+\chi_{2,2}(0,0)].$$

The two $\Xi$ functions of the $^1E_g$ term remain to be determined. These must be orthogonal to $\Xi(^1A_{1g}|(e_g)^2|M_S = 0$. Without difficulty one finds:

$$\Xi_1(^1E_g|(e_g)^2|M_S=0)=\chi_{1,2}(0,0),$$

$$\Xi_2(^1E_g|(e_g)^2|M_S=0)=\frac{1}{\sqrt{2}}[\chi_{1,1}(0,0)-\chi_{2,2}(0,0)].$$

The perturbation energies which are produced in the system without electron interaction by the ligand field (Step 1) are equal for all terms of the configuration $(e_g)^2$; this follows from (3–60) and (3–31c)

$$\Delta\tilde{E}_{\text{Lig.}}(^3A_{2g}|(e_g)^2)=\Delta\tilde{E}_{\text{Lig.}}(^1A_{1g}|(e_g)^2)=\Delta\tilde{E}_{\text{Lig.}}(^1E_g|(e_g)^2)=2\varepsilon_0+12Dq.$$

The perturbation energy which arises from the electron interaction will be calculated using the example of the $^3A_{2g}$ term. According to (3–61) all three components ($M_S = 1, 0, -1$) of the term experience the same perturbation. If the function $\Xi(^3A_{2g}|(e_g)^2|M_S = 1) = \Phi(\phi_1^+;\phi_2^+)$ is chosen for the calculation of the perturbation energy, then this becomes with consideration of rule (a), p. 249

$$\begin{aligned} \Delta\tilde{E}_{\text{el.}}(^3A_{2g}|(e_g)^2) &= (\Phi(\phi_1^+;\phi_2^+), 1/r_{12}\Phi(\phi_1^+;\phi_2^+)) \\ &= \iint \phi_1^*(1)\,\phi_2^*(2)\,1/r_{12}\phi_1(1)\,\phi_2(2)\,d\tau_1\,d\tau_2 \\ &-\iint \phi_1^*(1)\,\phi_2^*(2)\,1/r_{12}\,\phi_2(1)\,\phi_1(2)\,d\tau_1\,d\tau_2. \end{aligned}$$

With

$$\phi_1 = \psi_{n,2,0}$$
$$\phi_2 = \frac{1}{\sqrt{2}}(\psi_{n,2,2} + \psi_{n,2,-2})$$

and using the nomenclature introduced earlier, p. 249, one obtains

$$\begin{aligned}\Delta\tilde{E}_{\text{el.}}({}^3A_{2g}|(e_g)^2) = \frac{1}{2}\{&V(0;2;0;2) + V(0;2;0;-2)\\ &+ V(0;-2;0;2) + V(0;-2;0;-2)\\ &- V(0;2;2;0) - V(0;2;-2;0)\\ &- V(0;-2;2;0) - V(0;-2;-2;0)\}.\end{aligned}$$

The $V$ integrals may be reduced to integrals $F^k$ or $F_k$ or $A, B, C$ using equations (1–91, 92, 93, 95). Following the procedure explicitly explained in chapter 1 one obtains

$$\Delta\tilde{E}_{\text{el.}}({}^3A_{2g}|(e_g)^2) = A - 8B.$$

Likewise the perturbation energies

$$\Delta\tilde{E}_{\text{el.}}({}^1A_{1g}|(e_g)^2) = A + 8B + 4C,$$
$$\Delta\tilde{E}_{\text{el.}}({}^1E_g|(e_g)^2) = A + 2C.$$

arc found without difficulty.

In Table B.24 the values of all nonzero integrals

$$\tilde{V}(i;i';i'';i''') = \iint \phi_i^*(1)\phi_{i'}^*(2)\, 1/r_{12}\, \phi_{i''}(1)\phi_{i'''}(2)\, d\tau_1 d\tau_2$$

are tabulated for future reference, whereby the $\phi$ are the real $d$ functions defined on p. 369.

*Configuration* $(t_{2g})^2$. The possible antisymmetrized products are

$$\begin{array}{ll}\Phi(\phi_3^+;\phi_4^+),\ \Phi(\phi_3^+;\phi_5^+),\ \Phi(\phi_4^+;\phi_5^+) & M_S = 1\\ \left.\begin{array}{l}\Phi(\phi_3^+;\phi_3^-),\ \Phi(\phi_4^+;\phi_4^-),\ \Phi(\phi_5^+;\phi_5^-)\\ \Phi(\phi_3^+;\phi_4^-),\ \Phi(\phi_3^+;\phi_5^-),\ \Phi(\phi_4^+;\phi_5^-)\\ \Phi(\phi_3^-;\phi_4^+),\ \Phi(\phi_3^-;\phi_5^+),\ \Phi(\phi_4^-;\phi_5^+)\end{array}\right\} & M_S = 0\\ \Phi(\phi_3^-;\phi_4^-),\ \Phi(\phi_3^-;\phi_5^-),\ \Phi(\phi_4^-;\phi_5^-) & M_S = -1\end{array}$$

Because there are three antisymmetrized products for $M_S = 1$, nine for $M_S = 0$ and three for $M_S = -1$, it must hold that there are

*nine* triplet functions and
*six* singlet functions.

The three antisymmetrized products with $M_S = 1$ are assuredly triplet functions. With the aid of the lowering operator $\boldsymbol{S}_-$ the remaining triplet functions are obtained, in total:

$$\chi_{3,4}(1,1) \quad = \Phi(\phi_3^+;\phi_4^+)$$
$$\chi_{3,4}(1,0) \quad = \frac{1}{\sqrt{2}}[\Phi(\phi_3^+;\phi_4^-)+\Phi(\phi_3^-;\phi_4^+)]$$
$$\chi_{3,4}(1,-1) = \Phi(\phi_3^-;\phi_4^-)$$

$$\chi_{3,5}(1,1) \quad = \Phi(\phi_3^+;\phi_5^+)$$
$$\chi_{3,5}(1,0) \quad = \frac{1}{\sqrt{2}}[\Phi(\phi_3^+;\phi_5^-)+\Phi(\phi_3^-;\phi_5^+)]$$
$$\chi_{3,5}(1,-1) = \Phi(\phi_3^-;\phi_5^-)$$

$$\chi_{4,5}(1,1) \quad = \Phi(\phi_4^+;\phi_5^+)$$
$$\chi_{4,5}(1,0) \quad = \frac{1}{\sqrt{2}}[\Phi(\phi_4^+;\phi_5^-)+\Phi(\phi_4^-;\phi_5^+)]$$
$$\chi_{4,5}(1,-1) = \Phi(\phi_4^-;\phi_5^-)$$

Singlet functions are

$$\chi_{3,3}(0,0) \quad = \Phi(\phi_3^+;\phi_3^-)$$
$$\chi_{4,4}(0,0) \quad = \Phi(\phi_4^+;\phi_4^-)$$
$$\chi_{5,5}(0,0) \quad = \Phi(\phi_5^+;\phi_5^-)$$

and the three linear combinations (orthogonal to the triplet functions) are

$$\chi_{3,4}(0,0) \quad = \frac{1}{\sqrt{2}}[\Phi(\phi_3^+;\phi_4^-)-\Phi(\phi_3^-;\phi_4^+)]$$
$$\chi_{3,5}(0,0) \quad = \frac{1}{\sqrt{2}}[\Phi(\phi_3^+;\phi_5^-)-\Phi(\phi_3^-;\phi_5^+)]$$
$$\chi_{4,5}(0,0) \quad = \frac{1}{\sqrt{2}}[\Phi(\phi_4^+;\phi_5^-)-\Phi(\phi_4^-;\phi_5^+)]$$

The space parts of the $\chi$ functions induce a representation of $O_h$ containing the following irreducible components (cf. Table B.19):

$$T_{2g} \dot{\times} T_{2g} = A_{1g} \dot{+} E_g \dot{+} T_{1g} \dot{+} T_{2g}.$$

We shall first study the transformation behaviour of the space parts of the $\chi$ functions under the symmetry operations of $O_h$. With the aid of Table B.22 the results given in Table B.25 are obtained. It is shown that in each case three $\chi$ functions transform (into one another). The three function triads $\chi_{3,4}(1, M_S)$, $\chi_{3,5}(1, M_S)$, $\chi_{4,5}(1, M_S)$ with $M_S = 1, 0, -1$ transform in the same manner. As a comparison of the characters with the character table of $O_h$ shows, in each case a triad induces the irreducible representation

Table B.24. Values of the nonzero electron interaction integrals expressed in terms of the *Slater* parameters $F_0, F_2, F_4$ or the *Racah* parameters $A$, $B$, $C$

| | $a$ | $b$ | $c$ | $d$ | $\int\int a^*(1)\cdot b^*(2)\,1/r_{1,2}\,c(1)\cdot d(2)\cdot d\tau_1\cdot d\tau_2$ | |
|---|---|---|---|---|---|---|
| Coulomb integrals | $\phi_1$ | $\phi_1$ | $\phi_1$ | $\phi_1$ | $F_0+4F_2+36F_4$ | $A+4B+3C$ |
| | $\phi_2$ | $\phi_2$ | $\phi_2$ | $\phi_2$ | | |
| | $\phi_3$ | $\phi_3$ | $\phi_3$ | $\phi_3$ | | |
| | $\phi_4$ | $\phi_4$ | $\phi_4$ | $\phi_4$ | | |
| | $\phi_5$ | $\phi_5$ | $\phi_5$ | $\phi_5$ | | |
| | $\phi_1$ | $\phi_2$ | $\phi_1$ | $\phi_2$ | $F_0-4F_2+6F_4$ | $A-4B+C$ |
| | $\phi_1$ | $\phi_3$ | $\phi_1$ | $\phi_3$ | | |
| | $\phi_1$ | $\phi_4$ | $\phi_1$ | $\phi_4$ | $F_0+2F_2-24F_4$ | $1+2B+C$ |
| | $\phi_1$ | $\phi_5$ | $\phi_1$ | $\phi_5$ | | |
| | $\phi_2$ | $\phi_3$ | $\phi_2$ | $\phi_3$ | $F_0+4F_2-34F_4$ | $A+4B+C$ |
| | $\phi_2$ | $\phi_4$ | $\phi_2$ | $\phi_4$ | $F_0-2F_2-4F_4$ | $A-2B+C$ |
| | $\phi_2$ | $\phi_5$ | $\phi_2$ | $\phi_5$ | | |
| | $\phi_3$ | $\phi_4$ | $\phi_3$ | $\phi_4$ | | |
| | $\phi_3$ | $\phi_5$ | $\phi_3$ | $\phi_5$ | | |
| | $\phi_4$ | $\phi_5$ | $\phi_4$ | $\phi_5$ | | |
| exchange integrals | $\phi_1$ | $\phi_2$ | $\phi_2$ | $\phi_1$ | $4F_2+15F_4$ | $4B+C$ |
| | $\phi_1$ | $\phi_3$ | $\phi_3$ | $\phi_1$ | | |
| | $\phi_1$ | $\phi_4$ | $\phi_4$ | $\phi_1$ | $F_2+30F_4$ | $B+C$ |
| | $\phi_1$ | $\phi_5$ | $\phi_5$ | $\phi_1$ | | |
| | $\phi_2$ | $\phi_3$ | $\phi_3$ | $\phi_2$ | $35F_4$ | $C$ |
| | $\phi_2$ | $\phi_4$ | $\phi_4$ | $\phi_2$ | $3F_2+20F_4$ | $3B+C$ |
| | $\phi_2$ | $\phi_5$ | $\phi_5$ | $\phi_2$ | | |
| | $\phi_3$ | $\phi_4$ | $\phi_4$ | $\phi_3$ | | |
| | $\phi_3$ | $\phi_5$ | $\phi_5$ | $\phi_3$ | | |
| | $\phi_4$ | $\phi_5$ | $\phi_5$ | $\phi_4$ | | |
| | $\phi_1$ | $\phi_3$ | $\phi_4$ | $\phi_5$ | $\sqrt{3}F_2-5\sqrt{3}F_4$ | $\sqrt{3}B$ |
| | $\phi_1$ | $\phi_3$ | $\phi_5$ | $\phi_4$ | | |
| | $\phi_1$ | $\phi_2$ | $\phi_4$ | $\phi_4$ | | |
| | $\phi_1$ | $\phi_2$ | $\phi_5$ | $\phi_5$ | $-\sqrt{3}F_2+5\sqrt{3}F_4$ | $-\sqrt{3}B$ |
| | $\phi_1$ | $\phi_4$ | $\phi_3$ | $\phi_5$ | $-2\sqrt{3}F_2+10\sqrt{3}F_4$ | $-2\sqrt{3}B$ |
| | $\phi_1$ | $\phi_4$ | $\phi_2$ | $\phi_4$ | | |
| | $\phi_1$ | $\phi_5$ | $\phi_2$ | $\phi_5$ | $2\sqrt{3}F_2-10\sqrt{3}F_4$ | $2\sqrt{3}B$ |
| | $\phi_2$ | $\phi_3$ | $\phi_4$ | $\phi_5$ | $3F_2-15F_4$ | $3B$ |
| | $\phi_2$ | $\phi_3$ | $\phi_5$ | $\phi_4$ | $-3F_2+15F_4$ | $-3B$ |

$T_{1g}$. The nine functions mentioned describe a $^3T_{1g}$ term. Further $\chi_{3,4}(0, 0)$, $\chi_{3,5}(0, 0)$, $\chi_{4,5}(0, 0)$ prove to be functions of a $^1T_{2g}$ term. The three singlet functions $\chi_{3,3}(0, 0)$, $\chi_{4,4}(0, 0)$, $\chi_{5,5}(0, 0)$ induce a reducible representation, as the associated character system shows. The character system $\{3, 0, 3, 1, 1\}$ is (cf. character table $O_h$) the sums of the character systems of the irreducible representations $A_{1g}$ and $E_g$. This is in agreement with the conclusion drawn above that the totality of the $\chi$ functions describes in each case a term of the irreducible representation $A_{1g}$, $E_g$, $T_{1g}$, $T_{2g}$. While all $\chi$ functions with the exception of $\chi_{3,3}(0, 0)$, $\chi_{4,4}(0, 0)$, $\chi_{5,5}(0, 0)$ are already symmetry adapted (only $\chi$ functions of the same term transform into one another) another linear combination is to be constructed from the functions $\chi_{3,3}(0, 0)$, $\chi_{4,4}(0, 0)$, $\chi_{5,5}(0, 0)$ whose space part transforms as $A_{1g}$, and two linear combinations whose space parts transform as $E_g$.

We set the $^1A_{1g}$ function equal to:

$$\Xi(^1A_{1g}|(t_{2g})^2|M_S=0)=a\chi_{3,3}(0,0)+b\chi_{4,4}(0,0)+c\chi_{5,5}(0,0).$$

In agreement with the transformation behaviour of a $A_{1g}$ function invariance with respect to all symmetry operations of $O_h$ is to be required, e.g.

$$\begin{aligned} C_3[a\chi_{3,3}(0,0)+b\chi_{4,4}(0,0)+c\chi_{5,5}(0,0)] &= b\chi_{3,3}(0,0)+c\chi_{4,4}(0,0)+a\chi_{5,5}(0,0) \\ &= a\chi_{3,3}(0,0)+b\chi_{4,4}(0,0)+c\chi_{5,5}(0,0) \end{aligned}$$

It follows that

$$a=b,\ b=c,\ c=a, \quad \text{i.e.} \quad a=b=c.$$

and further

$$\Xi(^1A_{1g}|(t_{2g})^2|M_S=0)=a[\chi_{3,3}(0,0)+\chi_{4,4}(0,0)+\chi_{5,5}(0,0)].$$

$a$ is a normalization constant which must have the value $1/\sqrt{3}$ because the $\chi$ functions are orthonormalized:

$$\Xi(^1A_{1g}|(t_{2g})^2|M_S=0)=\frac{1}{\sqrt{3}}[\chi_{3,3}(0,0)+\chi_{4,4}(0,0)+\chi_{5,5}(0,0)].$$

The $^1E_g$ functions must be orthogonal to this $^1A_{1g}$ function. We set:

$$\Xi(^1E_g|(t_{2g})^2|M_S=0)=\alpha\chi_{3,3}(0,0)+\beta\chi_{4,4}(0,0)+\gamma\chi_{5,5}(0,0).$$

Orthogonality to $\Xi(^1A_{1g}|(t_{2g})^2|M_S = 0)$ is apparent when

$$\alpha+\beta+\gamma=0$$

holds.

Table B.25. Transformation behaviour of $\chi$ functions of the configuration $(t_{2g})^2$. Operations as in Table B.22

| | $E$ | $C_3$ | $C_2$ | $C_4$ | $C_2'$ | |
|---|---|---|---|---|---|---|
| $\chi_{3,4}(1, M_S)$ $M_S = -1, 0, +1$ | $\chi_{3,4}(1, M_S)$ | $-\chi_{3,5}(1, M_S)$ | $-\chi_{3,4}(1, M_S)$ | $-\chi_{3,5}(1, M_S)$ | $-\chi_{3,5}(1, M_S)$ | |
| $\chi_{3,5}(1, M_S)$ $M_S = -1, 0, +1$ | $\chi_{3,5}(1, M_S)$ | $-\chi_{4,5}(1, M_S)$ | $-\chi_{3,5}(1, M_S)$ | $\chi_{3,4}(1, M_S)$ | $-\chi_{3,4}(1, M_S)$ | $^3T_{1g}$ |
| $\chi_{4,5}(1, M_S)$ $M_S = -1, 0, +1$ | $\chi_{3,5}(1, M_S)$ | $\chi_{3,4}(1, M_S)$ | $\chi_{4,5}(1, M_S)$ | $\chi_{4,5}(1, M_S)$ | $-\chi_{4,5}(1, M_S)$ | |
| Character | 3 | 0 | $-1$ | 1 | $-1$ | |
| $\chi_{3,4}(0, 0)$ | $\chi_{3,4}(0, 0)$ | $\chi_{3,5}(0, 0)$ | $-\chi_{3,4}(0, 0)$ | $-\chi_{3,5}(0, 0)$ | $-\chi_{3,5}(0, 0)$ | |
| $\chi_{3,5}(0, 0)$ | $\chi_{3,5}(0, 0)$ | $\chi_{4,5}(0, 0)$ | $-\chi_{3,5}(0, 0)$ | $\chi_{3,4}(0, 0)$ | $-\chi_{3,4}(0, 0)$ | $^1T_{2g}$ |
| $\chi_{4,5}(0, 0)$ | $\chi_{4,5}(0, 0)$ | $\chi_{3,4}(0, 0)$ | $\chi_{4,5}(0, 0)$ | $-\chi_{4,5}(0, 0)$ | $\chi_{4,5}(0, 0)$ | |
| Character | 3 | 0 | $-1$ | $-1$ | 1 | |
| $\chi_{3,3}(0, 0)$ | $\chi_{3,3}(0, 0)$ | $\chi_{5,5}(0, 0)$ | $\chi_{3,3}(0, 0)$ | $\chi_{3,3}(0, 0)$ | $\chi_{3,3}(0, 0)$ | |
| $\chi_{4,4}(0, 0)$ | $\chi_{4,4}(0, 0)$ | $\chi_{3,3}(0, 0)$ | $\chi_{4,4}(0, 0)$ | $\chi_{5,5}(0, 0)$ | $\chi_{5,5}(0, 0)$ | $^1A_{1g} \dot{+} {^1E_g}$ |
| $\chi_{5,5}(0, 0)$ | $\chi_{5,5}(0, 0)$ | $\chi_{4,4}(0, 0)$ | $\chi_{5,5}(0, 0)$ | $\chi_{4,4}(0, 0)$ | $\chi_{4,4}(0, 0)$ | |
| Character | 3 | 0 | 3 | 1 | 1 | |

One of the $^1E_g$ functions is, for example, obtained, when $\alpha = 0$ and $\beta = -\gamma$ are assumed and normalization is carried out.

$$\Xi_1(^1E_g|(t_{2g})^2|M_S=0) = \frac{1}{\sqrt{2}}[\chi_{4,4}(0,0) - \chi_{5,5}(0,0)].$$

While there was still some freedom in the choice of this $\Xi$ function, the other $^1E_g$ function (except for the normalization constant) is here uniquely determined because it must be orthogonal to $\Xi(^1A_{1g}|(t_{2g})^2|M_S = 0)$ as well as to $\Xi(^1E_g|(t_{2g})^2|M_S = 0)$. The assumption

$$\Xi_2(^1E_g|(t_{2g})^2|M_S=0) = x\chi_{3,3}(0,0) + y\chi_{4,4}(0,0) + z\chi_{5,5}(0,0)$$

leads, together with the two orthogonality conditions mentioned, to the relations

$$x+y+z=0,$$
$$y-z=0.$$

From this it follows

$$y=z, \quad x=-2z,$$

i.e.,

$$\Xi_2(^1E_g|(t_{2g})^2|M_S=0) = z[-2\chi_{3,3}(0,0) + \chi_{4,4}(0,0) + \chi_{5,5}(0,0)]$$

or when normalized:

$$\Xi_2(^1E_g|(t_{2g})^2|M_S=0) = \frac{1}{\sqrt{6}}[2\chi_{3,3}(0,0) - \chi_{4,4}(0,0) - \chi_{5,5}(0,0)].$$

With this all 15 symmetry adapted $\Xi$ functions of the configuration $(t_{2g})^2$ are known.

In the first step of the perturbation calculation the energy perturbation produced by the ligand field is determined. The electron interaction is at first neglected. According to (3–60) the secular determinant formed using the 15 $\Xi$ basis functions is completely diagonal. The diagonal elements and thereby the perturbation energies are all equal:

$$\Delta\tilde{E}_{\text{Lig.}}(^3T_{1g}|(t_{2g})^2) = \Delta\tilde{E}_{\text{Lig.}}(^1T_{2g}|(t_{2g})^2) = \Delta\tilde{E}_{\text{Lig.}}(^1E_g|(t_{2g})^2)$$
$$= \Delta\tilde{E}_{\text{Lig.}}(^1A_{1g}|(t_{2g})^2) = 2(\varepsilon_0 - 4Dq) = 2\varepsilon_0 - 8Dq.$$

This complete degeneracy of all terms is in general removed when in step 2 of the perturbation calculation the electron interaction is considered. One finds then using the procedure illustrated for the case of the $(e_g)^2$ configuration the perturbation energies

$$\Delta\tilde{E}_{\text{el.}}(^3T_{1g}|(t_{2g})^2) = A - 5B,$$

$$\Delta\tilde{E}_{\text{el.}}\,({}^1T_{2g}|(t_{2g})^2) = A+B+2C\,,$$
$$\Delta\tilde{E}_{\text{el.}}\,({}^1E_{g}|(t_{2g})^2) = A+B+2C\,,$$
$$\Delta\tilde{E}_{\text{el.}}\,({}^1A_{1g}|(t_{2g})^2) = A+10B+5C\,.$$

The terms ${}^1T_{2g}$ and ${}^1E_g$ are (accidentally) degenerate for the strong field case.

*Configuration* $(e_g)^1(t_{2g})^1$. The following antisymmetrized products may then be written:

$$\left.\begin{matrix}\Phi(\phi_1^+;\phi_3^+),\ \Phi(\phi_1^+;\phi_4^+),\ \Phi(\phi_1^+;\phi_5^+)\\ \Phi(\phi_2^+;\phi_3^+),\ \Phi(\phi_2^+;\phi_4^+),\ \Phi(\phi_2^+;\phi_5^+)\end{matrix}\right\}\quad M_S=1$$

$$\left.\begin{matrix}\Phi(\phi_1^+;\phi_3^-),\ \Phi(\phi_1^+;\phi_4^-),\ \Phi(\phi_1^+;\phi_5^-)\\ \Phi(\phi_1^-;\phi_3^+),\ \Phi(\phi_1^-;\phi_4^+),\ \Phi(\phi_1^-;\phi_5^+)\\ \Phi(\phi_2^+;\phi_3^-),\ \Phi(\phi_2^+;\phi_4^-),\ \Phi(\phi_2^+;\phi_5^-)\\ \Phi(\phi_2^-;\phi_3^+),\ \Phi(\phi_2^-;\phi_4^+),\ \Phi(\phi_2^-;\phi_5^+)\end{matrix}\right\}\quad M_S=0$$

$$\left.\begin{matrix}\Phi(\phi_1^-;\phi_3^-),\ \Phi(\phi_1^-;\phi_4^-),\ \Phi(\phi_1^-;\phi_5^-)\\ \Phi(\phi_2^-;\phi_3^-).\ \Phi(\phi_2^-;\phi_4^-),\ \Phi(\phi_2^-;\phi_5^-)\end{matrix}\right\}\quad M_S=-1$$

From the fact that there are 6 antisymmetrized products with $M_S = 1$, 12 with $M_S = 0$ and 6 with $M_S = -1$ the existence of

18 triplet functions and
6 singlet functions

can be concluded. Functions of the type $\Phi(\phi_i^+;\phi_j^+)$, $i = 1, 2, j = 3, 4, 5$ are certainly triplet functions. With the aid of the lowering operator $\boldsymbol{S}_-$ the remaining triplet functions may be derived from these. One finds in this manner the following 18 triplet functions $\chi_{i,j}(1, M_S)$:

$$\begin{aligned}\chi_{i,j}(1,1) &= \Phi(\phi_i^+;\phi_j^+)\,,\\ \chi_{i,j}(1,0) &= \frac{1}{\sqrt{2}}[\Phi(\phi_i^+;\phi_j^-)+\Phi(\phi_i^-;\phi_j^+)]\,,\\ \chi_{i,j}(1,-1) &= \Phi(\phi_i^-;\phi_j^-)\,.\end{aligned}\qquad \begin{matrix}i=1,2\\ j=3,4,5\end{matrix}$$

The other six functions which can be constructed orthogonal to the triplet functions from the antisymmetrized products must be singlet functions:

$$\chi_{i,j}(0,0) = \frac{1}{\sqrt{2}}[\Phi(\phi_i^+;\phi_j^-)-\Phi(\phi_i^-;\phi_j^+)]\,.\quad i=1,2\,,\quad j=3,4,5$$

The space parts of the 24 $\chi$ functions induce a representation of the group $O_h$. The irreducible components are found when corresponding to the configuration under consideration $(e_g)^1(t_{2g})^1$ the product representation

$E_g \dot{\times} T_{2g}$ is formed and reduced. According to Table B.19 this is

$$E_g \dot{\times} T_{2g} = T_{1g} \dot{+} T_{2g}.$$

The transformation behaviour of the $\chi$ functions is—with consideration of Table B.22—summarized in Table B.26. Accordingly each of the four sets of functions induces

$$\chi_{1,3}(S,M_S),\ \chi_{1,4}(S,M_S),\ \chi_{1,5}(S,M_S),\ \chi_{2,3}(S,M_S),\ \chi_{2,4}(S,M_S),\ \chi_{2,5}(S,M_S)$$
$$S=1,\ M_S=1,0,-1$$
$$S=0,\ M_S=0$$

the same reducible representation of $O_h$ with the character system $\{6, 0, -2, 0, 0\}$. A comparison with the character table of the irreducible representation of $O_h$ shows that the irreducible components $T_{1g}$ and $T_{2g}$ are contained in the reducible representation. The $\chi$ functions of the three sets with $S = 1$ are suited for the description of two triplet terms, one $^3T_{1g}$ and one $^3T_{2g}$ term. The set of functions with $S = 0$ should be associated with the two singlet terms $^1T_{1g}$ and $^1T_{2g}$.

What is the form of the symmetry adapted linear combinations of the $\chi$ functions? Because the four sets of functions mentioned all transform in the same manner under the symmetry operations of $O_h$, the question may be answered quite generally by searching for the symmetry adapted linear combinations of a representative set of functions

$$\chi_{i,j}(S,M_S),\quad i=1,2,\quad j=3,4,5$$

and then introducing the special values of $S$ and $M_S$. To determine the symmetry adapted linear combinations we choose (for the sake of variety) the method of 'trial and error'. We make the trial assumption that $\chi_{1,3}(S, M_S)$ is already a symmetry adapted $\Xi$ function. As far as this holds one can write for it

$$\Xi_1(^{2S+1}T_x|\,(e_g)^1(t_{2g})^1|M_S) = \chi_{1,3}(S,M_S).$$

It is only unknown if the function belongs to $T_{1g}$ or to $T_{2g}$. For this reason the index on the representation symbol is left open. From Table B.26 it can be seen that $\chi_{1,3}(S, M_S)$ transforms into itself under the symmetry operations $E$, $C_2$, $C_2'$ and $C_4$ (except for the sign). $C_3$ generates from $\chi_{1,3}(S, M_S)$ the functions $-\sqrt{1/4}\chi_{1,5}(S, M_S)+\sqrt{3/4}\chi_{2,5}(S, M_S)$. We call them

$$\Xi_2(^{2S+1}T_x|\,(e_g)^1(t_{2g})^1|M_S) = -\sqrt{1/4}\,\chi_{1,5}(S,M_S) + \sqrt{3/4}\,\chi_{2,5}(S,M_S)$$

and look for the transformation behaviour of $\Xi_2$. With the aid of Table

Table B.26. Transformation behaviour of the $\chi$ functions of the configuration $(e_g)^1(t_{2g})^1$. Operations as in Table B.22

| | $E$ | $C_3$ | $C_2$ | $C_4$ | $C_2'$ | |
|---|---|---|---|---|---|---|
| $\chi_{1,3}(1, M_S)$ | $\chi_{1,3}(1, M_S)$ | $-\sqrt{1/4}\chi_{1,5}(1, M_S)+\sqrt{3/4}\chi_{2,5}(1, M_S)$ | $\chi_{1,3}(1, M_S)$ | $-\chi_{1,3}(1, M_S)$ | $\chi_{1,3}(1, M_S)$ | $^3T_{1g} \dot{+} {}^3T_{2g}$ |
| $\chi_{1,4}(1, M_S)$ | $\chi_{1,4}(1, M_S)$ | $-\sqrt{1/4}\chi_{1,3}(1, M_S)+\sqrt{3/4}\chi_{2,3}(1, M_S)$ | $-\chi_{1,4}(1, M_S)$ | $\chi_{1,5}(1, M_S)$ | $-\chi_{1,5}(1, M_S)$ | |
| $\chi_{1,5}(1, M_S)$ | $\chi_{1,5}(1, M_S)$ | $-\sqrt{1/4}\chi_{1,4}(1, M_S)+\sqrt{3/4}\chi_{2,4}(1, M_S)$ | $-\chi_{1,5}(1, M_S)$ | $-\chi_{1,4}(1, M_S)$ | $-\chi_{1,4}(1, M_S)$ | |
| $\chi_{2,3}(1, M_S)$ | $\chi_{2,3}(1, M_S)$ | $-\sqrt{3/4}\chi_{1,5}(1, M_S)-\sqrt{1/4}\chi_{2,5}(1, M_S)$ | $\chi_{2,3}(1, M_S)$ | $\chi_{2,3}(1, M_S)$ | $-\chi_{2,3}(1, M_S)$ | |
| $\chi_{2,4}(1, M_S)$ | $\chi_{2,4}(1, M_S)$ | $-\sqrt{3/4}\chi_{1,3}(1, M_S)-\sqrt{1/4}\chi_{2,3}(1, M_S)$ | $-\chi_{2,4}(1, M_S)$ | $-\chi_{2,5}(1, M_S)$ | $\chi_{2,5}(1, M_S)$ | |
| $\chi_{2,5}(1, M_S)$ | $\chi_{2,5}(1, M_S)$ | $-\sqrt{3/4}\chi_{1,4}(1, M_S)-\sqrt{1/4}\chi_{2,4}(1, M_S)$ | $-\chi_{2,5}(1, M_S)$ | $\chi_{2,4}(1, M_S)$ | $\chi_{2,4}(1, M_S)$ | |
| Character | 6 | 0 | $-2$ | 0 | 0 | |
| $\chi_{1,3}(0, 0)$ | $\chi_{1,3}(0, 0)$ | $-\sqrt{1/4}\chi_{1,5}(0, 0)+\sqrt{3/4}\chi_{2,5}(0, 0)$ | $\chi_{1,3}(0, 0)$ | $-\chi_{1,3}(0, 0)$ | $\chi_{1,3}(0, 0)$ | $^1T_{1g} \dot{+} {}^1T_{2g}$ |
| $\chi_{1,4}(0, 0)$ | $\chi_{1,4}(0, 0)$ | $-\sqrt{1/4}\chi_{1,3}(0, 0)+\sqrt{3/4}\chi_{2,3}(0, 0)$ | $-\chi_{1,4}(0, 0)$ | $\chi_{1,5}(0, 0)$ | $-\chi_{1,5}(0, 0)$ | |
| $\chi_{1,5}(0, 0)$ | $\chi_{1,5}(0, 0)$ | $-\sqrt{1/4}\chi_{1,4}(0, 0)+\sqrt{3/4}\chi_{2,4}(0, 0)$ | $-\chi_{1,5}(0, 0)$ | $-\chi_{1,4}(0, 0)$ | $-\chi_{1,4}(0, 0)$ | |
| $\chi_{2,3}(0, 0)$ | $\chi_{2,3}(0, 0)$ | $-\sqrt{3/4}\chi_{1,5}(0, 0)-\sqrt{1/4}\chi_{2,5}(0, 0)$ | $\chi_{2,3}(0, 0)$ | $\chi_{2,3}(0, 0)$ | $-\chi_{2,3}(0, 0)$ | |
| $\chi_{2,4}(0, 0)$ | $\chi_{2,4}(0, 0)$ | $-\sqrt{3/4}\chi_{1,3}(0, 0)-\sqrt{1/4}\chi_{2,3}(0, 0)$ | $-\chi_{2,4}(0, 0)$ | $-\chi_{2,5}(0, 0)$ | $\chi_{2,5}(0, 0)$ | |
| $\chi_{2,5}(0, 0)$ | $\chi_{2,5}(0, 0)$ | $-\sqrt{3/4}\chi_{1,4}(0, 0)-\sqrt{1/4}\chi_{2,4}(0, 0)$ | $-\chi_{2,5}(0, 0)$ | $\chi_{2,4}(0, 0)$ | $\chi_{2,4}(0, 0)$ | |
| Character | 6 | 0 | $-2$ | 0 | 0 | |

$M_S = -1, 0, +1$

B.26 we obtain

$$\begin{aligned}
E\Xi_2 &= \Xi_2,\\
C_3\Xi_2 &= -\sqrt{1/4}\chi_{1,4}-\sqrt{3/4}\chi_{2,4},\\
C_2\Xi_2 &= +\sqrt{1/4}\chi_{1,5}-\sqrt{3/4}\chi_{2,5} = -\Xi_2.\\
C_4\Xi_2 &= \sqrt{1/4}\chi_{1,4}+\sqrt{3/4}\chi_{2,4},\\
C_2'\Xi_2 &= \sqrt{1/4}\chi_{1,4}+\sqrt{3/4}\chi_{2,4}.
\end{aligned}$$

One obtains as a result either $\pm\Xi_2$ or $\pm[-\sqrt{1/4}\chi_{1,4}-\sqrt{3/4}\chi_{2,4}]$. The new functions is denoted

$$\Xi_3({}^{2S+1}T_x|(e_g)^1(t_{2g})^1|M_S) = -\sqrt{1/4}\chi_{1,4}(S, M_S)-\sqrt{3/4}\chi_{2,4}(S, M_S).$$

If the symmetry operations are applied to $\Xi_3$ we find

$$\begin{aligned}
E\Xi_3 &= \Xi_3,\\
C_3\Xi_3 &= \chi_{1,3} = \Xi_1,\\
C_2\Xi_3 &= \sqrt{1/4}\chi_{1,4}+\sqrt{3/4}\chi_{2,4} = -\Xi_3,\\
C_4\Xi_3 &= -\sqrt{1/4}\chi_{1,5}+\sqrt{3/4}\chi_{2,5} = \Xi_2,\\
C_2'\Xi_3 &= \sqrt{1/4}\chi_{1,5}-\sqrt{3/4}\chi_{2,5} = -\Xi_2.
\end{aligned}$$

The transformation behaviour of the functions $\Xi_1({}^{2S+1}T_x|(e_g)^1(t_{2g})^1|M_S)$, $\Xi_2({}^{2S+1}T_x|(e_g)^1(t_{2g})^1|M_S)$, $\Xi_3({}^{2S+1}T_x|(e_g)^1(t_{2g})^1|M_S)$ is summarized in Table B.27. One sees that these three functions transform only into one another and—as a comparison with the character table of $O_h$ shows—according to the irreducible representation $T_{2g}$. The ${}^3T_{2g}$ term is described by the nine symmetry adapted functions

$$\begin{aligned}
&\Xi_1({}^3T_{2g}|(e_g)^1(t_{2g})^1|M_S) = \chi_{1,3}(1, M_S),\\
&\Xi_2({}^3T_{2g}|(e_g)^1(t_{2g})^1|M_S) = -\sqrt{1/4}\,\chi_{1,5}(1, M_S) + \sqrt{3/4}\,\chi_{2,5}(1, M_S),\\
&\Xi_3({}^3T_{2g}|(e_g)^1(t_{2g})^1|M_S) = -\sqrt{1/4}\,\chi_{1,4}(1, M_S) - \sqrt{3/4}\,\chi_{2,4}(1, M_S),\\
&\qquad M_S = 1, 0, -1.
\end{aligned}$$

The three functions

$$\begin{aligned}
&\Xi_1({}^1T_{2g}|(e_g)^1(t_{2g})^1|0) = \chi_{1,3}(0,0),\\
&\Xi_2({}^1T_{2g}|(e_g)^1(t_{2g})^1|0) = -\sqrt{1/4}\,\chi_{1,5}(0,0) + \sqrt{3/4}\,\chi_{2,5}(0,0),\\
&\Xi_3({}^1T_{2g}|(e_g)^1(t_{2g})^1|0) = -\sqrt{1/4}\,\chi_{1,4}(0,0) - \sqrt{3/4}\,\chi_{2,4}(0,0)
\end{aligned}$$

belong to the ${}^1T_{2g}$ term.

The functions of the symmetry species $T_{1g}$ can be easily determined. The

Table B.27. Transformation behaviour of the $T_{2g}$ functions of the configuration $(e_g)^1(t_{2g})^1$. Operations as in Table B.22

| | $E$ | $C_3$ | $C_2$ | $C_4$ | $C_2'$ |
|---|---|---|---|---|---|
| $\Xi_1$ | $\Xi_1$ | $\Xi_2$ | $\Xi_1$ | $-\Xi_1$ | $\Xi_1$ |
| $\Xi_2$ | $\Xi_2$ | $\Xi_3$ | $-\Xi_2$ | $-\Xi_3$ | $-\Xi_3$ |
| $\Xi_3$ | $\Xi_3$ | $\Xi_1$ | $-\Xi_3$ | $\Xi_2$ | $-\Xi_2$ |
| Character | 3 | 0 | $-1$ | $-1$ | 1 |

$T_{1g}$ functions must be orthogonal to the $T_{2g}$ functions

$$\chi_{1,3}(S,M_S),\quad -\sqrt{1/4}\,\chi_{1,5}(S,M_S)+\sqrt{3/4}\,\chi_{2,5}(S,M_S),$$
$$-\sqrt{1/4}\,\chi_{1,4}(S,M_S)-\sqrt{3/4}\,\chi_{2,4}(S,M_S)$$

$\chi_{2,3}(S, M_S)$ is certainly such a function because it is not contained in the $T_{2g}$ functions:

$$\Xi_1(^{2S+1}T_{1g}|(e_g)^1(t_{2g})^1|M_S)=\chi_{2,3}(S,M_S).$$

Further one recognizes that the two following functions are orthogonal to all remaining $\Xi$ functions and also orthogonal to one another:

$$\Xi_2(^{2S+1}T_{1g}|(e_g)^1(t_{2g})^1|M_S)=-\sqrt{3/4}\,\chi_{1,5}(S,M_S)-\sqrt{1/4}\,\chi_{2,5}(S,M_S),$$
$$\Xi_3(^{2S+1}T_{1g}|(e_g)^1(t_{2g})^1|M_S)=-\sqrt{3/4}\,\chi_{1,4}(S,M_S)+\sqrt{1/4}\,\chi_{2,4}(S,M_S).$$

Therefore there are nine symmetry adapted $^3T_{1g}$ functions

$$\Xi_1(^3T_{1g}|(e_g)^1(t_{2g})^1|M_S)=\chi_{2,3}(1,M_S),$$
$$\Xi_2(^3T_{1g}|(e_g)^1(t_{2g})^1|M_S)=-\sqrt{3/4}\,\chi_{1,5}(1,M_S)-\sqrt{1/4}\,\chi_{2,5}(1,M_S),$$
$$\Xi_3(^3T_{1g}|(e_g)^1(t_{2g})^1|M_S)=-\sqrt{3/4}\,\chi_{1,4}(1,M_S)+\sqrt{1/4}\,\chi_{2,4}(1,M_S),$$
$$M_S=1,0,-1,$$

and three $^1T_{1g}$ functions

$$\Xi_1(^1T_{1g}|(e_g)^1(t_{2g})^1|0)=\chi_{2,3}(0,0),$$
$$\Xi_2(^1T_{1g}|(e_g)^1(t_{2g})^1|0)=-\sqrt{3/4}\,\chi_{1,5}(0,0)-\sqrt{1/4}\,\chi_{2,5}(0,0),$$
$$\Xi_3(^1T_{1g}|(e_g)^1(t_{2g})^1|0)=-\sqrt{3/4}\,\chi_{1,4}(0,0)+\sqrt{1/4}\,\chi_{2,4}(0,0).$$

In the ligand field all term energies of the configuration $(e_g)^1(t_{2g})^1$ are equal, when, as in step 1, the electron interaction is neglected, cf. (3–60):

$$\begin{aligned}\Delta\tilde{E}_{\text{Lig.}}({}^3T_{2g}|(e_g)^1(t_{2g})^1) &= \Delta\tilde{E}_{\text{Lig.}}({}^3T_{1g}|(e_g)^1(t_{2g})^1)\\ &= \Delta\tilde{E}_{\text{Lig.}}({}^1T_{2g}|(e_g)^1(t_{2g})^1) = \Delta\tilde{E}_{\text{Lig.}}({}^1T_{1g}|(e_g)^1(t_{2g})^1)\\ &= (\varepsilon_0 + 6Dq) + (\varepsilon_0 - 4Dq) = 2\varepsilon_0 + 2Dq.\end{aligned}$$

The perturbation energies produced by the electron interaction are calculated by following the procedure described in detail above and we obtain:

$$\begin{aligned}\Delta\tilde{E}_{\text{el.}}({}^3T_{2g}|(e_g)^1(t_{2g})^1) &= A - 8B,\\ \Delta\tilde{E}_{\text{el.}}({}^3T_{1g}|(e_g)^1(t_{2g})^1) &= A + 4B,\\ \Delta\tilde{E}_{\text{el.}}({}^1T_{2g}|(e_g)^1(t_{2g})^1) &= A + 2C,\\ \Delta\tilde{E}_{\text{el.}}({}^1T_{1g}|(e_g)^1(t_{2g})^1) &= A + 4B + 2C.\end{aligned}$$

Figure B.25 shows the term system which results for the strong-field case for $d^2$ configurations.

In the strong-field approximation the centre of gravity theorem is also valid, when it is applied to terms of the same configuration. For example one sees that the arithmetic mean (calculated taking the degrees of degeneracy into consideration) of the *splitting* energies of the $(e_g)^1(t_{2g})^1$ progeny terms is equal to zero. The arithmetic means of the $\Delta\tilde{E}_{\text{el.}}$ values is equal to the *common* energy $A$ of all terms:

$$\begin{aligned}\frac{1}{9+9+3+3}\,[9\Delta\tilde{E}_{\text{el.}}({}^3T_{2g}|(e_g)^1(t_{2g})^1) + 9\Delta\tilde{E}_{\text{el.}}({}^3T_{1g}|(e_g)^1(t_{2g})^1)\\ + 3\Delta\tilde{E}_{\text{el.}}({}^1T_{2g}|(e_g)^1(t_{2g})^1) + 3\Delta\tilde{E}_{\text{el.}}({}^1T_{1g}|(e_g)^1(t_{2g})^1)] = A.\end{aligned}$$

### 3.4.3. *Rigorous treatment*

As was shown at the beginning of the section concerning $d^N$ systems, in the rigorous treatment of a $d^N$ ion bound in a complex the secular problem (equation (3–44)) must be solved. In the limiting strong and weak field cases a rigorous treatment is not necessary. In both cases integrals in the complete secular determinant are neglected in the second step of the calculation.

In the weak-field treatment in the second step the perturbation calculation is carried out separately for each of the terms ${}^{2S+1}L$ obtained in the first step. Here the integrals $(\Theta_i, \boldsymbol{H}_{\text{Lig.}}\Theta_j)$ which are formed from functions $\Theta_i$ and $\Theta_j$ of *different* terms ${}^{2S+1}L$ are neglected. The omission of these integrals can be corrected by determining the coupling between the states

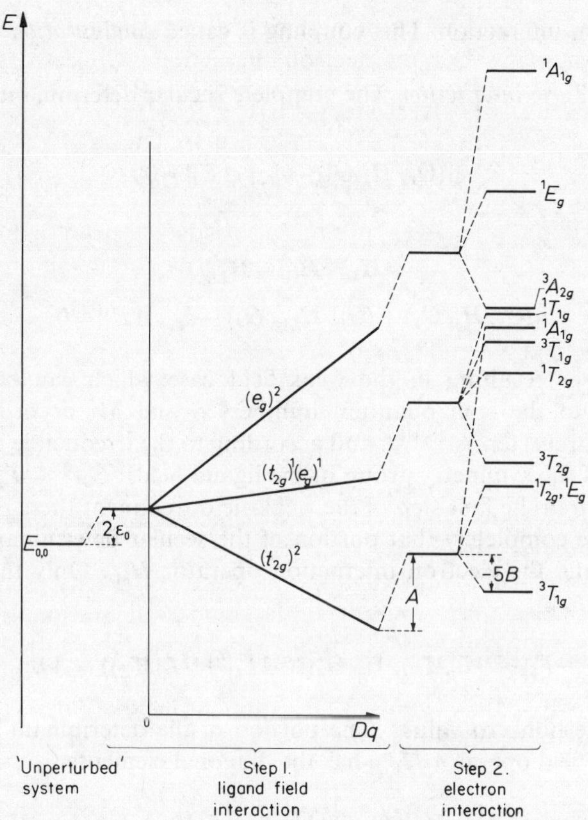


Figure B.25. Splitting diagram for octahedrally bound $d^2$ systems for the strong-field (without term interaction) with $C/B = 4$.

of different terms described by these $(\Theta_i, \boldsymbol{H}_{\text{Lig.}}\Theta_j)$ subsequently, using a perturbation calculation. The coupling which can be treated in this manner is called *term interaction.*

In the strong-field case the first step yields the terms $^{2S+1}\Gamma_i$ which are degenerate if they belong to the same configuration $\gamma_1^a\gamma_2^b\ldots$. The second step considers the electron interaction although it is limited to treating separately the terms arising from the same configuration. The electron interaction between states of different configurations is neglected as before. Starting with the results of the strong-field case the solutions of the complete problem (3–44) can be obtained by taking into consideration the coupling between states of different configurations which results from the electron interaction. This coupling is called *configuration interaction.*

3.4.3.1. *Term interaction.* The complete secular determinant (3–44) has the form

$$||(\Theta_r, \boldsymbol{H}_1\, \Theta_s) - \delta_{r,s}\, \Delta E|| = 0\,, \tag{3–44a}$$

or because of

$$\boldsymbol{H}_1 = \boldsymbol{H}_{\text{el.}} + \boldsymbol{H}_{\text{Lig.}}$$

$$||(\Theta_r, \boldsymbol{H}_{\text{el.}}\Theta_s) + (\Theta_r, \boldsymbol{H}_{\text{Lig.}}\,\Theta_s) - \delta_{r,s}\, \Delta E|| = 0\,. \tag{3–44b}$$

The states resulting in the weak-field case which can be classified according to the spin quantum numbers $S$ and $M_S$ according to the original (parent) terms $^{2S+1}L$ and according to the irreducible representations $\Gamma_i$ (of the symmetry group of the ligand field): $\Theta_r(^{2S+1}\Gamma_i|^{S+1}L|M_S)$. As is shown in the first step of the weak-field treatment these $\Theta$ functions diagonalize completely that portion of the secular determinant which is formed using the electron interaction operator $\boldsymbol{H}_{\text{el.}}$. Only the diagonal elements

$$(\Theta_r(^{2S+1}\Gamma_i|^{2S+1}L|M_S),\, \boldsymbol{H}_{\text{el.}}\Theta_r(^{2S+1}\Gamma_i|^{2S+1}L|M_S)) = \Delta E_{\text{el.}}(^{2S+1}L)$$

can assume nonzero values. The portion of the determinant containing the ligand field operator $\boldsymbol{H}_{\text{Lig.}}$ has the diagonal elements

$$(\Theta_r(^{2S+1}\Gamma_i|^{2S+1}L|M_S),\, \boldsymbol{H}_{\text{Lig.}}\Theta_r(^{2S+1}\Gamma_i|^{2S+1}L|M_S)) = \Delta E_{\text{Lig.}}(^{2S+1}\Gamma_i|^{2S+1}L)$$

and is in general only partially diagonalized by the $\Theta$ functions. According to the results of the second step (of the weak-field treatment) only those nondiagonal elements are with certainty zero which are formed with $\Theta$ functions of the same parent term $^{2S+1}L$. Nondiagonal elements containing $\Theta$ functions of *different* parent terms $^{2S+1}L = {}^{2S+1}L'$ were neglected.

For the calculation of these *off-diagonal elements* which have not yet been treated

$$(\Theta_r(^{2S+1}\Gamma_i|^{2S+1}L|M_S),\, \boldsymbol{H}_{\text{Lig.}}\Theta_{r'}(^{2S'+1}\Gamma_{i'}|^{2S'+1}L'|M'_S))$$

it should first be considered which of these integrals can possibly assume nonzero values. Because $\boldsymbol{H}_{\text{Lig.}}$ has no effect on the spin variables and because the $\Theta$ belong to irreducible representations of the symmetry group of $\boldsymbol{H}_{\text{Lig.}}$ all integrals vanish identically for which the following conditions do not hold

$$S = S'$$
$$M_S = M'_S$$
$$\Gamma_i = \Gamma_{i'}$$
$$r = r' \quad \text{(column index of } \Gamma_i)$$

If the $\Theta_r({}^{2S+1}\Gamma_i|{}^{2S+1}L|M_S)$ are arranged so that $\Theta$ functions with the same $\Gamma_i$ and the same $S$ stand as close as possible to one another in the secular determinant (3–44), then (3–44) becomes a step determinant. The subdeterminants lying along the diagonal which likewise contain $\Theta$ functions with the same $\Gamma_i$ and the same $S$ are to be set individually equal to zero. The roots of all of these subdeterminants are equal to the perturbation energies of the complete problem. These perturbation energies differ from those perturbation energies obtained in the weak field case by the *term interaction energies.*

We shall now consider these particular subdeterminants more closely. For this purpose it is assumed that from all of the terms of the free ion (in step 2 of the weak-field case) $k$ progeny terms ${}^{2S+1}\Gamma_i$ arise where the dimensionality of $\Gamma_i$ is equal to $g_i$. The $k$ progeny terms ${}^{2S+1}\Gamma_i$ are themselves composed of in total $k \times (2S+1) \times g_i$ states. $k \times (2S+1) \times g_i$ is then the order of the associated subdeterminant. If the $\Theta$ functions are so arranged when these subdeterminants are formulated that $\Theta$ functions with the same $M_S$ and the same index $r$ (which denotes that the function is in the $r$th column of the representation matrices of $\Gamma_i$) as far as possible stand next to one another, then the subdeterminant decomposes, because of the non-combination rule mentioned above for $\Theta$ functions with different $M_S$ values and different indices $r$ into $(2S+1) \times g_i$ determinants of order $k$. Each of these determinants contains $\Theta$ functions with the same index $r$ and the same $M_S$ value. The $\Theta$ functions in a determinant differ therefore only in their parent term ${}^{2S+1}L$. Further all of the $(2S+1) \times g_i$ determinants have the same roots because according to theorem 2b, p. 317 the equalities

$$(\Theta_r({}^{2S+1}\Gamma_i|{}^{2S+1}L|M_S), \mathbf{H}_{\text{Lig.}}\Theta_r({}^{2S+1}\Gamma_i|{}^{2S+1}L'|M_S))$$
$$= (\Theta_{r'}({}^{2S+1}\Gamma_i|{}^{2S+1}L|M'_S), \mathbf{H}_{\text{Lig.}}\Theta_{r'}({}^{2S+1}\Gamma_i|{}^{2S+1}L'|M'_S))$$

hold.

The solution functions which diagonalize these determinants are linear combinations, that is 'mixtures' of the appropriate $\Theta$ functions. To be able to characterize these functions and the perturbation energies and to distinguish among them a knowledge of all contributing parent terms is essential.

*Example of a $d^2$ ion in a ligand field with octahedral symmetry $O_h$.* We shall now calculate the term interaction energies for the example of octahedrally coordinated $d^2$ ions. In the weak field case the following original and progeny terms were found:

$$
\begin{array}{lllllll}
d^2: {}^1S \rightarrow & \boxed{{}^1A_{1g}} & & & & & \\
{}^1D \rightarrow & & & \boxed{{}^1E_g} & & + & \boxed{{}^1T_{2g}} \\
{}^1G \rightarrow & \boxed{{}^1A_{1g}} & + & \boxed{{}^1E_g} & + {}^1T_{1g} & + & \boxed{{}^1T_{2g}} \\
{}^3P \rightarrow & & & & \boxed{{}^3T_{1g}} & & \\
{}^3F \rightarrow & {}^3A_{2g} & & + & \boxed{{}^3T_{1g}} & + & {}^3T_{2g}
\end{array}
$$

Term interaction can occur only between terms which are enclosed in the same box (the same multiplicity and the same symmetry species).

*Term interaction* ${}^1A_{1g}|{}^1S, {}^1G$. Because ${}^1A_{1g}$ terms are nondegenerate, we have *one* subdeterminant of order 2. Using the functions (cf. Table B.16 and B.6)

$$
\begin{aligned}
K_1 = \Theta({}^1A_{1g}|{}^1S|0) &= \Psi(0,0,0,0) \\
&= \sqrt{1/5}[\Phi(2^+;-2^-) - \Phi(2^-;-2^+) - \Phi(1^+;-1^-) \\
&\quad + \Phi(1^-;-1^+) + \Phi(0^+;0^-)], \\
K_2 = \Theta({}^1A_{1g}|{}^1G|0) &= \sqrt{7/12}\Psi(4,0,0,0) + \sqrt{5/24}\Psi(4,4,0,0) \\
&\quad + \sqrt{5/24}\Psi(4,-4,0,0) \\
&= \sqrt{1/120}[\Phi(2^+;-2^-) - \Phi(2^-;-2^+) + 4\Phi(1^+;-1^-) \\
&\quad - 4\Phi(1^-;-1^+) + 6\Phi(0^+;0^-) + 5\Phi(2^+;2^-) \\
&\quad + 5\Phi(-2^+;-2^-)]
\end{aligned}
$$

and the abbreviations

$$H_{i,j} = (K_i, \boldsymbol{H}_1 K_j)$$

they are

$$
\begin{vmatrix}
H_{1,1} - \Delta E({}^1A_{1g}|{}^1S, {}^1G) & H_{1,2} \\
H_{2,1} & H_{2,2} - \Delta E({}^1A_{1g}|{}^1S, {}^1G)
\end{vmatrix} = 0
$$

and have the roots

$$\Delta E_{a,b}({}^1A_{1g}|{}^1S, {}^1G) = \frac{H_{1,1} + H_{2,2}}{2} \pm \frac{1}{2}\sqrt{(H_{1,1} - H_{2,2})^2 + 4H_{1,2}H_{2,1}}\,.$$

According to the results on pp. 254, 354, 357 the diagonal elements $H_{1,1}$ and $H_{2,2}$ have the values

$$
\begin{aligned}
H_{1,1} = (K_1, \boldsymbol{H}_1 K_1) &= (K_1, \boldsymbol{H}_{\text{el.}} K_1) + (K_1, \boldsymbol{H}_{\text{Lig.}} K_1) \\
&= \Delta E_{\text{el.}}({}^1S) + \Delta E_{\text{Lig.}}({}^1A_{1g}|{}^1S) \\
&= A + 14B + 7C + 2\varepsilon_0,
\end{aligned}
$$

$$H_{2,2}=(K_2,\ \mathbf{H}_1 K_2)=(K_2,\ \mathbf{H}_{\text{el.}}K_2)+(K_2,\ \mathbf{H}_{\text{Lig.}}K_2)$$
$$=\Delta E_{\text{el.}}(^1G)+\Delta E_{\text{Lig.}}(^1A_{1g}|^1G)$$
$$=A+4B+2C+2\varepsilon_0+4Dq.$$

One has for the nondiagonal elements after the linear combinations of the $\Phi$ given above have been substituted for the $K$ using the rules of p. 249

$$H_{2,1}=H_{1,2}=(K_1,\ \mathbf{H}_{\text{el.}}\ K_2)+(K_1,\ \mathbf{H}_{\text{Lig.}}\ K_2)=(K_1,\ \mathbf{H}_{\text{Lig.}}\ K_2)$$
$$=\frac{1}{5\sqrt{6}}[2(2|\boldsymbol{V}|2)-8(1|\boldsymbol{V}|1)+6(0|\boldsymbol{V}|0)$$
$$+5(2|\boldsymbol{V}|-2)+5(-2|\boldsymbol{V}|2)].$$

The values of the one-electron integrals $(m|\boldsymbol{V}|m')$ are taken from (3–50, 53) and this gives

$$H_{2,1}=H_{1,2}=\frac{1}{5\sqrt{6}}\ [2(\varepsilon_0+Dq)-8(\varepsilon_0-4Dq)+6(\varepsilon_0+6Dq)+10\cdot 5Dq]$$
$$=4\sqrt{6}Dq.$$

The two perturbation energies are then

$$\Delta E_{a,b}\ (^1A_{1g}|^1S,{}^1G)=A+9B+\frac{9}{2}C+2\varepsilon_0+2Dq$$
$$\pm\frac{1}{2}[(10B+5C-4Dq)^2+384Dq^2]^{1/2}.$$

Let us now discuss the energy expression

$$\Delta E_{a,b}=\frac{H_{1,1}+H_{2,2}}{2}+\frac{1}{2}\sqrt{(H_{1,1}-H_{2,2})^2+4H_{1,2}H_{2,1}}$$

in complete generality. Because $H_{1,2}=H_{2,1}$ is a measure of the term interaction, it appears reasonable to consider the magnitudes of $\Delta E_a$ and $\Delta E_b$ as functions of $H_{1,2}=H_{2,1}$. For vanishing term interaction, that is for $H_{1,2}=H_{2,1}=0$ the roots are $\Delta E_a=H_{1,1}$ and $\Delta E_b=H_{2,2}$. These are of course exactly the perturbation energies given by the weak-field treatment. With an increasing value of $H_{1,2}=H_{2,1}$ the energetically higher-lying term is labilized and the lower-lying term stabilized, cf. Figure B.26. Both terms 'repel one another'.

The relation for $\Delta E_{a,b}$ gives a hyperbola. The asymptotes of the hyperbola cross at the point $\frac{1}{2}(H_{1,1}+H_{2,2})$ and have the slopes $+1$ and $-1$.

*Term interaction* $^1E_g|^1D,{}^1G$. Here the $4\times 4$ subdeterminant decomposes into two $2\times 2$ determinants, each of which has two functions which transform as the same column of the irreducible representation $E_g$.

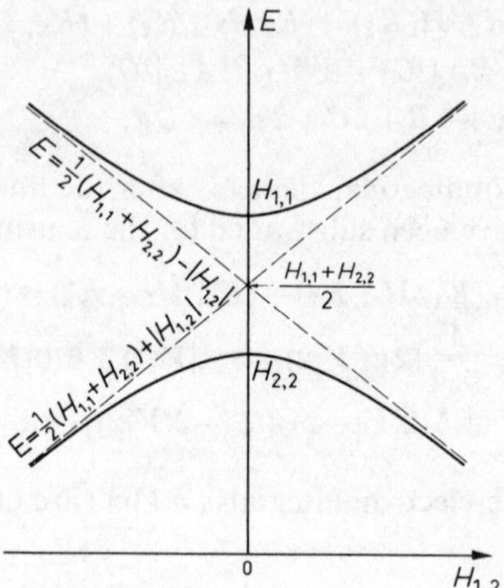


Figure B.26. 'Repulsion' of two terms as a result of term interaction.

Two such functions are according to Table B.16 and B.6:

$$K_1=\Theta_2(^1E_g|^1D|0)=\sqrt{1/2}[\Psi(2,2,0,0)+\Psi(2,-2,0,0)]$$
$$=\sqrt{1/14}[\sqrt{2}\Phi(2^+;0^-)-\sqrt{2}\Phi(2^-;0^+)-\sqrt{3}\Phi(1^+;1^-)$$
$$+\sqrt{2}\Phi(0^+;-2^-)-\sqrt{2}\Phi(0^-;-2^+)-\sqrt{3}\Phi(-1^+;-1^-)],$$

$$K_2=\Theta_2(^1E_g|^1G|0)=\sqrt{1/2}[\Psi(4,2,0,0)+\Psi(4,-2,0,0)]$$
$$=\sqrt{1/28}[\sqrt{3}\Phi(2^+;0^-)-\sqrt{3}\Phi(2^-;0^+)+\sqrt{8}\Phi(1^+;1^-)$$
$$+\sqrt{3}\Phi(0^+;-2^-)-\sqrt{3}\Phi(0^-;-2^+)+\sqrt{8}\Phi(-1^+;-1^-)].$$

The determinant for $K_1$ and $K_2$ is

$$\begin{vmatrix} H_{1,1}-\Delta E(^1E_g|^1D,\,^1G) & H_{1,2} \\ H_{2,1} & H_{2,2}-\Delta E(^1E_g|^1D,\,^1G) \end{vmatrix}=0.$$

According to pp. 254, 356 f. the diagonal elements have the values

$$H_{1,1}=(K_1\,\mathbf{H}_{\text{el.}}\,K_1)+(K_1,\,\mathbf{H}_{\text{Lig.}}\,K_1)$$
$$=\Delta E_{\text{el.}}(^1D)+\Delta E_{\text{Lig.}}(^1E_g|^1D)$$
$$=A-3B+2C+2\varepsilon_0+\frac{24}{7}Dq,$$
$$H_{2,2}=(K_2,\,\mathbf{H}_{\text{el.}}\,K_2)+(K_2,\,\mathbf{H}_{\text{Lig.}}\,K_2)$$
$$=\Delta E_{\text{el.}}(^1G)+\Delta E_{\text{Lig.}}(^1E_g|^1G)$$
$$=A+4B+2C+2\varepsilon_0+\frac{4}{7}Dq.$$

For the off-diagonal elements we obtain using the rules on p. 249

$$H_{1,2}=H_{2,1}=(K_2,\ \boldsymbol{H}_{\text{el.}}\,K_1)+(K_2,\ \boldsymbol{H}_{\text{Lig.}}\,K_1)=(K_2,\ \boldsymbol{H}_{\text{Lig.}}\,K_1)$$
$$=\sqrt{3/49}[2\,(2|\boldsymbol{V}|2)-4\,(1|\boldsymbol{V}|1)+2\,(0|\boldsymbol{V}|0)$$
$$+(2|\boldsymbol{V}|-2)+(-2|\boldsymbol{V}|2)]$$

and with consideration of (3–50, 53)

$$H_{1,2}=H_{2,1}=\frac{40\sqrt{3}}{7}Dq.$$

The perturbation energies are then

$$\Delta E_{a,b}(^1E_g|^1D,\,^1G)=A+\frac{1}{2}B+2C+2\varepsilon_0+2Dq$$
$$\pm\frac{1}{2}\left[\left(-7B+\frac{20}{7}Dq\right)^2+\frac{19200}{49}Dq^2\right]^{\frac{1}{2}}.$$

*Term interaction* $^1T_{2g}|^1D,\,^1G$. The $6\times 6$ determinant decomposes into three $2\times 2$ determinants, one for each column of the representation matrices of $T_{2g}$. We choose the following function pair for the calculation of the perturbation energies (cf. Table B.16 and B.6)

$$K_1=\Theta_2(^1T_{2g}|^1D|0)=\sqrt{1/2}[\Psi(2,2,0,0)-\Psi(2,-2,0,0)]$$
$$=\sqrt{1/14}[\sqrt{2}\Phi(2^+;0^-)-\sqrt{2}\Phi(2^-;0^+)-\sqrt{3}\Phi(1^+;1^-)$$
$$\sqrt{2}\Phi(0^+;-2^-)+\sqrt{2}\Phi(0^-;-2^+)+\sqrt{3}\Phi(-1^+;-1^-)],$$

$$K_2=\Theta_2(^1T_{2g}|^1G|0)=\sqrt{1/2}[\Psi(4,2,0,0)-\Psi(4,-2,0,0)]$$
$$=\sqrt{1/28}[\sqrt{3}\Phi(2^+;0^-)-\sqrt{3}\Phi(2^-;0^+)+\sqrt{8}\Phi(1^+;1^-)$$
$$-\sqrt{3}\Phi(0^+;-2^-)+\sqrt{3}\Phi(0^-;-2^+)-\sqrt{8}\Phi(-1^+;-1^-)].$$

The diagonal elements appearing in the determinant

$$\begin{vmatrix} H_{1,1}-\Delta E(^1T_{2g}|^1D,\,^1G) & H_{1,2} \\ H_{2,1} & H_{2,2}-\Delta E(^1T_{2g}|^1D,\,^1G) \end{vmatrix}=0$$

have according to pp. 254, 348 ff. the values

$$H_{1,1}=(K_1,\ \boldsymbol{H}_{\text{el.}}\,K_1)+(K_1,\ \boldsymbol{H}_{\text{Lig.}}\,K_1)$$
$$=\Delta E_{\text{el.}}(^1D)+\Delta E_{\text{Lig.}}(^1T_{2g}|^1D)$$
$$=A-3B+2C+2\varepsilon_0-\frac{16}{7}Dq,$$

$$H_{2,2}=(K_2,\ \boldsymbol{H}_{\text{el.}}\,K_2)+(K_2,\ \boldsymbol{H}_{\text{Lig.}}\,K_2)$$
$$=\Delta E_{\text{el.}}(^1G)+\Delta E_{\text{Lig.}}(^1T_{2g}|^1G)$$
$$=A+4B+2C+2\varepsilon_0-\frac{26}{7}Dq.$$

For the off-diagonal elements one obtains with consideration of the rules on p. 249 and of equations (3–50, 53)

$$
\begin{aligned}
H_{1,2} = H_{2,1} &= (K_2, \boldsymbol{H}_{\text{el.}} K_1) + (K_2, \boldsymbol{H}_{\text{Lig.}} K_1) = (K_2, \boldsymbol{H}_{\text{Lig.}} K_1) \\
&= \sqrt{3/49}\,[2(2|\boldsymbol{V}|2) - 4(1|\boldsymbol{V}|1) + 2(0|\boldsymbol{V}|0) \\
&\quad - (2|\boldsymbol{V}|-2) - (-2|\boldsymbol{V}|2)] \\
&= -\frac{20\sqrt{3}}{7} Dq.
\end{aligned}
$$

The total perturbation energies are thus

$$
\begin{aligned}
\Delta E_{a,b}(^1T_{2g}|^1D, {}^1G) = A + \frac{1}{2} B + 2C + 2\varepsilon_0 - 3Dq \\
\pm \frac{1}{2}\left[\left(-7B + \frac{10}{7} Dq\right)^2 + \frac{4800}{49} Dq^2\right]^{\frac{1}{2}}
\end{aligned}
$$

*Term interaction* $^3T_{1g}|^3P, {}^3F$. Because of the multiplicity $2S+1 = 3(M_S = 1, 0, -1)$ the associated subdeterminant is of degree $2 \times 3 \times 3 = 18$. It decomposes into nine $2 \times 2$ determinants. Pairs of functions belonging to the same column of $T_{1g}$ and having the same $M_S$ values compose the determinant. For the function pair (cf. Table B.16 and B.6)

$$
\begin{aligned}
K_1 = \Theta_2(^3T_{1g}|^3P|1) &= \Psi(1,0,1,1) \\
&= -\sqrt{4/5}\,\Phi(2^+; -2^+) + \sqrt{1/5}\,\Phi(1^+; -1^+), \\
K_2 = \Theta_2(^3T_{1g}|^3F|1) &= \Psi(3,0,1,1) \\
&= \sqrt{1/5}\,\Phi(2^+; -2^+) + \sqrt{4/5}\,\Phi(1^+; -1^+)
\end{aligned}
$$

the diagonal elements are (cf. pp. 254, 356)

$$
\begin{aligned}
H_{1,1} &= (K_1, \boldsymbol{H}_{\text{el.}} K_1) + (K_1, \boldsymbol{H}_{\text{Lig.}} K_1) \\
&= \Delta E_{\text{el.}}(^3P) + \Delta E_{\text{Lig.}}(^3T_{1g}|^3P) \\
&= A + 7B + 2\varepsilon_0, \\
H_{2,2} &= (K_2, \boldsymbol{H}_{\text{el.}} K_2) + (K_2, \boldsymbol{H}_{\text{Lig.}} K_2) \\
&= \Delta E_{\text{el.}}(^3F) + \Delta E_{\text{Lig.}}(^3T_{1g}|^3F) \\
&= A - 8B + 2\varepsilon_0 - 6Dq.
\end{aligned}
$$

The off-diagonal elements have the value (cf. p. 249 and (3–50, 53))

$$
\begin{aligned}
H_{1,2} = H_{2,1} &= (K_2, \boldsymbol{H}_{\text{el.}} K_1) + (K_2, \boldsymbol{H}_{\text{Lig.}} K_1) = (K_2, \boldsymbol{H}_{\text{Lig.}} K_1) \\
&= -\frac{4}{5}(2|\boldsymbol{V}|2) + \frac{4}{5}(1|\boldsymbol{V}|1) \\
&= -4Dq.
\end{aligned}
$$

The determinant

$$\begin{vmatrix} H_{1,1}-\Delta E(^3T_{1g}|^3P,{}^3F) & H_{1,2} \\ H_{2,1} & H_{2,2}-\Delta E(^3T_{1g}|^3P,{}^3F) \end{vmatrix} = 0$$

has the roots

$$\Delta E_{a,b}(^3T_{1g}|^3P,{}^3F) = A - \frac{1}{2}B + 2\varepsilon_0 - 3Dq \pm \frac{1}{2}[(15B+6Dq)^2 + 64Dq^2]^{1/2}.$$

The consequences of term interaction for the term energies are illustrated in the left-hand side of Figure B.27 for octahedrally coordinated $d^2$ ions.

3.4.3.2. *Configuration interaction.* The functions for the strong-field case $\Xi_r(^{2S+1}\Gamma_i|\gamma_1^a\gamma_2^b\ldots|M_S)$ were selected in the following manner:

1. The $\Xi$ diagonalize completely that part of the total secular determinant (3–44) formed with the ligand field operator $\boldsymbol{H}_{\text{Lig.}}$ whereby the diagonal elements [cf. equation (3–60)] have the values

$$(\Xi_r(^{2S+1}\Gamma_i|\gamma_1^a\gamma_2^b\ldots|M_S), \boldsymbol{H}_{\text{Lig.}}\Xi_r(^{2S+1}\Gamma_i|\gamma_1^a\gamma_2^b\ldots|M_S)) = \Delta\tilde{E}_{\text{Lig.}}(\gamma_1^a\gamma_2^b\ldots)$$

2. The part of (3–44) arising from the electron interaction operator $\boldsymbol{H}_{\text{el.}}$ with the diagonal elements (cf. p. 368)

$$(\Xi_r(^{2S+1}\Gamma_i|\gamma_1^a\gamma_2^b\ldots|M_S), \boldsymbol{H}_{\text{el.}}\Xi_r(^{2S+1}\Gamma_i|\gamma_1^a\gamma_2^b\ldots|M_S)) = \Delta\tilde{E}_{\text{el.}}(^{2S+1}\Gamma_i|\gamma_1^a\gamma_2^b\ldots)$$

is in general only partially diagonalized by the $\Xi$ functions. Only the off-diagonal elements $(\Xi_r, \boldsymbol{H}_{\text{el.}}\Xi_s)$ are with certainty zero when the $\Xi$ functions contain the same configuration $\gamma_1^a\gamma_2^b\ldots$. Nondiagonal elements $(\Xi_r, \boldsymbol{H}_{\text{el.}}\Xi_s)$ with $\Xi$ functions of *different* configurations do not in general vanish. In the strong-field case these elements were, however, set equal to zero, which is of course not allowed in a rigorous treatment. It is exactly these nondiagonal elements which determine the degree of *configuration interaction.*

The off-diagonal elements in question

$$(\Xi_r(^{2S+1}\Gamma_i|\gamma_1^a\gamma_2^b\ldots|M_S), \boldsymbol{H}_{\text{el.}}\Xi_{r'}(^{2S'+1}\Gamma_{i'}|\gamma_1^{a'}\gamma_2^{b'}\ldots|M'_S))$$

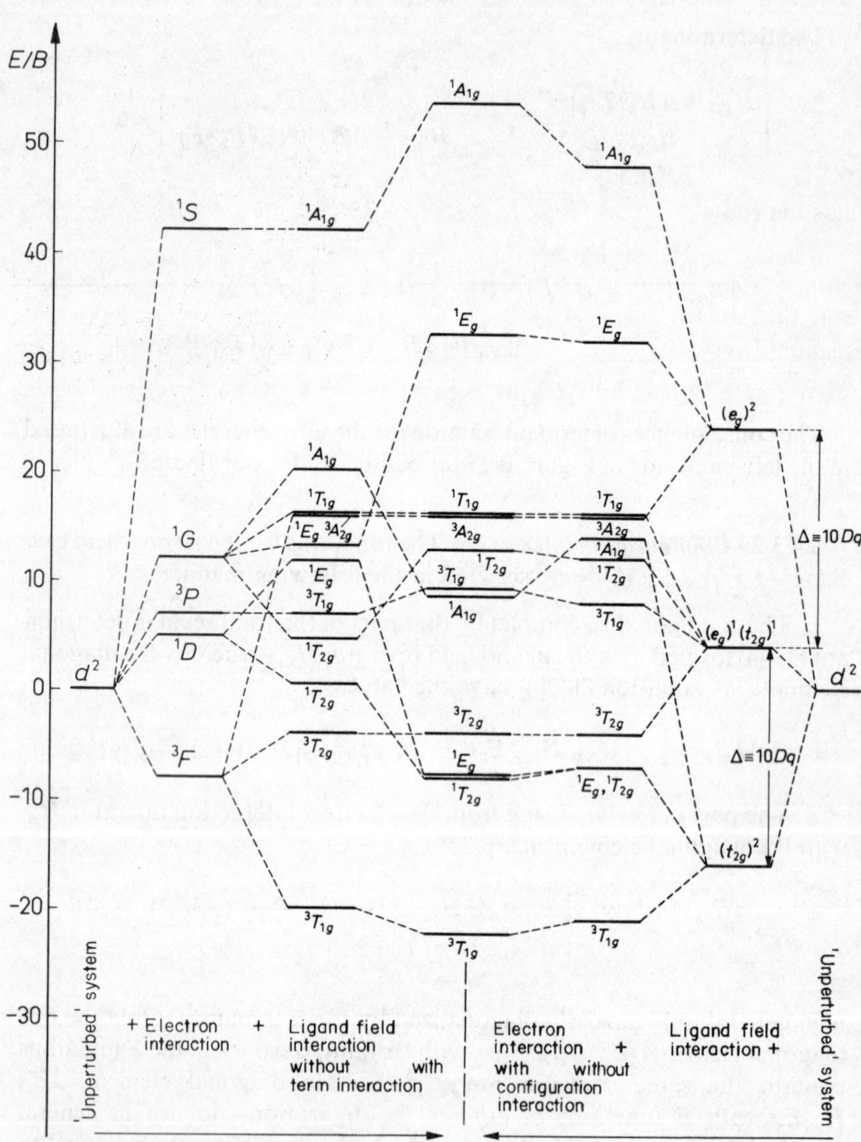


Figure B.27. Term splitting of an octahedrally coordinated $d^2$ system with $C/B = 4$ and $\Delta \equiv 10\,Dq = 20B$. (The shifts by $A$ and $2\varepsilon_0$ which are equally large for all terms are not considered.)

can apparently only then assume nonzero values when

$$S=S'$$
$$M_S=M'_S$$
$$\Gamma_i=\Gamma_{i'}$$
$$r=r' \quad \text{(column index of } \Gamma_i)$$

are satisfied.

If the complete secular determinant is so formulated that the $\Xi$ functions with the same $S$ and same $\Gamma_i$ are as close to one another as possible, then equation (3–44) decomposes into subdeterminants arranged along the diagonal of (3–44). The subdeterminants which are all individually set equal to zero contain in all cases functions with the same $S$ and the same $\Gamma_i$. Their roots are the perturbation energies of the complete problem. They differ from the perturbation energies of the strong-field treatment by the *configuration interaction* energies.

The determination of the roots of the subdeterminants can often be greatly simplified. Let us assume that the $k$ terms ${}^{2S+1}\Gamma_i$ (let the dimensionality of $\Gamma_i$ be $g_i$) arise from all the configurations $\gamma_1^a\gamma_2^b\ldots, \gamma_1^{a'}\gamma_2^{b'}\ldots$. The strong-field treatment gives $k\times(2S+1)\times g_i$ functions $\Xi$ to describe these $k\times(2S+1)\times g_i$ states. These $\Xi$ are the basis functions for the associated subdeterminant formed with the perturbation operator $\boldsymbol{H}_1 = \boldsymbol{H}_{\text{el.}}+\boldsymbol{H}_{\text{Lig.}}$. If the subdeterminant is formulated so that the $\Xi$ functions with the same $M_S$ and the same index $r$ (= index of the columns of the representation matrices of $\Gamma_i$) stand as close to one another as possible, then the subdeterminant decomposes into single determinants in which in each case only $\Xi$ functions of the same $M_S$ and the same $r$ are contained*. $(2S+1)\times g_i$ new determinants of order $k$ arise from the subdeterminant of order $k\times(2S+1)\times g_i$. All these new determinants which arise from the same subdeterminant have the same roots because of (cf. p. 317)

$$(\Xi_r({}^{2S+1}\Gamma_i|\gamma_1^a\gamma_2^b\ldots|M_S), \boldsymbol{H}_{\text{el.}}\Xi_r({}^{2S+1}\Gamma_i|\gamma_1^{a'}\gamma_2^{b'}\ldots|M_S)) =$$
$$=(\Xi_{r'}({}^{2S+1}\Gamma_i|\gamma_1^a\gamma_2^b\ldots|M'_S), \boldsymbol{H}_{\text{el.}}\Xi_{r'}({}^{2S+1}\Gamma_i|\gamma_1^{a'}\gamma_2^{b'}\ldots|M'_S))$$

Because the states which arise from configuration interaction can no longer be associated with one single configuration, the contributing configurations must be given to completely characterize and label the perturbation energies.

*Example of a $d^2$ ion in a ligand field of octahedral symmetry $O_h$*: We shall now study the influence of configuration interaction on the perturbation energies using the example of octahedrally coordinated $d^2$ ion. It is known

* As is explained above nondiagonal elements with two $\Xi$ functions always vanish when both $\Xi$ functions do not have the same values of $M_S$ and $r$.

from the strong-field treatment that the following configurations and progeny terms exist:

$$
d^2: \quad \begin{array}{rl}
(e_g)^2 \rightarrow & \boxed{{}^1A_{1g}} + {}^3A_{2g} + \boxed{{}^1E_g} \\
(e_g)^1(t_{2g})^1 \rightarrow & {}^1T_{1g} + \boxed{{}^3T_{1g}} + \boxed{{}^1T_{2g}} + {}^3T_{2g} \\
(t_{2g})^2 \rightarrow & \boxed{{}^1A_{1g}} + \boxed{{}^1E_g} + \boxed{{}^3T_{1g}} + \boxed{{}^1T_{2g}}
\end{array}
$$

Configuration interaction occurs only between terms in the same box (the same multiplicity and the same symmetry species).

*Configuration interaction* ${}^1A_{1g}|(e_g)^2, (t_{2g})^2$. Here we have $g_1 = 1$ and $2S+1 = 1$. There is then only one $2\times 2$ subdeterminant and this has the functions (cf. pp. 374, 378)

$$
\begin{aligned}
K_1 = \Xi({}^1A_{1g}|(e_g)^2|0) &= \sqrt{1/2}[\chi_{1,1}(0,0)+\chi_{2,2}(0,0)] \\
&= \sqrt{1/2}\,[\Phi(\phi_1^+;\phi_1^-)+\Phi(\phi_2^+;\phi_2^-)], \\
K_2 = \Xi({}^1A_{1g}|(t_{2g})^2|0) &= \sqrt{1/3}\,[\chi_{3,3}(0,0)+\chi_{4,4}(0,0)+\chi_{5,5}(0,0)] \\
&= \sqrt{1/3}\,[\Phi(\phi_3^+;\phi_3^-)+\Phi(\phi_4^+;\phi_4^-)+\Phi(\phi_5^+;\phi_5^-)].
\end{aligned}
$$

It is specifically

$$
\begin{vmatrix}
H_{1,1}-\Delta E({}^1A_{1g}|(e_g)^2, (t_{2g})^2) & H_{1,2} \\
H_{2,1} & H_{2,2}-\Delta E({}^1A_{1g}|(e_g)^2, (t_{2g})^2)
\end{vmatrix} = 0
$$

with

$$H_{i,j} = (K_i, \boldsymbol{H}_1 K_j).$$

The diagonal elements are (cf. pp. 366, 375, 378)

$$
\begin{aligned}
H_{1,1} &= (K_1, \boldsymbol{H}_1 K_1) = (K_1, \boldsymbol{H}_{\text{Lig.}} K_1) + (K_1, \boldsymbol{H}_{\text{el.}} K_1) \\
&= \Delta\tilde{E}_{\text{Lig.}}((e_g)^2) + \Delta\tilde{E}_{\text{el.}}({}^1A_{1g}|(e_g)^2) \\
&= 2\varepsilon_0 + 12Dq + A + 8B + 4C, \\
H_{2,2} &= (K_2, \boldsymbol{H}_1 K_2) = (K_2, \boldsymbol{H}_{\text{Lig.}} K_2) + (K_2, \boldsymbol{H}_{\text{el.}} K_2) \\
&= \Delta\tilde{E}_{\text{Lig.}}((t_{2g})^2) + \Delta\tilde{E}_{\text{el.}}({}^1A_{1g}|(t_{2g})^2) \\
&= 2\varepsilon_0 - 8Dq + A + 10B + 5C.
\end{aligned}
$$

The nondiagonal elements are obtained using the rules on p. 249 and Table B.24

$$
\begin{aligned}
H_{1,2} = H_{2,1} &= (K_2, \boldsymbol{H}_1 K_1) = (K_2, \boldsymbol{H}_{\text{Lig.}} K_1) + (K_2, \boldsymbol{H}_{\text{el.}} K_1) \\
&= (K_2, \boldsymbol{H}_{\text{el.}} K_1) \\
&= \sqrt{1/6}\,[\tilde{V}(\phi_1;\phi_1;\phi_3;\phi_3) + \tilde{V}(\phi_2;\phi_2;\phi_3;\phi_3) \\
&\quad + \tilde{V}(\phi_1;\phi_1;\phi_4;\phi_4) + \tilde{V}(\phi_2;\phi_2;\phi_4;\phi_4) \\
&\quad + \tilde{V}(\phi_1;\phi_1;\phi_5;\phi_5) + \tilde{V}(\phi_2;\phi_2;\phi_5;\phi_5)] \\
&= \sqrt{1/6}\,(12B + 6C).
\end{aligned}
$$

With this the total perturbation energies are

$$\Delta E_{a,b}(^1A_{1g}|(e_g)^2, (t_{2g})^2) = A + 9B + \frac{9}{2}C + 2\varepsilon_0 + 2Dq$$
$$\pm \frac{1}{2}[(-2B - C + 20Dq)^2$$
$$+ \frac{2}{3}(12B + 6C)^2]^{1/2}$$

*Configuration interaction* $^1E_g|(e_g)^2, (t_{2g})^2$. The subdeterminant is of the fourth order and decomposes into two $2\times 2$ determinants. The one is formed with the $\Xi$ functions which transform as the first column of the representation matrices of $E_g$, (cf. pp. 374, 378).

$$K_1 = \Xi_1(^1E_g|(e_g)^2|0) = \chi_{1,2}(0,0)$$
$$= \sqrt{1/2}[\Phi(\phi_1^+;\phi_2^-) - \Phi(\phi_1^-;\phi_2^+)],$$
$$K_2 = \Xi_1(^1E_g|(t_{2g})^2|0) = \sqrt{1/2}[\chi_{4,4}(0,0) - \chi_{5,5}(0,0)]$$
$$= \sqrt{1/2}[\Phi(\phi_4^+;\phi_4^-) - \Phi(\phi_5^+;\phi_5^-)].$$

The other $2\times 2$ determinant has the $\Xi$ functions which transform as the second column of the $E_g$ representation matrices:

$$\Xi_2(^1E_g|(e_g)^2|0) = \sqrt{1/2}[\chi_{1,1}(0,0) - \chi_{2,2}(0,0)]$$
$$= \sqrt{1/2}[\Phi(\phi_1^+;\phi_1^-) - \Phi(\phi_2^+;\phi_2^-)],$$
$$\Xi_2(^1E_g|(t_{2g})^2|0) = \sqrt{1/6}[2\chi_{3,3}(0,0) - \chi_{4,4}(0,0) - \chi_{5,5}(0,0)]$$
$$= \sqrt{1/6}[2\Phi(\phi_3^+;\phi_3^-) - \Phi(\phi_4^+;\phi_4^-) - \Phi(\phi_5^+;\phi_5^-)].$$

It must be tested from case to case if the indexing of the $\Xi$ functions and thereby the assignment to the columns of the representation matrices is correct. In consideration of the calculation of configuration interaction the indexing of the $\Xi$ functions was already carried out correctly earlier. The correctness of the indexing can be demonstrated by determining the transformation behaviour of the $\Xi$ functions given above with the help of the Tables B.23 and B.25:

| | | $E$ | $C_3$ | $C_2$ | $C_4$ | $C_2'$ |
|---|---|---|---|---|---|---|
| $(e_g)^2$: | $\Xi_1$ | $\Xi_1$ | $-\frac{1}{2}\Xi_1 + \frac{\sqrt{3}}{2}\Xi_2$ | $\Xi_1$ | $-\Xi_1$ | $-\Xi_1$ |
| | $\Xi_2$ | $\Xi_2$ | $-\frac{\sqrt{3}}{2}\Xi_1 - \frac{1}{2}\Xi_2$ | $\Xi_2$ | $\Xi_2$ | $\Xi_2$ |

| | | $E$ | $C_3$ | $C_2$ | $C_4$ | $C_2'$ |
|---|---|---|---|---|---|---|
| $(t_{2g})^2$: | $\Xi_1$ | $\Xi_1$ | $-\frac{1}{2}\Xi_1+\frac{\sqrt{3}}{2}\Xi_2$ | $\Xi_1$ | $-\Xi_1$ | $-\Xi_1$ |
| | $\Xi_2$ | $\Xi_2$ | $-\frac{\sqrt{3}}{2}\Xi_1-\frac{1}{2}\Xi_2$ | $\Xi_2$ | $\Xi_2$ | $\Xi_2$ |

The tables show that both $\Xi_1$ functions transform in the same way, that is, according to the same column of the transformation matrices. This is also true of the two $\Xi_2$ functions*.

We shall use the basis $K_1, K_2$ to calculate the total perturbation energy

$$\begin{vmatrix} H_{1,1}-\Delta E({}^1E_g|(e_g)^2, (t_{2g})^2) & H_{1,2} \\ H_{2,1} & H_{2,2}-\Delta E({}^1E_g|(e_g)^2, (t_{2g})^2) \end{vmatrix}=0.$$

According to pp. 366, 375, 378 we have

$$\begin{aligned} H_{1,1} &= (K_1, \boldsymbol{H}_{\text{Lig.}}K_1)+(K_1, \boldsymbol{H}_{\text{el.}}K_1) \\ &= \Delta\tilde{E}_{\text{Lig.}}((e_g)^2)+\Delta\tilde{E}_{\text{el.}}({}^1E_g|(e_g)^2) \\ &= 2\varepsilon_0+12Dq+A+2C, \\ H_{2,2} &= (K_2, \boldsymbol{H}_{\text{Lig.}}K_2)+(K_2, \boldsymbol{H}_{\text{el.}}K_2) \\ &= \Delta\tilde{E}_{\text{Lig.}}((t_{2g})^2)+\Delta\tilde{E}_{\text{el.}}({}^1E_g|(t_{2g})^2) \\ &= 2\varepsilon_0-8Dq+A+B+2C. \end{aligned}$$

Using the rules on p. 249 and Table B.24 one obtains further

$$\begin{aligned} H_{1,2}=H_{2,1} &= (K_2, \boldsymbol{H}_{\text{Lig.}}K_1)+(K_2, \boldsymbol{H}_{\text{el.}}K_1)=(K_2, \boldsymbol{H}_{\text{el.}}K_1) \\ &= 2\sqrt{3}B. \end{aligned}$$

With this the total perturbation energies become

$$\begin{aligned} \Delta E_{a,b}({}^1E_g|(e_g)^2, (t_{2g})^2) = A+\frac{1}{2}B+2C+2\varepsilon_0+2Dq \\ \pm\frac{1}{2}[(-B+20Dq)^2+48B^2]^{1/2}. \end{aligned}$$

*Configuration interaction* ${}^1T_{2g}|(e_g)^1(t_{2g})^1, (t_{2g})^2$. The subdeterminant is of the order $2\times1\times3 = 6$ and decomposes into three $2\times2$ determinants. The

---

* If we had determined that the transformation behaviour of both $(t_{2g})^2$ functions does not go as the same columns as the transformation behaviour of $(e_g)^2$ functions we would only need to form two linear combinations of the two $(t_{2g})^2$ functions which have the desired property, that is, transform as the same column as the $(e_g)^2$ function. We could have then proceeded further using these new $(t_{2g})^2$ functions.

associated function pairs are (according to pp. 382, 387)

$$K_1 = \Xi_1({}^1T_{2g}|(e_g)^1(t_{2g})^1|0) = \chi_{1,3}(0,0)$$
$$= \sqrt{1/2}\,[\Phi(\phi_1^+;\phi_3^-) - \Phi(\phi_1^-;\phi_3^+)],$$
$$K_2 = \Xi_1({}^1T_{2g}|(t_{2g})^2|0) = \chi_{4,5}(0,0)$$
$$= \sqrt{1/2}\,[\Phi(\phi_4^+;\phi_5^-) - \Phi(\phi_4^-;\phi_5^+)],$$
$$\Xi_2({}^1T_{2g}|(e_g)^1(t_{2g})^1|0) = -\sqrt{1/4}\,\chi_{1,5}(0,0) + \sqrt{3/4}\,\chi_{2,5}(0,0)$$
$$= \sqrt{1/8}\,[-\Phi(\phi_1^+;\phi_5^-) + \Phi(\phi_1^-;\phi_5^+) + \sqrt{3}\Phi(\phi_2^+;\phi_5^-) - \sqrt{3}\Phi(\phi_2^-;\phi_5^+)],$$
$$\Xi_2({}^1T_{2g}|(t_{2g})^2|0) = \chi_{3,4}(0,0)$$
$$= \sqrt{1/2}\,[\Phi(\phi_3^+;\phi_4^-) - \Phi(\phi_3^-;\phi_4^+)],$$
$$\Xi_3({}^1T_{2g}|(e_g)^1(t_{2g})^1|0) = -\sqrt{1/4}\,\chi_{1,4}(0,0) - \sqrt{3/4}\,\chi_{2,4}(0,0)$$
$$= -\sqrt{1/8}\,[\Phi(\phi_1^+;\phi_4^-) - \Phi(\phi_1^-;\phi_4^+) + \sqrt{3}\Phi(\phi_2^+;\phi_4^-) - \sqrt{3}\Phi(\phi_2^-;\phi_4^+)],$$
$$\Xi_3({}^1T_{2g}|(t_{2g})^2|0) = \chi_{3,5}(0,0)$$
$$= \sqrt{1/2}\,[\Phi(\phi_3^+;\phi_5^-) - \Phi(\phi_3^-;\phi_5^+)].$$

Following the procedure explained above we have

$$H_{1,2} = 2\sqrt{3}\,B$$

and with this

$$\Delta E_{a,b}({}^1T_{2g}|(e_g)^1(t_{2g})^1,(t_{2g})^2) = A + \frac{1}{2}B + 2C + 2\varepsilon_0 - 3Dq \pm \frac{1}{2}[(B-10Dq)^2 + 48B^2]^{1/2}.$$

*Configuration interaction* ${}^3T_{1g}|(e_g)^1(t_{2g})^1,(t_{2g})^2$. Here the subdeterminant is of order $2\times 3\times 3 = 18$. Because functions with different $M_S$ values cannot combine with one another with reference to $\boldsymbol{H}_1$ the subdeterminant decomposes at first into three $6\times 6$ determinants. If as a basis $\Xi$ functions are selected which transform as the same column of the matrix representation, then from every $6\times 6$ determinant three $2\times 2$ determinants result which are in each case formed with pairs of $\Xi$ functions having the same $M_S$ value and which transform as the same column of the representation matrices of $T_{1g}$. According to pp. 350, 386 there are the following function pairs when $M_S = 1$:

$$K_1 = \Xi_1({}^3T_{1g}|(e_g)^1(t_{2g})^1|1) = \chi_{2,3}(1,1) = \Phi(\phi_2^+;\phi_3^+),$$
$$K_2 = \Xi_1({}^3T_{1g}|(t_{2g})^2|1) = \chi_{4,5}(1,1) = \Phi(\phi_4^+;\phi_5^+),$$

$$\Xi_2(^3T_{1g}|(e_g)^1(t_{2g})^1|1) = -\sqrt{3/4}\chi_{1,5}(1,1) - \sqrt{1/4}\chi_{2,5}(1,1)$$
$$= -\sqrt{3/4}\Phi(\phi_1^+;\phi_5^+) - \sqrt{1/4}\Phi(\phi_2^+;\phi_5^+),$$
$$\Xi_2(^3T_{1g}|(t_{2g})^2|1) = \chi_{3,4}(1,1) = \Phi(\phi_3^+;\phi_4^+),$$
$$\Xi_3(^3T_{1g}|(e_g)^1(t_{2g})^1|1) = -\sqrt{3/4}\chi_{1,4}(1,1) + \sqrt{1/4}\chi_{2,4}(1,1)$$
$$= -\sqrt{3/4}\Phi(\phi_1^+;\phi_4^+) + \sqrt{1/4}\Phi(\phi_2^+;\phi_4^+),$$
$$\Xi_3(^3T_{1g}|(t_{2g})^2|1) = \chi_{3,5}(1,1) = \Phi(\phi_3^+;\phi_5^+).$$

When $M_S = 0$ and $M_S = -1$ there are corresponding pairs of functions. The roots of each $2\times 2$ determinant which is formed with any one of these function pairs are because of

$$H_{1,2} = 6B$$

$$\Delta E_{a,b}(^3T_{1g}|(e_g)^1(t_{2g})^1,(t_{2g})^2) = A - \frac{1}{2}B + 2\varepsilon_0 - 3Dq \pm \frac{1}{2}[(-9B-10Dq)^2 + 144B^2]^{1/2}.$$

The perturbation energies for octahedrally bound $d^2$ ions are shown in Figure B.27. The term positions for the strong-field case are sketched in the right-hand side of the diagram. As one proceeds towards the middle, the consequences of configuration interaction are taken into account. In the left-hand side of the figure the results for the weak-field case are given and as one proceeds toward the middle the corrections for term interaction are included. As can be seen from the figure and as an inspection of the calculated perturbation energies shows the *same term energies* are obtained independent of if one

(a) begins with the weak-field treatment and considers term interaction or

(b) starts from the strong-field case and considers configuration interaction.

For the $^3T_{1g}$ terms e.g., we obtained (p. 395) using the weak-field case and term interaction:

$$\Delta E_{a,b}(^3T_{1g}|^3P,^3F) = A - \frac{1}{2}B + 2\varepsilon_0 - 3Dq \pm \frac{1}{2}[(15B+6Dq)^2 + 64Dq^2]^{1/2}$$

and above (using the strong-field case and configuration interaction):

$$\Delta E_{a,b}(^3T_{1g}|(e_g)^1(t_{2g})^1,(t_{2g})^2) = A - \frac{1}{2}B + 2\varepsilon_0 - 3Dq \pm \frac{1}{2}[(-9B-10Dq)^2 + 144B^2]^{1/2}.$$

The two expressions differ only in the *form* of the factor under the square root sign. They are, however, in both cases the same, $225\,B^2+180\,BDq+100\,Dq^2$.

This behaviour, which holds not only for the $d^2$ ions treated here as examples, is sometimes overlooked. It is however apparent. In both cases one proceeds from the solutions $\Phi$ of the unperturbed system (3–41) and determines the roots of the complete secular determinant (3–44) using the $\Phi$ basis. It has of course no effect on the values of the roots if the secular determinant is formulated using the $\Phi$ themselves or with a set of linearly independent linear combinations of the $\Phi$ functions. The $\Theta_r({}^{2S+1}\Gamma_i|{}^{2S+1}L|M_S)$ as well as the $\Xi_r({}^{2S+1}\Gamma_i|\gamma_1^a\gamma_2^b\ldots|M_S)$ are such sets of functions. That in general different results are obtained in the strong-field case than in the weak-field case should not however be unexpected. In both approximations certain nondiagonal elements in the secular determinant (3–44) are arbitrarily set equal to zero; in the strong-field case there are other elements than in the weak-field case. Solely herein lies the difference of the results obtained by the two approximation methods.

For the sake of clarity the results obtained above for octahedrally coordinated $d^2$ systems are collected in Table B.28, 29, 30, and 31. From Tables B.29 and B.31 one sees that the nondiagonal elements of the same matrix have identical values. Further one sees that the nondiagonal elements of the secular determinants (Table B.29) used in the calculation of the term interaction energy are multiples of $Dq$, while in the calculation

Table B.28. Weak-field case. Term energies of octahedrally bound $d^2$ systems

| Term of the free ion | $\Delta E_{\text{el.}}$ | Term of the ion in the ligand field | $\Delta E_{\text{Lig.}}$ |
|---|---|---|---|
| ${}^3F$ | $A-8B$ | ${}^3A_{2g}$ | $2\varepsilon_0+12\,Dq$ |
| | | ${}^3T_{1g}$ | $2\varepsilon_0-6\,Dq$ |
| | | ${}^3T_{2g}$ | $2\varepsilon_0+2\,Dq$ |
| ${}^3P$ | $A+7B$ | ${}^3T_{1g}$ | $2\varepsilon_0$ |
| ${}^1S$ | $A+14B+7C$ | ${}^1A_{1g}$ | $2\varepsilon_0$ |
| ${}^1D$ | $A-3B+2C$ | ${}^1E_g$ | $2\varepsilon_0+\frac{24}{7}\,Dq$ |
| | | ${}^1T_{2g}$ | $2\varepsilon_0-\frac{16}{7}\,Dq$ |
| ${}^1G$ | $A+4B+2C$ | ${}^1A_{1g}$ | $2\varepsilon_0+4\,Dq$ |
| | | ${}^1E_g$ | $2\varepsilon_0+\frac{4}{7}\,Dq$ |
| | | ${}^1T_{1g}$ | $2\varepsilon_0+2\,Dq$ |
| | | ${}^1T_{2g}$ | $2\varepsilon_0-\frac{26}{7}\,Dq$ |

Table B.29.

$^1A_{1g}$

$$\begin{array}{c|cc|} & ^1S & ^1G \\ & A+14B+7C+2\varepsilon_0 & 4\sqrt{6}Dq \\ & 4\sqrt{6}Dq & A+4B+2C+2\varepsilon_0+4Dq \end{array}$$

$^1E_g$

$$\begin{array}{c|cc|} & ^1D & ^1G \\ & A-3B+2C+2\varepsilon_0+\frac{24}{7}Dq & \frac{40}{7}\sqrt{3}Dq \\ & \frac{40}{7}\sqrt{3}Dq & A+4B+2C+2\varepsilon_0+\frac{4}{7}Dq \end{array}$$

$^1T_{2g}$

$$\begin{array}{c|cc|} & ^1D & ^1G \\ & A-3B+2C+2\varepsilon_0-\frac{16}{7}Dq & \frac{20}{7}\sqrt{3}Dq \\ & \frac{20}{7}\sqrt{3}Dq & A+4B+2C+2\varepsilon_0-\frac{26}{7}Dq \end{array}$$

$^3T_{1g}$

$$\begin{array}{c|cc|} & ^3F & ^3P \\ & A-8B+2\varepsilon_0-6Dq & -4Dq \\ & -4Dq & A+7B+2\varepsilon_0 \end{array}$$

| $^1T_{1g}$ | $^3T_{2g}$ | $^3A_{2g}$ |
|---|---|---|
| $^1G$ | $^3F$ | $^3F$ |
| $A+4B+2C+2\varepsilon_0+2Dq$ | $A-8B+2\varepsilon_0+2Dq$ | $A-8B+2\varepsilon_0+12Dq$ |

of the configuration interaction energy only the electron interaction parameters $B$ and $C$ appear in the nondiagonal elements.

Matrix elements obtained using the weak-field method with consideration of term interaction have been published by L. E. Orgel (*J. Chem. Physics*, **23**, 1004 (1955)) for $d^2$, $d^8$ and $d^5$ and by R. Finkelstein and J. H. van Vleck (*J. Chem. Physics*, **8**, 790 (1940)) for $d^3$ and $d^7$ (octahedral symmetry). These authors give the matrix elements in an abbreviated notation in units of $Dq$, e.g., $^1A_{1g}$ for the $d^2$ problem:

$$\begin{array}{|cc|} ^1S & ^1G \\ 0 & 4\sqrt{6} \\ 4\sqrt{6} & 4 \end{array}$$

Table B.30. Strong-field case. Term energies of octahedrally bound $d^2$ systems

| Ligand field configuration | $\Delta\tilde{E}_{\text{Lig.}}$ | Term of the ion in the ligand field | $\Delta\tilde{E}_{\text{el.}}$ |
|---|---|---|---|
| $(t_{2g})^2$ | $2\varepsilon_0 - 8Dq$ | $^3T_{1g}$ | $A-5B$ |
| | | $^1A_{1g}$ | $A+10B+5C$ |
| | | $^1E_g$ | $A+B+2C$ |
| | | $^1T_{2g}$ | $A+B+2C$ |
| $(t_{2g})^1(e_g)^1$ | $2\varepsilon_0 + 2Dq$ | $^3T_{2g}$ | $A-8B$ |
| | | $^3T_{1g}$ | $A+4B$ |
| | | $^1T_{2g}$ | $A+2C$ |
| | | $^1T_{1g}$ | $A+4B+2C$ |
| $(e_g)^2$ | $2\varepsilon_0 + 12Dq$ | $^3A_{2g}$ | $A-8B$ |
| | | $^1A_{1g}$ | $A+8B+4C$ |
| | | $^1E_g$ | $A+2C$ |

A comparison with the corresponding data in Table B.29 shows that for the diagonal elements the equal shifts for all terms of $2\varepsilon_0$ and $A$ have been omitted. They do not play a rôle for the optical and magnetic properties. Further the energy difference between the terms $^1S$ and $^1G$ of the free ion is not taken into account. It is as can be seen from Table B.28 $10\,B+5\,C$.

Y. Tanabe and S. Sugano (*J. Phys. Soc.* (*Japan*), **9**, 753, 766 (1954)) have given the matrix elements for $d^2$ to $d^8$ obtained using the strong-field method and considering configuration interaction. They also use an abbreviated notation, e.g., for $^1A_{1g}$ in the $d^2$ problem:

$$\begin{array}{cc} (t_{2g})^2 & (e_g)^2 \\ \left| \begin{array}{cc} 10\,B+5C-10\,Dq & \sqrt{6}(2\,B+C) \\ \sqrt{6}(2\,B+C) & 8\,B+4C+10\,Dq \end{array} \right| \end{array}$$

The equal shift $A+2\varepsilon_0$ for all terms in the diagonal elements is omitted. Further the zero point of the energy was assigned to the configuration $(t_{2g})^1(e_g)^1$ so that $\Delta\tilde{E}_{\text{Lig.}}$ assumes the values $-10\,Dq$ for $(t_{2g})^2$ and $+10\,Dq$ for $(e_g)^2$.

For the matrix elements obtained for the strong-field and weak-field cases and for their relationships to one another the reader is referred to C. K. Jørgensen, *Absorption Spectra and Chemical Bonding in Complexes*, Pergamon Press, 1962, pp. 81–84.

Table B.31. Matrix elements for a rigorous treatment of octahedrally bound $d^2$ systems (with configuration interaction). The secular determinants given in section 3.4.3.2 are obtained by adding the term $-\Delta E(^{2S+1}\Gamma|(e_g)^a(t_{2g})^b, (e_g)^{a'}(t_{2g})^{b'})$ to the diagonal elements and setting the determinants equal to zero.

$$^1A_{1g}$$

$$\begin{array}{cc} (t_{2g})^2 & (e_g)^2 \end{array}$$

$$\begin{vmatrix} 2\varepsilon_0-8Dq+A+10B+5C & \sqrt{6}(2B+C) \\ \sqrt{6}(2B+C) & 2\varepsilon_0+12Dq+A+8B+4C \end{vmatrix}$$

$$^1E_g$$

$$\begin{array}{cc} (t_{2g})^2 & (e_g)^2 \end{array}$$

$$\begin{vmatrix} 2\varepsilon_0-8Dq+A+B+2C & 2\sqrt{3}B \\ 2\sqrt{3}B & 2\varepsilon_0+12Dq+A+2C \end{vmatrix}$$

$$^1T_{2g}$$

$$\begin{array}{cc} (t_{2g})^2 & (t_{2g})^1(e_g)^1 \end{array}$$

$$\begin{vmatrix} 2\varepsilon_0-8Dq+A+B+2C & 2\sqrt{3}B \\ 2\sqrt{3}B & 2\varepsilon_0+2Dq+A+2C \end{vmatrix}$$

$$^3T_{1g}$$

$$\begin{array}{cc} (t_{2g})^2 & (t_{2g})^1(e_g)^1 \end{array}$$

$$\begin{vmatrix} 2\varepsilon_0-8Dq+A-5B & 6B \\ 6B & 2\varepsilon_0+2Dq+A+4B \end{vmatrix}$$

| $^1T_{1g}$ | $^3T_{2g}$ | $^3A_{2g}$ |
|---|---|---|
| $(t^2{}_g)^1(e_g)^1$ | $(t_{2g})^1(e_g)^1$ | $(e_g)^2$ |
| $2\varepsilon_0+2Dq+A+4B+2C$ | $2\varepsilon_0+2Dq+A-8B$ | $2\varepsilon_0+12Dq+A-8B$ |

# 4. The molecular orbital theory as an extension of the ionic ligand field theory

## 4.1. The method

The extended electrostatic model upon which the ligand field theory is based and which was extensively treated in Part A, section 1.12, constitutes a very rough approximation. To be sure it allows numerous properties of transition metal complexes to be described at least qualitatively accurately. However, it is in general not capable of explaining detailed properties of such complexes. It is to be noted that the range of applicability of this model becomes more restricted the less *ionic* character the central ion-ligand bonds show. Certain properties of transition metal complexes which are typical *covalent* compounds can be understood only in a limited way or not at all using the ionic ligand field theory. These observations are explained by considering the nature of the extended ionic model. Here a complex ion is treated as a system whose electrons belong either exclusively to the central ion or to the ligands. In other words, the electrons are localized. This postulation specifically excludes covalent bonds and even covalent bond contributions. Therefore, it is not surprising that this theory, originally conceived essentially for transition metal complexes having predominantly ionic bonds, loses its usefulness to the degree to which covalent bond contributions gain on importance.

In order to arrive at an all-embracing description of transition metal complexes the ionic ligand field theory must first of all be extended so that the appearance of covalent bonds is taken into consideration. In a theory extended in this manner the ionic ligand field theory is a limiting case which is reached when the covalent bond contributions can be neglected in comparison to the ionic contributions. The MO theory offers the possibility of such an extension. The conception underlying this theory is that the individual valence electrons in a complex ion do not move either exclusively in orbitals of the central ion or of the ligands, as is assumed in the ionic ligand field theory; rather, the valence electrons are permitted to move over the entire complex ion. The requirement for localization of the valence electrons, as is characteristic for the ionic ligand field theory, is removed.

In order to formulate the molecular orbitals of a system of valence electrons in a complex ion it is useful to the first rough approximation to use a *one-electron scheme.* One considers that in the zeroth approximation the valence electrons move independently from one another in a potential $V$ that arises first of all from the charges on the core of the central ion and the ligands and secondly from an average charge distribution of the valence electrons. The *Hamiltonian* for the motion of a *single valence* electron is accordingly (in atomic units):

$$H = -\frac{1}{2}\Delta + V. \tag{4–1}$$

If the system consists of a central ion and $L$ ligands, the potential $V$ can be divided advantageously into single additive terms:

$$V = V_Z + \sum_{\lambda=1}^{L} V_\lambda + V'. \tag{4–2}$$

$V_Z$ is defined here so that $H_Z = -\frac{1}{2}\Delta + V_Z$ is the *Hamiltonian* for the motion of one valence electron in the *effective potential of the isolated central ion.* The corresponding wave functions and the associated energy eigenvalues are $\psi_i$ and $\varepsilon_{Z,i}$:

$$H_Z \psi_i = \varepsilon_{Z,i} \psi_i. \tag{4–3}$$

In the same sense $H_\lambda = -\frac{1}{2}\Delta + V_\lambda$ is the *Hamiltonian* for the motion of one valence electron in the *effective potential of the isolated λth ligand* and has the eigenfunctions $\varphi_{\lambda,j}$ and the eigenvalues $\varepsilon_{\lambda,j}$:

$$H_\lambda \varphi_{\lambda,j} = \varepsilon_{\lambda,j}\, \varphi_{\lambda,j}. \tag{4–4}$$

$V'$ is the so-called *shielding potential* that tends to zero with increasing separation between the components of the complex.

If the complex components are located very far apart, then the functions $\psi_1, \ldots, \psi_i, \ldots; \varphi_{1,1}, \ldots; \ldots; \varphi_{\lambda,1}, \ldots, \varphi_{\lambda,j}, \ldots; \ldots; \varphi_{L,1}, \ldots, \varphi_{L,j}, \ldots$ are also solutions of the Schrödinger equation with the complete *Hamiltonian* (4–1) and have the eigenvalues $\varepsilon_{Z,i}$ or $\varepsilon_{\lambda,j}$. It is reasonable to use this set of functions $\psi_i$ and $\varphi_{\lambda,j}$ as the starting basis for an approximate solution of the *Schrödinger* equation for the case that the separations of the complex components have finite values, in particular the empirically determined values. In order to take into consideration the possibility that the valence electrons move over the entire complex ion, one can assume *linear combinations* of the $\psi_i$ and $\varphi_{\lambda,j}$ for the one-electron functions:

$$\Psi = \sum_{i=1}^{I} c_i \psi_i + \sum_{\lambda=1}^{L} \sum_{j=1}^{J_\lambda} k_{\lambda,j}\, \varphi_{\lambda,j}. \tag{4–5}$$

We have used in this trial solution $I$ central ion functions and $J_\lambda$ functions of the $\lambda th$ ligand:

$$\psi_1, \psi_2, \ldots, \psi_I; \varphi_{1,1}, \ldots, \varphi_{1,J_1}; \varphi_{2,1}, \ldots, \varphi_{2,J_2}; \ldots; \varphi_{L,1}, \ldots, \varphi_{L,J_L}. \qquad (4\text{–}6)$$

Which functions these are will be discussed later using an example. The assumption (4–5) is called a *linear variation assumption** because the undetermined coefficients $c_i$ and $k_{\lambda,j}$ can be determined using a variation calculation, which leads to a secular determinant of the order $I + \sum_{\lambda=1}^{L} J_\lambda = g$ (corresponding to the number of basis functions (4–6)):

$$\begin{vmatrix} (\psi_1, \boldsymbol{H}\psi_1) - (\psi_1, \psi_1)\varepsilon & (\psi_2, \boldsymbol{H}\psi_2) - (\psi_1, \psi_2)\varepsilon \ldots & (\psi_1, \boldsymbol{H}\varphi_{L,J_L}) - (\psi_1, \varphi_{L,J_L})\varepsilon \\ (\psi_2, \boldsymbol{H}\psi_1) - (\psi_2, \psi_1)\varepsilon & (\psi_2, \boldsymbol{H}\psi_2) - (\psi_2, \psi_2)\varepsilon \ldots & \cdot \\ \cdot & \cdot \quad\quad \cdot & \cdot \\ \cdot & \cdot \quad\quad \cdot & \cdot \\ \cdot & \cdot \quad\quad \cdot & \cdot \\ (\varphi_{L,J_L}, \boldsymbol{H}\psi_1) - (\varphi_{L,J_L}, \psi_1)\varepsilon & \cdot & \ldots (\varphi_{L,J_L}, \boldsymbol{H}\varphi_{L,J_L}) - (\varphi_{L,J_L}, \varphi_{L,J_L})\varepsilon \end{vmatrix} = 0. \qquad (4\text{–}7)$$

The roots $\varepsilon$ of this determinant are the possible energies which a valence electron can assume.

The energies $\varepsilon$ can also be determined when one begins with a basis of functions different from (4–6). Under the condition that the functions (4–6) be linearly independent (which can always be fulfilled), sets of in each case $g$ linearly independent linear combinations of the old basis functions (4–6) can be constructed. We shall call the functions of one of these sets

$$\Xi_1, \Xi_2, \ldots, \Xi_g \qquad (4\text{–}8)$$

These $g$ $\Xi$ functions can be considered to be new basis functions and a linear combination of them can be made:

$$\Phi = \sum_{i=1}^{g} a_i \Xi_i. \qquad (4\text{–}9)$$

The secular determinant which follows from a variation calculation is then

$$\|H_{ij} - S_{ij}\varepsilon\| = 0, \qquad i,j = 1,2,\ldots,g \qquad (4\text{–}10)$$

---

* For the example of complexes the term LCAO (linear combination of atomic orbitals) assumption is somewhat misleading. The $\varphi_{\lambda j}$ are one-electron functions of the ligands which, when we are concerned with polyatomic systems, are no longer atomic orbitals, but ligand molecular orbitals.

with

$$H_{ij} \equiv (\Xi_i, \boldsymbol{H}\Xi_j) \quad \text{and} \quad S_{ij} \equiv (\Xi_i, \Xi_j).$$

The roots ε of this secular determinant are the same as the roots ε of the determinant (4–7).

The transition from a basis (4–6) to a basis (4–8) can bring great advantages. If the functions of the $\Xi$ basis are chosen so that they transform as the *irreducible representations of the symmetry group of the Hamiltonian* $\boldsymbol{H}$, then the theorems 1 and 2, p. 317, can be applied. These give information as to when the integrals which appear in the secular determinant (4–10) vanish or are the same. The secular problem can usually be greatly simplified in this manner.

The procedure of setting up symmetry-adapted $\Xi$ functions will be demonstrated for the case of octahedral complex ions. We consider a complex ion consisting of a transition metal central ion and six identical ligands occupying the vertices of a regular octahedron. The system thus has the symmetry $O_h$ and the *Hamiltonian* $\boldsymbol{H}$ is invariant under all symmetry operations of the group $O_h$. We have at our disposal as basis functions of the type (4–6) the functions of the central ion

$$\psi_1, \psi_2, \ldots, \psi_I$$

and of the ligands

$$\begin{array}{l} \varphi_{1,1}, \varphi_{1,2}, \ldots, \varphi_{1,J_1}, \\ \varphi_{2,1}, \varphi_{2,2}, \ldots, \varphi_{2,J_2}, \\ \cdots\cdots\cdots\cdots \\ \varphi_{6,1}, \varphi_{6,2}, \ldots, \varphi_{6,J_6} \end{array}$$

whereby $J_1 = J_2 = J_3 = J_4 = J_5 = J_6$ holds because it was assumed that the six ligands are identical. It can also be seen that under the symmetry operations of the group $O_h$ respectively only central ion functions and only ligand functions can transform into one another. It is therefore reasonable to construct the symmetry-adapted $\Xi$ basis (4–8) so that it consists of $\Xi$ functions which arise from suitable linear combinations of only central ion functions and of such $\Xi$ functions which consist of ligand functions only.

## 4.2. Symmetry-adapted functions

We consider first the transformation behaviour of the ligand functions $\varphi_{\lambda,j}$. The set of the $\varphi_{\lambda,j}$ of a ligand can be subdivided into $\varphi_{\lambda,j}$ functions which allow δ bonds with the central ion and into those which are suited for π, δ bonds, etc.

All those $\varphi_{\lambda,j}$ functions which are rotationally symmetric about the line of centres ligand-central ion can produce $\sigma$ *bonds.* If the ligand is e.g., an atomic ion, then $s$, $p_z$, $d_{z^2}$ ... ligand functions are suited for $\sigma$ bonds, assuming that the line of centres coincides with the $z$ axis of a cartesian coordinate system centred in the ligand (cf. Figure B.28). The $\varphi$ functions of the $\lambda$th ligand which can induce $\sigma$ bonds are denoted with $z_\lambda$. In Figure B.29 they are symbolized for octahedral complexes with arrows marked with $z_\lambda$. The $z$ arrows express the rotational symmetry about the line of centres.

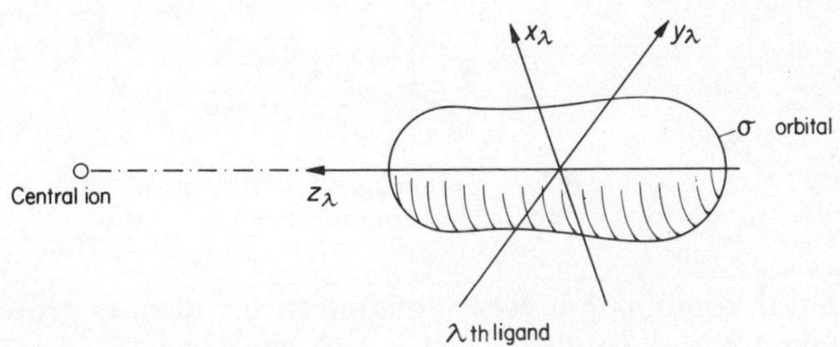


Figure B.28. A $\sigma$ orbital.

Which representation of the group $O_h$ is induced by a basis of six functions $z_1, z_2, z_3, z_4, z_5, z_6$? To answer this question one determines first the characters, that is, the diagonal sums of the transformation matrices, associated with the symmetry operations. It is here sufficient to determine how many $z_\lambda$ are brought into themselves by the individual symmetry operations, because only such $z_\lambda$ contribute to the diagonal sums, in each case exactly $+1$. The identity operation $E$, e.g., brings each of the six $z$ functions into itself; the associated character $\chi(E)$ is thus equal to 6. The rotation $C_4$ about the $z$ axis of the octahedron brings only the two $z$ functions $z_3$ and $z_4$ into themselves: $\chi(C_4) = 2$, etc. The characters determined in this manner for the representation of the group $O_h$ induced by the six $z_\lambda$ are summarized here:

| $R$ | $E$ | $8C_3$ | $3C_2$ | $6C_4$ | $6C_2'$ | $i$ | $8iC_3$ | $3iC_2$ | $6iC_4$ | $6iC_2$ |
|---|---|---|---|---|---|---|---|---|---|---|
| $\chi(R)$ | 6 | 0 | 2 | 2 | 0 | 0 | 0 | 4 | 0 | 2 |

With the aid of the reduction formula (2–4b) and with consideration of the character table for $O_h$, p. 479, one finds that the irreducible representations $A_{1g}$, $E_g$ and $T_{1u}$ of the group $O_h$ are contained in this representation.

Symmetry-adapted linear combinations of the six $z_\lambda$ can be easily formulated on the basis of plausibility arguments.

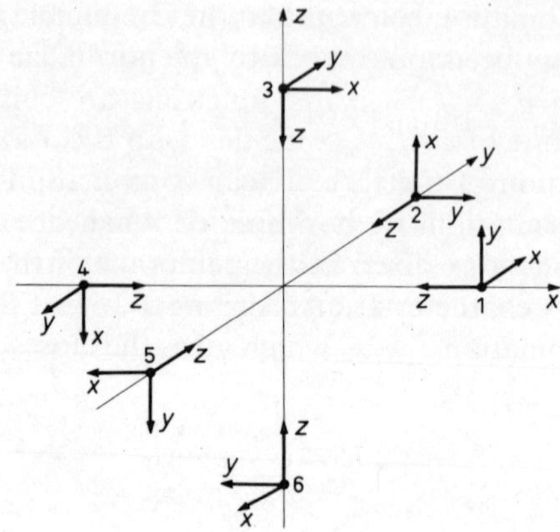


Figure B.29. $\sigma$ and $\pi$ ligand orbitals in an octahedral complex ion.

The linear combination corresponding to the identity representation $A_{1g}$ is found without difficulty when one considers that an $s$ function centred at the midpoint of the octahedron also transforms as $A_{1g}$. The functional behaviour of such a spherically symmetric $s$ function is apparently approximated most closely by the following linear combination of the $z_\lambda$ (cf. Figure B.30):

$$z_1 + z_2 + z_3 + z_4 + z_5 + z_6 .$$

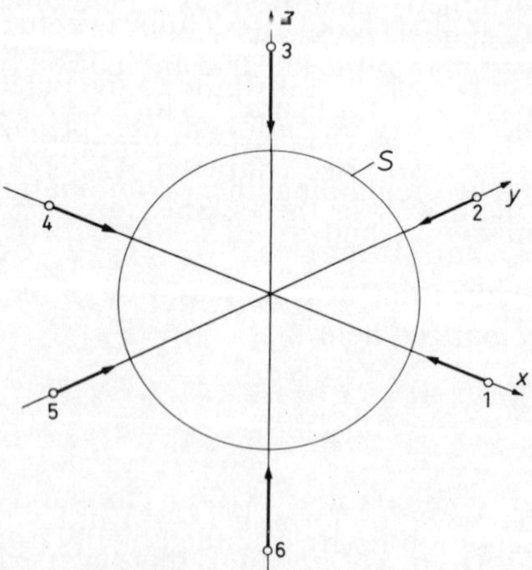


Figure B.30. $\sigma$ ligand orbitals of symmetry species $A_{1g}$ and a central ion $s$ orbital.

The linear combination belonging to the irreducible representation $A_{1g}$ is then, after having been multiplied* by the normalization factor†

$$a_{1g} = \sqrt{1/6}\,(z_1 + z_2 + z_3 + z_4 + z_5 + z_6) = \Xi_{12}.$$

The determination of the three linear combinations belonging to the irreducible representation $T_{1u}$ is simplified when one considers that the three $p_x, p_y, p_z$ functions (fixed at the midpoint of the octahedron) also have the symmetry character $T_{1u}$. According to Figure B.31a it is apparently the linear combination $z_1 - z_4$ which gives the closest approximation to

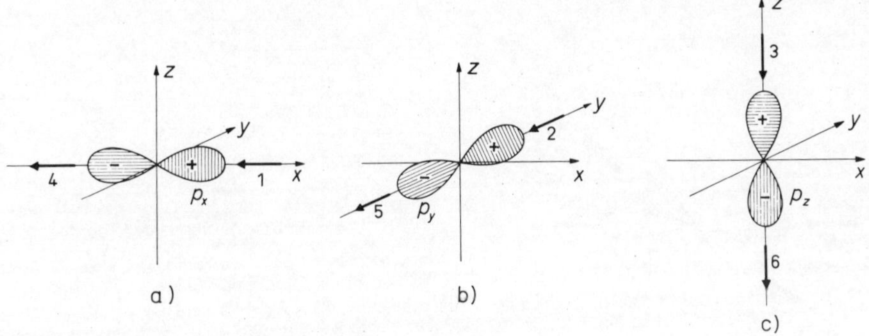


Figure B.31. Ligand $\sigma$ orbitals of species $T_{1u}$ and central ion $p$ orbitals.

the function $p_x$. (The $yz$ plane is a nodal plane for the function $p_x$ so that the $z$ functions lying in this plane cannot appear in the linear combination. $z_1$ goes with a plus sign into the linear combination because $p_x$ is positive on the positive $x$ axis, where the ligand 1 is located. Correspondingly $z_4$ has a negative sign in the linear combination.) From the Figure B.31b and c it can be seen that the two remaining linear combinations corresponding to $T_{1u}$ have the form $z_2 - z_5$ and $z_3 - z_6$. Normalized to 1 the three $T_{1u}$ functions are:

$$t_{1u}^{(x)} = \sqrt{1/2}\,(z_1 - z_4) = \Xi_{14},$$
$$t_{1u}^{(y)} = \sqrt{1/2}\,(z_2 - z_5) = \Xi_{17},$$
$$t_{1u}^{(z)} = \sqrt{1/2}\,(z_3 - z_6) = \Xi_{20}.$$

The two linear combinations with $\sigma$ functions which induce the irreducible representation $E_g$ are obtained finally from a consideration of

---

* The normalization factor is (as in the following) determined under the assumption that the overlap of the $z_\lambda$ is vanishingly small.

† The $\Xi$ functions have been numbered for convenience in later studies.

the behaviour of two functions for which it is known that they transform as $E_g$: $d_{z^2}$ and $d_{x^2-y^2}$ functions centred at the midpoint of the octahedron. Figure B.32(a) indicates the form $z_1 - z_2 + z_4 - z_5$ for the linear combination corresponding to the function $d_{x^2-y^2}$, or normalized to one this linear combination is:

$$e_g^{(x^2-y^2)} = \frac{1}{2}(z_1 - z_2 + z_4 - z_5) = \Xi_2 .$$

The linear combination $e_g^{(z^2)}$ cannot be found quantitatively from the corresponding Figure B.32(b). From the variation of the sign of the $d_{z^2}$

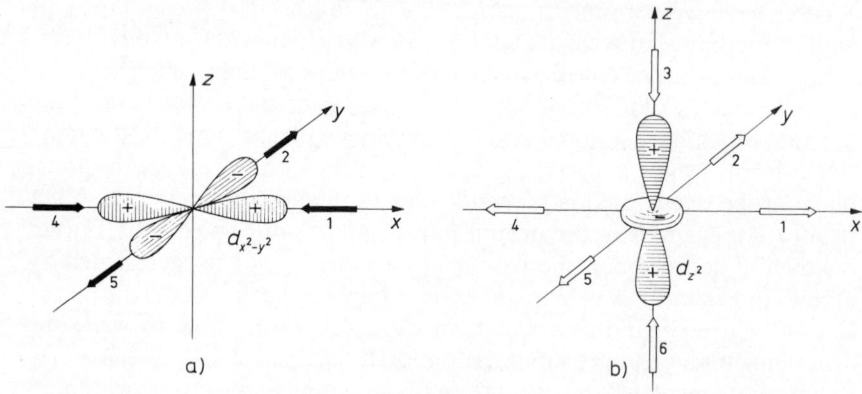


Figure B.32. Ligand $\sigma$ orbitals of species $E_g$ and central ion $d$ orbitals.

function one learns that $z_1, z_2, z_4, z_5$ appear with the same sign and that this must differ from the sign of $z_3$ and $z_6$. Nothing can be said concerning the magnitude of the coefficients $z_\lambda$ in the linear combination purely on the basis of the form of the $d_{z^2}$ function. One can, however, make use of the fact that the linear combination sought $e_g^{(z^2)}$ must be orthogonal to the remaining five linear combinations $a_{1g}$, $t_{1u}^{(x)}$, $t_{1u}^{(y)}$, $t_{1u}^{(z)}$, $e_g^{(x^2-y^2)}$, which are already known. If one writes first of all the general expression

$$e_g^{(z^2)} = c_1 z_1 + c_2 z_2 + c_3 z_3 + c_4 z_4 + c_5 z_5 + c_6 z_6 ,$$

then the relations of the coefficients $c_1, \ldots, c_6$ to one another are obtained from the five orthogonality conditions

$$\begin{aligned}
(a_{1g}, e_g^{(z^2)}) &= \sqrt{1/6}\,(c_1 + c_2 + c_3 + c_4 + c_5 + c_6) = 0 ,\\
(t_{1u}^{(x)}, e_g^{(z^2)}) &= \sqrt{1/2}\,(c_1 - c_4) = 0 ,\\
(t_{1u}^{(y)}, e_g^{(z^2)}) &= \sqrt{1/2}\,(c_2 - c_5) = 0 ,
\end{aligned}$$

$$(t_{1u}^{(z)}, e_g^{(z^2)}) \quad = \sqrt{1/2}\,(c_3 - c_6) = 0\,,$$
$$(e_g^{(x^2-y^2)}, e_g^{(z^2)}) \quad = \sqrt{1/4}\,(c_1 - c_2 + c_4 - c_5) = 0\,.$$

The five equations yield

$$c_1 = c_2 = -\frac{1}{2}\,c_3 = c_4 = c_5 = -\frac{1}{2}\,c_6\,.$$

The linear combination $e_g^{(z^2)}$ has then the (normalized) form

$$e_g^{(z^2)} = \frac{1}{2\sqrt{3}}\,(z_1 + z_2 - 2z_3 + z_4 + z_5 - 2z_6) = \Xi_4\,.$$

After having determined the linear combinations of the ligand orbitals with $\sigma$ bonding character, we shall study which symmetry adapted linear combinations can be constructed from the ligand orbitals, which lead to $\pi$ *bonds*. Under 'ligand orbitals with $\pi$ bond character' one understands those orbitals which have a nodal plane coinciding with the bond axis. $\pi$ functions always appear in pairs. The functions of a pair differ in that the nodal plane of the one orbital is perpendicular to the nodal plane of the other orbital. An example is the orbital pair $p_x$ and $p_y$ in Figure B.33. For an octahedral complex ion the pairs of $\pi$ orbitals $x_\lambda, y_\lambda$ are represented by arrows in Figure B.29, which are oriented perpendicular to the particular line of centres ligand-central ion in question. To be able to write the symmetry adapted linear combinations of the twelve $x_\lambda, y_\lambda$ $(\lambda = 1, 2, \ldots, 6)$ one studies first which representation of the symmetry group $O_h$ is induced by the basis of the twelve $x_\lambda, y_\lambda$ and which irreducible components are contained in the representation. The characters $\chi$ of the representation induced by $x_\lambda, y_\lambda$ are obtained by applying one symmetry operation from each class to the $x_\lambda, y_\lambda$ and taking the diagonal sum (spur) of the associated transformation matrices*. The identity operation $E$, e.g., leaves all twelve $x_\lambda, y_\lambda$ in their original position: $\chi(E) = 12$. According to the usual procedure one obtains the following characters:

| $R$ | $E$ | $8C_3$ | $3C_2$ | $6C_4$ | $6C_2'$ | $i$ | $8iC_3$ | $3iC_2$ | $6iC_4$ | $6iC_2'$ |
|---|---|---|---|---|---|---|---|---|---|---|
| $\chi(R)$ | 12 | 0 | −4 | 0 | 0 | 0 | 0 | 0 | 0 | 0 |

The irreducible representations $T_{1g}$, $T_{1u}$, $T_{2g}$ and $T_{2u}$ are contained in this representation, as an application of the reduction formula (2–4b)

* For the case of $\sigma$ orbitals $z_\lambda$ we could restrict ourselves in the determination of the characters to counting the number of $z_\lambda$ orbitals which do not change their positions under the influence of the individual symmetry operations. For the $\pi$ orbitals simply counting is not sufficient because the case can occur where a $\pi$ orbital remains on the same ligand after an operation but changes its sign. Of course then this function contributes the value $-1$ instead of $+1$ to the sum.

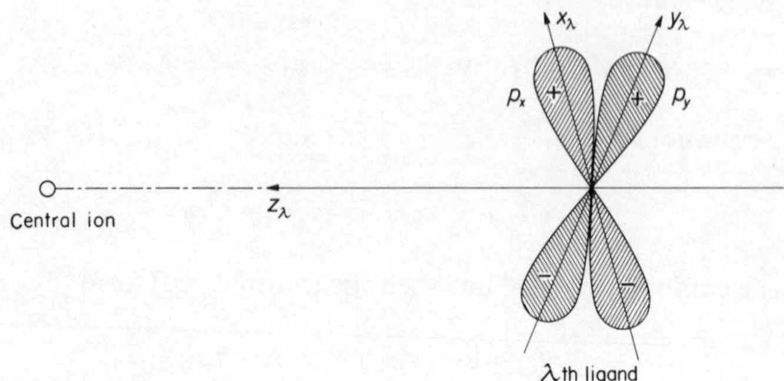


Figure B.33. Diagram of two $\pi$ orbitals.

shows. The associated linear combinations can be found using plausibility arguments, as was the case for the $\sigma$ functions.

Functions having the irreducible representations (symmetry species) $T_{1u}$ are obtained by making a linear combination of $x_\lambda, y_\lambda$ so that the orbitals $p_x, p_y, p_z$ (which also transform as $T_{1u}$) are approximated as closely as possible. From Figure B.34(a), (b), (c), one finds the following three (normalized) $\pi$ linear combinations (cf. the selection of the coordinate systems centred in the ligands shown in Figure B.29):

$$t_{1u}^{(x)} = \sqrt{1/4}\,(y_2 + x_3 - x_5 - y_6) = \Xi_{15},$$
$$t_{1u}^{(y)} = \sqrt{1/4}\,(x_1 + y_3 - y_4 - x_6) = \Xi_{18},$$
$$t_{1u}^{(z)} = \sqrt{1/4}\,(y_1 + x_2 - x_4 - y_5) = \Xi_{21}.$$

The linear combinations which transform as $T_{2g}$ can be easily found from a consideration of the functions $d_{xy}, d_{xz}, d_{yz}$ which likewise have the symmetry character $T_{2g}$. According to Figures 35(a), (b), (c), we have

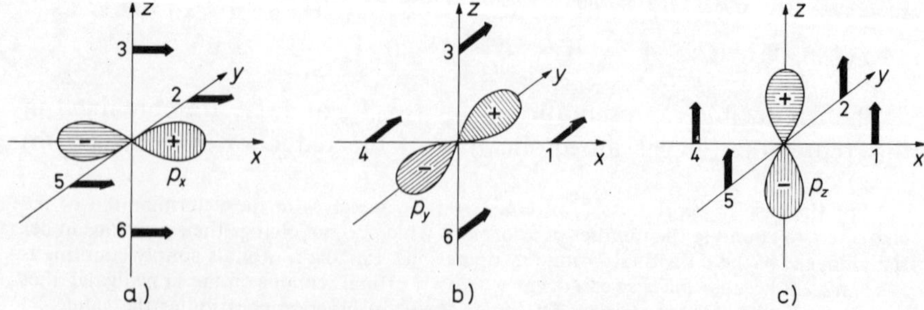


Figure B.34. Ligand $\pi$ orbitals of species $T_{1u}$ and central ion $p$ orbitals.

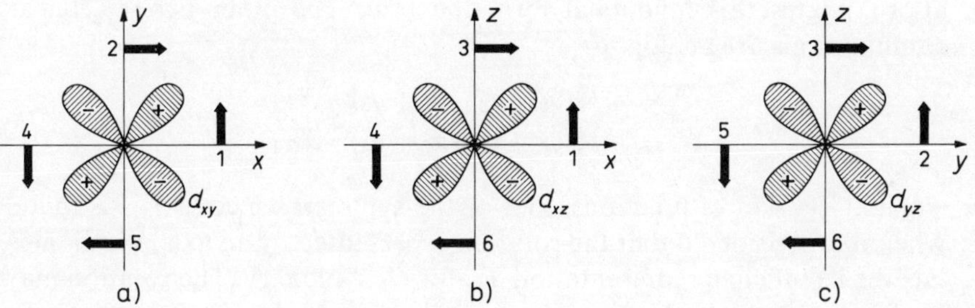


Figure B.35. Ligand $\pi$ orbitals of species $T_{2g}$ and central ion $d$ orbitals.

three linear combinations:

$$t_{2g}^{(xy)} = \sqrt{1/4}\,(x_1 + y_2 + y_4 + x_5) = \Xi_6\,,$$
$$t_{2g}^{(xz)} = \sqrt{1/4}\,(y_1 + x_3 + x_4 + y_6) = \Xi_8\,,$$
$$t_{2g}^{(yz)} = \sqrt{1/4}\,(x_2 + y_3 + y_5 + x_6) = \Xi_{10}\,.$$

The linear combinations of the irreducible representation $T_{2u}$ can be found without difficulty when one considers that according to Table B.13 these are three $f$ functions ($l = 3$) which transform as $T_{2u}$. One of these three is proportional to $Y_3^2 + Y_3^{-2} \sim z(x^2 - y^2)$. Its functional behaviour is shown schematically in Figure B.36. It is seen that the linear combination

$$t_{2u}^{(1)} = \sqrt{1/4}\,(y_1 - x_2 - x_4 + y_5) = \Xi_{22}$$

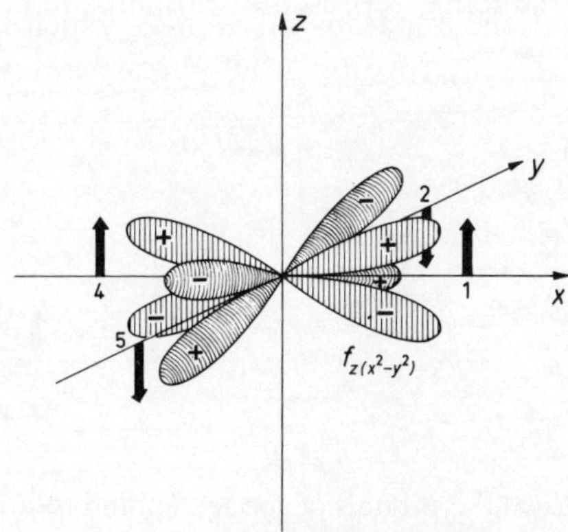


Figure B.36. Ligand $\pi$ functions of the species $T_{2u}$ and a central ion $f$ orbital.

approximates this functional form the best. The other two $T_{2u}$ linear combinations are analogous

$$t_{2u}^{(2)} = \sqrt{1/4}(y_2 - x_3 - x_5 + y_6) = \Xi_{23},$$
$$t_{2u}^{(3)} = \sqrt{1/4}(y_3 - x_1 - x_6 + y_4) = \Xi_{24}.$$

The last triad of functions, that of the symmetry species $T_{1g}$, is found when it is considered that the rotation represented by an axial vector also has the irreducible representation $T_{1g}$ for $O_h$ symmetry. The rotation may be decomposed into three components, that is, into the rotations about the $x$, $y$ and $z$ axes. In the Figure B.37(a), (b), (c) these rotational components are shown symbolically. One finds the corresponding linear combinations of the ligand $\pi$ orbitals from the figures:

$$t_{1g}^{(x)} = \sqrt{1/4}(-x_2 + y_3 - y_5 + x_6) = \Xi_{25},$$
$$t_{1g}^{(y)} = \sqrt{1/4}(y_1 - x_3 + x_4 - y_6) = \Xi_{26},$$
$$t_{1g}^{(z)} = \sqrt{1/4}(-x_1 + y_2 - y_4 + x_5) = \Xi_{27}.$$

The symmetry-adapted linear combinations of ligand functions with $\delta$ or higher bond characters can be determined according to the procedure shown. Besides the procedure used here, determination of the "correct" linear combinations of ligand functions by a comparison with central ion functions of known symmetry behaviour, there is also the (often quite complicated) method of computing the linear combinations with the aid of projection operators*.

In Table B.32 (columns 3 and 4) the linear combinations of the ligand functions with $\sigma$ and $\pi$ bond character determined above for octahedral systems (symmetry $O_h$) are summarized. The first column contains the symbols for the irreducible representations. In the column "central ion

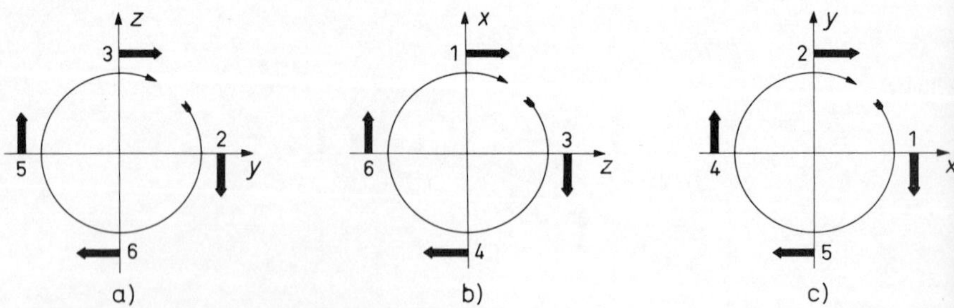


Figure B.37. Ligand $\pi$ orbitals of species $T_{1g}$ and rotations symbols.

* Cf. J. C. Slater: *Quantum Theory of Molecules and Solids*, Vol. 1, pp. 172 ff. and 323 ff. McGraw-Hill Book Company, Inc., New York, San Francisco, Toronto, London, 1963.

orbitals" the central ion functions are given which have the same *symmetry behaviour* as the linear combinations in the same row. Functions in the same row do not have the same irreducible representation, rather they transform as the same column of the appropriate representation matrices. For example, the linear combination $\sqrt{1/2}(z_1 - z_4)$ and $\sqrt{1/4}(y_2 + x_3 - x_5 - y_6)$ transform under all symmetry operations of the group $O_h$ as the same column of the representation matrices of $T_{1u}$ as the central ion orbital $p_x$. The symbols given in square brackets, e.g., [$g, \ldots$], indicate which central ion orbitals with $l > 2$ ($l$ = orbital angular momentum quantum number) also have the symmetry species of the appropriate line in the table*. The presence of only square brackets in the second column indicates that in that case no central ion orbital of suitable symmetry exists which has a smaller orbital angular momentum quantum number than that of the central ion function given in the square bracket. For example there are no $s$, $p$ and $d$ orbitals for the irreducible representation $T_{2u}$. Only $f$ (and higher) functions show such symmetry behaviour.

## 4.3. Energies

The energies which a valence electron can assume within the one-electron scheme are the same as the roots $\varepsilon$ of the secular determinant (4–10). The order of the secular determinant is equal to the number of the basis functions (4–6) and thereby also to the number of $\Xi$ functions (4–8), which are used in the linear variation calculation (4–9). We have not yet determined which and how many basis functions should be used. This decision is of course determined by the purpose of the calculation. In the theory of transition metal complexes principally electronic states of lower energy are of interest, that is the ground state and the lower-lying excited states. The description of these states will of course have the most success when the basis consists of such functions $\psi_i$ and $\varphi_{\lambda,j}$ which represent states of low energy in the isolated central ion or ligand.

In an isolated central ion of the first transition metal series the most stable one-electron states still available for valence electrons are the five $3d$, the $4s$, and the three $4p$ states with the energies $\varepsilon_{Z,3d} < \varepsilon_{Z,4s} < \varepsilon_{Z,4p}$. It is to be expected from these nine central ion states that of the one-electron states of the central ion they contribute the most intensively to the stable molecular orbitals of the complexes. For the example of regular octahedral complex ions the symmetry-adapted central ion $d$ functions are already

*Although an explicit assignment of $s$, $p$ and $d$ functions to the linear combinations is given, for functions with $l > 2$ only the symbols for $l$ have been entered into the table, without indication of the specific form of the central ion orbitals.

Table B.32. Assignment of central ion functions and ligand functions of the same symmetry species for octahedral complex ions (symmetry $O_h$), cf. Figure B.29.

| Species | Central ion functions | Ligand functions: σ type | Ligand functions: π type |
|---|---|---|---|
| $A_{1g}$ | $s = \Xi_{11}$ [g, ...] | $\sqrt{1/6}(z_1+z_2+z_3+z_4+z_5+z_6)$ | |
| $T_{1u}$ | $p_x = \Xi_{13}$ | $\sqrt{1/2}(z_1-z_4) = \Xi_{14}$ | $\sqrt{1/4}(y_2+x_3+x_5-y_6) = \Xi_{15}$ |
| | $p_y = \Xi_{16}$ | $\sqrt{1/2}(z_2-z_5) = \Xi_{17}$ | $\sqrt{1/4}(x_1+y_3-y_4-x_6) = \Xi_{18}$ |
| | $p_z = \Xi_{19}$ [f, ...] | $\sqrt{1/2}(z_3-z_6) = \Xi_{20}$ | $\sqrt{1/4}(y_1+x_2-x_4-y_5) = \Xi_{21}$ |
| $E_g$ | $d_{x^2-y^2} = \Xi_1$ | $\sqrt{1/4}(z_1-z_2+z_4-z_5) = \Xi_2$ | |
| | $d_{z^2} = \Xi_3$ [g, ...] | $\sqrt{1/12}(z_1+z_2-2z_3+z_4+z_5-2z_6) = \Xi_4$ | |
| $T_{2g}$ | $d_{xy} = \Xi_5$ | | $\sqrt{1/4}(x_1+y_2+y_4+x_5) = \Xi_6$ |
| | $d_{xz} = \Xi_7$ | | $\sqrt{1/4}(y_1+x_3+x_4+y_6) = \Xi_8$ |
| | $d_{yz} = \Xi_9$ [g, ...] | | $\sqrt{1/4}(x_2+y_3+y_5+x_6) = \Xi_{10}$ |
| $T_{2u}$ | [f, ...] | | $\sqrt{1/4}(y_1-x_2-x_4+y_5) = \Xi_{22}$ |
| | | | $\sqrt{1/4}(y_2-x_3-x_5+y_6) = \Xi_{23}$ |
| | | | $\sqrt{1/4}(y_3-x_1-x_6+y_4) = \Xi_{24}$ |
| $T_{1g}$ | [g, ...] | | $\sqrt{1/4}(-x_2+y_3-y_5+x_6) = \Xi_{25}$ |
| | | | $\sqrt{1/4}(y_1-x_3+x_4-y_6) = \Xi_{26}$ |
| | | | $\sqrt{1/4}(-x_1+y_2-y_4+x_5) = \Xi_{27}$ |

known from section 3.4.2, p. 369: $3d_{x^2-y^2} = \Xi$ and $3d_{z^2} = \Xi_3$, species $E_g$, as well as $3d_{xy} = \Xi_5$, $3d_{xz} = \Xi_7$ and $3d_{zy} = \Xi_9$ to the irreducible representation $T_{2g}$. The three central ion $p$ functions $4p_x = \Xi_{13}$, $4p_y = \Xi_{16}$, $4p_z = \Xi_{19}$ have the symmetry character $T_{1u}$, while the central ion function $4s = \Xi_{11}$ transforms as $A_{1g}$. In the left-hand part of Figure B.38 the energy levels of these functions for an isolated central ion are shown schematically*.

The right-hand side of Figure B.38 contains a schematic representation of the one-electron terms of the symmetry-adapted ligand functions with $\sigma$ and $\pi$ bond character. One-electron states of ligand systems with $\delta$ bond character are only indicated†.

The energetically lower-lying one-electron MO's should accordingly be described quite well by linear combinations of the $\Xi$ in which besides

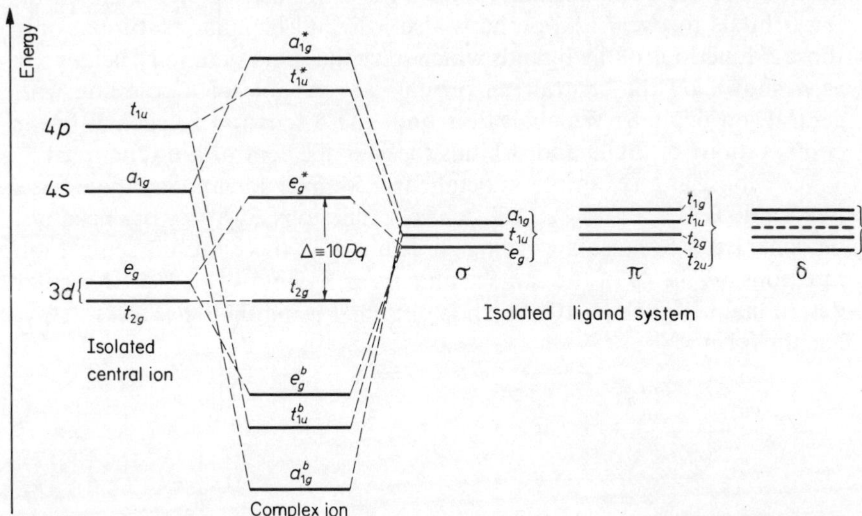


Figure B.38. Diagram of the coupling between central ion states and ligand $\sigma$ orbitals in octahedral complex ions. (Coupling with ligand $\pi$ and $\delta$ states has been neglected.)

* Energy levels which are enclosed in brackets have equal energy. They are shown in Figure B.38 as separated for the sake of clarity only.

† In Figure B.38 the three $\sigma$ terms as well as the four $\pi$ terms have been set in brackets to indicate degeneracy. This degeneracy is of course only rigorously satisfied when not only the central ion is removed from the complex ion, but when the separations of the six ligands from the midpoint of the octahedron are very great. For isolated ligand systems with normal ligand-octahedron midpoint separations the degeneracy, for example that of the three $\sigma$ orbitals, is in general more or less strongly removed. For the further consideration it therefore suffices to a good approximation to consider the terms of the isolated system which are enclosed in brackets as being degenerate.

central ion functions $3d$, $4s$, $4p$, low-lying $\sigma, \pi, \delta, \ldots$ orbitals are contained as well. If the components of the complex which are first thought of as being separated and whose energies are represented in the left- and right-hand portions of Figure B.38 are brought together to form a complex, a new coupling between the complex components appears, which is expressed in the off-diagonal elements of the secular determinant. According to experience this coupling decreases with increasing bonding character of the participating ligand orbitals. $\sigma$ functions usually extend predominantly in the direction of the central ion and produce in this manner maximum overlap. Correspondingly the coupling between ligand $\sigma$ orbitals and central ion orbitals is the greatest.

Ligand $\pi$ orbitals couple in general more weakly with the central ion states than do ligand $\sigma$ orbitals, etc. In the normal case it is therefore adequate in the linear combinations of the $\Xi$ besides the $3d$, $4s$, $4p$ central ion orbitals to use $\sigma$ and perhaps also $\pi$ ligand orbitals. Naturally only those $\Xi$ functions of the ligands which have the same symmetry behaviour as is shown by the central ion orbitals are capable of a coupling with central ion orbitals. We shall thus limit the discussion to such $\Xi$ linear combinations of the ligand orbitals (besides the central ion functions).

We choose as an example an octahedral complex ion and use the $\Xi$ basis (see Table B.32): $\Xi_1, \Xi_2, \ldots, \Xi_{21}$. The $\pi$ functions $\Xi_{22}, \ldots, \Xi_{27}$ need not be considered because according to Table B.32 there are no central ion functions $3d$, $4s$, $4p$ of the same symmetry behaviour. The $21 \times 21$ secular determinant of type (4–10) which is obtained using the basis $\Xi_1, \ldots, \Xi_{21}$ has the form

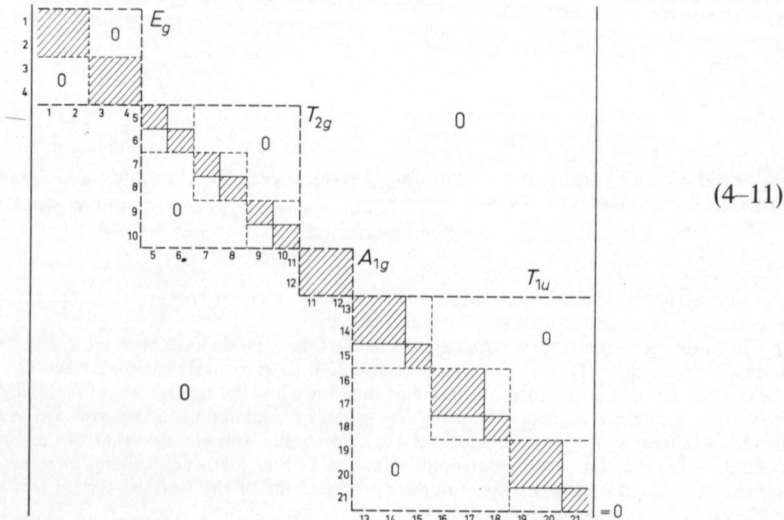


(4–11)

Because according to theorems 1a and 2a, p. 317 the integrals $(\Xi_i, \boldsymbol{H}\Xi_j)$ and $(\Xi_i, \Xi_j)$ vanish when $\Xi_i$ and $\Xi_j$ belong to different irreducible representations of the symmetry group of $\boldsymbol{H}$, in this case the group $O_h$, all elements lying outside of the dashed lines in (4–11) are zero. Using the theorems 1a and 2a, p. 317, according to which functions of the same irreducible representation which transform as *different columns* of the representation matrices cannot combine, all elements outside of the finely dashed lines can be set equal to zero. The $21 \times 21$ determinant becomes in this manner a block determinant. The following subdeterminants, which are individually to be set equal to zero, remain along the diagonal: two $2 \times 2$ determinants of irreducible representation $E_g$, three $2 \times 2$ determinants of the irreducible representation $T_{2g}$, one $2 \times 2$ determinant of the irreducible representation $A_{1g}$ and three $3 \times 3$ determinants of the irreducible representation $E_g$. If there are several subdeterminants for a given symmetry species, e.g., two for the species $E_g$, then it suffices to find the roots of one subdeterminant because according to theorem 2b, p. 317, the remaining subdeterminants of the same species have the same roots.

Besides these rigorous symmetry-determined statements, which lead to a reduction of the $21 \times 21$ determinant to determinants of lower order we shall make a further simplification which does not follow from the symmetry but from the fact that the coupling of $\pi$ functions with the central ion functions and with $\sigma$ functions is small (compared to the central ion $\sigma$ coupling)*. This will be carried out after we have seen how to proceed when we wish to consider $\pi$ orbitals in addition to $\sigma$ orbitals. Under this assumption it is permissible to set all of the integrals $(\Xi_i, \boldsymbol{H}\Xi_j)$ and $(\Xi_i, \Xi_j)$ equal to zero in which one of the $\Xi$ functions is a $\pi$ function, the other, however, a central ion function or a $\sigma$ function. With this the $21 \times 21$ determinant is reduced to the shaded determinants along the diagonal.

We shall treat one of the $E_g$ determinants as being representative of the two-dimensional subdeterminants:

$$\begin{vmatrix} H_{1,1}-\varepsilon & H_{1,2}-S_{1,2}\,\varepsilon \\ H_{2,1}-S_{2,1}\varepsilon & H_{2,2}-\varepsilon \end{vmatrix} = 0\,. \tag{4–12}$$

The two roots may be expressed in the form:

$$\varepsilon_{1,2}(E_g) = X \pm Y$$

where (for $H_{1,2} \neq 0$) the smaller root in the energy diagram lies lower than $H_{1,1}$ and $H_{2,2}$. The larger root, however, lies higher than $H_{1,1}$ and $H_{2,2}$ (cf. Figure B.39). States whose energies are stabilized by virtue of coupling are designated as *bonding* states, those which are labilized as *antibonding*

* Cf. Part A, p. 106.

states. If the states which result are denoted by their irreducible representation, then it is conventional to add a star to the upper right-hand corner of the antibonding term symbol and to add a '*b*' to that of the bonding state. The state with the higher energy in Figure B.39 is accordingly denoted as the $e_g^*$ state and the bonding state as the $e_g^b$ state. Were the integrals $H_{1,2} = H_{2,1}$ and $S_{1,2} = S_{2,1}$ which determine the magnitude of the coupling zero, then the off-diagonal elements of the determinant (4–12) would vanish and we would have the solutions $\varepsilon_1 = H_{1,1}$ and $\varepsilon_2 = H_{2,2}$. In this special case the two states are designated as nonbonding states because coupling, which is necessary for the appearance of a bond, is not present.

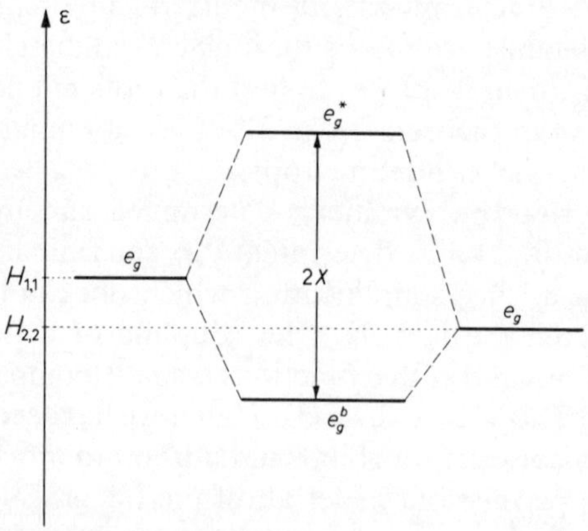


Figure B.39. Coupling between two states of the same species ($e_g$).

For an *ab initio* calculation of the two energies $\varepsilon_1$ and $\varepsilon_2$ it is necessary to determine explicitly the integrals $H_{1,1}, H_{1,2} = H_{2,1}, H_{2,2}$ and $S_{1,2} = S_{2,1}$. While the overlap integrals $S_{ij}$ can be evaluated without difficulty, enormous barriers are met if we wish to determine the values of $H_{ij}$. These arise from the fact that the potential expressions $V_z$, $V_\lambda$, $V'$ contained in the Hamiltonian are not known sufficiently well and that integrals $(\Xi_i, H\Xi_j)$ with ligand $\Xi$ functions are many-centre integrals*, the evaluation

* We may write for the integral $H_{1,2}$ in an octahedral complex

$$(\Xi_1, H\,\Xi_2) = (d_{x^2-y^2}, H\, {}^1/_2[z_1 - z_2 + z_4 - z_5])$$
$$= {}^1/_2\{(d_{x^2-y^2}, H\,z_1) - (d_{x^2-y^2}, H\,z_2) + (d_{x^2-y^2}, H\,z_4) - (d_{x^2-y^2}, H\,z_5)\}.$$

of which is extraordinarily difficult. Usually then these integrals $H_{ii}$ and $H_{ij}$ are treated as parameters. It is conventional to set energy integrals with two identical indices, $H_{ii}$, equal to the ionization energies of the complex ion being considered. (This procedure is, however, questionable because the functions $\Xi_i$ employed in the integrals $H_{ii} = (\Xi_i, \boldsymbol{H}\Xi_i)$ do not represent MO's. The $H_{ii}$ are rather the expectation values of the energy of an electron which moves in a central ion orbital or in an orbital of the ligand system.) Energy integrals with two different indices, $H_{ij}$ with $i \neq j$, are usually estimated using an approximation given by Mulliken*. The essential and plausible assumption underlying this relation is that $H_{ij}$ is proportional to the appropriate overlap integral $S_{ij}$ as well as to the arithmetic mean of the energies $H_{ii}$ and $H_{jj}$:

$$H_{ij} \approx F \cdot S_{ij} \frac{H_{ii} + H_{jj}}{2}$$

The numerical factor $F$ is likewise adjusted to experimental data, such as spectroscopically determined term separations. $F$ is accordingly set $\approx 2$. The overlap integrals $S_{ij}$ are calculated explicitly, whereby the central ion functions and ligand functions which appear are approximated by *Hartree–Fock* or by *Slater* functions. Thus the calculation of all integrals in the secular determinant is reduced to the calculation of overlap integrals. One must of course always keep the numerous approximations in mind which were made purely in order to simplify the evaluation of the integrals when attempting to estimate the accuracy of an LCAO-calculation carried out according to the procedure explained above.

In the central portion of Figure B.38 the terms of the symmetry species $E_g$ which arise as a result of the coupling between the central ion and the ligands are schematically shown, both the bonding $e_g^b$ term and the anti-bonding $e_g^*$ term. The progeny terms of the remaining original terms for octahedral complexes are obtained analogously to the case of the $E_g$ terms.

---

If the explicit form of $\boldsymbol{H}$ is used, (4–1) with (4–2), then the first integral in braces is

$$(d_{x^2-y^2}, \boldsymbol{H}\, z_1) = (d_{x^2-y^2}, [-{}^1\!/_2\Delta + V_Z + V' + V_1] z_1) + (d_{x^2-y^2}, [V_2 + V_3 + V_4 + V_5 + V_6] z_1).$$

The last integral can be represented as a sum of five integrals of the form $(d_{x^2-y^2}, V_i z_1)$, $i = 2, 3, 4, 5, 6$. These are obviously many-centre integrals, in particular three-centre integrals. For example, the integral $(d_{x^2-y^2}, V_2 z_1)$ contains the function $d_{x^2-y^2}$, which is defined with reference to the position of the central ion, and further the potential $V_2$ and the ligand function $z_1$, which is defined with reference to the positions of the second and the first ligands.

* R. S. Mulliken: *J. Chim. Phys.*, **46**, 497, 675 (1949); M. Wolfsberg and L. Helmholtz: *J. Chem. Physics*, **20**, 837 (1952).

There is no coupling for the species $T_{2g}$ according to the secular determinant (4–11) for $\sigma$ bonding (but vanishing $\pi$ bonding). Only the shaded one-dimensional subdeterminants remain. The central ion $t_{2g}$ state is therefore nonbonding (cf. central portion of Figure B.38). Accordingly for a $\sigma$ bond the $A_{1g}$ and $T_{1u}$ states each lead to a bonding and to an antibonding state.

If we consider an octahedral complex with a $d^1$ central ion and with ligands whose outer shells are closed, then for the case of predominantly $\sigma$ bonding one must start from the central portion of the term scheme depicted in Figure B.38 when the ground state and the low-lying excited states of the complex ion are to be determined. As is usual in the electron schemes, the ground state of the system is obtained by filling the terms according to their degree of degeneracy from the bottom with the available valence electrons. For the case of $\sigma$ bonding each of the six ligands contributes two electrons, and the $d^1$ central ion one $d$ electron to the valence system of the complex ion. The resulting ground state is according to Figure B.38 a $t_{2g}$ state, the first excited state an $e_g^*$ state, in agreement with the results of the ionic ligand field theory. The energy difference between the $t_{2g}$ and the $e_g^*$ state is then set equal to $\Delta \equiv 10\,Dq(\text{oct})$. Completely analogous considerations are carried out when the central ion contributes $N$ $d$ electrons to the valance electron system. A detailed discussion of this problem has already been given in Part A, section 1.12.

After having treated in detail the methods of determining the one-electron energies within the framework of the LCAO theory, we shall now turn our attention to the question of the general form of the wave functions corresponding to the LCAO energies. This is found by determining the coefficients $a_i$ in the linear combinations (4–9). As a typical example we shall treat once again the $e_g^*$ state of octahedral complex ions. From the secular determinant (4–12) it is seen that for the roots $\varepsilon_1$ and $\varepsilon_2$ linear combinations of the form

$$\Phi_1 = a_1 \Xi_1 + b_1 \Xi_2,$$
$$\Phi_2 = a_2 \Xi_1 + b_2 \Xi_2,$$

come into consideration. The coefficient ratios $a_1/b_1$ and $a_2/b_2$ are obtained from one of the equations*

$$\begin{aligned} a_i(H_{1,1}-\varepsilon_i) \qquad &+ b_i(H_{1,2}-S_{1,2}\varepsilon_i) = 0, \\ a_i(H_{1,2}-S_{1,2}\varepsilon_i) &+ b_i(H_{2,2}-\varepsilon_i) \qquad = 0. \end{aligned} \qquad i = 1,2$$

which were derived using a variational calculation.

* For the case of a two term linear variational function it is of no consequence which of these two secular equations is used. The first equation yields

$$\frac{a_i}{b_i} = -\frac{H_{1,2}-S_{1,2}\varepsilon_i}{H_{1,1}-\varepsilon_i}.$$

If the coefficient ratios are substituted into the functions $\Phi_1, \Phi_2$, then these are completely determined except for a constant factor, which is conventionally chosen so that the $\Phi$ functions are normalized to one:

$$(\Phi_1, \Phi_1) = (\Phi_2, \Phi_2) = 1 .$$

Interesting conclusions concerning the type of bond between the central ion and the ligands can be drawn from the magnitude of the coefficients $a_i$ and $b_i$. If for example the coefficient of the central ion function $\Xi_1 = d_{x^2-y^2}$ is equal to one (The coefficient of the ligand functions must then certainly be equal to one.), this means that the valence electrons described by $\Phi$ are localized on the central ion. In this case the assumption of ionic bonding which is made in the ionic ligand field theory is rigorously satisfied. Equality of the absolute values of the coefficients of $\Xi_1$ and $\Xi_2$ means that the valence electrons are distributed equally over the central ion and the ligands. Under these conditions the case treated by Pauling is realized: Rigorous covalent binding is present.

If in particular the $\Xi$ are orthogonal (vanishing overlap) and normalized to one, i.e., $(\Xi_i, \Xi_j) = 0$ for $i \neq j$ and $(\Xi_i, \Xi_i) = 1$, then the coefficients $a_i$ and $b_i$ in the $\Phi$ functions have the special form:

$$a_1 = \alpha, \quad b_1 = \sqrt{1-\alpha^2},$$
$$a_2 = \sqrt{1-\alpha^2}, \quad b_2 = -\alpha.$$

The case of rigorous covalent binding is then achieved when $\alpha = \sqrt{1/2}$:

$$a_1 = b_1 = \sqrt{1/2},$$
$$a_2 = -b_2 = \sqrt{1/2}.$$

Rigorous ionic bonding is present when $\alpha = 1$:

$$a_1 = 1, \quad b_1 = 0,$$
$$a_2 = 0, \quad b_2 = -1.$$

(cf. Part A, p. 101).

---

From the second equation it follows

$$\frac{a_i}{b_i} = -\frac{H_{2,2} - \varepsilon_i}{H_{2,1} - S_{2,1}\varepsilon_i}.$$

That the right-hand sides of these equations are truly identical can be seen when $\varepsilon = \varepsilon_i$ is substituted into the secular determinant (4–12) and the determinant is expanded

$$(H_{1,1} - \varepsilon_i)(H_{2,2} - \varepsilon_i) = (H_{1,2} - S_{1,2}\varepsilon_i)(H_{2,1} - S_{2,1}\varepsilon_i),$$

or

$$\frac{H_{2,2} - \varepsilon_i}{H_{2,1} - S_{2,1}\varepsilon_i} = \frac{H_{1,2} - S_{1,2}\varepsilon_i}{H_{1,1} - \varepsilon_i}.$$

Up to this point we have discussed the valence electron systems of complex ions in the one-electron LCAO scheme only. The interaction between the valence electrons was only approximately treated by the introduction of effective potentials, such as the shielding potential $V'$. From the theory of free atoms for example it is known that the electron interaction can have a very strong influence on the term energies. One could in principle consider the electron interaction starting from the one-electron system. For this one needs only to form the antisymmetrized products of the LCAO functions and, using this basis, to carry out a perturbation calculation with the complete many-electron *Hamiltonian* in which the electron interaction operators $1/r_{ij}$ appear explicitly. This attempt, however, meets enormous difficulties: One is led to many-centre integrals, which, with the exception of a few especially simple types, have not yet been evaluated. Recently attempts to carry out an absolute calculation of the energies in the many-electron scheme of the MO theory have been made*. Because of the difficulties explained above, however, the investigations which have been published cannot claim to be absolute calculations. They use many assumptions and approximations, the validity of which has not been proven.

* Cf. e.g., R. G. Shulman and S. Sugano: *Phys. Rev. Letters*, **7**, 157 (1961).

# 5. Spin–orbit coupling and magnetic properties

In the study of free atomic ions and of transition metal ions bound in complexes we have considered up until now only *Coulomb*-type forces. Magnetic coupling was neglected. In this chapter we shall treat first of all one type of magnetic coupling, *spin–orbit* coupling. Further the effect of *external magnetic fields* on the energy states of free and complex-bound atomic ions will also be considered. The results obtained in this manner will be applied in the last section to the calculation of magnetic susceptibilities. We shall limit the discussion in this chapter to the presentation of the theory of *one-electron systems* and to extensive treatment of several selected examples.

## 5.1. Spin–orbit coupling in free atoms and atomic ions

While treating free atomic one-electron systems in chapter 1, it was shown that the *Schrödinger* equation (1–1) allows only an approximate description of the states of motion of an electron. The magnetic moment of the electron is proportional to its intrinsic angular momentum, the spin. Further an electron with nonzero *orbital angular momentum* (as a charge revolving about an atomic nucleus) likewise generates a magnetic moment. This orbital magnetic moment and the magnetic spin moment interact: *spin–orbit coupling* is the result. The operator which represents the energy of the spin–orbit coupling can be derived within the framework of relativistic quantum mechanics and for an electron moving in a central field with the potential $V(r)$ has the form

$$\boldsymbol{H}_{SO} = \xi(r)\, \boldsymbol{l} \cdot \boldsymbol{s} \tag{5–1}$$

with

$$\xi(r) = -\frac{e}{2m^2c^2}\,\frac{1}{r}\,\frac{\partial V(r)}{\partial r}\,. \tag{5–2}$$

$\boldsymbol{l}$ and $\boldsymbol{s}$ are here the orbital angular momentum and spin operators*. $c$ is the velocity of light and $m$ the mass of the electron.

---

* $\boldsymbol{l} \cdot \boldsymbol{s} = l_x s_x + l_y s_y + l_z s_z$ is the direct product of $\boldsymbol{l}$ and $\boldsymbol{s}$.

With consideration of the spin–orbit coupling the extended *Hamiltonian*

$$\boldsymbol{H} = -\frac{\hbar^2}{2m}\Delta - e\,V(r) + \xi(r)\,\boldsymbol{l}\cdot\boldsymbol{s}. \tag{5–3}$$

arises from (1–1). Exact solutions of the *Schrödinger* equation with the *Hamiltonian* (5–3) cannot be given. One therefore attempts to take account of the influence of the spin–orbit coupling by treating it as a perturbation of the unperturbed system with the *Hamiltonian*

$$\boldsymbol{H}_0 = -\frac{\hbar^2}{2m}\Delta - eV(r) \tag{1–1}$$

The solutions for the unperturbed system are known from chapter 1. The energy eigenvalues can be classified according to the principal quantum number $n$ and the orbital angular momentum quantum number $l$: $E_{n,l}$. For every $E_{n,l}$ there are $2(2l+1)$ (degenerate) states, which can be described by the functions*

$$\begin{aligned} \psi_{n,l,m_l,m_s} &= R_{n,l}(r)\cdot Y_{l,m_l}(\theta,\phi)\cdot\chi_{m_s}(\sigma) \\ m_l &= -l,\ -l+1,\ldots,l-1,l \\ m_s &= -1/2,\ +1/2 \end{aligned} \tag{5–4}$$

or by a set of $2(2l+1)$ linearly independent linear combinations of these $\psi$ functions. The term system of the unperturbed system is schematically shown in the left-hand portion of Figure B.40. The numbers in brackets give the degrees of degeneracy $2(2l+1)$.

The consequences of the perturbation of the unperturbed system produced by the spin–orbit coupling are obtained approximately by solving the secular problem of the total *Hamiltonian* (5–3). Either all the functions

$$\psi_{n,l,m_l,m_s} \qquad \begin{aligned} n &= 1,2,3,\ldots \\ l &= 0,1,\ldots,n-1 \\ m_l &= -l,\ldots,+l \\ m_s &= -1/2,\ +1/2 \end{aligned}$$

or a set of linearly independent combinations of these may be used as basis functions for the secular problem.

---

*According to pp. 220 ff. if holds that $\chi_{+\frac{1}{2}} = \alpha$ and $\chi_{-\frac{1}{2}} = \beta$.

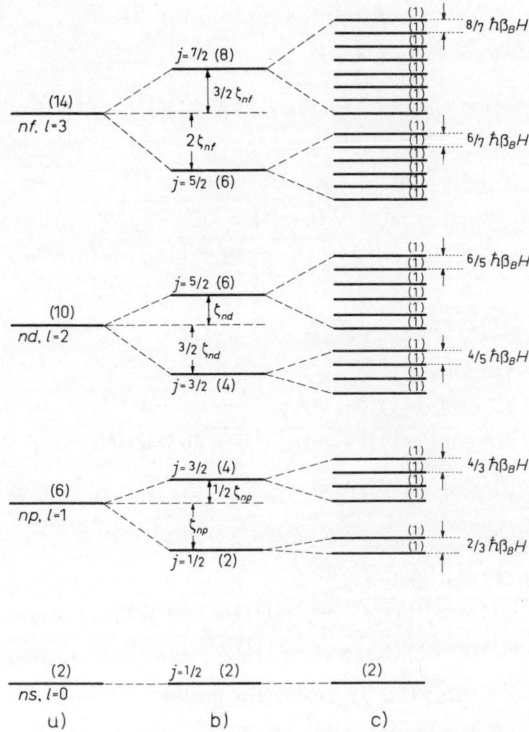


Figure B.40. Term system of an atomic one-electron system (schematic): (a) free system without spin–orbit coupling; (b) free system with spin–orbit coupling; (c) systems with spin–orbit coupling in a homogeneous magnetic field of strength $H$.

First of all we shall formulate the secular determinant using the basis of the $\psi_{n,l,m_l,m_s}$.

$$\|(\psi_{n',l',m_l',m_s'}[\boldsymbol{H}_0+\boldsymbol{H}_{SO}]\psi_{n,l,m_l,m_s})-\delta_{n',n}\delta_{l',l}\delta_{m_l',m_l}\delta_{m_s',m_s}E\|=0. \tag{5–5}$$

Because they are solutions of the unperturbed problem the $\psi$ functions diagonalize that portion of the determinant containing $H_0$ completely:

$$(\psi_{n',l',m_l',m_s'}\boldsymbol{H}_0\psi_{n,l,m_l,m_s})=\delta_{n',n}\delta_{l',l}\delta_{m_l',m_l}\delta_{m_s',m_s}E_{n,l}. \tag{5–6}$$

What can be said about the integrals formed with $\boldsymbol{H}_{SO}$?

$$(\psi_{n',l',m_l',m_s'}\boldsymbol{H}_{SO}\psi_{n,l,m_l,m_s})=(\psi_{n',l',m_l',m_s'}\xi(r)\,\boldsymbol{l}\cdot\boldsymbol{s}\,\psi_{n,l,m_l,m_s})$$

Because of the following equations and cf. pp. 219 ff.

$$\boldsymbol{l}\cdot\boldsymbol{s}=\boldsymbol{l}_x\,\boldsymbol{s}_x+\boldsymbol{l}_y\,\boldsymbol{s}_y+\boldsymbol{l}_z\,\boldsymbol{s}_z$$
$$=\boldsymbol{l}_z\,\boldsymbol{s}_z+\frac{1}{2}(\boldsymbol{l}_x+i\,\boldsymbol{l}_y)(\boldsymbol{s}_x-i\,\boldsymbol{s}_y)+\frac{1}{2}(\boldsymbol{l}_x-i\,\boldsymbol{l}_y)(\boldsymbol{s}_x+i\,\boldsymbol{s}_y)$$
$$\boldsymbol{l}_z\psi_{n,l,m_l,m_s}=\hbar m_l\psi_{n,l,m_l,m_s}$$
$$\boldsymbol{s}_z\psi_{n,l,m_l,m_s}=\hbar m_s\psi_{n,l,m_l,m_s}$$
$$(\boldsymbol{l}_x\pm i\,\boldsymbol{l}_y)\psi_{n,l,m_l,m_s}=\hbar\sqrt{(l\pm m_l+1)(l\mp m_l)}\,\psi_{n,l,m_l\pm1,m_s}$$
$$(\boldsymbol{s}_x\pm i\,\boldsymbol{s}_y)\psi_{n,l,m_l,m_s}=\hbar\sqrt{(1/2\pm m_s+1)(1/2\mp m_s)}\,\psi_{n,l,m_l,m_s\pm1}$$

we have

$$\boldsymbol{H}_{SO}\psi_{n,l,m_l,m_s}=\xi(r)\,\boldsymbol{l}\cdot\boldsymbol{s}\,\psi_{n,l,m_l,m_s}$$
$$=\xi(r)\hbar^2[m_l m_s\psi_{n,l,m_l,m_s}$$
$$+1/2\sqrt{(l+m_l+1)(l-m_l)}\sqrt{(1/2-m_s+1)(1/2+m_s)}\,\psi_{n,l,m_l+1,m_s-1}$$
$$+1/2\sqrt{(l-m_l+1)(l+m_l)}\sqrt{(1/2+m_s+1)(1/2-m_s)}\,\psi_{n,l,m_l-1,m_s+1}]\,.$$

With consideration of the form $\psi_{n,l,m_l,m_s} = R_{n,l}Y_{l,m_l}\chi_{m_s}$ this becomes

$$(\psi_{n',l',m_l',m_s'},\boldsymbol{H}_{SO}\,\psi_{n,l,m_l,m_s})=\hbar^2(R_{n',l'},\xi(r)\,R_{n,l})\times$$
$$\times\{m_l m_s(Y_{l',m_l'},Y_{l,m_l})(\chi_{m_s'},\chi_{m_s})$$
$$+1/2\sqrt{(l+m_l+1)(l-m_l)}\sqrt{(1/2-m_s+1)(1/2+m_s)}(Y_{l',m_l'},Y_{l,m_l+1})(\chi_{m_s'},\chi_{m_s-1})$$
$$+1/2\sqrt{(l-m_l+1)(l+m_l)}\sqrt{(1/2+m_s+1)(1/2-m_s)}(Y_{l',m_l'},Y_{l,m_l-1})(\chi_{m_s'},\chi_{m_s+1})\}$$

or because the $Y_{l,m_l}$ and the $\chi_{m_s}$ are orthogonal

$$(\psi_{n',l',m_l',m_s'},\boldsymbol{H}_{SO}\,\psi_{n,l,m_l,m_s})=\hbar^2(R_{n',l'},\xi(r)\,R_{n,l})\times$$
$$\times\,\delta_{l',l}\{m_l m_s\,\delta_{m_l',m_l}\,\delta_{m_s',m_s} \tag{5–7}$$
$$+1/2\sqrt{(l+m_l+1)(l-m_l)}\sqrt{(1/2-m_s+1)(1/2+m_s)}\,\delta_{m_l',m_l+1}\,\delta_{m_s',m_s-1}$$
$$+1/2\sqrt{(l-m_l+1)(l+m_l)}\sqrt{(1/2+m_s+1)(1/2-m_s)}\,\delta_{m_l',m_l-1}\,\delta_{m_s',m_s+1}\}.$$

One sees that the integral for $\psi$ functions with $n' \neq n$ does not necessarily vanish, although this is the case for $l' \neq l$. Now it is known from experiments that the separations between terms of states with different principal quantum numbers $n' \neq n$ but the same orbital angular momentum quantum numbers $l' = l$ are in general large compared to the spin–orbit coupling energies. The error is then not large when all nondiagonal elements in (5–5) which are formed with functions $\psi_{n',l}$ and $\psi_{n,l}$ (two such widely separated states with $n' \neq n$) are set equal to zero. The complete secular determinant (5–5) decomposes into subdeterminants arranged along the diagonal and which in each case contain only $\psi$ functions with the same principal quantum number $n$ and the same orbital angular momentum quantum number $l$, that is $\psi$ functions of the same energy eigenvalue $E_{n,l}$ of the unperturbed system. The subdeterminant of such a function basis (5–4) is of order $2(2l+1)$ and with consideration of (5–6) and (5–7) has the form

$$\|(\psi_{n,l,m_l',m_s'}, \boldsymbol{H}_{SO}\,\psi_{n,l,m_l,m_s}) - \delta_{m_l',m_l}\delta_{m_s',m_s}(E-E_{n,l})\|=0$$
$$m_l', m_l = -l, -l+1, \ldots, l-1, l \qquad (5\text{–}8)$$
$$m_s', m_s = -1/2, +1/2$$

From (5–7) it can be seen that the integrals which appear there can be nonzero only if either

$$m_l' = m_l, \quad m_s' = m_s$$

or
$$m_l' = m_l + 1, \quad m_s' = m_s - 1$$

or
$$m_l' = m_l - 1, \quad m_s' = m_s + 1$$

hold. In summary one may write for the three pairs of conditions the single condition

$$m_l' + m_s' = m_l + m_s \qquad (5\text{–}9)$$

With consideration of this condition it would in principle be possible to determine without calculation which integrals in (5–8) must be zero and then to calculate the remaining integrals, to substitute these into (5–8), and finally to obtain the roots $E$ of (5–8).

We shall, however, also consider a somewhat more elegant procedure for determining the roots. Determination of the roots is of course the easiest when the secular determinant is in completely diagonal form. With the condition (5–9) and equation (5–7) we recognize that with the exception of the case $l = 0$ the $\psi_{n,l,m_l,m_s}$ themselves do not diagonalize the secular determinant (5–8). Rather linear combinations of the $\chi_{n,l,m_l,m_s}$ tend to bring (5–8) into diagonal form. With consideration of only the condition (5–9) it can be said that the solution functions are linear combinations which in each case contain only functions $\psi_{n,l,m_l,m_s}$ for which the sum of the quantum numbers $m_l$ and $m_s$ are equal:

$$a\,\psi_{n,l,m_l,m_s} + b\,\psi_{n,l,m_l',m_s'} + c\,\psi_{n,l,m_l'',m_s''} + \cdots$$

with

$$m_l + m_s = m_l' + m_s' = m_l'' + m_s'' = \cdots = m_j\,.$$

Further it is known that only the two $m_s$ values $+1/2$ and $-1/2$ can appear. For a given $m_j$ only two $m_l$ values may be given because of

$$m_j = m_l + 1/2 = m_l' - 1/2\,,$$

$m_l = m_j - 1/2$ for $m_s = +1/2$ and $m_l' = m_j + 1/2$ for $m_s' = -1/2$. The linear combinations given above, again with the exception of the case $l = 0$, consist in each case of two $\psi$ functions

$$F=a(\psi_{n,l,m_j-1/2,1/2}+x\,\psi_{n,l,m_j+1/2,-1/2}). \qquad (5\text{–}10)$$

$a$ and $x$ are constants which have not yet been determined.

If it is possible to select the $F$ functions so that they are (orthogonal) eigenfunctions of $\boldsymbol{l}\cdot\boldsymbol{s}$, then the secular determinant formed using these functions is certainly diagonal. The $F$ are eigenfunctions of $\boldsymbol{l}\cdot\boldsymbol{s}$ when the coefficient $x$ is determined so as to give:

$$\begin{aligned}\boldsymbol{l}\cdot\boldsymbol{s}\,a(\psi_{n,l,m_j-1/2,1/2}+x\,\psi_{n,l,m_j+1/2,-1/2})\\ =K\cdot a(\psi_{n,l,m_j-1/2,1/2}+x\,\psi_{n,l,m_j+1/2,-1/2}),\end{aligned}$$

whereby $K$ is a constant which is the eigenvalue of $\boldsymbol{l}\cdot\boldsymbol{s}$. If $\boldsymbol{l}\cdot\boldsymbol{s}$ is applied to $F$ we have

$$\begin{aligned}\boldsymbol{l}\cdot\boldsymbol{s}\,F=&\left[\boldsymbol{l}_z\boldsymbol{s}_z+\frac{1}{2}(\boldsymbol{l}_x+i\,\boldsymbol{l}_y)(\boldsymbol{s}_x-i\,\boldsymbol{s}_y)+\frac{1}{2}(\boldsymbol{l}_x-i\,\boldsymbol{l}_y)(\boldsymbol{s}_x+i\,\boldsymbol{s}_y)\right]\times\\ &\times[a(\psi_{n,l,m_j-1/2,1/2}+x\,\psi_{n,l,m_j+1/2,-1/2})]\\ =&\hbar^2a\{(m_j-1/2)\cdot 1/2\,\psi_{n,l,m_j-1/2,1/2}\\ &+1/2\sqrt{(l+m_j-1/2+1)(l-m_j+1/2)}\sqrt{(1/2-1/2+1)(1/2+1/2)}\times\\ &\times\psi_{n,l,m_j+1/2,-1/2}\\ &+1/2\sqrt{(l-m_j+1/2+1)(l+m_j-1/2)}\sqrt{(1/2+1/2+1)(1/2-1/2)}\times\\ &\times\psi_{n,l,m_j-3/2,3/2}\\ &+x\cdot(m_j+1/2)(-1/2)\psi_{n,l,m_j+1/2,-1/2}\\ &+x\cdot 1/2\sqrt{(l+m_j+1/2+1)(l-m_j-1/2)}\sqrt{(1/2+1/2+1)(1/2-1/2)}\times\\ &\times\psi_{n,l,m_j+3/2,-3/2}\\ &+x\cdot 1/2\sqrt{(l-m_j-1/2+1)(l+m_j+1/2)}\sqrt{(1/2-1/2+1)(1/2+1/2)}\times\\ &\times\psi_{n,l,m_j-1/2,1/2}\}\\ =&\hbar^2a\{[(m_j-1/2)\cdot 1/2+x\,1/2\sqrt{(l+m_j+1/2)(l-m_j+1/2)}]\times\\ &\times\psi_{n,l,m_j-1/2,1/2}\\ &+[x(m_j+1/2)(-1/2)+1/2\sqrt{(l+m_j+1/2)(l-m_j+1/2)}]\times\\ &\times\psi_{n,l,m_j+1/2,-1/2}\}.\end{aligned}$$

By rearranging and applying the abbreviation

$$W=\sqrt{(l+m_j+1/2)(l-m_j+1/2)}$$

we obtain further

$$\begin{aligned}\boldsymbol{l}\cdot\boldsymbol{s}\,F=&\hbar^2[(m_j-1/2)\,1/2+x\cdot 1/2\,W][a(\psi_{n,l,m_j-1/2,1/2}\\ &+\frac{x(m_j+1/2)(-1/2)+1/2\,W}{(m_j-1/2)\cdot 1/2+x\cdot 1/2\,W}\,\psi_{n,l,m_j+1/2,-1/2})].\end{aligned} \qquad (5\text{–}11)$$

A comparison with the original function $F$ shows that $F$ is an eigenfunction of $\boldsymbol{l}\cdot\boldsymbol{s}$ only if the coefficient of $\psi_{n,l,m_j+1/2,-1/2}$ is equal to $x$:

$$\frac{x(m_j+{}^1\!/_2)(-{}^1\!/_2)+{}^1\!/_2\,W}{(m_j-{}^1\!/_2)\cdot{}^1\!/_2+x\cdot{}^1\!/_2\,W}=x.$$

This equation is quadratic in $x$

$$x^2+2\,\frac{m_j}{W}\,x-1=0$$

and has the roots

$$x_{1,2}=-\frac{m_j}{W}\pm\frac{1}{2W}\sqrt{4m_j^2+4W^2}$$

or (with consideration of the explicit form of $W$)

$$x_{1,2}=\frac{-m_j\pm(l+{}^1\!/_2)}{\sqrt{(l+m_j+{}^1\!/_2)(l-m_j+{}^1\!/_2)}},$$

$$x_1=\sqrt{\frac{l-m_j+{}^1\!/_2}{l+m_j+{}^1\!/_2}},\qquad x_2=-\sqrt{\frac{l+m_j+{}^1\!/_2}{l-m_j+{}^1\!/_2}}. \tag{5–12}$$

From the linear assumption (5–10) two eigenfunctions of $\boldsymbol{l}\cdot\boldsymbol{s}$ result for a given $l$ and $m_j$

$$F_i=a_i(\psi_{n,l,m_j-1/2,1/2}+x_i\psi_{n,l,m_j+1/2,-1/2}),\qquad i=1,2.$$

The undetermined constants $a_i$ are determined from the normalization conditions

$$1=(F_i,F_i)=a_i^2(1+x_i^2)$$

to be

$$a_i=\sqrt{\frac{1}{1+x_i^2}}$$

and one obtains using (5–12)

$$F_1=\sqrt{\frac{l+m_j+{}^1\!/_2}{2l+1}}\,\psi_{n,l,m_j-1/2,1/2}+\sqrt{\frac{l-m_j+{}^1\!/_2}{2l+1}}\,\psi_{n,l,m_j+1/2,-1/2},$$

$$F_2=\sqrt{\frac{l-m_j+{}^1\!/_2}{2l+1}}\,\psi_{n,l,m_j-1/2,1/2}-\sqrt{\frac{l+m_j+{}^1\!/_2}{2l+1}}\,\psi_{n,l,m_j+1/2,-1/2}. \tag{5–13}$$

The corresponding eigenvalues are according to (5–11)

$$(ls)_i=\hbar^2[(m_j-{}^1\!/_2)\,{}^1\!/_2+x_i\,{}^1\!/_2\sqrt{(l+m_j+{}^1\!/_2)(l-m_j+{}^1\!/_2)}\,].$$

With consideration of (5–12) this becomes

$$(ls)_1 = \frac{1}{2}\,\hbar^2 l,$$
$$(ls)_2 = -\frac{1}{2}\,\hbar^2 (l+1). \tag{5–14}$$

The $F_i$ are not only eigenfunctions of $\boldsymbol{l}\cdot\boldsymbol{s}$ but are also eigenfunctions of $\boldsymbol{l}^2$ and $\boldsymbol{s}^2$*,

$$\boldsymbol{l}^2 F_i = \hbar^2 l(l+1) F_i,$$
$$\boldsymbol{s}^2 F_i = \hbar^2 \frac{1}{2}\left(\frac{1}{2}+1\right) F_i.$$

Because, as can be easily shown, the relation $\boldsymbol{ls}F_i = \boldsymbol{sl}F_i$ holds, the $F_i$ are also eigenfunctions of the operator

$$\boldsymbol{l}^2 + \boldsymbol{ls} + \boldsymbol{sl} + \boldsymbol{s}^2 = (\boldsymbol{l}+\boldsymbol{s})^2 = \boldsymbol{j}^2,$$

that is of the operator $\boldsymbol{j}^2$ which represents the *square of the resultants of the orbital and spin angular momenta*

$$\boldsymbol{j}^2 F_i = \hbar^2 j_i (j_i + 1) F_i.$$

In particular with the aid of (5–14) we obtain

$$\boldsymbol{j}^2 F_1 = (\boldsymbol{l}^2 + 2\,\boldsymbol{ls} + \boldsymbol{s}^2) F_1 = \hbar^2 \left[ l(l+1) + 2\cdot\frac{1}{2}\,l + \frac{1}{2}\left(\frac{1}{2}+1\right)\right] F_i$$
$$= \hbar^2 \left[ l^2 + 2l + \frac{3}{4}\right] F_1 = \hbar^2 \left[ l + \frac{1}{2}\right]\left[\left(l+\frac{1}{2}\right)+1\right] F_1,$$

i.e.,

$$\boldsymbol{j}^2 F_1 = \hbar^2 j_1 (j_1+1) F_1 \quad \text{with} \quad j_1 = l + {}^1/_2. \tag{5–15a}$$

Correspondingly we have

$$\boldsymbol{j}^2 F_2 = \hbar^2 j_2 (j_2+1) F_2 \quad \text{with} \quad j_2 = l - {}^1/_2.$$

Further it is seen that the $F_i$ are eigenfunctions of the operator $\boldsymbol{j}_z = \boldsymbol{l}_z + \boldsymbol{s}_z$ which represents the *z component of the resultant of the orbital and spin angular momenta*

$$\boldsymbol{j}_z F_1 = \hbar m_j F_1, \quad m_j = -j_1, -j_1+1, \ldots, j_1 - 1, j_1,$$
$$\boldsymbol{j}_z F_2 = \hbar m_j F_2, \quad m_j = -j_2, -j_2+1, \ldots, j_2 - 1, j_2. \tag{5–16}$$

* The $\psi$ functions appearing in an $F$ function have equal $l$ and equal $s$ values.

$m_j = m_l + m_s$ is called *the total magnetic angular momentum quantum number*. The fact that the $F_i$ are eigenfunctions of the operators $\boldsymbol{l}^2, \boldsymbol{s}^2, \boldsymbol{j}^2, \boldsymbol{j}_z$ leads one to classify the $F_i$ according to the eigenvalues of these operators. Here the following notation is conventional. The appropriate principal quantum number $n$ and the letter symbol for the orbital angular momentum quantum number are set in brackets after the function symbol.

The multiplicity is written in the upper left-hand corner (here $s = 1/2$ and therefore $2S+1 = 2$), in the upper right-hand corner the quantum number $m_j$ of $\boldsymbol{j}_z$ and to the lower right the quantum number $j$ of $j^2$: $F(n^{2S+1}L_j^{m_j})$. For the functions (5–13) we then write:

$$\begin{aligned} F(n^2L^{m_j}_{l+1/2}) &= \sqrt{\frac{l+m_j+{}^1\!/_2}{2l+1}}\,\psi_{n,\,l,\,m_j-1/2,\,1/2} + \sqrt{\frac{l-m_j+{}^1\!/_2}{2l+1}}\,\psi_{n,l,m_j+1/2,\,-1/2}\,, \\ F(n^2L^{m_j}_{l-1/2}) &= \sqrt{\frac{l-m_j+{}^1\!/_2}{2l+1}}\,\psi_{n,\,l,\,m_j-1/2,\,1/2} - \sqrt{\frac{l+m_j+{}^1\!/_2}{2l+1}}\,\psi_{n,\,l,\,m_j+1/2,\,-1/2}\,. \end{aligned} \tag{5–17}$$

Let us return to the calculation of the spin–orbit coupling energy. Because the (orthogonal) functions (5–17) are eigenfunctions of $\boldsymbol{l}\cdot\boldsymbol{s}$, the secular determinant formed using them and $\boldsymbol{H} = \boldsymbol{H}_0 + \boldsymbol{H}_{SO}$ is completely diagonal. The roots are

$$\begin{aligned} E(n^2L^{m_j}_{l+1/2}) &= (F(n^2L^{m_j}_{l+1/2}), [\boldsymbol{H}_0 + \xi(r)\,\boldsymbol{l}\cdot\boldsymbol{s}]\,F(n^2L^{m_j}_{l+1/2})) \\ &= E_{n,\,l} + \frac{1}{2}\hbar^2\,l\,(F(n^2L^{m_j}_{l+1/2}), \xi(r)\,F(n^2L^{m_j}_{l+1/2})) \\ &= E_{n,\,l} + \frac{1}{2}\hbar^2\,l\left\{\frac{l+m_j+{}^1\!/_2}{2l+1} + \frac{l-m_j+{}^1\!/_2}{2l+1}\right\}\times \\ &\quad \times (R_{n,\,l}(r), \xi(r)\,R_{n,\,l}(r))\,, \end{aligned}$$

for $j_1 = 1+1/2$ because of (5–6) and (5–14), or with the abbreviation

$$\zeta_{n,\,l} = \hbar^2(R_{n,\,l}(r), \xi(r)\,R_{n,\,l}(r)) > 0\,, \tag{5–18}$$

$$\begin{aligned} E(n^2L^{m_j}_{l+1/2}) &= E_{n,\,l} + {}^1\!/_2\,l\,\zeta_{n,\,l}, \\ |m_j| &\leq j_1 = l + {}^1\!/_2. \end{aligned} \tag{5–19a}$$

Analogously, for $j_2 = l - 1/2$ we have

$$\begin{aligned} E(n^2L^{m_j}_{l-1/2}) &= E_{n,\,l} - {}^1\!/_2(l+1)\,\zeta_{n,\,l}, \\ |m_j| &\leqq j_2 = l - {}^1\!/_2. \end{aligned} \tag{5–19b}$$

The magnitude $\zeta_{n,l}$ defined in (5–18) is called the *one-electron spin–orbit coupling constant.*

One sees that the $2(2l+1)$ states which are degenerate for the case of vanishing spin–orbit coupling and have the energy $E_{n,l}$ under the influence of the spin–orbit coupling, split energetically into two terms, a *term doublet.* The splitting term $n^2L_{l+1/2}$ with $j_1 = l+1/2$ consists of $2j_1+1 = 2l+2$ states which differ in their $m_j$ values and is labilized as a result of the spin–orbit coupling by the energy $(1/2)l\zeta_{n,l}$*.

The other splitting term $n^2L_{l-1/2}$ (with $j_2 = l-1/2$) is stabilized by $-1/2(l+1)\zeta_{n,l}$ consists of a total of $2j_2+1 = 2j$ states. For $d$ electrons in particular it holds according to (5–19) that

$$E(n^2D_{5/2}) = E_{nd} + \zeta_{nd}, \qquad m_j = -5/2, -3/2, -1/2, +1/2, +3/2, +5/2,$$
$$E(n^2D_{3/2}) = E_{nd} - 3/2\zeta_{nd}, \quad m_j = -3/2, -1/2, +1/2, +3/2. \tag{5–20}$$

The situation is schematically represented in the central portion of Figure B.40.

If the spherically symmetric potential $V(r)$ in which the electron moves is a *Coulomb*-type potential $V(r) = Ze/r$, then the spin–orbit coupling constant $\zeta_{n,l}$ with consideration of (5–2, 18) and using the hydrogen-like radial functions $R_{n,l}(r)$. Cf. (1–12) can be given in the following form:

$$\zeta_{n,l} = \frac{Ze^2\hbar^2}{2m^2c^2a_0^3}\int_0^\infty r^{-3}R_{n,l}^2(r)r^2dr = \frac{e^2\hbar^2}{2m^2c^2a_0^3}\frac{Z^4}{n^3l(l+1/2)(l+1)}.$$

For $3d$ electrons we have specifically

$$\zeta_{3d} \approx 1{,}4 \cdot 10^{-2}Z^4\ \mathrm{cm}^{-1},$$

and for $4d$ electrons

$$\zeta_{4d} \approx 6{,}1 \cdot 10^{-3}Z^4\ \mathrm{cm}^{-1}.$$

## 5.2. Spin–orbit coupling of $d^1$ ions in ligand fields of octahedral symmetry

According to section 5.1 the energy of the spin–orbit coupling of an electron moving in a *spherically symmetric* potential is represented by the operator

$$\boldsymbol{H}_{SO} = \xi(r)\,\boldsymbol{l}\cdot\boldsymbol{s}. \tag{5–1}$$

The spin–orbit coupling operator for an electron in fields of lower symmetry, for example, in an octahedral environment, has a different form. However, the following assumption is made in the treatment of $d$ electron systems bound in complexes: in the ligand field theory the coupling of the orbital angular momentum of a central ion $d$ electron with its own spin angular momentum is also approximately described by the operator (5–1). One attempts to correct for inaccuracies possibly accompanying this assumption by treating the quantities $\zeta_{nd}$ (cf. 5–18) appearing in the spin–orbit coupling energies as parameters which are adjusted to experimental data, e.g., to spectroscopically determined term separations.

In the following we shall study the consequences of spin–orbit coupling for a $d^1$ ion bound in a complex. Starting from the states of a free $d^1$ ion without spin–orbit coupling two ways of proceeding are open:

A. In the first step the spin–orbit coupling in the free ion is considered. The resulting states can be classified according to the quantum numbers $l, s, j, m_j$ (cf. section 5.1). In the second step one studies the changes which are produced in these states by the influence of the ligand field.

B. The first step gives the states of the ion in the ligand field without spin–orbit coupling (cf. section 3.3) and in the second step spin–orbit coupling is taken into consideration.

If the ligand field is octahedral, then the total *Hamiltonian* and thereby also $\boldsymbol{H}_{SO}$ is invariant under all symmetry transformations of the double group $O'$ (cf. section 2.6). The states which an electron in an octahedral environment can assume under consideration of spin–orbit coupling can thus be classified according to the irreducible representations of the *double group* $O'$. For the procedures A and B given above one obtains the following group theoretical results:

A. According to section 5.1, p. 438, a free $d^1$ ion has under effective spin–orbit coupling six states (term $^2D_{5/2}$) with $j_1 = l+s = 5/2$ ($m_j = -5/2, -3/2, -1/2, +1/2, +3/2, +5/2$) and four states (term $^2D_{3/2}$) with $j_2 = l-s = 3/2$ ($m_j = -3/2, -1/2, +1/2, +3/2$). The six $^2D_{5/2}$ wave functions and the four $^2D_{3/2}$ wave functions form in each case a basis for a representation of the octahedral double group $O'$. The characters of the representations can be determined with the aid of formula (2–58). The results have been entered into Table B.33. A comparison with the character table for $O'$, p. 480, shows that the representation for $j_2 = 3/2$ is the irreducible representation $\Gamma_8$. The representation for $j_1 = 5/2$ is reducible. Application of the reduction formula (2–4b) leads to the result that the irreducible representations $\Gamma_7$ and $\Gamma_8$ are each contained once in the $j_1$ representation. We have then

$$\Gamma(j_1 = 5/2) = \Gamma_7 \dotplus \Gamma_8,$$
$$\Gamma(j_2 = 3/2) = \Gamma_8,$$

and therefore expect that under the influence of an octahedral field the term $^2D_{3/2}$ does not split, but that the $^2D_{5/2}$ term on the other hand splits

Table B.33. Character table for case A

| | $E$ | $R$ | $4C_3$ $4C_3^2R$ | $4C_3R$ $4C_3^2$ | $3C_2$ $3C_2R$ | $3C_4$ $3C_4^3R$ | $3C_4R$ $3C_4^3$ | $6C_2'$ $6C_2'R$ |
|---|---|---|---|---|---|---|---|---|
| $\chi_{j-3/2}$ | 4 | $-4$ | $-1$ | 1 | 0 | 0 | 0 | 0 |
| $\chi_{j-5/2}$ | 6 | $-6$ | 0 | 0 | 0 | $-\sqrt{2}$ | $\sqrt{2}$ | 0 |

into two terms having the irreducible representation $\Gamma_7$ or $\Gamma_8$. The results are shown schematically in the right-hand side of Figure B.41.

B. Under the influence of an octahedral ligand field the tenfold degenerate (fivefold orbitally degenerate × twofold spin degeneracy) $^2D$ term of a $d^1$ ion with vanishing spin–orbit coupling splits into a fourfold $^2E_g$ term and a sixfold $^2T_{2g}$ term (cf. section 3.3). The wave functions of the $^2E_g$ term and the $^2T_{2g}$ term are products of the space functions and the spin functions. The representation symbols $E_g$ and $T_{2g}$ of the simple group $O_h$ refer to the transformation behaviour of the space parts alone. The complete wave functions (space part × spin part) of the $^2E_g$ term and those of the $^2T_{2g}$ term induce in each case a representation of the double group $O'$. These representations are equal to the product representations (cf. section 2.5) of space and spin functions.

Let us consider first the $^2E_g$ term. A comparison of the characters of the $E_g$ representation (of the simple group, cf. p. 479) with the character systems of the double group $O'$ (cf. p. 480) shows that the $E_g$ representation which is characteristic of the space parts of the wave functions, corresponds

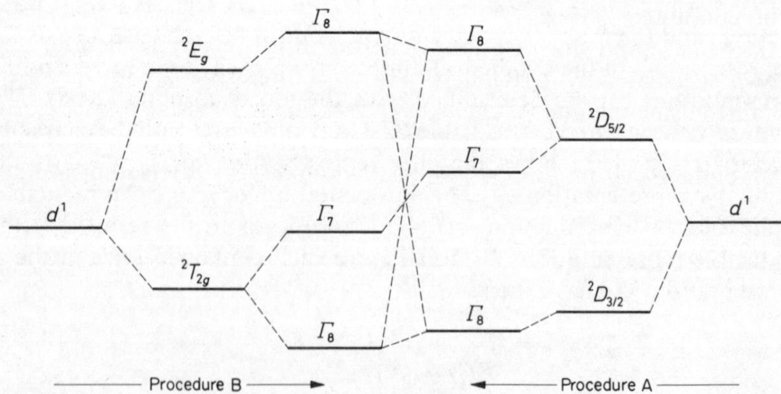


Figure B.41. Diagram of the term splitting of an octahedrally coordinate $d^1$ ion with spin–orbit coupling.

to the irreducible representation $\Gamma_3$ of the double group. The characters are given in the second row of the Table B.34. The spin functions of the first electron ($s = 1/2$) induce likewise a representation of the double group $O'$, whose character is obtained by substituting $j = 1/2$ into the relation (2–58). The results have been entered into the third row of Table B.34. The character system is the same as the character system of the irreducible representation $\Gamma_6$ of $O'$ (cf. p. 480). From the facts that the space parts transform as $\Gamma_3$ and the spin parts as $\Gamma_6$ and that the complete wave functions are products of the space and spin parts, it can be concluded that the $^2E_g$ functions induce a representation of $O'$ which is equivalent to the *product representation* $\Gamma_3 \dot{\times} \Gamma_6$. With the aid of Table B.19 it is seen that $\Gamma_3 \dot{\times} \Gamma_6 = \Gamma_8$. Symbolically one can then write

$$\Gamma(e_g) \dot{\times} \Gamma(s = 1/2) = \Gamma_3 \dot{\times} \Gamma_6 = \Gamma_8.$$

The $^2E_g$ term then cannot split under the influence of spin–orbit coupling.

The investigation of the $^2T_{2g}$ term is carried out completely analogously. The space parts of the $^2T_{2g}$ functions belong to the irreducible representation $T_{2g}$ of the (simple) group $O_h$ which corresponds to the irreducible representation $\Gamma_5$ of the double group $O'$ (comparison of the characters). As has already been shown the spin parts ($s = 1/2$) transform as $\Gamma_6$. The $^2T_{2g}$ functions themselves induce therefore the representation $\Gamma_5 \dot{\times} \Gamma_6$ of $O'$. According to Table B.19, the irreducible components $\Gamma_7$ and $\Gamma_8$ are each contained once:

$$\Gamma(t_{2g}) \dot{\times} \Gamma(s = 1/2) = \Gamma_5 \dot{\times} \Gamma_6 = \Gamma_7 \dot{+} \Gamma_8.$$

The sixfold term $^2T_{2g}$ can therefore split into a twofold term of symmetry $\Gamma_7$ and a fourfold term of symmetry $\Gamma_8$ under the influence of the spin–orbit coupling.

In the left-hand side of Figure B.41 the results given by method B are shown schematically. It is interesting to compare these with the results obtained using method A. One sees that the term $\Gamma_7$ in the one case arises from the $^2T_{2g}$ term and in the other case from the $^2D_{5/2}$ term. The assignment of the $\Gamma_8$ terms, thus cannot be unambiguously carried out. Without

Table B.34. Character table for case B

| | $E$ | $R$ | $4C_3$<br>$4C_3^2R$ | $4C_3R$<br>$4C_3^2$ | $3C_2$<br>$3C_2R$ | $3C_4$<br>$3C_4^3R$ | $3C_4R$<br>$3C_4^3$ | $6C_2'$<br>$6C_2'R$ |
|---|---|---|---|---|---|---|---|---|
| $\chi_{Eg} = \chi_{\Gamma_3}$ | 2 | 2 | $-1$ | $-1$ | 2 | 2 | 0 | 0 |
| $\chi_{j=1/2}$ | 2 | $-2$ | 1 | $-1$ | 0 | $\sqrt{2}$ | $-\sqrt{2}$ | 0 |
| $\chi_{T_{2g}} = \chi_{\Gamma_5}$ | 3 | 3 | 0 | 0 | $-1$ | $-1$ | $-1$ | 1 |

further consideration it cannot be seen from the scheme, e.g., if the $^2E_g$ progeny term $\Gamma_8$ is to be assigned to the term $\Gamma_8$ from $^2D_{5/2}$ to $\Gamma_8$ from $^2D_{3/2}$ or to a 'mixture' of $\Gamma_8(^2D_{5/2})$ and $\Gamma_8(^2D_{3/2})$. (The last mentioned possibility will prove to be the correct one.) Such uncertainty is also met for the assignment of the $^2T_{2g}$ progeny term $\Gamma_8$ to the terms $\Gamma_8(^2D_{5/2})$ and $\Gamma_8(^2D_{3/2})$.

We shall now proceed to look for symmetry adapted wave functions. The possibility of assigning the $\Gamma_7(^2T_{2g})$ term unambiguously to the $^2D_{5/2}$ progeny term $\Gamma_7$ simplifies the problem. One can conclude that the wave functions of the $\Gamma_7$ term as well as the eigenfunctions of $\boldsymbol{j}^2$ must have the quantum number $j = 5/2$ and can be formulated with the aid of the $T_{2g}$ space parts. For a one-electron $d$ system according to (5–17), there are the following wave functions with $j = 5/2$:

$$F(n^2D_{5/2}^{m_j}) = \sqrt{\frac{5/2+m_j}{5}}\,\psi_{n,2,m_j-1/2,1/2} + \sqrt{\frac{5/2-m_j}{5}}\,\psi_{n,2,m_j+1/2,-1/2},$$
$$m_j = -5/2, \ldots, +5/2,$$

or with use of the notation

$$\psi_{n,2,m_j\pm1/2,\pm1/2} = d_{m_j\pm1/2}\begin{Bmatrix}\alpha\\ \beta\end{Bmatrix},$$
$$F(n^2D_{5/2}^{m_j}) = \sqrt{\frac{5/2+m_j}{5}}\,d_{m_j-1/2}\alpha + \sqrt{\frac{5/2-m_j}{5}}\,d_{m_j+1/2}\beta. \tag{5–21}$$

Because the irreducible representation $\Gamma_7$ has the dimensionality 2, it must be possible to construct two linear combinations of the symmetry species $\Gamma_7$ from the six $F$ functions. The general form of these two symmetry adapted linear combinations can be seen from Table B.16 (case $j = 5/2$):

$$\Gamma_7: \sqrt{1/6}\,(j=5/2,\ m_j=+5/2) - \sqrt{5/6}\,(j=5/2,\ m_j=-3/2),$$
$$\sqrt{1/6}\,(j=5/2,\ m_j=-5/2) - \sqrt{5/6}\,(j=5/2,\ m_j=+3/2).$$

In our special case we have $(j, m_j) = F(n^2D_j^{m_j})$. With this the two symmetry adapted $\Gamma_7$ functions have the form*

$$\Omega_1(\Gamma_7|t_{2g}) = \sqrt{1/6}\,F(n^2D_{5/2}^{-5/2}) - \sqrt{5/6}\,F(n^2D_{5/2}^{+3/2}),$$
$$\Omega_2(\Gamma_7|t_{2g}) = \sqrt{1/6}\,F(n^2D_{5/2}^{+5/2}) - \sqrt{5/6}\,F(n^2D_{5/2}^{-3/2}).$$

---

* We include $t_{2g}$ as well as $\Gamma_7$ in the brackets after the function symbol $\Omega$ because on the basis of the assignment explained above the symmetry-adapted $\Gamma_7$ states arise unambiguously from the $T_{2g}$ term.

With consideration of (5–21) we can write for example for $\Omega_2$

$$\begin{aligned}\Omega_2(\Gamma_7|t_{2g}) &= \sqrt{1/6}\,\{d_2\alpha - \sqrt{5}[\sqrt{1/5}\,d_{-2}\alpha + \sqrt{4/5}\,d_{-1}\beta]\} \\ &= \sqrt{1/3}\,\{\sqrt{1/2}\,(d_2 - d_{-2})\alpha - \sqrt{2}\,d_{-1}\beta\},\end{aligned}$$

and with the use of the functions defined in (3–29) this becomes

$$\Omega_2(\Gamma_7|t_{2g}) = \sqrt{1/3}\,\{t_{2g}^0\alpha - \sqrt{2}\,t_{2g}^-\beta\}, \tag{5–22a}$$

and correspondingly

$$\Omega_1(\Gamma_7|t_{2g}) = \sqrt{1/3}\,\{-t_{2g}^0\beta - \sqrt{2}\,t_{2g}^+\alpha\}. \tag{5–22b}$$

The four $\Gamma_8$ functions which likewise follow from ${}^2T_{2g}$ must be orthogonal to both $\Omega(\Gamma_7|t_{2g})$ functions. Such a set of functions is obviously given by

$$\begin{aligned}\Omega_1(\Gamma_8|t_{2g}) &= \sqrt{1/3}\,\{-\sqrt{2}\,t_{2g}^0\beta + t_{2g}^+\alpha\}, \\ \Omega_2(\Gamma_8|t_{2g}) &= \sqrt{1/3}\,\{-\sqrt{2}\,t_{2g}^0\alpha - t_{2g}^-\beta\}, \\ \Omega_3(\Gamma_8|t_{2g}) &= t_{2g}^-\alpha, \\ \Omega_4(\Gamma_8|t_{2g}) &= -t_{2g}^+\beta.\end{aligned} \tag{5–23}$$

Four (orthogonal) $\Gamma_8$ functions formed with the $E_g$ space parts must now be found. Four such functions are*

$$\begin{aligned}\Omega_1(\Gamma_8|e_g) &= e_g^b\beta, \\ \Omega_2(\Gamma_8|e_g) &= -e_g^b\alpha, \\ \Omega_3(\Gamma_8|e_g) &= e_g^a\beta, \\ \Omega_4(\Gamma_8|e_g) &= -e_g^a\alpha.\end{aligned} \tag{5–24}$$

Now that we have sets of symmetry-adapted functions, the term energies of an octahedrally bound $d^1$ ion with inclusion of spin–orbit coupling can be easily computed. The perturbation operator is made up additively of the ligand field operator $\boldsymbol{H}_{\text{Lig.}}$ and the spin–orbit coupling operator $\boldsymbol{H}_{SO}$: $\boldsymbol{H}_{\text{pert}} = \boldsymbol{H}_{\text{Lig.}} + \boldsymbol{H}_{SO}$. The complete secular determinant formed using the ten $\Omega$ functions gives the term energies relative to the energy of a free $d^1$ ion without spin–orbit coupling. Because $\boldsymbol{H}_{\text{pert}}$ is invariant under all symmetry operations of the double group $O'$ according to theorem 2a, p. 317, only those $\Omega$ functions can combine with one another with respect to $\boldsymbol{H}_{\text{pert}}$ which have the same irreducible representation of $O'$. The $10 \times 10$ determinant decomposes accordingly into a $2 \times 2$ determinant (for $\Gamma_7$) and an $8 \times 8$ determinant (for $\Gamma_8$). Because according to theorem 2a, p. 317, a

---

* A set of four $\Omega(\Gamma_8|e_g)$ functions can of course be given without difficulty. We have chosen and indexed the $\Omega(\Gamma_8|e_g)$ functions so that they transform as the same columns of the representation matrices of $\Omega(\Gamma_8|t_{2g})$.

non-combination rule is in effect for functions which transform as different columns of the same irreducible representation, a further decomposition of the two determinants takes place. The $\Gamma_7$ determinant decomposes into two $1\times 1$ determinants, the $\Gamma_8$ determinant into four $2\times 2$ determinants. According to theorem 2b, p. 317, the two $\Gamma_7$ determinants and the four $\Gamma_8$ determinants have in each case the same roots:

$$E(\Gamma_7)=(\Omega_i(\Gamma_7|t_{2g}),\, \boldsymbol{H}_{\text{pert}}\, \Omega_i(\Gamma_7|t_{2g}))\,, \qquad i=1,2 \tag{5–25}$$

$$\begin{vmatrix} (\Omega_i(\Gamma_8|t_{2g}),\, \boldsymbol{H}_{\text{pert}}\, \Omega_i(\Gamma_8|t_{2g}))-E(\Gamma_8) & (\Omega_i(\Gamma_8|t_{2g}),\, \boldsymbol{H}_{\text{pert}}\, \Omega_i(\Gamma_8|e_g)) \\ (\Omega_i(\Gamma_8|e_g),\, \boldsymbol{H}_{\text{pert}}\, \Omega_i(\Gamma_8|t_{2g})) & (\Omega_i(\Gamma_8|e_g),\, \boldsymbol{H}_{\text{pert}}\, \Omega_i(\Gamma_8|e_g))-E(\Gamma_8) \end{vmatrix}=0\,.$$
$$i=1,2,3,4 \tag{5–26}$$

According to (3–31c, d) we have for the partial integrals with $\boldsymbol{H}_{\text{Lig.}}$*:

$$\begin{aligned} &(\Omega_i(\Gamma_7|t_{2g}),\, \boldsymbol{H}_{\text{Lig.}}\, \Omega_i(\Gamma_7|t_{2g}))=(\Omega_i(\Gamma_8|t_{2g}),\, \boldsymbol{H}_{\text{Lig}}\, \Omega_i(\Gamma_8|t_{2g}))=\varepsilon_0-4Dq,\\ &(\Omega_i(\Gamma_8|e_g),\ \boldsymbol{H}_{\text{Lig.}}\ \Omega_i(\Gamma_8|e_g)) \ =\varepsilon_0+6Dq,\\ &(\Omega_i(\Gamma_8|t_{2g}),\, \boldsymbol{H}_{\text{Lig.}}\ \Omega_i(\Gamma_8|e_g)) \ =0. \end{aligned} \tag{5–27}$$

In order to evaluate the integrals with $\boldsymbol{H}_{SO}$ we study first the effect of $\boldsymbol{l}\cdot\boldsymbol{s}$ on the $\Omega$ functions. Because of

$$\boldsymbol{l}\cdot\boldsymbol{s}=\boldsymbol{l}_z\boldsymbol{s}_z+{}^1\!/_2(\boldsymbol{l}_x+i\boldsymbol{l}_y)(\boldsymbol{s}_x-i\boldsymbol{s}_y)+{}^1\!/_2(\boldsymbol{l}_x-i\boldsymbol{l}_y)(\boldsymbol{s}_x+i\boldsymbol{s}_y)$$

and with consideration of p. 430 and (3–29) we have

$$\begin{aligned} \boldsymbol{l}\cdot\boldsymbol{s}\,\Omega_1(\Gamma_8|e_g)&=\boldsymbol{l}\cdot\boldsymbol{s}e_g^b\beta=\boldsymbol{l}\cdot\boldsymbol{s}\sqrt{1/2}\,(d_2+d_{-2})\beta\\ &=\hbar^2\sqrt{1/2}\,(2d_2-2d_{-2})(-1/2)\beta\\ &\quad+{}^1\!/_2\hbar^2\sqrt{1/2}\,[\sqrt{(2-2+1)(2+2)}\,d_1+\sqrt{(2+2+1)(2-2)}\,d_{-3}]\alpha\\ &=\hbar^2\{-\sqrt{1/2}\,(d_2-d_{-2})\beta+\sqrt{1/2}\,d_1\alpha\}\\ &=\sqrt{1/2}\,\hbar^2\{-\sqrt{2}\,t_{2g}^0\beta+t_{2g}^+\alpha\}=\sqrt{3/2}\,\hbar^2\,\Omega_1(\Gamma_8|t_{2g}), \end{aligned}$$

and correspondingly

$$\begin{aligned} &\boldsymbol{l}\cdot\boldsymbol{s}\,\Omega_i(\Gamma_8|e_g)\ =\sqrt{3/2}\,\hbar^2\Omega_i(\Gamma_8|t_{2g}), \qquad i=2,3,4\\ &\boldsymbol{l}\cdot\boldsymbol{s}\,\Omega_i(\Gamma_8|t_{2g})=-{}^1\!/_2\hbar^2\Omega_i(\Gamma_8|t_{2g})+\sqrt{3/2}\,\hbar^2\Omega_i(\Gamma_8|e_g), \qquad i=1,2,3,4\\ &\boldsymbol{l}\cdot\boldsymbol{s}\,\Omega_i(\Gamma_7|t_{2g})=\hbar^2\Omega_i(\Gamma_7|t_{2g}), \qquad i=1,2. \end{aligned}$$

Using the definition of $\zeta_{nd}$ (5–18) the integrals with $\boldsymbol{H}_{SO}=\xi(r)\boldsymbol{l}\cdot\boldsymbol{s}$ are obtained

$$(\Omega_i(\Gamma_7|t_{2g}),\, \boldsymbol{H}_{SO}\, \Omega_i(\Gamma_7|t_{2g}))=\zeta_{nd}, \qquad i=1,2$$

* The validity of (5–27) is easily seen when one considers that the $\Omega$ are *normalized linear combinations* of $t_{2g}$ or of $e_g$ functions.

$$\begin{aligned}
&(\Omega_i(\Gamma_8|t_{2g}),\ \boldsymbol{H}_{SO}\,\Omega_i(\Gamma_8|t_{2g})) = -{}^1/_2\zeta_{nd}, \quad i=1,2,3,4\\
&(\Omega_i(\Gamma_8|e_g),\ \boldsymbol{H}_{SO}\,\Omega_i(\Gamma_8|e_g)) = 0, \quad i=1,2,3,4\\
&(\Omega_i(\Gamma_8|t_{2g}),\ \boldsymbol{H}_{SO}\,\Omega_i(\Gamma_8|e_g)) = (\Omega_i(\Gamma_8|e_g),\ \boldsymbol{H}_{SO}\Omega_i(\Gamma_8|t_{2g}))\\
&\qquad = \sqrt{3/2}\,\zeta_{nd}, \quad i=1,2,3,4.
\end{aligned} \tag{5–28}$$

If (5–27) and (5–28) are substituted into (5–25) the energy of the $\Gamma_7$ term becomes:

$$E(\Gamma_7) = \varepsilon_0 - 4Dq + \zeta_{nd}. \tag{5–29}$$

Analogously the energies of the two $\Gamma_8$ terms are obtained from

$$\begin{vmatrix} \varepsilon_0 - 4Dq - {}^1/_2\zeta_{nd} - E(\Gamma_8) & \sqrt{3/2}\,\zeta_{nd} \\ \sqrt{3/2}\,\zeta_{nd} & \varepsilon_0 + 6Dq - E(\Gamma_8) \end{vmatrix} = 0, \tag{5–30}$$

with the roots

$$E_{a,b}(\Gamma_8) = \varepsilon_0 + Dq - \frac{1}{4}\zeta_{nd} \pm \frac{1}{2}(10Dq + {}^1/_2\zeta_{nd})\cdot\sqrt{1 + \frac{6\zeta_{nd}^2}{(10Dq + {}^1/_2\zeta_{nd})^2}}.$$

In order to obtain a connection to the energy expressions for vanishing spin–orbit coupling we can also write

$$\begin{aligned}
E_a(\Gamma_8) &= \varepsilon_0 + 6Dq + \sqrt{3/2}\,\zeta_{nd}\cdot U,\\
E_b(\Gamma_8) &= \varepsilon_0 - 4Dq - {}^1/_2\zeta_{nd} - \sqrt{3/2}\,\zeta_{nd}\cdot U,
\end{aligned} \tag{5–31}$$

with

$$U = \frac{10Dq + {}^1/_2\zeta_{nd}}{\sqrt{6}\,\zeta_{nd}}\left\{\sqrt{1 + \frac{6\zeta_{nd}^2}{(10Dq + {}^1/_2\zeta_{nd})^2}} - 1\right\}. \tag{5–32}$$

The results obtained can be interpreted in the following manner: Under the influence of spin–orbit coupling the three terms $\Gamma_7, \Gamma_8, \Gamma_8$ arise from $^2T_{2g}$ and $^2E_g$. The progeny term $\Gamma_7$ arises unambiguously from the $^2T_{2g}$ term and is according to (5–29) labilized by $\zeta_{nd}$ in comparison to its parent term. Because an unambiguous assignment is not possible for the two $\Gamma_8$ progeny terms, it can be concluded that the nondiagonal elements in (5–30) do not vanish. This combination of $\Omega(\Gamma_8|e_g)$ with $\Omega(\Gamma_8|t_{2g})$ states with reference to $\boldsymbol{H}_{SO}$ shows that both $\Gamma_8$ progeny terms arise from a *mixture* of $E_g$ states with $^2T_{2g}$ states.

If in (5–30) the nondiagonal elements and thus the mixture of $^2E_g$ with $^2T_{2g}$ terms are neglected, then the energies become:

$$\begin{aligned}
E_a(\Gamma_8) &= \varepsilon_0 + 6Dq,\\
E_b(\Gamma_8) &= \varepsilon_0 - 4Dq - {}^1/_2\zeta_{nd}.
\end{aligned} \tag{5–33}$$

In this case the $\Gamma_8$ progeny term with the energy $E_a(\Gamma_8)$ can be unambiguously assigned to the $^2E_g$ term and analogously the $\Gamma_8$ progeny term with the energy $E_b(\Gamma_8)$ to the $^2T_{2g}$ term. The $\Gamma_8$ term from $^2E_g$ remains completely unaffected by the spin–orbit coupling and only the $\Gamma_8$ term which arises from $^2T_{2g}$ is stabilized by the energy $-\frac{1}{2}\zeta_{nd}$. In this case it is also precise to say that the $^2T_g$ term splits under the influence of spin–orbit coupling into a $\Gamma_7$ term and a $\Gamma_8$ term with the separation $|+\zeta_{nd}|+|-\frac{1}{2}\zeta_{nd}| = \frac{3}{2}\zeta_{nd}$. As will be shown in the following the conditions which allow the neglect of nondiagonal elements are given when the spin–orbit coupling parameter $\zeta_{nd}$ is very much smaller than the $^2E_g$-$^2T_{2g}$ separation 10 $Dq$. The term system which approximately holds for this limiting case, that of $d^1$ systems of the first series, is shown under (b) in Fig. B.42.

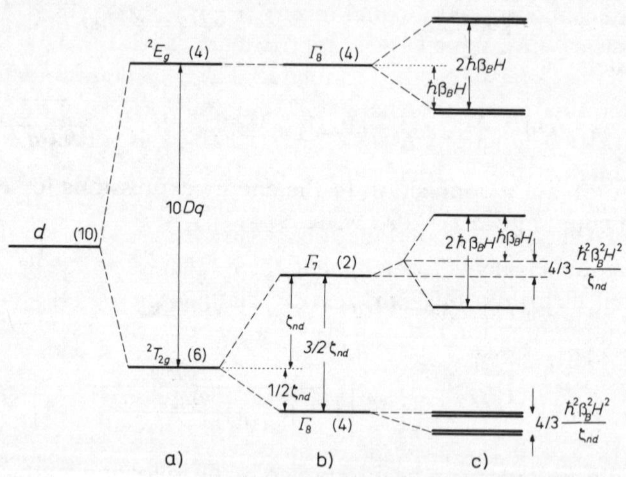


Figure B.42. Diagram of the term splitting of an octahedrally coordinated $d^1$ ion: (a) without spin–orbit coupling; (b) with weak spin–orbit coupling; (c) with weak spin–orbit coupling in a 'weak' magnetic field of field strength $H$.

If $\zeta_{nd} \ll 10\,Dq$, i.e., $\zeta_{nd}/10\,Dq \ll 1$, then we are allowed to use for the root in the expression (5–32) because of

$$\frac{6\,\zeta_{nd}^2}{(10\,Dq+{}^1/_2\zeta_{nd})^2}=x\ll 1$$

and because of

$$(1+x)^{1/2}\approx 1+{}^1/_2x\,,\qquad [x\ll 1]$$

we can write:

$$\sqrt{1+\frac{6\zeta_{nd}^2}{(10Dq+{}^1/_2\zeta_{nd})^2}}\approx 1+{}^1/_2\,\frac{6\zeta_{nd}^2}{(10Dq+{}^1/_2\zeta_{nd})^2}\,.$$

With this we obtain:

$$U \approx {}^1/_2 \frac{\sqrt{6}\,\zeta_{nd}}{10Dq + {}^1/_2\zeta_{nd}} \approx \sqrt{\frac{3}{2}}\,\frac{\zeta_{nd}}{10Dq}.$$

Bearing this condition in mind, we obtain the following approximate expressions for the energies (5–31)

$$E_a(\Gamma_8) \approx \varepsilon_0 + 6Dq + {}^3/_2(\zeta_{nd}^2/10Dq) \approx \varepsilon_0 + 6Dq,$$
$$E_b(\Gamma_8) \approx \varepsilon_0 - 4Dq - {}^1/_2\zeta_{nd} - {}^3/_2(\zeta_{nd}^2/10Dq) \approx \varepsilon_0 - 4Dq - {}^1/_2\zeta_{nd}. \qquad (5\text{–}34)$$

These are, however, exactly the energies which result from (5–30) when the off-diagonal elements are set equal to zero.

As was the case for the energies, approximate expressions for the wave functions of the $\Gamma_8$ terms can be formulated. The $2 \times 2$ determinants (5–30) for the determination of the energies $E_a(\Gamma_8)$ and $E_b(\Gamma_8)$ are in each case formed with two functions $\Omega_i(\Gamma_8|t_{2g})$, $\Omega_i(\Gamma_8|e_g)$. The solution functions which diagonalize the $2 \times 2$ determinant will thus always be two orthogonal linear combinations of associated function pairs $\Omega_i(\Gamma_8|t_{2g})$, $\Omega_i(\Gamma_8|e_g)$. In order for two such linear combinations actually to be solution functions the nondiagonal elements formed using them must of course be zero and the diagonal clcments must be equal to the energies $E_a(\Gamma_8)$ or $E_b(\Gamma_8)$. For the solution function of the energy $E_a(\Gamma_8)$ one assumes

$$\Psi_{a,i}(\Gamma_8) = \sqrt{\frac{1}{1+a^2}}\,[\Omega_i(\Gamma_8|e_g) + a\Omega_i(\Gamma_8|t_{2g})],$$

where $\sqrt{1(1+a^2)}$ is the normalization constant. With consideration of (5–27, 28) the diagonal element becomes:

$$\varepsilon_0 + 6Dq + \sqrt{3/2}\,\zeta_{nd}U = \frac{1}{1+a^2}\,[\varepsilon_0 + 6Dq + 2a\sqrt{3/2}\,\zeta_{nd} + a^2(\varepsilon_0 - 4Dq - {}^1/_2\zeta_{nd})].$$

This diagonal element must be equal to $E_a(\Gamma_8)$ (cf. 5–31)

$$(\Psi_{a,i}(\Gamma_8), \boldsymbol{H}_{\text{pert}}\,\Psi_{a,i}(\Gamma_8)) = \frac{1}{1+a^2}\,[\varepsilon_0 + 6Dq + 2a\sqrt{3/2}\,\zeta_{nd} + a^2(\varepsilon_0 - 4Dq - {}^1/_2\zeta_{nd})].$$

By rearranging terms this becomes

$$a^2(10Dq + {}^1/_2\,\zeta_{nd} + \sqrt{3/2}\,\zeta_{nd}U) - 2a\sqrt{3/2}\,\zeta_{nd} + \sqrt{3/2}\,\zeta_{nd}U = 0. \qquad (5\text{–}35)$$

For the case $\zeta_{nd} \ll 10\,Dq$ it holds approximately that

$$10Dq + {}^1/_2\,\zeta_{nd} + \sqrt{3/2}\,\zeta_{nd}U \approx 10Dq$$

and

$$U \approx \sqrt{3/2}\,(\zeta_{nd}/10\,Dq).$$

(5–35) becomes then

$$a^2 - 2a[\sqrt{3/2}\,(\zeta_{nd}/10\,Dq)] + [\sqrt{3/2}\,(\zeta_{nd}/10\,Dq)]^2 = 0,$$

or

$$[a - \sqrt{3/2}\,(\zeta_{nd}/10\,Dq)]^2 = 0.$$

That is

$$a = \sqrt{3/2}\,(\zeta_{nd}/10\,Dq).$$

Because for $\zeta_{nd} \ll 10\,Dq$ we may set $1 + a^2 \approx 1$ we obtain finally the following approximate expression for the wave functions of the energy $E_a(\Gamma_8)$:

$$\Psi_{a,i}(\Gamma_8) \approx \Omega_i(\Gamma_8|e_g) + \sqrt{3/2}\,(\zeta_{nd}/10\,Dq)\,\Omega_i(\Gamma_8|t_{2g}). \qquad i = 1,2,3,4$$

Using the same procedure the approximate expressions for the wave functions for the energy $E_b(\Gamma_8)$ are found:

$$\Psi_{b,i}(\Gamma_8) \approx \Omega_i(\Gamma_8|t_{2g}) - \sqrt{3/2}\,(\zeta_{nd}/10\,Dq)\,\Omega_i(\Gamma_8|e_g). \qquad i = 1,2,3,4$$

It must finally be emphasized that the complete energy expressions (5–31) would naturally also have been obtained if one had not started with the functions $\Omega$, but with any other set of ten linearly independent $d$ functions, e.g., from the $F(n^2D^{m_j}_{5/2})$ and $F(n^2D^{m_j}_{3/2})$ which describe the states of a free $d^1$ system for a nonzero spin–orbit coupling.

## 5.3. Zeeman effect in free atoms and atomic ions

If an atom or atomic ion is subjected to a magnetic field then the atomic electrons by virtue of their magnetic orbital and spin moments can interact with the field. The quantum mechanical operator which represents the energy of interaction between an atomic electron and a homogeneous magnetic field of the strength $\vec{H}$ has the form*

$$\boldsymbol{H}_M = \beta_B \vec{H} \cdot (\boldsymbol{l} + 2\boldsymbol{s}). \qquad (5\text{–}35)$$

Here $\beta_B$ is the *Bohr magneton* (= $0{\cdot}9273 \times 10^{-20}$ erg/gauss). The factor 2

---

* According to classical electrodynamics the interaction between a magnetic field of strength $\vec{H}$ and a magnetic dipole $\vec{\mu}$ is equal to

$$-\vec{H} \cdot \vec{\mu} = -(H_x\,\mu_x + H_y\,\mu_y + H_z\,\mu_z).$$

before the spin operator $\boldsymbol{s}$ arises from the anomalous spin magnetic moment.

In section 5.1 it was shown that the states of a free atomic one-electron system under consideration of spin–orbit coupling can be classified according to the quantum numbers $n$, $l$, $j$, $m_j$. The states belonging to a given principal quantum number $n$ and a given orbital angular momentum quantum number $l$ can be divided into two terms $n^2L_{l+1/2}$ and $n^2L_{l-1/2}$ with the energies (cf. (5–19)).

$$E(n^2L_{l+1/2}) = E_{n,l} + {}^1/_2 l\zeta_{nl}, \tag{5–36a}$$

and

$$E(n^2L_{l-1/2}) = E_{n,l} - {}^1/_2 (l+1)\zeta_{nl}, \tag{5–36b}$$

and the wave functions $F(n^2L^{m_j}_{l+1/2})$ and $F(n^2L^{m_j}_{l-1/2})$, (cf. (5–17)). The energy separation between the terms of the same doublet is in general small compared to the separations of different doublets.

We shall now study what consequences an external magnetic field of strength $\vec{H}$ has for the states $F(n^2L^{m_j}_{l+1/2})$ and $F(n^2L^{m_j}_{l-1/2})$ of a doublet. Because of the relatively large separation of different doublets we assume that it is legitimate to study a single doublet term isolated from the others. The energies of the states which an atomic one-electron system can assume under the influence of a magnetic field $\vec{H}$ are obtained approximately for a given $n$ and $l$ by solving the secular determinant having the $(2l+2)+(2l) = 4l+2$ functions $F(n^2L^{m_j}_{l+1/2})$ and $F(n^2L^{m_j}_{l-1/2})$ as a basis and the total *Hamiltonian*

$$\boldsymbol{H} = \boldsymbol{H}_0 + \boldsymbol{H}_{SO} + \boldsymbol{H}_M. \tag{5–37}$$

It is convenient and in no way reduces the generality when the $z$ axis of the coordinate system is chosen so that it coincides with the direction of the field $\vec{H}$. We can then write for (5–37) with consideration of (5–35):

$$\boldsymbol{H} = \boldsymbol{H}_0 + \boldsymbol{H}_{SO} + \beta_B H(\boldsymbol{l}_z + 2\boldsymbol{s}_z), \tag{5–38}$$

where $H$ is the absolute value of $\vec{H}$. The choice of the functions $F(n^2L^{m_j}_{l+1/2})$ and $F(n^2L^{m_j}_{l-1/2})$ as the basis of the secular problem is advantageous in so far as these functions will already diagonalize that portion of the secular determinant formed with $\boldsymbol{H}_0 + \boldsymbol{H}_{SO}$ and yields the diagonal elements (cf. (5–36)):

$$(F(n^2L^{m_j}_{l+1/2}), [\boldsymbol{H}_0 + \boldsymbol{H}_{SO}]F(n^2L^{m_j}_{l+1/2})) = E_{n,l} + {}^1/_2 l\zeta_{nl},$$
$$(F(n^2L^{m_j}_{l-1/2}), [\boldsymbol{H}_0 + \boldsymbol{H}_{SO}]F(n^2L^{m_j}_{l-1/2})) = E_{n,l} - {}^1/_2 (l+1)\zeta_{nl}. \tag{5–39}$$

We need only to evaluate the integrals with the operator $\boldsymbol{H}_M = \beta_B H(\boldsymbol{l}_z + 2\boldsymbol{s}_z)$ which appear in the elements of the secular determinant. For this one applies first $\boldsymbol{H}_M$ to the functions defined in (5–17) $F(n^2L^{m_j}_{l\pm 1/2})$ and uses in this case the validity of

$$\boldsymbol{l}_z\psi_{n,\,l,\,m_l,\,m_s} = m_l\hbar\psi_{n,\,l,\,m_l,\,m_s},$$
$$\boldsymbol{s}_z\psi_{n,l,\,m_l,\,m_s} = m_s\hbar\psi_{n,\,l,\,m_l,\,m_s},$$

$$\begin{aligned}\boldsymbol{H}_M F(n^2L^{m_j}_{l\pm1/2}) &= \beta_B H(\boldsymbol{l}_z+2\boldsymbol{s}_z)\left[\sqrt{\frac{l\pm m_j+{}^1\!/_2}{2l+1}}\,\psi_{n,\,l,\,m_j-1/2,1/2}\right.\\ &\qquad\left.\pm\sqrt{\frac{l\mp m_j+{}^1\!/_2}{2l+1}}\,\psi_{n,\,l,\,m_j+1/2,-1/2}\right]\\ &= \beta_B H\hbar\left[[(m_j-{}^1\!/_2)+2\cdot{}^1\!/_2]\sqrt{\frac{l\pm m_j+{}^1\!/_2}{2l+1}}\,\psi_{n,\,l,\,m_j-1/2,1/2}\right.\\ &\qquad\left.\pm[(m_j+{}^1\!/_2)+2(-{}^1\!/_2)]\sqrt{\frac{l\mp m_j+{}^1\!/_2}{2l+1}}\,\psi_{n,\,l,\,m_j+1/2,-1/2}\right]\\ &= \beta_B H\hbar\left[(m_j+{}^1\!/_2)\sqrt{\frac{l\pm m_j+{}^1\!/_2}{2l+1}}\,\psi_{n,\,l,\,m_j-1/2,1/2}\right.\\ &\qquad\left.\pm(m_j-{}^1\!/_2)\sqrt{\frac{l\mp m_j+{}^1\!/_2}{2l+1}}\,\psi_{n,\,l,\,m_j+1/2,-1/2}\right].\end{aligned}$$

Thus the integrals which are formed with $F$ functions of the same $j$ value become

$$\begin{aligned}(F(n^2L^{m'_j}_{l\pm1/2}),\,\boldsymbol{H}_M F(n^2L^{m_j}_{l\pm1/2})) &= \beta_B H\hbar\cdot\delta_{m_j,\,m'_j}\left[(m_j+{}^1\!/_2)\frac{l\pm m_j+{}^1\!/_2}{2l+1}\right.\\ &\qquad\left.+(m_j-{}^1\!/_2)\frac{l\mp m_j+{}^1\!/_2}{2l+1}\right]\\ &= \delta_{m_j,\,m'_j}\beta_B H\hbar\cdot m_j\frac{2l+1\pm1}{2l+1},\end{aligned} \tag{5–40}$$

and for integrals with $F$ functions of different $j$ values we obtain

$$\begin{aligned}(F(n^2L^{m'_j}_{l\mp1/2}),\,\boldsymbol{H}_M F(n^2L^{m_j}_{l\pm1/2})) &= \beta_B H\hbar\,\delta_{m_j,\,m'_j}\times\\ &\quad\times\left[(m_j+{}^1\!/_2)\sqrt{\frac{(l\mp m_j+{}^1\!/_2)(l\pm m_j+{}^1\!/_2)}{(2l+1)^2}}\right.\\ &\qquad\left.-(m_j-{}^1\!/_2)\sqrt{\frac{(l\mp m_j+{}^1\!/_2)(l\pm m_j+{}^1\!/_2)}{(2l+1)^2}}\right]\\ &= \delta_{m_j,\,m'_j}\beta_B H\hbar\sqrt{\frac{(l\pm m_j+{}^1\!/_2)(l\mp m_j+{}^1\!/_2)}{(2l+1)^2}}\end{aligned} \tag{5–41}$$

In any case only those $F$ functions can combine with one another with reference to $\boldsymbol{H}_M$ which have the same $m_j$ value. The decomposition of the secular determinant of order $4l+2$ into subdeterminants of lower order is based on this fact.

First of all it must be considered that for $m_j$ values $l+1/2$ and $-(l+1/2)$ there is always only one $F$ function, in particular for $j = l+1/2$:

$F(n^2L_{l+1/2}^{l+1/2})$ or $F(n^2L_{l+1/2}^{-(l+1/2)})$. These functions can not combine with any other $F$ functions because of $\delta_{m_j,m'_j}$ in (5–40) and (5–41). For this reason the complete secular determinant decomposes into the two $1\times 1$ determinants with the roots (cf. (5–39, 40))

$$\begin{aligned} E_M(n^2L_{l+1/2}^{l+1/2}) &= (F(n^2L_{l+1/2}^{l+1/2}), [H_0+H_{SO}+H_M]F(n^2L_{l+1/2}^{l+1/2})) \\ &= E_{n,l}+{}^1/_2 l\zeta_{nl}+\beta_B H\hbar(l+{}^1/_2)\frac{2l+1+1}{2l+1} \\ &= E_{n,l}+{}^1/_2 l\zeta_{nl}+(l+1)\hbar\beta_B H, \end{aligned} \tag{5–42a}$$

or

$$E_M(n^2L_{l+1/2}^{-(l+1/2)}) = E_{n,l}+{}^1/_2 l\zeta_{nl}-(l+1)\hbar\beta_B H, \tag{5–42b}$$

and the $4l\times 4l$ subdeterminants formed with all of the remaining functions. Of these residual functions there always exist for every $m_j$ value ($|m_j| < l+1/2$) exactly two functions: $F(n^2L_{l+1/2}^{m_j})$, $F(n^2L_{l-1/2}^{m_j})$. The $4l\times 4l$ subdeterminant decomposes because of $\delta_{m_l,m'_l}$ thus into $2l$ $2\times 2$ determinants

$$\begin{vmatrix} (F(n^2L_{l+1/2}^{m_j}), HF(n^2L_{l+1/2}^{m_j}))-E_M & (F(n^2L_{l+1/2}^{m_j}), HF(n^2L_{l-1/2}^{m_j})) \\ (F(n^2L_{l-1/2}^{m_j}), HF(n^2L_{l+1/2}^{m_j})) & (F(n^2L_{l-1/2}^{m_j}), HF(n^2L_{l-1/2}^{m_j}))-E_M \end{vmatrix}=0, \quad |m_j|<l+1/2. \tag{5–43}$$

With consideration of (5–39, 40, 41) and using the abbreviation $4E = E_M - E_{n,l}$ we can write

$$\begin{vmatrix} \frac{\hbar\beta_B H}{2l+1}\cdot m_j(2l+2)+{}^1/_2 l\zeta_{nl}-\Delta E & \frac{\hbar\beta_B H}{2l+1}\sqrt{(l+m_j+{}^1/_2)(l-m_j+{}^1/_2)} \\ \frac{\hbar\beta_B H}{2l+1}\sqrt{(l+m_j+{}^1/_2)(l-m_j+{}^1/_2)} & \frac{\hbar\beta_B H}{2l+1} m_j\cdot 2l-{}^1/_2(l+1)\zeta_{nl}-\Delta E \end{vmatrix}=0.$$

The roots are

$$\begin{aligned} \Delta E_\pm = E_M^\pm - E_{n,l} &= \hbar\beta_B H m_j - {}^1/_4\zeta_{nl} \\ &\pm {}^1/_4[4\hbar^2\beta_B^2H^2+8\hbar\beta_B H m_j\zeta_{nl}+(2l+1)^2\zeta_{nl}^2]^{1/2}. \end{aligned} \tag{5–44}$$

Because accordingly, in contrast to the field-free case, the energies depend upon the quantum number $m_j$, in general a splitting of the terms present in the field-free system $n^2L_{l+1/2}$ or $n^2L_{l-1/2}$ will appear under the influence of a homogeneous magnetic field*.

It is often of interest for the spectroscopist to consider two limiting cases. The first occurs when the *energetic effects of the external field are*

* The terms $n^2L_j$ of the field-free system are of course composed of degenerate states $F(n^2L_j^m j)$ with (different) $m_j = -j, \ldots, +j$.

*small compared to the spin–orbit coupling constant* $\zeta_{nl}$, that is the case

$$\hbar\beta_B H \ll \zeta_{nl}.$$

If we write for (5–44)

$$E_M^{\pm} = E_{n,l} + \hbar\beta_B H m_j - {}^1\!/_4\zeta_{nl} \pm {}^1\!/_4(2l+1)\zeta_{nl}\sqrt{1 + \frac{8m_j}{(2l+1)^2}\frac{\hbar\beta_B H}{\zeta_{nl}} + \frac{4}{(2l+1)^2}\left(\frac{\hbar\beta_B H}{\zeta_{nl}}\right)^2},$$

and neglect the quadratic term in $\hbar\beta_B H/\zeta_{nl}$, which is very small compared to 1 because $\hbar\beta_B H \ll \zeta_{nl}$, in the radical and expand the remaining portion of the radical using the approximation

$$(1+x)^{1/2} \approx 1 + {}^1\!/_2 x, \qquad [x \ll 1],$$

then this becomes

$$\begin{aligned} E_M^{\pm} &= E_{n,l} + \hbar\beta_B H m_j - {}^1\!/_4\zeta_{nl} \pm {}^1\!/_4(2l+1)\zeta_{nl}\left[1 + {}^1\!/_2\frac{8m_j}{(2l+1)^2}\frac{\hbar\beta_B H}{\zeta_{nl}}\right] \\ &= E_{n,l} - {}^1\!/_4[1 \mp (2l+1)]\zeta_{nl} + \hbar\beta_B H m_j \frac{2l+1\pm1}{2l+1}, \ |m_j| < l + 1/2. \end{aligned} \tag{5–45}$$

When we add the energy expressions (5–42) without change, then we obtain

$$E_M^{+} = E_{n,l} + {}^1\!/_2 l\zeta_{nl} + \hbar\beta_B H m_j \frac{2l+2}{2l+1}, m_j = -(l+1/2), \ldots, +(l+1/2), \tag{5–46a}$$

$$E_M^{-} = E_{n,l} - {}^1\!/_2(l+1)\zeta_{nl} + \hbar\beta_B H m_j \frac{2l}{2l+1}, m_j = -(l-1/2), \ldots, +(l-1/2). \tag{5–46b}$$

These results can be interpreted in the following manner: In the weakfield case, $\hbar\beta_B H \ll \zeta_{nl}$, the terms $n^2L_{l+1/2}$ and $n^2L_{l-1/2}$ of the field-free system split into $2l+2$ or $2l$ equally distant simple (nondegenerate) terms. The splittings are symmetric with reference to the term position in the field-free case. Two neighbouring splitting terms arising from $n^2L_{l+1/2}$ are separated by $\hbar\beta_B H(2l+2)/(2l+1)$, the separation of two neighbouring $n^2L_{l-1/2}$ progeny terms is $\hbar\beta_B H(2l)/(2l+1)$. The splittings are thus proportional to the field strength and depend only upon the orbital angular momentum quantum number $l$ and therefore not upon the principal quantum number $n$ and the nature of the spherically symmetric potential (arising from the atomic nucleus or atomic core) in which the electron moves.

The results are shown schematically in the right-hand side of Figure B.40.

The factor $(2l+1\pm1)(2l+1)$ which appears in (5–45) is called the *Landé* g factor and is a special form of the general expression

$$g=1+\frac{J(J+1)-L(L+1)+S(S+1)}{2J(J+1)},$$

obtained when the substitutions $J=j=l\pm1/2$, $L=l$, $S=s=1/2$ are made.

The other limiting case is obtained when *the energetic consequences of the external field are large in comparison to the effects of the spin–orbit coupling*, that is, for the case

$$\hbar\beta_B H \gg \zeta_{nl}.$$

Approximate expressions for the energies for this limiting case are obtained when in (5–44) terms with $(\zeta_{n,l}/\hbar\beta_B H)^2$ are neglected compared to 1. Together with (5–42) one then finds

$$\begin{aligned} E_M^+ &= E_{n,l}+\hbar\beta_B H(m_j+{}^1\!/_2)+{}^1\!/_2(m_j-{}^1\!/_2)\zeta_{nl}, \\ &\qquad |m_j|<l+1/2 \quad\text{and}\quad m_j=l+1/2, \\ E_M^- &= E_{n,l}+\hbar\beta_B H(m_j-{}^1\!/_2)-{}^1\!/_2(m_j+{}^1\!/_2)\zeta_{nl}, \\ &\qquad |m_j|<l+1/2 \quad\text{and}\quad m_j=-(l+1/2). \end{aligned} \tag{5–47}$$

The effects which appear for the case $\hbar\beta_B H \ll \zeta_{nl}$ are denoted summarily as the *Paschen–Back effect*.

## 5.4. Zeeman effect of octahedrally bound $d^1$ ions

In this section we shall study what consequences an external homogeneous magnetic field has for the states of an octahedrally bound $d^1$ ion. We shall, however, not attempt to solve this problem in complete generality. Rather we shall restrict ourselves to treating those systems in which the *splitting*, 10 *Dq*, *produced by the ligand field is large compared to the spin–orbit coupling constant* $\zeta_{nd}$ *and also compared to the energetic effects of the magnetic field*. Examples of such systems are octahedrally bound $d^1$ ions of the first transition metal series which are subjected to magnetic fields of strengths normally produced in the laboratory ($H \lesssim 15{,}000$ gauss).

The perturbation operator which represents the perturbation energy for a $d^1$ ion in an octahedral ligand field under the influence of spin–orbit coupling when a homogeneous magnetic field of strength $\vec{H}$ is applied is

$$\boldsymbol{H}_{\text{pert}} = \boldsymbol{H}_{\text{Lig.}}+\boldsymbol{H}_{SO}+\boldsymbol{H}_M. \tag{5–48}$$

The operators $\boldsymbol{H}_{\text{Lig.}}$ and $\boldsymbol{H}_{SO}$ are defined in sections 3.3 and 5.2. The magnetic field operator is (cf. p. 448)

$$\boldsymbol{H}_M=\beta_B\vec{H}\cdot(\boldsymbol{l}+2\boldsymbol{s}). \tag{5–49}$$

The partial operator $\boldsymbol{H}_{\text{Lig.}}+\boldsymbol{H}_{SO}$ of $\boldsymbol{H}_{\text{pert}}$ is, as has already been explained in section 5.2, invariant under all symmetry operations of the octahedral double group $O'$. With the addition of the magnetic field operator $\boldsymbol{H}_M$ the symmetry is reduced, in particular the axial symmetry which is determined by the direction of $\vec{H}$. From Figure B.9 it can, however, be seen that it must be physically irrelevant if $\vec{H}$ is oriented parallel to the $x$, $y$ or $z$ coordinate axis. The three coordinate axes are in this sense equivalent; we have isotropy. Without being arbitrary one can choose the direction of $\vec{H}$ along any one of the equivalent coordinate axes. We shall select the $z$ axis and then have $\boldsymbol{H}_M$ in the form

$$\boldsymbol{H}_M=\beta_B H(\boldsymbol{l}_z+2\boldsymbol{s}_z). \tag{5–50}$$

$H$ is here the absolute value of $\vec{H}$.

We shall now determine the states of a $d$ electron which are compatible with the perturbation operator (5–48). The perturbation energies are obtained approximately when the secular determinant with the perturbation operator (5–48) is solved using as basis functions ten linearly independent $d$ functions (space part × spin part). It is useful to choose the ten functions $\Omega_i(\Gamma_7|t_{2g})$, $\Omega_i(\Gamma_8|t_{2g})$, $\Omega_i(\Gamma_8|e_g)$ defined in equations (5–22, 23, 24).

The $\Omega$ functions diagonalize the portion of the secular determinant formed with $\boldsymbol{H}_{\text{Lig.}}$. Here the functions $\Omega_i(\Gamma_8|e_g)$ describe the labelized $E_g$ term with the ligand field energy $\varepsilon_0+4\,Dq$. The functions $\Omega_i(\Gamma_7|t_{2g})$ and $\Omega_i(\Gamma_8|t_{2g})$ belong to the stabilized $T_{2g}$ term with the ligand field energy $\varepsilon_0-4\,Dq$, cf. p. 444. If, however, the $E_g-T_{2g}$ separation (equal to 10 $Dq$) is very large compared to the perturbation energies produced by $\boldsymbol{H}_{SO}$ and $\boldsymbol{H}_M$, then in the complete secular determinant (with the total perturbation operator $\boldsymbol{H}_{\text{pert}}=\boldsymbol{H}_{\text{Lig.}}+\boldsymbol{H}_{SO}+\boldsymbol{H}_M$) all of the off-diagonal elements which are formed with $E_g$ functions $\Omega_i(\Gamma_8|e_g)$ and $T_{2g}$ functions $\Omega_i(\Gamma_7|t_{2g})$ or $\Omega_i(\Gamma_8|t_{2g})$ can be set equal to zero:

$$(\Omega_i(\Gamma_8|e_g),\ \boldsymbol{H}_{\text{pert}}\ \Omega_{i'}(\Gamma_7|t_{2g}))=(\Omega_i(\Gamma_8|e_g),\ \boldsymbol{H}_{\text{pert}}\ \Omega_{i''}(\Gamma_8|t_{2g}))=0.$$
$$i=1,2,3,4,\quad i'=1,2,\quad i''=1,2,3,4.$$

This means physically that the spin–orbit coupling and the effect of the magnetic field are looked upon as small perturbations which are superposed upon the comparatively large perturbation produced by the ligand field*. The 10 × 10 secular determinant thus decomposes into a 4 × 4

* For the case of very weak spin–orbit coupling, $\zeta_{nd} \ll 10\,Dq$, it was shown explicitly in section 5.2, pp. 445 ff. that setting the $E_g-T_{2g}$ off-diagonal elements equal to zero in actuality leads to the energies which one obtains when the complete solutions are determined and then, according to the assumptions made above the terms with $\zeta_{nd}/10\,Dq$ are neglected as very small compared to one. Analogous results can be proven for the case treated here.

determinant with the functions $\Omega_i(\Gamma_8|e_g)$, $i = 1, 2, 3, 4$ and a $6\times 6$ determinant with the functions $\Omega_i(\Gamma_7|t_{2g})$, $i = 1, 2$ and $\Omega_i(\Gamma_8|t_{2g})$, $i = 1, 2, 3, 4$.

We consider first the $4\times 4$ determinant

$$||(\Omega_i(\Gamma_8|e_g), [\boldsymbol{H}_{\text{Lig.}} + \boldsymbol{H}_{SO} + \boldsymbol{H}_M]\Omega_{i'}(\Gamma_8|e_g)) - \delta_{i,i'}E(e_g)|| = 0.$$
$$i, i' = 1,2,3,4 \tag{5–51}$$

For the integrals which appear here one obtains because of (cf. (5–27))

$$(\Omega_i(\Gamma_8|e_g), \boldsymbol{H}_{\text{Lig.}}\Omega_{i'}(\Gamma_8|e_g)) = \delta_{i,i'}(\varepsilon_0 + 6Dq),$$

and because of (cf.(5–28))

$$\begin{aligned}(\Omega_i(\Gamma_8|e_g), \boldsymbol{H}_{SO}\ \Omega_{i'}(\Gamma_8|e_g)) &= 0,\\ (\Omega_i(\Gamma_8|e_g), [\boldsymbol{H}_{\text{Lig.}} + \boldsymbol{H}_{SO} + \boldsymbol{H}_M]\Omega_{i'}(\Gamma_8|e_g)) &= \\ = \delta_{i,i'}(\varepsilon_0 + 6Dq) &+ (\Omega_i(\Gamma_8|e_g), \boldsymbol{H}_M\Omega_{i'}(\Gamma_8|e_g)).\end{aligned} \tag{5–52}$$

For the evaluation of the integrals which are still unknown and which contain $\boldsymbol{H}_M$ we apply first of all $\boldsymbol{H}_M = \beta_B H(\boldsymbol{l}_z + 2\boldsymbol{s}_z)$ to the $\Omega_i(\Gamma_8|e_g)$

$\boldsymbol{H}_M = \beta_B H(\boldsymbol{l}_z + 2\boldsymbol{s}_z)$ auf die $\Omega_i(\Gamma_8|e_g)$ an:

$$\begin{aligned}\beta_B H(\boldsymbol{l}_z + 2\boldsymbol{s}_z)\Omega_1(\Gamma_8|e_g) &= \beta_B H(\boldsymbol{l}_z + 2\boldsymbol{s}_z)e_g^b\beta = \beta_B H(\boldsymbol{l}_z + 2\boldsymbol{s}_z)\sqrt{1/2}(d_2 + d_{-2})\beta\\ &= \beta_B H\hbar[\sqrt{1/2}(2d_2 - 2d_{-2})\beta\\ &\qquad + 2\sqrt{1/2}(d_2 + d_{-2})(-1/2)\beta]\\ &= \beta_B H\hbar[2\sqrt{1/2}(d_2 - d_{-2})\beta - \sqrt{1/2}(d_2 + d_{-2})\beta]\\ &= \beta_B H\hbar[2t_{2g}^0\beta - e_g^b\beta],\\ \beta_B H(\boldsymbol{l}_z + 2\boldsymbol{s}_z)\Omega_2(\Gamma_8|e_g) &= \beta_B H\hbar[-2t_{2g}^0\alpha - e_g^b\alpha],\\ \beta_B H(\boldsymbol{l}_z + 2\boldsymbol{s}_z)\Omega_3(\Gamma_8|e_g) &= \beta_B H\hbar(-e_g^a\beta),\\ \beta_B H(\boldsymbol{l}_z + 2\boldsymbol{s}_z)\Omega_4(\Gamma_8|e_g) &= \beta_B H\hbar(-e_g^a\alpha).\end{aligned}$$

With this, e.g., we have

$$\begin{aligned}(\Omega_1(\Gamma_8|e_g), \boldsymbol{H}_M\Omega_1(\Gamma_8|e_g)) &= \beta_B H\hbar(e_g^b\beta, [2t_{2g}^0\beta - e_g^b\beta])\\ &= -\hbar\beta_B H,\end{aligned}$$

and correspondingly

$$\begin{aligned}&(\Omega_2(\Gamma_8|e_g), \boldsymbol{H}_M\Omega_2(\Gamma_8|e_g)) = (\Omega_4(\Gamma_8|e_g), \boldsymbol{H}_M\Omega_4(\Gamma_8|e_g)) = +\hbar\beta_B H,\\ &(\Omega_3(\Gamma_8|e_g), \boldsymbol{H}_M\Omega_3(\Gamma_8|e_g)) = -\hbar\beta_B H,\\ &(\Omega_i(\Gamma_8|e_g), \boldsymbol{H}_M\Omega_{i'}(\Gamma_8|e_g)) = 0, \qquad i \neq i'.\end{aligned}$$

With consideration of (5–52) when in addition the abbreviation $\Delta E(e_g) = E(e_g)-(\varepsilon_0+6\,Dq)$ is used, we obtain for the determinant (5–51):

$$\begin{vmatrix} -\hbar\beta_B H-\Delta E(e_g) & 0 & 0 & 0 \\ 0 & +\hbar\beta_B H-\Delta E(e_g) & 0 & 0 \\ 0 & 0 & -\hbar\beta_B H-\Delta E(e_g) & 0 \\ 0 & 0 & 0 & +\hbar\beta_B H-\Delta E(e_g) \end{vmatrix} = 0.$$

This has the two double roots

$$\begin{aligned} E_+(e_g) &= \varepsilon_0+6Dq+\hbar\beta_B H, \\ E_-(e_g) &= \varepsilon_0+6Dq-\hbar\beta_B H. \end{aligned} \tag{5–53}$$

Thus the fourfold $^2E_g$ term splits under the influence of the magnetic field into two twofold terms with the separation $2\hbar\beta_B H$. The splitting is proportional to the field strength $\vec{H}$ (cf. Figure B.42).

Three types of integrals appear in the $6\times 6$ determinant with the $T_{2g}$ functions $\Omega_i(\Gamma_7|t_{2g})$ and $\Omega_i(\Gamma_8|t_{2g})$, those with two $\Gamma_7$ functions, with two $\Gamma_8$ functions and 'mixed' integrals with one $\Gamma_7$ and one $\Gamma_8$ functions. With consideration of (5–48), (5–27) and (5–28) one obtains for these

$$\begin{aligned} (\Omega_i(\Gamma_7|t_{2g}),\, \boldsymbol{H}_{\text{pert.}}\Omega_{i'}(\Gamma_7|t_{2g})) &= \delta_{i,i'}(\varepsilon_0-4Dq+\zeta_{nd}) \\ &\quad + (\Omega_i(\Gamma_7|t_{2g}),\, \boldsymbol{H}_M\Omega_{i'}(\Gamma_7|t_{2g})), \\ (\Omega_i(\Gamma_8|t_{2g}),\, \boldsymbol{H}_{\text{pert.}}\Omega_{i'}(\Gamma_8|t_{2g})) &= \delta_{i,i'}(\varepsilon_0-4Dq-{}^1\!/_2\zeta_{nd}) \\ &\quad + (\Omega_i(\Gamma_8|t_{2g}),\, \boldsymbol{H}_M\Omega_{i'}(\Gamma_8|t_{2g})), \\ (\Omega_i(\Gamma_7|t_{2g}),\, \boldsymbol{H}_{\text{pert.}}\Omega_{i'}(\Gamma_8|t_{2g})) &= (\Omega_i(\Gamma_7|t_{2g}),\, \boldsymbol{H}_M\Omega_{i'}(\Gamma_8|t_{2g})). \end{aligned}$$

The $\boldsymbol{H}_M$ integrals can be evaluated using the same procedure as in the case of $4\times 4$ determinants, by first applying $\boldsymbol{H}_M = \beta_B H(\boldsymbol{l}_z+2\boldsymbol{s}_z)$ to the function from the right and then multiplying this result with the function $\Omega^*$ from the left and finally integrating. We thus find

$$\begin{aligned} &(\Omega_1(\Gamma_7|t_{2g}),\, \boldsymbol{H}_M\Omega_1(\Gamma_7|t_{2g})) = -(\Omega_2(\Gamma_7|t_{2g}),\, \boldsymbol{H}_M\Omega_2(\Gamma_7|t_{2g})) = \hbar\beta_B H, \\ &(\Omega_1(\Gamma_7|t_{2g}),\, \boldsymbol{H}_M\Omega_2(\Gamma_7|t_{2g})) = 0, \\ &(\Omega_i(\Gamma_8|t_{2g}),\, \boldsymbol{H}_M\Omega_{i'}(\Gamma_8|t_{2g})) = 0, \quad i,i'=1,2,3,4 \\ &(\Omega_1(\Gamma_7|t_{2g}),\, \boldsymbol{H}_M\Omega_1(\Gamma_8|t_{2g})) = (\Omega_2(\Gamma_7|t_{2g}),\, \boldsymbol{H}_M\Omega_2(\Gamma_8|t_{2g})) = -\sqrt{2}\hbar\beta_B H, \\ &(\Omega_i(\Gamma_7|t_{2g}),\, \boldsymbol{H}_M\Omega_{i'}(\Gamma_8|t_{2g})) = 0 \quad \text{except for} \quad i=i'=1 \quad \text{and} \quad i=i'=2. \end{aligned}$$

The $6\times 6$ determinant becomes then

$$
\begin{array}{cccccc}
\Omega_1(\Gamma_7|t_{2g}) & \Omega_1(\Gamma_8|t_{2g}) & \Omega_2(\Gamma_8|t_{2g}) & \Omega_2(\Gamma_8|t_{2g}) & \Omega_3(\Gamma_8|t_{2g}) & \Omega_4(\Gamma_8|t_{2g})
\end{array}
$$

$$
\left|\begin{array}{cc:cc:c:c}
\zeta_{nd} + \hbar\beta_{\mathrm{B}}H - \Delta E(t_{2g}) & -\sqrt{2}\hbar\beta_{\mathrm{B}}H & 0 & 0 & 0 & 0 \\
-\sqrt{2}\hbar\beta_{\mathrm{B}}H & -{}^{1}/{}_{2}\,\zeta_{nd} - \Delta E(t_{2g}) & 0 & 0 & 0 & 0 \\
\hdashline
0 & 0 & \zeta_{nd} - \hbar\beta_{\mathrm{B}}H - \Delta E(t_{2g}) & -\sqrt{2}\hbar\beta_{\mathrm{B}}H & 0 & 0 \\
0 & 0 & -\sqrt{2}\hbar\beta_{\mathrm{B}}H & -{}^{1}/{}_{2}\,\zeta_{nd} - \Delta E(t_{2g}) & 0 & 0 \\
\hdashline
0 & 0 & 0 & 0 & -{}^{1}/{}_{2}\,\zeta_{nd} - \Delta E(t_{2g}) & 0 \\
\hdashline
0 & 0 & 0 & 0 & 0 & -{}^{1}/{}_{2}\,\zeta_{nd} - \Delta E(t_{2g})
\end{array}\right| = 0. \qquad (5\text{–}54)
$$

when the abbreviation

$$\Delta E(t_{2g}) = E(t_{2g}) - (\varepsilon_0 - 4Dq)$$

is used.

This decomposes into two $2 \times 2$ determinants and two $1 \times 1$ determinants. The latter have the identical roots

$$E_{1,2}(t_{2g}) = \varepsilon_0 - 4Dq - {}^1\!/_2\zeta_{nd}. \tag{5–55}$$

If the $2 \times 2$ determinant in the upper left in (5–54) is set equal to zero and solved for powers of $\Delta E(T_{2g})$ then one obtains

$$\Delta E^2(t_{2g}) - \Delta E(t_{2g}) \cdot ({}^1\!/_2\zeta_{nd} + \hbar\beta_B H) - {}^1\!/_2\zeta_{nd}^2 - {}^1\!/_2\zeta_{nd}\hbar\beta_B H - 2\hbar^2\beta_B^2 H^2 = 0,$$

with the roots

$$\begin{aligned}\Delta E_{3,4}(t_{2g}) &= {}^1\!/_4\zeta_{nd} + {}^1\!/_2\hbar\beta_B H \\ &\quad \pm {}^1\!/_2\sqrt{({}^1\!/_2\zeta_{nd} + \hbar\beta_B H)^2 + 2\zeta_{nd}^2 + 2\zeta_{nd}\hbar\beta_B H + 8\hbar^2\beta_B^2 H^2} \\ &= {}^1\!/_4\zeta_{nd} + {}^1\!/_2\hbar\beta_B H \\ &\quad \pm {}^3\!/_4\zeta_{nd}\sqrt{1 + {}^4\!/_3(\hbar\beta_B H/\zeta_{nd}) + 4(\hbar\beta_B H/\zeta_{nd})^2}\,.\end{aligned}$$

For

$${}^4\!/_3(\hbar\beta_B H/\zeta_{nd}) + 4(\hbar\beta_B H/\zeta_{nd})^2 \leqslant 1$$

because of

$$(1+x)^{1/2} = 1 + {}^1\!/_2 x - {}^1\!/_8 x^2 + \cdots, \qquad [x^2 \leqslant 1],$$

the square root in the expression for $\Delta E_{3,4}(t_{2g})$ can be expanded in a series of increasing powers of $\hbar\beta_B H/\zeta_{nd}$:

$$\begin{aligned}\Delta E_{3,4}(t_{2g}) = {}^1\!/_4\zeta_{nd} + {}^1\!/_2\hbar\beta_B H \pm {}^3\!/_4\zeta_{nd}\{1 + {}^1\!/_2[{}^4\!/_3(\hbar\beta_B H/\zeta_{nd}) + \\ + 4(\hbar\beta_B H/\zeta_{nd})^2] - {}^1\!/_8[{}^4\!/_3(\hbar\beta_B H/\zeta_{nd}) + 4(\hbar\beta_B H/\zeta_{nd})^2]^2 + \cdots\}.\end{aligned}$$

Neglecting the higher than square terms in $\hbar\beta_B H/\zeta_{nd}$ we obtain

$$\begin{aligned}\Delta E_{3,4}(t_{2g}) = {}^1\!/_4\zeta_{nd} + {}^1\!/_2\hbar\beta_B H \pm {}^3\!/_4\zeta_{nd}\{1 + {}^2\!/_3(\hbar\beta_B H/\zeta_{nd}) + \\ + 2(\hbar\beta_B H/\zeta_{nd})^2 - {}^2\!/_9(\hbar\beta_B H/\zeta_{nd})^2\},\end{aligned}$$

or with consideration of the abbreviation $\Delta E(t_{2g}) = E(t_{2g}) - (\varepsilon_0 - 4\,Dq)$

$$\begin{aligned}E_3(t_{2g}) &= \varepsilon_0 - 4Dq + \zeta_{nd} + \hbar\beta_B H + {}^4\!/_3(\hbar^2\beta_B^2 H^2/\zeta_{nd}), \\ E_4(t_{2g}) &= \varepsilon_0 - 4Dq - {}^1\!/_2\zeta_{nd} - {}^4\!/_3(\hbar^2\beta_B^2 H^2/\zeta_{nd}).\end{aligned} \tag{5–56}$$

The other $2 \times 2$ determinant in (5–54) yields analogous roots

$$\begin{aligned}E_5(t_{2g}) &= \varepsilon_0 - 4Dq - {}^1\!/_2\zeta_{nd} - {}^4\!/_3(\hbar^2\beta_B^2 H^2/\zeta_{nd}), \\ E_6(t_{2g}) &= \varepsilon_0 - 4Dq + \zeta_{nd} - \hbar\beta_B H + {}^4\!/_3(\hbar^2\beta_B^2 H^2/\zeta_{nd}).\end{aligned} \tag{5–57}$$

The solutions (5–55, 56, 57) show that the sixfold ${}^2T_{2g}$ term for nonzero spin–orbit coupling and under the influence of a magnetic field of strength $H$ splits into four terms, in particular into a twofold term with the energy

$$E_1(t_{2g}) = E_2(t_{2g}) = \varepsilon_0 - 4Dq - {}^1\!/_2\zeta_{nd}, \tag{5–58a}$$

a twofold term with the energy

$$E_4(t_{2g}) = E_5(t_{2g}) = \varepsilon_0 - 4Dq - {}^1\!/_2\zeta_{nd} - {}^4\!/_3(\hbar^2\beta_B^2 H^2/\zeta_{nd}), \tag{5–58b}$$

and two onefold terms with the energies

$$E_6(t_{2g}) = \varepsilon_0 - 4Dq + \zeta_{nd} - \hbar\beta_B H + {}^4\!/_3(\hbar^2\beta_B^2 H^2/\zeta_{nd}), \tag{5–58c}$$

$$E_3(t_{2g}) = \varepsilon_0 - 4Dq + \zeta_{nd} + \hbar\beta_B H + {}^4\!/_3(\hbar^2\beta_B^2 H^2/\zeta_{nd}). \tag{5–58d}$$

Because in the field-free case $H = 0$, the ${}^2T_{2g}$ term is already split into a twofold term with the energy $\varepsilon_0 - 4\,Dq + \zeta_{nd}$ and a fourfold term with the energy $\varepsilon_0 - 4\,Dq - \frac{1}{2}\zeta_{nd}$ as a result of the spin–orbit coupling, the consequences of the effect of the magnetic field can be expressed by saying: A homogeneous magnetic field leads to a splitting of the fourfold spin–orbit coupling term (energy $\varepsilon_0 - 4\,Dq - \frac{1}{2}\zeta_{nd}$) into two twofold terms and to a splitting of the twofold spin–orbit coupling term (energy $\varepsilon_0 - 4\,Dq + \zeta_{nd}$) into two onefold terms. These relationships are shown schematically in Figure B.42.

## 5.5. Magnetic susceptibility

Let us consider an atomic or molecular system with the term energies* $E_n^{(0)}$, $n = 0, 1, 2, \ldots$. If we subject this system to the influence of a magnetic field of strength $\vec{H}(|\vec{H}| = H)$ the term energies will then in general be changed as is shown in sections 5.3 and 5.4 for specific examples and the existing degeneracies partially or completely removed. The new energy states can be formally expressed in a power series in $H$†

$$E_{n,m} = E_n^{(0)} + HE_{n,m}^{(1)} + H^2E_{n,m}^{(2)} + \cdots \tag{5–59}$$

With the double indexing of the energies $E_{n,m}$ it can be seen that the term present in the field-free case $H = 0$ and having the energy $E_n^{(0)}$ can decompose into states with the energies $E_{n,1}, E_{n,2}, \ldots, E_{n,m}$‡. $m$ is in this sense a magnetic quantum number.

---

* $n$ is a term index which does not need to be the principal quantum number $n$.

† Cf. e.g., J. H. van Vleck, *Electric and Magnetic Susceptibilities*, Oxford University Press, Oxford and New York, 1932.

‡ Not all energies $E_{n,1}\ E_{n,2} \cdots$ need be different. That is, the degeneracy need not be completely removed.

In a direct application of the corresponding relation in classical electrodynamics, the magnetic moment of an atomic or molecular system, that is, in a state with the energy $E_{n,m}$ measured in the direction of the magnetic field is equal to

$$\mu^{\parallel}_{n,m} = -\frac{\partial E_{n,m}}{\partial H}, \tag{5–60}$$

or with consideration of (5–59)

$$\mu^{\parallel}_{n,m} = -E^{(1)}_{n,m} - 2HE^{(2)}_{n,m} - \cdots \tag{5–61}$$

If a large number, say $N_L \approx 6\times 10^{23}$ (*Loschmidt* number) of similar atomic or molecular systems are present, one can starting from the microscopic magnitude $\mu$, arrive at conclusions concerning the macroscopic magnitude of the molar magnetic polarization $P$. If the totality of the systems has the absolute temperature $T$, then the distribution of the systems over the energy levels $E_{n,m}$ is regulated by the *Boltzmann* relation.

According to the principle for the formation of statistical average values, the molar magnetic polarization $P_{\parallel}$ (parallel to $\vec{H}$) is given by

$$P_{\parallel}(H) = N_L \frac{\sum\limits_{n,m} \mu^{\parallel}_{n,m} e^{-\frac{E_{n,m}}{kT}}}{\sum\limits_{n,m} e^{-\frac{E_{n,m}}{kT}}}. \tag{5–62}$$

With consideration of (5–59) and (5–61) we can write the above as

$$P_{\parallel}(H) = N_L \frac{\sum\limits_{n,m} (-E^{(1)}_{n,m} - 2HE^{(2)}_{n,m} - \cdots).\exp\left[-\frac{1}{kT}(E^{(0)}_n + HE^{(1)}_{n,m} + H^2E^{(2)}_{n,m} + \cdots)\right]}{\sum\limits_{n,m} \exp\left[-\frac{1}{kT}(E^{(0)}_n + HE^{(1)}_{n,m} + H^2E^{(2)}_{n,m} + \cdots)\right]}$$

$$= N_L \frac{\sum\limits_{n,m} (-E^{(1)}_{n,m} - 2HE^{(2)}_{n,m} - \cdots)\exp\left[-\frac{1}{kT}(HE^{(1)}_{n,m} + H^2E^{(2)}_{n,m} + \cdots)\right] e^{-\frac{E^{(0)}_n}{kT}}}{\sum\limits_{n,m} \exp\left[-\frac{1}{kT}(HE^{(1)}_{n,m} + H^2E^{(2)}_{n,m} + \cdots)\right] e^{-\frac{E^{(0)}_n}{kT}}}.$$

Making use of the relation

$$e^{-a} = 1 - a + \ldots,$$

this becomes

$$P_{\parallel}(H) = N_L \frac{\sum\limits_{n,m} (-E^{(1)}_{n,m} - 2HE^{(2)}_{n,m} - \cdots)\left(1 - \frac{HE^{(1)}_{n,m}}{kT} - \frac{H^2E^{(2)}_{n,m}}{kT} - \cdots\right) e^{-\frac{E^{(0)}_n}{kT}}}{\sum\limits_{n,m}\left(1 - \frac{HE^{(1)}_{n,m}}{kT} - \frac{H^2E^{(2)}_{n,m}}{kT} - \cdots\right) e^{-\frac{E^{(0)}_n}{kT}}}. \tag{5–63}$$

If the brackets in the numerator are expanded and if terms in the numerator and denominator of second and higher order in $H$ are neglected, we obtain

$$P_{\parallel}(H) = N_L \frac{\sum\limits_{n,m}\left(-E^{(1)}_{n,m} - 2HE^{(2)}_{n,m} + \frac{H(E^{(1)}_{n,m})^2}{kT}\right) e^{-\frac{E^{(0)}_n}{kT}}}{\sum\limits_{n,m}\left(1 - \frac{HE^{(1)}_{n,m}}{kT}\right) e^{-\frac{E^{(0)}_n}{kT}}}.$$

The polarization $P_{\parallel}$ must of course be zero for vanishing field strength $H = 0$, i.e.,

$$P_{\parallel}(0) = N_L \frac{\sum\limits_{n,m} -E^{(1)}_{n,m} e^{-\frac{E^{(0)}_n}{kT}}}{\sum\limits_{n,m} e^{-\frac{E^{(0)}_n}{kT}}} = 0,$$

or

$$\sum_{n,m} -E^{(1)}_{n,m} e^{-\frac{E^{(0)}_n}{kT}} = 0.$$

With this one finds finally for the molar magnetic polarization

$$P_{\parallel}(H) = N_L \frac{H\sum\limits_{n,m}\left[\frac{(E^{(1)}_{n,m})^2}{kT} - 2E^{(2)}_{n,m}\right] e^{-\frac{E^{(0)}_n}{kT}}}{\sum\limits_{n,m} e^{-\frac{E^{(0)}_n}{kT}}}.$$

The molar magnetic susceptibility $\chi_{\parallel}$ follows from this according to the relation

$$\chi_{\parallel} = \frac{P_{\parallel}}{H}$$

giving

$$\chi_{\parallel} = N_L \frac{\sum\limits_{n,m} \left[ \frac{(E^{(1)}_{n,m})^2}{kT} - 2E^{(2)}_{n,m} \right] e^{-\frac{E^{(0)}_n}{kT}}}{\sum\limits_{n,m} e^{-\frac{E^{(0)}_n}{kT}}}. \tag{5–64}$$

It is remarkable that the susceptibility calculated in this manner is independent of the field strength $H$. The reason for this is that above terms of the second and higher order in $H$ were neglected, an approximation which is certainly also applicable for strong magnetic fields generated in the laboratory*.

As an example of the application of the theory, the magnetic susceptibility of an octahedrally bound $d^1$ ion will be calculated. As in the preceding section, let us again assume the case applicable for light transition metal ions, that is, we shall assume that the ligand field splitting 10 $Dq$ is large compared to the spin–orbit coupling constant $\zeta_{nd}$ : 10 $Dq \gg \zeta_{nd}$. Further, we wish to limit our discussion to temperatures for which $kT$ is not substantially larger than $\zeta_{nd}$ : $kT \lesssim \zeta_{nd} \ll 10\,Dq$. Under these conditions it is adequate to use only those terms in the sums (5–64) which arise from $T_{2g}$ states†. In the preceding section the term energies resulting from $T_{2g}$ were already calculated (cf. (5–58)). In order to be consistent with the notation used here, we shall relabel several quantities. If the four states which are degenerate for vanishing field strength $H$ and have the energy $\varepsilon_0 - 4\,Dq - \frac{1}{2}\zeta_{nd}$ are given the index $n = 1$ with reference to (5–59), that is

$$E_{1,1} = E_1(t_{2g}),$$
$$E_{1,2} = E_2(t_{2g}),$$
$$E_{1,3} = E_4(t_{2g}),$$
$$E_{1,4} = E_5(t_{2g}),$$

and both states with the energy $\varepsilon_0 - 4\,Dq + \zeta_{nd}$ for $H = 0$ are given the index $n = 2$

$$E_{2,1} = E_6(t_{2g}),$$
$$E_{2,2} = E_3(t_{2g}),$$

* The dependency of $\chi$ upon the field strength which appears for extraordinarily intense magnetic fields can be easily calculated by considering terms of the 2nd, 3rd, etc. order in (5–63) corresponding to growing field strength and then using $\chi_{\parallel} = P_{\parallel}/H$.

† For 10 $Dq \gg kT$ the $E_g$ states give only very small contributions to the sums because of the exponential factors in (5–64).

then one obtains from a comparison of the coefficients from (5–58) with (5–59) the coefficients

$$\begin{aligned}
E_1^{(0)} &= \varepsilon_0 - 4\,Dq - {}^1/_2\zeta_{nd},\\
E_{1,1}^{(1)} &= 0, \qquad E_{1,1}^{(2)} = 0,\\
E_{1,2}^{(1)} &= 0, \qquad E_{1,2}^{(2)} = 0,\\
E_{1,3}^{(1)} &= 0, \qquad E_{1,3}^{(2)} = -{}^4/_3(\hbar^2\beta_B^2/\zeta_{nd}),\\
E_{1,4}^{(1)} &= 0, \qquad E_{1,4}^{(2)} = -{}^4/_3(\hbar^2\beta_B^2/\zeta_{nd}),\\
E_2^{(0)} &= \varepsilon_0 - 4\,Dq + \zeta_{nd},\\
E_{2,1}^{(1)} &= -\hbar\beta_B, \; E_{2,1}^{(2)} = {}^4/_3(\hbar^2\beta_B^2/\zeta_{nd}),\\
E_{2,2}^{(1)} &= +\hbar\beta_B, \; E_{2,2}^{(2)} = {}^4/_3(\hbar^2\beta_B^2/\zeta_{nd}).
\end{aligned} \tag{5–65}$$

We are then in a position to calculate $\chi$ directly using the formula (5–64):

$$\begin{aligned}
\chi = N_L \cdot \Bigg\{ & \left[\left(\frac{(E_{1,1}^{(1)})^2}{kT} - 2E_{1,1}^{(2)}\right) + \left(\frac{(E_{1,2}^{(1)})^2}{kT} - 2E_{1,2}^{(2)}\right) + \left(\frac{(E_{1,3}^{(1)})^2}{kT} - 2E_{1,3}^{(2)}\right) + \right.\\
& \left. + \left(\frac{(E_{1,4}^{(1)})^2}{kT} - 2E_{1,4}^{(2)}\right)\right] e^{-\frac{E_1^{(0)}}{kT}} + \\
& + \left[\left(\frac{(E_{2,1}^{(1)})^2}{kT} - 2E_{2,1}^{(2)}\right) + \left(\frac{(E_{2,2}^{(1)})^2}{kT} - 2E_{2,2}^{(2)}\right)\right] e^{-\frac{E_2^{(0)}}{kT}} \Bigg\} \times \\
& \times \left\{ 4\,e^{-\frac{E_1^{(0)}}{kT}} + 2\,e^{-\frac{E_2^{(0)}}{kT}} \right\}^{-1}
\end{aligned}$$

If (5–65) is substituted into (5–64) then the common factor of all the exponential functions $e^{-\varepsilon_0 - 4Dq/kT}$ can be cancelled out and one obtains

$$\begin{aligned}
\chi &= N_L \frac{\left[2\left(\frac{0^2}{kT} - 2\cdot 0\right) + 2\left(\frac{0^2}{kT} + 2\frac{4}{3}\frac{\hbar^2\beta_B^2}{\zeta_{nd}}\right)\right] e^{\frac{\zeta_{nd}}{2kT}} + 2\left(\frac{\hbar^2\beta_B^2}{kT} - 2\frac{4}{3}\frac{\hbar^2\beta_B^2}{\zeta_{nd}}\right) e^{-\frac{\zeta_{nd}}{kT}}}{4\,e^{\frac{\zeta_{nd}}{2kT}} + 2\,e^{-\frac{\zeta_{nd}}{kT}}}\\
&= N_L \frac{\hbar^2\beta_B^2}{3kT} \frac{8 + \left(3\frac{\zeta_{nd}}{kT} - 8\right) e^{-\frac{3}{2}\frac{\zeta_{nd}}{kT}}}{\frac{\zeta_{nd}}{kT}\left(2 + e^{-\frac{3}{2}\frac{\zeta_{nd}}{kT}}\right)}
\end{aligned}$$

or with the abbreviation $\zeta_{nd}/kT = x$

$$\chi = N_L \frac{\hbar^2\beta_B^2}{3kT} \frac{8 + (3x - 8)\,e^{-\frac{3}{2}x}}{x\left(2 + e^{-\frac{3}{2}x}\right)}.$$

With the aid of the relation (cf. Part A, p. 118)

$$\chi = N_L \mu_{\text{eff.}}^2 / 3kT$$

from this the relation for the *effective magnetic moment*

$$\mu_{\text{eff.}} = \hbar \beta_B \left\{ \frac{8 + (3x - 8)\, e^{-\frac{3}{2}x}}{x\left(2 + e^{-\frac{3}{2}x}\right)} \right\}^{1/2}$$

is obtained. A discussion of this expression was given in Part A, chapter 2.

PART C

# Appendix

# 1. Mathematical terms and aids

In this portion of the appendix several mathematical terms and theorems will be discussed which are of use in solving the problems treated in the text. No claim can be made to completeness or to rigour in the sense of the mathematician.

### (a) Functions

The (physically meaningful) functions which appear in quantum mechanics are *regular*. A function $f(x)$ is regular* in the domain $a \leqq x \leqq b$ when

the function is *single-valued*, i.e., when $f(x)$ has exactly one value for each value of $x$ (in the domain),

the function is *continuous* (except for a finite number of $x$ values in the domain for which $f(x)$ can become infinite) and when

the integral of the square of the absolute value of $f(x)$ taken over the domain has a *finite value*†.

$$\int_a^b |f(x)|^2 \, \mathrm{d}x = \int_a^b f^*(x) f(x) \, \mathrm{d}x .$$

* Consider as a typical example a function of *one* (independent) variable. The functions arising in the quantum mechanics of particles are of course functions of several independent variables. A function which describes the state of $N$ electrons is a function of the $3N$ spatial coordinates $x_i$, $y_i$, $z_i$ $(i = 1, 2, \ldots, N)$ and the $N$ spin coordinates $\sigma_i$ $(i = 1, 2, \ldots, N)$. The 'domain' is in this case

$$\left.\begin{aligned} -\infty \leq x_i \leq +\infty \\ -\infty \leq y_i \leq +\infty \\ -\infty \leq z_i \leq +\infty \\ \sigma_i = -{}^1/_2, +{}^1/_2. \end{aligned}\right. \quad i = 1, 2, \ldots, N$$

The considerations for $f(x)$ can be extended without difficulty to functions of several variables.

† $f^*(x)$ denotes the complex conjugate function to $f(x)$.

*Example*: $f(x) = u(x) + iv(x)$, $u(x)$ and $v(x)$ are real functions.
$f^*(x) = u(x) - iv(x)$.

*Example:* $f(x) = e^{imx}$, domain $0 \leqq x \leqq 2\pi$. It is apparent that the (complex) function $e^{imx} = \cos mx + i \sin mx$ is single-valued and continuous in this domain. Further we have

$$\int_0^{2\pi} (e^{imx})^* e^{imx} dx = \int_0^{2\pi} e^{-imx} e^{imx} dx = \int_0^{2\pi} dx = 2\pi.$$

$e^{imx}$ is thus regular in the domain $0 \leqq x \leqq 2\pi$.

A function $f(x)$ is *normalized to one* in the domain $a \leqq x \leqq b$ if

$$\int_a^b f^*(x) f(x)\, dx = 1$$

holds.

*Example:* Which value must the (real) constant $N$ assume in the expression $f(x) = N \cdot e^{imx}$ so that $f(x)$ is normalized to one in the domain $0 \leqq x \leqq 2\pi$?

$$\int_0^{2\pi} (N \cdot e^{imx})^* (N \cdot e^{imx})\, dx = N^2 \int_0^{2\pi} e^{-imx} e^{imx} dx = N^2 \int_0^{2\pi} dx = N^2 \cdot 2\pi.$$

$N^2 \times 2\pi = 1$ holds when the *normalization constant* $N$ is:

$$N = \pm 1/\sqrt{2\pi}.$$

Two functions $f(x)$ and $g(x)$ are *orthogonal** (to one another) in the domain $a \leqq x \leqq b$ if

$$\int_a^b f^*(x) g(x)\, dx = \int_a^b g^*(x) f(x)\, dx = 0$$

$$f(x) = e^{ix},\ g(x) = e^{i2x}, \qquad 0 \leq x \leq 2\pi\,.$$

$$\int_0^{2\pi} (e^{ix})^* e^{i2x}\, dx = \int_0^{2\pi} e^{-ix} e^{2ix} dx = \int_0^{2\pi} e^{ix} dx = \frac{1}{i}\, e^{ix}\Big|_0^{2\pi} = \frac{1}{i}(1-1) = 0$$

holds.

*Example:* $f(x) = e^{ix}$, $g(x) = e^{i2x}$, domain $0 \leqq x \leqq 2\pi$.

$$\int_a^b f^*(x) g(x)\, dx = \int_a^b g^*(x) f(x)\, dx = 0$$

---

* The term *orthogonality* comes from vector analysis. If two vectors in space $\vec{r}$ and $s$ are perpendicular to one another, then the scalar product $\vec{r} \cdot \vec{s}$ is by definition equal to zero, $\vec{r} \cdot \vec{s} = 0$, and one says $\vec{r}$ and $\vec{s}$ are *orthogonal.* These concepts can be formally applied to functions in that the functions are considered as 'vectors' in a (possibly many-dimensional) function 'space', the so-called *Hilbert space.* The vanishing of the integral

$$\int_a^b f^*(x)\, g(x)\, dx$$

is interpreted by saying that the functions $f(x)$ and $g(x)$ are 'perpendicular to one another in *Hilbert* space' or that they are 'orthogonal.'

$$f(x)=e^{ix},\ g(x)=e^{i2x},\ \text{Gebiet } 0\leq x\leq 2\pi\,.$$

$$\int_0^{2\pi}(e^{ix})^* e^{i2x}\,dx=\int_0^{2\pi}e^{-ix}e^{2ix}\,dx=\int_0^{2\pi}e^{ix}\,dx=\frac{1}{i}\,e^{ix}\Big|_0^{2\pi}=\frac{1}{i}\,(1-1)=0.$$

### (b) Operators

An *operator* $\boldsymbol{a}$ is defined as a rule for obtaining a new function $F(x)$ from a given function $f(x)$:

$$\boldsymbol{a}\,f(x)=F(x)\,.$$

1. *Example:* Operator $\boldsymbol{a}$ = d/dx, function $f(x) = x^n$

$$\boldsymbol{a}f(x)=\frac{d}{dx}\,x^n=n\cdot x^{n-1},$$

i.e., the new function is $F(x) = n \cdot x^{n-1}$.

2. *Example:*

$$\boldsymbol{a}=\frac{\partial^2}{\partial x^2}+\frac{\partial^2}{\partial y^2}+\frac{\partial^2}{\partial z^2},\quad f(x,y,z)=\sin\alpha x\cdot\sin\beta y\cdot\sin\gamma z.$$

$$\begin{aligned}\boldsymbol{a}f(x,y,z)&=\sin\beta y\cdot\sin\gamma z\cdot\frac{\partial^2}{\partial x^2}\sin\alpha x\\&\quad+\sin\alpha x\cdot\sin\gamma z\cdot\frac{\partial^2}{\partial y^2}\sin\beta y\\&\quad+\sin\alpha x\cdot\sin\beta y\cdot\frac{\partial^2}{\partial z^2}\sin\gamma z\\&=\sin\beta y\cdot\sin\gamma z\cdot(-\alpha^2)\sin\alpha x\\&\quad+\sin\alpha x\cdot\sin\gamma z\cdot(-\beta^2)\sin\beta y\\&\quad+\sin\alpha x\cdot\sin\beta y\cdot(-\gamma^2)\cdot\sin\gamma z\\&=(-\alpha^2-\beta^2-\gamma^2)\cdot\sin\alpha x\cdot\sin\beta y\cdot\sin\gamma z\\&=-(\alpha^2+\beta^2+\gamma^2)f(x,y,z).\end{aligned}$$

Further the *sum of two operators* $\boldsymbol{a}$ and $\boldsymbol{b}$ is defined by the equation

$$(\boldsymbol{a}+\boldsymbol{b})f(x)=\boldsymbol{a}f(x)+\boldsymbol{b}f(x),$$

and the *product of two operators* $\boldsymbol{a}$ and $\boldsymbol{b}$ by

$$\boldsymbol{a}\,\boldsymbol{b}f(x)=\boldsymbol{a}[\boldsymbol{b}f(x)].$$

While the order of two operators $\boldsymbol{a}$ and $\boldsymbol{b}$ in a sum is always reversible

$$(\boldsymbol{a}+\boldsymbol{b})f(x)=(\boldsymbol{b}+\boldsymbol{a})f(x),$$

the order of the operators in a product is in general not irrelevant, i.e., we have

$$\boldsymbol{a}\,\boldsymbol{b}f(x)\neq\boldsymbol{b}\,\boldsymbol{a}f(x).$$

Then one says that $\boldsymbol{a}$ and $\boldsymbol{b}$ do not commute. In the special case

$$\boldsymbol{a}\,\boldsymbol{b}f(x)=\boldsymbol{b}\,\boldsymbol{a}f(x)$$

one speaks of *commuting* operators $\boldsymbol{a}$ and $\boldsymbol{b}$.

*Example:* $\boldsymbol{a}$ = d/dx; $\boldsymbol{b}$ is an operator which when applied to a function $f(x)$ gives as result the function $f(x)$ multiplied by the independent variable $x$

$$\boldsymbol{a}\,\boldsymbol{b}\neq\boldsymbol{b}\,\boldsymbol{a},$$

$\boldsymbol{a}$ and $\boldsymbol{b}$ thus do not commute

$$\boldsymbol{b}f(x)=x\cdot f(x).$$

$$\boldsymbol{a}\,\boldsymbol{b}f(x)=\boldsymbol{a}[\boldsymbol{b}f(x)]=\boldsymbol{a}[x\cdot f(x)]=\frac{\mathrm{d}}{\mathrm{d}x}[x\cdot f(x)]=f(x)+x\cdot f'(x),$$

$$\boldsymbol{b}\,\boldsymbol{a}f(x)=\boldsymbol{b}[\boldsymbol{a}f(x)]=\boldsymbol{b}[f'(x)]=x\cdot f'(x)$$

because $\boldsymbol{ab}$ has a different effect than does $\boldsymbol{ba}$.

An operator $\boldsymbol{a}$ is *linear* if

$$\boldsymbol{a}[c_1f_1(x)+c_2f_2(x)]=c_1\,\boldsymbol{a}f_1(x)+c_2\,\boldsymbol{a}f_2(x)$$

holds. (This relation is not trivial. If $\boldsymbol{a}$ had, e.g., the meaning 'extraction of the square root, $\boldsymbol{a} = \sqrt{\ }$,' then the relation above would of course not be satisfied and $\boldsymbol{a}$ would not be a nonlinear operator.)

A linear operator $\boldsymbol{a}$ is *Hermitian* when the following equality holds for two regular functions $f$ and g

$$\int f^*(\boldsymbol{a}g)\,\mathrm{d}x = \int g(\boldsymbol{a}f)^*\,\mathrm{d}x$$

*Example:* Operator $\boldsymbol{a} = i(\mathrm{d}/\mathrm{d}x)$, domain $-\infty \leqq x \leqq +\infty$.

In order that the two functions $f(x)$ and $g(x)$ be regular and normalizable for the given domain, i.e., that the integrals

$$\int_{-\infty}^{+\infty} f^*(x)f(x)\,\mathrm{d}x \quad \text{and} \quad \int_{-\infty}^{+\infty} g^*(x)g(x)\,\mathrm{d}x$$

remain finite, then the functions must of necessity be zero at infinity:

$$f(-\infty)=f(+\infty)=g(-\infty)=g(+\infty)=0.$$

The left-hand side of the above differential equation for *Hermitian* operators assumes the following form when partial integration is carried out

$$\int_{-\infty}^{+\infty} f^*(x)\left(i\frac{\mathrm{d}}{\mathrm{d}x}g(x)\right)\mathrm{d}x = if^*(x)g(x)\Big|_{x=-\infty}^{x=+\infty} - \int_{-\infty}^{+\infty}\left(i\frac{\mathrm{d}}{\mathrm{d}x}f^*(x)\right)g(x)\,\mathrm{d}x.$$

The first term of the result vanishes because the functions vanish at the limits of integration. The second term can be rewritten:

$$-\int_{-\infty}^{+\infty}\left(i\frac{\mathrm{d}}{\mathrm{d}x}f^*(x)\right)g(x)\mathrm{d}x = \int_{-\infty}^{+\infty} g(x)\left(-i\frac{\mathrm{d}}{\mathrm{d}x}\right)f^*(x)\mathrm{d}x = \int_{-\infty}^{+\infty} g(x)\left(i\frac{\mathrm{d}}{\mathrm{d}x}\right)^* f^*(x)\,\mathrm{d}x.$$

This is equal to the right-hand side of the defining equation. Thus $\boldsymbol{a} = i(\mathrm{d}/\mathrm{d}x)$ is Hermitian.

It must be emphasized that *Hamiltonians* are *Hermitian operators.*

**(c) Eigenvalue equations**

If an operator $\boldsymbol{a}$ acts on a function $f(x)$ (in other words, if $\boldsymbol{a}$ is applied to $f(x)$) and if as the result the same function $f(x)$ is obtained, except for a constant factor,

$$\boldsymbol{a}f(x) = Af(x),$$

then $f(x)$ is an *eigenfunction* of the operator $\boldsymbol{a}$ having the *eigenvalue* $A$. The equation itself is an *eigenvalue equation.*

*Example:* Consider the operator $\boldsymbol{a} = (\mathrm{d}^2/\mathrm{d}x^2)$ operating on all functions which vanish at the points $x = 0$ and $x = \pi$. One eigenfunction of $\boldsymbol{a}$ is $f(x) = \sin mx$, where $m$ is an integer. The corresponding eigenvalue is $-m^2$;

$$\frac{\mathrm{d}^2}{\mathrm{d}x^2}\sin mx = -m^2 \sin mx.$$

If there are several linearly independent eigenfunctions which have the same eigenvalue one speaks of *degeneracy.* The number of different eigenfunctions having the same eigenvalue is the *degree of degeneracy.*

*Theorem* A.1. If for a given linear operator $A$ there are two or more eigenfunctions $f_1, f_2, \ldots$ having the same eigenvalue $A$, then every *linear combination* of these functions is also an eigenfunction having the eigenvalue $A$, i.e., if

$$\boldsymbol{a} f_1 = A f_1, \qquad \boldsymbol{a} f_2 = A f_2, \ldots,$$

then

$$\boldsymbol{a}(c_1 f_1 + c_2 f_2 + \cdots) = A(c_1 f_1 + c_2 f_2 + \cdots).$$

*Theorem* A.2. The eigenvalues $A$ of a *Hermitian* operator $\boldsymbol{a}$ are *real*

$$\boldsymbol{a} f = A f,$$
$$A^* = A.$$

*Theorem* A.3. The eigenfunctions of a *Hermitian* operator $\boldsymbol{a}$ belonging to different eigenvalues $A_1$ and $A_2$

$$\begin{aligned} \boldsymbol{a} f_1 &= A_1 f_1, \\ \boldsymbol{a} f_2 &= A_2 f_2, \end{aligned} \qquad A_1 \neq A_2,$$

are *orthogonal* (with respect to the given domain of the variables)

$$\int f_1^* f_2 \, \mathrm{d}x = \int f_2^* f_1 \, \mathrm{d}x = 0.$$

The eigenfunctions of a *Hermitian* operator having the same eigenvalue (degenerate case) may always be chosen so that they are orthogonal. (Construction of appropriate linear combinations.)

*Theorem* A.4: If two operators $\boldsymbol{a}$ and $\boldsymbol{b}$ commute, then there exists a set of functions which are eigenfunctions of both operators simultaneously.

The converse: If a (complete) set of (orthogonal) functions exists so that they are eigenfunctions of two operators $\boldsymbol{a}$ and $\boldsymbol{b}$, then $\boldsymbol{a}$ and $\boldsymbol{b}$ commute.

*Theorem* A.5: If two *Hermitian* operators $\boldsymbol{a}$ and $\boldsymbol{b}$ commute and if $f_1$ and $f_2$ are eigenfunctions of $\boldsymbol{a}$ corresponding to different eigenvalues $A_1$ and $A_2$

$$\begin{aligned} \boldsymbol{a} f_1 &= A_1 f_1, \\ \boldsymbol{a} f_2 &= A_2 f_2, \\ \boldsymbol{a}\boldsymbol{b} &= \boldsymbol{b}\boldsymbol{a}, \end{aligned} \qquad A_1 \neq A_2$$

then

$$\int f_1^* \, \boldsymbol{b} f_2 \, \mathrm{d}x = \int f_2^* \, \boldsymbol{b} f_1 \, \mathrm{d}x = 0$$

holds (non-combination rule).

### (d) Matrices

An array of numbers of the form

$$\begin{pmatrix} 3 & 7 & 1 & 0 \\ 1 & 2 & 9 & 4 \\ 7 & 5 & 3 & 8 \\ 13 & -1 & 4 & 0 \end{pmatrix}$$

is called a *matrix*. According to the general definition a *matrix* is a rectangular array of $n \times m$ numbers $a_{ij}$ called *elements*.

$$\begin{pmatrix} a_{1,1} & a_{1,2} & a_{1,3} & \cdots & a_{1,n} \\ a_{2,1} & a_{2,2} & a_{2,3} & \cdots & a_{2,n} \\ a_{3,1} & a_{3,2} & a_{3,3} & \cdots & a_{3,n} \\ \vdots & \vdots & \vdots & & \vdots \\ a_{m,1} & a_{m,2} & a_{m,3} & \cdots & a_{m,n} \end{pmatrix}.$$

The elements $a_{i,1}, a_{i,2}, \ldots a_{i,n}$ constitute the $i$th row, the elements $a_{1,j}$, $a_{2,j}, a_{3,j}, \ldots, a_{m,j}$ the $j$th column of the matrix. $n$ gives the number of columns, $m$ the number of rows. Matrices are denoted in this book with bold face type, e.g. **A**.

A matrix consisting of equally many rows and columns, $n = m$, and thus containing $n^2$ elements is called a *square matrix*.

$$\mathbf{A} = \begin{pmatrix} a_{1,1} & a_{1,2} & a_{1,3} & \cdots & a_{1,n} \\ a_{2,1} & a_{2,2} & a_{2,3} & \cdots & a_{2,n} \\ a_{3,1} & a_{3,2} & a_{3,3} & \cdots & a_{3,n} \\ \vdots & \vdots & \vdots & & \vdots \\ a_{n,1} & a_{n,2} & a_{n,3} & \cdots & a_{n,n} \end{pmatrix}.$$

$n$ is the *dimensionality* of the square matrix **A**. The elements of a square matrix labeled with two equal indices are called *diagonal elements* of the matrix because they are arranged along the *principal diagonal* which runs from the upper left to the lower right. The sum of the diagonal elements is the *spur, trace,* or *character* of the matrix **A**;

$$\text{Spur } \mathbf{A} = a_{1,1} + a_{2,2} + a_{3,3} + \cdots + a_{n,n}.$$

*Example*: The matrix given at the beginning of this section is a square matrix having the dimensionality $n = 4$. Its diagonal elements are $a_{1,1} = 3$, $a_{2,2} = 2$, $a_{3,3} = 3$, $a_{4,4} = 0$. Their sum gives the spur $a_{1,1} + a_{2,2} + a_{3,3} + a_{4,4} = 3+2+3+0 = 8$.

If all diagonal elements of a square matrix are 1 and all nondiagonal elements zero, then one speaks of the unit matrix **E**,

$$\mathbf{E} = \begin{pmatrix} 1 & 0 & 0 & 0 \cdots 0 \\ 0 & 1 & 0 & 0 \cdots 0 \\ 0 & 0 & 1 & 0 \cdots 0 \\ 0 & 0 & 0 & 1 \cdots 0 \\ \vdots & \vdots & \vdots & \vdots \quad \vdots \\ 0 & 0 & 0 & 0 \cdots 1 \end{pmatrix}$$

Two matrices

$$\mathbf{A} = \begin{pmatrix} a_{1,1} & a_{1,2} & \cdots & a_{1,n} \\ a_{2,1} & a_{2,2} & \cdots & a_{2,n} \\ \vdots & & & \vdots \\ a_{m,1} & a_{m,2} & \cdots & a_{m,n} \end{pmatrix} \quad \text{and} \quad \mathbf{B} = \begin{pmatrix} b_{1,1} & b_{1,2} & \cdots & b_{1,n} \\ b_{2,1} & b_{2,2} & \cdots & b_{2,n} \\ \vdots & & & \\ b_{m,1} & b_{m,2} & \cdots & b_{m,n} \end{pmatrix}$$

are *equal* if their corresponding elements are equal, i.e.*, $\mathbf{A} = \mathbf{B}$, if $a_{i,j} = b_{i,j}$, $i = 1, 2, \ldots, m$, $j = 1, 2, \ldots, n$.

For matrices the following composition *rules* are defined: The *sum of two matrices** **A** (with the elements $a_{i,j}$) and **B** (with the elements $b_{i,j}$) is defined as the matrix **C**, whose elements $c_{i,j}$ are equal to the sum of the corresponding elements from **A** and **B**

$$\mathbf{A}+\mathbf{B} = \mathbf{C}, \quad c_{i,j} = a_{i,j}+b_{i,j}, \quad i = 1, 2, \ldots, m, \quad j = 1, 2, \ldots, n.$$

*Example:*

$$\mathbf{A} = \begin{pmatrix} 3 & 7 & 1 & 0 \\ 1 & 2 & 9 & 4 \\ 7 & 5 & 3 & 8 \\ 13 & -1 & 4 & 0 \end{pmatrix}, \quad \mathbf{B} = \begin{pmatrix} 7 & 0 & 11 & -3 \\ 4 & 5 & 9 & 1 \\ 9 & -1 & 7 & 4 \\ 8 & 6 & 0 & 2 \end{pmatrix}$$

$$\mathbf{C} = \mathbf{A}+\mathbf{B} = \begin{pmatrix} 3+7 & 7+0 & 1+11 & 0-3 \\ 1+4 & 2+5 & 9+9 & 4+1 \\ 7+9 & 5-1 & 3+7 & 8+4 \\ 13+8 & -1+6 & 4+0 & 0+2 \end{pmatrix} = \begin{pmatrix} 10 & 7 & 12 & -3 \\ 5 & 7 & 18 & 5 \\ 16 & 4 & 10 & 12 \\ 21 & 5 & 4 & 2 \end{pmatrix}$$

The *product of a matrix* **A** *with a number s* is obtained by multiplying each element $a_{i,j}$ of **A** with $s$:

$$s\mathbf{A} = \begin{pmatrix} sa_{1,1} & sa_{1,2} & \cdots & sa_{1,n} \\ sa_{2,1} & sa_{2,2} & \cdots & sa_{2,n} \\ \vdots & \vdots & & \vdots \\ sa_{m,1} & sa_{m,2} & \cdots & sa_{m,n} \end{pmatrix} = \mathbf{A}\, s.$$

---

* The requirement is of course that **A** and **B** have equally many ($m$) rows and equally many ($n$) columns.

The *product of two matrices* **A** ($m$ rows, $n$ columns) and **B** ($m'$ rows and $n'$ columns) $\mathbf{AB} = \mathbf{C}$ is defined only for the case where the number $n$ of columns of **A** is equal to the number $m'$ of rows of **B**, $n = m'$ (or for square matrices: if **A** and **B** have equal dimensionalities). If the number of rows of **A** is equal to $m$ and the number of columns of **B** is $n'$, then the product **C** is a matrix with $m$ rows and $n'$ columns. The elements $c_{i,j}$ of the product matrix **C** are calculated according to the rule:

$$c_{i,j} = \sum_{k=1}^{n} a_{i,k} b_{k,j}$$

$$= a_{i,1} b_{1,j} + a_{i,2} b_{2,j} + a_{i,3} b_{3,j} + \cdots + a_{i,n} b_{n,j}$$

*Example:*

$$\mathbf{A} = \overset{n=2}{\begin{pmatrix} 7 & 4 \\ 6 & -3 \end{pmatrix}}_{m=2} \qquad \mathbf{B} = \overset{n'=4}{\begin{pmatrix} -1 & 8 & 9 & 3 \\ 0 & 5 & 10 & 2 \end{pmatrix}}_{m'=2}$$

The product matrix $\mathbf{AB} = \mathbf{C}$ has $m = 2$ rows and $n = 4$ columns

$$\mathbf{C} = \begin{pmatrix} c_{1,1} & c_{1,2} & c_{1,3} & c_{1,4} \\ c_{2,1} & c_{2,2} & c_{2,3} & c_{2,4} \end{pmatrix},$$

$$c_{1,1} = 7(-1) + 4 \cdot 0 = -7, \qquad c_{2,1} = 6(-1) + (-3)0 = -6,$$
$$c_{1,2} = 7 \cdot 8 + 4 \cdot 5 = 76, \qquad c_{2,2} = 6 \cdot 8 + (-3)5 = 33,$$
$$c_{1,3} = 7 \cdot 9 + 4 \cdot 10 = 103, \qquad c_{2,3} = 6 \cdot 9 + (-3)10 = 24,$$
$$c_{1,4} = 7 \cdot 3 + 4 \cdot 2 = 29, \qquad c_{2,4} = 6 \cdot 3 + (-3)2 = 12,$$

$$\mathbf{C} = \begin{pmatrix} -7 & 76 & 103 & 29 \\ -6 & 33 & 24 & 12 \end{pmatrix}.$$

The example shows clearly why the rule for the formation of matrix products is meaningful only if the number of columns of **A** is equal to the number of rows of **B**.

If two matrices **A** and **B** are square matrices and have the same dimensionality, then besides the product **AB** one can also give the product **BA**. These two matrix products are in general different, i.e., matrix multiplication is in general *noncommutative*

$$\mathbf{AB} \neq \mathbf{BA}$$

The order of the 'factors' is important.

*Example:*

$$\mathbf{A}=\begin{pmatrix}7 & 4\\ 6 & -3\end{pmatrix},\ \mathbf{B}=\begin{pmatrix}1 & 1\\ 2 & 0\end{pmatrix}$$

$$\mathbf{AB}=\begin{pmatrix}7 & 4\\ 6 & -3\end{pmatrix}\begin{pmatrix}1 & 1\\ 2 & 0\end{pmatrix}=\begin{pmatrix}15 & 7\\ 0 & 6\end{pmatrix},\ \mathbf{BA}=\begin{pmatrix}1 & 1\\ 2 & 0\end{pmatrix}\begin{pmatrix}7 & 4\\ 6 & -3\end{pmatrix}=\begin{pmatrix}13 & 1\\ 14 & 8\end{pmatrix},$$

i.e., $\mathbf{AB} \neq \mathbf{BA}$.

If the product **BA** of two square matrices **A** and **B** yields the unit matrix **E**,

$$\mathbf{BA} = \mathbf{E},$$

then **B** is the reciprocal of **A** and is denoted by $\mathbf{A}^{-1}$. The reciprocal of a square matrix **A** is defined only if the determinant of **A**, $|\mathbf{A}|$, is non-zero, that is, if **A** is *nonsingular*.

*Example:*

$$\mathbf{A}=\begin{pmatrix}7 & 4\\ 6 & -3\end{pmatrix},\quad \mathbf{A}^{-1}=\begin{pmatrix}\frac{1}{15} & \frac{4}{45}\\ \frac{2}{15} & -\frac{7}{45}\end{pmatrix},$$

$$\mathbf{A}^{-1}\mathbf{A}=\begin{pmatrix}\frac{1}{15} & \frac{4}{45}\\ \frac{2}{15} & -\frac{7}{45}\end{pmatrix}\begin{pmatrix}7 & 4\\ 6 & -3\end{pmatrix}=\begin{pmatrix}1 & 0\\ 0 & 1\end{pmatrix}=\mathbf{E},$$

$$|A|=\begin{vmatrix}7 & 4\\ 6 & -3\end{vmatrix}=7(-3)-6\cdot 4=-45\neq 0.$$

Matrices find application in describing *transformations.* Many examples are given in the text, cf. pp. 271 ff.

### (e) Secular determinants

If a quantum mechanical system is subjected to a *weak perturbation* (a so-called *secular perturbation**) then in general the energies of the possible states of the system are changed. The energy changes are found from the roots of the corresponding *secular equation.*

* The term *secular* (lat. *saec(u)lum* = age, generation, century) in the sense used here comes from classical mechanics, in particular from celestial mechanics. Weak perturbations of planetary orbits, e.g., appear as small changes of these orbits. An example is the forced precessional motion, the period of which is very large compared to the period of revolution of the planet (about the sun), and is of a 'secular' order of magnitude. A direct analogy with the stationary states handled here is not possible; the relationship to classical mechanics becomes apparent with a consideration of *time-dependent* quantum mechanical problems.

One starts with the *unperturbed* system having the *Hamiltonian* $H_0$ and the *Schrödinger* equation

$$H_0\psi = E_0\psi,$$

whose solutions are already known. Let there be $\eta_i$ different (linearly independent) eigenfunctions $\psi_{i,1}, \psi_{i,2}, \ldots, \psi_{i,n_i}$ having the eigenvalue $E_{0,i}$, i.e., let the degree of degeneracy of the state $E_{0,i}$ be $\eta_i$

$$\underbrace{\psi_{1,1}, \psi_{1,2}, \ldots, \psi_{1,\eta_1}}_{E_{0,1}(\eta_1\text{-fold})} \quad \underbrace{\psi_{2,1}, \psi_{2,2}, \ldots, \psi_{2,\eta_2}}_{E_{0,2}(\eta_2\text{-fold})} \ldots \underbrace{\psi_{i,1}, \psi_{i,2}, \ldots, \psi_{i,\eta_i}}_{E_{0,i}(\eta_i\text{-fold})} \ldots$$

If the system is subjected to a *perturbation* represented by the *perturbation operator* $H_1$, then two cases can be distinguished.

1. The energetic effects of the perturbation are small compared to the energy difference $E_{0,i} - E_{0,i+1}$ between neighbouring states of the *unperturbed system.* In this case the energies of the possible states of the perturbed system are obtained by using the complete *Hamiltonian* $H_0 + H_1$ to set up a *secular determinant* separately for each degenerate family of functions (basis of functions) and then determining their roots $E$. The secular determinant for the degenerate basis of functions $\psi_{i,1}$, $\psi_{i,2}, \ldots, \psi_i, \eta_i$ which (in the unperturbed system) corresponds to the energy $E_{0,i}$ is

$$\begin{vmatrix} H_{1,1}^{(i,i)} - S_{1,1}^{(i,i)}E^{(i)} & H_{1,2}^{(i,i)} - S_{1,2}^{(i,i)}E^{(i)} \cdots & H_{1,\eta_i}^{(i,i)} - S_{1,\eta_i}^{(i,i)}E^{(i)} \\ H_{2,1}^{(i,i)} - S_{2,1}^{(i,i)}E^{(i)} & H_{2,2}^{(i,i)} - S_{2,2}^{(i,i)}E^{(i)} \cdots & H_{2,\eta_i}^{(i,i)} - S_{2,\eta_i}^{(i,i)}E^{(i)} \\ \vdots & \vdots & \vdots \\ H_{\eta_i,1}^{(i,i)} - S_{\eta_i,1}^{(i,i)}E^{(i)} & H_{\eta_i,2}^{(i,i)} - S_{\eta_i,2}^{(i,i)}E^{(i)} \cdots & H_{\eta_i,\eta_i}^{(i,i)} - S_{\eta_i,\eta_i}^{(i,i)}E^{(i)} \end{vmatrix} = 0,$$

with

$$H_{r,s}^{(i,i)} = (\psi_{i,r}, [H_0 + H_1]\psi_{i,s}) \quad \text{(energy integral)}$$

$$S_{r,s}^{(i,i)} = (\psi_{i,r}, \psi_{i,s}) \quad \text{(overlap integral)}$$

The $\eta_i$ roots $E_k^{(i)}$, $k = 1, 2, \ldots, \eta_i$ are found by expanding the determinant in a polynomial in $E^{(i)}$ and finding the zeros of this polynomial*.

2. The energetic effects of a perturbation are of the same order of magnitude as the energy differences between neighbouring states of the unperturbed system. In these cases *one* secular determinant must be

---

* This problem is simplified because complete computer programs for the determination of the roots of secular determinants are almost always available. One then needs only to evaluate the integrals $H_{r,s}^{(ii)}$ and $S_{r,s}^{(i,i)}$ individually. For a procedure for expanding determinants see pp. 477 ff.

constructed in principle using all solution functions of the unperturbed system. If one is interested in a certain energy interval, let us say the ground state, then it suffices when the secular determinant is *abbreviated*. To achieve this purpose only those $\psi$ functions are used which describe states (in the unperturbed system) in the energy interval of interest. If, e.g., the energies in the unperturbed system are $E_{0,1} < E_{0,2} < E_{0,3} \ldots$, and the energy changes of the low-lying states are to be determined, then in any case the functions corresponding to $E_{0,1}$ are used when formulating the secular determinant and, further, depending upon the desired accuracy, the functions corresponding to $E_{0,2}$ to $E_{0,3}$, etc. are used in addition. The determination of the roots (= energies of the perturbed system) follows likewise according to the procedure sketched here.

A common feature of secular determinants constructed using real functions is that they are *symmetric* about the principal diagonal (running from the upper left to the lower right): Elements lying in mirror image positions about the principal diagonal are equal:

$$H_{r,s} = H_{s,r}, \qquad S_{r,s} = S_{s,r}.$$

If a (secular) determinant has the form

$$\left|\begin{array}{cccccccc}
\times & \times & & & & & & \\
\times & \times & & & & 0 & & \\
 & & \times & \times & \times & & & \\
 & & \times & \times & \times & & & \\
 & & \times & \times & \times & & & \\
 & & & & & \times & & \\
 & & 0 & & & & \times & \times \\
 & & & & & & \times & \times
\end{array}\right|$$

i.e., if there are square areas (blocks) along the principal diagonals outside of which all elements are zero, then one speaks of step or block determinants. Setting the individual blocks of the block determinant equal to zero is equivalent to setting the entire block determinant equal to zero, e.g.,

$$|A| = \left|\begin{array}{cccc}
|A_1| & & 0 & \\
 & |A_2| & & \\
 & & |A_3| & \\
0 & & & |A_4|
\end{array}\right| = 0, \qquad \begin{array}{l} |A_1| = 0, \\ |A_2| = 0, \\ |A_3| = 0, \\ |A_4| = 0. \end{array}$$

For every element of a determinant

$$|A| = \begin{vmatrix} a_{1,1} & a_{1,2} & \cdots & a_{1,n} \\ a_{2,1} & a_{2,2} & \cdots & a_{2,n} \\ \vdots & \vdots & & \vdots \\ a_{n,1} & a_{n,2} & \cdots & a_{n,n} \end{vmatrix}$$

of order $n$ a *minor* of order $n-1$ is defined. The minor corresponding to the element $a_{i,j}|A_{i,j}|$ is obtained by omitting in $|A|$ the $i$th row and the $j$th column:

$$|A_{i,j}| = \begin{vmatrix} a_{1,1} & a_{1,2} & \cdots & a_{1,j-1} & a_{1,j+1} & \cdots & a_{1,n} \\ a_{2,1} & a_{2,2} & \cdots & a_{2,j-1} & a_{2,j+1} & \cdots & a_{2,n} \\ \vdots & \vdots & & \vdots & \vdots & & \vdots \\ a_{i-1,1} & a_{i-1,2} & \cdots & a_{i-1,j-1} & a_{i-1,j+1} & \cdots & a_{i-1,n} \\ a_{i+1,1} & a_{i+1,2} & \cdots & a_{i+1,j-1} & a_{i+1,j+1} & \cdots & a_{i+1,n} \\ \vdots & \vdots & & \vdots & \vdots & & \vdots \\ a_{n,1} & a_{n,2} & \cdots & a_{n,j-1} & a_{n,j+1} & \cdots & a_{n,n} \end{vmatrix}.$$

*Example*:

$$|A| = \begin{vmatrix} a & b & c \\ d & e & f \\ g & h & i \end{vmatrix},$$

$$|A_{2,3}| = \begin{vmatrix} a & b \\ g & h \end{vmatrix}.$$

A determinant $|A|$ can be expanded according to the minors (of the elements) of a row from $|A|$ or of a column from $|A|$. The expansion for the $i$th row of $|A|$ is given by

$$|A| = a_{i,1}(-1)^{i+1}|A_{i,1}| + a_{i,2}(-1)^{i+2}|A_{i,2}| + \cdots$$
$$\cdots + a_{i,j}(-1)^{i+j}|A_{i,j}| + \cdots + a_{i,n}(-1)^{i+n}|A_{i,n}|,$$

and the expansion for the $j$th column by

$$|A| = a_{1,j}(-1)^{1+j}|A_{1,j}| + a_{2,j}(-1)^{2+j}|A_{2,j}| + \cdots$$
$$\cdots + a_{i,j}(-1)^{i+j}|A_{i,j}| + \cdots + a_{n,j}(-1)^{n+j}|A_{n,j}|.$$

Because the minors appearing in the equation above can themselves be expanded in terms of their minors, a determinant can be represented as a sum of products of the $n$th degree of its elements (when necessary by repeated expansion):

*Example*:

$$|A| = \begin{vmatrix} a & b & c \\ d & e & f \\ g & h & i \end{vmatrix}.$$

Expansion according to the first row gives

$$\begin{aligned}|A| &= a(-1)^{1+1}|A_{1,1}| + b(-1)^{1+2}|A_{1,2}| + c(-1)^{1+3}|A_{1,3}| \\ &= a|A_{1,1}| - b|A_{1,2}| + c|A_{1,3}|\end{aligned}$$

with

$$|A_{1,1}| = \begin{vmatrix} e & f \\ h & i \end{vmatrix}, \; |A_{1,2}| = \begin{vmatrix} d & f \\ g & i \end{vmatrix}, \; |A_{1,3}| = \begin{vmatrix} d & e \\ g & h \end{vmatrix}.$$

The three minors can also be expanded in terms of their minors (in each case according to the first row) as follows:

$$\begin{aligned}|A_{1,1}| &= e(-1)^{1+1}|i| + f(-1)^{1+2}|h| = e \cdot i - f \cdot h, \\ |A_{1,2}| &= d(-1)^{1+1}|i| + f(-1)^{1+2}|g| = d \cdot i - f \cdot g, \\ |A_{1,3}| &= d(-1)^{1+1}|h| + e(-1)^{1+2}|g| = d \cdot h - e \cdot g.\end{aligned}$$

Substitution into the expression for $|A|$ yields finally

$$\begin{aligned}|A| &= a(e \cdot i - f \cdot h) - b(d \cdot i - f \cdot g) + c(d \cdot h - e \cdot g) \\ &= a \cdot e \cdot i - a \cdot f \cdot h - b \cdot d \cdot i + b \cdot f \cdot g + c \cdot d \cdot h - c \cdot e \cdot g.\end{aligned}$$

# 2. Character tables of the point groups of most importance for the ligand field theory

| Mulliken's Nomenclature: $O_h$ | $E$ | $8C_3$ | $3C_2$ | $6C_4$ | $6C_2'$ | $i$ | $8iC_3$ | $3iC_2$ | $6iC_4$ | $6iC_2'$ | Bethe's Nomenclature |
|---|---|---|---|---|---|---|---|---|---|---|---|
| $A_{1g}$ | 1 | 1 | 1 | 1 | 1 | 1 | 1 | 1 | 1 | 1 | $\Gamma_1$ |
| $A_{1u}$ | 1 | 1 | 1 | 1 | 1 | −1 | −1 | −1 | −1 | −1 | |
| $A_{2g}$ | 1 | 1 | 1 | −1 | −1 | 1 | 1 | 1 | −1 | −1 | $\Gamma_2$ |
| $A_{2u}$ | 1 | 1 | 1 | −1 | −1 | −1 | −1 | −1 | 1 | 1 | |
| $E_g$ | 2 | −1 | 2 | 0 | 0 | 2 | −1 | 2 | 0 | 0 | $\Gamma_3$ |
| $E_u$ | 2 | −1 | 2 | 0 | 0 | −2 | 1 | −2 | 0 | 0 | |
| $T_{1g}$ | 3 | 0 | −1 | 1 | −1 | 3 | 0 | −1 | 1 | −1 | $\Gamma_4$ |
| $T_{1u}$ | 3 | 0 | −1 | 1 | −1 | −3 | 0 | 1 | −1 | 1 | |
| $T_{2g}$ | 3 | 0 | −1 | −1 | 1 | 3 | 0 | −1 | −1 | 1 | $\Gamma_5$ |
| $T_{2u}$ | 3 | 0 | −1 | −1 | 1 | −3 | 0 | 1 | 1 | −1 | |

| Mulliken's Nomenclature: $D_{4h}$ | $E$ | $2C_4$ | $C_2$ | $2C_2'$ | $2C_2''$ | $i$ | $2S_4$ | $\sigma_h$ | $2\sigma_v$ | $2\sigma_d$ | Bethe's Nomenclature |
|---|---|---|---|---|---|---|---|---|---|---|---|
| $A_{1g}$ | 1 | 1 | 1 | 1 | 1 | 1 | 1 | 1 | 1 | 1 | $\Gamma_1$ |
| $A_{1u}$ | 1 | 1 | 1 | 1 | 1 | −1 | −1 | −1 | −1 | −1 | |
| $A_{2g}$ | 1 | 1 | 1 | −1 | −1 | 1 | 1 | 1 | −1 | −1 | $\Gamma_2$ |
| $A_{2u}$ | 1 | 1 | 1 | −1 | −1 | −1 | −1 | −1 | 1 | 1 | |
| $B_{1g}$ | 1 | −1 | 1 | 1 | −1 | 1 | −1 | 1 | 1 | −1 | $\Gamma_3$ |
| $B_{1u}$ | 1 | −1 | 1 | 1 | −1 | −1 | 1 | −1 | −1 | 1 | |
| $B_{2g}$ | 1 | −1 | 1 | −1 | 1 | 1 | −1 | 1 | −1 | 1 | $\Gamma_4$ |
| $B_{2u}$ | 1 | −1 | 1 | −1 | 1 | −1 | 1 | −1 | 1 | −1 | |
| $E_g$ | 2 | 0 | −2 | 0 | 0 | 2 | 0 | −2 | 0 | 0 | $\Gamma_5$ |
| $E_u$ | 2 | 0 | −2 | 0 | 0 | −2 | 0 | 2 | 0 | 0 | |

| Mulliken's Nomenclature: $C_{4v}$ | $E$ | $2C_4$ | $C_2$ | $2\sigma_v$ | $2\sigma_d$ | Bethe's Nomenclature |
|---|---|---|---|---|---|---|
| $A_1$ | 1 | 1 | 1 | 1 | 1 | $\Gamma_1$ |
| $A_2$ | 1 | 1 | 1 | −1 | −1 | $\Gamma_2$ |
| $B_1$ | 1 | −1 | 1 | 1 | −1 | $\Gamma_3$ |
| $B_2$ | 1 | −1 | 1 | −1 | 1 | $\Gamma_4$ |
| $E$ | 2 | 0 | −2 | 0 | 0 | $\Gamma_5$ |

| Mulliken's Nomenclature $T_d$ | $E$ | $8C_3$ | $3C_2$ | $6S_4$ | $6\sigma_d$ | Bethe's Nomenclature |
|---|---|---|---|---|---|---|
| $A_1$ | 1 | 1 | 1 | 1 | 1 | $\Gamma_1$ |
| $A_2$ | 1 | 1 | 1 | $-1$ | $-1$ | $\Gamma_2$ |
| $E$ | 2 | $-1$ | 2 | 0 | 0 | $\Gamma_3$ |
| $T_1$ | 3 | 0 | $-1$ | 1 | $-1$ | $\Gamma_4$ |
| $T_2$ | 3 | 0 | $-1$ | $-1$ | 1 | $\Gamma_5$ |

| Mulliken's Nomenclature $D_3$ | $E$ | $2C_3$ | $3C_2$ | Bethe's Nomenclature |
|---|---|---|---|---|
| $A_1$ | 1 | 1 | 1 | $\Gamma_1$ |
| $A_2$ | 1 | 1 | $-1$ | $\Gamma_2$ |
| $E$ | 2 | $-1$ | 0 | $\Gamma_3$ |

| Mulliken's Nomenclature $O'$ | $E$ | $R$ | $4C_3$ $4C_3^2R$ | $4C_3R$ $4C_3^2$ | $3C_2$ $3C_2R$ | $3C_4$ $3C_4^3R$ | $3C_4R$ $3C_4$ | $6C'$ $6C'_2R$ | Bethe's Nomenclature |
|---|---|---|---|---|---|---|---|---|---|
| $A_1$ | 1 | 1 | 1 | 1 | 1 | 1 | 1 | 1 | $\Gamma_1$ |
| $A_2$ | 1 | 1 | 1 | 1 | 1 | $-1$ | $-1$ | $-1$ | $\Gamma_2$ |
| $E$ | 2 | 2 | $-1$ | $-1$ | 2 | 0 | 0 | 0 | $\Gamma_3$ |
| $T_1$ | 3 | 3 | 0 | 0 | $-1$ | 1 | 1 | $-1$ | $\Gamma_4$ |
| $T_2$ | 3 | 3 | 0 | 0 | $-1$ | $-1$ | $-1$ | 1 | $\Gamma_5$ |
| $E'$ | 2 | $-2$ | 1 | $-1$ | 0 | $\sqrt{2}$ | $-\sqrt{2}$ | 0 | $\Gamma_6$ |
| $E''$ | 2 | $-2$ | 1 | $-1$ | 0 | $-\sqrt{2}$ | $\sqrt{2}$ | 0 | $\Gamma_7$ |
| $G'$ | 4 | $-4$ | $-1$ | 1 | 0 | 0 | 0 | 0 | $\Gamma_8$ |

* One also finds the notation $E_{1/2}$ for $E'$, $E_{5/2}$ for $E''$, $G$ for $G'$. The various systems of notation for the completely cubic group are summarized by R. S. Knox and A. Gold: *Symmetry in the Solid State*, p. 82. W. A. Benjamin, Inc., New York and Amsterdam, 1964.

| Mulliken's Nomenclature $C_{2v}$ | $E$ | $C_2$ | $\sigma_v$ | $\sigma'_v$ |
|---|---|---|---|---|
| $A_1$ | 1 | 1 | 1 | 1 |
| $A_2$ | 1 | 1 | $-1$ | $-1$ |
| $B_1$ | 1 | $-1$ | 1 | $-1$ |
| $B_2$ | 1 | $-1$ | $-1$ | 1 |

# 3. The most important symmetries of complex ions

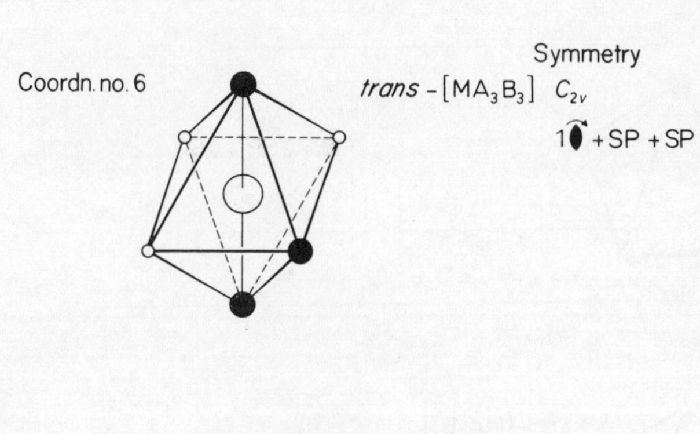


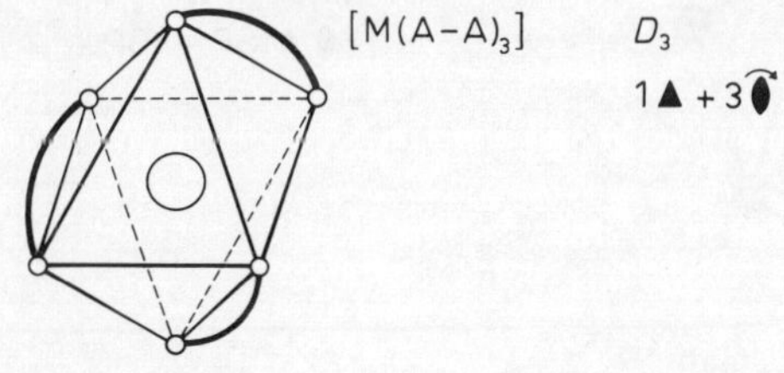


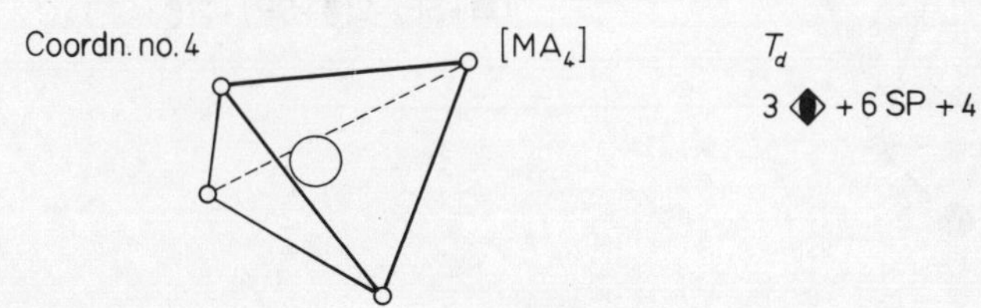

Symmetry

Coordn. no. 6

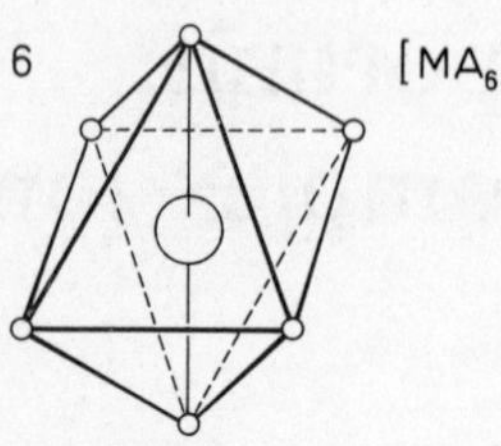

$[MA_6]$ $O_h$

3(■ + SP) + 6(⬮ + SP) + 4 ▲ + C

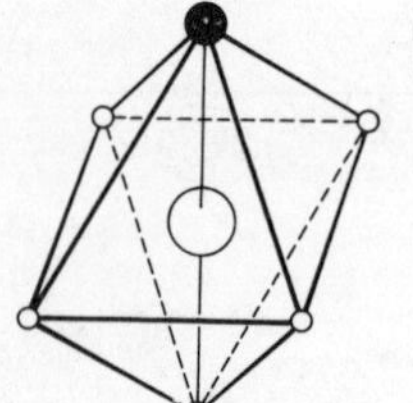

$[MA_5B]$ $C_{4v}$

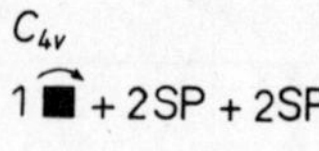


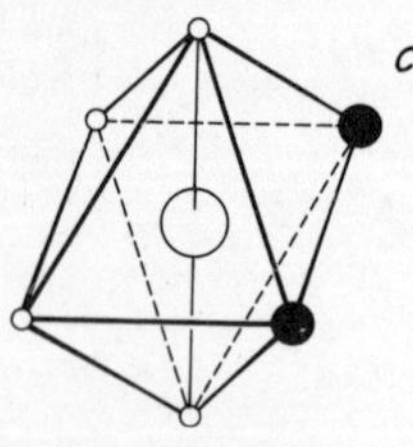

*cis* $-[MA_4B_2]$ $C_{2v}$

1 ⬮ + 1SP + 1SP

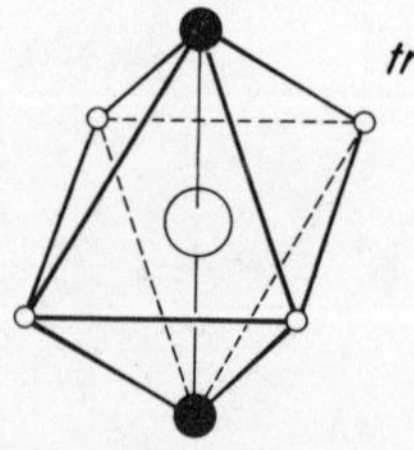

*trans* $-[MA_4B_2]$ $D_{4h}$

1(■ + SP) + 2(⬮ + SP) + 2(⬮ + SP) + C

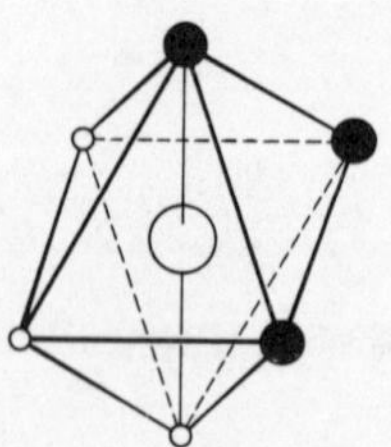

*cis* $-[MA_3B_3]$ $C_{3v}$

1 ▲ + 3

SP = Symmetry plane (mirror plane)
⬮ = 2-fold rotation axis
▲ = 3-fold " "
■ = 4-fold " "
◆ = 4-fold rotation-reflexion axis
⬢ = 6-fold " " "
C = Centre of symmetry
⌒ = Polar axis

Parentheses( )around symbols denote that the axis and symmetry plane are perpendicular to one another

# 4. Hybridization of central ion orbitals in the valence bond theory

In Part A, section 1.3 the treatment of the ground states of complex ions by means of the *valence bond theory* was discussed. According to Sidgwick (*J. Amer. Chem. Soc.*, **53** 1367, 3225 (1931) and Pauling (*The Nature of the Chemical Bond*, pp. 92 ff., Cornell University Press, Ithaca, New York, 1960) the ground state of a complex ion is described by the valence state obtained when per ligand one electron of the lone electron pair is transferred to the central ion so that as a result of the interaction of a transferred electron with the electron remaining on the ligand a bond arises. Depending upon the original charge of the central ion certain (unoccupied) orbitals of the central ion are available for the transferred ligand electrons. For energetic reasons the central ion *d*, *s* and *p* orbitals are to be considered most likely to accept ligand electrons, cf. pp. 120 ff., in particular one *s* and three *p* orbitals ($p_x, p_y, p_z$) as well as several *d* orbitals, the number of which, as discussed in section 1.3, is limited by the number of *d* electrons already present on the central ion.

Pauling has suggested a procedure by which the spatial structure of a complex ion can usually be determined for a given coordination number and for a given type and number of empty central ion orbitals. The basic idea is that a particular structure can be realized only when *K equivalent orthonormalized linear combinations* (*hybrid functions*) can be formed whose maximal spatial extension points in the direction of the *K* ligands (having the positions corresponding to the assumed structure of the complex ion). Square planar complex ions for example should appear only when four hybrid functions can be constructed from the empty central ion orbitals having maxima in the direction of the corners of a square (with the central ion at its centre).

If sets of equivalent hybrid orbitals for several different structures can be given for an assumed coordination number and type and number of empty central ion orbitals (e.g., $K = 4$, tetrahedron and square planar structure), then according to Pauling that structure is the most stable for which the *bond strength* is the greatest. We take as a measure of the bond strength the value that the angular dependent portion of a hybrid function has along the line of centres central ion-ligand. It is conventional to set the bond strength of an s function equal to 1. In the following three examples will be treated.

### (a) Regular octahedron ($Kz. = 6$)

Which *s*, *p* and *d* orbitals are necessary to construct six equivalent hybrid functions which extend in the directions of the vertices of a regular octahedron? The hybrid orbitals which produce a bond with the 1st, 2nd, ..., 6th ligand (cf. Figure B.9) will be denoted with $\pi_1, \pi_2, \ldots, \pi_6$. For their determination we shall take a general linear combination with one *s*, three *p* and five *d* orbitals.

$$\pi_i = \alpha_i s + \beta_i p_x + \gamma_i p_y + \delta_i p_z \qquad i = 1,2,3,4,5,6$$
$$+ \varepsilon_i d_{z^2} + \zeta_i d_{x^2-y^2} + \eta_i d_{xy} + \vartheta_i d_{xz} + \varkappa_i d_{yz}.$$

The function $\pi_1$, which is to extend essentially along the positive *x* axis must be invariant under all the symmetry operations of the octahedron in which the *x* axis does not change its position. One of these operations is the reflection $\sigma_{xy}(x \to x, y \to y, z \to -z)$. Application of $\sigma_{xy}$ to $\pi_1$ yields:

$$\sigma_{xy}\pi_1 = \alpha_1 s + \beta_1 p_x + \gamma_1 p_y - \delta_1 p_z + \varepsilon_1 d_{z^2} + \zeta_1 d_{x^2-y^2} + \eta_1 d_{xy}$$
$$- \vartheta_1 d_{xz} - \varkappa_1 d_{yz}.$$

Because of the invariance requirement $\sigma_{xy}\pi_1 = \pi_1$ it follows from a comparison of coefficients

$$\delta_1 = 0, \quad \vartheta_1 = 0, \quad \varkappa_1 = 0.$$

Completely analogously the invariance requirement $\sigma_{xz}\pi_1 = \pi_1$ can be satisfied when further

$$\gamma_1 = 0, \quad \eta_1 = 0$$

holds. The other symmetry operations of the octahedron which leave the position of the *x* axis unchanged do not yield further information concerning the remaining undetermined coefficients $\alpha_1, \beta_1, \varepsilon_1, \zeta_1$. The function $\pi_1$ thus has the general form*:

$$\pi_1 = \alpha_1 s + \beta_1 p_x + \varepsilon_1 d_{z^2} + \zeta_1 d_{x^2-y^2}.$$

By means of a coordinate transformation the remaining five functions $\pi_2, \pi_3, \ldots, \pi_6$ can be generated. The function $\pi_2$, which has its maximum

* Because the functions $p_y$, $p_z$, $d_{xy}$, $d_{xz}$, $d_{yz}$ have nodal planes containing the *x* axis they cannot contribute to a bond along the *x* axis.

extension along the positive $y$ axis and is equivalent to $\pi_1$ is obtained by applying a coordination transformation to $\pi_1$ which transforms the $x$ axis into the $y$ axis. The rotation $C_4^1(z)$ about the $z$ axis is such a transformation. It gives

$$C_4^1(z)\,\pi_1 = \pi_2 = \alpha_1 s + \beta_1 p_y + \varepsilon_1 d_{z^2} - \zeta_1 d_{x^2-y^2}.$$

The remaining functions are obtained in a completely analogous manner:

$$\begin{aligned}
C_4^1(y)\,\pi_1 &= \pi_3 = \alpha_1 s + \beta_1 p_z + \varepsilon_1\left(-\frac{1}{2}\,d_{z^2} + \frac{\sqrt{3}}{2}\,d_{x^2-y^2}\right) + \\
&\quad + \zeta_1\left(\frac{\sqrt{3}}{2}\,d_{z^2} + \frac{1}{2}\,d_{x^2-y^2}\right) \\
&= \alpha_1 s + \beta_1 p_z + \left(-\frac{1}{2}\,\varepsilon_1 + \frac{\sqrt{3}}{2}\,\zeta_1\right) d_{z^2} + \left(\frac{\sqrt{3}}{2}\,\varepsilon_1 + \frac{1}{2}\,\zeta_1\right) d_{x^2-y^2}, \\
\sigma_{yz}\,\pi_1 &= \pi_4 = \alpha_1 s - \beta_1 p_x + \varepsilon_1 d_{z^2} + \zeta_1 d_{x^2-y^2}, \\
C_4^3(z)\,\pi_1 &= \pi_5 = \alpha_1 s - \beta_1 p_y + \varepsilon_1 d_{z^2} - \zeta_1 d_{x^2-y^2}, \\
C_4^3(y)\,\pi_1 &= \pi_6 = \alpha_1 s - \beta_1 p_z + \left(-\frac{1}{2}\,\varepsilon_1 + \frac{\sqrt{3}}{2}\,\zeta_1\right) d_{z^2} \\
&\quad + \left(\frac{\sqrt{3}}{2}\,\varepsilon_1 + \frac{1}{2}\,\zeta_1\right) d_{x^2-y^2}.
\end{aligned}$$

The undetermined coefficients follow from the orthogonality and normalization conditions $(\pi_i, \pi_j) = \delta_{ij}$:

$$\begin{aligned}
&(\pi_1, \pi_2) = (\pi_1, \pi_5) = (\pi_2, \pi_4) = (\pi_4, \pi_5) = \alpha_1^2 + \varepsilon_1^2 - \zeta_1^2 = 0, \\
&(\pi_1, \pi_3) = (\pi_1, \pi_6) = (\pi_3, \pi_4) = (\pi_4, \pi_6) = \alpha_1^2 - {}^1\!/_2\,\varepsilon_1^2 + \sqrt{3}\,\varepsilon_1\zeta_1 + {}^1\!/_2\,\zeta_1^2 = 0, \\
&(\pi_1, \pi_4) = (\pi_3, \pi_6) = \alpha_1^2 - \beta_1^2 + \varepsilon_1^2 + \zeta_1^2 = 0, \\
&(\pi_2, \pi_3) = (\pi_2, \pi_6) = (\pi_3, \pi_5) = (\pi_5, \pi_6) = \alpha_1^2 - {}^1\!/_2\,\varepsilon_1^2 - {}^1\!/_2\,\zeta_1^2 = 0, \\
&(\pi_i, \pi_i) = \alpha_1^2 + \beta_1^2 + \varepsilon_1^2 + \zeta_1^2 = 1, \qquad i = 1,2,3,4,5,6.
\end{aligned}$$

From these five equations we obtain

$$\alpha_1 = 1/\sqrt{6}, \quad \beta_1 = 1/\sqrt{2}, \quad \varepsilon_1 = -1/\sqrt{12}, \quad \zeta_1 = 1/2.$$

Hence the six octahedral hybridized orbitals are

$$\begin{aligned}
\pi_1 &= \frac{1}{\sqrt{6}}\,s + \frac{1}{\sqrt{2}}\,p_x - \frac{1}{\sqrt{12}}\,d_{z^2} + \frac{1}{2}\,d_{x^2-y^2}, \\
\pi_2 &= \frac{1}{\sqrt{6}}\,s + \frac{1}{\sqrt{2}}\,p_y - \frac{1}{\sqrt{12}}\,d_{z^2} - \frac{1}{2}\,d_{x^2-y^2}, \\
\pi_3 &= \frac{1}{\sqrt{6}}\,s + \frac{1}{\sqrt{2}}\,p_z + \frac{1}{\sqrt{3}}\,d_{z^2}, \\
\pi_4 &= \frac{1}{\sqrt{6}}\,s - \frac{1}{\sqrt{2}}\,p_x - \frac{1}{\sqrt{12}}\,d_{z^2} + \frac{1}{2}\,d_{x^2-y^2},
\end{aligned}$$

$$\pi_5 = \frac{1}{\sqrt{6}}\, s - \frac{1}{\sqrt{2}}\, p_y - \frac{1}{\sqrt{12}}\, d_{z^2} - \frac{1}{2}\, d_{x^2-y^2},$$

$$\pi_6 = \frac{1}{\sqrt{6}}\, s - \frac{1}{\sqrt{2}}\, p_z + \frac{1}{\sqrt{3}}\, d_{z^2}.$$

Because all six $\pi$ orbitals are built up from the orbitals $s$, $p_x$, $p_y$, $p_z$, $d_{z^2}$, $d_{x^2-y^2}$, that is from one $s$, three $p$ and two $d$ orbitals, one speaks of a *$d^2sp^3$ hybrid**.

The bond strength is equal for all $\pi$ orbitals because these are equivalent and differ only in their spatial orientation. We shall calculate the bond strength of the $d^2sp^3$ hybrid functions of the octahedron using the example of the orbital

$$\pi_1 = \frac{1}{\sqrt{6}}\, s + \frac{1}{\sqrt{2}}\, p_x - \frac{1}{\sqrt{12}}\, d_{z^2} + \frac{1}{2}\, d_{x^2-y^2}.$$

The angular dependent parts of the orbitals appearing are

$$(s) = \frac{1}{2\sqrt{\pi}},$$

$$(p_x) = \frac{1}{2\sqrt{\pi}}\sqrt{3}\sin\theta\cdot\cos\phi,$$

$$(d_{z^2}) = \frac{1}{2\sqrt{\pi}}\,\frac{\sqrt{5}}{2}\,(3\cos^2\theta - 1),$$

$$(d_{x^2-y^2}) = \frac{1}{2\sqrt{\pi}}\,\frac{\sqrt{15}}{2}\,\sin^2\theta\cos 2\phi.$$

Along the $x$ axis ($\theta = \pi/2$, $\phi = 0$) they have the values

$$(s)_x = \frac{1}{2\sqrt{\pi}},$$

$$(p_x)_x = \frac{1}{2\sqrt{\pi}}\sqrt{3},$$

$$(d_{z^2})_x = -\frac{1}{2\sqrt{\pi}}\,\frac{\sqrt{5}}{2},$$

$$(d_{x^2-y^2})_x = \frac{1}{2\sqrt{\pi}}\,\frac{\sqrt{15}}{2}.$$

* The symbol used here $d^2sp^3$ must not be confused with the symbols for electron configurations. Here we state which functions constitute the *linear combinations* which themselves represent one-electron states. In the case of the electron configuration symbols we are concerned with shorthand notations which are applicable to many-electron systems when they contain several function symbols. The (many-electron) wave functions belonging to such an electron configuration are (antisymmetrical) *products* of the one-electron functions appearing in the symbol.

The *bond strength* $B(f)$ of a function $f$ in a certain direction is according to definition equal to the value of the angular part of the function multiplied by $2\sqrt{\pi}$.

$$B_x(s) = 1,$$
$$B_x(p_x) = \sqrt{3},$$
$$B_x(d_{z^2}) = -\sqrt{5}/2,$$
$$B_x(d_{x^2-y^2}) = \sqrt{15}/2.$$

With this one obtains for the bond strength of the orbital $\pi_1$ in the $x$ direction:

$$B_x(\pi_1) = \frac{1}{\sqrt{6}} \cdot 1 + \frac{1}{\sqrt{2}} \cdot \sqrt{3} - \frac{1}{\sqrt{12}}\left(-\frac{\sqrt{5}}{2}\right) + \frac{1}{2}\frac{\sqrt{15}}{2} = 2{\cdot}92.$$

The bond strengths for the remaining five $\pi$ orbitals are equally large in the corresponding directions.

The angular dependence of the hybrid orbital $\pi_1$ is shown in Figure C.1.

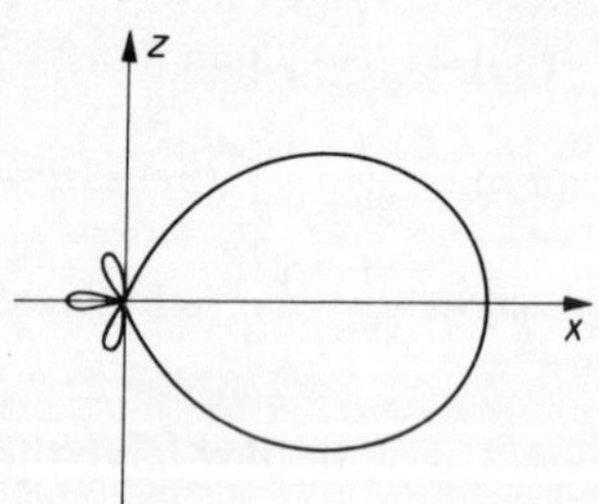


Figure C.1. The angular dependence of the octahedral $d^2sp^3$ hybrid orbital $\pi_1$ (after L. Pauling, *op. cit.*, p. 95).

**(b) Square planar structure** ($Kz. = 4$)

We assume the following form for the hybrid orbital $\pi_1$ which points to ligand one (on the positive $x$ axis (cf. Figure C.2))

$$\pi_1 = \alpha s + \beta p_x + \gamma p_y + \delta p_z + \varepsilon d_{z^2} + \zeta d_{x^2-y^2} + \eta d_{xy} + \vartheta d_{xz} + \varkappa d_{yz}.$$

From the invariance conditions

$$\sigma_{xz}\pi_1 = \pi_1 \quad \text{and} \quad \sigma_{xy}\pi_1 = \pi_1$$

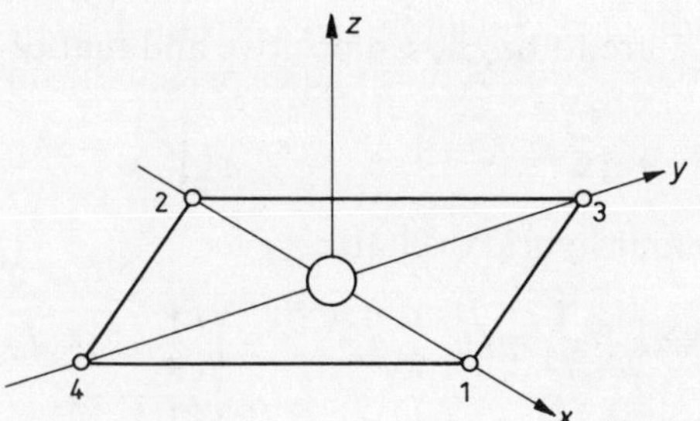


Figure C.2. Square planar coordination.

it follows

$$\gamma = 0, \eta = 0, \varkappa = 0 \quad \text{or} \quad \delta = 0, \vartheta = 0,$$

i.e.

$$\pi_1 = \alpha s + \beta p_x + \varepsilon d_{z^2} + \zeta d_{x^2-y^2}.$$

With the aid of the coordinate transformation $x \to -x, y \to y, z \to z$

$$\pi_2 = \alpha s - \beta p_x + \varepsilon d_{z^2} + \zeta d_{x^2-y^2}$$

is obtained.

The coordinate transformations $x \to y$, $y \to -z$, $z \to z$ as well as $x \to -y$, $y \to x$, $z \to z$ lead to

$$\pi_3 = \alpha s + \beta p_y + \varepsilon d_{z^2} - \zeta d_{x^2-y^2},$$

or

$$\pi_4 = \alpha s - \beta p_y + \varepsilon d_{z^2} - \zeta d_{x^2-y^2}.$$

The orthogonality and normalization conditions

$$(\pi_1, \pi_2) = (\pi_3, \pi_4) = \alpha^2 - \beta^2 + \varepsilon^2 + \zeta^2 = 0,$$
$$(\pi_1, \pi_3) = (\pi_1, \pi_4) = (\pi_2, \pi_3) = (\pi_2, \pi_4) = \alpha^2 + \varepsilon^2 - \zeta^2 = 0,$$
$$(\pi_1, \pi_1) = (\pi_2, \pi_2) = (\pi_3, \pi_3) = (\pi_4, \pi_4) = \alpha^2 + \beta^2 + \varepsilon^2 + \zeta^2 = 1$$

do not allow a complete determination of the coefficients $\alpha$, $\beta$, $\varepsilon$, $\zeta$. One obtains

$$\beta^2 = \frac{1}{2}, \quad \zeta^2 = \frac{1}{4}, \quad \varepsilon^2 = \frac{1}{4} - \alpha^2.$$

In order to obtain the maximum bond strength because of

$$B_x(p_x) = \sqrt{3} > 0, \ B_x(d_{z^2}) = -\frac{\sqrt{5}}{2} < 0, \ B_x(d_{x^2-y^2}) = \frac{\sqrt{15}}{2} > 0$$

the signs of $\beta$ and $\zeta$ are to be chosen positive and that of $\varepsilon$ negative:

$$\beta = +\frac{1}{\sqrt{2}}, \quad \zeta = +\frac{1}{2}, \quad \varepsilon = -\sqrt{\frac{1}{4} - \alpha^2}\,.$$

Thus we have the four hybrid orbitals:

$$\begin{aligned}
\pi_1 &= \alpha s + \frac{1}{\sqrt{2}}\, p_x + \frac{1}{2}\, d_{x^2-y^2} - \sqrt{\frac{1}{4} - \alpha^2}\; d_{z^2},\\
\pi_2 &= \alpha s - \frac{1}{\sqrt{2}}\, p_x + \frac{1}{2}\, d_{x^2-y^2} - \sqrt{\frac{1}{4} - \alpha^2}\; d_{z^2},\\
\pi_3 &= \alpha s + \frac{1}{\sqrt{2}}\, p_y - \frac{1}{2}\, d_{x^2-y^2} - \sqrt{\frac{1}{4} - \alpha^2}\; d_{z^2},\\
\pi_4 &= \alpha s - \frac{1}{\sqrt{2}}\, p_y - \frac{1}{2}\, d_{x^2-y^2} - \sqrt{\frac{1}{4} - \alpha^2}\; d_{z^2}.
\end{aligned}$$

Two boundary cases are of interest:

Case $\alpha = 1/2$:

$$\begin{aligned}
\pi_1 &= \frac{1}{2}\, s + \frac{1}{\sqrt{2}}\, p_x + \frac{1}{2}\, d_{x^2-y^2},\\
\pi_2 &= \frac{1}{2}\, s - \frac{1}{\sqrt{2}}\, p_x + \frac{1}{2}\, d_{x^2-y^2},\\
\pi_3 &= \frac{1}{2}\, s + \frac{1}{\sqrt{2}}\, p_y - \frac{1}{2}\, d_{x^2-y^2},\\
\pi_4 &= \frac{1}{2}\, s - \frac{1}{\sqrt{2}}\, p_y - \frac{1}{2}\, d_{x^2-y^2}.
\end{aligned}$$

In this case the orbitals $s, p_x, p_y, d_{x^2-y^2}$ appear and we have a *$dsp^2$ hybridization*. The bond strengths 2·69 are obtained using the procedure described above for the octahedron.

Case $\alpha = 0$:

$$\begin{aligned}
\pi_1 &= \frac{1}{\sqrt{2}}\, p_x + \frac{1}{2}\, d_{x^2-y^2} - \frac{1}{2}\, d_{z^2},\\
\pi_2 &= -\frac{1}{\sqrt{2}}\, p_x + \frac{1}{2}\, d_{x^2-y^2} - \frac{1}{2}\, d_{z^2},\\
\pi_3 &= \frac{1}{\sqrt{2}}\, p_y - \frac{1}{2}\, d_{x^2-y^2} - \frac{1}{2}\, d_{z^2},\\
\pi_4 &= -\frac{1}{\sqrt{2}}\, p_y - \frac{1}{2}\, d_{x^2-y^2} - \frac{1}{2}\, d_{z^2}.
\end{aligned}$$

Here we have *$d^2p^2$ hybridization* with the bond strength 2·75.

**(c) Regular tetrahedron** (*Kz.* = 4)

For the hybrid orbital which extends principally to ligand 1 (cf. Figure B.12) we assume:

$$\pi_1 = \alpha s + \beta p_x + \gamma p_y + \delta p_z + \varepsilon d_{z^2} + \zeta d_{x^2-y^2} + \eta d_{xy} + \vartheta d_{xz} + \varkappa d_{yz}.$$

$\pi_1$ must be invariant with regard to reflection in the plane defined by the $z$ axis and the line of centres between the ligands 1 and 3, $x \to y, y \to x, z \to z$. (The position of the axis central ion ligand 1 remains unchanged):

$$\sigma_{z,1-3}\,\pi_1 = \pi_1.$$

From a comparison of coefficients it follows

$$\zeta = 0, \;\gamma = \beta, \;\vartheta = \varkappa.$$

A further symmetry operation with regard to which $\pi_1$ must be invariant because the position of the axis central ion-ligand 1 is not affected is the reflection in the plane defined by the $x$ axis and the line of centres 1–2, $x \to x, y \to z, z \to y$,

$$\sigma_{x,1-2}\,\pi_1 = \pi_1.$$

Comparing the coefficients yields

$$\varepsilon = 0, \;\delta = \gamma, \;\eta = \vartheta.$$

With this we have

$$\pi_1 = \alpha s + \beta(p_x + p_y + p_z) + \eta(d_{xy} + d_{xz} + d_{yz}).$$

The $\pi$ orbitals pointing to the ligands 2, 3, 4 are obtained by the appropriate coordinate transformation, i.e., from $x \to x, y \to -y, z \to -z$

$$\pi_2 = \alpha s + \beta(p_x - p_y - p_z) + \eta(-d_{xy} - d_{xz} + d_{yz}),$$

from $x \to -x, y \to -y, z \to z$

$$\pi_3 = \alpha s + \beta(-p_x - p_y + p_z) + \eta(d_{xy} - d_{xz} - d_{yz}),$$

from $x \to -x, y \to y, z \to -z$

$$\pi_4 = \alpha s + \beta(-p_x + p_y - p_z) + \eta(-d_{xy} + d_{xz} - d_{yz}).$$

The orthogonality and normalization conditions

$$(\pi_1, \pi_2) = (\pi_1, \pi_3) = (\pi_1, \pi_4) = (\pi_2, \pi_3) = (\pi_2, \pi_4) = (\pi_3, \pi_4)$$
$$= \alpha^2 - \beta^2 - \eta^2 = 0,$$

$$(\pi_1, \pi_1) = (\pi_2, \pi_2) = (\pi_3, \pi_3) = (\pi_4, \pi_4)$$
$$= \alpha^2 + 3\beta^2 + 3\eta^2 = 1,$$

yield

$$\alpha^2 = \frac{1}{4} \quad \text{and} \quad \eta^2 = \frac{1}{4} - \beta^2.$$

In order to obtain the maximum bond strength because of*

$$B_1(s) = 1 > 0, \; B_1(d_{xy}) = B_1(d_{xz}) = B_1(d_{yz}) = \sqrt{15}/3 > 0$$

the positive roots for $\alpha$ and $\beta$ are to be chosen:

$$\alpha = \frac{1}{2}, \quad \eta = +\sqrt{\frac{1}{4} - \beta^2}.$$

Thus the $\pi$ orbitals have the form

$$\pi_1 = \frac{1}{2}\, s + \beta(p_x + p_y + p_z) \; + \sqrt{\frac{1}{4} - \beta^2}\; (d_{xy} + d_{xz} + d_{yz}),$$
$$\pi_2 = \frac{1}{2}\, s + \beta(p_x - p_y - p_z) \; + \sqrt{\frac{1}{4} - \beta^2}\; (-d_{xy} - d_{xz} + d_{yz}),$$
$$\pi_3 = \frac{1}{2}\, s + \beta(-p_x - p_y + p_z) + \sqrt{\frac{1}{4} - \beta^2}\; (d_{xy} - d_{xz} - d_{yz}),$$
$$\pi_4 = \frac{1}{2}\, s + \beta(-p_x + p_y - p_z) + \sqrt{\frac{1}{4} - \beta^2}\; (-d_{xy} + d_{xz} - d_{yz}).$$

Similarly as for a square planar arrangement two limiting cases can be distinguished.

Case $\beta = 1/2$:

$$\pi_1 = {}^1/_2(s + p_x + p_y + p_z),$$
$$\pi_2 = {}^1/_2(s + p_x - p_y - p_z),$$
$$\pi_3 = {}^1/_2(s - p_x - p_y + p_z),$$
$$\pi_4 = {}^1/_2(s - p_x + p_y - p_z).$$

This case describes *$sp^3$ hybridization* with the bond strength 2·00. The angular dependence of the hybrid function $\pi_1$ is shown in Figure C.3.

Case $\beta = 0$:

$$\pi_1 = {}^1/_2(s + d_{xy} + d_{xz} + d_{yz}),$$
$$\pi_2 = {}^1/_2(s - d_{xy} - d_{xz} + d_{yz}),$$
$$\pi_3 = {}^1/_2(s + d_{xy} - d_{xz} - d_{yz}),$$
$$\pi_4 = {}^1/_2(s - d_{xy} + d_{xz} - d_{yz}).$$

* The coordinates of the ligand 1 are given by $\sin\theta = \sqrt{2/3}$ and $\phi = \pi/4$.

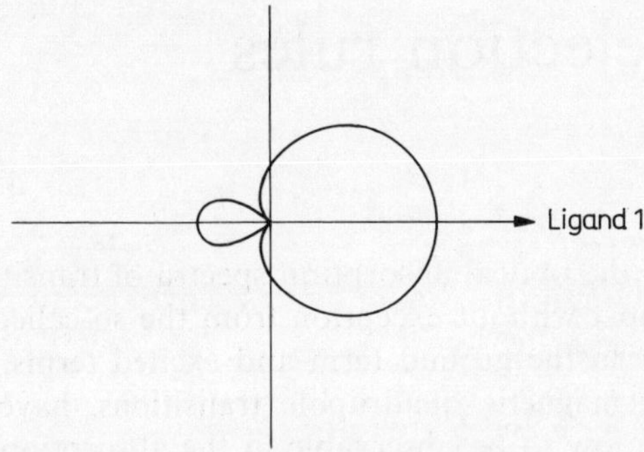


Figure C.3. The angular dependence of the tetrahedral $sp^3$ hybrid orbital $\pi_1$ (after L. Pauling, *op. cit.*, p. 82).

This is a case of $d^3s$ *hybridization*. The bond strength is calculated to be 2·44.

It is interesting to compare the results obtained for the tetrahedron and for square planar coordination. In both cases the coordination number $Kz$ is 4. If we have a central ion with one empty $d$ orbital, one empty $s$ orbital and three empty $p$ orbitals (Example $Ni^{2+}$), then for tetrahedral structure $sp^3$ hybrids with the bond strength 2·00 or for square planar coordination $dsp^2$ hybrids with the bond strength 2·69 can be constructed. According to *Pauling*'s concepts the structure with the greatest bond strength should be the most stable; thus in the present case the square planar coordination.

# 5. Selection rules

The bands in the optical absorption spectra of transition metal complexes arise almost without exception from the so-called *electric dipole transitions* between the ground term and excited terms. Other sorts of transitions, e.g., magnetic quadrupole transitions, have in general an intensity far too low to be observable in the absorption spectrum. The intensity with which an electric dipole transition between two states $\Psi_a \rightarrow \Psi_e$ can take place is essentially determined by the *transition moment* $P$:

$$P = (\Psi_a, \vec{\mu}_{el} \Psi_e).$$

Here $\vec{\mu}_{el}$ is the 'dipole vector' which has the same symmetry behaviour as the position vector $\vec{r} = ix + jy + kz$. An electric dipole transition is only possible when the transition moment has a non-zero value.

The functions $\Psi_k$ of a molecular entity can be written approximately as products of three functions*:

$$\Psi_k = \psi_{k,\text{ space}} \cdot \psi_{k,\text{ spin}} \cdot \phi_{k,\text{ vib.}}$$

$\psi_{k,\text{ space}}$ and $\psi_{k,\text{ spin}}$ describe the state of the electron system. $\psi_{k,\text{ vib.}}$ gives information concerning the vibrational state of the molecule.

Because the dipole vector $\vec{\mu}_{el}$ is a vector in coordinate space and therefore has no effect on the spin coordinates we may write for $P$:

$$P = (\psi_{a,\text{ spin}}, \psi_{e,\text{ spin}})(\psi_{a,\text{space}}\phi_{a,\text{ vib.}}, \vec{\mu}_{el}\psi_{e,\text{ space}}\phi_{e,\text{ vib.}}).$$

As a consequence of the orthogonality of the spin functions the integral over the spin functions can be nonzero only if

$$\psi_{a,\text{ Spin}} = \psi_{e,\text{ Spin}}$$

holds. The spin quantum numbers $S$ for the original and the final states must therefore be equal. In other words: electric dipole transitions can only occur between terms of the same multiplicity (for vanishing spin–orbit coupling).

---

* A necessary condition for the factorization given above is that the spin–orbit coupling be vanishingly small and that the Born–Oppenheimer approximation (*Ann. Physik*, **84** 457 (1927)) be applicable. Those portions of the complete function which describe the total state of motion of the molecule (rotation, translation) were not included.

For $P$ to be nonzero it must further hold that

$$(\psi_{a,\,\mathrm{space}}\phi_{a,\,\mathrm{vib.}},\, \vec{\mu}_{el}\psi_{e,\,\mathrm{space}}\phi_{e,\,\mathrm{vib.}}) \neq 0.$$

For the discussion of this integral we shall treat first the special (hypothetical) case where the molecule does not vibrate and the nuclei are fixed in their equilibrium positions. In this case the given integral is reduced to the form

$$(\psi_{a,\,\mathrm{space}},\, \vec{\mu}_{el}\psi_{e,\,\mathrm{space}}).$$

This integral can only have a nonzero value when the integrand or portions of the integral are invariant with respect to all operations of the symmetry group of $\vec{\mu}_{el}$, cf. Part B, section 2.5, p. 316 f. Because $\vec{\mu}_{el}$ has the same symmetry behaviour as the position vector $\vec{r}$, the rotation reflection group of the sphere is applicable. When the invariance of the integrand is violated for only one operation from the rotation reflection group of the sphere, then the integral must have the value zero.

In the ligand field theory it is conventional when judging the invariance properties of integrands of the above type to limit the discussion to those symmetry operations which compose the symmetry group of the ligand system and are only a subset of the symmetry operations of the complete rotation reflection group of the sphere. As will be shown using several examples, this procedure is very convenient but must be applied with due caution.

If a molecular structure has for example the symmetry $O_h$, then the electron space coordinate functions $\psi_{a,\,\mathrm{space}}$ and $\psi_{e,\,\mathrm{space}}$ can be assigned to certain irreducible representations of $O_h$, in general say $\Gamma_{(a)}$ or $\Gamma_{(e)}$. The dipole vector $\vec{\mu}_{el}$ transforms as $\vec{r}$ under the operations of $O_h$ as $T_{1u}$. With consideration of the results obtained in Part B, section 2.5, p. 316 f., the integral in $(\psi_{a,\,\mathrm{space}},\, \vec{\mu}_{ee}\psi_{e,\,\mathrm{space}})$ transforms according to the representation $\Gamma$ of $O_h$ which is equal to the product

$$\Gamma = \Gamma(a) \dot{\times} T_{1u} \dot{\times} \Gamma(e).$$

In order for the integral having the integrands mentioned to have a nonzero value, $\Gamma$ must contain the identity representation $A_{1g}$ of $O_h$ as an irreducible component. The product representation must thus be *even*. Because $T_{1u}$ is *odd*, *even* $\Gamma$ can be obtained only when $\Gamma(a)$ and $\Gamma(e)$ represent different parities. That is, either the original state $\psi_{a,\,\mathrm{space}}$ is even and the final state $\psi_{e,\,\mathrm{space}}$ is odd or vice versa. *Transitions* $g \to g$ and $u \to u$ *are forbidden* (LaPorte Rule). Such forbidden transitions are called 'LaPorte forbidden.'

$d \to d$ transitions ($g \to g$) for a fixed nuclear configuration should not occur as electric dipole radiation in octahedral complexes.

The LaPorte Rule for $d \rightarrow d$ transitions (with fixed nuclei) is by no means limited to systems with a centre of inversion, although an uncritical application of the formalism would appear to lead to such a result. Consider for example systems with tetrahedral symmetry $T_d$. Because of the absence of a centre of inversion the irreducible representations are not differentiated here by the indices g and *u*. $\psi_{el}$ transforms as $T_2$. If, e.g., the original *d* state belongs to the species *E* and the final *d* state to the species $T_2$, it holds for the representation of the integrand

$$E \dot{\times} T_2 \dot{\times} T_2 = A_1 \dot{+} A_2 \dot{+} 2E \dot{+} 2T_1 \dot{+} 2T_2.$$

This contains the identity representation $A_1$. This does not, however, mean that the electric dipole transition $E \rightarrow T_2$ in the *d* electron system is allowed. In the given representation product it can be seen that the integrand (or a portion of it) is invariant under the operations of the group $T_d$.

It was already shown above that the invariance of the integrand must hold for all operations of the rotation-reflection group of the sphere and thus also for inversion. The LaPorte rule must also be considered in this case. It follows that also in $T_d$ systems pure $d \rightarrow d$ electronic transitions $(g \rightarrow g)$ are forbidden. (As long as the nuclear system is at rest.) For tetrahedral complexes, however, as opposed to systems with a centre of inversion, the peculiarity arises that as a result of the absence of a centre of inversion in general no pure *d* states are present. *p* state contributions can be mixed in with low-lying *d* states. As a result the integrals $(\psi_{a,\text{space}}, \vec{\mu}_{el}\psi_{e,\text{space}})$ have nonzero values for tetrahedral complexes.

For vanishing spin–orbit coupling and for a fixed nuclear system the following selection rules hold for electronic dipole transitions:

(a) *Transitions between terms of different multiplicity* ($\Delta S \neq 0$) *are forbidden.*

(b) *Pure d → d transitions are* (*LaPorte*) *forbidden.* For systems lacking a centre of inversion the transition prohibition can be lifted by the *intermixing of states* of odd parity (e.g. *p*, *f*, ...) with the *d* state.

These selection rules do not hold rigorously for transition metal complexes for the following reasons:

For (a) as a result of *spin–orbit coupling* the prohibition for $\Delta S \neq 0$ is in general removed. The intensities of the *intercombination bands* which then appear grow with increasing spin–orbit coupling but are very weak compared to the intensities of spin–allowed transitions.

For (b) the assumption of a nuclear system completely at rest does not hold. As a result of *vibrational motion* the prohibition for pure $d \rightarrow d$ transitions is removed to a certain degree. The observed intensities in centrosymmetric systems are in general weaker than in systems without a centre of inversion, for in the latter case electric dipole transitions are possible because of the intermixing of states of odd parity, even without coupling with vibrational motion.

# 6. Atomic units

The quantities which appear in atomic and molecular physics can be expressed in so-called *atomic units (au)*. These are obtained when the following units of charge, mass and length are employed:

the charge of the electron $e$,
the mass of the electron $m$,
the *Bohr* radius $a_0 = \hbar^2/me^2$

In the following Table C.1 the units obtained in this manner for the most important quantities are summarized (after H. Hartmann, *Theorie der chemischen Bindung*, Springer-Verlag, Berlin–Göttingen–Heidelberg 1954).

Table C.1. Atomic Units (au)

| Quantity | Unit | Numerical values of important quantities in atomic units (au) |
|---|---|---|
| charge | $e$ | |
| mass | $m$ | |
| length | $a_0$ | |
| energy | $e^2/a_0$ | ionization energy of H atom $= \frac{1}{2}$ au |
| angular momentum | $\hbar$ | angular momentum of the electron about an axis (in H atom) = 1, 2, 3, ... au. |
| force | $e^2/a_0^2$ | force between nucleus and electron with a separation of $a_0 = 1$ au in H atom. |
| velocity | $e^2/\hbar$ | velocity of the electron in the first orbit of the *Bohr* theory = 1 au. |
| time | $a_0\hbar/e^2$ | period of revolution of the electron in the first *Bohr* orbit = 2 au. |
| electric field strength | $e/a_0^2$ | field strength at a distance $a_0$ from H nucleus = 1 au. |
| electric potential | $e/a_0$ | potential at a distance $a_0$ from the H nucleus = 1 au. |

# Bibliography

A. General complex chemistry, reviews
B. Introductory books on the quantum theory of the chemical bond
C. Theory of free atoms and ions
D. Group theory
E. Ligand field theory, monographs
F. Review articles on ligand field theory
G. Applications of ligand field theory
   1. Magnetochemistry
   2. Optical spectra
   3. Thermodynamical stability
   4. Ionic radii
   5. Stereochemistry
   6. Reaction kinetics
   7. Jahn-Teller effect
H. Computational aids
I. Miscellaneous

In the following a number of important monographs and original papers have been cited which should enable the reader to increase his knowledge of the ligand field theory. This bibliography is a summary of the literature, selected for didactical reasons and for usefulness, and no claim to completeness can be made.

## A. General complex chemistry, review articles

1. WERNER, A.: Neuere Anschauungen auf dem Gebiet der anorganischen Chemie (Neu bearbeitet von Paul PFEIFFER). Friedr. Vieweg u. Sohn, Braunschweig 1923. (Klassische Darstellung des Begründers der Koordinationslehre.) (Classical treatment by the founder of the coordination theory.)

2. WEINLAND, R.: Einführung in die Chemie der Komplex-Verbindungen. 2. Aufl., Ferd. Enke, Stuttgart 1924. (Ausführliche Übersicht über die klassischen Komplexverbindungen.) (Detailed summary of the classical complexes.)

3. HEIN, F.: Chemische Koordinationslehre. Verlag S. Hirzel, Leipzig 1950. (Darstellung des Gesamtgebietes der Koordinationschemie auf klassischer Grundlage; präparative Gesichtspunkte stehen im Vordergrund.) (Treatment of the entire subject of coordination chemistry on a classical basis. The emphasis is on preparative chemistry.)

4. GRINBERG, A. A.: Einführung in die Chemie der Komplexverbindungen. VEB Verlag Technik, Berlin 1955. Übersetzung der russischen 2. Aufl., Moskau 1951. (Darstellung des Gesamtgebietes unter spezieller Berücksichtigung der Chemie der Komplexe der Platinmetalle; die klassischen Gesichtspunkte stehen im Vordergrund.) German translation of the 2nd Russian edition. Moscow, 1951. (A treatment of the entire subject with a special consideration of the chemistry of the complexes of platinum metals; classical considerations are in the foreground.)

5. BAILAR, J. C. jr.: The Chemistry of the Coordinate Compounds. ACS Monograph No. 131. Reinhold Publ. Corp., New York; Chapman & Hall Ltd., London 1956. (Ausgezeichnete Monographie, die auch moderne physikalisch-chemische Gesichtspunkte mit berücksichtigt; umfangreiche Literaturangaben.) (A fine monograph which also considers the modern physical-chemical viewpoint; comprehensive bibliography.)

6. GRADDON, D. P.: An Introduction to Co-ordination Chemistry. Pergamon Press, Oxford—London—New York—Paris 1961. (Knappe, einführende Darstellung, die die modernen physikalisch-chemischen Gesichtspunkte berücksichtigt.) (A terse, introductory presentation which takes the modern physical chemical point of view into account.)

7. MARTELL, A. E. u. M. CALVIN: Chemistry of the Metal Chelate Compounds. Prentice-Hall, New York 1952. Deutsche Übersetzung Verlag Chemie, Weinheim/Bergstraße 1958. (Ausgezeichnete Übersicht über die Chemie der Metallchelate; moderne physikalisch-chemische und bindungstheoretische Gesichtspunkte sind jedoch nicht berücksichtigt; zahlreiche Literaturangaben.) (A fine survey of the chemistry of metal chelates; modern physical chemical and bond theoretical point of view are however, not considered; numerous references to the literature.)

8. ZEISS, H.: Organometallic Chemistry. ACS Monograph No. 147. Reinhold Publ. Corp., New York; Chapman & Hall Ltd., London 1960.
(Behandelt in Beiträgen zahlreicher Autoren die verschiedenen metallorganischen Verbindungen; umfangreiche Bibliographie.)
(In contributions by numerous authors the various metal-organic compounds are treated; a large bibliography.)

9. JØRGENSEN, C. K.: Inorganic Complexes. Academic Press Inc., New York—London 1963.
(Die Komplexverbindungen werden nach dem Konzept der anorganischen Chromophore $MX_6$, $MX_4$ usw. behandelt. Die Kapitel sind nach der Art der Liganden X geordnet, X = Halogene, N, O, S usw. Die Diskussion erfolgt u. a. auf der Grundlage der Ligandenfeldtheorie; ausführliche Bibliographie, 1176 Literaturzitate.)
(Complexes are treated according to the concept of the inorganic chromophores $MX_6$, $MX_4$ etc. The chapters are arranged according to the nature of the ligands X, X = halogens, N, O, S, etc. The discussion proceeds, among other things, along the lines of the ligand field theory; a detailed bibliography with 1176 literature references.)

10. BASOLO, F. u. R. C. JOHNSON: Coordination Chemistry. The Chemistry of Metal Complexes. W. A. Benjamin Inc., New York—Amsterdam 1964.
(Elementare moderne Darstellung, leicht lesbar. Darin auch Kapitel über Ligandenfeldtheorie.)
(An elementary modern presentation, easily readable. A chapter on the ligand field theory is also contained.)

## B. Introductory books on the quantum theory of the chemical bond

1. PAULING, L. u. E. B. WILSON: Introduction to Quantum Mechanics with Application to Chemistry. McGraw-Hill, New York–London 1935.
(Ausgezeichnete Einführung speziell für den Chemiker.)
(A fine introduction, especially for the chemist.)

2. KAUZMANN, W.: Quantum Chemistry. Academic Press Inc., New York 1957.
(Elementare einführende Darstellung, speciell für den Chemiker geeignet.)
(Elementary introductory presentation, suited especially for the chemist.)

3. EYRING, H., J. WALTER u. G. E. KIMBALL: Quantum Chemistry. John Wiley & Sons, New York; Chapman & Hall Ltd., London 1954.
(Führt weiter als 1. und 2., jedoch schwieriger zu lesen; enthält einen Abschnitt über Gruppentheorie.)
(Goes farther than 1 and 2, but is more difficult to read; contains a section on group theory.)

4. HARTMANN, H.: Theorie der chemischen Bindung auf quantentheoretischer Grundlage. Springer-Verlag, Berlin–Göttingen–Heidelberg 1954.
(Knappe Darstellung, Schwierigheitsgrad über 1., 2. und 3. hinausgehend; der spezielle Teil, der die chemischen Anwendungen bringt, ist besonders wertvoll.)
(A terse presentation, more difficult than 1, 2, and 3; the specialized part which presents the chemical applications is especially valuable.)

5. HELLMANN, H.: Einführung in die Quantenchemie. Franz Deuticke Verlag, Leipzig 1937.

6. COULSON, C. A.: Valence. 2nd edition. Oxford Univ. Press, London 1961. (Leicht lesbare qualitative Einführung.) (An easily readable, qualitative introduction.)

7. PAULING, L.: The Nature of the Chemical Bond. 3rd edition. Cornell Univ. Press, Ithaca, N.Y. 1960. (Darin speziell eine ausführliche Darstellung der Valenzstrukturtheorie und ihrer Anwendung auf Übergangsmetallkomplexe sowie der Resonanz.) (This contains a detailed treatment of the valence bond theory and its application to transition metal complexes and a treatment of resonance.)

## C. The theory of free atoms and ions

1. FINKELNBURG, W.: Einführung in die Atomphysik, 9./10. Aufl. Springer-Verlag, Berlin–Göttingen–Heidelberg 1964. (Bringt die experimentellen und die theoretischen Grundlagen in elementarer Form.) (Presents the experimental and theoretical foundation in elementary form.) English translation.

2. CONDON, E. U. u. G. H. SHORTLEY: The Theory of Atomic Spectra. Cambridge University Press, Cambridge 1959. (Standardwerk; bringt die Theorie in strenger Darstellung.) (Standard text and reference; presents the theory in a rigorous form.)

3. SLATER, J. C.: Quantum Theory of Atomic Structure. Vol. I and II. McGraw-Hill, New York–Toronto–London 1960. (Sehr eingehende Darstellung mit umfangreicher Bibliographie.) (A very detailed presentation with an extensive bibliography.)

4. SOMMERFELD, A.: Atombau und Spektrallinien, Bd. 1 und 2. Friedr. Vieweg & Sohn, Braunschweig 1939. (Die klassische deutschsprachige Darstellung.) (The classical German language presentation.)

5. MOORE, C. E.: Atomic Energy Levels. Nat. Bur. Standards Cir. **467.** Vol. 1 1949; Vol. 2 1952; Vol. 3 1959. (Zusammenstellung von Energiezuständen freier Ionen, entnommen aus der Analyse von Emissionsspektren.) (A summary of the energy states of free ions taken from the analysis of emission spectra.)

Vgl. auch die entsprechenden Abschnitte in den unter B. angeführten Werken. (Cf. The appropriate sections in the references listed under B.)

## D. Group theory

1. WILSON, E. B., J. C. DECIUS u. P. C. CROSS: Molecular Vibrations. McGraw-Hill, New York 1955. (Enthält die Charakterentafeln für alle im Bereich der Molekularphysik wichtigen Gruppen.)

(Contains the character tables of all of the groups of importance in molecular physics.)

2. WIGNER, E.: Gruppentheorie. Friedr. Vieweg & Sohn, Braunschweig 1931. Engl. Übersetzung: Academic Press Inc., New York 1959.
3. COTTON, F. A.: Chemical Applications of Group Theory. Interscience Publ. Inc., New York–London 1963.
   (Darstellung ist speziell auf die Belange des an der Ligandenfeld- und M.-O.-Theorie interessierten Chemikers abgestellt; leicht lesbar.)
   (The presentation is suited especially to the needs of the chemist interested in the ligand field and MO theory; easily readable.)
4. TINKHAM, M.: Group Theory and Quantum Mechanics. McGraw-Hill, New York—San Francisco—Toronto—London 1964.
5. HEINE, V.: Group Theory in Quantum Mechanics. Pergamon Press, Oxford—London—New York—Paris 1964.

Vgl. auch unter B. 3., Kapitel X.
(Cf. also under B.3, Chapter X.)

## E. Ligand field theory, monographs

1. BALLHAUSEN, C. J.: Introduction to Ligand Field Theory. McGraw-Hill, London–New York 1962.
   (Ausgezeichnete klare Darstellung der Kristall- und Ligandenfeldtheorie; die mathematischen Formulierungen sind z.T. ausführlich angegeben, einzelne instruktive Beispiele werden explizit durchgerechnet. Ein Anhang gibt eine Übersicht über die elektronische Struktur ausgewählter anorganischer Komplexe an Hand der Diskussion der optischen und magnetischen Eigenschaften. Umfangreiche Bibliographie.)
   (A very fine clear presentation of the crystal and ligand field theory; the mathematical formulations are given in detail and selected instructive examples are explicitly calculated. An appendix gives a survey of the electronic structure of specific organic complexes on the basis of a discussion of their optical and magnetic properties. A comprehensive bibliography.)
2. GRIFFITH, J. S.: The Theory of Transition-Metal Ions. Cambridge Univ. Press, Cambridge 1962.
   (Strenge, anspruchsvolle Darstellung in der Art von CONDON u. SHORTLEY »The Theory of Atomic Spectra«; zahlreiche Tabellen und Übersichten im Anhang, Bibliographie wichtiger Arbeiten.)
   (A rigorous, demanding presentation in the style of Condon and Shortley 'The Theory of Atomic Spectra'; numerous tables and summaries are included in the appendix; a bibliography of the important papers.)
3. JØRGENSEN, C. K.: Absorption Spectra and Chemical Bonding in Complexes. Pergamon Press, Oxford–London–New York–Paris 1962.
   (Diskussion der Absorptionsspektren von Übergangsmetallen, behandelt auch Elektronenübergangsspektren; Tabellen von Spektren mit Bandenzuordnungen, umfangreiche Literaturzusammenstellung; für denjenigen, der sich mit den Spektren von Übergangsmetallkomplexen beschäftigt, unentbehrlich.)
   (A discussion of the absorption spectra of transition metals, also treats

electron transfer spectra; tables of spectra with band assignments, comprehensive bibliography; essential for those interested in the spectra of transition metal complexes.)

4. JØRGENSEN, C. K.: Orbitals in Atoms and Molecules. Academic Press Inc., New York–London 1962.
(Stellt eine wertvolle Ergänzung zu 3. dar; Versuch einer einheitlichen Beschreibung von kugelsymmetrischen Atomen bzw. Ionen sowie niedriger symmetrischen Komplexen auf der Basis von Elektronenkonfigurationen und Einelektroneneigenfunktionen; stark durch den individuellen Stil des Autors geprägt, enthält eine Fülle von interessanten Gedanken und Anregungen, die schwerlich an anderer Stelle zu finden sind.)
(This is a valuable supplement to 3; an attempt to give a unified description of spherically symmetric atoms and ions as well as complexes of lower symmetry on the basis of electron configurations and one-electron eigenfunctions; strongly influenced by the individual style of the author, contains a wealth of interesting ideas and stimulating suggestions which are difficult to find in any other place.)

5. ORGEL, L. E.: An Introduction to Transition-Metal Chemistry, Ligand Field Theory. Methuen & Co. Ltd., London 1960.
(Einführung in die Ligandenfeldtheorie auf elementarer Grundlage, ohne mathematischen Formalismus. Im zweiten Teil des Buches werden auch u. a. Carbonyle, Dicyclopentadienylkomplexe u. ähnliche Verbindungen diskutiert.)
(An introduction to the ligand field theory on an elementary basis without mathematical formalism. In the second part of the book carbonyls, dicyclopentadienyl complexes and similar compounds are discussed.)

6. BERSUKER, I. B. u. A. V. ABLOV: Khimicheskaya svyasi v komplexnyth sojedinenya. Chemische Bindung in Komplexverbindungen (in russischer Sprache). Akademie der Wissenschaften der Moldauischen S.S.R. Kishinev 1962.
(Chemical bond in complexes (in Russian). Academy of Sciences of the Moldau S.S.R., Kishinev 1962.)
(Eingführung, die in der Art der Darstellung in Teilen etwas weiter geht als die von ORGEL; 378 Literaturzitate.)
(Introduction, which in the manner of presentation in part goes farther than that of Orgel; 378 literature references.)

7. DUNN, T. M., D. S. MCCLURE u. R. C. PEARSON: Some Aspects of Crystal Field Theory. Harper & Row Publ., New York 1965.
(Knappe Darstellung der theoretischen Grundlagen, der Anwendung auf die optischen Absorptionsspektren, auf das magnetische Verhalten sowie Erläuterung der Kristallfeldstabilisierung und reaktionskinetischer Aspekte; zahlreiche Literaturzitate.)
(A concise presentation of the theoretical principles and their application to optical absorption spectra and magnetic behaviour as well as an explanation of crystal field stabilization and reaction kinetic aspects; numerous references to the literature.)

8. FIGGIS, B. N.: Introduction to Ligand Fields. Interscience Publ., New York–London–Sydney 1966.

(Einführung in die Ligandenfeldtheorie auf elementarer Grundlage mit besonderer Berücksichtigung der Anwendungen. Leicht lesbar. Die Teile über die magnetischen Eigenschaften von Übergangsmetallkomplexverbindungen sind besonders gut gelungen, zahlreiche Literaturhinweise.)
(An introduction to the ligand field theory on an elementary basis with special emphasis on applications. Easily readable. The sections on magnetic properties of transition metal complexes are especially good, numerous literature references.)

*Werke speziell über M.O.-Theorie:*

9. BALLHAUSEN, C. J. u. H. B. GRAY: Molecular Orbital Theory (An Introduction Lecture Note and Reprint Volume). W. A. Benjamin Inc., New York–Amsterdam 1964.
(Gibt eine gute Einführung in die M.O.-Theorie, enthält im Nachdruck eine große Zahl instruktiver Originalarbeiten zur allgemeinen Theorie und über Übergangsmetallkomplexe.)

*Specialized books on the MO theory*

(Gives a good introduction to the MO theory, contains in reprint form a large number of instructive original papers on the general theory and on transition metal complexes.)

10. GRAY, H. B.: Electrons and Chemical Bonding. W. A. Benjamin Inc., New York–Amsterdam 1965.
(M.O.-Theorie in elementarer Darstellung.)
(An elementary presentation of MO theory.)

## F. Review articles on ligand field theory

1. MOFFITT, W. u. C. J. BALLHAUSEN: Quantum Theory. Annu. Rev. physic. Chem. 7 (1956) 107.
(Einführung in die Kristallfeldtheorie, schwache und starke Felder, spektrochemische Serie, Spin-Bahn-Kopplung, oktaed rische und Felder geringerer Symmetrie, spektroskopische Splitting-Faktoren, *Jahn-Teller*-Effekt, optische Drehung, Literatur bis 1955, 109 Zitate.)
(Introduction to crystal field theory, strong and weak fields, spectrochemical series, spin-orbit coupling, octahedral and fields of lower symmetry, spectroscopic splitting factors, *Jahn-Teller* Effect, optical rotation, literature up to 1955, 109 references.)

2. BALLHAUSEN, C. J.: Elektrontilstande i Komplexser af 1. Overgangsgruppe (En Studie i Krystalfelt Teori). Dissertation Kopenhagen 1958. (Literatur bis 1957, 106 Zitate.)
(Literature up to 1957, 106 references.)

3. BALLHAUSEN, C. J.: Theories of Bonding in Coordination Compounds. Advances in the Chemistry of the Coordination Compounds, Proceedings of the sixth Conference on Coordination Chemistry, p. 3. Macmillan Co., New York 1961.
(Diskussion der Theorien der Molekülzustände, der Valenzstrukturtheorie sowie der Ligandenfeldtheorie.)
(Discussion of the molecular orbital, valence bond and ligand field theories.)

4. Jørgensen, C. K.: Absorption Spectra of Complexes with Unfilled *d*-Shells. Proc. 10th Solvay Conf., p. 355. Stoops, Brüssel 1956.
(Abriß der Theorie, Zusammenstellung der Diagonalelemente für $O_h$-Symmetrie für Ionen $d^1$ bis $d^9$, Methode des starken Feldes, Energiezustände freier Ionen ausgedrückt als Summe von *Racah*-Parametern, Zusammenstellung von Absorptionsspektren oktaedrischer $d^1$- bis $d^9$-Übergangsmetallkomplexe, Tabelle von *Dq*-Werten, 160 Literaturzitate.)
(Summary of the theory, tabulation of the diagonal elements for $O_h$ symmetry for ions $d^1$ to $d^9$, strong-field method, energy states of free ions expressed as sums of *Racah* parameters, tabulation of absorption spectra of $d^1$ to $d^9$ octahedral transition metal complexes, tables of *Dq* values, 160 literature references.)

5. Jørgensen, C. K.: Energy Levels of Complexes and Gaseous Ions. Dissertation, Jul. Gjellerups Forlag, Kopenhagen 1957.
(Übersicht über die Ligandenfeldtheorie, auch Komplexe der Seltenen Erden behandelt, 128 Literaturzitate.)
(Survey of the ligand field theory, complexes of the rare earths are also treated, 128 literature references.)

6. Jørgensen, C. K.: Spectroscopy of Transition-Group Complexes. Advances chem. Physics **5** (1963) 33, Interscience Publ., New York–London 1963.
(Ausgezeichnete Übersicht über die Ligandenfeldtheorie, Lanthaniden und 5*f*-Elemente eingeschlossen zahlreiche Tabellen, z.B. über Absorptionsspektren mit Angabe der Bandenzuordnung, Werte der nephelauxetischen Parameter $\beta_{55}$ und $\beta_{35}$, Termenergien für $O_h$-Symmetrie für Komplexe der Konfiguration $d^2$ bis $d^8$ unter Berücksichtigung der Nichtdiagonalelemente auf der Grundlage der Methode des starken Feldes; ausgedrückt als a $Dq+$ b$B+$c$C-$e$B^2/Dq$, Literatur bis 1961, 270 Zitate.)
(A very fine survey of the ligand field theory, lanthanide and 5*f* elements; numerous tables are included, e.g., on absorption spectra with band assignments, values of the nephelauxetic parameters $\beta_{55}$ and $\beta_{35}$ term energies for $O_h$ symmetry, for complexes of the configuration $d^2$ to $d^8$ with consideration of nondiagonal elements on the basis of the strong-field method; expressed as $aDq+bB+cC-eB^2/Dq$, literature up to 1961, 270 references.)

7. Runciman, W. A.: Absorption and Fluorescence Spectra of Ions in Crystals. Rep. Progr. Physics **21** (1958) 30.
(Optische Spektren in Kristallen mit Übergangsmetallionen sowie mit Ionen der Seltenen Erden; auch die Emissionsspektren werden behandelt; zahlreiche Literaturhinweise.)
(Optical spectra of crystals with transition metal ions and with ions of the rare earths; also emission spectra are treated; numerous literature references.)

8. Dunn, T. M.: The Visible and Ultra-Violet Spectra of Complex Compounds, in: Modern Coordination Chemistry (Principles and Methods) by J. Lewis and R. G. Wilkins, S. 230. Interscience Publ., New York–London 1960.
(Ausgezeichnete und prägnante Zusammenstellung der Theorie der Spektren sowie Diskussion experimenteller Befunde, umfangreiche Literaturzusammenstellung, 204 Zitate.)
(A very fine, stimulating summary of the theory of the spectra as well as a discussion of experimental results, comprehensive survey of the literature, 204 references.)

9. DUNN, T. M.: Appraisal of Experiment and Theory in the Spectra of Complexes. Proc. 7th Int. Conf. Coordination Chemistry, London. Butterworths, London 1963. (vgl. auch Pure and Applied Chemistry, Vol. 6, No. 1).
(Abriß der Kristallfeldtheorie, Kristallspektren gemessen mit polarisiertem Licht, Rotationsdispersion, Zirkulardichroismus, Elektronenspinresonanz, *Zeeman*-Effekt und magn. Daten, Intensität, *Jahn-Teller*-Effekt; 112 Literaturzitate.)
(Summary of the crystal field theory, crystal spectra measured with polarized light, circular dichroism, electron spin resonance, *Zeeman* effect and magnetic data, intensity, *Jahn-Teller* effect; 112 literature references.)

10. MCCLURE, D. S.: Electronic Spectra of Molecules and Ions in Crystals, in: Solid State Physics, Vol. 9, p. 399. Academic Press Inc., New York–London 1959.
(Ausführliche Darstellung des experimentellen Materials über Festkörperspektren von Übergangsmetall- sowie auch von Verbindungen der Seltenen Erden, Abriß der theoretischen Grundlagen, Abschnitte über Molecular-Orbital-Theorie sowie Charge-Transfer-Spektren, Angabe der Matrixelemente für $O_h$-Symmetrie auf der Grundlage des starken Feldes für $d^2$ bis $d^8$-Ionen, umfangreiche Literaturzitate.)
(Detailed presentation of the experimental material on solid state spectra of transition metal and rare earth complexes, summary of the theory, sections on MO theory and charge-transfer spectra, tabulation of the matrix elements for $O_h$ symmetry on the basis of the strong-field method for $d^2$ to $d^8$ ions, comprehensive literature references.)

11. Ions of the Transition Elements. Discuss. Faraday Soc. **26** (1959).
(Zusammenstellung einer Reihe von Vorträgen mit ausführlichen Diskussionsbemerkungen; gibt einen ausgezeichneten Überblick über die Probleme im Bereich der Spektroskopie, des magnetischen Verhaltens usw. der Übergangsmetallverbindungen.)
(A series of lectures with detailed comments on the discussion; gives a very fine survey of the problems in the field of spectroscopy, magnetic behaviour, etc. of the transition metal complexes.)

12. GRIFFITH, J. S. u. L. E. ORGEL: Ligand-Field Theory. Quart. Rev. (chem. Soc. London) **11** (1957) 381.
(Grundzüge der Theorie, Molecular-Orbital-Theorie, Frage der Bindung in Komplexen, Magnetisches Kriterium für die Bindung, Diskussion der thermodynamischen Eigenschaften, Ligandenfeld-Theorie und *Pauling*sche Theorie, 41 Literaturzitate.)
(Theoretical basis, MO theory, question of the formation of complexes, magnetic criterion for bonding, discussion of the thermodynamic properties, ligand field theory and Pauling's theory, 41 literature references.)

13. ORGEL, L. E.: Some Application of Crystal-Field Theory to the Problems in Transition-Metal Chemistry. Proc. 10th Solvay conf., p. 289. Stoops, Brüssel 1956.
(Hydratationswärmen, Stabilität von Komplexionen in Lösung, Gitterenergien, relative Stabilität oktaedrischer und tetraedrischer Komplexe, Charge-Transfer-Spektren, Ionenkomplexe und kovalente Komplexe, 40 Literaturzitate.)
(Heats of hydration, stability of complex ion in solution, lattice energies, relative stability of octahedral and tetrahedral complexes, charge-transfer

spectra, ionic complexes and covalent complexes, 40 literature references.)

14. SUTTON, L. E.: Theory of Bonding in Metal Complexes. Chemistry of the Co-ordinate Compounds, Symposium Rome 1957, p. 24. Pergamon Press, New York–London 1958.
vgl. auch J. inorg. nucl. Chem. **8** (1958) 24.
(Cf. also *J. inorg. nucl. Chem.* **8**, 24 (1958).

15. HARTMANN, H., u. H. L. SCHLÄFER: Farbe und Konstitution von Komplexverbindungen der Übergangsmetalle. Angew. Chem. **66** (1954) 768.
(Elementare qualitative Darstellung der Theorie im Hinblick auf die Deutang der Lichtabsorptionseigenschaften.)
(Elementary qualitative presentation of the theory with reference to the interpretation of light-absorption properties.)

16. HARTMANN, H. u. H. L. SCHLÄFER: Zur Frage der Bindungsverhältnisse bei Komplexverbindungen. Angew. Chem. **70** (1958) 155.
(Diskussion, inwieweit auf Grund der Ergebnisse physikalischer oder chemischer Versuchsresultate Aussagen über den Bindungszustand möglich sind. Die Konsequenzen, die sich aus dem erweiterten elektrostatischen Modell ergeben, werden diskutiert. Die Bedeutung des Variationsprinzips für die Diskussion von Bindungsproblemen wird erläutert.)
(Discussion of in how far it is possible to make statements concerning the state of bonding on the basis of physical or chemical experimental results. The consequences of the extended electrostatic model are discussed. The importance of the variation principle for the discussion of bonding problems is explained.)

17. DORAIN, P. B.: Symmetry in Inorganic Chemistry. Addison-Wesley Publ. Inc., Reading, Mass.–Palo Alto–London 1965. (Kap. 4, S. 76ff.: Crystal-Field Theory.)
(Chapter 4, pp. 76ff: Crystal field theory.)

18. JØRGENSEN, C. K.: Recent Progress in Ligand Field Theory. Structure and Bonding **1** (1966) 3. Springer-Verlag, Berlin–Heidelberg–New York.
(Ligandenfeldtheorie und Interpretation von Spektren, Erweiterung der Ligandenfeldtheorie durch Betrachtung von molecular orbitals, angular-overlap-Modell, Nephelauxetische Parameter, 49 Literaturzitate.)
(Ligand field theory and interpretation of spectra, extension of the ligand field theory by a consideration of molecular orbitals, angular overlap model, nephelauxetic parameters, 49 literature references.)

## Angular overlap model

19. MCCLURE, D.: In Advances in the Chemistry of the Coordination Compounds. Macmillan, New York 1961, p. 498.

20. SCHÄFFER, C. E. and C. K. JØRGENSEN: Molecular Physics **9** (1966) 401.
Dan. Mat. Fys. Medd. **32** Nr. 13 (1965).

21. JØRGENSEN, C. K.: J. Physique **26** (1965) 825.

22. SCHÄFFER, C. E.: Theoret. chim. Acta (Berl.) **4** (1966), 166.
Proc. Roy. Soc. **A297** (1967), 96.

23. JØRGENSEN, C. J., R. POPPARLARDO and H. H. SCHMIDTKE: J. Chem. Phys. **39** (1963), 1422.

24. SCHMIDTKE, H. H.: *Electronic Absorption Spectroscopy*. Physical Methods in Advanced Inorganic Chemistry.
Ed. by H. A. O. Hill and P. Day. Interscience Publ., London–New York–Sydney 1968, p. 107. 127 references. Spectra of transition metal complexes and ligand field theory.

25. JØRGENSEN, C. K.: Recent Progress in Ligand Field Theory. Structure and Bonding **1** (1966), 3.
Ligand field theory and interpretation of spectra, extension of the ligand field theory by consideration of molecular orbitals. Angular overlap model and the influence of Madelung potentials on Wolfsberg-Helmholz calculations. 48 references.

26. KOIDE, S. and K.-J. GONDAIRA: Recent Developments in the Theory of Transition Metal Ions.
Prog. Theor. Physics Supplement No. 40 (1967) 160.
Review of the Japanese contributions from Kotani's school, magnetic properties, summary of the $d^n$ energy levels, covalency effects, interaction with phonon and magnon systems. 53 references.

27. LEVER, A. B. P.: The Electronic Spectra of Tetragonal Metal Complexes. Analysis and Significance. Coord. Chem. Revs. **3** (1968) 119.

28. The Crystal Field Splitting Parameter *Dq*, Calculation and Significance.
Advances in Chemistry Series, Nr. 62. "Werner Centennial" 1966, 430.
A method for the rapid determination of $Dq < \beta$ spin-allowed transitions of high-spin $d^2$, $d^3$, $d^7$ and $d^8$ complexes. Method for the determination of *Dq* for cubic molecules, 50 references.

## G. Applications of the ligand field theory

### *G.1. Magnetochemistry*

*Monographs*

1. VAN VLECK, J. H.: The Theory of Electric and Magnetic Susceptibilities. Oxford Univ. Press, London 1932.
(Strenge, anspruchsvolle Darstellung der Theorie, unentbehrlich für alle, die magnetochemisch arbeiten, zahlreiche Literaturhinweise.)
(Rigorous, challenging treatment of the theory, indispensable for all those working in the field of magnetochemistry, numerous references to the literature.)

2. SELWOOD, P. W.: Magnetochemistry, 2nd edit. Interscience Publ. Inc., New York–London 1956.
(Gibt eine gute Übersicht über das experimentelle Tatsachenmaterial, nur einige Grundzüge der Theorie werden behandelt, ausführliche Bibliographie.)
(Gives a good survey of the experimental results available, only a few main themes of the theory are treated, comprehensive bibliography).

3. Bates, L. F.: Modern Magnetism, 4th edit. Cambridge Univ. Press, Cambridge 1961.
(Ausführliche Darstellung der physikalischen Grundlagen des Magnetismus; die chemischen Anwendungen sind nur am Rande behandelt.)
(Detailed presentation of the physical basis of magnetism; chemical applications are treated only in passing.)

4. Klemm, W.: Magnetochemie. Akademische Verlagsgesellschaft, Leipzig 1936.
(Darstellung der experimentellen Grundlagen sowie der chemischen Anwendungen; auch heute noch als erste Einführung in die experimentellen Gegebenheiten zu empfehlen.)
(Presentation of the experimental basis as well as the chemical applications; still today to be recommended as a first introduction to the experimental situation.)

5. Stoner, E. C.: Magnetism and Matter. Methuen & Co. Ltd., London 1934.
(Darstellung der Grundlagen des Magnetismus.)
(Presentation of the fundamentals of magnetism.)

6. Bhatnagar, S. S. u. K. N. Mathur: Physical Principles and Applications of Magnetochemistry. Macmillan & Co., London 1935.

7. Earnshaw, A.: Introduction to Magnetochemistry. Academic Press, London and New York 1968.
Gives a short and well written summary of the present state of the interpretation of measurements of magnetic susceptibilities of coordination compounds in connection with ligand field theory, chemical bonding and stereochemistry. 112 pages.

*Important review articles on the magnetic behaviour of transition metal complexes*

1. Nyholm, R. S.: Magnetism and the Shape of Transition Metal Complexes. Rec. chem. Progr. **19**, No. 2 (1958) 45.
(Darstellung der Grundlagen des Magnetismus bei chemischen Verbindungen, ausführliche Diskussion der verschiedenen Typen von Paramagnetismus, Ligandenfeldtheorie und Magnetismus bei Übergangsmetallverbindungen, Bahnmomentbeitrag und Stereochemie, Kotanische Theorie.)
(Presentation of the fundamentals of magnetism for chemical compounds, detailed discussion of various types of paramagnetism, ligand field theory and magnetism of transition metal compounds, orbital moment contribution and stereochemistry, Kotani's theory.)

2. Nyholm, R. S.: Complex Compounds of the Transition Metals. Proc. 10th Solvay conf., p. 225. Stoops, Brüssel 1956.
(Valenzstrukturtheorie, Ligandenfeldtheorie, Diskussion des magnetischen Verhaltens auf der Grundlage der Ligandenfeldtheorie, Bahnmomentbeitrag und Stereochemie, Übersicht über die Valenzzustände, die Stereochemie und das magnetische Verhalten von Übergangsmetallkomplexen, 87 Literaturzitate.)
(Valence bond theory, ligand field theory, discussion of magnetic behaviour on the basis of the ligand field theory, orbital moment contribution and stereochemistry, survey of the valence states, the stereochemistry and the magnetic behaviour of transition metal complexes, 87 literature references.)

3. Nyholm, R. S.: Magnetism and Stereochemistry. J. Proc. Roy. Soc. New South Wales **89** (1955) 8.
(Physikalische Grundlagen, Para- und Diamagnetismus, Anwendung paramagnetischer Suszeptibilitätsmessungen auf chem. Probleme: 1. Bestimmung der Wertigkeit, 2. Frage nach der Bindung in Metallkomplexen, 3. Stereochemie, Bahnmomentbeiträge bei Ionen- und kovalenten Komplexen.)
(Physical fundamentals, para- and diamagnetism, application of paramagnetic susceptibilities to chemical problems; (1) Determination of valence, (2) Question of the bond in metal complexes, (3) Stereochemistry, orbital moment contributions for ionic and covalent complexes.)

4. Nyholm, R. S.: Magnetism and Inorganic Chemistry. Quart. Rev. (chem. Soc. London) **7** (1953) 377.
(Grundlagen des Para- und Diamagnetismus, Anwendung auf chem. Probleme, Diskussion der Verbindungen aus den verschiedenen Gruppen des periodischen Systems der Elemente, Bahnmomentbeitrag und Stereochemie, Paramagnetische Resonanzabsorption, 138 Literaturzitate.)
(Principles of para- and diamagnetism, application to the chemical problems, discussion of compounds of the various groups of the periodic system of the elements, orbital moment contribution and stereochemistry, paramagnetic resonance absorption, 138 references to the literature.)

5. Nyholm, R. S.: Magnetochemistry. J. inorg. nucl. Chem. **8** (1958) 401. Chemistry of the Co-ordinate Compounds, Symposium Rome 1957, p. 401. Pergamon Press, London–New York 1958.
(Übersicht über die Grundlagen, Diskussion des magnetischen Verhaltens von Übergangsmetallverbindungen auf der Grundlage der Ligandfeldtheorie, Kotanische Theorie, magnetische Suszeptibilität und Valenzprobleme, Bindungsverhältnisse und Stereochemie.)
(Survey of the principles, discussion of the magnetic behaviour of transition metal compounds on the basis of the ligand field theory, Kotani's theory, magnetic susceptibility and valence problems, bonding and stereochemistry.)

6. Cossee, P.: Magnetische Eigenschappen van Ionen van Overgangselementen (in holländischer Sprache). Chem. Weekblad **58**, 28 (1962) 337.
(In the Dutch language.)

7. Figgis, B. N. u. J. Lewis: The Magnetochemistry of Complex Compounds; in: Modern Coordination Chemistry (Principles and Methods) by J. Lewis and R. G. Wilkins, p. 400. Interscience Publ. Inc., New York–London 1960.
(Elementare Theorie des Magnetismus, experimentelle Methoden zur Messung der magnetischen Suszeptibilität, Paramagnetismus; magnetische Eigenschaften freier Ionen, Ligandenfelder, Bahnmomentbeitrag, low-spin-Komplexe der ersten, zweiten und dritten Übergangsmetallreihe, Bahnmomentbeiträge bei nicht oktaedrischen Komplexverbindungen, Temperaturabhängigkeit der Suszeptibilität, antiferromagnetische Wechselwirkungen, Actiniden, magnetische Momente bei Verbindungen der zweiten und dritten Übergangsmetallreihe; 120 Literaturzitate.)
(Elementary theory of magnetism, experimental methods of measuring the magnetic susceptibility, paramagnetism; magnetic properties of free ions, ligand fields, orbital moment contribution, low-spin complexes of the first, second and third transition metal series, orbital moment contributions for

nonoctahedral complexes, temperature dependence of the susceptibility, antiferromagnetic interaction, actinides, magnetic moments of compounds of the second and the third transition metal series; 120 literature references.)

8. KOIDE, S. u. OGUCHI, T.: Theories on the Magnetic Properties of Compounds. Advances chem. Physics V., p. 189, ed. by I. PRIGOGINE. Interscience Publ. Inc., New York–London 1963.
(Ausführliche Darstellung der Theorie, paramagnetische Suszeptibilität und Kristallfeldtheorie, *Jahn-Teller*-Effekt, kovalente Effekte, Austauschwechselwirkung, 103 Literaturzitate.)
(Detailed presentation of the theory, paramagnetic susceptibility and crystal field theory, *Jahn-Teller* effect, covalent effects, exchange interaction, 103 references to the literature.)
9. FIGGIS, B. N. u. J. LEWIS: The Magnetic Properties of Transition Metal Complexes, Progr. inorg. Chem. (Interscience) **6** (1964) 37.
(Grundlagen des Paramagnetismus bei Koordinationsverbindungen: Allgemeines, freie Ionen, Ligandenfelder, magnetische Anisotropie, Gleichgewicht magnetisch normaler und magnetisch anomaler Systeme, Antiferromagnetismus. Übersicht über magnetische Daten vom Übergangsmetallkomplexen, 670 Literaturzitate. Hervorragende, klar geschriebene Übersicht, die die modernen Gesichtspunkte berücksichtigt.)
(Fundamentals of paramagnetism of coordination compounds: general theory, free ions, ligand fields, magnetic anisotropy, equilibrium of high-spin and low-spin systems, antiferromagnetism. Survey of the magnetic data for transition metal complexes, 670 references to the literature. Superb, clearly written survey which considers the modern points of view.)

*Important original papers from the field of magnetochemistry*

1. VAN VLECK, J. H.: Valence Strength and the Magnetism of Complex Salts, J. chem. Physics **3** (1935) 807.
2. HOWARD, J. B.: The Principal Magnetic Susceptibilities of $K_3$ [$Fe(CN)_6$]. J. chem. Physics **3** (1935) 813.
3. SCHLAPP, R. u. W. G. PENNEY: Influence of Crystalline Fields on the Susceptibilities of Salts of Paramagnetic Ions. I. Physic. Rev. **41** (1932) 194; II. Physic. Rev. **42** (1932) 666.
4. VAN VLECK, J. H. u. W. G. PENNEY: The Theory of the Paramagnetic Rotation and Susceptibility in Manganous and Ferric Salts. Philos. Mag. **17** (1934) 961.
5. STEVENS, K. W. H.: On the Magnetic Properties of Covalent $XY_6$ Complexes. Proc. Roy. Soc. **A 219** (1953) 542.
6. GRIFFITH, J. H. E., J. OWEN u. I. H. WARD: Paramagnetic Resonance in Palladium and Platinum Group Compounds. Proc. Roy. Soc. **A 219** (1953) 526.
7. OWEN, J.: The Colours and Magnetic Properties of Hydrated Iron Group Salts and Evidence for Covalent Bonding. Proc. Roy. Soc. **A 227** (1955) 183.
8. OWEN, J.: The Magnetic Evidence for Charge Transfer in Octahedral Complexes. Discuss. Faraday Soc. **19** (1955) 127.
9. VAN VLECK, J. H.: The Significance of the Results of Microwave Spectroscopy to the Theory of Magnetism. Ann. New York Academy of Sciences **55** Art. 5 (1952) 928.

10. Bleaney, B., K. D. Bowers u. D. J. E. Ingram: Paramagnetic Resonance in Diluted Copper Salts. I. Hyperfine Structure in Diluted Copper Tutton Salts. Proc. Roy. Soc. **A 228** (1955) 147.

11. Bleaney, B., K. D. Bowers u. R. S. Trenam: II. Salts with Trigonal Symmetry. Proc. Roy. Soc. **A 228** (1955) 157.

12. Bleaney, B., K. D. Bowers u. M. H. L. Pryce: III. Theory and Evaluation of the Nuclear Electric Quadrupole Moments of $^{63}Cu$ and $^{65}Cu$. Proc. Roy. Soc. **A 228** (1955) 166.

13. Abragam, A. u. M. H. L. Pryce: Theory of the Nuclear Hyperfine Structure of Paramagnetic Resonance Spectra in Crystals. Proc. Roy. Soc. **A 205** (1951) 135.

14. Griffith, J. H. E. u. J. Owen: Complex Hyperfine Structure in Microwave Spectra of Covalent Iridium Compounds. Proc. Roy. Soc. **A 226** (1954) 96.

15. Abragam, A. u. M. H. L. Pryce: The Theory of Paramagnetic Resonance in Hydrated Cobalt Salts. Proc. Roy. Soc. **A 206** (1951) 173.

16. Griffith, J. H. E. u. J. Owen: Paramagnetic Resonance in the Nickel Tutton Salts. Proc. Roy. Soc. **A 213** (1952) 459.

17. Abragam, A. u. M. H. L. Pryce: The Theory of the Nuclear Hyperfine Structure of Paramagnetic Resonance Spectra in the Copper Tutton Salts. Proc. Roy. Soc. **A 206** (1951) 164.

18. Kotani, M.: On the Magnetic Moment of Complex Ions. J. phys. Soc. (Japan) **4** (1949) 293.

19. Griffith, J. S.: Octahedral Complexes. I. Magnetic Susceptibilities in the First Transition Series. Trans. Faraday Soc. **54** (1958) 1109.

20. Figgis, B. N.: Magnetic Properties of Transition Metal Ions in Asymmetric Ligand Fields. 1. Cubic Fields $A$ and $E$ terms. Trans. Faraday Soc. **56** (1960) 1553.
2. Cubic Fields $^2T_2$ Terms. Trans. Faraday Soc. **57** (1961) 198.
3. The Behaviour of Some $t_{2g}^5$ Complexes. Trans. Faraday Soc. **57** (1961) 204.

21. Figgis, B. N.: Magnetic Properties of Spin-Free Transition Series Complexes. Nature (London) **182** (1958) 1568.

22. Figgis, B. N., J. Lewis, R. S. Nyholm u. R. D. Peacock: The Magnetic Properties of Some $d\varepsilon^3$, $d\varepsilon^4$ and $d\varepsilon^5$ Configurations. Discuss. Faraday Soc. **26** (1958) 103.

23. Dunn, T. M.: Spin-Orbit Coupling in the First and Second Transition Series. Trans. Faraday Soc. **57** (1961) 1441.

24. Kamimura, H.: On the Magnetic Moment and g-Value of Complex Ions. J. phys. Soc. (Japan) **11** (1956) 1171.

25. Figgis, B. N., J. Lewis u. F. Mabbs: The Magnetic Properties of Vanadium (III) Complexes. J. chem. Soc. (London) (1960) 2480.

26. Earnshaw, A. u. J. Lewis: Polynuclear Compounds. I. Magnetic Properties of some Binuclear Complexes. J. chem. Soc. (London) (1961) 396.

27. Figgis, B. N., J. Lewis u. F. Mabbs: The Magnetic Properties of Some $d^3$-Complexes. J. chem. Soc. (London) (1961) 3138.

28. EARNSHAW, A., B. N. FIGGIS, J. LEWIS u. R. D. PEACOCK: The Magnetic Properties of Some $d^4$-Complexes. J. chem. Soc. (London) (1961) 3132.

29. FIGGIS, B. N. u. C. M. HARRIS: Magnetochemistry III. Compounds of Bivalent Copper. J. chem. Soc. (London) (1959) 855.

30. BALLHAUSEN, C. J. u. A. D. LIEHR: Some Comments on Anomalous Magnetic Behaviour of Certain Ni (II) Complexes. J. Amer. chem. Soc. **81** (1959) 538.

## *G.2. Optical spectra*

(Cf. also the review articles under F.1–10, 15, 16 and 18.)

*Several important original papers on ligand field theory and optical spectra*

1. ILSE, F. E. u. H. HARTMANN: Termsysteme elektrostatischer Komplexe der Übergangsmetalle mit einem *d*-Elektron. Untersuchungen über die Absorptionsspektren von Komplexverbindungen, 1. Mitteilung. Z. physik. Chem. **197** (1951) 239. (Diskussion des $d^1$-Problems, $Ti^{3+}$.)
   (Discussion of the $d^1$ problem, $Ti^{3+}$.)

2. ILSE, F. E. u. H. HARTMANN: Termsystem eines Ions mit zwei *d*-Außenelektronen im elektrischen Feld oktaedrischer Symmetrie. Farbe und Konstitution von Komplexverbindungen. 3. Mitteilung. Z. Naturforsch. **6 a** (1951) 751. (Diskussion des $d^2$-Problems, $V^{3+}$.)
   (Discussion of the $d^2$ problem, $V^{3+}$.)

3. HARTMANN, H. u. H. L. SCHLÄFER: Über die Absorptionsspektren elektrostatischer Komplexionen dreiwertiger Übergangselemente mit oktaedrischer Symmetrie. Farbe und Konstitution von Komplexverbindungen, 5. Mitteilung. Z. Naturforsch. **6 a** (1951) 760.
   ($Ti^{3+}$, $V^{3+}$, $Mn^{3+}$ und $Fe^{3+}$, vereinfachte Theorie der Spektren.)
   ($Ti^{3+}$, $V^{3+}$, $Mn^{3+}$ and $Fe^{3+}$, simplified theory of spectra.)

4. TANABE, Y. u. S. SUGANO: On the Absorption Spectra of Complex Ions I and II. J. phys. Soc. (Japan) **9** (1954) 753, 766.
   (Termdiagramme $\frac{E}{B} = f\left(\frac{Dq}{B}\right)$ sowie Matrixelemente für oktaedrische Komplexe der Konfiguration $d^2$ bis $d^8$, Methode des starken Feldes mit Konfigurationswechselwirkung.)
   (Term diagrams $E/B = f(Dq/B)$ as well as matrix elements for octahedral complexes of the configuration $d^2$ to $d^8$, Strong-field method with configuration iteration.)

5. ORGEL, L. E.: Spectra of Transition Metal Complexes. J. chem. Physics **23** (1955) 1004.
   (Methode des schwachen Feldes, Diagramm $E = f(Dq)$, $d^2$, $d^8$, $d^3$, $d^7$, $d^5$ oktaedrische Symmetrie.)
   (Weak-field method, diagram $E = f(Dq)$, $d^2$, $d^8$, $d^3$, $d^7$, $d^5$ octahedral symmetry.)

6. ORGEL, L. E.: The Effects of Crystal Fields on the Properties of Transition Metal Ions. J. chem. Soc. (London) (1952) 4756.
   ($d^1$, $d^2$, $d^3$, $d^6$, schwache Felder, $O_h$-Symmetrie.)
   ($d^1$, $d^2$, $d^3$, $d^6$, weak fields, $O_h$ symmetry.)

7. ORGEL, L. E.: Band Widths in the Spectra of Manganous and Other Transition-Metal Complexes. J. chem. Physics **23** (1955) 1824.
(Diskussion der Halbwertsbreiten, speziell bei $[Mn(H_2O)_6]^{2+}$.)
(Discussion of half-widths, especially for $[Mn(H_2O)_6]^{2+}$.)

8. ORGEL, L. E.: Electronic Structures of Transition Metal Complexes. J. chem. Physics **23** (1955) 1819
(Interpretation von $Dq$, Werte von $F_2$ und $F_4$ für verschiedene Metallionen.)
(Interpretation of $Dq$, values of $F_2$ and $F_4$ for various metallic ions.)

9. FINKELSTEIN, R. u. J. H. VAN VLECK: On the Energy Levels of Chrome Alum. J. chem. Physics **8** (1940) 790.
(Matrixelemente für $d^3$, $O_h$-Symmetrie, Rechnung nach der Methode des schwachen Feldes mit Termwechselwirkung für die Quartett- und Dublett-Terme, Interpretation des »Chrom-Dubletts« als Übergang $^4A_{2g} \rightarrow {}^2E_g$.)
(Matrix elements for $d^3$, $O_h$ symmetry, calculation using the weak-field method with term interaction for the quartet and doublet terms, interpretation of the "chromium doublet" as the transition $^4A_{2g} \rightarrow {}^2E_g$.)

10. BJERRUM, J., C. J. BALLHAUSEN u. C. K. JØRGENSEN: Studies on Absorption Spectra.
I. Results of Calculation on the Spectra and Configuration of Copper (II) Ion. Acta chem. scand. **8** (1954) 1275.

11. BALLHAUSEN, C. J.: II. Theory of Copper (II) Spectra. Kgl. Danske Vidensk. Selsk. mat. fysiske Medd. **29** Nr. 4 (1954).

12. JØRGENSEN, C. K.: IV. Some New Transition Group Bands of Low Intensity. Acta chem. scand. **8** (1954) 1502.

13. BALLHAUSEN, C. J.: V. The Spectra of Nickel (II) Complexes. Kgl. Danske Vidensk. Selsk. mat. fysiske Medd. **29** Nr. 8 (1955).

14. JØRGENSEN, C. K.: VIII. Three and More $d$-Electrons in Cubic Crystal Fields. Acta chem. scand. **9** (1955) 116.

15. BALLHAUSEN, C. J. u. C. K. JØRGENSEN: IX. The Spectra of Cobalt (II) Complexes. Acta chem. scand. **9** (1955) 397.

16. BALLHAUSEN, C. J. u. C. K. JØRGENSEN: X. $d$-Electrons in Crystal Fields of Different Symmetry. Kgl. Danske Vidensk. Selsk. mat. fysiske Medd. **29** Nr. 14 (1955).

17. LIEHR, A. u. C. J. BALLHAUSEN: Intensities in Inorganic Complexes. Physic. Rev. **106** (1957) 1161.

18. LIEHR, A. D. u. C. J. BALLHAUSEN: Complete Theory of Ni (II) and V (III) in Cubic Crystalline Fields, Octahedral Complexes. Ann. Physics **6** (1959) 134.
(Behandlung mit Spin-Bahn-Kopplung.)
(Treatment with spin-orbit coupling.)

19. BALLHAUSEN, C. J. u. A. D. LIEHR: Intensities in Inorganic Complexes III. Octahedral Complexes of Ni (II) and V (III). Mol. Physics **2** (1959) 123.

20. JØRGENSEN, C. K.: Some Problems of Intensity of Absorption Bands and Chemical Bonding. Acta chem. scand. **9** (1955) 405.

21. BALLHAUSEN, C. J.: On Some Intensity Problems in the Absorption Bands of Complex Ions. Acta chem. scand. **9** (1955) 821.

22. BASOLO, F., C. J. BALLHAUSEN u. J. BJERRUM: Absorption Spectra of Geometrical Isomers of Hexa Coordinated Complexes. Acta chem. scand. 9 (1955) 810. (Spektren von [Co(III)$A_6$]- sowie trans- und cis-[Co(III)$A_4B_2$].) (Spectra of [Co(III)$A_6$] and trans- and cis-[Co(III)$A_4B_2$].)

23. HARTMANN, H. u. H.-H. KRUSE: Über die Absorptionsspektren der Komplexverbindungen des dreiwertigen Chroms. Z. phys. Chem. (Neue Folge) **5** (1955) 9. (Termsystem für $O_h$, $D_{4h}$, $C_{4v}$ und $C_2$ Symmetrie.) (Term system for $O_h$, $D_{4h}$, $C_{4v}$ and $C_2$ symmetry.)

24. GRIFFITH, J. S. u. L. E. ORGEL: Spectra of Transition Metal Complexes of the Type $[Co(NH_3)_5X]^{2+}$. J. chem. Soc. (London) (1956) 4981.

25. JØRGENSEN, C. K.: Comparative Ligand Field Studies. III. The Decrease of the Integrals $F^k$ For Manganese (II) Complexes, Evidence for Partly Covalent Bonding. Acta chem. scand. **11** (1957) 53.

26. JØRGENSEN, C. K.: The Interelectronic Repulsion Integrals and the Purity of Molecular Orbital Configurations, as Derived from Absorption Spectra of Transition Group Complexes. Acta chem. scand. **12** (1958) 903.

27. HARTMANN, H. u. H. FISCHER-WASELS: Über die Änderungen des Termsystems des $Ni^{++}$-Ions beim Einbau in tetraedrische und tetragonale Komplexe. Z. phys. Chem. (Neue Folge) **4** (1955) 20.

28. HARTMANN, H. u. C. FURLANI: Über die Zuordnung der Banden bei den Spektren komplexer Vanadin(III)-Verbindungen. Z. phys. Chem. (Neue Folge) **9** (1956) 162.

*Nephelauxetic effect*

1. SCHÄFFER, C. E. u. C. K. JØRGENSEN: The Nephelauxetic Series of Ligands Corresponding to Increasing Tendency of Partly Covalent Bonding. J. inorg. nucl. Chem. **8** (1958) 143; Chemistry of the co-ordinate compounds, Symposium Rom 1957, p. 143. Pergamon Press, London–New York 1958.

2. JØRGENSEN, C. K.: The Nephelauxetic Series; Progr. inorg. Chem. (Interscience) **4** (1962) 73. (Ausführlicher Reviewartikel, viele Tabellen, $\beta_{35}$ und $\beta_{55}$ für zahlreiche Komplexionen, 104 Literaturzitate.)

Siehe auch unter E.3, E.4, F.6, F.18.

(Detailed review article, many tables, and for numerous complex ions, 104 literature references.)

(Cf. also under E.3, E.4, F.6, F.18.)

## *G.3. Thermodynamic stability of transition metal complexes (ligand field stabilization)*

*(Ligand field stabilization)*

1. GEORGE, P. u. D. S. MCCLURE: The Effect of Inner Orbital Splitting on the Thermodynamic Properties of Transition Metal Compounds and Coordination Complexes. Progr. inorg. Chem. (Interscience) **1** (1959) 381. (Zusammenfassende Darstellung, 36 Literaturzitate.) (A general survey, 36 literature references.)

2. siehe auch unter F.11 und F.12.
See also F.11 and F.12.

3. JØRGENSEN, C. K. u. J. BJERRUM: Crystal Field Stabilization of First Transition Group Complexes. Acta chem. scand. **9** (1955) 180.

4. GEORGE, P., D. S. MCCLURE, J. S. GRIFFITH u. L. E. ORGEL: Ligand Field Theory and the Stability of Transition Metal Compounds. J. chem. Physics **24** (1956) 1269.

5. GRIFFITHS, J. S.: On the Stabilities of Transition Metal Complexes. Theory of Energies. J. inorg. nucl. Chem. **2** (1956) 1.

Vgl. auch die entsprechenden Abschnitte in G.6, Zitat 1.
(Cf. also the corresponding sections in G.6, Ref. 1.)

## *G.4. Ionic radii*

1. VAN SANTEN, J. H. u. J. S. VAN WIERINGEN: Some Remarks on the Ionic Radii of Iron-Group Elements. The influence of the Crystalline Field. Recueil Trav. chim. Pays-Bas **71** (1952) 240.
(Grundlegende Arbeit über die kubischen Ionenradien ($O_h$, $T_d$) der Übergangsmetallionen der Reihe $d^1$–$d^{10}$, Diskussion an Hand der $d\varepsilon^n d\gamma^{N-n}$ Konfigurationen.)
(A fundamental paper on the cubic ionic radii ($O_h$, $T_d$) of the transition metal ions of the series $d^1$–$d^{10}$, discussion on the basis of the $d\varepsilon^n d\gamma^{N-n}$ configurations.)

2. HUSH, N. S. u. M. H. PRYCE: Influence on the Crystal-Field Potential on Interionic Separation in Salts of Divalent Iron-Group Ions. J. chem. Physics **28** (1958) 244.
($d^1$–$d^9$ Ionen, Gang der Radien.)
($d^1$–$d^9$ ions, range of the radii.)

3. HUSH, N. S. u. M. H. L. PRYCE: Radii of Transition Ions in Crystal Fields. J. chem. Physics **26** (1957) 143.

4. SCHÄFFER, C. E.: Ligand Field Theory and the Stabilization of Ions. Lecture at »Internationale Ferienkurse über Theoretische Chemie«, Konstanz 1962.
(Außer den Ionenradien werden die Kristallfeldstabilisierungsenergien behandelt.)
(Besides the ionic radii the crystal field stabilization energies are also treated.)

5. LIEHR, A. D.: »Ionic Radii«, Spin-Orbit Coupling and the Geometrical Stability of Inorganic Complexes. Bell System techn. J. **89** (1960) 1617.
(Ionenradien, Einfluß der Spin-Bahn-Kopplung, 30 Literaturzitate.)
(Ionic radii, influence of spin-orbit coupling, 30 literature references.)

## *G.5. Stereochemistry of transition metal complexes*

1. DUNITZ, J. D. u. L. E. ORGEL: Stereochemistry of Ionic Solids. Adv. Inorg. Chem. Radiochem. **2** (1960) 1.
(Ligandenfeldtheoretische Gesichtspunkte in Kap. IV, S. 12.)
(Ligand field theoretical points of view in Chap. IV, p. 12.)

2. NYHOLM, R. S.: The Stereochemistry of Complex Compounds. Progress in Stereochemistry I, hsg. von W. KLYNE. Butterworths, London 1954.

(S. 322, darin Abschnitt über Ligandenfeldtheorie und Stereochemie.)
(P. 322, section on ligand field theory and stereochemistry.)

3. Gillespie, R. J. u. R. S. Nyholm: The Stereochemistry of Inorganic Molecules and Complex Ions. Progress in Stereochemistry II, S. 261. hsg. von P. B. D. de la Mare u. W. Klyne. Butterworths, London 1958.
(Abschnitt S. 288 über Ligandenfeldtheorie und Stereochemie.)
(Section on p. 288 on ligand field theory and stereochemistry.)

4. Dunitz, J. D. u. L. E. Orgel: Electronic Properties of Transition-Metal Oxides I. Distortion from Cubic Symmetry. J. phys. chem. Solids **3** (1957) 20.
(Diskussion der Abweichungen von der kubischen Symmetrie bei bestimmten Übergangsmetalloxiden auf der Basis der Ligandenfeldtheorie unter Berücksichtigung des *Jahn-Teller*-Theorems.)
(Discussion of the deviations from cubic symmetry for certain transition metal oxides on the basis of the ligand field theory with consideration of the *Jahn-Teller* theorem.)

5. Dunitz, J. D. u. L. E. Orgel: Electronic Properties of Transition-Metal Oxides II. Cation Distribution amongst Octahedral and Tetrahedral Sites. J. phys. chem. Solids **3** (1957) 318.

Eine gute Übersicht über die Strukturen anorganischer Verbindungen gibt
A good survey of the structures of inorganic compounds is given by

6. Wells, A. F.: Structural Inorganic Chemistry, 3rd ed., Oxford University Press, London 1962.
(Abschnitt über Ligandenfeldtheorie, S. 86 ff.)
(Section on ligand field theory, pp. 86 ff.)

7. Nyholm, R. S.: Stereochemistry of Complex Halides of the Transition Metals. Croatica chem. Acta **33** (1961) 157.
(Diskussion auf der Grundlage der Kristallfeldtheorie.)
(Discussion on the basis of the crystal field theory.)

## *G.6. Reactions with transition metal complexes*

(*Kinetic stability*)

1. Basolo, F. u. R. G. Pearson: Mechanisms of Inorganic Reactions. A Study of Metal Complexes in Solution. John Wiley & Sons, New York–London 1958.
(Ausführliche monographische Darstellung, in der die Bedeutung ligandenfeldtheoretischer Gesichtspunkte bezüglich der Reaktionskinetik behandelt wird; zahlreiche Tabellen, ausgedehnte Bibliographie; für den an reaktionskinetischen Problemen interessierten Komplexchemiker unentbehrlich.)
(Detailed monographic presentation in which the importance of the ligand field theoretical point of view with reference to reaction kinetics is treated; numerous tables, extensive bibliography; indispensable for the chemist interested in the problems of the reaction kinetics of complexes.)

2. Pearson, R. G.: Crystal Field Theory and Substitution Reactions of Metal Ions. J. chem. Educat. **38** (1961) 164.
(Übersichtsartikel, knappe Darstellung der wichtigsten Grundlagen.)
(Review article, concise presentation of the most important principles.)

3. Dowden, D. A. u. D. Wells: A Crystal Field Interpretation of some Activity Patterns. Proceedings 2ème Congrès International de Catalyse, Paris 1960, S. 1499. Edition Tèchnipes, Paris 1960.
(Heterogene Katalyse und Kristallfeldtheorie bei Katalysatoren, die Übergangsmetallionen enthalten.)
(Heterogeneous catalysis and crystal field theory of catalysts containing transition metal ions.)

### *G.7. Jahn-Teller effect*

1. Jahn, H. A. u. E. Teller: Proc. Roy. Soc. (London) **A 161** (1937) 220.
2. Jahn, H. A.: Proc. Roy. Soc. (London) **A 164** (1938) 117.
3. Öpik, U. u. M. H. L. Pryce: Proc. Roy. Soc. (London) **A 238** (1957) 425.
4. Ballhausen, C. J.: *Jahn-Teller* Configurational Instability of Inorganic Complexes. Lecture at »Internationale Ferienkurse für theoretische Chemie«, Konstanz 1962.
(50 Literaturzitate, auch dynamischer Effekt behandelt.)
(50 literature references, also the dynamic effect is treated.)
5. Bersuker, I. B.: Restricted Motions in Transition Metal Complexes. Optics and Spectroscopy **11** (1961) 172.
6. Avvakumov, V. I.: Distortion of the Complexes $[Cr(H_2O)_6]^{3+}$ and $[Ni(H_2O)_6]^{2+}$ and Splitting of the Spin Levels of $Cr^{3+}$ and $Ni^{2+}$ by the *Jahn-Teller*-Effect. Optics and Spectroscopy **13** (1962) 332.

vgl. auch entsprechende Kapitel in E. 1–6.
(Cf. also the corresponding chapters of E. 1–6.)

7. Liehr, A. D.: Topological Aspects of the Conformational Stability Problem. I. Degenerate Electronic States; II. Nondegenerate Electronic States. J. physic. Chem. **67** (1963) 389, 471.
8. Liehr, A. D.: The Coupling of Vibrational and Electronic Motions in Degenerate Electronic States of Inorganic Complexes. I. II. Progr. inorg. Chem. (Interscience) **3** (1961) 281; **4** (1962) 455.

vgl. auch Annu. Rev. physic. Chem. **13** (1962) 41.
(Cf. also Ann. Rev. phys. chem. **13** (1962) 41.)

## H. Computational aids

1. Furlani, C.: Auto-funzioni di alcune configurazioni $d^n$ in accoppiamento Russell-Sanders a loro proprietà di transformazione. Università degli studi di Trieste. Instituto di Chimica N. 14, 1957.
(Eigenfunktionen $\Psi$ ($L$, $S$, $M_L$, $M_S$) für $d^1$ bis $d^9$, Angabe der Termlagen der Ionen als Funktion der *Slater*-Integrale $F_2$ und $F_4$. Transformationsverhalten für die Symmetrien $O_h$ bzw. $T_d$, $D_{4h}$ bzw. $C_{4v}$, $D_{\infty h}$ bzw. $C_{\infty v}$.)
(Eigenfunctions $\Psi$ ($L$, $S$, $M_L$, $M_S$) for $d^1$ to $d^9$, tabulation of the term positions of the ions as a function of the *Slater* integrals $F_2$ and $F_4$. Transformation behaviour of the symmetries $O_h$ or $T_d$, $D_{4h}$ or $C_{4v}$, $D_h$ or $C_v$.)
2. Hartmann, H. u. E. König: Matrixelemente des Ligandenfeldpotentials in Komplexverbindungen der Übergangsmetalle. Z. physik. Chem. (Neue Folge) **28** (1961) 425.

Matrixelemente $(m_1|V|m'_1)$ für $m_1 = 2, 1, 0, -1, -2)$.
(Matrix elements $(m_1|V|m'_1)$ for $m_1 = 2, 1, 0, -1, -2)$.)

3. Ballhausen, C. J. u. E. M. Ancom: Kgl. Danske Vidensk. Selsk. mat. fysiske Medd. **31** Nr. 9 (1958).
$$\left(\text{Zahlenwerte der Integrale } G_k = \int_0^\infty R^2_{\text{nd,nd}}(r)\frac{r^k_<}{r^{k+1}_>}r^2\,dr.\right)$$
$$\left(\text{Numerical values of the integrals } G_k = \int_0^\infty R^2_{\text{nd,nd}}(r)\frac{r^k_<}{r^{k+1}_>}r^2\,dr.\right)$$

4. Kruglyak, Yu. A.: Rechenmethoden in der Quantenchemie. IV. Radialintegrale in der Ligandenfeldtheorie. Zhur. Strukt. Khim. **5** (1964) 765 (Russ.).

## I. Miscellaneous

1. Schläfer, H. L.: Photochemie von Übergangsmetallkomplexen, insbesondere Chrom(III)-Komplexen. Z. physik. Chem. (Neue Folge) **11** (1957) 65; Acta chim. Acad. Sci. hung. **18** (1959) 375; Z. Elektrochem., Ber. Bunsenges. physik. Chem. **64** (1960) 887; J. physic. Chem. **69** (1965) 2201.

2. Porter, G. B. u. H. L. Schläfer: Lumineszenz von Übergangsmetallverbindungen. Ber. Bunsenges. physik. Chem. **68** (1964) 316.
(Zusammenfassender Artikel, 50 Literaturzitate.)
Z. physik. Chem. (Neue Folge) **37** (1963) 109; **38** (1963) 227; **40** (1964) 280.
(Review article with 50 literature references.)

# Subject Index


Absorption Spectra
of dissolved halogen ions, interpretation as charge-transfer spectra 5
of various free ligands 8, 9
of complex ions of low symmetry 34 ff
of a transition metal complex ion schematic 4, 5
Absorption spectra of selected compounds, graphical representation, (unless otherwise indicated, they are solution spectra)
$Cr^{3+}/\alpha-Al_2O_3$
Ruby crystalline spectrum 45
$[CrCl_3py_3]$ reflectance spectrum 12
$[Cr(C_2O_4)_5]^{3-}$ 6
$[Cr(H_2O)_6]^{3+}$ 42
$[Cr(NH_3)_6]^{3+}$ 35
$[Cren_3]^{3+}$ 35, 43
$[Cr(NH_3)_5X]^{2}+X = Cl, Br, J$ 37, 38
$[Cren_2Cl_2]^{+}$ *cis-* and *trans-* 39
$[Co(H_2O)_6]^{2+}$ 86
$[CoCl_4]^{2-}$ 86
$[Coen_3]^{3+}$ 16, 92
$[Co(NH_3)_6]^{3+}$ 16, 92
$[Co(NH_3)_5X]^{2+}X = F, Cl, Br, J$ 111
$NH_4Fe(SO_4)_2 . 12H_2O$ crystal spectrum 49
Ligands, free
$Cl^-, Br^-, J^-, ClO_4^-, C_2O_4^2, H_2O, NH_3$, en, $SO_4^{2-}, S_2O_6^{2-}, NO_3^-$ 8
Pyridin 8
$MnCl_2 . 4H_2O$ crystal spectrum 46
$[Mn(H_2O)_6]^{2+}$ 46, 93
$MnO_4^-$ 12
$K_3[MoCl_6]$ reflectance spectra 11
$(NH_4)_2[MoCl_5(H_2O)]$ reflectance spectra 11
$[MoCl_3py_3]$ reflectance spectrum 12
$[Ti(H_2O)_6]^{3+}$ 10
$[Ti(Co(NH_2)_2)_6]J_3$ crystalline spectrum 30
reflectance spectrum 31
$TiCl_3 . 6H_2O$ remission spectrum 32
$[TlCl_3py_3]$ reflectance spectrum 12
$CsV(SO_4)_2 . 12H_2O$, $NH_4V(SO_4)_2 . 12H_2O$ crystalline spectra 10
Activation energy 181 ff
and Arrhenius relation 185
Ligand field contribution to
$S_{N1}$ mechanism 183
$S_{N2}$ mechanism 184
Angular momentum 132, 217 ff
Anistropy
of the absorption of complex compounds of low symmetry
of the *g* factor 174
of susceptibility 140
Antiferromagnetic interactions 138
Archimedean (tetragonal) antiprism, symmetry $D_{4d}$ 15, 63, 64
Arrhenius relation for the reaction rate constant 185
Atomic units 29, 498
Atomic orbitals, in M.O. considerations 97 ff

Back donation 114
Binding criterion
magnetic 124, 130, 188
Bipyramid
pentagonal as a transition state for an $S_{N2}$ mechanism 181 ff
trigonal as a transition state for an $S_{N1}$ mechanism 181
Bohr magneton (cf. magneton)
Bohr radius 98
Boltzmann distribution 135, 138, 193, 460
Bond
electrovalent 13, 18, 124, 186 ff
covalent 14 f, 18, 82, 83 f, 121, 130, 186 ff

Bonding of complexes 186 ff
bonds
  $\pi$ 17, 82, 84, 100 ff, 105 ff, 114, 171, 195, 416 ff
  $d\pi \rightarrow p\pi$ 82
  $p\pi \rightarrow d\pi$ 82, 107, 108
  $\sigma$ 82, 84, 100 ff, 114, 130, 411 ff
  orbitals of the central ion and orbitals of the ligands
bond strength 485, 488, 490, 493
Born–Oppenheimer approximation 494

Centre of gravity rule 60, 63, 75, 360, 361
Central field 207
Central field approximation 222 ff, 242
Central field potential 224, 310, 408
Central ion bands 4, 116
Character of a matrix 472, 473
Character tables for the point symmetry groups
  $O_h$, $D_{4h}$, $C_{4u}$, Td, $C_3$, O′, $C_{2u}$ 479–480
Charge distribution, spherically symmetric 142, 150, 151, 153, 159, 161
Charge, effective 23
Charge-transfer bands (cf. electronic transition bands)
Charge-transfer spectra (cf. electronic transition spectra)
Chemical shift 200
Chromium doublet 22, 43
Compounds, mentioned
  Acetyl acetonate 200
  Acetyl complexes 204
  Alums
    chromium (III) alum 22, 43, 45, 73, 203
    iron (III) alum 49
    titanium (III) alum 173
    vanadium (III) alum 9, 10, 203
  $[Al(C_2O_4)_3]^{3-}$ 57
  Arsine complexes 106
  Carbonyls 9, 11, 186, 187, 204
  $Cr(CO)_6$ 186
  $[Cr\ dip_3]^+$ 178
  $[Cr(CN)_6]^{4-}$ 178
  $[Cr\ dip_3]^{2+}$ 178
  $CrF_2$ 172
  $[Cr(H_2O)_6]^{2+}$ 149
  $[Cr(NH_3)_6]^{2+}$ 149
  Chromium alum 22, 43, 45, 73
  $Cr^{3+}/\alpha\text{–}Al_2O_3$ (ruby) 43, 45, 46
  $[Cr\ Cl_3\ py_3]$ 9, 12, 34
  $[Cr(C_2O_4)_3]^{3-}$ 5, 6, 7, 35, 42
  $[Cr\ en_3]^{3+}$ 35–37, 174
  $[Cr\ tn_3]^{3+}$ 174, 268
  $[Cr\ en_2Cl_2]^+$ *cis-* and *trans-* 37 ff, 86
  $[Cr(H_2O)_6]^{3+}$ 29, 42, 72, 74, 109
  $[Cr\ malonate_3]^{3-}$ 42
  $[Cr(CO(NH_2)_2)_6]^{3+}$ 264
  $[Cr(NH_3)_6]^{3+}$ 29, 35–37, 85, 186, 264
  $[Cr(NH_3)_5X]^{2+}$ X = Cl, Br 36 ff, 266
  $[Cr(NH_3)_4X_2]^+$ *cis-* and *trans-* X = Cl, Br 38
  $CrO_4^{2-}$ 117
  $[CoCl_4]^{2-}$ 85, 86, 96, 267
  $(pyH)_2[CoCl_4]$ 136
  $[CoCl_2, 2(C_2H_5)_3P]$ 136
  $[CoCl_2py_2]$ 136
  $[Co\ dip_3](ClO_4)_2$ 136
  $(pyH)_2[CoBr_4]$ 136
  $[Co(H_2O)_6]^{2+}$ 86, 123
  $[Co(H_2O)_6]Cl_2$ 136
  $[Co(H_2O)_6](ClO_4)_2 . 6H_2O$ 136
  $[Co(NH_3)_6]^{2+}$ 86
  $[Co(NH_3)_6](ClO_4)_2$ 136
  $[Co\ py_6](ClO_4)_2$ 136
  $Hg[Co(NCS)_4]$ 136
  $[Co(NCS)_4]^{2-}$ 267
  $[Co\ en_3]^{3+}$ 2, 16, 92, 149, 268
  $[Co\ en_2Cl_2]^+$ *cis-* and *trans-* 87, 88
  $[CoF_6]^{3-}$ 50, 123, 126, 129, 178, 264
  $[Co(H_2O)_6]^{3+}$ 50, 125, 149, 176, 188
  $[Co(NH_3)_6]^{3+}$ 2, 14, 15, 16, 40, 50, 87, 91, 92, 119, 127, 149, 176, 184
  $[Co(NH_3)_5Cl]^{2+}$ 2, 110, 111
  $[Co(NH_3)_5Br]^{2+}$ 2, 266
  $[Co(NH_3)_5H_2O]^{3+}$ 2
  $[Co(NH_3)_5N_3]^{2+}$ 112, 118
  $[Co(NH_3)_4Cl_2]^+$ *cis-* and *trans-* 2, 87, 265, 269
  *trans-* $[Co(NH_3)_4Br_2]^+$ 265, 275
  $[Co(NH_3)_4(N_3)_2]^+$ 112, 118
  $[Co(NH_3)_4(NO_2)_2]^+$ *cis-* and *trans-* 2
  *trans-* $[Co(NH_3)_3(H_2O)_3]^{3+}$ 269
  $[Co(NH_3)_3(N_3)_3]$ 112, 118
  $[Co(C_2O_4)_3]^{3-}$ 268
  $[CuCl_4]^{2-}$ 85, 267
  $[Cu\ en_2(H_2O)_2]^{2+}$ 156
  $[Cu\ X_2py_2]X$ = Cl, Br 110
  $CuF_2$ 172

$NaCuF_3$ 172
$K_2CuF_4$ 172
$[Cu(H_2O)_6]^{2+}$ 79, 141
$[Cu(H_2O)_6]\,SiF_6$ 174
$[Cu(NH_3)_5H_2O]^{2+}$ 266
$[Cu(CN_4]^{2-}$ 267
$CuSO_4 . 5H_2O$ 79
Cyano complexes 9, 14, 15, 50, 82, 120, 127, 140, 177, 178, 194, 264, 265, 266
$\alpha,\alpha'$-dipyridyl complexes 7, 11, 178, 204
$[Fe(CN)_6]^{4-}$ 50, 127, 129, 178
$[Fe\ dip_3]^{2+}$ 178, 188
$[FeF_6]^{4-}$ 195
$[Fe(H_2O)_6]^{2+}$ 50, 173
$[Fe(NH_3)_6]^{2+}$ 50, 123, 127, 129
$[Fe\ phen_3]^{2+}$ 178
iron alum 49
$[Fe(CN)_6]^{3-}$ 14, 119, 140, 178, 264
$[FeF_6]^{3-}$ 127
$[FeCl_4]^{-}$ 159
$[FeCl_6]^{3-}$ 159
$[Fe\ dip_3]^{3+}$ 178
$[Fe(H_2O)_6]^{3+}$ 46 ff, 87, 123, 149, 158
$[Fe(NH_3)_6]^{3+}$ 149
$[Fe\ phen_3]^{3+}$ 178
$H_2$ 194
$[IrCl_6]^{2-}$ 194, 196
$[IrF_6]^{2-}$ 203
$[\overset{III}{Ir}\ X_3py_3]$ 110
$MnCl_2 . 4H_2O$ 46
$[MnF_6]^{4-}$ 195, 202
$[Mn(CN)_6]^{4-}$ 178
$[Mn(H_2O)_6]^{2+}$ 1, 46, 89, 91, 149, 202
$[Mn(NCS)_6]^{4-}$ 202
$[Mn(NH_3)_6]^{2+}$ 149
$[MnF_6]^{2-}$ 72
$[MnF_6]^{3-}$ 172
$MnF_3$ 172
$[Mn(H_2O)_6]^{3+}$ 29
$MnO_4^-$ 110
$K_3(MoCl_6)$ 11
$[MoCl_6]^{3-}$ 72, 264
$(NH_4)_2[MoCl_5(H_2O)]$ 9, 11
$[MoCl_3py_3]$ 12, 110
$[Mo(CN)_8]^{4-}$ 15
$[Mo(CN)_8]^{3-}$ 195
NaCl 186
$(Ph_3MeAs)_2[NiCl_4]$ 137
$[Ni(CN)_4]^{2-}$ 176, 177, 188, 265
$[Ni\ en_3]^{2+}$ 157
$[Ni(H_2O)_6]^{2+}$ 177, 196
$NaNiO_2$ 173
$NiSO_4 . 7H_2O$ 87
Nitrosyl complexes 9, 11
$[OsF_6]^{2-}$ 203
$K_2[OsCl_6]$ 140
o-phenanthroline complexes 110, 178
phosphine complexes 106
$[PtCl_4]^{2-}$ 117, 265
$[Pt(CN)_4]^{2-}$ 265
$[PtX_2py_2]$ 110
$[PtF_6]^{2-}$ 203
$[ReF_6]^{2-}$ 203
$TiCl_3 . 6H_2O$ 30, 31, 32
$[Ti(CO(NH_2)_2)_6]^{3+}$ 30 ff
$[Ti(H_2O)_6]^{3+}$ 9, 10, 29, 31, 71, 85, 108, 173, 264
$[Ti(H_2O)_4Cl_2]^{+}$ 31
$CsTi(SO_4)_2 . 12H_2O$ 182
$[TlCl_3py_3]$ 9, 12
Tutton salts 203
$[V\ dip_3]^{-}$ 7
$[V\ dip_3]$ 7
$[V\ dip_3]^{+}$ 7, 78
$[V(H_2O)_6]^{2+}$ 72
vanadium (III) alums 10
$[V(H_2O)_6]^{3+}$ 9, 10, 29
$[Zn(NH_3)_4]^{2+}$ 159
Configuration interaction 52, 69 ff, 76, 128, 146, 147, 162, 183, 184, 188, 395
example $d^2$ ion in the ligand field of symmetry $O_h$ 397 ff
Coordinates
relation between Cartesian and polar coordinates 207, 208
Core electrons 227
Correlation diagrams 68
for $d^3$ in the octahedral ligand field (only quartet terms) 68–69
Correlation schemes of the irreducible representations of the groups $O_h$, $T_d$, $D_3$, $D_{4h}$, $D_{4v}$, $C_{2v}$, induced by the basis functions $\psi_{n,l,ml} l = 0, 1, 2, 3, 4, 5$ 304
Correlation diagrams for subgroups of $O_h$ and $D_{4h}$ 285
Coulomb integral 98, 377

Coulomb interaction
classical 143
Coulomb potential 209
Covalency
central field 83–84
symmetry restricted 83–84
Covalent complex model according to Pauling 13 ff, 16, 18, 121 ff, 186–187, 188
quantitative treatment 484 ff
Covalence (cf. compound, covalent)
Covering operations (cf. symmetry operations)
Croceo salts 2
Cross over point 50, 129, 130, 150
Crystal field 17
Crystal field model (cf. electrostatic complex model, extended)
Crystal field stabilization energy (CFSE) (cf. ligand field stabilization energy (LFSE))
Crystal field theory 17
Cube 57, 61, 62, 341
Curie constant 137
Curie law 137
Curie–Weiss law 137

*d* orbitals 21, 53 ff
analytic expressions for the five real 55, 214
graphical representation of the five angular dependent parts 53, 217
photos of models of the five angular dependent parts 50
real transformation behaviour for symmetry operations of the *O* group 380
$d^1$ systems in ligand fields of various symmetries (quantitative treatment) 333 ff
$D_{4h}$ (tetragonally distorted octahedron; square planar configuration) 343 ff
$O_h$ (octahedron) 336 ff
(cube) 341 ff
$T_d$ (tetrahedron) 342
*d* state, one-electron
splitting in the ligand field 53 ff, 333 ff
of octahedral symmetry ($O_h$) 56 ff, 336 ff
of tetrahedral symmetry ($T_d$) 61, 62, 342, 343
weak and strong tetragonal field (elongated and compressed octahedral square planar arrangement) ($D_{4h}$) 64, 65, 343 ff
arrangement of the ligands at the corners of a cube ($O_h$) 62, 63, 349
tetragonal Archimedean antiprism ($D_{4h}$) 64
square planar arrangement of ligands ($D_{4h}$) 64, 65
splitting for a square pyramid (Kz 5) and pentagonal bi-pyramid (Kz 7) 183
$d\varepsilon$ and $d\gamma$ orbitals 53 ff
$d \rightarrow d$ bands (cf. ligand field bands and central ion bands)
$^2D$ term
splitting for $O_h$, $D_3$, $D_{4h}$ and $D_{2h}$ symmetry 31, 33
$D_3$ symmetry for $Cr^{3+}$ complexes
additional splitting compared to $O_h$ symmetry 35
point group 267
$D_{4h}$ symmetry
*d* one electron orbitals 65, 343 ff
$^2D$ state, splitting 33
point group 264, 265
trans $[CrA_4B_2]$ 40
degeneracy 469
accidental 310
Degree of degeneracy 25–26, 67, 213, 228, 229, 232, 323, 438
Determinants 475 ff
Diagonal elements of a matrix 471
Diagonal sum rule 245
Diagram of potential curves for the explanation of the Jahn–Teller theorem 166
Dipole
electric 13, 23, 56
point dipole 329 ff
magnetic 131
Dipole moment of ligands 13, 22–23
permanent 13
effective 23
induced 13
Dipole radiation transitions
electric 85, 494

magnetic 85, 192
Dissociation equilibria 155
Distortion of the coordinate octahedron as a consequence of the Jahn–Teller effect 33, 34, 164 ff
possible for the $\varepsilon_g$ case 168
Double groups 317 ff, 439
nomenclature 319
Durchdringungskomplexe 187

Eigenfunction 469
Eigenvalue 469
Eigenvalue problems (equations) 469 ff
Electric moment of higher order 142
Electronegativity 83, 114, 152
optical according to Jørgensen 114
Electrovalence (cf. Bond, electrovalent)
Electron configuration as treated in the central field approximation 226 ff
Electron delocalization 83, 97, 106, 141, 187, 194 ff, 196 ff, 201 ff
Electron delocalization parameter according to STEVENS 141
Electron density distribution 190 ff
Electron diffraction 172
Electron interaction integrals (Coulomb and exchange integrals) expressed by $F_0$, $F_2$, $E_4$ and $A$, $B$, and $C$ 377
Electron interaction parameter (cf. Racah-parameter $A$, $B$, and $C$ and Slater radial integrals $F_0$, $F_2$, and $F_4$)
Electron pairs
long 114, 119, 186, 190
for low and high spin complexes 145–147
number of electron pairs with parallel spin for octahedral complexes $d^1$ to $d^9$ 149
Electron pair bond 14, 121, 130, 186–187, 484
Electron spin (cf. spin)
Electron spin resonance measurements (cf. paramagnetic resonance measurements)
Electron interaction 51 ff, 109, 146, 185, 243 ff, 324, 348, 360, 362, 367 ff
Electronic transitions 108, 116
possible for complexes (schematic representation)
Electronic transition bands 4 ff, 109 ff
classic electrostatic treatment 112
possible cases of charge-transfer bands for octahedral microsymmetry 114
Electrostatic model for complexes
classical according to Kossel and Magnus 13, 16, 18, 120, 143
Extended (electrostatic ligand field model) 15 ff, 321 ff
Emission spectra in the gas phase of metal ions 17, 77, 82, 93
Energy integral 475
Energy of formation for complexes 13
Equilibrium between high and low spin complexes in the vicinity of the point of intersection 150
Equilibrium constants 176
formation and dissociation constants 154, 176, 177
Eulerian angle 287, 288
Exchange energy 147
average for electron pairs 148, 149
Exchange integrals 377
Exchange interaction 147
Extinction 9
Extinction coefficient
molar decadic 9

Formation constants
net reaction constant 155
individual, stepwise 155
for $[Men_3]^{2+}$ complexes M = Mn, Fe, Co, Ni, Cu, Zn 156
Formation equilibria 155
Flavo salts 2
Force constants 188
Franck–Condon principle 89
Free energy of reaction of dissolved complex ions 154 ff, 176
Free ions
term systems 17
theory of free atoms and ions 206 ff
one electron systems 207 ff
many electron systems 222 ff

g factor (cf. Landé-factor)
Gray–Wills method 238 ff

Ground state 1, 23, 27, 118, 126, 138, 162
and orbital moment contribution 131 ff
ground states of octahedral complex ions for high and low-spin compounds 118
ground states and ligand field stabilization energy 23, 143
$T_2$ ground terms $\mu_{eff} = f(T)$ according to Figgis 140
and Jahn–Teller theorem 33, 162 ff, 172 ff
multiplet splitting and magnetic behaviour 131
multiplicity and magnetic behaviour 1, 23, 128
term splitting of the ground states of the ions $Ti^{3+}$ to $Cu^{2+}$ 27
Group
definition 260
class of a group 260 ff
order of a group 260
symmetry group 260
special symmetry groups 260, 261 ff
$C_{2v}$ 268
$C_{4v}$ 265
$D_3$ 267
$D_{4h}$ 264
$O_h$ 263
$R_{3i}$ 262
$T_d$ 266
subgroup 261, 269
Group theory 22, 24, 38, 255 ff
Gaussian curves
for the reduction of absorption bands into partial bands 32, 174

Half width of bands
definition 34
of $d \rightarrow d$ bands 89 ff
and occupation numbers of the one electron states 90, 91
of the rare earths 97
for $[Mn(H_2O)_6]^{2+}$ 93
Half life 176, 177
Heat of formation 108
Heat of formation of dissolved complex ions 154–157
Heat of hydration 150–152
as a function of the atomic number for bivalent and trivalent ions 151
Heat of reaction 155, 156
High-spin complexes 13, 119, 163, 178, 183, 184, 188
Hilbert space 466
"Holes" in the *d* shell 27, 232
equivalence of electrons and holes 237
Homomorphism 275
Hund's rule for determination of the ground state 20
Hyperfine splitting (cf. paramagnetic resonance)
Hybridization coefficient for M.O. calculations 101, 105, 113, 196, 427
Hybridization of complexes 14 ff, 178

Individual, stepwise (consecutive) complex formation constants (cf. formation constants)
Infrared spectra 188
Intra ligand bands 4, 7 ff, 115, 116
Inner ligand transitions (cf. intra ligand bands)
Inner orbital complexes 125, 177, 178
Intensity of $d \rightarrow d$ bands 84 ff, 204
for tetrahedral and octahedral coordination 85–86
for *cis-trans*-isomers 86
Intercombination bands 44, 45 ff, 84, 87, 89 ff, 174, 498
intensity of intercombination bands and spin orbital coupling 96
Intercombination spectra
pure 46 ff
Intrinsic angular momentum (cf. spin)
Inversion symmetry 85, 497
Ionic radii 151, 152, 160
of the transition metals of the iron series 160 ff
as a function of the atomic number for octahedral coordination 161
for tetrahedral coordination 162
Isomers, *cis-*, *trans-*
spectroscopic differentiation of $[CrA_4B_2]$ and $[CoA_4B_2]^+$ $A = NH_3$, $B =$ Cl, Br, J 36 ff, 80
Intensity relations 40, 86
Isomorphism 275

Jahn–Teller theorem 33, 34, 152, 156, 162, 164 ff, 185
and orbital moment contribution 135
diagram of potential curves for the explanation of 166
dynamic J. T. effect 33, 165, 172
effect on $Dq$ values 79, 80
experimental evidence 172 ff
for low spin complexes 149
for high spin $d^6$ complexes 50
normal vibration for the J.T. distortion of an octahedron 165, 167
J.T. stabilization for tetrahedrally distorted octahedral ligand arrangement 171
and stereochemistry of $d^4$ and $d^5$ complexes 162
for $[TiA_6]$ complexes 33, 34

Kepler problem 209 ff
Kotani diagram for $d\varepsilon^1$, $d\varepsilon^2$, $d\varepsilon^3$, $d\varepsilon^4$ and $d\varepsilon^5$ 139
calculation for $d\varepsilon^1$ 459 ff
Kotani's theory for $\mu_{eff} = f(T)$ 138–140
Kramer's degeneracy 175
Kramer's doublets 96
Kronecker symbol 209, 226
Kubelka–Munk function 9

Laguerre polynomials 212
Lambert Beer law 9
Landé g factor 132, 192
Lanthanide contraction 160
Laplace operator 207, 222
Laporte rule 29, 85, 496
Lattice energy 152–153, 204
for metal halogenide $MX_2$ as a function of the atomic number 153
Lattice field (Madelung) 17
(LCAO–MO) method 97 ff, 409 ff
Legendre polynomials, associated 209
Ligand bands 4 ff
Ligand exchange reactions 176 ff, 187
Ligand fields 324 ff
definition 17
weak 51 ff, 126, 145, 323
strong 51 ff, 127, 130, 144, 146, 323
strength of the ligand field and magnetic behaviour 51 ff, 126, 127, 188
Ligand field bands (cf. $d \rightarrow d$ and central ion bands)
Ligand field components
of low symmetry and magnetic behaviour 138, 140
axially symmetric and magnetic behaviour (Figgis theory) 141
Ligand field configuration 66 ff, 364
energy in the zeroth approximation
for octahedral symmetry
$d\varepsilon^n\, d\gamma^{N-n}$ 66
for tetrahedral symmetry
$d\gamma^n\, d\varepsilon^{N-n}$ 66
and increase of the $E = f(Dq)$ curves in the Orgel diagram 90
and relation to the half widths of the $d \rightarrow d$ bands 89 ff
Ligand field stabilization 142 ff
for cubic complex ions 142 ff
and formation tendency of the low spin complexes 147
additional through Jahn–Teller distortion 152
contribution to the heat of hydration 150–152
contribution to the lattice energy 152, 153
contribution to the enthalpy of formation for complex equilibria of formation 154–158
for $O_h$ and $T_d$ symmetry 158
and ionic radii 160 ff
Ligand field stabilization energy (LFSE) 23, 142 ff, 150 ff
in the strong field approximation
for $O_h$ symmetry 146
for $T_d$ symmetry 146
in the weak field approximation
for $O_h$ symmetry 145
for $T_d$ symmetry 145
difference between the LFSE for tetrahedral and for octahedral ligand arrangements (cf. site preference energy)
for octahedral and tetrahedral symmetry, in relation to $Dq$ (octahedral) 160
for octahedral and square pyramidal coordination as well as contributions to the activation

energy for an $S_{N1}$ mechanism 183
for octahedral and pentagonal bipyramidal coordination as well as contributions to the activation energy for an $S_{N2}$ mechanism 184
Ligand field strength
critical 48, 129, 149, 188
calculation of the critical ligand field strength 149
Ligand field strength parameters
Determination of $Dq$ 76–77
values of $Dq$ for Aquo complexes 78
for various transition metal complex ions 81
definition of the parameter $\Delta \equiv 10\,Dq$ for cubic symmetry 59 ff, 106
ligand field parameter for $D_{4h}$ symmetry $Dq, Ds, Dt$ 66, 346
Terms contributing to $Dq$ 82, 108
explicit expressions for $\Delta \equiv 10\,Dq$ for $O_h$ symmetry (point charges or point dipoles)
rules for the magnitude of $Dq$ 77 ff
$\Delta \equiv 10\,Dq$ in the M.O. model 106
Ligand potential
spherically symmetric part $V_k(r)$ 24, 25, 59, 60, 143, 331, 332
non-spherically symmetric part $V_0$ $(x, y, z)$ 59, 60, 143, 331, 332
Lower level complexes (cf. inner orbital complexes)
Low-spin complexes 14, 120 ff, 147 ff, 178, 183, 184, 188
explanation according to Pauling 120–125
explanation according to the ligand field theory 126 ff
in the one-electron scheme 126–128
in the many electron term system 128–131
explanation within the M.O. theory 130
criterion for the appearance of 149
stabilization and formation tendency 147–150
octahedral radii of 162–163
LS coupling (cf. Russell–Saunders coupling)
Luteo salts 2

Magnetic behaviour
of octahedral $d^6$ complexes in the one-electron scheme 127
in the many electron term system 128
Magnetic moment
effective 119 ff, 139
of complexes of the first transition metal series 120
of complexes of the higher transition metal series 139, 140
of octahedral and tetrahedral cobalt (II) and nickel (II) complexes 136–137
temperature dependence according to Kotani's theory 138–140, 459 ff
Magnetic moment of the nucleus (cf. nuclear spin and nuclear moment)
Magneton, Bohr 118, 131, 192
Many centre integrals 424, 428
Many electron problem according to the strong field method 65 ff, 362 ff, 395 ff
Many electron systems
$d^N$ systems (free ions) 222 ff
$d^N$ systems in ligand fields 347 ff
Matrices 471 ff
character or trace 471
product of two 473
square 471
sum of two 472
Matrix elements
for the weak field method, according to Orgel 70, 71, 144, 348 ff, 388 ff
for the strong field method, according to Tanabe and Sugano 71, 72, 77, 362 ff, 395 ff
for a $d^2$ system, symmetry $O_h$
according to the weak-field method with term interaction 404
according to the strong field method with configuration interaction 406
Matrix multiplication 272 ff, 288, 289, 473

Mesomerism 186
Metal oxidation bands 110, 114, 116
Metal reduction bands 110, 111, 114, 116
Microsymmetry 17
$M_L$, $M_S$ tables 231, 235
M.O. energy diagrams 99 ff, 419 ff
for octahedral complex ions
only $\sigma$ bonds 105, 421
consideration of $\pi$ bonds of the $t_{2g}$ ligand states with the $t_{2g}$ central ion states 107
Molecular orbitals, M.O.s
for a complex ion with $O_h$ symmetry $\sigma$ and $\pi$ bonds 104, 420
Molecular orbital theory 17, 34, 84, 97 ff, 109, 113, 130, 161, 171, 187, 189, 190, 407 ff
antibonding 98 ff, 171
bonding 98 ff, 101
non-bonding 98 ff, 161, 171, 424
Mössbauer effect 201
Multiplet splitting 19, 94, 97, 131, 429 ff
Multiplication of symmetry operations 257 ff
Multiplication table 258
for the symmetry group $C_{3v}$ 259
for transformation matrices 274
Multiplicity 18, 128
Hund's principle of maximum multiplicity 119, 123, 124
change in multiplicity of the ground state and magnetic behaviour 127 ff

Na–D line 19
Nephelauxetic series 82 ff
of the ligands 83
of the central ions 83
nephelauxetic parameter $\beta$ 83, 202
Nephelauxetic effect 82 ff, 202
relativistic 96, 141, 202
Net reaction constant for complex equilibria 155
Neutron diffraction 172, 191
Noncombination theorem 244, 317, 470
Normal complexes 187
Normal coordinates 165
Normal vibrations (modes) 29, 85, 87
in the Jahn–Teller theorem 166 ff
of a $XY_6$ molecule (point group $O_h$) 167
Normal $d \rightarrow d$ bands (cf. spin allowed bands)
Normalization 466
Nuclear configuration
totally symmetric, non-totally symmetric 164–166
Nuclear $g$ factor 197
Nuclear magneton 197
Nuclear moment, magnetic 131, 194 ff
Nuclear quadrupole moment 199
Nuclear resonance, magnetic 191, 196 ff.
Nuclear spin 131, 195 ff
Nuclear spin quantum number I 194, 196, 197
Number of electrons, effective according to Sidgwick 14

Octahedron, regular 340
elongated, compressed 32, 64–66, 168 ff, 265, 344, 345
with system of coordinates inscribed in a cube 57, 263
non uniformly elongated 265
Octahedron group, $O_h$ 263, 264
One-electron functions
atomic 20, 207 ff
Operators 467 ff
Orbital angular momentum quantum number $l$, $L$ 19, 131, 192, 194, 212, 218, 235–237
magnetic $m_l$ 55, 212, 218
$M_L$ 230, 235, 236
Orbital angular momentum $\vec{l}\,\vec{L}$ 19, 131, 215 ff, 232 ff
Orbitals, angular part $Y(\theta, \phi)$ 22, 23, 53 ff
graphical representation of the five $d$ orbitals 53, 217
of three $p$ orbitals 98, 217
of the $s$ orbital 98, 217
photos of models of the angular parts of the $s$, $p$, and $d$ angular orbitals 54
$d$ function, behaviour upon rotation, qualitative discussion of the orbital angular momentum contribution to the magnetic moment 134
Orbital degeneracy, degree of orbital degeneracy

of ground terms and orbital moment 26, 226 f
contribution to $\mu_{eff}$ 136
Orbitals
six equivalent for $O_h$ symmetry 121
of ligands for $O_h$ symmetry 104
of the central ion for $O_h$ symmetry 104
of the central ion and of the ligands for octahedral complex ions for pure $\sigma$ bonding 102, 103
Orbital magnetism 131 ff
Orbital moment, magnetic
suppression 132 ff
for octahedral high- and low-spin complexes 134, 135
for tetrahedral complexes 135
Orbital moment contribution to the magnetic moment and stereo-chemistry 131 ff
Orgel diagram 70, 71, 73, 74, 75
for $d^5 Mn^{2+}$ 70
and half width of the absorption bands 91
Orthogonality 466
Oscillator strength 87
of intercombination bands and spin-orbit coupling 96
Outer orbital complexes 125, 177, 178
Overlap 100, 106, 424
Overlap charge 190
Overlap integrals 101, 116, 424, 475

*p* orbitals 14, 21, 85, 97 ff, 121 ff, 177, 178, 484 ff
representation of the three *p* orbitals 96, 217
photos of three-dimensional models of the three *p* orbitals 54
Packing effect in the crystal lattice 172, 173
Paramagnetic resonance 141, 172, 174, 192
hyperfine structure of the 194 ff
Paramagnetic susceptibility
temperature dependence 137 ff
for $d^1$ according to Kotani (quantitative) 459 ff
Parity, 29, 85 ff, 110, 301, 495
Parity prohibition 85, 174, 495
Paschen–Back effect 453
Pauli principle 52, 66, 93, 126, 130, 181, 226 ff
Pauling's theory of coordination complexes 13 ff, 120 ff, 484 ff
interpretation of the low spin magnetic behaviour according to 120 ff
Perturbation calculation 22, 25, 48, 243, 309 ff, 322, 333, 474 ff
Perturbation potential for octahedral symmetry
$V = V_k(r) + V_0(x, y, z)$ (Reduction into spherically symmetrical and non spherically symmetrical parts) 58 ff, 332, 333
for other symmetries 332
Phase factor 208
Point charge 17, 18, 109, 120, 329 ff
Point dipole 17, 18, 109, 120, 329 ff
Polarization 3, 80
Polarizability 13, 23
Polarized light in the study of the band assignments 87
Possible distribution of two electrons between the $d\varepsilon$ states 67
Potential barrier for the Jahn–Teller effect 166, 174
Potential surfaces in the Jahn–Teller effect of $[Cu(H_2O)_6][SiF_6]$ 174
Potential curves and half widths 90, 91
Potentials of various ligand systems 332 ($O_h$ [octahedron, cube], $T_d$, $D_{4h}$, $D_{4d}$)
Praseo salts 2
Pre-exponential factor 185
Principal quantum number *n* 213
Product representation 85, 312 ff, 363, 364
irreducible components of the representation products $\Gamma_i \times \Gamma_j$ of the double group O′ 315
for the evaluation of quantum mechanical interaction integrals 314–316
Product functions 225, 226
Projection operators 305
Purpureo salts 2
Pyramids
square as transition state for an $S_{N1}$ dissociation mechanism 181 ff

bipyramids, pentagonal as transition state for an $S_{N2}$ mechanism 181 ff
trigonal 181

Quadrupole radiation transitions, electric 85

Racah parameters *A*, *B* and *C* of the electron interaction 18, 71
reduction of *B* 73, 74, 77, 82 ff, 202, 203
*B* values of various free metal ions 77
*B* values of various transition metal complex ions 81
spin pairing energy as a function of *B* and *C* 147
$B_{55}$, $B_{35}$ 202
electron interaction integrals as a function of the Racah parameters 377
Radial function *R*(*r*) 21, 22, 53, 56, 207, 209 ff, 219
Radio carbon 176
Raman spectra 188
Rare earths 97, 324
Reaction rate constant 176
relation to energy of activation and pre-exponential factor 185
Reciprocity between electrons and holes 27
Redox reactions, photochemical 111
Reduction formula of the theory of representations 281 ff
Reduction of an absorption band into partial bands 32, 174
Reflectance spectroscopy 3, 9
Remission spectroscopy (cf. reflectance spectroscopy)
Reflection in a plane of symmetry 256, 257, 262
Representation of groups 270 ff
equivalent representation 276
notation of irreducible representations according to Mulliken and Bethe 26, 281, 479, 480
character of a representation 277
character system 277
character tables 280, 281, 479, 480
definition of a representation 276
dimensionality of a representation 276
reduction of representations 278 ff
reducible and irreducible 279, 280, 281 ff
ambiguous representations 141
Roseo salts 2
Rotary reflection (cf. improper rotation) 262
Rotation reflection group of the sphere 262, 310
Ruby lines $R_1$ and $R_2$ 45
Rule of the average environment 79
Rule of Irving and Williams 157
Rules
phenomenological of Tsuchida and Shibato concerning absorption spectra of complexes 2
non crossing rule 74
for term diagrams 74 ff
Russell–Saunders coupling (LS coupling) 18 ff, 235 ff
Rydberg spectra 115, 116

*s* orbital 214
orbitals, graphical representation 98, 217
photo of a three-dimensional model of the orbital 54
Secular determinants 243, 323, 474 ff
Selection rules for optical absorption spectra 494 ff
Shells
*K*, *L*, *M*, *N* (principal shell) 213, 227
subshells 214, 227
Shifting (raising and lowering) operators for the angular momentum 220, 233
spin shifting operators 221, 234
Site preference energy 159
Slater determinants
Heisenberg–Slater functions (HS functions) 66, 225, 226
Slater functions $R_{n_e}(r)$ 22, 56
Slater's method for determining the term systems of free ions 18, 223 ff
Slater's radial integrals $F_0$, $F_2$ and $F_4$ 18, 71, 202, 249
electron interaction integrals as a function of $F_0$, $F_2$ and $F_4$ 377

$S_{N1}$ mechanism, $S_{N2}$ mechanism (cf. substitution reactions)
Spectrochemical series 76 ff, 188
of the ligands 78
of the central ions 79
and half widths of the spin allowed $d \rightarrow d$ bands 92
and magnetic behaviour 149, 150
Spin = electron spin 19, 20, 220 ff, 232 ff
Spin orbital coupling 20, 87, 93 ff, 97, 136, 138, 323, 429 ff, 494, 498
indication in the absorption spectra for tetrahedral coordination 96
and intensity for intercombination bands 96
for the interpretation of the magnetic moments 138 ff, 438 ff
in free atoms and atomic ions 429 ff
of $d^1$ ions in ligand fields of octahedral symmetry 438 ff
Spin orbital coupling constant 93 ff, 136
definition of $\zeta_{n,l}$ and $\lambda$ 93, 438
values of $\zeta_{n,l}$ for various free ions 95
reduction for complex ions (relativistic, nephelauxetic effect) 96, 141, 196, 202, 203
definition of the relativistic nephelauxetic parameter $\beta^*$ 97
of the halogens Cl, Br and J and intensity of the charge-transfer bands $\pi \rightarrow dz^2$ for $[Co(NH_3)_5X]$ 111
and interpretation of the magnetic properties 138 ff, 459 ff
Spin eigenfunctions 221
Spin allowed Laporte forbidden bands 85 ff, 91
Spin free complexes (cf. high-spin complexes)
Spin magnetism 131 ff, 136
relation for $\mu_{eff}$ for pure spin magnetism 119, 131
Spin moments
effective in B.M. as a function of the number of the unpaired electrons 119
influencing by interaction with excited states through spin orbital coupling 136
Spin pairing 119, 126, 131, 146–149
Spin pairing energy $P$ 146–147
average $\pi$ pro $10\,Dq$ ligand field stabilization 147–149
Coulomb and exchange part of $\pi$ 148
$P$ and $\pi$ as function of $B$ and $C$ 147
Spin paired complexes (cf. low-spin complexes)
Spin quantum number, $s$, $S$ 19, 20, 131
magnetic spin quantum number $m_s$, $M_s$ 67, 222, 230, 235, 236
Spin forbidden bands (cf. intercombination bands)
Spin states for $(t_2)^2$ 67
Spherical harmonics $Y_L$, $m_l$ 56, 208 ff, 218, 299 ff, 326
characters of representation matrices, induced by spherical harmonics or by atomic orbitals 303
expansion of $1/|\vec{R} - \vec{r}|$ in spherical harmonics 324
real orthonormalized linear combinations of spherical harmonics for $l = 0, 1, 2,$
symmetry adapted linear combinations of the spherical harmonics for the symmetry groups
$O$ 305 f
$D_{4h}$ 307
$D_3$ 308
Tables for $l = 0, 1, 2, 3, 4,$ 210, 211
transformation behaviour under symmetry operations 299 ff
Splitting factor
spectroscopic 194
reduction of the orbital magnetic moment contribution to the spectroscopic splitting factor 196
Square planar complexes 13, 14, 63, 64, 121, 123, 127, 347, 488 ff
Stability
thermodynamic 154 ff, 176, 187, 188
kinetic 176 ff, 187, 188
relative of tetrahedral and octahedral high spin complexes 158–159
Stark effect, innercomplex 20, 21 ff, 321 ff
State of solution of dissolved complex ions 2, 3
Stereochemistry 14, 122, 131 ff, 484 ff
of $d^4$ and $d^9$ complexes 161

stereochemical considerations and conditions 188, 203
and hybridization for octahedral, tetrahedral and square planar configuration 14, 121 ff, 484 ff
Strong field method 51 ff, 69, 70, 362 ff, 395 ff
example: $d^2$ ion in the ligand field of the symmetry
$O_h$ without configuration interaction 369 ff
with configuration interaction 397 ff
Subdeterminant = Minor 477
Substitution reactions 176 ff
mechanism for $S_{N1}$, $S_{N2}$ 179–181
Symmetry centre (inversion centre) 85, 174, 204, 261 ff, 496
Symmetry of the ligand field 17, 21
cubic 24 ff, 336 ff
lower 343 ff
the most important symmetries of complex ions 481–483
Symmetry axis 262
multiplicity of 262
Symmetry plane 262
Symmetry groups 255 ff
symmetry group of the Hamilton operator 293 ff
Symmetry operations 255 ff
symbols for 261 ff

Tanabe–Sugano diagram 71–73
for $d^3$ 72
for $d^6$ 74
and magnetic behaviour 129
Temperature dependence
of the intensity of the $d \rightarrow d$ bands for $NiSO_4 . 7H_2O$ 87 f
of the paramagnetic susceptibility 118, 137 ff, 459 ff
Term diagrams 70 ff
rules for 74–75
Term energies
of free ions for $d^N$ $N = 2, 3, 4, 5$ or $8, 7, 6$ 254
for octahedral symmetry in the approximation of the weak field for the ground terms of the configurations $d^1 - d^9$ 75
for $d^2$ ions, symmetry $O_h$, for all terms 403, 405
for a $d^3$ ion, only $^4F$, weak field case 361
Term energy as a function of the Racah parameters $A$, $B$ and $C$ 254
Term interaction 42 ff, 48, 51, 52, 68 ff, 75, 76, 128, 145, 185, 388 ff, 399 ff
example: $d^2$ ion in the ligand field of symmetry $O_h$ 390 ff
Term inversion
for the transition from octahedral to tetrahedral symmetry 25 ff
for the transition from electrons to "holes" 27
Term nomenclature
for free ions (cf. Russell–Saunders coupling)
for cubic ($O_h$ and $T_d$) symmetry acording to Bethe and Mulliken 26
for atomic one-electron states 214
for the octahedral double group 319
Terms of free ions 237
belonging to the configuration $d^N$ ($N = 1$ to $9$)
Term splitting 24 ff
in cubic ligand fields 24–27
of the ground states of the ions $Ti^{3+}$ to $Cu^{2+}$ for $O_h$ symmetry 27
of a $^2D$ state with consideration of the spin orbital coupling for $O_h$ symmetry 95, 438 ff
of a one-electron $d$ state for $O_h$ symmetry, with consideration of the spin orbital coupling 141, 438 ff
and influence of a magnetic field 141, 446
of a $^2D$ state for $D_3$, $O_h$, $D_{4h}$ and $D_{2h}$ symmetry 33
of a $d$ state for $D_{4h}$ symmetry 64 ff, 343 ff
of the sextet and quartet states of a $d^5$ ion for $O_h$ symmetry 87
of the low-lying singlet terms of $Co^{3+}$ for the transition of $O_h$ to $D_{4h}$ 48
for a $d^2$ system. $O_h$ symmetry, weak field case, without term interaction 351 ff
strong-field case without configuration interaction 369 ff

with term of configuration interaction 388 ff, 397 ff
Term system
of the free $Cr^{3+}$ ion 21
of $Cr^{3+}$ in the octahedral ligand field
only quartet terms 41
with doublet terms 44
of $[Mn(H_2O)_6]^{2+}$ 46, 70
octahedral cobalt (III) and iron (II) complexes, simplified 50
for $d^3$ ($O_h$) symmetry 72
for $d^6$ ($O_h$) symmetry 74
for chromium (III) complex ions of symmetry $C_{4v}$, $D_{4h}$, and $C_{2v}$
for titanium (III) complexes of symmetry $O_h$, $D_3$, $D_{4h}$ and $D_{2h}$ 33
for $d^2$ ($O_h$) symmetry 360
weak-field case 360
strong-field case 387
with term and configuration interaction 396
M.O. term system for $O_h$ symmetry 105, 107, 421
Term system of an atomic one electron system for Coulomb or non-Coulomb potential 213
of the $(nd)^2$ configuration
(a) in the central field approximation 247
(b) with consideration of the electron interaction 247
for $d^1$ systems
octahedron, cube, tetrahedron 63–64, 339
$D_{4h}$ symmetry 65
$D_{4d}$ symmetry 64
Term overlapping 48 ff, 129
in the Tanabe–Sugano diagram for $d^6$ 73, 129
consequences for absorption spectra 49, 50
consequences for magnetic behaviour 129–130, 188
Tetrahedron 15, 62 ff, 86 ff, 123, 158 ff, 267, 268, 342, 491 ff
with coordinate system, inscribed in a cube 62, 267
Tetrahedron group $T_d$ 266, 267
Thermomagnetic properties of transition metal complexes and ligand field stabilization energy 150 ff
Time reversal 175
Total angular momentum $\vec{j}\,\vec{J}$ 19, 20, 131
quantum number $j$, $J$ 19, 20, 131, 192
magnetic total angular momentum quantum number $m_j$, $M_j$ 192, 437
Total degree of degeneracy 67, 68
for $d\varepsilon$ and $d\gamma$ 59
Total intrinsic angular momentum (cf. spin)
Transformations of coordinates 286 ff
Translational vector 85
Transition moment 494
Transition state
for $S_{N1}$ and $S_{N2}$ mechanisms 181

Valence bond method
according to Pauling for the description of complex ions 13 ff, 186, 189, 484 ff
for the interpretation of the low spin behaviour 120 ff
critic or the valence bond method in relation to complexes 125
and interpretation of the magnetic behaviour according to Taube 177
for octahedral arrangement (quantitative treatment) 485 ff
for square planar arrangement (quantitative treatment) 488 ff
Valence eigenfunctions
directional properties 121, 484 ff
Valence electrons 227
Variation, linear 409
Variation principle of quantum mechanics 188 ff
Vileo salts 2

Wavefunctions
for the Kepler problem 212 ff
complex and real 214–215
Weak-field method 51, 52, 69, 348 ff, 388 ff
example: $d^2$ ion in the ligand field of the symmetry $O_h$ without term interaction 351 ff
with term interaction 371 ff
$d^2$ ion only $^4F$ term 361
Weiss' constant 137

Zeeman effect
in free atoms and atomic ions 448 ff
in $d^1$ ions in the octahedral ligand field 453 ff
Zeeman energies, first and second order 138
Zeeman splitting 91, 197, 198, 448 ff, 454 ff